STATISTICS
for Management and Economics
THIRD EDITION

William Mendenhall
University of Florida

James E. Reinmuth
University of Oregon

DUXBURY PRESS
North Scituate, Massachusetts

Library of Congress Cataloging in Publication Data

Mendenhall, William.
  Statistics for management and economics.

  Includes index.
  1. Statistics. I. Reinmuth, James E., 1940-joint author. II. Title.
HA29.M533   1978   519.5   77-22411
ISBN 0-87872-142-8

Duxbury Press
A Division of Wadsworth Publishing Company, Inc.

Statistics for Management and Economics, 3d edition was edited and
prepared for composition by Carol Beal. Interior design was provided by
Sandra Rigney and the cover was designed by Oliver Kline.

L.C. Cat. Card No.: 77-22411
ISBN 0-87872-142-8
Printed in the United States of America

1 2 3 4 5 6 7 8 9 - 82 81 80 79 78

Contents

four

PROBABILITY   75

five

RANDOM VARIABLES AND PROBABILITY DISTRIBUTIONS   121

six

THREE USEFUL DISCRETE PROBABILITY DISTRIBUTIONS   145

seven

THE NORMAL PROBABILITY DISTRIBUTION   187

eight

nine

ten

eleven

twelve

thirteen

fourteen

fifteen

sixteen

seventeen

eighteen

Preface

The desire to update and polish the content of the second edition of *Statistics for Management and Economics* has prompted us to prepare this third edition. Our experience with the second edition, coupled with valuable advice from the book's users, has helped us prepare what we hope to be a much improved book with a number of timely features. We hope that these additions and innovations will enhance the book's pedagogical effectiveness and will prove appealing to users. At the same time we have endeavored to retain the spirit of the first two editions by presenting a unified, streamlined treatment of the subject. To this end inference is explained in the first chapter as the objective of statistics and the introduction to every subsequent chapter discusses the role of the new material in making inferences and measuring their reliabilities.

The following changes are, in our view, the main improvements in this third edition:

1. A new chapter 12, on "Multiple Regression," has been added. With the advent of computers, this topic has acquired a new importance for the business decision maker and we feel that it deserves the reasonably thorough treatment we have given it. Chapter 12 offers a *computer-oriented approach* to multiple regression analysis, as most exercises and examples in it are solved by using commonly available multiple regression programs.
2. Because of its growing usefulness to modern business, especially in accounting and in marketing, we have added a new chapter 16 on "Survey Sampling."
3. The chapter on "Quality Control," which appeared as chapter 11 in the first two editions, has been omitted from the present edition. However, some exercises reflecting the uses of quality control in business have been added to chapters 8 and 9 of the third edition so that the student still receives exposure to this important topic.
4. The total number of exercises has been greatly increased, especially because in this edition exercises have been added *at the end of most sections.* This is an addition to the large number of exercises included in a supplementary exercises section *at the end of each chapter.* These latter exercises

are inclusive in character and relate some or all of the techniques within the corresponding chapter.

5. We have made a strong effort to provide this revision with a distinctly applied flavor. To this end many exercises and examples have been taken from *real life instances,* which were extracted from real studies published in recent issues of business and economics journals.

6. We have also attempted to demonstrate the breadth of applicability of statistical methodology in management and economics. To this end most exercises in this edition have been indexed according to their primary area of application. While some exercises are purely instructional, most have been drawn from the functional areas of business and administration. The coding used to indicate the area of application is as follows:

Actg	Accounting
Fin	Finance
Mgt	Management
Mktg	Marketing
Pdn	Production
PA	Public Administration
RE	Real Estate
Trn	Transportation

7. At the end of most chapters you will find another new feature: "Experiences With Real Data." These consist of student (or, in some cases, class) projects associated with realistic experimental situations. They provide experimental verification of statistical concepts or involve students in a live sampling or data analysis situation. The purpose of these "Experiences" is to teach by experimentation and to make statistics more real to the student.

8. Last, but by no means least, we have made numerous small but to our mind significant changes throughout the text. They make it easier, simpler, and clearer and also provide more motivation for the student. Most have originated in student suggestions and in advice given to us by the second edition's users. We believe that they make this edition a much better teaching and learning tool.

The authors are grateful to the editorial staff of Duxbury Press, especially to Alexander Kugushev and Carol Beal, for assistance and cooperation in the preparation of this book. Thanks are due to a number of colleagues and friends of both authors for their helpful comments. Special thanks are due M. James Dunn for assistance in reading rough drafts of the original manuscript and to Jack Butler for his assistance in the preparation of chapter 15. Thanks are also due P. L. Claypool, Oklahoma State University; D. James Croft, University of Utah; George Heitmann and Roger Pfaffenberger, Pennsylvania State University; and William Whiston, University of Massachusetts, for their helpful reviews of the manuscript, and to D. R. Cox, A. Hald, S. M. Selby, D. L. Burkholder, R. A. Wilcox, and C. W. Dunnett for their kind permission to use tables reprinted in the appendix. Finally, we acknowledge the constant encouragement and assistance of our wives and children and their partnership in this writing endeavor.

William Mendenhall
James E. Reinmuth

STATISTICS
for Management
and Economics

chapter objectives

GENERAL OBJECTIVE The purpose of this chapter is to identify the nature of statistics, its objective, and how it plays an important role in business and economics.

SPECIFIC OBJECTIVES

1. To answer the question, "What is statistics?"
>> Sections 1.1, 1.2, 1.3

2. To identify statistical inference as the objective of modern statistics.
>> Section 1.3

3. To identify the contributions that statistics can make to business decision making.
>> Section 1.4

4. To define the basic concepts used in statistics.
>> Sections 1.1 through 1.5

chapter one

What Is Statistics

1.1 Illustrative Statistical Problems

What is statistics? How does it function? How does it help to solve certain business problems? Rather than attempting a definition at this point, let us examine several problems that might come to the attention of the business statistician. From these business problems we can select the essential elements of a statistical problem.

Statistics plays an important role in accounting. For example, auditing the inventory of a large hospital can be very costly and time-consuming. Instead of counting and pricing the thousands of items in stock, a sample of items can be selected from the complete inventory. The value of these items actually in stock can be compared with the values shown in the hospital's records, and an estimate can be constructed of the ratio between the total value of supplies on hand to those shown in the hospital records.

Or consider the example of a small business that must decide each morning how many units of a perishable commodity should be stocked. This decision is based on the observed sales of the commodity over a recent period of time. Based on this sample data and the costs for a shortage or overstock, the manager decides on the inventory level to be used for that day.

Another example of the use of statistics is the sampling inspection of purchased items in a manufacturing plant. On the basis of an inspection, each lot of incoming goods must either be accepted or be rejected and returned to the supplier. The inspection might involve drawing a sample of 10 items from each lot and recording the number of defectives. The decision to accept or reject the lot would then be based on the number of defective items observed.

Similarly, the production of a manufacturing plant depends on many factors unique to the type of manufacturing plant under consideration. By observing sample data collected on these factors and the production over a period of time, we can construct a prediction equation relating production to the observed factors. Then prediction of the plant's future production for a given set of factor conditions can be made by substituting values of the factors into the prediction equation. Methods of identifying the important factors needed for the prediction equation as well as a method for assessing the error of prediction will be discussed in subsequent chapters.

Marketing research offers another example of a prediction problem. A representative sample of customers is selected and each person is asked to give an opinion concerning a manufacturer's product. From the data obtained in this opinion survey, the market analyst must decide whether a sufficient demand exists for the product. If the demand exists, the analyst must select the package design, the best selling price, and the market area. All these questions can be answered from information derived from the sample survey data.

These problems illustrate that business statistics involves the use of sample data to predict, estimate, and, ultimately, make business decisions. As you will see subsequently, **modern statistics offers a variety of analytical procedures to aid the business statistician in making decisions in the presence of uncertainty.** Of course, we are not implying here that uncertainties exist only within a business context or that modern statistics applies only to business. However, our emphasis in this text is to show the applications of statistical techniques to business problems. Other areas of application are suggested through the exercises at the end of each chapter.

1.2 The Population and the Sample

The examples above are varied in nature and complexity, but each involves prediction or decision making or both. In addition, each involves sampling. A specified number of items (objects or bits of information)—a **sample**—are drawn from a much larger body of data called the **population**. Note that the word "population" refers to data and not to people. Recording the daily sales for a perishable commodity gives a sample of all possible daily levels of demand (the population) that have occurred in the past or may occur in the future. In the sampling inspection problem, we assume that each sample of 10 items is a representative sample of the lot (the population) from which it was selected. The market researcher draws a sample of opinions from the statistical population that represents the entire potential market for his product.

> **Definition**
>
> A **population** is the set representing all measurements of interest to the sample collector.

In all the previous examples, we were primarily interested in the population. However, in most instances it is impractical or far too costly to observe or measure every element in the entire population. So we select a sample that we hope is a small-scale representation of the underlying population. The sample may be of immediate interest, but ultimately we are interested in describing the population from which the sample is drawn.

> **Definition**
>
> A **sample** is a subset of measurements selected from the population of interest.

As used by most people, the word "sample" has two meanings. It can refer to the set of objects on which measurements are to be taken or it can refer to the objects themselves. A similar double use could be made of the word "population."

For example, you read in the newspapers that a Gallup poll was based on a sample of 1823 people. In this use of the word "sample," the objects selected for the sample are clearly people. Presumably each person is interviewed on a particular question, and that person's response represents a single item of data. The collection of data corresponding to the people represents a sample of data.

In a study of sample survey methods, it is important to distinguish between the objects measured and the measurements themselves. To experimenters the objects measured are called "experimental units." The sample survey statistician calls them "elements of the sample."

To avoid a proliferation of terms, we will use the word "sample" in its everyday meaning. Most of the time we will be referring to the set of measurements made on the experimental units (elements of the sample). If occasionally we use the term in referring to a collection of experimental units, the context of the discussion will clarify the meaning.

1.3   The Parts of a Statistical Problem

The objective of statistics and the parts of a statistical problem are now defined and discussed.

Objective of Statistics

The **objective of statistics** is to make inferences (predictions, decisions) about certain characteristics of a population based on information contained in a sample.

How will we achieve this objective? We will find that every statistical problem involves five parts. By successfully completing each part, we can achieve the objective of statistics.

The first and most important part of a statistical problem is a clear specification of the question to be answered and of the population of data that is related to the question.

The second part of a statistical problem is deciding how the sample will be selected. This is called the *design of the experiment* or the *sampling procedure*. This second part is important because data cost money and time. In fact, it is not unusual for a business survey to cost $50,000 to $500,000, and the costs of many technological experiments can run into the millions. And what do these experiments and surveys produce? Numbers on a sheet of paper, or, in brief, information. So planning the experiment is important. Including too many observations in the sample is often costly and wasteful; including too few is also unsatisfactory. In addition, the way the sample is selected will often affect the amount of information per observation. Thus a good sampling design can sometimes reduce the costs of data collection to one-tenth or as little as one-hundredth of the cost of another sampling design.

The third part of a statistical problem involves the analysis of the sample data. No matter how much information the data contain about the practical question, you must use the appropriate method of data analysis to extract the information from the data.

The fourth part of a statistical problem is using the sample data to make an inference about the population. As you will subsequently learn, many different procedures can be employed to make an estimate or decision about some characteristic of a population or to predict the value of some member of the population. For example, 10 different methods might be available to forecast a company's sales but one procedure might be much more accurate than another. Therefore, you will wish to employ the best inference-making procedure when you use sample data to make an estimate or decision about a population or a prediction about some member of a population.

The final part of a statistical problem identifies what is perhaps the most important contribution of statistics to business decision making. It answers the question, "How good is this inference?" To illustrate, suppose someone conducts a statistical survey for you and estimates that your company's product will gain 34% of the market next year. We hope that you will not be satisfied with this inference but will ask, "How accurate is the estimate?" Of what value is the estimate without a measure of its reliability? Is the estimate accurate to within 1%, 5%, or 20%? Is it reliable enough to be used in setting production goals? As you will subsequently learn, statistical estimation and decision-making and prediction procedures enable you to calculate a measure of goodness for every inference. Consequently, in a practical inference-making situation, every inference should be accompanied by a measure that tells you how much faith you can place in the inference.

Parts of a Statistical Problem

1. A clear definition of the population of interest.
2. The design of the experiment or the sampling procedure.
3. The collection and analysis of data.
4. The procedure for making inferences about the population based on sample information.
5. The provision of a measure of goodness (reliability) for the inference.

1.4 The Statistician and Business Decision Making

How does statistics contribute to business decision making? We can use the parts of a statistical problem to identify the major contributions of statistics to business decision making.

One of the major occupations and contributions of the business statistician is in planning surveys and experiments—buying a specified quantity of information at minimum cost. As an example, a knowledge of market conditions for both raw materials and a company's products is an essential prerequisite to making business decisions. The price of this information, the cost of the survey or experiment to assess market conditions, can be greatly reduced by a good sampling design.

A second contribution concerns parts 3 and 4 of the statistical problem. A business statistician can aid in the selection of an appropriate method of data analysis and of a good inference-making procedure.

The most important contribution of business statisticians, however, is in providing a measure of goodness (i.e., reliability) to go with each inference. The measure of goodness tells you how much faith you can place in the inference. For example, if you were a businessperson deciding whether to make a capital investment, you would like to know the future course of the economy. A forecast would be helpful but only if you knew whether the forecast was reliable. An unreliable forecast could lead you to an erroneous and costly decision. Statistical procedures address this problem, providing a measure of goodness for every decision, estimate, or prediction.

1.5 Summary

Statistics is a branch of science concerned with the design of experiments or sampling procedures, the analysis of data, and the procedures for making inferences about a population of measurements from information contained in a sample. The statistician is concerned with developing and using procedures for design analysis, and for inference making, which will provide the best inferences for a given expenditure. In addition to making the best inference, he is concerned with providing a quantitative measure of the goodness of the inference-making procedure. Also, it is essential that the business statistician effectively communicate his experimental conclusions to management.

1.6 A Note to the Reader

We have stated the objective of statistics and, we hope, have answered the question, What is statistics? The remainder of this text is devoted to the development of the basic concepts involved in statistical methodology. In other words, we wish to explain how statistical techniques actually work and why they work.

Statistics is a theory of information in which applied mathematics plays a major role. Most of the fundamental rules (called *theorems* in mathematics) are developed and based on a knowledge of the calculus or higher mathematics. Since this is an introductory text, we omit proofs except where they can be easily derived. Where concepts or theorems are intuitively reasonable, we will attempt to give an informal explanation. Hence we will attempt to convince you with the aid of examples and intuitive arguments rather than with rigorous mathematical derivations.

As in any other business technique, common sense must be employed when applying statistical methods to the solution of a real problem. Since inferences are made about a population from sample data, the inferences are meaningful only if the sample is selected from the population of interest. For example, if we proposed a study to find the average height of all students at a particular university, we would not select our entire sample from the members of the basketball team. The basketball players should be represented within a sample in approximately the same proportion that they exist within the entire university. Another common misuse of statistics is in making comparisons. The National Safety Council claims that on a holiday weekend there are about twice as many highway deaths as on an ordinary weekend.

We might conclude that driving on a holiday weekend is much less safe than driving on an ordinary weekend. But this is true only if the traffic volume is about the same for all weekends—an unrealistic assumption. It is important that we view all statistics and sets of data with a critical eye and apply common sense and intuition about the problem to our decision format before arriving at a conclusion. The conclusions that apply to a particular problem are unique to that problem and seldom apply to other or related problems.

As you read through the text, refer occasionally to this chapter and review the objective of statistics, its role in business research, and the parts of a statistical problem. Each of the following chapters should, in some way, be directed toward answering the questions posed here. Each is essential to completing the overall picture of the role of statistics in business, management, and economics.

Experiences with Real Data

In any experiment involving the analysis of data, it is important to have a clear picture of the composition of the sample and the population. Often samples and populations are confused, or they are defined so loosely that experimental results are difficult to interpret. Occasionally such errors lead to absurd or otherwise meaningless results.

To clarify the notion of the sample and population, we will consider three realistic activities, which may be conducted individually or by groups of students.

1. Visit the research library of your college or university and select a research journal appropriate to your field. For advice in the selection of an appropriate journal, consider the following guide:

Field	Journals
Accounting	*Accounting Review*
	Journal of Accounting Research
Economics	*Quarterly Journal of Economics*
	American Economics Review
Finance	*Journal of Finance*
	Journal of Financial and Quantitative Analysis
Management	*Academy of Management Journal*
	Management Science
Marketing	*Journal of Marketing*
	Journal of Marketing Research

After selecting a journal, choose an article that deals with experimentation involving a sample survey. The article need not contain the actual data set used in the experiment. Now discuss the research objective and the procedure by which the data were obtained. Define and identify the experimental units, the sample, and the population. Recall the objective of statistics. Explain how statistical inference about the population may help to answer the researcher's questions.

2. Consumer surveys or opinion polls, conducted by such firms as A. C. Nielsen, George Gallup, and Louis Harris, are printed in almost every Sunday newspaper.

Find one such published survey or poll and discuss the objective of the study and the procedure employed to obtain the sample data. Define and identify the experimental units, the sample, and the population. Explain how the reported findings represent an inference about a population.

Preceding election day during each presidential election year, we are continually bombarded by numerous candidate preference surveys. What special problems confront pollsters in the selection of experimental units and in the definition of the sample and population in preelection surveys—problems that are not factors in consumer or public opinion surveys? (Hint: Is the pollster concerned only with those who will actually vote on election day? If so, how can he identify potential voters from nonvoters?) How can the pollster accommodate these problems?

3. Select a busy traffic intersection near your school and set up an experiment to count the daily traffic flow through that intersection. Define the population to include any class or type of vehicles you choose. Define the experimental units and the sample, and devise a procedure for the selection of the sample data. What special problems were encountered in designing the experiment, defining the experimental units, the sample, and the population? How did the statement of your research objective assist your efforts in obtaining a clear image of the experimental units, the sample, and the population?

References

AAKER, D. A. *Multivariate Analysis in Marketing: Theory and Applications.* Belmont, Calif.: Wadsworth, 1971.

Careers in Statistics. American Statistical Association and the Institute of Mathematical Statistics, 1974.

FOURAKER, L., and S. SIEGEL. *Bargaining and Group Decision Making.* New York: McGraw-Hill, 1963.

HUFF, D., and I. GEIS. *How to Lie with Statistics.* New York: Norton, 1954.

MANSFIELD, E. "Entry, Gibrat's Law, Innovation and the Growth of Firms." *American Economic Review* 53 (1962): 1023-1051.

MARKOWITZ, H. *Portfolio Analysis.* New York: Wiley, 1959.

NETER, J. "Some Applications of Statistics for Auditing." *Journal of the American Statistical Association* 47 (1952): 6-25.

REICHMANN, W. J. *Use and Abuse of Statistics.* London: Methuen & Company Ltd., 1961.

SEMON, T.; R. COHEN; S. RICHMOND; and J. STOCK. "Sampling in Marketing Research." *Journal of Marketing* 23 (1959): 263-274.

SIELAFF, T. J. *Statistics in Action.* San Jose, Calif.: Lansford Press, 1963.

TANUR, J. M., et al. *Statistics: A Guide to the Unknown.* San Francisco: Holden-Day, 1972.

VANCE, L. L. "Review of Developments in Statistical Sampling for Accountants." *Accounting Review* 35 (1960): 19-29.

chapter objectives

GENERAL OBJECTIVE Because functional notation and summation notation are used throughout this text, the objective of this chapter is to review (or introduce) these two important mathematical notations.

SPECIFIC OBJECTIVES

1. To develop an understanding of functional notation. This notation will be used in the subsequent discussion of summation notation.
Section 2.2

2. To develop an understanding of the meaning of summation symbols.
Sections 2.3, 2.4

3. To show how summation notation is used to provide instructions for the summation of numbers. Summation notation will be used in the summing of sample data in chapter 3 and subsequent chapters.
Section 2.4

4. To present three useful theorems that can be used to find simplified expressions for summation formulas that appear in chapter 3 and subsequent chapters.
Section 2.5

chapter two

Useful Mathematical Notation

2.1 Introduction

As you will learn subsequently, the statistical analysis of data requires numerous arithmetical operations which are too cumbersome to describe in words. The solution to this difficulty is to express the operations in terms of one or more formulas. Then, to analyze a given set of sample data, you substitute the sample measurements into the appropriate set of formulas.

One of the most common operations in a statistical analysis of data is the process of summation (addition). Consequently, we need a symbol to instruct the reader to sum the sample measurements or perhaps a set of numbers computed from the sample measurements. This symbol, called summation notation, will be very familiar to some readers and entirely new to others.

If you are unfamiliar with the notation or need a review, it is suggested that you read this chapter and work the exercises at the ends of the sections. If you are familiar with summation notation, the chapter may be omitted.

To understand summation notation, you should be familiar with two topics from elementary algebra: functional notation and the notion of a sequence. Functional notation is important because we will often wish to sum some function of the sample measurements (to give a simple example, their squares). Thus the function of the sample measurements to be summed will be given as a formula in functional notation. Sequences are important because we must order (identify by arranging in order) the sample measurements so that we can give instructions on which measurements are to be summed.

At first glance this chapter may seem more mathematical than necessary.

Please bear with us. You will need an understanding of this notation if you are to understand the formulas that will appear in subsequent chapters.

2.2 Functional Notation

Let us consider two sets* of elements (objects, numbers, or anything we want to use) and their relation to one another. Let the symbol x represent an element of the first set and y an element of the second. We could specify many different rules defining **relationships** between x and y. For instance, if our elements are people, we have rules for determining whether x and y, two persons, are first cousins. Or suppose that x and y are integers taking values 1, 2, 3, 4, ... For some reason we might wish to say that x and y are "related" if $x = y + 2$.

Now let us direct our attention to a specific relationship useful in mathematics—a **functional relation** between x and y.

Definition

A **function** consists of two sets of elements and a specified correspondence between the elements of the sets such that for each element in the first set there corresponds exactly one element in the second set.

A function may be exhibited as a collection of ordered pairs of elements, written (x, y), where x represents an element from the first set and y represents an element from the second. Most often x and y are variables taking numerical values. The two sets of elements represent all the possible numerical values that x and y may take, and the rule defining the correspondence between them is usually an equation. For example, if we state that x and y are real numbers and that

$$y = x + 2$$

then y is a function of x. Assigning a value to x (that is, choosing an element of the set of all real numbers), there corresponds exactly one value of y. When $x = 1$, then $y = 3$. When $x = -4$, then $y = -2$, and so on.

The area A of a circle is related to the radius r by the formula

$$A = \pi r^2$$

Note that if we assign a value to r, the value of A can be determined from the formula. Hence A is a function of r. Likewise, the circumference C of a circle is a function of the radius.

As a third example of a functional relation, consider a classroom

*A set is a collection of specific things.

containing 20 students. Let x represent a specific body in the classroom and let y represent a name. Is y a function of x? To answer the question we examine x and y in light of the definition. We note that each body has a name y attached to it. When x is specified, y will be uniquely determined. According to our definition, y is a function of x. Note that the defining rule for functional relations does not require that x and y take numerical values. We will encounter an important functional relation of this type in chapter 4.

Having defined a functional relation, we may now turn to **functional notation**. Mathematical writing frequently uses the same phrase or refers to a specific object many times in the course of a discussion. Unnecessary repetition wastes the reader's time, takes up valuable space, and is cumbersome to the writer. Hence the mathematician resorts to mathematical symbolism, which is, in some respects, a type of mathematical shorthand. Rather than state, "The area of a circle A is a function of the radius r," the mathematician writes $A(r)$. The expression

$$y = f(x)$$

tells us that y is a function of the variable appearing in parentheses, namely x. Note that this expression does not tell us the specific functional relation existing between y and x.

Consider this function of x:

$$y = 3x + 2$$

In functional notation, this is written as

$$f(x) = 3x + 2$$

It is understood that $y = f(x)$.

Functional notation is especially advantageous when we wish to indicate the value y of the function when x takes a specific value, say $x = 2$. For the preceding example we see that when $x = 2$,

$$y = 3x + 2 = 3(2) + 2 = 8$$

Rather than write this, we use the simpler notation

$$f(2) = 8$$

Similarly, $f(5)$ is the value of the function when $x = 5$:

$$f(5) = 3(5) + 2 = 15 + 2 = 17$$

Thus $f(5) = 17$.

To find the value of the function when x is any value, say $x = c$, we substitute the value of c for x in the equation and obtain

$$f(c) = 3c + 2$$

Example 2.1 Consider this function of x:

$$g(x) = \frac{1}{x} + 3x$$

a. Find $g(4)$.

$$g(4) = \tfrac{1}{4} + 3(4) = .25 + 12 = 12.25$$

b. Find $g(0)$.

$$g(0) = \tfrac{1}{0} + 3(0)$$

Division by 0 is mathematically undefined, so we say that $g(0)$ does not exist; that is, there is no $g(0)$.

c. Find $g(1/a)$.

$$g\left(\frac{1}{a}\right) = \left(\frac{1}{1/a}\right) + 3\left(\frac{1}{a}\right) = a + \frac{3}{a} = \frac{a^2 + 3}{a}$$

Example 2.2 Let $p(y) = (1 - a)a^y$. Find $p(2)$ and $p(3)$.

$$p(2) = (1 - a)a^2$$
$$p(3) = (1 - a)a^3$$

Example 2.3 Let $f(y) = 4$. Find $f(2)$ and $f(3)$.

$$f(2) = 4 \qquad f(3) = 4$$

Note that $f(y)$ always equals 4, regardless of the value of y. Hence when a value of y is assigned, the value of the function is determined and equals 4.

Example 2.4 Let $f(x) = x^2 + 1$ and $g(x) = x + 2$. Find $f[g(x)]$. We find $f[g(x)]$ by substituting $g(x)$ for x in the function $f(x)$. Thus

$$f[g(x)] = [g(x)]^2 + 1 = (x + 2)^2 + 1$$
$$= x^2 + 4x + 4 + 1 = x^2 + 4x + 5$$

For instance,

$$f[g(2)] = 2^2 + (4)(2) + 5 = 17$$

Exercises

2.1. For $f(y) = y - 2$, find each of the following:
a. $f(0)$ b. $f(1)$ c. $f(-2)$
d. $f(1/2)$ e. $f(a + 3)$ f. $f(1 - a)$

2.2. For $f(y) = (y^2 - 1)/y$, find each of the following:
a. $f(2)$ b. $f(-3)$ c. $f(-1)$
d. $f(x)$ e. $f(a - 1)$

2.3. For $g(y) = (1/y) - 1$, find $g(1)$ and $g(2)$.

2.4. For $p(x) = 3^x - 2$, find $p(0)$, $p(1)$, and $p(2)$.

2.5.   If $f(i) = y_i$, find $f(1)$, $f(2)$, and $f(3)$.

2.6.   If $f(y) = y^2 - 2y + 1$ and $g(y) = 1 - y$, find each of the following:

a. $f(-2)$ b. $g(1/2)$ c. $f(a + 1)$ d. $f[g(y)]$

2.3 Numerical Sequences

In statistics we are concerned with samples consisting of sets of measurements. There is a first measurement, a second, and so on. Introducing the notation of a mathematical sequence at this point provides us with a simple notation for discussing data and, at the same time, supplies our first practical application of functional notation. This will, in turn, be used in the summation notation to be introduced in section 2.4.

A set of objects a_1, a_2, a_3, a_4, ..., ordered in the sense that we can identify the first member a_1 of the set, the second a_2, and so on, is called a **sequence**. Most often, a_1, a_2, a_3, ..., called **elements of the sequence**, are numbers, but this is not a requirement. For example, the numbers

$$1, 5, 4, 8, 7, 11, 10, ...$$

form a sequence moving from left to right. Note that the elements are ordered only in their position in the sequence and need not be ordered in magnitude. Likewise, in some card games we are interested in a sequence of cards; for example,

$$10, jack, queen, king, ace$$

Although nonnumerical sequences are of interest, we will be concerned solely with numerical sequences. Specifically, data obtained in a sample from a population will be regarded as a sequence of measurements.

Since sequences will form an important part of subsequent discussions, let us turn to a shortcut method of writing sequences by using functional notation. Since the elements of a sequence are ordered in position in the sequence, it seems natural to attempt to write a formula for a typical element of the sequence as a function of its position. For example, consider the sequence

$$3, 4, 5, 6, 7, ...$$

where each element in the sequence is one greater than the preceding element. Let y be the position variable for the sequence, so that y can take values 1, 2, 3, 4,.... Then we might write a formula for the element in position y as

$$f(y) = y + 2$$

Thus the first element in the sequence is in position $y = 1$, and $f(1)$ is

$$f(1) = 1 + 2 = 3$$

Likewise, the second element is in position $y = 2$, and the second element of the sequence is

$$f(2) = 2 + 2 = 4$$

A brief check convinces us that this formula works for all elements of the sequence. Note that finding a proper formula (function) is a matter of trial and error and requires a bit of practice.

Example 2.5 For the sequence

$$1, 4, 9, 16, 25, \ldots$$

find a formula expressing a typical element in terms of a position variable y. We note that each element is the square of the position variable; hence

$$f(y) = y^2$$

Example 2.6 The formula for the typical element of the following sequence is not so obvious:

$$0, 3, 8, 15, 24, 35, \ldots$$

The typical element is

$$f(y) = y^2 - 1$$

Readers (and this includes the authors) prefer consistency in the use of mathematical notation. Unfortunately, this is not always practical. Writers are limited by the number of symbols available and also by the desire to make their notation consistent with some other texts on the subject. Hence x and y are very often used in referring to a variable, but we could just as well use i, j, k, or z. For instance, suppose we wish to refer to a set of measurements and we denote the measurements as a variable y. We might write this sequence of measurements as

$$y_1, y_2, y_3, y_4, y_5, \ldots$$

using a subscript to denote a particular element in the sequence. We are now forced to choose a new position variable (since y has been used). Suppose that we use the letter i. Then we could write a typical element as

$$f(i) = y_i$$

Note that, as previously, $f(1) = y_1$, $f(2) = y_2$, and so on.

Exercises

2.7. Let y be the position variable for a sequence and let the formula for a typical element be $3y - 1$. Give the first four elements of the sequence.

2.8. Let y be the position variable for a sequence and let the formula for a typical element be $y^2 + 2$. Give the first four elements of the sequence.

2.9. For the sequence

$$2, 3, 4, 5, 6, \dots$$

give a formula for a typical element of the sequence as a function of y, the position variable.

2.4 Summation Notation

As we will observe in chapter 3, in analyzing statistical data we often work with sums of numbers and hence we need a simple notation for indicating a sum. For instance, consider the sequence of numbers

$$1, 2, 3, 4, 5, \dots$$

and suppose we wish to discuss the sum of the squares of the first four numbers of the sequence. Using summation notation this would be written as

$$\sum_{y=1}^{4} y^2$$

Interpreting the summation notation is relatively easy. The Greek letter Σ (capital sigma), corresponding to "S" in the English alphabet (the first letter in the word "sum"), tells us to **sum elements of a sequence**. A typical element of the sequence is given to the right of the summation symbol, and the position variable, called the **variable of summation**, is shown beneath. For our example, y^2 is the typical element, y is the variable of summation, and the implied sequence is

$$1, 4, 9, 16, 25, 36, \dots .$$

Which element of the sequence should appear in the sum? The position of the first element in the sum is indicated below the summation sign, the last, above. The sum would include all elements proceeding in order from the first to last. Thus in our example the sum includes the sum of the elements beginning with the first and ending with the fourth:

$$\sum_{y=1}^{4} y^2 = 1 + 4 + 9 + 16 = 30$$

Example 2.7

$$\sum_{y=2}^{4} (y - 1) = (2 - 1) + (3 - 1) + (4 - 1) = 6$$

Example 2.8

$$\sum_{x=2}^{5} 3x = 3(2) + 3(3) + 3(4) + 3(5) = 42$$

We emphasize that the typical element is a function only of the variable of summation. If other symbols appear in the formula for the typical element, they should be regarded as constants. That is, they will not change from one term of a sequence to another.

Example 2.9

$$\sum_{i=1}^{3} (y_i - a) = (y_1 - a) + (y_2 - a) + (y_3 - a)$$

In this example note that i is the variable of summation and that it appears as a subscript in the typical element. The symbols y and a appear in the same manner in each of the three elements entering into the sum.

Example 2.10

$$\sum_{i=1}^{2} (x - i + 1) = (x - 1 + 1) + (x - 2 + 1) = 2x - 1$$

Observe that i is the variable of summation and that therefore x is a constant.

Example 2.11

$$\left(\sum_{y=2}^{4} y\right) - 1 = (2 + 3 + 4) - 1 = 8$$

Note the difference between example 2.7 and example 2.11. The quantity $(y - 1)$ is the typical element in example 2.7, whereas y is the typical element in example 2.11.

Summation notation will be used throughout this text to write formulas for the analysis of sample data. Although at this point you may not understand why we wish to calculate some of the quantities, the reasons for their use will become apparent in succeeding chapters. The following example shows how we will use summation notation in chapter 3.

Example 2.12 Suppose that you randomly select $n = 5$ electric utility companies from the totality (the population) in the United States. Their dividend rates, as a percentage of their stock prices per share, are 8.2, 6.8, 6.4, 8.1, and 7.3. Represent these five dividend rates symbolically as $y_1 = 8.2$, $y_2 = 6.8$, $y_3 = 6.4$, $y_4 = 8.1$, and $y_5 = 7.3$.

a. Find $\sum_{i=1}^{n} y_i$.
  b. Find $\dfrac{\sum_{i=1}^{n} y_i}{n}$.
  c. Find $\sum_{i=1}^{n} y_i^2$.

Solution Recall that the number of measurements in the sample is $n = 5$. Then we have the following calculations:

a. $\displaystyle\sum_{i=1}^{n} y_i = \sum_{i=1}^{5} y_i = 8.2 + 6.8 + 6.4 + 8.1 + 7.3 = 36.8$

b. $\displaystyle\frac{\sum_{i=1}^{n} y_i}{n} = \frac{\sum_{i=1}^{5} y_i}{5} = \frac{36.8}{5} = 7.36$

c. $\displaystyle\sum_{i=1}^{n} y_i^2 = \sum_{i=1}^{5} y_i^2 = (8.2)^2 + (6.8)^2 + (6.4)^2 + (8.1)^2 + (7.3)^2 = 273.34$

Exercises

2.10.   Evaluate the following summations:

a. $\displaystyle\sum_{y=0}^{5} (y - 4)$ b. $\displaystyle\sum_{y=2}^{6} (y^2 - 5)$ c. $\displaystyle\sum_{i=1}^{4} (y_i - 2)$ d. $\displaystyle\sum_{i=1}^{3} (y + 2i)$

2.11.   Suppose y is the position variable for a sequence and the formula for a typical element is $y^2 - 1$.
a. Write the first four elements of the sequence.
b. Use summation notation to write an expression for the sum of the first four terms of the sequence.
c. Find the sum of part b.

2.12.   Suppose y is the position variable for a sequence and the formula for a typical element is $(y - 3)^2$.
a. Write the first five elements of the sequence.
b. Use summation notation to write an expression for the sum of the first five terms of the sequence.
c. Find the sum of part b.

2.5   Useful Theorems Relating to Sums

Consider the summation

$$\sum_{y=1}^{3} 5$$

The typical element is 5 and it does not change. The sequence is, therefore,

$$5, 5, 5, 5, \ldots$$

and

$$\sum_{y=1}^{3} 5 = 5 + 5 + 5 = 15$$

> **Theorem 2.1**
>
> Let c be a constant (an element that does not involve the variable of summation) and y be the variable of summation. Then
>
> $$\sum_{y=1}^{n} c = nc$$

Proof

$$\sum_{y=1}^{n} c = c + c + c + c + \cdots + c$$

where the sum involves n elements. Hence

$$\sum_{y=1}^{n} c = nc$$

Example 2.13

$$\sum_{y=1}^{4} 3a = 4(3a) = 12a$$

Example 2.14

$$\sum_{i=1}^{3} (3x - 5) = 3(3x - 5)$$

Note that i is the variable of summation.

There is one other important point to consider when applying theorem 2.1. As the theorem is stated, the sum begins with the first term of the sequence (at $y = 1$). If you wish to begin the sum with other than the first term, say at $y = a$, then you would have

$$\sum_{y=a}^{n} c = (n - a + 1)c$$

A second theorem is illustrated by using example 2.8. We note that 3 is a common factor in each term. Therefore,

$$\sum_{y=2}^{5} 3y = 3(2) + 3(3) + 3(4) + 3(5)$$

$$= 3(2 + 3 + 4 + 5) = 3 \sum_{y=2}^{5} y$$

It would appear that the summation of a constant times a variable is equal to the constant times the summation of the variable.

Theorem 2.2

Let c be a constant. Then

$$\sum_{i=1}^{n} cy_i = c \sum_{i=1}^{n} y_i$$

Proof

$$\sum_{i=1}^{n} cy_i = cy_1 + cy_2 + cy_3 + \cdots + cy_n$$

$$= c(y_1 + y_2 + \cdots + y_n) = c \sum_{i=1}^{n} y_i$$

The constant c need only be a term whose value does not depend on the variable of summation. For instance, in the expression

$$\sum_{y=1}^{4} x^2 y$$

the term x^2 is not a function of y, the variable of summation. Therefore, it can be considered a constant. Thus

$$\sum_{y=1}^{4} x^2 y = x^2 \sum_{y=1}^{4} y = x^2(1 + 2 + 3 + 4) = 10x^2$$

Theorem 2.3

$$\sum_{i=1}^{n} (x_i + y_i + z_i) = \sum_{i=1}^{n} x_i + \sum_{i=1}^{n} y_i + \sum_{i=1}^{n} z_i$$

Proof

$$\sum_{i=1}^{n} (x_i + y_i + z_i) = x_1 + y_1 + z_1 + x_2 + y_2 + z_2 + x_3 + y_3$$

$$+ z_3 + \cdots + x_n + y_n + z_n$$

Regrouping, we have

$$\sum_{i=1}^{n} (x_i + y_i + z_i)$$
$$= (x_1 + x_2 + \cdots + x_n) + (y_1 + y_2 + \cdots + y_n) + (z_1 + z_2 + \cdots + z_n)$$
$$= \sum_{i=1}^{n} x_i + \sum_{i=1}^{n} y_i + \sum_{i=1}^{n} z_i$$

In words, we say that the summation of a typical element which is itself a sum of a number of terms is equal to the sum of the summations of the terms.

Theorems 2.1, 2.2, and 2.3 can be used jointly to simplify summations. Consider the following examples.

Example 2.15

$$\sum_{x=1}^{3} (x^2 + ax + 5) = \sum_{x=1}^{3} x^2 + \sum_{x=1}^{3} ax + \sum_{x=1}^{3} 5 = \sum_{x=1}^{3} x^2 + a \sum_{x=1}^{3} x + 3(5)$$
$$= (1 + 4 + 9) + a(1 + 2 + 3) + 15 = 6a + 29$$

Example 2.16

$$\sum_{i=1}^{4} (x^2 + 3i) = \sum_{i=1}^{4} x^2 + \sum_{i=1}^{4} 3i = 4x^2 + 3 \sum_{i=1}^{4} i$$
$$= 4x^2 + 3(1 + 2 + 3 + 4) = 4x^2 + 30$$

Exercises

Utilize theorems 2.1, 2.2, and 2.3 to simplify and evaluate the following summations.

2.13. $\displaystyle\sum_{x=1}^{4} 3x$ 2.14. $\displaystyle\sum_{i=1}^{4} x_i$ 2.15. $\displaystyle\sum_{x=1}^{4} (x^2 + 2)$

2.16. $\displaystyle\sum_{y=2}^{5} (y^2 - 3y + 2)$ 2.17. $\displaystyle\sum_{i=1}^{4} (2y_i - 5)$ 2.18. $\displaystyle\sum_{i=1}^{n} (y_i - 3)^2$

2.6 Summary

Two types of mathematical notation have been presented: functional notation and summation notation. The former is used in summation notation to express the typical element as a function of the variable of summation. Other uses for functional notation will be discussed in chapter 4. Summation notation will be employed in chapter 3 and succeeding chapters.

Supplementary Exercises

2.19. For $f(y) = 4y + 3$, find each of the following:
a. $f(0)$ b. $f(1)$ c. $f(2)$ d. $f(-1)$
e. $f(-2)$ f. $f(a^2)$ g. $f(-a)$ h. $f(1 - y)$

2.20. For $f(y) = (y - 2)^2$, find each of the following:
a. $f(2)$ b. $f(-3)$ c. $f(-1)$
d. $f(x)$ e. $f(a - 1)$

2.21. For $f(x) = x^2 - x + 1$, find the following:
a. $f(-2)$ b. $f(a + b)$

2.22. If $g(y) = (y^2 - 1)/(y + 1)$, find the following:
a. $g(1)$ b. $g(-1)$ c. $g(4)$ d. $g(-a^2)$

2.23. If $p(x) = (1 - a)^x$, find $p(0)$ and $p(1)$.

2.24. If $f(x) = 3x^2 - 3x + 1$ and $g(x) = x - 3$, find the following:
a. $f(1/2)$ b. $g(-3)$ c. $f(1/x)$ d. $f[g(x)]$

2.25. If $h(x) = 2$, find the following:
a. $h(0)$ b. $h(1)$ c. $h(2)$

Utilize theorems 2.1, 2.2, and 2.3 to simplify and evaluate the following summations.

2.26. $\displaystyle\sum_{y=1}^{3} y^3$ 2.27. $\displaystyle\sum_{y=1}^{3} 6$

2.28. $\displaystyle\sum_{x=2}^{3} (1 + 3x + x^2)$ 2.29. $\displaystyle\sum_{i=1}^{5} (x^2 + 2i)$

2.30. $\displaystyle\sum_{y=0}^{5} (x^2 + y^2)$ 2.31. $\displaystyle\sum_{x=0}^{2} (x^3 + 2ix)$

2.32. $\displaystyle\sum_{x=1}^{4} (x + xy^2)$ 2.33. $\displaystyle\sum_{i=1}^{2} (y_i - i)$

2.34. $\displaystyle\sum_{i=1}^{n} (y_i - a)$ 2.35. $\displaystyle\sum_{i=1}^{n} (y_i - a)^2$

Using the following set of measurements, calculate the sums in exercises 2.36 through 2.41.

i	1	2	3	4	5	6	7	8	9	10	11	12	13
y_i	3	12	10	-6	0	11	2	-9	-5	8	-7	4	-5

2.36. $\displaystyle\sum_{i=1}^{13} 2y_i$ 2.37. $\displaystyle\sum_{i=1}^{13} (2y_i - 5)$ 2.38. $\displaystyle\sum_{i=3}^{10} y_i^2$

2.39. $\displaystyle\sum_{i=1}^{10} (y_i^2 + y_i)$ 2.40. $\displaystyle\sum_{i=1}^{13} (y_i - 2)^2$ 2.41. $\displaystyle\sum_{i=1}^{13} y_i^2 - \frac{1}{13}\left(\sum_{i=1}^{13} y_i\right)^2$

Verify the following identities. Each identity is a shortcut formula that will be used in later chapters. The symbols $\bar{x}$ and $\bar{y}$ appearing in these identities have the following definitions:

$$\bar{x} = \frac{\sum_{i=1}^{n} x_i}{n} \qquad \bar{y} = \frac{\sum_{i=1}^{n} y_i}{n}$$

2.42. $\displaystyle\sum_{i=1}^{n} (y_i - \bar{y})^2 = \sum_{i=1}^{n} y_i^2 - \frac{\left(\sum_{i=1}^{n} y_i\right)^2}{n}$

2.43. $\displaystyle\sum_{i=1}^{n} (x_i - \bar{x})(y_i - \bar{y}) = \sum_{i=1}^{n} x_i y_i - \frac{\left(\sum_{i=1}^{n} x_i\right)\left(\sum_{i=1}^{n} y_i\right)}{n}$

2.44. $\displaystyle\frac{\sum_{i=1}^{n} (y_i - \bar{y})^2}{n - 1} = \frac{1}{n - 1}\left[\sum_{i=1}^{n} y_i^2 - \frac{1}{n}\left(\sum_{i=1}^{n} y_i\right)^2\right]$

2.45. $\displaystyle\frac{\sum_{i=1}^{n} (x_i - \bar{x})(y_i - \bar{y})}{\sum_{i=1}^{n} (x_i - \bar{x})^2} = \frac{n\sum_{i=1}^{n} x_i y_i - \left(\sum_{i=1}^{n} x_i\right)\left(\sum_{i=1}^{n} y_i\right)}{n\sum_{i=1}^{n} x_i^2 - \left(\sum_{i=1}^{n} x_i\right)^2}$

2.46.   Suppose the numbers 5, 1, 6, 2, 1, 3 represent a random sample of $n = 6$ measurements from a population. Let $y_1 = 5$, $y_2 = 1$, $y_3 = 6$, $y_4 = 2$, $y_5 = 1$, and $y_6 = 3$.

a. Find $\displaystyle\sum_{i=1}^{n} y_i$. \qquad b. Find $\displaystyle\sum_{i=1}^{n} y_i^2$.

2.47.   Use the results of exercise 2.46 to calculate the quantity indicated in exercise 2.42.

2.48.   A random sampling by a department store of monthly inventory losses due to theft gave the following measurements (in thousands of dollars): 3.9, 3.8, 4.6, 3.2, 4.3, 4.1, and 3.8. Let $y_1 = 3.9$, $y_2 = 3.8$, ... , $y_7 = 3.8$.

a. Find $\displaystyle\sum_{i=1}^{n} y_i$. \qquad b. Find $\displaystyle\sum_{i=1}^{n} y_i^2$.

2.49.   Use the results of exercise 2.48 to calculate the quantity indicated in exercise 2.44.

chapter objectives

GENERAL OBJECTIVES Methods for describing sets of data are needed so that (1) you will better understand data sets and thus more easily construct theories about the phenomena they represent and (2) you will be able to phrase an inference (make descriptive statements) about a population based on sample data. Consequently, the objective of this chapter is to find a simple compact method for describing a set of data.

SPECIFIC OBJECTIVES

1. To present a variety of graphical methods for describing data, along with their advantages and possible shortcomings.
                    Sections 3.2 through 3.4

2. To develop the notion of a relative frequency distribution (or relative frequency histogram).
                    Section 3.2

3. To present important numerical descriptive methods, particularly measures of central tendency and variation.
                    Sections 3.6, 3.7

4. To provide you with a method for interpreting a standard deviation so that you can construct a mental picture of the frequency distribution for a set of data.
                    Section 3.8

5. To present an easy way to calculate the standard deviation of a set of measurements.
                    Section 3.9

chapter three

Describing Sets of Measurements

3.1 Introduction

You will recall that the objective of modern statistics is to make inferences about a large body of data, a population, based on information contained in a sample. But one difficulty arises: How do we phrase the inferences? How do we describe a set of measurements, whether they be the sample or the population? If the population were before us, how would we describe this large set of measurements?

Numerous texts have been devoted to the methods of descriptive statistics—that is, the methods of describing sets of numerical data. These methods can be categorized essentially as graphical methods and numerical methods. In this text we will restrict our discussion to a few graphical and numerical methods that are useful not only for descriptive purposes but also for statistical inference. Some common numerical methods have been excluded from our discussion to preserve continuity. Such methods are redundant in the presence of high-speed electric desk calculators and electronic computers and would contribute little, if any, to the main objective of our study. If you are interested in descriptive statistics not contained within this chapter, refer to the references listed at the end of the chapter.

A remark is necessary at this point. The graphical methods that follow can be applied either to a set of population measurements or to sample measurements, without making a specific distinction as to which case is involved. Numerical descriptive measures also apply to both population and sample measurements, but different symbols are used to indicate whether the measure was obtained from the set of population measurements or from a sample set.

3.2 Frequency Distributions

It seems natural to introduce appropriate graphical and numerical methods for describing sets of data by considering a set of real data. Individuals and organizations attempt to maintain an investment portfolio that provides for a maximum of return at tolerable levels of risk. One measure of the potential return and inherent risk of a security is its price-earnings ratio. Generally, securities with a low price-earnings ratio are preferred to those with a high ratio. The data presented in table 3.1 represent the price-earnings ratios for 25 different common stocks.

Table 3.1 *Price-earnings ratios for 25 common stocks*

20.5	19.5	15.6	24.1	9.9
15.4	12.7	5.4	17.0	28.6
16.9	7.8	23.3	11.8	18.4
13.4	14.3	19.2	9.2	16.8
8.8	22.1	20.8	12.6	15.9

A cursory examination of table 3.1 reveals that the largest price-earnings ratio in the sample is 28.6 and the smallest is 5.4. You might ask how the other 23 price-earnings ratios are distributed over the interval from 5.4 to 28.6. To answer this question we divide the interval into an arbitrary number of equal subintervals, the number being determined by the amount of data available. (As a rule of thumb, the number of subintervals chosen would range from 5 to 20; the larger the amount of data available, the more subintervals employed.) For the preceding data we might use the subintervals 5.00 to 8.99, 9.00 to 12.99, 13.00 to 16.99, ... Note that the points dividing the subintervals have been chosen so that it is impossible for a measurement to fall on the point of division, thus eliminating any ambiguity in allocating a particular measurement. The subintervals, called *classes* in statistical language, form cells, or pockets. We wish to determine the manner in which

Table 3.2 *Relative frequencies for the 25 price-earnings ratios*

CLASS, i	CLASS BOUNDARIES	TALLY	CLASS FREQUENCY, f_i	CLASS RELATIVE FREQUENCY
1	5.00–8.99	|||	3	3/25
2	9.00–12.99	|||||	5	5/25
3	13.00–16.99	||||| ||	7	7/25
4	17.00–20.99	|||||		6	6/25
5	21.00–24.99	|||	3	3/25
6	25.00–28.99			1	1/25
Totals			25	1

the measurements are distributed in the classes. A tally of the data from table 3.1 is presented in table 3.2.

The 25 measurements fall in one of six classes, which, for purposes of identification, we will number from 1 to 6. The identification number appears in the first column of table 3.2 and the corresponding class boundaries are given in the second column. The third column of the table is used for the tally, a mark entered opposite the appropriate class for each measurement falling in the class. For example, 3 of the 25 measurements fall in class 1, 5 in class 2, and so on. The number of measurements falling in a particular class, say class i, is called the *class frequency* and is designated by the symbol f_i. The class frequency is given in the fourth column of table 3.2. The last column of the table presents the fraction of the total number of measurements falling in each class. We call this the class relative frequency. If we let n represent the total number of measurements—for instance, in our example $n = 25$—then the relative frequency for the ith class is f_i divided by n:

$$\text{relative frequency} = \frac{f_i}{n}$$

The resulting tabulation can be presented graphically in the form of a *frequency histogram*, as in figure 3.1. In a frequency histogram, rectangles are constructed over each class interval, their height being proportional to the number of measurements (class frequency) falling in each class interval. When we look at the frequency histogram, we can see at a glance how the price-earnings ratios are distributed over the interval.

It is often more convenient to modify the frequency histogram by plotting class relative frequency rather than class frequency. A relative frequency histogram is presented in figure 3.2. Statisticians rarely make a distinction between the frequency histogram and the relative frequency histogram and

Figure 3.1 *Frequency histogram*

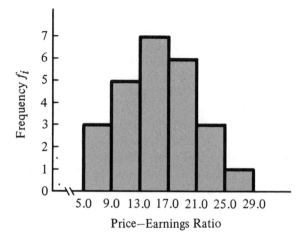

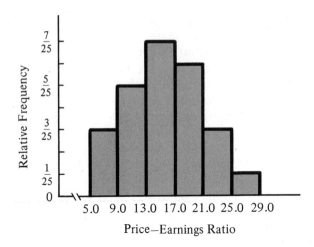

Figure 3.2 *Relative frequency histogram*

refer to either as a frequency histogram or simply a histogram. If corresponding values of frequency and relative frequency are marked along the vertical axes of the graphs, then the frequency and relative frequency histograms are identical (compare figures 3.1 and 3.2).

Although we are interested in describing the set of $n = 25$ measurements, we are much more interested in the population from which the sample was drawn. We might view the 25 price-earnings ratios as a representative sample drawn from a population consisting of all the firms whose stock was listed by the New York stock exchange in 1976. If we had the price-earnings ratios for all the firms in our population, we could construct a population relative frequency histogram. However, gathering the data from the entire population may be very costly and time-consuming. Thus a representative sample may be the best, and only, information we have to describe the population.

Let us consider the relative frequency histogram for the sample in greater detail. What fraction of the firms had price-earnings ratios equal to 17.0 or greater? Checking the relative frequency histogram, we see that the fraction involves all classes to the right of 17.0. Using table 3.2 we see that 10 companies had price-earnings ratios greater than or equal to 17.0. Hence the fraction is $\frac{10}{25}$, or 40%. We note that this is also the percentage of the total area of the histogram in figures 3.1 and 3.2 that is to the right of 17.0.

Suppose that we write each of the 25 price-earnings ratios on a piece of paper, place them in a hat, and then draw one piece of paper from the hat. What is the chance that this paper contains a price-earnings value greater than or equal to 17.0? Since 10 of the 25 pieces of paper are marked with numbers greater than or equal to 17.0, we say that we have 10 chances out of 25. Or we might say that the probability is $\frac{10}{25}$. You have undoubtedly encountered the word "probability" in ordinary conversation, and we are content to defer a definition of it and a discussion of its significance until chapter 4.

Let us now look at the population from which the sample was drawn. What fraction of the firms in the population of all New York stock exchange firms for 1976 had price-earnings ratios greater than or equal to 17.0? If we possessed the relative frequency histogram for the population, we could give the exact answer to this question. Unfortunately, we are forced to make an inference from our sample information. Our estimate for the true population fraction or proportion, based on the sample information, would likely be $\frac{10}{25}$, or 40%. Without knowledge of the population relative frequency histogram, we would infer that the population histogram is similar to the sample histogram and that approximately 40% of the price-earnings ratios in the population are greater than or equal to 17.0. Most likely this estimate would differ from the true population percentage. We will examine the magnitude of this error in chapter 8.

The relative frequency histogram is often called a frequency distribution because it shows the manner in which the data are distributed along the abscissa (horizontal axis) of the graph. We note that the rectangles constructed above each class are subject to two interpretations. They represent the fraction of observations falling in a given class. Also, if a measurement is drawn from the data, a particular class relative frequency is also the chance or probability that the measurement will fall in that class. The most significant feature of the sample frequency histogram is that it provides information about the population frequency histogram that describes the population. We would expect the two frequency histograms, sample and population, to be similar. Such is the case. The degree of resemblance will increase as more and more data are added to the sample. If the sample were enlarged to include the entire population, the sample and population would be the same and the histograms would be identical.

In the preceding discussion we showed you how you could construct a frequency distribution for the price-earnings ratio data of table 3.1, and we explained how such a distribution could be interpreted. Before concluding this topic we will summarize the principles you should employ in constructing a frequency distribution for a set of data. (See the box on the next page.)

To illustrate the use of these principles, suppose we wish to group the 36 incomes of the employees of a small firm into five classes, where the smallest income is $5,500 and the largest is $29,500. Applying our rule, we have

$$\text{class width} = \frac{\$29,500 - \$5,500}{5} = \$4,800$$

It might be more convenient to use $5,000 as our class width. The class boundaries would then be

$5,000-$9,999
$10,000-$14,999
$15,000-$19,999
$20,000-$24,999
$25,000-$29,999

Principles for Constructing a Frequency Distribution

1. **Determine the number of classes.** It is usually best to have from 5 to 20 classes. The larger the amount of data available, the more the classes that should be employed. If the number of classes is too small, we might be concealing important characteristics of the data by grouping. If we have too many classes, empty classes may result and the distribution will be meaningless. The number of classes should be determined from the amount of data present and the uniformity of the data. A small sample would require fewer classes.

2. **Find the range and determine the class width.** As a general rule for finding the class width, divide the difference between the largest and the smallest measurements by the number of classes desired and add enough to the quotient to arrive at a convenient figure for class width. All classes, with the possible exception of the smallest and largest classes, should be of equal width. This allows us to make uniform comparisons of the class frequencies.

3. **Locate the class boundaries.** Begin with the lowest class so that you include the smallest measurement. Then add the remaining classes. Class boundaries should be chosen so that it is impossible for a measurement to fall on a boundary.

Here none of our principles for constructing a frequency distribution is violated. However, if our data were widely dispersed, we might wish to make the lower or upper class open-ended. For instance, if the president's $100,000 salary is included as the 37th income value, our largest class would be $25,000 and up. If we were to apply the class width rule, empty classes would occur and would confuse the interpretation of our results.

Exercises

3.1.   A city housing commission would like to investigate the rental rate structure of a city. Comment on the effect of using each of the following class interval widths to classify $n = 100$ monthly rental rates selected from apartments throughout the city if these monthly rates range from $120 to $550:

$$\$.50, \$1.00, \$10.00, \$25.00, \$50.00, \$100.00$$

3.2.   Three different designations have been proposed for the classification of the hourly wages of cabinetmakers in a certain New England state (see the table). Criticize the use of each designation.

DESIGNATION I	DESIGNATION II	DESIGNATION III
$ 0–3.00	$ 0–3.50	$ 0–under 3.00
3.00–6.00	3.51–7.00	4.00–under 7.00
6.00–9.00	7.01–10.00	8.00–under 11.00
9.00–12.00	10.01–15.00	12.00–under 15.00
12.00–15.00	over 15.00	over 15.00

3.3.   The Federal Trade Commission (FTC) has issued an order prohibiting automobile manufacturers from making unsubstantiated gasoline mileage claims for their vehicles.* The FTC now requires manufacturers to base advertised gasoline mileage ratings on many test runs conducted under a variety of highway, traffic, and driver conditions in order to best simulate actual driving conditions. In test runs of $n = 50$ cars, an automobile manufacturer records the following gasoline mileage ratings for the company's subcompact model:

27.9	29.3	31.8	22.5	34.2
34.2	32.7	26.5	26.4	31.6
35.6	31.0	28.0	33.7	32.0
28.5	27.5	29.8	31.2	28.7
30.0	28.7	33.2	30.5	27.9
31.2	29.5	28.7	23.0	30.1
30.5	31.3	24.9	26.8	29.9
28.7	30.4	31.3	32.7	30.3
33.5	30.5	30.6	35.1	28.6
30.1	30.3	29.6	31.4	32.4

a. Construct a relative frequency histogram for these data using five classes of equal width.

b. The automobile manufacturing company has designed this subcompact to achieve a gasoline mileage rating of at least 30 mpg. Give the fraction of cars in the sample that achieve the company's test standard.

3.4.   During times of economic recession, dentists and physicians tend to incur special problems with account collection. The number of days since billing for 30 accounts receivable of a physician is shown below.

17	57	10	35	26	3
21	11	7	72	5	86
6	20	105	40	14	42
12	32	28	13	19	28
45	8	19	21	38	20

a. Construct a relative frequency histogram for these data.

b. Accounts at least 6 weeks (42 days) delinquent are considered uncollectible and are turned over to a collection agency. What fraction of the 30 physician's accounts should be turned over to the collection agency?

3.3 Other Graphical Methods

Data collected from different time periods or geographical areas often are best presented by using statistical tables, charts, or pictograms. The principles behind the construction and use of each are discussed in this section.

A statistical table is a classified or subdivided frequency distribution comparing the frequencies or the relative frequencies for samples drawn from two or more different populations. The populations might correspond to different time periods, different geographical areas, different but related firms, different areas within a firm, and so on. Within each sample the classifications must be the same so that we can perform a meaningful cross analysis of the data.

FTC News Summary, 15 August 1975.

Bar charts, line charts, and ratio charts are designed to serve as visual summaries of our data. Usually line and ratio charts are plots of points tracing a firm's profits, sales, or productivity, or their change over time. Bar charts and pictograms are pictorial frequency histograms. Many other types of graphical and pictorial methods are useful for the business statistician, but time limits our discussion within this chapter.

Table 3.3 *Analysis of employees for a small manufacturing firm*

	1974	1975	1976	1977
Total Number of Employees	100	115	110	150
College Graduates	40	44	42	58
Male	37	40	39	48
Female	3	4	3	10
High School Graduates	60	71	68	92
Male	50	58	56	69
Female	10	13	12	23

Table 3.3, an example of a statistical table, shows the breakdown of the employees of a small midwestern manufacturing firm for the years 1974 through 1977. The classifications for each of the four years are the same, so we can make meaningful comparisons among the years. Entries in the table give the total number of employees who are college or high school graduates for each year. Notice that these two numbers add to the total number of employees for a given year. The table also gives a breakdown of college and high school graduates into the numbers of males and females.

Percentages are also used as entries in statistical tables. Such tables should be examined with care because comparisons between corresponding entries for different samples can be misleading if the number of measurements differ from sample to sample. For example, two corresponding entries in a table might both show 50%, but the first entry might represent 100 out of a total of 200 and the second might represent 1 out of a total of 2. The first entry (based on a total of 200) is certainly more meaningful than the second (based on a total of 2). Consequently, the two percentages should be visually compared with caution.

Figure 3.3 shows how a bar chart could be used to display pictorially some of the employee data of table 3.3. Bar charts are not ordinarily as finely subdivided as a classification table since the extra partitions tend to clutter the appearance of a chart. The intention is to produce a chart that is easy to read and that provides a quick analysis of the data.

A bar chart can be constructed in other ways had we chosen to do so. We could have illustrated the same information by drawing three separate rectangles (bars) for each year, showing separately the number of male employees, the number of female employees, and the total number of employees. The type of bar chart employed is not important as long as the chart is factual and easy to interpret.

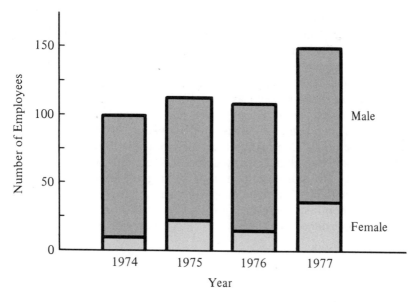

Figure 3.3 *Bar graph*

Figure 3.4 *New housing starts and new available rental units in the United States from 1965 through 1974*

Source: U.S. Department of Commerce, Census Bureau.

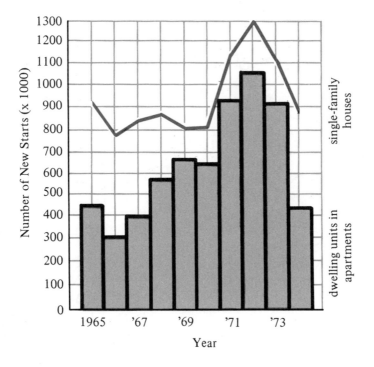

Figure 3.4 illustrates the use of a bar chart and a line chart together. The bars indicate the number of rental units built each year and the line indicates the number of houses built each year for the years 1965 through 1974. Note the growth in both new rental units and new housing starts over the years 1965 to 1972 and the decline in the prerecession years of 1973 and 1974. Many readers, such as loan companies, commercial banks, and real estate agencies, would be interested not so much in the exact number of rental units and homes built as in the relation of the total of one to the total of the other. Contrasting representations like those shown in figure 3.4 (bar chart and line chart) clearly depict this relation.

Bar charts are most useful in representing the total amount of some quantity for each of a given number of years or for each of a group of categories. In contrast, **pie charts** are useful in showing how a single total quantity is apportioned to a group of categories. For example, figure 3.5

Figure 3.5 *Principal commodities carried by water, 1972*

Source: Army Corps of Engineers, *Waterborne Commerce of the United States*, part 5, 1972.

Total Commerce

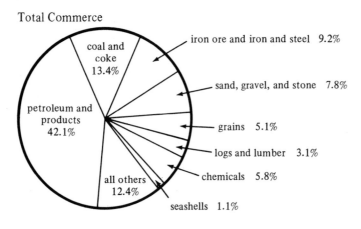

Foreign Commerce   coal and coke   10.5%   Domestic Commerce

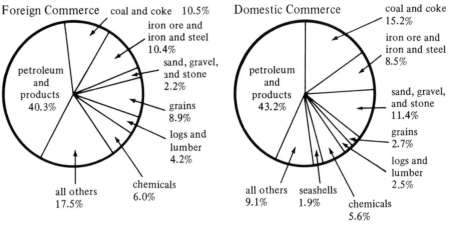

shows the breakdown of the principal commodities transported by water in 1972 into the respective product categories. The pie chart at the top shows the breakdown of total commerce transported by water. The chart clearly shows that petroleum was the major product transported by water (42.1% of the total) while grains represented only a small portion (5.1%) of the total. Similar interpretations can be made for the two pie charts showing the breakdowns for foreign commerce and domestic commerce. The primary usefulness of the pie chart is that it allows you to see quickly how much of a total is represented by each subdivision of the total.

A pie chart is easy to construct if you remember that the total pie contains 360 degrees and that this angle must correspond to 100% of the total that is represented. For example, consider the Total Commerce pie chart of figure 3.5. Calculate the angular portion of the pie that should be allocated to petroleum and products. Since this portion represents 42.1% of the total commerce transported by water, it must be allocated 42.1% of the total 360 degrees of the "pie." This angular portion is

$$\frac{42.1\%}{100\%} (360 \text{ degrees}) = 151.56 \text{ degrees}$$

The portions of the pie chart assigned to each of the other products are calculated in a similar manner. The total of the pie-shaped portions corresponding to the various components covers 360 degrees of the pie chart.

The main purpose of any chart is to give a quick, easy-to-read-and-interpret pictorial representation of data. The type of chart or graphical presentation used and the format of its construction are incidental to its main purpose. A well-designed graphical presentation can effectively communicate the data's message in a language readily understood by almost everyone.

Exercises

3.5. The data in the accompanying table represent the personal consumption expenditures, by major categories, in the United States for selected years from 1929 through 1969 (in billions of dollars):

	YEAR			
EXPENDITURE CATEGORY	1929	1949	1959	1969
total durables	9.2	24.6	44.3	90.0
total nondurables	37.7	94.6	146.6	245.8
total services	30.3	54.6	120.3	241.6

Source: U.S. Department of Commerce, Office of Business Economics, *Survey of Current Business,* July 1970.

a. Construct a bar graph to represent these data.

b. Use the chart to discuss changes in the three different expenditure categories over the period from 1929 through 1969.

3.6. The estimated costs, in cents per mile, of operating a subcompact 1974 automobile for 10 years or 100,000 miles are shown below. Construct a pie chart to represent these estimated operating costs.

Depreciation	2.36
Repairs and Maintenance	2.12
Gasoline	1.82
Insurance	1.47
Garaging, Parking	1.96
Taxes and Fees	.92

Source: U.S. Department of Transportation, Federal Highway Administration, *Cost of Operating an Automobile*, April 1974.

 3.7.   The following distribution shows the number of shareholders of a corporation according to the number of shares owned:

SHARES	SHAREHOLDERS
under 1000	486
1000–1999	372
2000–2999	210
3000–3999	117
4000–4999	43
5000–5999	27
6000 and over	13

a. Construct a relative frequency histogram for these data.
b. Construct a bar chart to depict these data.

3.4 Cheating with Charts

Although graphical descriptive techniques are very useful for describing data, the figures must be interpreted with care. It is very easy to construct a figure that may lead an unsuspecting reader to the wrong conclusions. For example, one of the simplest methods to lead a reader astray is to shrink or stretch the axes of a graph.

To illustrate, suppose that the number of near-fatal collisions between aircraft per month at a major airport is recorded as 13, 14, 14, 15, and 15 for the period January through May. If you want this growth to appear small (perhaps you represent the Civil Aeronautics Administration), you show the results using the frequency polygon of figure 3.6. The growth is apparent, but it does not appear to be very great. If you want the growth to appear large (perhaps you belong to the Citizens' Safety Group), look at the graph of the same data in figure 3.7. The vertical axis is stretched and does not include 0. Note the impression of a substantial rise that is indicated by the steeper slope.

Another way to achieve the same effect—to decrease or increase a slope—is to stretch or shrink the horizontal axis. Of course, you are sometimes limited in the amount of shrink or stretch you can apply and still achieve

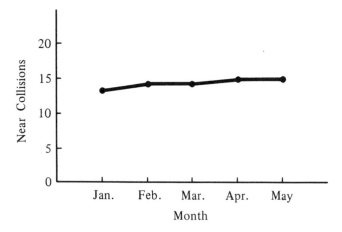

Figure 3.6 *Number of near collisions per month*

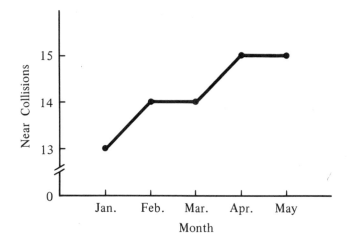

Figure 3.7 *Number of near collisions per month*

a picture that appears reasonable to the viewer. For example, you could not shrink or stretch the horizontal axes of figures 3.6 and 3.7 very much because of the limited number of data points ($n = 5$).

Shrinking or stretching axes to increase the slopes in bar graphs, histograms, frequency polygons, or other figures usually catches the hasty reader off guard; the distortions are apparent only if you look closely at the axes. The important point, however, is that increases or decreases in responses are judged large or small depending on the arbitrary importance to the observer of the changes, not on the slopes shown in graphic representations.

The preceding examples provide a simple illustration of how the truth may be distorted, accidently or purposely, using graphical descriptive methods. To protect yourself, carefully examine graphs and charts and note the scales of measurement. Check to see whether the axes are broken and ask yourself whether the observed changes in the variable described are meaningful from a practical point of view. For example, are you interested in the numerical or the percentage changes in the variable from one condition to another?

To summarize, draw your conclusions with extreme caution. Remember that charts and graphs are sometimes constructed to create a certain illusion and to tell a story that is far from the truth.

3.5 Numerical Descriptive Methods

Graphical methods are extremely useful for conveying a rapid general description of collected data and for presenting data. This statement supports, in many respects, the saying that a picture is worth a thousand words. There are, however, limitations to the use of graphical techniques for describing and analyzing data. For instance, suppose we wish to discuss our data before a group of people and must describe the data verbally. Unable to present the histogram visually, we would be forced to use other descriptive measures that would convey to the listeners a mental picture of the histogram.

A second and not so obvious limitation of the histogram and other graphical techniques is that they are difficult to use for purposes of statistical inference. Presumably we use the sample histogram to make inferences about the shape and position of the population histogram which describes the population and is unknown to us. Our inference is based on the correct assumption that some degree of similarity exists between the two histograms, but we are then faced with the problem of measuring the degree of similarity. We know when two figures are identical, but this situation is not likely to occur in practice. If the sample and population histograms differ, how can we measure the degree of difference or, expressing it positively, the degree of similarity? To be more specific, we might wonder about the degree of similarity between the histogram in figure 3.2 and the frequency histogram for the population of price-earnings ratios from which the sample was drawn. Although these difficulties are not insurmountable, we prefer to seek other descriptive measures which readily lend themselves for use as predictors of the shape of the population frequency distribution.

The limitations of the graphical method of describing data can be overcome by the use of numerical descriptive measures. Thus we would like to use the sample data to calculate a set of numbers that will convey to the statistician a good mental picture of the frequency distribution and will be useful in making inferences concerning the population.

Definition

Numerical descriptive measures computed from population measurements are called **parameters**; those computed from sample measurements are called **statistics**.

3.6 Measures of Central Tendency

In constructing a mental picture of the frequency distribution for a set of measurements, we would most likely envision a histogram similar to that shown in figure 3.2 for the data on price-earnings ratios. One of the first descriptive measures of interest is a **measure of central tendency,** that is, a measure of the center of the distribution. We note that the price-earnings data ranged from a low of 5.4 to a high of 28.6, the center of the histogram being located in the vicinity of 16.0. Let us now consider some definite rules for locating the center of a distribution of data.

One of the most common and useful measures of central tendency is the arithmetic average of a set of measurements. This is also often referred to as the *arithmetic mean,* or simply the *mean,* of a set of measurements.

Definition

The **arithmetic mean** of a set of n measurements $y_1, y_2, y_3, ..., y_n$ is equal to the sum of the measurements divided by n.

Recall that we are always concerned with both the sample and the population, each of which has a mean. To distinguish between the two, we will use the symbol $\bar{y}$ (y-bar) for the mean of the sample and μ (Greek letter mu) for the mean of the population. Since the n sample measurements are denoted by the symbols, $y_1, y_2, y_3, ..., y_n$, a formula for the sample mean is as given in the box.

Sample Mean

$$\bar{y} = \frac{\sum\limits_{i=1}^{n} y_i}{n}$$

Checking, you will find that $\bar{y}$ falls near the middle of the set of sample measurements from which $\bar{y}$ is computed.

Example 3.1 Find the mean of the measurements 2, 9, 11, 5, 6.

Solution Substituting the measurements into the formula, we have

$$\bar{y} = \frac{\sum\limits_{i=1}^{n} y_i}{n} = \frac{2 + 9 + 11 + 5 + 6}{5} = 6.6$$

We have seen that $\bar{y}$ is used to locate the center of a set of sample measurements. A more important use of $\bar{y}$ is as an estimator (predictor) of the value of the unknown population mean μ. For example, the mean of the sample given in table 3.1 is

$$\bar{y} = \frac{\sum_{i=1}^{n} y_i}{n} = \frac{400}{25} = 16.0$$

Note that this falls approximately in the center of the set of sample measurements. The mean of the entire population of price-earnings ratios is unknown to us, but if we were to estimate its value, our estimate of μ would be 16.0.

A second measure of central tendency is the *median*.

Definition

The **median** of a set of n measurements $y_1, y_2, y_3, ..., y_n$ is defined to be the value of y that falls in the middle when the measurements are arranged in order of magnitude. If the number of measurements is even, we choose the median as the value of y halfway between the two middle measurements.

Example 3.2 Consider the sample measurements

$$9, 2, 7, 11, 14$$

Arranging the measurements in order of magnitude, 2, 7, 9, 11, 14, we see that 9 is the median.

Example 3.3 Consider the sample measurements

$$9, 2, 7, 11, 14, 6$$

Arranged in order of magnitude, 2, 6, 7, 9, 11, 14, we choose the median as the value halfway between 7 and 9, which is 8.

Our rule for locating the median may seem a bit arbitrary in the case of an even number of measurements. But recall that we calculate the sample median either for descriptive purposes or as an estimator of the population median. If it is used for descriptive purposes, we may be as arbitrary as we please. If it is used as an estimator of the population median, "the proof of the pudding is in the eating." A rule for locating the sample median is poor or good depending on whether it tends to give a poor or good estimate of the population median.

A third measure sometimes used as a measure of central tendency is the *mode*.

Definition

The **mode** of a set of n measurements y_1, y_2, y_3, ..., y_n is defined to be the value of y occurring with the greatest frequency.

Example 3.4 Consider the sample measurements

$$9, 2, 7, 11, 14, 7, 2, 7$$

The value 7 occurs three times, 2 occurs twice, and the others, once each. Thus 7 is the mode of our sample measurements.

The mode is not a widely used measure of central tendency but it is useful in business planning for identifying those product sizes in greatest demand. For example, a shirt or dress manufacturer is interested in the sizes in greatest demand. Similarly, in scheduling the production of a drug, a manufacturer is interested in the drug potency most commonly prescribed by physicians. All these measurements are best described by the mode.

The mode, the y-value with the greatest frequency, might appear as shown in figure 3.8. (Note that the relative frequency histogram corresponding to a large quantity of data often will appear, for all practical purposes, as a smooth curve, as shown in figure 3.8.)

The relationships among the mean (μ), median (Md), and mode (Mo) can be seen by examining figure 3.9. For a symmetric frequency distribution (one for which values of the variable that are equidistant from the mean occur with equal frequency) such as shown in figure 3.9(a), the values of the mean, median, and mode are identical. If the distribution is skewed to the left (negative skew), the mean, median, and mode are aligned as shown in figure 3.9(b). If skewed to the right (positive skew), the mean, median, and mode appear as shown in figure 3.9(c).

The mean measures the "center of gravity" of a set of data and is

Figure 3.8 *Locating the mode for a frequency distribution*

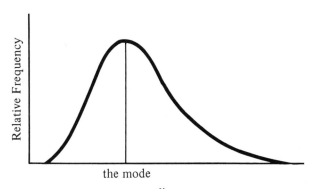

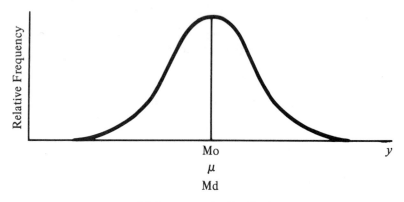

(a) A symmetric distribution

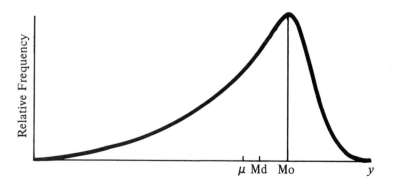

(b) Distribution skewed to the left

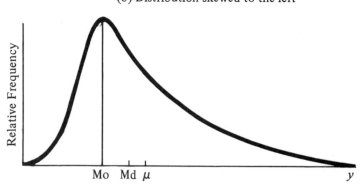

(c) Distribution skewed to the right

Figure 3.9 *Relationships among the mean (μ), median (Md), and mode (Mo)*

consequently influenced by extreme values. This property of the mean can be observed in figures 3.9(b) and 3.9(c). If the distribution is skewed to the left [figure 3.9(b)], the mean shifts to the left of the mode. If the distribution is skewed to the right, the mean shifts to the right of the mode. The greater the skewness, as measured by the preponderance of extreme values lying on one side of the mode, the greater is the shift of the mean in that direction.

The median ignores extreme values, except to take into account their location with respect to the middle value in an array.

Statistical inference is generally easier when using the mean. Because we will be concerned primarily with statistical inference in the following chapters, and because the sample mean is most widely used for this purpose, we will confine our attention in the subsequent discussions to the mean as a measure of central tendency.

Exercises

3.8. The following data represent the number of work stoppages per day due to machine breakdowns for 10 consecutive workdays in a food processing plant:

$$2, 3, 0, 5, 4, 3, 1, 3, 5, 2$$

Find the mean, median, and modal number of daily work stoppages in the plant during this two-week period.

3.9. Which measure of central tendency would be most useful in each of the following instances?

a. The production manager for a manufacturer of glass jars, who is concerned about the proper jar size to manufacture, has ample data on jar sizes ordered by customers. Would the mean, median, or modal jar size be of most value to the manager?

b. The sales manager for a quality furniture manufacturer is interested in selecting the regions most likely to purchase his firm's products. Would he be most interested in the mean or median family income in prospective sales areas?

c. A security analyst is interested in describing the daily market price change of the common stock of a manufacturing company. Only rarely does the market price of the stock change by more than one point, but occasionally the price will change by as many as four points in one day. Should the security analyst describe the daily price change of the stock in terms of the mean, median, or modal daily market price change?

3.7 Measures of Variability

Having located the center of a distribution of data, our next step is to provide a measure of the **variability** or **dispersion** of the data. Consider the two distributions shown in figure 3.10. Both distributions are located with a center at $y = 4$, but there is a vast difference in the variability of the measurements about the mean for the two distributions. The measurements in figure 3.10(a) vary from 3 to 5; in figure 3.10(b) the measurements vary from 0 to 8.

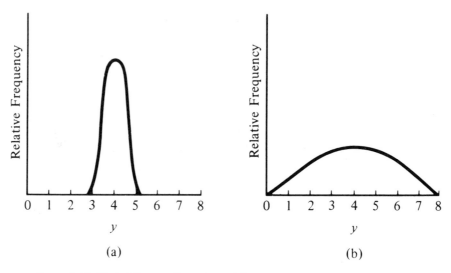

Figure 3.10 *Variability or dispersion of data*

Variation is a very important characteristic of data. For example, if we are manufacturing bolts, excessive variation in the bolt diameter would imply a high percentage of defective product. On the other hand, if we are using an examination to discriminate between good and poor accountants, we would be most unhappy if the examination always produced test grades with little variation, since this would make discrimination very difficult indeed.

In addition to the practical importance of variation in data, it is obvious that a measure of this characteristic is necessary to the construction of the mental image of the frequency distribution. Numerous measures of variability exist, and we will discuss a few of the most important.

The simplest measure of variation is the *range*.

Definition

The **range** of a set of n measurements $y_1, y_2, y_3, ..., y_n$ is defined to be the difference between the largest and smallest measurements.

For our price-earnings ratios in table 3.1, we note that the measurements vary from 5.4 to 28.6. Hence the range is $(28.6 - 5.4) = 23.2$.

Unfortunately, the range is not completely satisfactory as a measure of variation. Consider the two distributions in figure 3.11. Both distributions have the same range, but the data of figure 3.11(b) are more variable than the data of figure 3.11(a).

To overcome this limitation of the range, we introduce **quartiles** and **percentiles.** Remember that if we specify an interval along the y-axis of our histogram, the percentage of area under the histogram lying above the interval

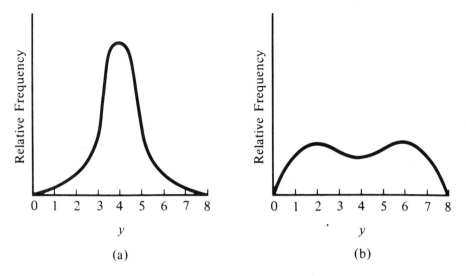

Figure 3.11 *Distribution with equal ranges and unequal variability*

is equal to the percentage of the total number of measurements falling in that interval. Since the median is the middle measurement when the data are arranged in order of magnitude, the median would be the value of y such that half the area of the histogram would lie to its left, half to the right. Similarly, we define *quartiles* as values of y that divide the area of the histogram into quarters.

Definition

Let $y_1, y_2, \ldots, y_n$ be a set of n measurements arranged in order of magnitude. The **lower quartile** is a value of y that exceeds $\frac{1}{4}$ of the measurements and is less than the remaining $\frac{3}{4}$. The second quartile is the median. The **upper quartile** (third quartile) is a value of y that exceeds $\frac{3}{4}$ of the measurements and is less than $\frac{1}{4}$.

Locating the lower quartile on the histogram in figure 3.12, we note that $\frac{1}{4}$ of the area lies to the left of the lower quartile, $\frac{3}{4}$ to the right. The upper quartile is the value of y such that $\frac{3}{4}$ of the area lies to the left, $\frac{1}{4}$ to the right.

For small quantities of data, the quartiles may fall between two measurements, thereby admitting the possibility of many numbers that would satisfy the definition given above. Arbitrary rules are available for locating specific values of the quartiles in this case. However, we will omit them from our discussion because you probably would not bother to calculate quartiles for

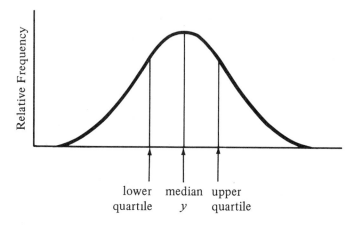

Figure 3.12 *Location of quartiles*

small quantities of data; you would use the range instead. For large quantities of data, the quartiles will either be located exactly or will fall in such small intervals between two observations that their exact location will be of little practical importance. Since we are primarily interested in interpreting quartiles for given sets of measurements, the preceding definition satisfies our needs.

For some applications, particularly those in which you have very large quantities of data, it is preferable to use *percentiles*.

Definition

Let $y_1, y_2, ..., y_n$ be a set of n measurements arranged in order of magnitude. The **pth percentile** is a value of y such that at most p percent of the measurements are less than the value y and at most $(100 - p)$ percent are greater.

For example, the 90th percentile for a set of data is a value of y that

Figure 3.13 *Dot diagram*

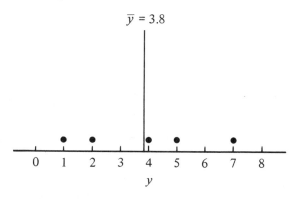

exceeds 90% of the measurements and is less than 10%. Just as in the case of quartiles, 90% of the area of the histogram lies to the left of the 90th percentile.

The range possesses simplicity in that it can be expressed as a single number. Quartiles and percentiles, on the other hand, provide more information about data location and variation, but several numbers must be given to provide an adequate description. Can we find a measure of variability expressible as a single number but more sensitive than the range?

Consider, as an example, the sample measurements 5, 7, 1, 2, 4. We can depict these data graphically, as in figure 3.13, by showing the measurements as dots falling along the y-axis. Figure 3.13 is called a **dot diagram.**

Calculating the mean as the measure of central tendency, we obtain

$$\bar{y} = \frac{\sum\limits_{i=1}^{n} y_i}{n} = \frac{19}{5} = 3.8$$

and we locate $\bar{y}$ on the dot diagram. We can now view variability in terms of distance between each dot (measurement) and the mean $\bar{y}$. If the distances are large, we can say that the data are more variable than if the distances were small. More explicitly, we define the **deviation** of a measurement from its mean to be the quantity $(y_i - \bar{y})$. Note that measurements to the right of the mean produce positive deviations, and those to the left, negative deviations. The values of y and the deviations for our example are shown in the first and second columns of table 3.4.

Table 3.4 *Computation of* $\sum\limits_{i=1}^{n} (y_i - \bar{y})^2$

y_i	$(y_i - \bar{y})$	$(y_i - \bar{y})^2$	y_i^2
5	1.2	1.44	25
7	3.2	10.24	49
1	−2.8	7.84	1
2	−1.8	3.24	4
4	.2	.04	16
Totals 19	0	22.80	95

If we now agree that deviations contain information on variation, our next step is to construct a formula, based on the deviations, that will provide a good measure of variation. As a first possibility we might choose the average of the deviations. Unfortunately, this will not work because some of the deviations are positive, some are negative, and the sum is always 0 (unless round-off errors have been introduced into the calculations). This can be shown by using our summation theorems (see chapter 2) as follows: Given n measurements $y_1, y_2, ..., y_n$,

$$\sum_{i=1}^{n} (y_i - \bar{y}) = \sum_{i=1}^{n} y_i - \sum_{i=1}^{n} \bar{y} = \sum_{i=1}^{n} y_i - n\bar{y}$$

$$= \sum_{i=1}^{n} y_i - n \frac{\sum_{i=1}^{n} y_i}{n} = \sum_{i=1}^{n} y_i - \sum_{i=1}^{n} y_i = 0$$

Note that the deviations in the second column of table 3.4 sum to zero.

You may have observed an easy solution to this problem. Why not calculate the average of the absolute values* of the deviations? This method has, in fact, been employed as a measure of variability, but it tends to be unsatisfactory for purposes of statistical inference. We prefer overcoming the difficulty caused by the sign of the deviations by working with the sum of their squares,

$$\sum_{i=1}^{n} (y_i - \bar{y})^2$$

For a fixed number of measurements, when this quantity is large, the data will be more variable than when it is small.

Definition

The **variance of a population** of N measurements $y_1, y_2, \ldots, y_N$ is defined to be the average of the square of the deviations of the measurements about their mean μ. The population variance is denoted by σ^2 (σ is the lowercase Greek letter sigma) and is given by the formula

$$\sigma^2 = \frac{1}{N} \sum_{i=1}^{N} (y_i - \mu)^2$$

Note that we use N to denote the number of measurements in the population and n to denote the number of measurements in the sample.

Typically we do not have all the population measurements available and must be satisfied with sample measurements selected from the population. Thus we must use the variance of a *sample,* as defined on the next page.

For example, we may calculate the variance for the set of $n = 5$ sample measurements presented in table 3.4. The square of the deviation of each measurement is recorded in the third column of table 3.4. Adding, we obtain

$$\sum_{i=1}^{5} (y_i - \bar{y})^2 = 22.80$$

---

*The absolute value of a number is its magnitude, ignoring its sign. For example, the absolute value of -2, represented by the symbol $|-2|$, is 2. The absolute value of 2, that is, $|2|$, is 2.

Definition

The **variance of a sample** of n measurements $y_1, y_2, ..., y_n$ is defined to be the sum of the squared deviations of the measurements about their mean $\bar{y}$ divided by $(n - 1)$. The sample variance is denoted by s^2 and is given by the formula

$$s^2 = \frac{1}{n-1} \sum_{i=1}^{n} (y_i - \bar{y})^2$$

The sample variance is

$$s^2 = \frac{1}{n-1} \sum_{i=1}^{n} (y_i - \bar{y})^2 = \frac{22.80}{4} = 5.70$$

You may wonder about the apparent inconsistency in the definitions of the population and sample variances. Recall that we use the sample mean $\bar{y}$ as an estimator of the population mean μ. Although it was not specifically stated, we wished to convey the impression that the sample mean provides a good estimate of μ. In the same vein, it might seem reasonable to assume

$$s'^{2} = \frac{1}{n} \sum_{i=1}^{n} (y_i - \bar{y})^2$$

would provide a good estimate of the population variance σ^2, based on a set of sample measurements. However, it can be shown that for small samples (n small), s'^2 tends to underestimate σ^2, and the sample variance s^2 provides better estimates of σ^2 than does s'^2. Note that s^2 and s'^2 differ only in the denominators and that when n is large, s'^2 and s^2 will be approximately equal. In later chapters we will have numerous occasions for use of an estimator of the population variance σ^2. In all our calculations we will use s^2 rather than s'^2.

At this point you may be understandably disappointed with the practical significance attached to variance as a measure of variability. Large variances imply a large amount of variation, but this statement only permits comparison of several sets of data. When we attempt to say something specific concerning a single set of data, we are at a loss. For example, what can be said about the variability of a set of data with a variance equal to 100? The question cannot be answered with the facts we have. We will remedy this situation by introducing a new definition, and, in section 3.7, a theorem and a rule.

Definition

The **standard deviation** of a set of n sample measurements $y_1, y_2, y_3, ..., y_n$ is equal to the positive square root of the variance.

The variance is measured in terms of the square of the original units of measurement. If the original measurements are in inches, the variance is expressed in square inches. Taking the square root of the variance, we obtain the standard deviation, which, most happily, returns our measure of variability to the original units of measurement.

Sample Standard Deviation

$$s = \sqrt{s^2} = \sqrt{\dfrac{\sum\limits_{i=1}^{n}(y_i - \bar{y})^2}{n-1}}$$

The population standard deviation is σ. As an aid for remembering sample standard deviation, note that the symbol s is the first letter in the word "standard."

Now that we have defined the standard deviation, you might wonder why we bothered to define the variance in the first place. Actually, both the variance and the standard deviation play an important role in statistics, a fact that you must accept on faith at this stage of our discussion.

3.8 On the Practical Significance of the Standard Deviation

We now introduce an interesting and useful theorem developed by the Russian mathematician Tchebysheff. Proof of the theorem is not difficult, but we omit it from our discussion.

Tchebysheff's Theorem

Given a number k greater than or equal to 1 and a set of n measurements $y_1, y_2, ..., y_n$, at least $(1 - 1/k^2)$ of the measurements lie within k standard deviations of their mean.

Tchebysheff's theorem applies to any set of measurements, and for purposes of illustration, we could refer either to the sample or to the population. We will use the notation appropriate for populations, but you should realize that we could just as easily use the mean and the standard deviation for the sample.

The idea involved in Tchebysheff's theorem is illustrated in figure 3.14. An interval is constructed by measuring a distance of $k\sigma$ on either side of

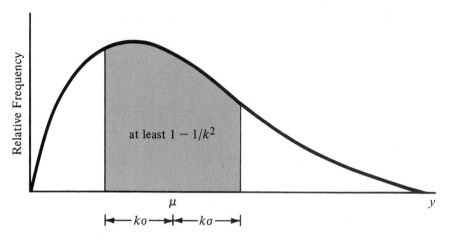

Figure 3.14 *Illustrating Tchebysheff's theorem*

the mean μ. Note that the theorem is true for any number we choose for k as long as it is greater than or equal to 1. Then, computing the fraction $(1 - 1/k^2)$, we see that Tchebysheff's theorem states that at least that fraction of the total number n of measurements lies in the constructed interval.

Let us choose a few numerical values for k and compute $(1 - 1/k^2)$ (see table 3.5). When $k = 1$, the theorem states that at least $1 - 1/(1)^2 = 0$ of the measurements lie in the interval from $(\mu - \sigma)$ to $(\mu + \sigma)$, a most unhelpful and uninformative result. However, when $k = 2$, we observe that at least $1 - 1/(2)^2 = 3/4$ of the measurements lie in the interval from $(\mu - 2\sigma)$ to $(\mu + 2\sigma)$. At least $8/9$ of the measurements lie within three standard deviations of the mean, that is, in the interval from $(\mu - 3\sigma)$ to $(\mu + 3\sigma)$. Although $k = 2$ and $k = 3$ are very useful in practice, k need not be an integer. For example, the fraction of measurements falling within $k = 2.5$ standard deviations of the mean is at least $1 - 1/(2.5)^2 = .84$.

To apply Tchebysheff's theorem to describe sample data, it is proper to use s' rather than s (s', defined in section 3.7, is a quantity slightly smaller than s) for use in constructing intervals about the mean. We will ignore this minute point because it is of no practical importance. The theorem always holds true when s is used instead of s'. Second, s and s' are nearly equal when n is large (and satisfactory for descriptive purposes when n is as small as 10). Finally, note that we are interested primarily in describing populations, not samples. Examples describing small sets of sample measurements are

Table 3.5 *Illustrative values of $(1 - 1/k^2)$*

k	$(1 - 1/k^2)$
1	0
2	3/4
3	8/9

presented solely to demonstrate the use of Tchebysheff's theorem.

Example 3.5 The mean and variance of a sample of $n = 25$ measurements are 75 and 100, respectively. Use Tchebysheff's theorem to describe the distribution of measurements.

Solution We are given $\bar{y} = 75$ and $s^2 = 100$. The standard deviation is $s = \sqrt{100} = 10$. The distribution of measurements is centered about $\bar{y} = 75$, and Tchebysheff's theorem states that:

1. At least $\frac{3}{4}$ of the 25 measurements lie in the interval $(\bar{y} \pm 2s)$
   $= [75 \pm 2(10)]$, that is, 55 to 95.
2. At least $\frac{8}{9}$ of the measurements lie in the interval $(\bar{y} \pm 3s)$
   $= [75 \pm 3(10)]$, that is, 45 to 105.

We emphasize the "at least" in Tchebysheff's theorem because the theorem is very conservative, applying to *any* distribution of measurements. In most situations the fraction of measurements falling in the specified interval will exceed $(1 - 1/k^2)$.

We now state a rule that describes accurately the variability of a particular bell-shaped distribution and describes reasonably well the variability of other mound-shaped distributions of data. The frequent occurrence of mound-shaped and bell-shaped distributions of data in nature—hence the applicability of our rule—leads us to call it the Empirical Rule.

The Empirical Rule

Given a distribution of measurements that is approximately bell-shaped (see figure 3.15), the interval

$(\mu \pm \sigma)$ contains approximately 68% of the measurements

$(\mu \pm 2\sigma)$ contains approximately 95% of the measurements

$(\mu \pm 3\sigma)$ contains all or almost all the measurements

---

Figure 3.15 *Normal distribution*

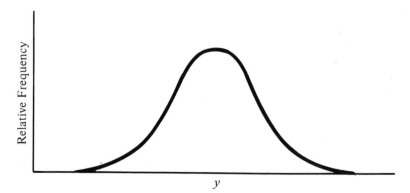

The bell-shaped distribution shown in figure 3.15 is commonly known as the normal distribution and will be discussed in detail in chapter 7. The point we wish to make here is that the Empirical Rule is extremely useful and provides an excellent description of variation for many types of data.

Example 3.6 A time study is conducted to determine the length of time necessary to perform a specified operation in a manufacturing plant. The length of time necessary to complete the operation is measured for each of $n = 40$ workmen. The mean and standard deviation are found to be 12.8 and 1.7, respectively. Describe the sample data by using the Empirical Rule.

Solution To describe the data we calculate the intervals

$$(\bar{y} \pm s) = (12.8 \pm 1.7) \qquad \text{or} \qquad 11.1 \text{ to } 14.5$$

$$(\bar{y} \pm 2s) = [12.8 \pm 2(1.7)] \qquad \text{or} \qquad 9.4 \text{ to } 16.2$$

$$(\bar{y} \pm 3s) = [12.8 \pm 3(1.7)] \qquad \text{or} \qquad 7.7 \text{ to } 17.9$$

According to the Empirical Rule, we expect approximately 68% of the measurements to fall in the interval from 11.1 to 14.5; 95% in the interval from 9.4 to 16.2; and all or almost all in the interval from 7.7 to 17.9.

If we doubt that the distribution of measurements is mound-shaped or wish, for some other reasons, to be conservative, we can apply Tchebysheff's theorem and be absolutely certain of our statements. Tchebysheff's theorem tells us that at least $\frac{3}{4}$ of the measurements fall in the interval from 9.4 to 16.2 and at least $\frac{8}{9}$ in the interval from 7.7 to 17.9.

Before leaving this topic we might wonder how well the Empirical Rule applies to the price-earnings ratios of table 3.1. We will show in section 3.9 that the mean and standard deviation for the $n = 25$ measurements is $\bar{y} = 16.0$ and $s = 5.6$. The appropriate intervals are calculated and the number of measurements falling in each interval recorded. The results are shown in table 3.6, with k in the first column and the interval $(\bar{y} \pm ks)$ in the second column, using $\bar{y} = 16.0$ and $s = 5.6$. The frequency, or number of measurements falling in each interval, is given in the third column and the relative frequency in the fourth column. Note that the relative frequency histogram for this data, figure 3.2, is not bell-shaped. Yet the percentages falling in the intervals $(\bar{y} \pm s)$, $(\bar{y} \pm 2s)$, and $(\bar{y} \pm 3s)$ agree reasonably well with the Empirical Rule.

Table 3.6 *Frequency of measurements lying within k standard deviations; data of table 3.1*

k	INTERVAL, $(\bar{y} \pm ks)$	FREQUENCY IN INTERVAL	RELATIVE FREQUENCY
1	10.4–21.6	16	.64
2	4.8–27.2	24	.96
3	−.8–32.8	25	1.00

Exercises

3.10.   Most graduate schools of business require that applicants submit a score on the GMAT (Graduate Management Admissions Test) examination, which is administered by the Educational Testing Service. Since 1965, GMAT scores have averaged 480, with a standard deviation of 100. Past evidence shows the distribution of GMAT scores to be mound-shaped (in fact, roughly bell-shaped).

a. What fraction of scores would you expect to find in the interval from 380 to 580?

b. What fraction of scores would you expect to find in the interval 280 to 680?

c. A certain prestigious graduate business school automatically accepts all applicants whose GMAT scores exceed 680. Approximately what fraction of all those taking the GMAT examination would qualify for admission under this criterion?

3.11.   Quality control techniques are used to monitor the quality of a manufacturing process to ensure uniformity and consistency in product output. A quality control engineer for a glass bottle manufacturing company, seeking to establish product quality standards at a time when the manufacturing process was known to be in control, randomly selected $n = 30$ bottles from the manufacturing process and recorded their weights. He found the average weight to be 8.2 ounces and the standard deviation to be 1 ounce. Describe the distribution of weights when the process is in control.

a. Use Tchebysheff's theorem.

b. Use the Empirical Rule. Is the Empirical Rule likely to be applicable in this case? Explain.

3.12.   Refer to exercise 3.11 and suppose the quality control engineer had selected a sample of $n = 4$ instead of $n = 30$ bottles from the manufacturing process. Would the Empirical Rule have been suitable for describing the bottle weights? Explain.

3.9 A Short Method for Calculating the Variance

The calculation of the variance and standard deviation of a set of measurements is no small task, regardless of the method employed, but it is particularly tedious if we proceed, according to the definition, by calculating each deviation individually as shown in table 3.4. We will use the data of table 3.4 to illustrate a shorter method of calculation. The tabulations are presented in table 3.7 in two columns, the first containing the individual

Table 3.7 *Table for simplified calculations of $\sum_{i=1}^{n}(y_i - \bar{y})^2$*

y_i	y_i^2
5	25
7	49
1	1
2	4
4	16
Totals 19	95

measurements and the second containing the squares of the measurements. We now calculate

$$\sum_{i=1}^{n} y_i^2 - \frac{\left(\sum_{i=1}^{n} y_i\right)^2}{n} = 95 - \frac{(19)^2}{5} = 95 - \frac{361}{5} = 95 - 72.2 = 22.8$$

Notice that this result is exactly equal to the sum of squares of the deviations

$$\sum_{i=1}^{n} (y_i - \bar{y})^2$$

given in the third column of table 3.4. Of course, this is no accident. The sum of squares of the deviations is always equal to the formula shown in the box.

Shortcut Formula for Calculating the Sum of Squares of Deviations

$$\sum_{i=1}^{n} (y_i - \bar{y})^2 = \sum_{i=1}^{n} y_i^2 - \frac{\left(\sum_{i=1}^{n} y_i\right)^2}{n}$$

This formula can be derived using the summation theorems of chapter 2 (see exercise 2.42).

We call this formula the shortcut method of calculating the sums of squares of deviations, which are needed in the formulas for the variance and standard deviation. Comparatively speaking, it is short because it eliminates all the subtractions required for calculating the individual deviations. A second and not so obvious advantage is that it tends to give better computational accuracy than the method utilizing the deviations. The beginning statistics student frequently finds the variance that he or she has calculated different from the answer in the text. This is usually caused by rounding off numbers in the computations. We suggest that rounding off be held at a minimum, since it may seriously affect the result of computation of the variance. A third advantage of the shortcut method is that it is especially suitable for calculators, some of which accumulate

$$\sum_{i=1}^{n} y_i \quad \text{and} \quad \sum_{i=1}^{n} y_i^2$$

simultaneously.

Before leaving this topic, we will calculate the standard deviation for the $n = 25$ price-earnings ratios of table 3.1. Using table 3.1 you may verify the following calculations:

$$\sum_{i=1}^{n} y_i = 400 \quad \text{and} \quad \sum_{i=1}^{n} y_i^2 = 7154.02$$

Then using the shortcut formula, we have

$$\sum_{i=1}^{n} (y - \bar{y})^2 = \sum_{i=1}^{n} y_i^2 - \frac{\left(\sum_{i=1}^{n} y_i\right)^2}{n} = 7154.02 - \frac{(400)^2}{25}$$

$$= 7154.02 - 6400 = 754.02$$

It follows that the standard deviation is (correct to one decimal place)

$$s = \sqrt{\frac{\sum_{i=1}^{n} (y_i - \bar{y})^2}{n - 1}} = \sqrt{\frac{754.02}{24}} = \sqrt{31.4175} = 5.6$$

Example 3.7 Calculate $\bar{y}$ and s for the measurements 85, 70, 60, 90, and 81.

Solution

$$\bar{y} = \frac{386}{5} = 77.2$$

y_i	y_i^2
85	7,225
70	4,900
60	3,600
90	8,100
81	6,561
386	30,386

$$\sum_{i=1}^{n} (y_i - \bar{y})^2 = \sum_{i=1}^{n} y_i^2 - \frac{\left(\sum_{i=1}^{n} y_i\right)^2}{n}$$

$$= 30,386 - \frac{(386)^2}{5}$$

$$= 30,386 - 29,799.2$$

$$= 586.8$$

Thus

$$s = \sqrt{\frac{\sum_{i=1}^{n} (y_i - \bar{y})^2}{n - 1}} = \sqrt{\frac{586.8}{4}} = \sqrt{146.7} = 12.1$$

Exercises

3.13.   Calculate $\bar{y}$, s^2, and s for the following data:

$$2, 1, 3, 0, 4, 2, 1, 3$$

Use the shortcut formula to calculate the sum of squares of deviations.

3.14.   Calculate $\bar{y}$, s^2, and s for the following data:

4, 1, 0, 3, 6, 4

Use the shortcut formula to calculate the sum of squares of deviations.

3.15.   Calculate $\bar{y}$, s^2, and s for the following data:

0, −2, −5, −3, 2, −4, 0, 3, 0, −1

3.16.   Refer to the daily work stoppages indicated in exercise 3.8 and compute the variance of daily work stoppages. What is the standard deviation of daily work stoppages?

3.17.   Residential properties in a common region often have very different mortgage interest rates, as these rates depend on prevailing economic conditions at the time of the purchase of the property. A sample of 25 properties in a certain residential suburb provided the following mortgage interest rates on loans outstanding on the properties (all listings are percentages):

5.2	6.0	7.5	8.0	5.0
7.9	6.6	9.2	7.4	6.5
8.0	9.0	7.3	7.5	7.0
4.6	8.5	5.5	9.3	9.5
6.5	7.5	6.5	8.1	8.2

a. Construct a relative frequency histogram for these data.
b. Calculate the mean, median, and modal interest rate for these properties.
c. Calculate s^2 and s for these data.

3.18.   Refer to exercise 3.17 and find the 50th percentile of the distribution of interest rates. Find the 90th percentile. Interpret these two values you have computed. Now find the lower quartile and upper quartile of the distribution of interest rates and interpret these values.

3.19.   Refer to exercise 3.17. Describe the distribution of interest rates.
a. Use Tchebysheff's theorem.
b. Use the Empirical Rule. Would you expect the Empirical Rule to be suitable for describing these data? Explain.
c. Calculate the percentage of measurements lying within one, two, and three standard deviations of the mean. How do these percentages compare with those given in the Empirical Rule? Those that might be expected according to Tchebysheff's theorem?

3.20.   To comply with FTC regulations on truth in advertising, an oil refinery has conducted $n = 10$ test runs involving the use of an experimental gasoline additive to their regular brand of gasoline. The increases in gasoline mileage noted on the 10 test runs were as follows (in miles per gallon):

1.5, 3.4, 0.2, 1.6, 2.0, 1.8, 2.9, 3.7, 0.4, 2.6

a. Calculate $\bar{y}$, the average mileage increase due to the gasoline additive.
b. Using the shortcut formula, calculate s^2 and s from these data.

3.10 Estimating the Mean and Variance for Grouped Data (Optional)

Often the only data available for analysis are listed in the form of a frequency histogram. Company reports often list data only in terms of class frequencies; governmental and news media sources usually use some type of a bar chart to display pertinent data. In such cases we may not know the exact values of the measurements falling within the class intervals. When this situation occurs, there is no way to compute the exact values of the sample mean and variance.

There is, however, a method for approximating the mean $\bar{y}$ and the variance s^2 when only grouped data are available. **This method is based on the assumption that the midpoint of each class in the grouped frequency classification is approximately equal to the arithmetic mean of the measurements contained within that class.** The midpoint of a particular class i is denoted by the symbol m_i. Now suppose that the midpoints do actually equal the mean of the measurements within their respective classes. Then for a particular class i, if we multiply m_i by f_i, the frequency within class i, we obtain the total of the measurements within class i. Summing the class totals then gives us the total of all measurements contained within the frequency distribution and $\bar{y}$ can be found by taking this sum total over the total number of measurements n in the usual manner. Naturally, the accuracy of the approximated mean obtained by using class midpoints depends heavily on the degree to which the midpoints accurately reflect the arithmetic mean of the measurements contained within each respective class. Usually such approximations are quite reliable, especially when the class frequencies f_i are of sufficient size to guarantee a rather even "coverage" of measurements over each class. For approximating the variance s^2 when only grouped data are available, we follow a procedure that generalizes the shortcut formula for the computation of s^2 introduced in section 3.9.

Mean and Variance for Grouped Data

If data are grouped according to frequency of occurrence in each of k nonoverlapping classes, then the mean $\bar{y}$ and variance s^2 of the measurements contained within the groupings are approximated by

$$\bar{y} \approx \frac{\sum_{i=1}^{k} f_i m_i}{n}$$

$$s^2 \approx \frac{\sum_{i=1}^{k} f_i m_i^2 - \left[\left(\sum_{i=1}^{k} f_i m_i\right)^2 \Big/ n\right]}{n-1}$$

where m_i is the midpoint of class i and f_i is the frequency of measurements within class i. (Note: The symbol $\approx$ means "is approximately equal to.")

Table 3.8 summarizes the computations necessary to compute $\bar{y}$ and s^2 from the frequency distribution of $n = 25$ price-earnings ratios for 25 common stocks shown in table 3.2. The computations required by the grouped

Table 3.8 *Class frequencies and class midpoints for the 25 price-earnings ratios listed in table 3.2*

CLASS, i	CLASS BOUNDARIES	f_i	m_i	$f_i m_i$	$f_i m_i^2$
1	5.00–8.99	3	7	21	147
2	9.00–12.99	5	11	55	605
3	13.00–16.99	7	15	105	1575
4	17.00–20.99	6	19	114	2166
5	21.00–24.99	3	23	69	1587
6	25.00–28.99	1	27	27	729
Totals		25		391	6809

data formulas given in the box are greatly simplified when the data are organized as in table 3.8. Using these formulas the mean can be approximated from the grouped data as

$$\bar{y} \approx \frac{\sum_{i=1}^{6} f_i m_i}{25} = \frac{391}{25} = 15.64$$

The approximation to the variance of the measurements if found by computing

$$s^2 \approx \frac{\sum_{i=1}^{6} f_i m_i^2 - \left[\left(\sum_{i=1}^{6} f_i m_i\right)^2 \Big/ 25\right]}{24} = \frac{6809 - [(391)^2/25]}{24}$$

$$= \frac{693.76}{24} = 28.91$$

Since the approximation to the variance of the $n = 25$ price-earnings ratios is 28.91, the approximate standard deviation is

$$s \approx \sqrt{28.91} = 5.377$$

In sections 3.6 and 3.9 we found the actual mean and standard deviation of the ungrouped sample of the $n = 25$ price-earnings ratios to be

$$\bar{y} = 16 \quad \text{and} \quad s = 5.6$$

Thus the approximations obtained from the grouped frequency distribution of the price-earnings ratios (table 3.2) appear to be satisfactory approximations to the values of $\bar{y}$ and s calculated from the ungrouped data.

Although it was mentioned in section 3.2 that the classes should be of equal width, the classes need not be of equal width to apply the grouped data mean and variance formulas. All that must be assumed is that the class midpoints are approximately equal to the arithmetic mean of the measurements within the classes. The grouped data procedures are not applicable, however, in the case where one or more of the classes are open-ended (one endpoint is located at either $+\infty$ or $-\infty$). In such cases it becomes impossible to find class midpoints for the open-ended classes.

Exercises

3.21.   Refer to exercise 3.3 and use the computing formulas for grouped data to approximate the mean and variance of gasoline mileage based on the relative frequency histogram derived in exercise 3.3a. (Hint: To find the class frequency f_i for each class, multiply the class relative frequency by 50.)

3.22.   Refer to exercise 3.17 and use the computing formulas for grouped data to approximate the mean and variance of the mortgage interest rates based on the frequency distribution derived in exercise 3.17a. How do these approximations compare with $\bar{y}$ and s^2 found in exercises 3.17b and 3.17c, respectively?

3.23.   An accountant for a large retail store is interested in estimating the average accounts receivable balance for the store's 10,000 credit customers. The frequency distribution shown in the table was constructed from a sample of $n = 100$ accounts selected at random from the store's accounts receivable files.

ACCOUNT BALANCE	FREQUENCY
$ 0 and under 20	10
20 and under 40	15
40 and under 60	40
60 and under 80	22
80 and under 100	13
	100

a. Use the computing formulas for grouped data to approximate the mean and standard deviation of the account balances.

b. Why are the values found in part a approximations to $\bar{y}$ and s and not, in fact, the true values for $\bar{y}$ and s for the data represented in the analysis?

c. Approximately what percentage of the company's accounts receivable have balances below $30?

d. Approximately how many of the 10,000 accounts have account balances exceeding $98?

3.11 Linear Transformations of Data (Optional)

Often when we wish to make comparisons between two or more sets of data, we find that some of the data sets are listed in a measuring system different from the others. A union official may wish to compare the retirement incomes of the employees of two automobile manufacturing plants, one in Detroit and one in Paris. A buyer is interested in comparing the gasoline economy of a British car, rated in kilometers per imperial gallon, to the economy of an American car, rated in miles per gallon. In each case comparisons cannot be made unless we transform all the sets of data to a common scale of measure. What we must do is find the mathematical relationship between one of the data sets, which will be used as a reference (called a reference data set), and all the other sets. Then applying the known mathematical relationship among the various data sets, we transform all the other sets of data to the scale of measure of the reference set.

If we are interested in comparing the data sets by comparing their means, variances, or standard deviations, based on a common scale of measurement, it is not necessary that we first transform each measurement to the common scale. For instance, if the union official mentioned above samples the retirement incomes of 50 Parisian automobile manufacturing employees, it is not necessary that he convert each of the 50 incomes from units of francs to dollar units before computing the numerical descriptive measures in dollar units. Knowing the mean, variance, and standard deviation in francs, he can convert them to their equivalent values in units of dollars by applying the following theorem, which we will call the *coding theorem*. Its proof is omitted but can be constructed from the definitions given earlier for the mean and variance.

Theorem 3.1   The Coding Theorem

Suppose that $\bar{y}$ and s_y^2 are the mean and variance of a set of n measurements $y_1, y_2, ..., y_n$. If we transform each measurement y_i by the linear transformation $x_i = a + by_i$, where a and b are any real numbers, then the mean of the transformed data is

$$\bar{x} = a + b\bar{y}$$

and the transformed variance is

$$s_x^2 = b^2 s_y^2$$

Similarly, the transformed standard deviation is

$$s_x = |b| s_y$$

We have stated the coding theorem for the sample mean and variance. However, the theorem applies as well for the transformed population mean and variance.

Example 3.8   In 1973 the employees of the Volkswagen assembly plant in Munich earned an average salary of 15,200 marks, with a standard deviation of 1,000 marks. To compare these West German wages to those of American automobile factory workers, we must transform the mean and standard deviation from marks to dollars. At the current rate of exchange, 1 mark = .39 dollars. Thus in the formula for $\bar{x}$, $a = b$ and $b = .39$, and we have

$$\bar{x}(\text{dollars}) = .39\,\bar{y}(\text{marks}) = .39(15,200) = 5,928$$

$$s_x(\text{dollars}) = .39\,s_y(\text{marks}) = .39(1,000) = 390$$

Since the rate of exchange relating West German marks to American dollars involves only a multiplicative constant, the addition constant from our theorem is assumed to be zero. Likewise, if our data are coded only by adding a constant to each measurement, the multiplicative constant becomes 1. In the latter case the variance of the transformed data remains unchanged.

Example 3.9 The daily high temperatures from Stockholm, Sweden, for the month of April had a mean of 10 degrees Celsius and a variance of 50 degrees Celsius squared. For comparative purposes we must convert to Fahrenheit degrees by the following transformation: Fahrenheit degrees $= 32 + (\frac{9}{5})$ degrees. Applying the coding theorem, we have

$$\bar{x}(\text{Fahrenheit}) = 32 + (\tfrac{9}{5})\cdot 10 = 50 \text{ degrees Fahrenheit}$$

$$s_x^2(\text{Fahrenheit}) = (\tfrac{9}{5})^2\cdot 50 = 162 \text{ degrees Fahrenheit squared}$$

Exercises

3.24.   For the sample measurements 27.995, 27.998, 28.005, 28.003, 28.004, code the measurements by subtracting an appropriate constant from each measurement. Then use the coding theorem to find the sample mean and variance of the measurements.

3.25.   Code the following sample measurements by multiplying by an appropriate constant: .029, .030, .028, .031, .032. Then use the coding theorem to find the sample mean and variance of the measurements.

3.26.   In 1975 the U.S. Department of Commerce performed a study that showed that four-person families spent an average of $218 per month on food, with a standard deviation of about $32. Use the coding theorem to find the mean and standard deviation of daily expenditures on food for four-person families in 1975. (Assume there are 30 days in a month.)

3.27.   The conversion from our current system of measurement to the metric system will create the necessity for us to familiarize ourselves with metric conversion factors and the metric standards. Use the coding theorem to convert the following descriptive statistics to their metric equivalents.

a. In a test run, 25 new Ford Pintos attained an average gasoline mileage of 27.8 mpg, with a standard deviation of 4.4 mpg. Express these results in terms of kilometers per gallon. (1 kilometer = .622 miles.)

b. Express the descriptive statistics in part a in terms of kilometers per liter. (1 liter = 1.056 quarts.)

c. A commercial fishing cannery in Astoria, Oregon, reports that the salmon caught by commercial fleets average 9.8 pounds per fish, with a variance of 17.3. Find the mean and standard deviation of salmon caught by commercial fleets in the Astoria region in kilograms. (1 kilogram = 2.205 pounds.)

d. A large diversified timber products company has found that topped 30-year-old Douglas fir trees provide logs that average 31 feet 6 inches in length, with a standard deviation of 19 feet 2 inches. Describe the length of logs obtained from topped 30-year-old Douglas fir trees in meters. (1 meter = 39.37 inches.)

3.12 Summary

The objective of a statistical study is to make inferences about a characteristic of a population based on information contained in a sample. Since populations are sets of data, we first need to consider ways to phrase an inference about a set of measurements. This latter point constituted the objective of our study in chapter 3.

Methods for describing sets of measurements fall into one of two categories: graphical methods and numerical methods. The relative frequency histogram is an extremely useful graphical method for characterizing a set of measurements. Other methods are useful as long as they provide a complete, easy-to-read-and-interpret summary of the data. Numerical descriptive measures are numbers that attempt to provide a mental image of the frequency distribution. We have restricted the discussion to measures of central tendency and variation, the most useful of which are the mean and standard deviation. Although the mode is not generally a measure of central tendency, its importance in characterizing demand levels should be noted by the business statistician. While the mean possesses intuitive significance, the standard deviation is meaningful only when used in conjunction with Tchebysheff's theorem and the Empirical Rule. The objective of sampling is the description of the population from which the sample was obtained. This is accomplished by using the sample mean $\bar{y}$ and the sample variance s^2 as estimators of the population mean μ and the variance σ^2.

Many descriptive methods and numerical measures have been presented in this chapter, but these are only a small percentage of the methods that might have been discussed. In addition, many special computational techniques usually found in elementary texts have been omitted. This choice is necessitated by the limited time available in an elementary course and because the advent and common use of electronic computers have minimized the importance of special computational formulas. But, more important, the inclusion of such techniques would tend to detract from and obscure the main objective of modern statistics and this text—statistical inference.

Supplementary Exercises

3.28. Conduct the following experiment: Toss a coin 10 times and record y, the number y of heads observed. Repeat this process $n = 50$ times, thus providing 50 values of y. Construct a relative frequency histogram for these data.

3.29. One researcher notes that differing interpretations to such factors as the price range to consider, when to purchase, the choice of neighborhood, home styles, and the availability of mortgage lenders cause the duration of the active search period for home buyers to be quite variable.* The following data represent the duration

*D. J. Hempel, "The Role of the Real Estate Broker in the Home Buying Process," Center for Real Estate and Urban Economic Studies, University of Connecticut (Storrs, Conn., 1969).

of active search (in weeks) of 25 home buyers in a certain city:

15	17	7	15	20
5	3	19	10	3
11	10	4	8	13
9	15	6	2	8
12	1	2	13	4

 a. Construct a relative frequency histogram for these data.

 b. What does this graphic description of the data tell you about the lengths of search time for new home buyers?

  3.30. The data in the accompanying table show the estimated gross proceeds from corporate nonconvertible bond offerings and whether the offerings were publicly offered or privately placed (cash sales in millions of dollars).

YEAR	TOTAL	PUBLICLY OFFERED	PRIVATELY PLACED
1965	$12,456	$ 4,389	$8,067
1966	13,695	6,262	7,434
1967	17,470	10,883	6,597
1968	14,103	8,068	6,032
1969	14,310	9,636	4,671
1970	27,658	23,121	4,540
1971	26,490	19,777	6,714
1972	23,502	14,940	8,562
1973	20,286	12,404	7,882
1974	31,075	24,886	6,188

Source: Securities and Exchange Commission, *Statistical Bulletin* (Washington, D.C., August 1975).

Construct a bar chart like the one in figure 3.3 to depict these data.

  3.31. The following $n = 6$ measurements represent the number of workdays missed due to illness during the past year by 6 employees of an insurance claims office:

$$3, 8, 4, 10, 5, 6$$

Calculate $\bar{y}$, s^2, and s for these data.

  3.32. The following $n = 7$ measurements represent the gain or loss in the daily closing price of a security for 7 consecutive market days:

$$-1, 3, 4, 1, 0, -4, -3$$

Calculate $\bar{y}$, s^2, and s for these data.

  3.33. Calculate $\bar{y}$, s^2, and s for the data in exercise 3.28.

  3.34. Calculate $\bar{y}$, s^2, and s for the data in exercise 3.29.

  3.35. Refer to the histogram constructed in exercise 3.28 and find the fraction of measurements lying in the interval $(\bar{y} \pm 2s)$. (Use the results of exercise 3.33.) Are these results consistent with Tchebysheff's theorem? Is the frequency histogram of exercise 3.28 relatively mound-shaped? Does the Empirical Rule adequately describe the variability of the data in exercise 3.28?

3.36.   Repeat the instructions of exercise 3.35 but use the interval $(\bar{y} \pm s)$.

3.37.   Refer to the data in exercise 3.29. Find the fraction of measurements falling in the intervals $(\bar{y} \pm s)$ and $(\bar{y} \pm 2s)$. Do these results agree with Tchebysheff's theorem and the Empirical Rule?

3.38.   The distribution shown in the accompanying table shows the ton-miles of shipments by the five primary modes of transportation in the United States for 1968.

MODE OF TRANSPORTATION	TON-MILES (BILLIONS)
rail (express and mail)	756.8
motor carriers (freight only)	396.3
water (including Great Lakes)	287.0
pipeline	332.3
airlines	2.9

Source: 84th ICC Annual Report to Congress, 1970.

Construct a pie chart showing the percentage of total ton-miles of shipments in the United States in 1968 for each mode of transportation.

3.39.   The distribution shown in the accompanying table represents the total advertising expenditures in the United States in 1970 by class of media (in millions of dollars).

MEDIA CLASS	AMOUNT SPENT
newspapers	$4936
magazines	1061
television	2853
radio	1128
direct mail	2548
point-of-purchase display	839
other	941

Source: Advertising Age, 7 June 1971.

Construct a pie chart to represent these data.

3.40.   Give appropriate class boundaries for classifying each of the following sets of measurements into 10 classes:

a. Number of weeks of ownership of an automatic washer before first necessary major repair. Data are available from 964 purchasers, with times before repair ranging from 0 weeks to 113 weeks.

b. The annual amount spent on life insurance premiums by the 583 employees of a manufacturing company. The premium values range from $0 to $467.

c. The annual salaries of all 1280 employees of a large pulp and paper processing firm. The salaries, including those of the company president and other executives, range from $6,850 to $85,000.

d. The number of customers entering a local discount department store during each of the past 100 days. The data range from 92 arrivals to 471 arrivals per day.

3.41.   Find the range for the data in exercise 3.29. Find the ratio of the range to s. If we possessed a large amount of data having a bell-shaped distribution, the range would be expected to equal how many standard deviations? (Note that this provides a rough check for the computation of s.)

3.42.   The rule of thumb for estimating s from the range found in exercise 3.41 can be extended to smaller amounts of data obtained in sampling from a bell-shaped distribution. Thus the calculated value for s should not be far different from the range divided by the appropriate ratio found in the accompanying table.

Number of Measurements	5	10	25
Expected Ratio of Range to s	2.5	3	4

a. For the data in exercise 3.17, estimate s as suggested. Compare this estimate with the calculated s.

b. For the data in exercise 3.29, estimate s as suggested. Compare this estimate with the calculated s.

3.43.   Differences in promotional spending among firms reflect differences in how sales respond to communications efforts. For example, pharmaceuticals and breakfast food manufacturers generally spend a much greater amount on promotion than, say, textile firms or manufacturers of primary metals. A survey of 10 pharmaceutical companies reveals the following expenditures on product promotion as a percentage of gross sales revenue:

$$21, 18, 25, 26, 25, 20, 24, 19, 28, 22$$

a. Observe the data and guess the value for s by use of the range approximation method.

b. Calculate $\bar{y}$ and s and compare with the range approximation of part a.

3.44.   From the following data, a student calculated s to be .263. On what grounds might we doubt his accuracy? What is the correct value (nearest hundredth)?

17.2	17.1	17.0	17.1	16.9
17.0	17.1	17.0	17.3	17.2
17.1	17.0	17.1	16.9	17.0
17.1	17.3	17.2	17.4	17.1

3.45.   A machining operation produces bolts with an average diameter of .51 inches and a standard deviation of .01 inch. If the distribution of bolt diameters is bell-shaped (approximately normal), what fraction of total production would possess diameters falling in the interval from .49 to .53 inches?

3.46.   Refer to exercise 3.45. Suppose that the bolt specifications required a diameter equal to (.5 ± .02) inches. Bolts not satisfying this requirement are considered to be defective. If the machining operation functioned as described in exercise 3.45, what fraction of total production would result in defective bolts?

3.47.   Television commercials on a certain television station average 35 seconds in length of air time, with a standard deviation of 5 seconds. If the distribution of air time for commercials on the television station is mound-shaped, approximately what fraction of total commercials would last from 30 to 40 seconds?

3.48.   Recent studies have shown that television commercials lasting longer than 40 seconds are excessively long and ineffective in communicating the intended message. What fraction (approximately) of the television commercials of the television station described in exercise 3.47 could be considered excessively long?

3.49.   The yield to maturity of industrial bonds depends on the issuing firm's bond rating and the state of the economy at the time of issue. For the quarter ending 31 March 1975, Longbrake found the average yield to maturity of industrial bonds issued during the quarter to be 8.55, with a standard deviation of yield of .70.* Assume the distribution of yields on industrial bonds is mound-shaped.

a. Describe the distribution of industrial bond yields during the first quarter of 1975.

b. During this same period the average yield to maturity for bank-issued bonds was 9.95. Approximately what percentage of the industrial bonds exceeded the bank bond average?

3.50.   A recent study by the Highway Loss Data Institute reported that the average loss payment per insurance claim by automobile owners during the first half of 1974 averaged $495, with a standard deviation of $75.† Assume the distribution of loss payment per claim is mound-shaped.

a. Describe the distribution of loss payments per claim during this period.

b. Approximately what fraction of loss payments exceeded $570 during this period?

3.51.   The frequency distribution shown in the table portrays the amount of property taxes paid during the previous year by each of a certain city's 442 residential property owners.

AMOUNT PAID (DOLLARS)	FREQUENCY
under 100	27
100–199	85
200–299	217
300–399	81
400–499	32

a. Use the computing formulas for grouped data to approximate the mean and standard deviation of the amount of property taxes paid by each of the city's residential property owners.

b. Use the Empirical Rule to approximate the proportion of the city's residential property owners who paid from $156 to $347 in property taxes during the past year.

3.52.   The issue of property tax relief is of vital concern to all municipal administrators. Suppose administrators from the city discussed in exercise 3.51 have decided to offer property tax relief to those residential property owners whose past year's property taxes were above the 84th percentile. Use the Empirical Rule to approximate the minimum amount of property taxes necessary to qualify for tax relief in the city.

3.53.   Refer to exercise 3.29 and use the computing formulas for grouped data to approximate the mean and variance of the search time for new home buyers. How do these approximations compare with $\bar{y}$ and s^2 computed in exercise 3.34?

3.54.   Code the data in exercise 3.29 by subtracting 10 from each observation. Calculate the sample mean and variance, then use the coding theorem to find $\bar{y}$ and s^2.

3.55.   Code the data in exercise 3.17 by first subtracting 4.6 from each

---

*W. A. Longbrake, "Financial Management of Banks and Bank Holding Companies," *Financial Management*, Winter 1974.
† *Money*, September 1974.

observation and then multiplying by 10. Calculate the sample mean and variance, then use the coding theorem to find $\bar{y}$ and s^2.

$ 3.56.   Carpeting is commonly priced in the United States in terms of the cost per square yard. A carpet manufacturer wishes to market a carpet in Great Britain, which sells for $8.50 per square yard in the United States. What should the manufacturer's selling price be in dollars per square meter if she expects to have an equivalent selling price for the carpet in both the United States and Britain? (1 square meter = 10.76 square feet.)

3.57.   Refer to exercise 3.56 and give the selling price in British pounds sterling per square meter if the manufacturer expects to have an equivalent selling price for the carpet in both the United States and Britain. (1 British pound sterling = 1.70 U.S. dollars.)

$ 3.58.   The three major U.S. automobile manufacturers have each enjoyed enormous success by equipping many of their cars with a proven and efficient V-8 engine with approximately 290 cubic inches of displacement. If the United States converts to the metric standard, what would be the advertised displacement of this engine in cubic centimeters? (1 inch = 2.54 centimeters.)

3.59.   The daily low temperatures during the month of May in Yakima, Washington, average 50 degrees Fahrenheit, with a standard deviation of 18 degrees Fahrenheit. What is the average daily low temperature during the month of May in Yakima in degrees Celsius?

 3.60.   Yakima, Washington, is located in the center of a very rich valley known for its production of apples, pears, and cherries. May is a very critical month in the development of these fruits on the trees and if, during this critical development period, the temperature drops below 0° Celsius, the young fruit can be damaged and an entire year's crop wiped out. Using the Empirical Rule find the approximate percentage of the days in May that are likely to be cold enough to place the Yakima Valley fruit crops in danger.

Experiences with Real Data

1. Select a business journal appropriate to your area (see Experiences with Real Data in chapter 1), a business periodical (such as *Business Week*, *Forbes*, *Fortune*), or a published government document and find an article that lists a set of experimental data. Select a sample of at least $n = 25$ observations on one of the variables represented in the data set.
    a. Define the population from which your sample was drawn.
    b. Construct a relative frequency histogram for these data.
    c. Calculate $\bar{y}$ and s for these data.
    d. Do these data appear to be mound-shaped and thereby satisfy the requirements of the Empirical Rule?
    e. What fraction of the observations lie within two standard deviations of $\bar{y}$? Three? Do these results agree with Tchebysheff's theorem? The Empirical Rule?

2. Conduct an experiment to estimate the mean and standard deviation of the daily sales volume at some local retail establishment. Choose a supermarket, a discount department store, a restaurant, or perhaps your college bookstore and record the amount of each sale during five different 30-minute segments of the establishment's workday. Sale amounts can be determined by noting the total sales figure as it is tabulated on a cash register. Since most establishments have more than one checkout stand, this experiment may require more than one active student participant. (It may be wise to discuss the purposes of your experiment with the management of the retail establishment before undertaking your study.) At the day's end, construct a

relative frequency histogram to represent your recorded sales data.

    a. Use the computing formulas for grouped data to approximate the mean and standard deviation of the amount per sale.

    b. Find the fraction of observations in the intervals $(\bar{y} \pm s)$ and $(\bar{y} \pm 2s)$. Does the Empirical Rule provide a satisfactory description of the variability of the data?

    c. Knowing the number of hours the retail establishment is open on the day of your investigation, you can easily estimate their *total* daily sales revenue. If n is the number of sales observations gathered in your investigation, then $n\bar{y}$ is an estimate of the total sales revenue over your five 30-minute sample observation segments. Therefore, if the retail establishment is open for business 10 hours, the estimated total daily sales revenue is $4(n\bar{y})$, or, in general, if it is open T hours, the estimated total is $(T/2.5)(n\bar{y})$. (A measure of reliability of this estimate can be obtained but we defer discussion of this topic until later in the text.)

References

Cangelosi, V. E.; P. H. Taylor; and P. F. Rice. *Basic Statistics: A Real World Approach.* New York: West Publishing, 1976.

Clark, C. T., and L. L. Schkade. *Statistical Methods for Business Decisions.* Cincinnati: South-Western Publishing, 1969.

Mode, E. B. *Elements of Statistics.* 3d ed. Englewood Cliffs, N.J.: Prentice-Hall, 1961.

Spurr, W. A., and C. P. Bonini. *Statistical Analysis for Business Decisions.* Homewood, Ill.: Richard D. Irwin, 1967.

chapter objectives

GENERAL OBJECTIVE The objective of this chapter is to develop an understanding of the basic concepts of probability. These concepts will be used in subsequent chapters to make inferences and to evaluate the reliability of inferences about populations based on sample data.

SPECIFIC OBJECTIVES

1. To show the use of probability in making a decision based on sample data by giving a simple example.
Section 4.1

2. To present a model for the repetition of an experiment, and to explain how the model provides a straightforward mechanism for calculating the probability of an experimental outcome.
Section 4.2

3. To present a method for representing an event as a composition of two or more other events and then to give two laws of probability that you can use to find the probability of the composition.
Sections 4.3, 4.4, 4.5

4. To present Bayes's law, an adaptation of conditional probability that is used in business decision making. This technique will be explained in chapter 10.
Section 4.6

5. To present a discussion of subjective probability and to identify its relevance to business decision making.
Section 4.8

6. To introduce the concepts of a numerical event and of random variables.
Section 4.9

chapter four

Probability

4.1 Introduction

As stated in chapter 1, the objective of statistics is to make inferences about a population based on information contained in a sample. Since the sample provides only partial information about the population, we require a mechanism that will accomplish this objective. Probability is such a mechanism; it enables us to use the partial information contained in a set of sample data to infer the nature of the larger set of data, the population. How we use probability to make inferences is best illustrated by considering an example.

A manufacturer wishes to compare consumer preference for two packages (call them A and B) designed for his product. Suppose you assume that design A is preferred. To test this assumption 20 consumers are randomly selected, presented with both packages, and asked to select the design they prefer. Suppose that all 20 consumers indicate a preference for design B. What would you conclude about your assumption?

To answer this question we must make an inference from a sample of 20 responses to the population associated with all possible purchasers of the manufacturer's product. Let 1 denote a consumer favoring design A and 0 denote a consumer favoring design B. Then we wish to decide whether the fraction of 1s in the population of 1s and 0s is greater than .5. If it is, then the original assumption that design A is preferred may be taken as true. If none of the consumers preferred design A, we would conclude that the assumption is false. To see how we are actually using probability to reach this conclusion, let us examine our reasoning.

If, in fact, the assumption is true that consumers prefer design A, the fraction of all potential purchasers of the product must be greater than $\frac{1}{2}$, and we would expect that something near the same fraction would be observed in the sample of 20 responses. Instead, none of the 20 consumers preferred design A, a result that is highly improbable if at least half of all consumers prefer design A. So what do you decide? You either conclude that we have observed a very improbable sample, or else you conclude that our original assumption was false and that less than half of all consumers favor design A. Thus implicit in this decision process is a reliance on the notion of probability—particularly, the probability of the observed sample results.

In the preceding discussion we found that the sample results were so extremely contrary to the original assumption that a decision to reject the assumption could be made quite readily. Suppose, however, that the sample of 20 consumer responses indicated that 3 favored design A and 17 favored B. Would we still conclude that the sample is so improbable that we would reject the original assumption? Or suppose that 8 favored A and 12 favored B? What would we then conclude about the original assumption? To answer these questions we would want to know "how improbable" a particular sample result is. In other words, we would need to find the probability of obtaining a sample as extreme as that observed, given that the original assumption is true. Having determined this probability, we are in a position to decide whether the assumption is reasonable or should be rejected as untrue. Thus probability provides the necessary mechanism for making inferences about the population on the basis of sample evidence.

4.2 The Sample Space

Data are obtained either by observation of uncontrolled events in nature or by controlled experimentation in the laboratory. To simplify our terminology we seek a word that will apply to either method of data collection, and hence we define the term *experiment*.

Definition

An **experiment** is the process by which an observation (or measurement) is noted.

Note that the observation need not produce a numerical value. Typical examples of experiments are the following:

1. Recording the income of a factory worker.
2. Interviewing a buyer to determine brand preference for a particular product.

3. Recording the price of a security at a particular time.
4. Inspecting an assembly line to determine if more than the allowable number of defectives are being produced.
5. Recording the type and size of policy sold by an insurance salesperson.

In reality, a population is a set of observations associated with a set of experimental units that are of interest to the business person. Therefore, a population could conceptually be generated by the repetition of an experiment. For instance, we might be interested in the length of life of television tubes produced in a plant during the month of June. Testing a single tube until it fails and measuring the length of its life represents a single experiment. Repetition of the experiment for all tubes produced during this period generates the entire population. A sample includes the results of a group of experiments selected from the population.

Let us now direct our attention to a careful analysis of some experiments and the construction of a mathematical model for the population. A by-product of our development will be a systematic and direct approach to the solution of probability problems.

We begin by noting that each experiment may result in one or more outcomes, which we will call **events** and denote by capital letters. Consider the following experiment.

Example 4.1 Toss a die and observe the number appearing on the upper face. Some events would be

event A: observe an odd number
event B: observe a number less than 4
event E_1: observe a 1
event E_2: observe a 2
event E_3: observe a 3
event E_4: observe a 4
event E_5: observe a 5
event E_6: observe a 6

These events do not represent a complete listing of all possible events associated with the experiment but they suffice to illustrate a point. You will readily note a difference between events A and B and events E_1, E_2, E_3, E_4, E_5, and E_6. Event A will occur if event E_1, E_3, or E_5 occurs, that is, if we observe a 1, 3, or 5. Thus A could be *decomposed* into a collection of simpler events, namely, E_1, E_3, and E_5. Likewise, event B will occur if E_1, E_2, or E_3 occurs and could be viewed as a collection of smaller or simpler events. In contrast, we note that it is impossible to decompose events E_1, E_2, E_3, ... , E_6. Events E_1, E_2, ... , E_6 are called *simple events* and A and B are *compound events*.

Events E_1, E_2, ... , E_6 represent a complete listing of all simple events associated with example 4.1. An interesting property of simple events is readily apparent. **An experiment will result in one and only one of the simple events.**

> **Definition**
>
> An event that cannot be decomposed is called a **simple event**. Simple events are denoted by the symbol *E* with a subscript.

For instance, if a die is tossed, we will observe a 1, 2, 3, 4, 5, or 6, but we cannot possibly observe more than one of the simple events at the same time. Hence a list of simple events provides a breakdown of all possible outcomes of the experiment. For purposes of illustration consider the following examples.

Example 4.2 Toss a coin. The simple events are

event E_1:   observe a head
event E_2:   observe a tail

Example 4.3 Toss two coins. The simple events are

EVENT	COIN 1	COIN 2
E_1	head	head
E_2	head	tail
E_3	tail	head
E_4	tail	tail

It would be extremely convenient if we could construct a model for an experiment which could be portrayed graphically. We do this by creating a correspondence between simple events and a set of points. To each simple event we assign a point, called a **sample point**. Thus the symbol E_1 will

Figure 4.1 *Venn diagram for die tossing*

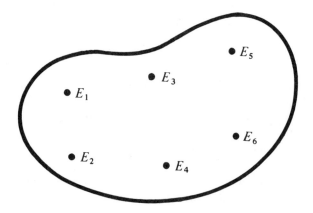

now be associated with either simple event E_1 or its corresponding sample point. The resulting diagram—the visual model—is called a **Venn diagram.**

Example 4.1 may be viewed symbolically in terms of the Venn diagram shown in figure 4.1. Six sample points are shown, corresponding to the six possible simple events enumerated in example 4.1. Likewise, a Venn diagram for the two-coin-toss experiment of example 4.3 represents an experiment that has four sample points.

Definition

The set of all sample points for an experiment is called the sample space and is represented by the symbol *S*. We say that *S* is the totality of all sample points.

What is an event in terms of the sample points? We recall that event *A* in example 4.1 occurs if any one of the simple events E_1, E_3, or E_5 occurs. That is, we observe event *A*, an odd number, if we observe a 1, 3, or 5. Event *B*, a number less than 4, occurs if E_1, E_2, or E_3 occurs. Thus if an event will occur only when one of a particular set of sample points occurs, the event is as clearly defined as if we had presented a verbal description of it. The event "observe E_1, E_3, or E_5" is obviously the same as the event "observe an odd number." This enables us to define a particular event as a specific collection of sample points.

Once again, how do you decide whether a sample point is included in a particular event? Check to see if the occurrence of the sample point implies the occurrence of the event. If it does, that sample point is included in the event. For example, in the die-tossing experiment, the sample point E_1 is in the event *A*, "observe an odd number," because if E_1 occurs, then *A* will occur.

Definition

An event is a specific collection of sample points.

Keep in mind that the preceding discussion refers to the outcome of a single experiment, that the performance of the experiment will result in the occurrence of one and only one sample point, and that an event will occur if any sample point in the event occurs.

An event can be represented on a Venn diagram by encircling the sample points in that event. Events *A* and *B* for the die-tossing problem are shown in figure 4.2. Note that points E_1 and E_3 are in both events *A* and *B* and that both *A* and *B* occur if either E_1 or E_3 occurs.

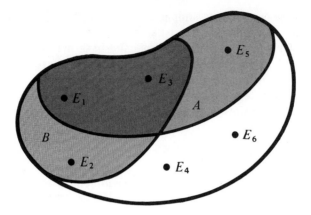

Figure 4.2 *Events A and B for die tossing*

Populations of observations are obtained by repeating an experiment a very large number of times. Some fraction of this very large number of experiments will result in E_1, another fraction in E_2, and so on. From a practical point of view, we think of the fraction of the population resulting in an event A as the probability of A. Putting it another way, **if an experiment is repeated a large number N of times and the event A is observed n_A times, the probability of A is**

$$P(A) = \frac{n_A}{N}$$

This practical view of the meaning of probability, a view held by most nonstatisticians, is called the **relative frequency concept of probability.**

In practice, the composition of the population is rarely known and hence the desired probabilities for various events are unknown. Mathematically speaking, we ignore this aspect of the problem and take the probabilities as given, thus providing a model for a real population. For instance, we would assume that a large population of die tosses for example 4.1 would yield

$$P(E_1) = P(E_2) = \cdots = P(E_6) = \tfrac{1}{6}$$

That is, we assume that the die is perfectly balanced and therefore that each of the numbers 1, 2, 3, 4, 5, and 6 should appear with approximately the same relative frequency in a long series of die tosses. Is there such a thing as a perfectly balanced die? Probably not, but we are inclined to think that the probability of the sample points would be so near $\tfrac{1}{6}$ that our assumption is quite valid for practical purposes and provides a good model for die tossing.

We complete our model for the population by adding the conditions given in the box.

To each point in the sample space we assign a number called the probability of E_i, denoted by the symbol $P(E_i)$, such that:

1. $0 \leq P(E_i) \leq 1$, for all i

2. $\displaystyle\sum_s P(E_i) = 1$

where the symbol $\displaystyle\sum_s$ means to sum the sample point probabilities over all sample points in S.

The two requirements placed on the probabilities of the sample points are necessary in order that the model conform to our relative frequency concept of probability. Thus we require that a probability be greater than or equal to 0 and less than or equal to 1 and that the sum of the probabilities over the entire sample space S be equal to 1. Furthermore, from a practical point of view, we would choose the $P(E_i)$ in a realistic way so that they would agree with the observed relative frequency of occurrence of the sample points.

Keeping in mind that a particular event is a specific collection of sample points, we can now state a simple rule for finding the probability of any event.

Definition

The **probability of an event** A is equal to the sum of the probabilities of the sample points in A.

Note that the definition agrees with our intuitive concept of probability.

Example 4.4 Calculate the probability of the event A for the die-tossing experiment of example 4.1.

Solution Event A, "observe an odd number," includes the sample points E_1, E_3, and E_5. Hence

$$P(A) = P(E_1) + P(E_3) + P(E_5) = \tfrac{1}{6} + \tfrac{1}{6} + \tfrac{1}{6} = \tfrac{1}{2}$$

Example 4.5 Calculate the probability of observing exactly one head in a toss of two coins.

Solution Construct the sample space, letting H represent a head and T a tail.

EVENT	FIRST COIN	SECOND COIN	$P(E_i)$
E_1	H	H	1/4
E_2	H	T	1/4
E_3	T	H	1/4
E_4	T	T	1/4

It seems reasonable to assign a probability of $\frac{1}{4}$ to each of the sample points. We are interested in

event A: observe exactly one head

Sample points E_2 and E_3 are in A. Hence

$$P(A) = P(E_2) + P(E_3) = \frac{1}{4} + \frac{1}{4} = \frac{1}{2}$$

Example 4.6 The personnel director of a company plans to hire two salespeople from a total of four applicants. Suppose he is completely incapable of correctly ranking the applicants according to their ability and, in effect, selects them at random.

a. What is the probability that he selects the two best candidates?

b. What is the probability that he selects at least one of the two best candidates?

Solution The experiment consists of selecting two applicants from the four available. Suppose that applicants vary in ability, and let 1, 2, 3, and 4 denote the applicants, with 1 and 2 representing the best and second best. Then the sample points in S are as shown in the table.

SAMPLE POINT	PAIR SELECTED	PROBABILITY
E_1	1, 2	1/6
E_2	1, 3	1/6
E_3	1, 4	1/6
E_4	2, 3	1/6
E_5	2, 4	1/6
E_6	3, 4	1/6

Since we would expect each of the six pairs to occur with approximately the same relative frequency in many repetitions of the experiment, the probability assigned to each sample point is $\frac{1}{6}$.

a. Define event A as "the personnel director selects the two best applicants." Since A will occur only if E_1 occurs, we have

$$P(A) = P(E_1) = \frac{1}{6}$$

b. Define event B as "the personnel director selects at least one of the two best applicants." Since B will occur if E_1, E_2, E_3, E_4, or E_5 occurs, we have

$$P(B) = P(E_1) + P(E_2) + P(E_3) + P(E_4) + P(E_5)$$

$$= \frac{1}{6} + \frac{1}{6} + \frac{1}{6} + \frac{1}{6} + \frac{1}{6} = \frac{5}{6}$$

Note the implications of the solutions to a and b. Since the personnel director's selection ability should be far better than a random selection, $P(A)$ and $P(B)$ should be less than the actual probabilities of selecting the "two best" and "at least one of the two best." A good personnel director should be able to make these selections with much higher probabilities.

As you can see, the procedure for calculating the probability of an event by summing the probabilities of the sample points requires the following steps:

Calculating the Probability of an Event:

Sample Point Approach

1. Define the experiment.
2. List the simple events associated with the experiment and test each to make certain that they cannot be decomposed. This defines the sample space S.
3. Assign reasonable probabilities to the sample points in S, making

   certain that $\sum_{S} P(E_i) = 1$.
4. Define the event of interest A as a specific collection of sample points. (A sample point is in A if A occurs when the sample point occurs. Test *all* sample points in S to locate those in A.)
5. Find $P(A)$ by summing the probabilities of the sample points in A.

Calculating the probability of an event by using the five-step procedure described in the box is systematic and leads to the correct solution if all the steps are correctly followed. A major source of error occurs in failing to define the experiment clearly (step 1) and then failing to specify simple events (step 2). A second source of error is the failure to assign valid probabilities to the sample points. The method becomes tedious (and, for all practical purposes, unmanageable) when the number of sample points in S is large, except in some special cases where sets of sample points are equiprobable. When this occurs, summation sometimes can be accomplished by using the counting rules of section 4.7.

The procedure we have described will enable you to construct a probabilistic model for a population—a model that possesses, in addition to elegance, a great deal of utility. The model provides us with a simple, logical, and direct method for calculating the probability of an event—or, if you like, the probability of a sample drawn from a theoretical population. Students familiar with probability problems will recognize the advantage of a systematic procedure for their solution. The disadvantages soon become apparent. Listing the sample points can be quite tedious, and you must be certain that none has been omitted. The total number of points in the sample space S may run into the millions.

We do not wish to pursue this point but simply mention that mathematical methods are available to simplify the counting procedure. A few useful theorems for counting sample points can be found in section 4.7 and in textbooks on combinatorial mathematics. We will study a second method for calculating the probability of an event in the following sections.

Exercises

4.1   An experiment involves tossing a single six-sided die. Specify the sample points in these events:

S: the sample space
B: observe a 4
B: observe an even number
C: observe a number less than 3

Assuming the die is balanced, calculate the probabilities of events A, B, and C by summing the probabilities of the appropriate sample points.

4.2.   Corporations must periodically decide whether or not to call in outstanding bonds and replace them with a less costly issue. At a time when prevailing interest rates are low, two corporations decide to exercise their call provision on outstanding bonds. Three investment banking firms have each submitted bids to underwrite each corporation's bond issue. One of the three firms will be selected to underwrite each of the two bond issues, and, since the underwriting firm is selected according to the lowest bid, one firm can possibly "win" both bond issues. Assume that none of the bids are identical.

a. List the sample points for the experiment.

b. Let A be the event that investment banking firm 1 wins at least one of the underwriting offers. List the sample points in A.

c. If the chance of any of the three investment banking firms submitting the lowest bid on an underwriting offer is the same for all three firms, find $P(A)$.

4.3.   A local school board consists of five residents of the community, two of whom are lawyers. The mayor plans to select two board members at random from the school board to form a subcommittee to meet with the local teachers' union on a collective bargaining issue. We are interested in the composition of the subcommittee.

a. Define the experiment.

b. List the sample points in S.

c. If all pairs of school board members have an equal chance of selection, what is the probability that the two lawyers will be selected to serve on the subcommittee?

d. What is the probability that the subcommittee will consist of at least one lawyer? No lawyers?

4.4.   In 1958 the U.S. Congress provided for the establishment of the Small Business Investment Companies (SBIC) to induce private investors to lend money to small business firms. Two small businesses in a certain metropolitan area containing four SBIC offices each seek to borrow money to supplement their working capital. Each small business randomly selects one of the four SBIC offices to apply for a loan and the result of their choices is observed.

    a. Define the experiment.

    b. List the sample points in *S*.

    c. Find the probability that both small businesses apply for a loan at the same SBIC office.

    d. Find the probability that the two small businesses each choose a different SBIC office.

4.3 Compound Events

Most events of interest in practical situations are compound events that require enumeration of a large number of sample points. Actually there is a second approach available for calculating the probability of events. This second method eliminates the necessity of listing the sample points and is therefore much less tedious and time-consuming. It is based on the classification of events, event relations, and two probability laws that will be discussed in sections 4.4 and 4.5.

Compound events, as the name suggests, are formed by a composition of two or more events. Composition takes place in one of two ways—a *union* or an *intersection*—or in a combination of the two.

Definition

Let *A* and *B* be two events in a sample space *S*. The **union of A and B** is the event containing all sample points in *A* or *B* or both. We denote the union of *A* and *B* by the symbol $A \cup B$.

Defined in ordinary terms, a union is the event that **either** event *A* or event *B* occurs **or** both *A* and *B* occur. For instance, in example 4.1, we had

$$\text{event } A: \quad E_1, E_3, E_5$$
$$\text{event } B: \quad E_1, E_2, E_3$$

The union $A \cup B$ is the collection of points E_1, E_2, E_3, and E_5. This is shown diagrammatically in figure 4.3.

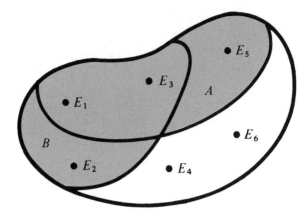

Figure 4.3 *Event A ∪ B in example 4.1; A ∪ B is represented by the shaded area*

Definition

Let *A* and *B* be two events in a sample space *S*. The **intersection of A and B** is the event composed of all sample points that are in both *A* and *B*. An intersection of events *A* and *B* is represented by the symbol *AB*. (Many authors use *A* ∩ *B*.)

The intersection *AB* is the event that **both** *A* and *B* occur. It would appear in a Venn diagram as the overlapping area between *A* and *B*. The intersection *AB* for example 4.1 is the event consisting of points E_1 and E_3. If either E_1 or E_3 occurs, both *A* and *B* occur. This is shown diagrammatically in figure 4.4.

Figure 4.4 *Event AB in example 4.1; AB is represented by the shaded area*

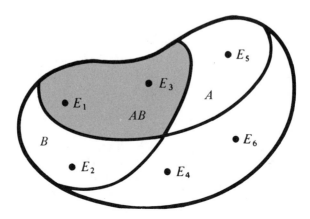

Example 4.7 Refer to the experiment of example 4.3, where two coins are tossed, and define events A and B as

event A: at least one head
event B: at least one tail

Define events A, B, AB, and $A \cup B$ as collections of sample points.

Solution Recall that the sample points for this experiment are

E_1: HH (head on first coin, head on second)
E_2: HT
E_3: TH
E_4: TT

The occurrence of sample points E_1, E_2, and E_3 implies and hence defines event A. The other events can be defined similarly:

event B: $\quad E_2, E_3, E_4$
event AB: $\quad E_2, E_3$
event $(A \cup B)$: $\quad E_1, E_2, E_3, E_4$

Note that $A \cup B = S$, which is the sample space, and thus is certain to occur.

Exercises

4.5. Consider the following experiment: Roll two fair six-sided dice and observe the total count on the upper faces of the two dice.
a. List the sample points in S.
b. List the sample points for the following events:

A: observe a 2
B: observe a 7
C: observe a total count less than 7
D: observe both A and C
E: observe both B and C
F: observe either A or C or both

c. Calculate the probabilities of events A, B, C, D, E, and F by summing the probabilities of the appropriate sample points.

4.6. A manufacturer of computing machinery has indicated that the monthly demand for the firm's small minicomputer ranges from one through seven. List the sample points in S. Then list the sample points in the following events:
a. A: two minicomputers are sold during a given month
b. B: less than four are sold
c. C: no more than five are sold
d. D: at least three are sold
e. AB
f. $A \cup B$
If we can assume that one demand level is as likely to occur as any other demand level, calculate the probabilities of events A, B, C, AB, and $A \cup B$ by summing the probabilities of the appropriate sample points.

4.7. An investor has the option of investing in two of four recommended securities. Unknown to the investor, only two of the securities will show a substantial profit within the next five years. Suppose the investor selects the two securities at random from among the four that have been recommended. List the sample points in S. Then list the sample points in the following events:

a. *A:* at least one of the profitable securities is selected
b. *B:* at least one of the unprofitable securities is selected
c. *AB*
d. *A* ∪ *B*
Construct a diagram like figure 4.3 to depict the event *A* ∪ *B*. Then construct a diagram like figure 4.4 to depict *AB*.

4.4 Event Relations

We will define three relations between events: complementary, independent, and mutually exclusive events. You will have many occasions to inquire whether two or more events bear a particular relationship to one another. The test for each relationship is inherent in the definition, as we will illustrate, and its use in the calculation of the probability of an event will become apparent in section 4.5.

Definition

The **complement of an event** *A* is the collection of all sample points that are in the sample space *S* but not in *A*. The complement of *A* is denoted by the symbol *Ā*.

Using a Venn diagram, complementary events appear as shown in figure 4.5.

Since we know that

Figure 4.5 *Complementary events*

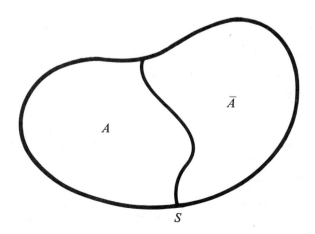

$$\sum_S P(E_i) = 1$$

we have

$$P(A) + P(\bar{A}) = 1$$

and so we get this useful equation for obtaining $P(A)$ when $P(\bar{A})$ is known or easily calculated:

$$P(A) = 1 - P(\bar{A})$$

Two events are often related in such a way that the probability of occurrence of one depends on whether the second has or has not occurred. For instance, suppose that one experiment consists in observing the weather on a specific day. Let A be the event "observe rain" and B be the event "observe an overcast sky." Events A and B are obviously related. The probability $P(A)$ of rain is not the same as the probability of rain given prior information that the day is cloudy. The probability of A, $P(A)$, is the fraction of the entire population of observations that result in rain. Now let us look only at the subpopulation of observations resulting in B, a cloudy day, and the fraction of these that result in A. This fraction, called the conditional probability of A given B, may equal $P(A)$ but we would expect the chance of rain, given that the day is cloudy, to be larger. The conditional probability of A, given that B has occurred, is denoted as

$$P(A|B)$$

where the vertical bar in the parentheses is read "given" and the events indicated to the right of the bar are the events that have occurred.

We define the conditional probabilities of B given A and A given B as shown in the box.

Definition

The conditional probability of B, given that A has occurred, is

$$P(B|A) = \frac{P(AB)}{P(A)}$$

The conditional probability of A, given that B has occurred, is

$$P(A|B) = \frac{P(AB)}{P(B)}$$

We can see that this definition of conditional probability is consistent with the relative frequency concept of probability by attaching some numbers to the probabilities in our weather example. Recall that A denotes rain on

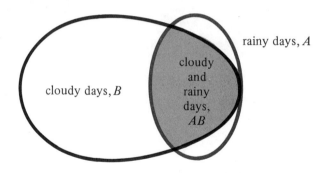

Figure 4.6 *Events A and B*

a given day; B denotes that the day is cloudy. Now suppose that 10% of all days are rainy and cloudy [that is, $P(AB) = .10$] and 30% of all days are cloudy [$P(B) = .30$].

The situation we have just described is graphically portrayed in figure 4.6. Each sample point in event B, denoted by the large egg-shaped area, is associated with a single cloudy day. Since 30% of all days will be cloudy, we can regard this area as .3. One-third (10%/30%) of these cloudy days will be rainy. These days are included in the shaded area, event AB. Hence if a single day is selected from the set of all days representing the population, what is the probability that we will select a rainy day, given that we know the day is cloudy? That is, what is $P(A|B)$?

Since we already know that the day is cloudy, we know that the sample point to be selected must fall in event B (figure 4.6). One-third of these days will result in rain. Hence the probability that we will select a rainy day is

$$P(A|B) = \tfrac{1}{3}$$

You can see that this result agrees with our definition for $P(A|B)$. That is,

$$P(A|B) = \frac{P(AB)}{P(B)} = \frac{.10}{.30} = \frac{1}{3}$$

Example 4.8 Calculate $P(A|B)$ for the die-tossing experiment described in example 4.1.

Solution Events A and B are defined as follows:

event A: observe an odd number (E_1, E_3, E_5)
event B: observe a number less than 4 (E_1, E_2, E_3)

Given that B has occurred, we are concerned only with sample points E_1, E_2, and E_3, which occur with equal frequency. Of these, E_1 and E_3 imply event A. Hence

$$P(A|B) = \tfrac{2}{3}$$

Or we could obtain $P(A|B)$ by substituting into the equation:

$$P(A|B) = \frac{P(AB)}{P(B)} = \frac{1/3}{1/2} = \frac{2}{3}$$

Note that $P(A|B) = \tfrac{2}{3}$ while $P(A) = \tfrac{1}{2}$, indicating that A and B are dependent on each other.

Definition

Two events A and B are said to be **independent** if either

$$P(A|B) = P(A) \qquad \text{or} \qquad P(B|A) = P(B)$$

Otherwise, the events are said to be **dependent**.

Expressing this definition verbally, two events are independent if the occurrence or nonoccurrence of one of the events does not change the probability of the occurrence of the other.

We note that if $P(A|B) = P(A)$, then $P(B|A)$ also equals $P(B)$. Similarly, if $P(A|B)$ and $P(A)$ are unequal, then $P(B|A)$ and $P(B)$ are unequal.

A third useful event relation was observed but not specifically defined in our discussion of simple events. Recall that an experiment can result in one and only one simple event. No two simple events can occur at exactly the same time. Two events A and B are said to be *mutually exclusive* if when one occurs, the other cannot occur. Another way to say this is to state that the intersection AB contains no sample points. It then follows that $P(AB) = 0$.

Definition

Two events A and B are **mutually exclusive** if the event AB contains no sample points.

Mutually exclusive events have no overlapping area in a Venn diagram (see figure 4.7).

Example 4.9 Are events A and B given in example 4.1 mutually exclusive? Are they complementary? Independent?

Solution

$$\text{event } A: \quad E_1, E_3, E_5$$
$$\text{event } B: \quad E_1, E_2, E_3$$

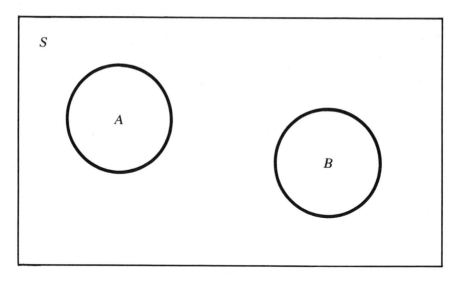

Figure 4.7 *Mutually exclusive events*

The event AB is the set of sample points in both A and B. You can see that AB includes points E_1 and E_3 and that $P(AB)$ is not equal to 0. Therefore, A and B are not mutually exclusive. They are not complementary because B does not contain all points in S that are not in A.

The test for independence lies in the definition. That is, we must check to see if $P(A|B) = P(A)$. From example 4.8, $P(A|B) = \frac{2}{3}$. Then, since $P(A) = \frac{1}{2}$,

$$P(A|B) \neq P(A)$$

and, by definition, events A and B are dependent.

Example 4.10 Experience has shown that a particular union-management contract negotiation has led to a contract settlement within a two-week period 50% of the time, that the union strike fund has been adequate to support a strike 60% of the time, and that both of these conditions have been satisfied 30% of the time. What is the probability of a contract settlement given that the union strike fund is adequate to support a strike? Is settlement of a contract within a two-week period dependent on whether the union strike fund is adequate to support a strike?

Solution We first define the following events:

event A: a contract settlement is reached within a two-week period
event B: the union strike fund is adequate to support a strike

Thus we wish to find $P(A|B)$.
We are given that $P(A) = .50$, $P(B) = .60$, and $P(AB) = .30$. Then

$$P(A|B) = \frac{P(AB)}{P(B)} = \frac{.30}{.60} = .50$$

To determine whether events A and B are dependent, we compare $P(A|B)$ and $P(A)$. Since

$$P(A|B) = P(A) = .50$$

then, by definition, events A and B are independent.

Exercises

4.8.   Refer to exercise 4.5 and suppose that the two dice are separately identifiable by color: one is red and the other green. Using the definitions in sections 4.2, 4.3, and 4.4, find the probability of each of the following events:

A: observe a 2 on the red die
B: observe an even number on the red die
C: observe a total count of 7 on both dice
D: observe a total count of 9 on both dice
E: $C|A$
F: $D|A$
G: $C|B$

4.9.   If it is known that the probability that daily demand for a product exceeds two units is $9/10$, what is the probability that the daily demand is less than three units?

4.10.   Oil drilling contractors typically conduct a seismic sounding of a parcel before deciding whether or not to drill. The detection of a closed structure in the terrain below the site is a hopeful sign, while the detection of no structure usually implies a lower probability of a successful drilling. The table represents the resulte of a very large number of seismic soundings on sites where drilling was actually conducted. Table entries are the approximate probabilities of occurrence of the four combinations of "structure type" and "drilling outcome."

	B_1, NO STRUCTURE	B_2, CLOSED STRUCTURE
A_1, *Unproductive Well*	.40	.10
A_2, *Productive Well*	.15	.35

Calculate the following probabilities and then describe each probability in terms of the foregoing problem:

$$P(A_1), P(\bar{A}_1), P(A_1|B_1), P(A_2|B_1), P(A_1B_2), P(B_2|A_2), P(A_2|A_1)$$

4.11.   Through a certain broker 100 stocks are available. Sixty are on the New York exchange; 30 are preferred stocks, of which 20 are preferred stocks on the New York exchange. A man goes to the broker and buys a stock from the New York exchange. (He is unaware that some stocks are preferred.) Given that he bought a stock from the New York exchange, what is the probability that it is preferred?

4.12.   A recent article described the use of computerized credit-scoring models as an aid in determining whether a financial institution should grant a loan to an applicant.* The studies show that 2% of all loans result in default at some time

*R. D. Edmister and G. G. Schlarbaum, "Credit Policy in Lending Institutions," *Journal of Financial and Quantitative Analysis*, June 1974.

before maturity and that the credit-scoring model is 84% accurate in correctly predicting good loans. If 90% of all applicants are scored favorably and granted a loan by the model, find the probability that an applicant will not default on a loan if he is granted a loan on the basis of the model.

4.5 Two Probability Laws and Their Use

As previously stated, a second approach to the solution of probability problems is based on the composition of events, event relations, and two probability laws, which we now state. The laws can be simply stated and accepted, as long as they are consistent with our model and with reality. The first, called the *multiplicative law* of probability, follows directly from the definition of conditional probability. It provides a formula for calculating the probability of an intersection.

The Multiplicative Law of Probability

Given two events A and B, the probability of the intersection AB is

$$P(AB) = P(A)P(B|A)$$
$$= P(B)P(A|B)$$

If A and B are independent, $P(AB) = P(A)P(B)$.

Here is an illustration of the multiplicative law.

Example 4.11 When receiving a shipment of goods from a supplier, the buyer typically conducts an inspection of the quality of the goods received. A discount store has received a lot of 100 portable television sets from a manufacturer. Unknown to the management of the discount store, 10 of the 100 television sets are defective. If 2 sets are randomly selected from the lot of 100 and then subjected to an exhaustive quality inspection, what is the probability that both will be defective?

Solution We first define the events:

event A: the first set is defective
event B: the second set is defective

Then AB is the event that both are defective, and

$$P(AB) = P(A)P(B|A)$$

$P(A) = .10$ since there are 10 defectives in the lot of 100. However, $P(B|A) = \frac{9}{99}$ since, after the first is selected and found to be defective, there are 99 sets remaining, of which 9 are defective. Thus

$$P(AB) = P(A)P(B|A) = \left(\frac{10}{100}\right)\left(\frac{9}{99}\right) = \frac{1}{110}$$

The second law of probability, called the *additive law*, applies to unions.

The Additive Law of Probability

The probability of a union $A \cup B$ is

$$P(A \cup B) = P(A) + P(B) - P(AB)$$

If A and B are mutually exclusive, then

$$P(AB) = 0 \quad \text{and} \quad P(A \cup B) = P(A) + P(B)$$

The additive law conforms to reality and our model. The sum $P(A)$ + $P(B)$ contains the sum of the probabilities of all sample points in $A \cup B$ but includes a double counting of the probabilities of all points in the intersection AB. Subtracting $P(AB)$ gives the correct result.

Example 4.12 A May 1975 issue of the *Wall Street Journal* reported that 40% of their subscribers regularly read *Time*, 32% read *U.S. News and World Report*, and 11% read both weekly news magazines. We define events A and B to be

> event A: a *WSJ* subscriber reads *Time*
> event B: a *WSJ* subscriber reads *U.S. News*

Find the probability of the events A, B, AB, and $A \cup B$.

Solution The experiment involves the selection of a single *WSJ* subscriber and a noting of the magazines he or she reads regularly. One sample point corresponds to the selection of each subscriber, and the sample points are equiprobable (equally likely to occur). Then since 40% of the *WSJ* subscribers read *Time*, 40% of the sample points in S are in the event A and

$$P(A) = .4$$

Similarly,

$$P(B) = .32$$

Since 11% of the *WSJ* subscribers read both magazines, we have

$$P(AB) = .11$$

Then

$$P(A \cup B) = P(A) + P(B) - P(AB) = .40 + .32 - .11 = .61$$

The use of the probability laws for calculating the probability of a compound event is less direct than the listing of sample points and it requires a bit of experience and ingenuity. The approach involves the expression of the event of interest as a union or intersection (or combination of both) of two or more events whose probabilities are known or easily calculated. This can often be done in many ways. The trick is to find the right combination, a task that requires considerable creativity in some cases. The usefulness of event relations is now apparent. If the event of interest is expressed as a union of mutually exclusive events, the probabilities of the intersection need not be known. If they are independent, we need not know the conditional probabilities to calculate the probability of an intersection. Example 4.13 illustrates the use of the probability laws and the technique described above.

Example 4.13 In spite of the demand for their services, competition among oil drilling contractors has increased. As a result, drilling contractors are more cost conscious and have tended to limit the number of holes they drill in any parcel of land. Suppose a drilling contractor has sufficient resources to drill up to two holes on a given parcel but will stop drilling with his first success. If the probability that he is successful on any given drill hole is .2, find the probability that the drilling contractor locates a productive well. Assume that the drilling outcomes are independent from one drill hole to another.

Solution Let us list the events which, when they occur, satisfy the objective of the problem.

A: a productive well is found within two trials
S_1: a productive well is found on the first trial
S_2: a productive well is found on the second trial
F_1: a dry well is found on the first trial
F_2: a dry well is found on the second trial

We see that event A occurs if

1. he strikes oil on the first trial, S_1, or
2. he hits a dry well on the first trial and strikes oil on the second, $F_1 S_2$.

Thus

$$A = S_1 \cup F_1 S_2$$

Note that these events are mutually exclusive. Then, since the events S_1 and $F_1 S_2$ are mutually exclusive, we have

$$P(A) = P(S_1 \cup F_1 S_2) = P(S_1) + P(F_1 S_2)$$

We know that

$$P(S_1) = P(S_2) = .2 \quad \text{and} \quad P(F_1) = P(F_2) = .8$$

and the outcomes "success" or "dry" at any trial are independent of the outcome at any other trial. Applying the multiplicative law, we have

$$P(F_1 S_2) = P(F_1)P(S_2) = (.8)(.2) = .16$$

Substituting we get

$$P(A) = P(S_1) + P(F_1 S_2) = .2 + .16 = .36$$

The steps for calculating the probability of an event using the event composition approach can be summarized as follows:

Calculating the Probability of an Event: Event Composition Approach

1. Define the experiment.
2. Clearly visualize the nature of the sample points. Identify a few to clarify your thinking.
3. Write an equation expressing the event of interest, say A, as a composition of two or more events using either or both of the two forms of composition (unions and intersections). Note that this equates point sets. Make certain that the event implied by the composition and event A represent the same set of sample points.
4. Apply the additive and multiplicative laws of probability to step 3 and find P(A).

Step 3 is the most difficult because we can form many compositions that will be equivalent to event A. The trick is to form a composition in which all the probabilities appearing in step 4 will be known. Thus we must visualize the results of step 4 for any composition and select the one for which the component probabilities are known.

The event composition approach to the calculation of the probability of an event is often more powerful than the method of section 4.2 because it avoids the necessity of actually listing the sample points in S and therefore can be used when the number of sample points is very large. However, it is less direct and requires considerably more creative ability.

Exercises

4.13. Buyers of large quantities of goods from a supplier often use sampling inspection schemes to judge the incoming quality of a supply. The supply of goods is accepted or rejected on the basis of results observed by administering tests to a few sample items selected from the supply. Suppose we know that an inspector for a food processing firm has accepted 98% of all good shipments and has incorrectly rejected 2% of good shipments. In addition, the inspector accepts 94% of all shipments and it is known that 5% of all shipments are of inferior quality.
   a. Find the probability that a shipment is rejected.
   b. Find the probability that a shipment is good.
   c. Find the probability that a shipment is good and that it is accepted.
   d. Find the probability that a shipment is of inferior quality and that it is accepted.

[Hint: Let A be the event that a shipment is accepted and G the event that a shipment is good. Then $P(A) = P(AG) + P(A\bar{G})$.]

    e. Find the probability that a shipment is accepted, given that it is of inferior quality.

    f. Find the probability that a shipment is rejected, given that it is good.

    4.14.   Television commercials are designed to appeal to the most likely viewing audience of the sponsored program. However, Ward notes that children often have a very low understanding of commercials, even of those designed to appeal especially to children.* Ward's studies show the following percentages of children who do or do not understand TV commercials, for the age groups listed:

		AGE	
	5–7	8–10	11–12
Do Not Understand (%)	55	40	15
Understand (%)	45	60	85

    An advertising agent has shown a television commercial to a 6-year-old and another to a 9-year-old child in a laboratory experiment to test their understanding of the commercials.

    a. What is the probability that the message of the commercial is understood by the 6-year-old child?

    b. What is the probability that both children demonstrate an understanding of the TV commercials?

    c. What is the probability that one or the other or both children demonstrate an understanding of the TV commercials?

    4.15.   In a recent study of the effects that tying agreements to a franchise have on the franchise system of distribution, a researcher contacted a large number of franchisees. In each case the researcher recorded whether the prices the franchisor charges for its goods to the franchisees are higher, about the same, or lower than the franchisee would pay if he were not obligated to buy his goods from the franchisor. The results obtained are as shown in the table.

PRICE LEVEL	PERCENTAGE
franchisor's prices are higher	47
franchisor's prices are about the same	28
franchisor's prices are lower	25

Source: Adapted from S. D. Hunt and J. R. Nevin, "Tying Agreements in Franchising," *Journal of Marketing*, July 1975, published by The American Marketing Association.

Use the percentages in the table as approximate probabilities.

    a. Find the probability that a randomly selected franchisee pays a higher price for goods due to the franchise agreement.

    b. Find the probability that in a randomly selected pair of franchisees, both pay a higher price for their goods due to their franchise-tying agreements.

    c. Find the probability that in a randomly selected pair of franchisees, only one pays a higher price for the goods due to the franchise-tying agreement.

    d. If three franchisees are selected at random, find the probability that at least

*S. Ward, "Children's Reactions to Commercials." Reprinted from the *Journal of Advertising Research.* © Copyright 1972 by the Advertising Research Foundation.

two pay a higher price for their goods due to their franchise-tying agreements.

4.16.   At the beginning of each month, a company decides whether to spend $100 or $200 on advertising for that month. Monthly decisions are independent of one another, and the two spending options are equiprobable.
a. What is the probability that in three consecutive months a total of more than $400 is spent on advertising?
b. What is the probability that $100 is spent in each of the three months?
[Note: The probability of the intersection of three events A, B, and C is given by the formula

$$P(ABC) = P(A)P(B|A)P(C|AB)$$

If A, B, and C are independent events, then we have $P(ABC) = P(A)P(B)P(C)$.]

4.17.   Refer to exercise 4.10 and find the following probabilities:
a. $P(A_1 \cup A_2)$
b. $P(A_1 \cup B_1)$
c. $P(A_1 \cup A_2 \cup B_2)$
d. $P(A_1 \cup A_2 | B_2)$
[Hint: The union of three events A, B, and C is given by the formula

$$P(A \cup B \cup C) = P(A) + P(B) + P(C) - P(AB) - P(AC) - P(BC) + P(ABC)$$

If A, B, and C are mutually exclusive events, then we have $P(A \cup B \cup C) = P(A) + P(B) + P(C)$.]

4.18.   Refer to exercise 4.10 and suppose a drilling contractor has conducted seismic soundings at three different sites, detecting a closed structure at two sites and an open structure at the third. If the contractor drills at all three sites, find the probability that he discovers exactly two productive wells. At least two productive wells. (Assume that the drilling outcomes at the three sites are independent of each other.)

4.19.   If a buyer rejects a supplier's shipment of goods on the basis of evidence gained from sampling inspection, the supplier, if he truly believes the shipment to be of acceptable quality, may insist that the buyer perform a second sampling inspection survey. Refer to exercise 4.13, and assume that the probability that a shipment is accepted is the same for an initial as well as a second inspection. Find the probability that a shipment of goods is accepted on the basis of a second sample inspection, given that it was rejected on an initial survey.

4.20.   A recent study reveals that about 90% of all American housewives believe that advertising should provide more information about the product and that manufacturers should be required to provide more clearly written warranties.* In a group of three housewives, find the probability that at least two are in favor of the "additional information" issue reported in the study. Assume that the responses for the three housewives are independent events.

4.6 Bayes's Law

In the usual sense of conditional probability, we seek the probability of some event A given that an event B has occurred. Usually we say that event A is an end effect for which event B is a possible cause and that B and A are sequenced, in that order, over time. For instance, we might

*N. Kangun, K. K. Cox, J. Higginbotham, and J. Burton, "Consumerism and Marketing Management," *Journal of Marketing*, April 1975.

be interested in the conditional probability $P(A|B)$, where A is the event "an insurance salesperson sells 15 policies" and B is the event "the salesperson contacts 40 clients." Suppose we know **after the fact** that the salesperson sold 15 policies last week but we do not know how many clients were contacted. That is, how can we find the probability that some particular event B is the "cause" among many possible causes of a known end effect A? Such probabilities are given by Bayes's law, which we state here without proof.

Bayes's Law

Let B be an event and $\bar{B}$ be its complement. If another event A occurs, then

$$P(B|A) = \frac{P(AB)}{P(A)} = \frac{P(A|B)P(B)}{P(A|B)P(B) + P(A|\bar{B})P(\bar{B})}$$

Our computed probability $P(B|A)$ is called the **posterior** probability of the event B given the information contained in event A. The unconditional probabilities $P(B)$ and $P(\bar{B})$ are called the **prior** probabilities of the events B and $\bar{B}$, respectively. In a sense Bayes's law is updating or revising the prior probability $P(B)$ by incorporating into the model the observed information contained within event A. For instance, if we want to find the probability, after the fact, that our salesperson contacted 40 clients last week, the proportion $P(B)$ of all work weeks the salesperson contacted 40 clients ignores the information that is contained in the known event A, the event that the salesperson sold 15 policies last week. However, the posterior probability $P(B|A)$ reflects both the prior and current information and provides a more efficient model for our problem.

Sometimes the conditional probabilities $P(A|B)$ and $P(A|\bar{B})$, or the unconditional probabilities $P(B)$ and $P(\bar{B})$, are not known exactly. When they are not known, empirical or subjective estimates are often used instead, with varying degrees of accuracy. This problem is explored in some detail in chapter 10.

Example 4.14 A department store is considering adopting a new credit management policy in an attempt to reduce the number of credit customers defaulting on their payments. The credit manager has suggested that, in the future, credit should be discontinued to any customer who has twice been a week or more late with his monthly installment payment. He supports his claim by noting that past credit records show that 90% of all those defaulting on their payments were late with at least two monthly payments.

Suppose from our own investigation we have found that 2% of all credit customers actually default on their payments and that 45% of those who have not defaulted have had at least two late monthly payments.

Find the probability that a customer with two or more late payments will actually default on his payments and, in light of this probability, criticize the credit manager's credit plan.

Solution Let the events L and D be defined as follows:

event L: a credit customer is two or more weeks late with at least two monthly payments

event D: a credit customer defaults on his payments

and let $\bar{D}$ denote the complement of event D. We seek the conditional probability

$$P(D|L) = \frac{P(DL)}{P(L)} = \frac{P(L|D)P(D)}{P(L|D)P(D) + P(L|\bar{D})P(\bar{D})}$$

From the information given within the problem description, we find that

$$P(D|L) = \frac{(.90)(.02)}{(.90)(.02) + (.45)(.98)} = \frac{.0180}{.0180 + .4410} = .0392$$

Therefore, if the credit manager's plan is adopted, the probability is only about .04, or the chances are only about 1 in 25, that a customer who loses his credit privileges would actually have defaulted on his payments. Unless management would consider it worthwhile to detect one prospective defaulter at the expense of losing 24 good credit customers, the credit manager's plan would be a poor business policy.

Exercises

4.21. Refer to exercise 4.13. Find the probability that a shipment is of acceptable quality, given that it has been accepted by the inspector.

4.22. Similar to the application discussed in exercise 4.13, sampling inspection plans are also used by a manufacturer of goods to monitor the ongoing quality of his manufactured items. This way he hopes to detect items of inferior quality and remove them from the lot before a shipment is sent to a buyer. Suppose that in a certain manufacturing plant, as items come to the end of a production line, an inspector chooses those items that are to go through a complete inspection. Ten percent of all items produced are defective, 60% of all defective items go through a complete inspection, and 20% of all good items go through a complete inspection. Given that an item is completely inspected, what is the probability that it is defective?

4.23. Each of two appliance stores (stores 1 and 2) sell each of two brands of washing machines (brands A and B). These are the only stores selling these machines. The probability that someone shops at Store 1 is 3/4. The probability that someone shopping at Store 1 buys Brand A is 1/3. The probability that someone shopping at Store 2 buys Brand A is 1/4. Given that someone bought a Brand A washing machine, what is the probability that it was purchased at Store 1?

4.24. One of the purposes of an audit is to detect the presence of material errors, procedural errors, or errors in judgment in the recording of accounting information. Suppose a CPA firm has been retained to conduct an audit of the accounting practices of a firm in which accounts are processed by two divisions, a wholesale division and a retail division. It is known that 70% of all accounts are retail accounts.

Furthermore, it is known that 10% of all retail accounts and 20% of all wholesale accounts contain some type of accounting error. If the auditor observes an accounting error in an account, find the probability that the account was processed by the firm's retail division.

4.7 Counting Sample Points (Optional)

The preceding sections cover the basic concepts of probability and provide a background that will help you to understand the role probability plays in making inferences. To illustrate the concepts we used only examples and exercises for which the total number of sample points in the sample space was small. This allowed you to list the sample points in S, to identify the sample points in the event of interest, and then to calculate its probability. Although these examples and exercises were adequate for our learning objective, most real life problems involve many more sample points. Consequently, we include this optional section for the student who wishes to improve his or her problem-solving ability. (Note: Developing the ability to solve problems requires practice.)

In this section we present some elementary counting rules that can be of assistance in solving probability problems that involve a large number of sample points. For example, suppose you are interested in the probability of an event A and you know that the sample points in S are equiprobable. Then

$$P(A) = \frac{n_A}{N}$$

where

n_A = the number of sample points in A

N = the number of sample points in S

Often we can use counting rules to find the values of n_A and N and thereby eliminate the necessity of listing the sample points in S.

We will present three elementary counting rules. The first, known as the *mn rule*, applies to situations where you seek the number of ways you can form pairs of objects, where the objects are selected, one each, from two different groups. For example, suppose that four companies have job openings in each of three areas: sales, manufacturing, and personnel. How many job opportunities are available to you? You can see that you have two sets of objects: companies (four) and types of jobs (three). Therefore, as shown in figure 4.8, there are three jobs for each of the four companies,

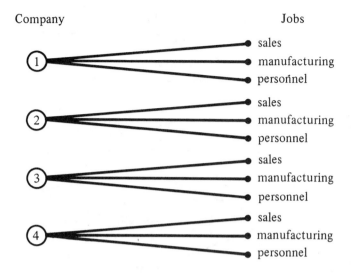

Figure 4.8 *Company-job combinations*

or (4)(3) = 12 possible company-job combinations.

The preceding example illustrates the use of the *mn* rule.

The *mn* Rule

With *m* elements a_1, a_2, ... , a_m and *n* elements b_1, b_2, ... , b_n, it is possible to form *mn* pairs that contain one element from each group.

Two dice are tossed. How many sample points are associated with the experiment?

Solution The first die can fall in one of six ways; that is, *m* = 6. Likewise, the second die can fall in *n* = 6 ways. The total number *N* of sample points is

$$N = mn = 6(6) = 36$$

As noted, the *mn* rule gives the number of pairs you can form in selecting one object from each of two groups. The rule can be extended to apply to triplets formed by selecting one object from each of three groups, quadruplets formed by selecting one object from each of four groups, and so on. The application to triplets is shown in figure 4.9. If you have *m* elements in the first group, *n* in the second, and *t* in the third, the total number of triplets that you can form, taking one object from each group, is equal to *mnt*, the number of branchings shown in figure 4.9.

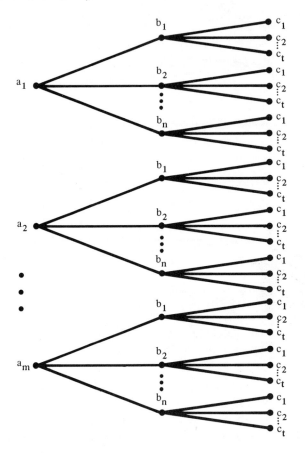

Figure 4.9 *Forming mnt triplets*

Example 4.16 How many sample points are in the sample space when three coins are tossed?

Solution Each coin can land in one of two ways. Hence

$$N = 2(2)(2) = 8$$

Example 4.17 A truck driver can take three routes in going from City A to City B, four from City B to City C, and three from City C to City D. If, in going from A to D, he must proceed A to B to C to D, how many possible A-to-D routes are available to him?

Solution Let

$$m = \text{the number of routes from A to B} = 3$$

$$n = \text{the number of routes from B to C} = 4$$

$t = $ the number of routes from C to D $= 3$

Then the total number of ways that you can construct a complete route, taking one subroute from each of the three groups (A to B), (B to C), and (C to D), is

$$mnt = (3)(4)(3) = 36$$

A second useful mathematical result is associated with orderings, or *permutations*. For instance, suppose that we have three books, b_1, b_2, and b_3. In how many ways can the books be arranged on a shelf, taking them two at a time? We enumerate, listing all combinations of two in the first column and a reordering of each in the second column:

COMBINATIONS OF TWO	REORDERING OF COMBINATIONS
$b_1 b_2$	$b_2 b_1$
$b_1 b_3$	$b_3 b_1$
$b_2 b_3$	$b_3 b_2$

The number of permutations is 6, a result easily obtained from the *mn* rule. The first book can be chosen in $m = 3$ ways, and, once selected, the second book can be chosen in $n = 2$ ways. The result is $mn = 6$.

In how many ways can three books be arranged on a shelf taking three at a time? Enumerating, we obtain

$$b_1 b_2 b_3 \qquad b_2 b_1 b_3 \qquad b_3 b_1 b_2$$
$$b_1 b_3 b_2 \qquad b_2 b_3 b_1 \qquad b_3 b_2 b_1$$

a total of 6. This again could be obtained easily by the extension of the *mn* rule. The first book can be chosen and placed in $m = 3$ ways. After choosing the first, the second can be chosen in $n = 2$ ways, and, finally, the third in $t = 1$ way. Hence the total number of ways is

$$N = mnt = 3 \cdot 2 \cdot 1 = 6$$

Definition

An ordered arrangement of r distinct objects is called a **permutation**. The number of ways of ordering n distinct (different) objects taken r at a time is designated by the symbol P_r^n.

As we have noticed, the number of permutations of n objects taken r at a time is equivalent to finding the number of ways of filling r positions with n distinct objects. The counting rule for finding P_r^n is as given in the box.

A Counting Rule for Permutations

$$P^n_r = n(n-1)(n-2) \cdots (n-r+1) = \frac{n!}{(n-r)!}$$

Recall that $n! = n(n-1)(n-2) \cdots (3)(2)(1)$ and $0! = 1$. Thus we have $4! = 4 \cdot 3 \cdot 2 \cdot 1 = 24$.

Example 4.18 Three lottery tickets are drawn from a total of 50. Assume that order is of importance. How many sample points are associated with the experiment?

Solution Since $n = 50$ and $r = 3$, the total number of sample points is

$$P^{50}_3 = \frac{50!}{47!} = 50(49)(48) = 117{,}600$$

Example 4.19 A piece of equipment is composed of five parts, which may be assembled in any order. A test is conducted to determine the length of time necessary for each order of assembly. If each order is tested once, how many tests must be conducted?

Solution Since $n = 5$ and $r = 5$, the total number of tests is

$$P^5_5 = \frac{5!}{0!} = 5(4)(3)(2)(1) = 120$$

The enumeration of the permutations of books in the previous discussion was performed in a systematic manner, first writing the combinations of n books taken r at a time and then writing the rearrangements of each combination. In many situations, ordering is unimportant because we are interested solely in the number of possible combinations. For instance, suppose that an experiment involves the selection of 5 men, a committee, from a total of 20 candidates. Then the simple events associated with this experiment correspond to the different combinations of men selected from the group of 20. How many simple events (different combinations) are associated with this experiment? Since order in a single selection is unimportant, permutations are not called for. We are interested in the number of *combinations* of $n = 20$ things taken $r = 5$ at a time.

Definition

The **number of combinations** of n objects taken r at a time is denoted by the symbol C^n_r. [Note: Some authors prefer the symbol $\binom{n}{r}$.]

The number of combinations of n objects taken r at a time can be found using the counting rule in the box.

A Counting Rule for Combinations

$$C_r^n = \frac{n!}{r!\,(n-r)!}$$

Example 4.20 A radio tube may be purchased from five suppliers. In how many ways can three suppliers be chosen from the five?

Solution

$$C_3^5 = \frac{5!}{3!\,2!} = \frac{(5)(4)}{2} = 10$$

The following example illustrates the use of the counting rules in the solution of a probability problem.

Example 4.21 Five manufacturers, of varying but unknown quality, produce a certain type of electronic tube. If we were to select three manufacturers at random, what is the chance that the selection would contain exactly two of the best three?

Solution Without enumerating the sample points, we would likely agree that each point, that is, any combination of three, would be assigned equal probability. If N points are in S, then each point receives probability

$$P(E_i) = \frac{1}{N}$$

Let n_A be the number of points in which two of the best three manufacturers are selected. Then the probability of including two of the best three manufacturers in a selection of three is

$$P = \frac{n_A}{N}$$

Our problem is to use the counting rules to find n_A and N.

Since order within a selection is unimportant and is unrecorded, each selection is a combination and hence

$$N = C_3^5 = \frac{5!}{3!\,2!} = 10$$

Determination of n is more difficult, but it can be obtained using the *mn* rule. Let a be the number of ways of selecting exactly two from the best three, or

$$a = C_2^3 = \frac{3!}{2!\,1!} = 3$$

and let b be the number of ways of choosing the remaining manufacturer from the two poorest, or

$$b = C_1^2 = \frac{2!}{1!\,1!} = 2$$

Then the total number of ways of choosing two of the best three in a selection of three is $n_A = ab = 6$.

Hence the probability P is

$$P = \frac{n_A}{N} = \frac{6}{10}$$

Students sometimes have difficulty deciding whether one of the counting rules will be applicable to a particular problem and, if so, deciding which one to use. The following clues may aid you in solving problems.

Diagnostic Clues for Applying the Counting Rules

1. One of three counting rules discussed in this section may be applicable in solving a probability problem if the sample point is identified by a fixed number, say r, of characteristics.
2. The *mn* rule may apply if the characteristics in part 1 are taken one each from *two or more sets*. In contrast, the permutation and combination rules apply to the selection of characteristics from a single set.
3. The combinations rule may apply if the characteristics are selected from a *single set* and reordering or rearranging the characteristics does not produce another sample point.
4. The permutations rule may apply if the characteristics are selected from a *single set* and each *arrangement* of a specific set of characteristics corresponds to a new sample point.

Exercises

Check each of the following exercises against the diagnostic clues.

4.25.   A traveler can take one of three airlines to San Francisco, and each airline operates four daily nonstop flights. If the selection of a particular airline-flight combination represents a sample point, how many characteristics are used to identify a sample point (see diagnostic clue 1)? How many sets are they selected from? Use the *mn* rule to give the number of airline-flight combinations available to the traveler. Construct a diagram similar to figure 4.8 to identify the airline-flight combinations.

4.26.   Use the *mn* rule to find the number of sample points associated with the experiment in exercise 4.2.

4.27.   A combination lock will open when the right combination of three digits is selected. Each digit can take a value from 0 to 9. If a particular combination of digits represents a sample point, how many characteristics are used to identify a sample point? How many sets are they selected from? Use the *mn* rule to give the number of possible combinations of digits.

4.28.   Ten persons enter a hospital clinic to be treated by three physicians. An outcome of this experiment is identified by a set of doctor-patient encounters (such as Doctor A sees Jones, Doctor A sees Smith, Doctor B sees Mucello, etc.). Use the *mn* rule to find the number of sample points for the experiment.

4.29.   The president, vice-president, secretary, and treasurer are to be selected from a group of 10 candidates. Use the permutations rule to give the number of ways the positions may be filled.

4.30.   A tennis team has 10 members, who may be assigned to play one of the six singles matches (the singles matches are ordered as no. 1 singles, no. 2 singles, etc.). Use the permutations rule to find how many different player-position assignments are available to the coach.

4.31.   Use the combinations rule to find the number of ways you can select two securities out of four. This will give you the total number of sample points for exercise 4.7.

4.32.   Five secretaries are selected from a group of 20 to form a secretarial pool. Use the combinations rule to give the total number of different pools that could be formed.

4.33.   An experiment consists in assigning 10 workmen to 10 different jobs. In how many different ways can the 10 men be assigned to the 10 jobs?

4.34.   Refer to exercise 4.33. Suppose there are only 4 different jobs available for the 10 men. In how many different ways can 4 men be selected from the 10 and assigned to the 4 jobs?

4.35.   A study is conducted to determine the attitudes of nurses in a hospital to various administrative procedures that are currently employed. If a sample of 10 nurses is selected from a total of 90, how many different samples can be selected? (Note that order within a sample is unimportant.)

4.36.   Buyers of television sets are offered a choice of one of three different styles. How many different outcomes could result if one customer makes a selection? Two customers? Ten?

4.37.   If 5 cards are to be selected, one after the other, in sequence from a 52-card deck, each card being replaced in the deck before the next draw, how many different selections are possible?

4.38.   Refer to exercise 4.37. Suppose that the 5 cards are drawn from the 52-card deck simultaneously and without replacement. How many different hands could be selected?

4.8 Subjective Probability

Probability can be viewed as a measure of one's belief in a particular outcome of an experiment. Thus a number is assigned to each sample point in such a way that

$$0 \le P(E_i) \le 1 \qquad \text{and} \qquad \sum_S P(E_i) = 1$$

Supposedly, the number assigned to each sample point E_i, $i = 1, 2, \ldots$, is a measure of the experimenter's belief that this sample point will occur when the experiment is conducted one time.

For many experiments we may view the probability of an event, the measure of our belief in its outcome, as the fraction of times the event will occur in a long series of repetitions of the experiment. This we referred to as the relative frequency concept of probability. Other experiments are not so easy to interpret. In particular, how do you view the probability of an event for an experiment that can be conducted only once because of the cost or the nature of the event? For example, the event "profit will rise next year after a corporate merger" is associated with an experiment that can never, under identical circumstances, be repeated. The raising or lowering of the prime interest rate at a particular time is also an experiment that can be conducted only once.

To answer the preceding question, we return to our opening statement. Probability can be viewed as a measure of one's belief in the outcome of a particular event. If we only consider experiments that can be repeated, we may wish to interpret the measure in terms of the relative frequency concept of probability. Then the sample point probabilities can be compared (as a check) with observed frequencies of occurrence of the sample points for a small number of trials. Experiments that cannot be repeated require a subjective evaluation of the probabilities of the sample points or, equivalently, the probability of an event. No opportunity will be available for an empirical check. These probabilities are usually specified by a person who is familiar with the experiment and who relies on his or her experience to determine a reasonably accurate probability for the event.

For example, a government economist is familiar with administration policies and current trends in the financial market. Thus he should be able to assign a reasonably accurate probability to the event that the prime interest rate will be raised (or lowered) at a particular time. (That economists are frequently unable to do this is indicated by the conflicting statements by experts regarding the probability or occurrence of various economic events.) Probabilities assigned to events that are acquired subjectively and based on "experience" are called *subjective probabilities* or, sometimes, personal probabilities.

How one determines the probabilities of the sample points or the probability of any event is irrelevant as long as the quantities so determined give reasonably accurate measures of the likelihood of the outcome of the events. Specifically, the accuracy of the results obtained by applying the laws of probability for calculating the probability of an event depends on the accuracies of the probabilities assigned to the sample points or to events appearing in the stated event composition. Incorrect assumptions lead to incorrect conclusions.

4.9 Random Variables

In section 4.2 we defined an experiment as the process by which an observation (or measurement) is obtained. Although the observations resulting from an experiment are not always numerically valued, most experiments produce data that are either quantitative or can be quantified by assigning numbers to represent categories. Thus we are particularly interested in experiments that result in numerically valued outcomes.

Suppose that the variable measured in an experiment is denoted by the symbol y. The variable y is called a *random variable* if the value that y assumes is a chance or random event.

Definition

A variable y is a **random variable** if the values that y assumes, corresponding to the various outcomes of an experiment, are chance or random events.

For example, consider the sampling of 20 consumers who are asked whether they prefer package design A or B. The number of consumers who indicate that they prefer design A may be regarded as a variable y that assumes any of the values 0, 1, 2, ... , 20. Each of these values corresponds to a particular experimental outcome, where the experiment consists of drawing a sample of 20 consumers' responses and noting how many indicate a preference for package design A. The variable y is a random variable since the value that y assumes is an outcome that cannot be predicted with absolute certainty in advance of the experiment; that is, the particular value that y takes is a chance or random event.

You may recall from our earlier discussion of this sampling problem (section 4.1) that the probabilities assumed by the random variable y are needed to make an inference about the preferences of all possible purchasers of the manufacturer's product. These probabilities, one associated with each value of y, are called the *probability distribution* for y. Probability distributions are the subject of chapter 5.

4.10 Summary

The theories of probability and statistics are concerned with samples drawn from a population. A probabilist assumes that the population is known and calculates the probability of observing a particular sample. A statistician assumes that the sample is known and, with the aid of probability, attempts to describe the frequency distribution of the population, which is unknown. Chapter 4 is concerned with the construction of a model, the sample space for the repetition of an experiment. Using this model we can find the probabilities of events associated with the experiment by one of two methods:

1. The summation of the probabilities of the sample points in the event of interest.
2. The joint use of event composition (compound events) and the laws of probability.

What use will be made of our ability to assess the probability of an event? Keep in mind the objective of statistics. We wish to make inferences about a population based on information contained in a sample. For us the experiment is the selection of a sample from a population and the event of interest is the particular sample outcome that we observe. Based on these sample results, we wish to deduce the nature of the population. The methods we will employ to accomplish this deduction (or inference as we call it) require that we know the probability of obtaining the observed sample.

Although we have not really explained how probability is used to make inferences, we gave an example in section 4.1 of how it was used to decide which package design was preferred by consumers. Rereading section 4.1 may help you to understand the role that probability plays in making inferences.

Supplementary Exercises

4.39. The board of county commissioners in a certain county is composed of two men and one woman. One commissioner is to be selected as commission chairman, another as secretary. Specify the sample points in the following events:

S: the sample space
A: the older man is selected as chairman
B: a man is selected as chairman
C: the woman is selected as secretary
D: events A and C occur
E: event B or event C or both occur

4.40. Refer to exercise 4.39. Suppose that each individual commissioner has an equal chance of being selected as either chairman or secretary. Compute the probabilities of the events A, B, C, D, and E by summing the probabilities of the appropriate sample points.

4.41. Patients arriving at a hospital outpatient clinic can select one of two stations for service. Suppose that physicians are randomly assigned to the stations and that the patients have no station preference. Three patients arrive at the clinic and their selection of stations is observed.

a. List the sample points for the experiment.

b. Let A be the event that each station receives at least one patient. List the sample points in A.

c. Make a reasonable assignment of probabilities to the sample points and find $P(A)$.

4.42. A retailer sells two styles of high-priced high-fidelity consoles that experience indicates are equal in demand. (Fifty percent of all potential customers prefer Style 1, 50% prefer Style 2.) If he stocks four of each, what is the probability that the first four customers seeking a console all purchase the same style?

a. Define the experiment.

b. List the sample points.

c. Define the event of interest A as a specific collection of sample points.

d. Assign probabilities to the sample points and find $P(A)$.

4.43. A large municipal bank chain has ordered five minicomputers to distribute among their five branches in a certain city. Unknown to the purchaser three of the five computers are defective. Before installing the computers an agent of the bank selects two of the five computers from the shipment, thoroughly tests them, and then classifies each as either defective or nondefective.

a. List the sample points for this experiment.

b. Let A be the event that the selection includes no defectives, and hence that the entire shipment is considered as acceptable. List the sample points in A.

c. Assign probabilities to the sample points and find $P(A)$.

4.44. As a result of fuel shortages due to the recent energy crisis, airlines have been forced to cancel many low capacity and duplicative flights. An airline company at present has four daily flights from New York to London. The first two flights are in the morning, the last two are in the afternoon. To conserve fuel the airline is going to cancel two flights. If the flights to be canceled are randomly chosen, what is the probability that a morning and an afternoon flight will still be available?

4.45. A small advertising firm consists of two men and one woman. The firm has two clients who are particularly difficult to deal with. To decide who sees the first client, one person is randomly selected from the three. The same procedure is followed for the second client.

a. Find the probability that both clients are served by the same person from the advertising firm.

b. Find the probability that both clients are served by men.

c. Find the probability that both the events described in parts a and b occur simultaneously.

4.46. Corporate bonds are rated A^+, A, B^+, B, or C, depending on the stability of the issuing firm. A novice bond buyer is unaware of the difference in corporate ratings and thus selects two firms at random from five. If each of the five firms has a different rating, what is the probability that she does not buy any bonds rated C? What is the probability that she buys only bonds rated A^+ or A?

4.47. One researcher has noted that affirmative action commitments by industrial organizations have led to an increase in the number of women in executive positions.* Suppose a company has vacancies for two positions of vice-president. The vacancies are to be filled by randomly selecting two people from a list of four

*S. Reynolds, "Women on the Line," *MBA*, February 1975.

candidates. On the list are two women and two men, all of whom have worked for the company for a long period of time.

   a. What is the probability that at least one of the women will be selected?

   b. What is the probability that neither woman will be selected?

   4.48.   A state highway department has contracted for the delivery of sand, gravel, and cement at a construction site. Due to other work commitments and labor force problems, contracting firms cannot always deliver items on the agreed delivery date. Based on past evidence the probabilities that sand, gravel, and cement will be delivered on the promised delivery dates by the contracting firms are .3, .6, and .8, respectively. Assume that the delivery or nondelivery of one material is independent of another.

   a. Find the probability that all three materials will be delivered on time.

   b. Find the probability that none of the three materials will be delivered on the promised delivery date.

   c. Find the probability that at least one of the materials will be delivered on time.

   4.49.   In a study to examine the relationship between social class and commercial bank credit card usage, researchers interviewed 1500 bank card holders and observed the results shown in the table. One respondent is selected at random from among the 1500 persons indicated in the table and the respondent's classification is noted.

| | NUMBER WHO USE CARD | |
SOCIAL CLASS	As a Convenience, A	For Installment Credit, B
upper, C	36	39
upper-middle, D	114	186
middle, E	174	426
lower-middle, F	72	228
lower, G	41	184

Source: H. L. Matthews and J. W. Slocum, Jr., "Social Class and Commercial Bank Credit Card Usage." Reprinted from the *Journal of Marketing*, January 1969, published by the American Marketing Association.

Let events A, B, C, D, E, F, and G be defined as labeled in the table. Find $P(A)$, $P(B)$, $P(E)$, $P(F)$, $P(AF)$, $P(BG)$, $P(AB)$, $P(A \cup F)$, $P(B \cup F)$, $P(A \cup B)$, and $P(A \cup C \cup D)$.

   4.50.   The accompanying table shows the location of all 200 stores in a trade association by city type and geographic region. One store is selected at random from the trade association to be used in testing the sales appeal of a new product. Calculate the following probabilities and then describe each probability in terms of the problem: $P(A_1)$, $P(B_3)$, $P(A_1 B_4)$, $P(B_1 | A_3)$, $P(A_2 \cup B_3)$, $P(B_1 \cup B_4)$, and $P(B_2 B_4)$.

| | GEOGRAPHIC REGION | | | |
CITY TYPE	East, B_1	South, B_2	Midwest, B_3	Far West, B_4
large, A_1	35	10	25	25
small, A_2	15	10	15	15
suburb, A_3	25	5	10	10

4.51.   Refer to exercises 4.39 and 4.40. Find $P(A|B)$, $P(A|C)$, and $P(B|C)$. Calculate the probabilities of the events AB, AC, BC, and $(A \cup B)$ by using the additive and multiplicative laws of probability. Are A and B independent? Mutually exclusive? Are B and C independent? Mutually exclusive?

4.52.   In a recent survey of 1700 companies, it was found that 49% of the firms reported that they perform extensive studies of advertising effectiveness, 61% conduct short-term sales forecasts, and 38% undertake both activities. Suppose a single firm is selected from the 1700 reporting firms. Define the events A and B as

> event A: the firm studies advertising effectiveness
> event B: the firm conducts short-term sales forecasts

Find $P(A)$, $P(B)$, $P(AB)$, $P(A \cup B)$, and $P(A|B)$.

4.53.   In the study referred to in exercise 4.52, it was also found that 64% of the firms undertake competitive product studies. Suppose one firm is selected at random from among those reported in the survey. Define the event C as "the firm undertakes competitive product studies."
  a. Find the probability that it does not undertake competitive product studies.
  b. Find the probability that it studies advertising effectiveness and undertakes competitive product studies but does no short-term sales forecasting. Assume that event C is independent of events A and B.
  c. Find the probability that it undertakes all three market research activities. Assume that event C is independent of events A and B.

4.54.   A company is going to temporarily hire two secretaries from a secretarial pool. One hundred secretaries are available. Twenty type under 40 words per minute, 50 type from 40 to 60 words per minute, and the rest type over 60 words per minute. Assume the two secretaries are randomly chosen from the secretarial pool.
  a. Find the probability that both secretaries type over 60 words per minute.
  b. Find the probability that the first types under 40 words per minute and the second types from 40 to 60 words per minute.
  c. Find the probability that the first types from 40 to 60 words per minute and the second types under 40 words per minute.
  d. Find the probability that one types under 40 words per minute and the other from 40 to 60 words per minute.

4.55.   A housewife is asked to rank four brands (A, B, C, D) of common household cleaner according to her preference, no. 1 being the cleaner she prefers best, and so on. Suppose the housewife really has no preference among the four brands and hence any ordering is equally likely to occur.
  a. What is the probability that Brand A is ranked first?
  b. What is the probability that C is first and D is second in the rankings?
  c. What is the probability that A is ranked either first or second?

4.56.   In its first annual report on environmental quality in New England, the regional office of the U.S. Environmental Protection Agency noted that only 30% of the region's solid waste disposal facilities meet EPA standards.* If three New England communities are selected at random, what is the probability that the solid waste disposal facilities in two of the communities fail to meet EPA standards? At least two fail to meet EPA standards? (Assume that each New England community has a single solid waste disposal facility and that the ability of a facility in one community to meet EPA standards is independent of the ability of a facility in any other community.)

*Environmental Protection Agency, *Environmental News*, August 1975.

4.57.   A salesman figures that the probability of his completing a sale during the first contact with a client is .4 but improves to .55 on the second contact if the client did not buy during the first contact. If the client does not buy on the first contact, the salesman will make one and only one return call to the client.

a. If the salesman contacts a client, calculate the probability that the client will buy.

b. If the salesman contacts a client, calculate the probability that the client will not buy.

4.58.   A shipping firm keeps two vehicles ready for local deliveries. Due to the demand on their time and the chance of mechanical failure, the probability that a specific vehicle will be available when needed is 9/10. The availability of one vehicle is independent of the other.

a. In the event of two large orders, what is the probability that both vehicles will be available?

b. What is the probability that neither will be available?

c. If one service call is placed, what is the probability that the delivery will be made (i.e., what is the probability that a vehicle will be available)?

4.59.   A certain article is visually inspected successively by two inspectors. When a defective article comes through, the probability that it gets by the first inspector is .1. Of those that do get past the first inspector, the second inspector will miss 5 out of 10. What fraction of the defectives will get by both inspectors?

4.60.   Some researchers suggest that guarantees should be simple, direct, and obvious in order to reduce perceived risks and encourage legitimate complaints.* A manufacturer of a sound movie camera reports that customer complaints are one of two types: those they regard as "defective products" and those they classify as "unsatisfactory products." Unsatisfactory products are those that do not meet consumer expectations. Currently, evidence suggests that 2% of the cameras are found to be defective by consumers, while 5% are judged to be unsatisfactory. If a camera is found to be defective, the probability of a complaint by a consumer is .9, while the probability of a complaint due to an unsatisfactory camera is .5. Given that a customer issues a complaint about a camera, what is the probability that the complaint is due to the product being judged unsatisfactory by the consumer? Would you then suggest that unsatisfactory products be covered under the product warranty?

4.61.   A company is going to buy two new desks for its office. The purchaser of the desks randomly chooses two from a total of five available. Although the purchaser does not know it, one desk has a major defect, two have minor defects, and two are in perfect condition.

a. What is the probability that both desks that are purchased are in perfect condition?

b. What is the probability that the desk with the major defect is purchased?

4.62.   A department store is going to run a sale on a particular item for one, two, or three days. The probability that the store chooses to run the sale for one day is .2, for two days is .3, and for three days is .5. The probabilities of selling all the items in stock if the sale is held one, two, or three days are .1, .7, or .9, respectively. If the store conducts a sale, what is the probability that all items in stock are sold by the store during the sale?

4.63.   A man takes either a bus or the subway to work, with probabilities .3 and .7, respectively. When he takes the bus, he is late on 30% of the days. When he takes the subway, he is late on 20% of the days. If the man is late for work on a particular day, what is the probability that he took the bus?

*C. L. Kendall and F. A. Russ, "Warranty and Complaint Policies: An Opportunity for Marketing Management," *Journal of Marketing*, April 1975.

4.64. A manufacturer has two machines that produce a certain product. Machine 1 produces 45% of the product and Machine 2 produces 55%. Machine 1 produces 10% defective items and Machine 2 produces 8% defective items. If a defective item is observed, what is the probability that it was produced by Machine 2?

4.65. An electronic fuse is produced by three production lines in a manufacturing operation. The fuses are costly, are quite reliable, and are shipped to suppliers in 100-unit lots. Because testing is destructive, most buyers of the fuses test only a small number before deciding to accept or reject lots of incoming fuses. All three production lines produce fuses at the same rate and normally produce only 2% defective fuses, which are randomly dispersed in the output. Unfortunately, production line 1 suffered mechanical difficulty and produced 5% defectives during the month of March. This situation became known to the manufacturer after the fuses had been shipped. A customer received a lot produced in March and tested three fuses. One failed. What is the probability that the lot came from one of the two other lines?

4.66. One study reported that 50% of all American families owe no installment debt, 23% owe installment debt not exceeding $500, while the remainder owe installment debt in excess of $500.* The author further notes that families with moderate incomes, from $7,500 to $10,000, are most likely to owe installment debt. In fact, he report. that 41% of all American families with incomes from $7,500 to $10,000 owe installment debt in excess of $500 while only 37% in this group owe no installment debt at all. If it is known that a certain family's income is within the range from $7,500 to $10,000, find the probability that they owe an installment debt exceeding $500.

4.67. A firm that manufactures precision metal components uses a special machine responsible for both the molding and finishing of the components. The machine is controlled by two electrical relay systems, one for the molding process and one for the finishing process. These two relay systems are connected in series so that the machine will work only if both electrical relay systems are in proper working order. Suppose the probability that a single relay system will operate satisfactorily is .99, and suppose that whether one will work is independent of whether the other will work. Find the probability that neither of the electrical relay systems will operate satisfactorily. That one of the relay systems will work. That both of the relay systems will work. What is the probability that the special machine will operate satisfactorily?

4.68. Refer to exercise 4.67 and assume that to increase the chances that the special machine will operate satisfactorily, a second pair of electrical relay systems, connected in series and designed to control, respectively, the molding and finishing processes, are installed. The special machine will now operate satisfactorily if one or the other of the pairs of electrical relay systems operates satisfactorily. Find the probability that neither of the pairs of relay systems works. That one of the pairs of relay systems works. That both pairs of relay systems work. What is the probability that the special machine will operate satisfactorily?

4.69. Each state delegates responsibility to its State Board of Public Accountancy for the administration of an annual CPA examination. The examination is in five parts, with two parts on accounting problems, and one each on auditing, business law, and theory of accounts. To be certified an applicant must pass all five parts. The State Board of Public Accountancy for the State of New York reports that past history shows the following percentages of applicants passing each part on their first attempt:[†]

*G. Katona, *The Mass Consumption Society* (New York: McGraw-Hill, 1964).

†R. G. Allyn, CPA, Executive Secretary, State Board of Public Accountancy, Albany, New York.

Part	account-ing 1	account-ing 2	audit-ing	business law	theory of accounts
Pass Percentage	.30	.30	.37	.35	.27

If success on one part of the examination is independent of success on any other part, what is the probability an applicant becomes certified on the first attempt to pass the CPA examination of the State of New York? Do you believe that this probability is a reasonable approximation to reality when you consider that a modest proportion of applicants pass all parts of the examination on their first attempt? Comment on the assumption of independence between parts of the test.

Experiences with Real Data

The use of noninstallment credit to pay for the costs of retail goods and services has expanded rapidly over the past 20 years. Many consumers find it a matter of convenience to pay by credit card or bank card, while some use such devices to delay payment for goods and services received until a later period when they expect to have more cash available. Most observers agree that the widespread use of noninstallment credit will eventually result in a society which has little, if any, dependence on cash.

To remain competitive department stores have begun to rely heavily on the use of noninstallment credit. Many offer their own in-house credit cards to preferred customers, while others accept only bank cards or credit cards offered by national credit service organizations such as Master Charge. The decision by a department store to offer or to not offer an in-house card depends on the degree to which noninstallment credit is used by its customers. Generally it is assumed that if a customer is more likely to purchase an item on noninstallment credit than to pay cash for the item, an in-house card is justified. If evidence indicates that a customer is more likely to pay cash for a purchase, most department stores contract with a local bank and offer noninstallment credit through the use of a bank card.

Select a department store in the vicinity of your college or university and conduct a study to estimate the probability that a customer of that store requests noninstallment credit when purchasing an item offered for sale in the store. Conduct your study by stationing yourself next to one checkout counter in the store and observe a single sale. Record whether the sale is a cash transaction or a noninstallment credit sale (credit card sale). Repeat the process for a total of 25 checkout counters. Estimate the probability that a customer of the store requests noninstallment credit by computing the fraction of sales transactions, of the $n = 25$ you observed, that were noninstallment credit transactions.

Observe the results of the other members of your class. Pool your results to estimate the probability a customer of the store requests noninstallment credit by computing the fraction of all observed sales transactions that were noninstallment sales transactions. Since this result is based on a much larger number of observations, it should be a much more accurate estimate of the desired probability than your estimate based on only $n = 25$ sales transactions. Note that you are unable to evaluate the accuracy of this estimate. We will explain how to evaluate the goodness of this type of estimate in chapter 8.

Save your data from this exercise: they will be required in exercise 8.81.

References

FELLER, W. *An Introduction to Probability Theory and Its Applications.* Vol. 1. 3d ed. New York: Wiley, 1968.

HYMANS, S. H. *Probability Theory with Applications to Econometrics and Decision-Making.* Englewood Cliffs, N.J.: Prentice-Hall, 1967.

KING, W. R. *Probability for Management Decisions.* New York: Wiley, 1968.

WOODROOFE, M. *Probability with Applications.* New York: McGraw-Hill, 1975.

chapter objectives

GENERAL OBJECTIVE In chapter 4 we gave an example to illustrate how probability is used in making inferences about a population based on information contained in a sample, and then we presented the concepts of probability that help us find the probability of experimental outcomes. Because most samples are measurements on random variables, we need to be able to find the probabilities associated with such measurements. The objective of this chapter is to distinguish between two types of random variables and to explain how to find the probability that a random variable will assume specific values.

SPECIFIC OBJECTIVES

1. To identify the role that random variables play in making inferences.
Section 5.1

2. To identify the two types of random variables—discrete and continuous—and to present probability models appropriate for each.
Sections 5.2, 5.3, 5.4

3. To define the expected value of a random variable and to identify this quantity as the mean of its probability distribution.
Section 5.5

4. To define and identify the variance of a random variable as an expectation and to explain how the expected value and variance of a random variable can be used to describe its probability distribution.
Section 5.6

chapter five

Random Variables and Probability Distributions

5.1 Random Variables: How They Relate to Statistical Inference

Recall that in section 4.2 we defined an **experiment** to be the process of collecting a measurement (or an observation). Most experiments yield a **numerical measurement** that varies from sample point to sample point in a random manner. Hence the measurement is called a random variable. Restating the definition presented in section 4.9, a variable *y* is a **random variable** if the value that *y* assumes is a chance or random event. The daily closing price of an industrial stock is a random event. Observing the number of defects on a piece of new furniture or recording the grade point average of a particular student are other examples of experiments yielding random numerical events.

The **population** associated with the experiment is generated when the experiment is repeated many times and a relatively large body of data is obtained. As previously noted, we never actually measure each member of the population, but we can certainly conceive of doing so. In lieu of this, we wish to obtain a small set of these measurements, called the **sample**, and we wish to use the information in the sample to **describe or make inferences about the population.**

We have stated that a measurement obtained from an experiment results in a specific value of the random variable of interest and represents a measurement drawn from a population. How can a single measurement or a larger sample of, say, *n* measurements be used to make inferences about the population of interest? With the consumer preference study of section 4.1 firmly in mind, we might **calculate the probability of the observed numerical** event, the sample, for a large set of possible populations and choose the set that gives the highest probability of observing the sample. We like to think that the method of inference described above appears reasonable and intuitively appealing, but note that it is not claimed to be the best, however "best" might be defined. (This procedure is the basis for one of the important methods for the statistical estimation of the parameters of a population and can be shown to provide "good" inferences in many situations.) We defer further discussion of inference until chapter 6. At this point it is sufficient to note that the procedure requires a knowledge of the probability associated with each value of the random variable. In other words, we need to know the probability distribution for the random variable, a distribution that represents the theoretical frequency histogram for the population of numerical measurements. The theory of probability presented in chapter 4 provides the mechanism for calculating these probabilities for some random variables.

5.2 Classification of Random Variables

Random variables are classified into two types: *discrete* or *continuous*.

Definition

A discrete random variable is one that can assume a countable number of values.

Countable means that you can associate the values that the random variable y can assume with the integers 1, 2, 3, 4, ... (that is, you can count them). The number of values that y can assume can be finite or can be infinite (because you can conceive of the count as proceeding and never ending).

The number y of errors that a mechanic can make in an assembly operation is a discrete random variable because y can assume a finite number of values, 0, 1, 2, 3, ..., N, where N is the total number of steps in the assembly operation. The number of years y until a corporation achieves \$1 billion in assets could conceivably be infinite, but nevertheless y is still a discrete random variable. In the latter example the number of values that y can assume could be infinitely many since we can imagine a count that could be continued indefinitely. Other examples of discrete random variables are the following:

1. The number of automobiles sold per month.
2. The number of accidents in a particular manufacturing plant for a given week.
3. The number of customers waiting at a supermarket checkout counter.

4. The number of television tubes produced in a given hour.

A continuous random variable is defined as follows.

Definition

A **continuous random variable** is one that can assume the infinitely many values corresponding to the points on a line interval.

For example, suppose that we measured the distance between a supplier and a buyer and represented this distance as the random variable y. If an exact measuring instrument could be used, each value of y could be viewed theoretically as a point on the line interval between the supplier and buyer. Thus we could associate each value of y with one of infinitely many points lying on a line interval. Other examples of continuous random variables are the following:

1. The length of time to complete an assembly operation in a manufacturing plant.
2. The amount of petroleum pumped per hour from a well.
3. The amount, in milligrams, of carbon monoxide in a cubic foot of air.
4. The amount of energy produced by a utility company in a given day.

The distinction between discrete and continuous random variables is an important one since different probability models are required for each. The probabilities associated with each value of a discrete random variable can be assigned so that the probabilities sum to 1. This is not possible with continuous random variables. Accordingly, we will consider the probability distributions for discrete and continuous random variables separately in the two sections that follow.

Exercises

5.1.   Identify the following as discrete or continuous random variables:
a. The number of defective transistors found in a shipment of 10,000 transistors received from a supplier.
b. The number of delinquent accounts in a department store at a particular time.
c. The length of time for an employee to complete a certain task observed in a time and motion study.
d. The number of policies sold by an insurance salesperson in a given week.
e. The gasoline consumption of a car driven over a specified 100-mile test run.

5.2.   Identify the following as discrete or continuous random variables:
a. The length of life of a light bulb observed during a life-testing experiment.
b. The gross revenue earned by a supermarket during a given day.
c. The market value of a publicly listed security on a particular day.

d. The tensile breaking strength, in pounds per square inch, of 1-inch-diameter steel cable.

e. The daily demand for electrical power by residential users in Redding, California.

5.3 Probability Distributions for Discrete Random Variables

The probability distribution for a discrete random variable can be represented by a formula, a table, or a graph that gives the probability associated with each value of the random variable. Since each value of the random variable y is a numerical event, we can apply the methods of chapter 4 to obtain the appropriate probabilities. Then, because one and only one value of y is assigned to each sample point in the sample space and because the values of y represent mutually exclusive events, it follows that the sum of the values of the probability distribution of y, summed over all values of y, is 1. That is,*

$$\sum_y p(y) = 1$$

Example 5.1 Consider an experiment that consists of tossing two coins, and let y be the number of heads observed. The sample points for this experiment with their respective probabilities are given in the table. Find the probability distribution for y.

SAMPLE POINT	COIN 1	COIN 2	$P(E_i)$	y
E_1	H	H	1/4	2
E_2	H	T	1/4	1
E_3	T	H	1/4	1
E_4	T	T	1/4	0

Solution We assign the value $y = 2$ to point E_1, $y = 1$ to point E_2, and so on. The probability of each value of y may be calculated by adding the probabilities of the sample points in that numerical event. The numerical event $y = 0$ contains one sample point, E_4; $y = 1$ contains two sample points, E_2 and E_3; and $y = 2$ contains one point, E_1. The values of y, with the respective probabilities, are given in table 5.1. Observe that

$$\sum_{y=0}^{2} p(y) = 1$$

*Note that the probability distribution of a random variable y is denoted by the symbol $p(y)$ [i.e., lowercase letter p is used]. In contrast, the probability of an event E is denoted by the symbol $P(E)$ [i.e., capital letter p]. We will make this distinction throughout the text.

Table 5.1 *Probability distribution for the number of heads when tossing two coins*

y	SAMPLE POINTS IN y	$p(y)$
0	E_4	1/4
1	E_2, E_3	1/2
2	E_1	1/4

$$\sum_{y=0}^{2} p(y) = 1$$

The probability distribution for y is shown graphically in figure 5.1

To illustrate how we use the probability distribution for a discrete random variable to make inferences, let us consider again the problem (introduced in section 4.1) of deciding whether package design A is preferred to design B. The experiment that was designed to answer this question consists of randomly selecting 20 consumers and asking each which package design he or she prefers. Then noting the number y of consumers in the sample favoring package design A, this question immediately arises: How small should this number be before we conclude that package design B is preferred to design A? To answer this question we need to find the probability of an observed sample result.

The easiest way to approach this problem is by way of analogy to a coin-tossing experiment. If the fraction of all consumers favoring package design A is at least $\frac{1}{2}$ (let us assume that it is exactly $\frac{1}{2}$), then observing the response of a randomly selected consumer is analogous to tossing a fair

Figure 5.1 *Probability histogram showing p(y) for example 5.1*

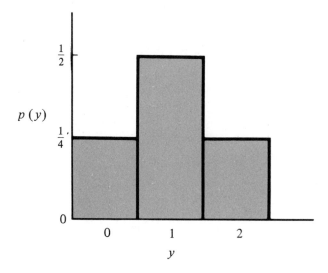

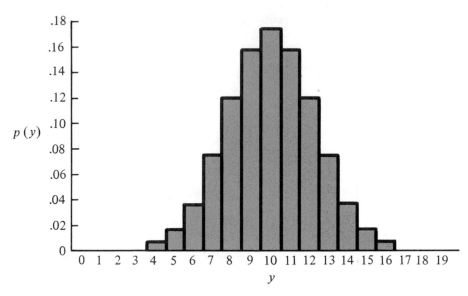

Figure 5.2 *Probability distribution for the number of consumers favoring package design A; n = 20 and p = $\frac{1}{2}$*

coin. Likewise, the random sampling of 20 consumers' responses is analogous to the tossing of 20 fair coins (an extension of the two-coin-toss experiment considered earlier). Thus the random variable y, the number of consumers favoring package design A, corresponds to the number of heads in a toss of 20 coins. For both experiments y can assume any of the values 0, 1, 2, ..., 20, with associated probabilities $p(0)$, $p(1)$, $p(2)$, ..., $p(20)$ given by the same probability distribution. Although the exact probability distribution for this experiment will be discussed in chapter 6, we present its graph here in figure 5.2.

From the graph of the probability distribution for y, we observe that the largest probability is that associated with the value $y = 10$—which is to be expected if half of all 20 consumers favor package design A. Moreover, we note that it is highly improbable that as few as $y = 3$ or even $y = 5$ consumers favor design A in the sample if the fraction in the population of all consumers is $\frac{1}{2}$. If $y = 0$, the result is so extremely improbable that we would conclude that the assumption is false. Thus if in our sample $y = 0$ or some very small value, we would infer that in the population of responses from consumers purchasing the manufacturer's product, less than $\frac{1}{2}$ favor package design A.

Exercises

5.3.   Suppose studies show that in a lower-income area of a certain city, only 20% of shoppers in food stores read unit-pricing labels before making a purchase decision. If $n = 2$ shoppers enter a food store in this lower-income area, calculate the probability distribution for y, the number of shoppers who take advantage of the unit-pricing labels. [Hint: Let F indicate the event that a shopper does not read the labels. Then $p(0)$ = P(both shoppers do not read the labels) = $P(F)P(F)$ = $(.8)^2$ = $.64$.] Construct a probability histogram for y.

5.4.   Simulate the experiment described in exercise 5.3 by using five marbles (you could use cards), four of one color and one of another. Draw one marble at random from the five, replace the marble, and repeat the process. Count the number y of times the different-colored marble appears in the two draws and record this observed value of y. The number y will correspond to the number of shoppers in a sample of two (corresponding to the two draws) who take advantage of the pricing labels. Repeat the dual drawing process 100 times and obtain 100 observed values of y. Construct a relative frequency histogram for the data and compare it with the population probability distribution you constructed in exercise 5.3.

5.5.   To save time and money but still provide safeguards on the quality of incoming goods, buyers of goods often inspect a portion of the shipment and judge the quality of the entire shipment on the basis of the observed quality of the sample. Suppose a buyer has received a shipment of four photocopy machines, two of which are defective. Two photocopy machines are selected at random from among the four and tested. Let y be the number of defective machines observed, where y = 0, 1, or 2. Find the probability distribution for y and express your results graphically as a probability histogram.

5.6.   Simulate the experiment described in exercise 5.5 by using four marbles (you could use cards), two of one color and two of another, to represent, respectively, the two good and the two defective photocopy machines. Place the marbles in a hat, mix, draw two, and record y, the number of defectives observed. Replace the marbles and repeat the process until a total of n = 100 observations on y have been recorded. Construct a relative frequency histogram for this sample and compare it with the population probability distribution you constructed in exercise 5.5.

5.7.   Competitive bids are required for the purchase of most services by governmental agencies. Suppose past experience has shown that a certain government contractor wins, on the average, three of every five contracts on which he submits a bid. Let y be the number of bid submissions until a contract is won by the contracting firm, and assume that successive bids are independent of one another.
   a. Find $p(1)$, $p(2)$, and $p(3)$.
   b. Give a formula for $p(y)$.
   c. Graph $p(y)$.

5.4 Continuous Random Variables and Their Probability Distributions

As indicated in section 5.2, continuous random variables can assume the infinitely many values corresponding to points on a line interval. However, it is not possible to assign a probability to each of these infinitely many points and have the probabilities sum to 1, as for discrete random variables. So a different approach is used to generate the probability distribution for a continuous random variable. The approach that we adopt uses the concept of a relative frequency histogram, such as that for the 25 price-earnings

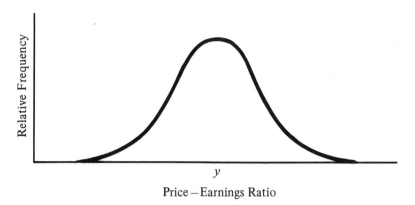

Figure 5.3 *Relative frequency histogram for a population*

ratios in figure 3.2. Recall that the width of the class interval in figure 3.2 was determined in accordance with the number of measurements involved. If more and more measurements are obtained, we might reduce the width of the class interval. The outline of the histogram would change slightly, for the most part becoming less and less irregular. When the number of measurements becomes very large and the intervals very small, the relative frequency histogram would appear, for all practical purposes, as a smooth curve, as shown in figure 5.3.

The relative frequency associated with a particular class in the population is the fraction of measurements in the population falling in that interval and also is the probability of drawing a measurement in that class. If the total area under the relative frequency histogram is adjusted to equal 1, areas under the frequency curve correspond to probabilities. In fact, this was the basis for the application of the Empirical Rule in chapter 3.

Let us construct a model for the probability distribution for a continuous random variable. Assume that the random variable *y* may take on any value on a real line, as in figure 5.3. We then distribute 1 unit of probability along the line, much as a person might distribute a handful of sand, each measurement in the population corresponding to a single grain. The probability, grains of sand or measurements, will pile up in certain places, and the result will be the probability distribution shown in figure 5.4. The depth or density of the probability, which varies with *y*, may be represented by a mathematical formula $f(y)$, called the probability distribution, or the probability density function, for the random variable *y*. The density function $f(y)$, represented graphically in figure 5.4, provides a mathematical model for the population relative frequency histogram that exists in reality. The total area under the curve $f(y)$ is equal to 1. The area lying above a given interval equals the probability that *y* will fall in that interval. Thus the probability that $a < y < b$ (*a* is less than *y* and *y* is less than *b*) is equal to the area under the density function between the two points *a* and *b*. This area is shown by the shading in figure 5.4.

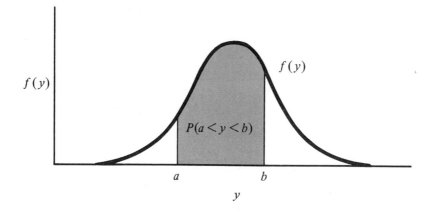

Figure 5.4 *The probability distribution f(y); P(a < y < b) is equal to the area under the curve shown by the shading*

A question remains. How do we choose the model—that is, the probability distribution $f(y)$—appropriate for a given physical situation? Fortunately we will find that most data collected on continuous random variables have mound-shaped frequency distributions, often very nearly bell-shaped. A probability model that provides a good approximation to such population distributions is the *normal probability distribution,* which we will study in detail in chapter 7.

The practice of adopting a model $f(y)$ that can only be expected to approximate the population relative frequency curve requires some comment. It is, of course, a practice that could lead to an invalid conclusion—if an inappropriate model is chosen out of poor judgment or insufficient knowledge of the phenomenon under study. On the other hand, using a density function $f(y)$ to approximate the population relative frequency distribution for a continuous random variable is a strategy that, when appropriately applied, has been found highly successful in scientific research. The equations, formulas, and various numerical expressions used in all the sciences are simply mathematical models that provide approximations to reality, the goodness of which are evaluated through experimental application. Hence a model is evaluated in terms of the results of its application—which are decisions or predictions in business situations. Do the resulting inferences fit in with the body of accumulated evidence? Are the deductions that follow from these inferences verified experimentally? If so, the model has proved its worth.

Exercises

5.8. Suppose that the sales y of a gasoline distributor has a uniform probability distribution, as shown in figure 5.5. Because of equipment limitations daily sales will never be less than 10,000 gallons per day and never greater than 50,000 gallons per day. Use the information in figure 5.5 to find the probability that the distributor sells these amounts:

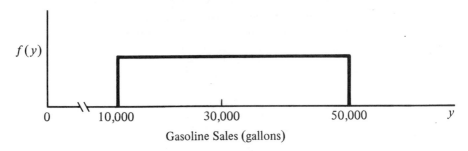

Figure 5.5 *A uniform probability distribution; exercise 5.8*

a. at least 40,000 gallons per day
b. between 20,000 and 30,000 gallons per day

     5.9.   Large pieces of equipment, which periodically have worn parts replaced and are subject to periodic maintenance checks and service, will frequently have a length of life y that has a *negative exponential probability distribution.* The probability distribution for such a piece of equipment is shown in figure 5.6. Use the information in figure 5.6 to find the approximate probability that one of these pieces of equipment, randomly selected from production and properly maintained, will last more than 100,000 operating hours. (Hint: Approximately what proportion of the total area under the curve is represented by the shaded area?)

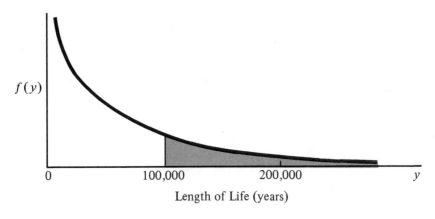

Figure 5.6 *A negative exponential probability distribution; exercise 5.9*

5.5 Mathematical Expectation

    The probability distributions described in sections 5.3 and 5.4 provide a model for the theoretical frequency distribution of a random variable. Hence these distributions must possess a mean, variance, standard deviation, and other descriptive measures associated with the theoretical populations that they represent. Recalling that both the mean and the variance are averages (sections 3.6 and 3.7), we will confine our attention to the problem of calculating the average value of a random variable defined over a theoretical population. This average is called the expected value of the random variable.

The method for calculating the population mean or expected value of a random variable can be more easily understood by considering an example. Let y be the number of heads observed in the toss of two coins, as shown in the table. For convenience, we also list $p(y)$ in the table.

y	0	1	2
$p(y)$	1/4	1/2	1/4

Let us suppose that the experiment is repeated a large number of times, say $n = 4,000,000$ times. Intuitively we would expect to observe approximately 1 million zeros, 2 million ones, and 1 million twos. Then the average value of y is

$$\frac{\text{sum of measurements}}{n} = \frac{(0)1,000,000 + (1)2,000,000 + (2)1,000,000}{4,000,000}$$

$$= \frac{(0)1,000,000}{4,000,000} + \frac{(1)2,000,000}{4,000,000} + \frac{(2)1,000,000}{4,000,000}$$

$$= (0)(\tfrac{1}{4}) + (1)(\tfrac{1}{2}) + (2)(\tfrac{1}{4})$$

Note that the first term in this sum is equal to $(0)p(0)$, the second is equal to $(1)p(1)$, and the third is equal to $(2)p(2)$. The average value of y, then, is

$$\sum_{y=0}^{2} yp(y) = 1$$

Of course, this result is not an accident. Thus it seems intuitively reasonable to define the expected value of y for a discrete random variable as given in the box.

Definition

Let y be a discrete random variable with probability distribution $p(y)$. Then $E(y)$, the **expected value of y,** is

$$E(y) = \sum_{y} yp(y)$$

where the elements are summed over all values of the random variable y.

Note that if $p(y)$ is an accurate description of the relative frequencies for a real population of data, then $E(y) = \mu$, the mean of the population. We will assume this to be true and let $E(y)$ be synonymous with μ.

The method for calculating the expected value of y for a continuous random variable is rather similar from an intuitive point of view, but, in practice, it involves the use of the calculus and is therefore beyond the scope of this text.

Example 5.2 In competitive bidding exercises a contractor will often submit a bid on a project if his analysis of the project and of the opposing bidders suggests that his expected return for submitting a bid exceeds some limiting amount. Suppose a contractor is considering a bid on a project that would return $50,000 if received. The cost of preparing a bid for the project is $5,000. The contractor feels that the probability that he will win the contract is .4. In addition, he will bid on all contracts with an expected return exceeding $12,000. Should he prepare a bid for the project?

Solution The contractor's net return y may take on one of two possible values. Either he will lose $5,000 (i.e., his return will be $-\$5,000$) or he will win $50,000 - \$5,000 = \$45,000$, with probabilities .6 and .4, respectively. The probability distribution for his net gain y is shown in the table.

y	$-\$5,000$	$\$45,000$
$p(y)$	.6	.4

The expected net return is

$$E(y) = \sum_y yp(y) = -\$5,000(.6) + \$45,000(.4) = \$15,000$$

Since $E(y) = \$15,000$ exceeds $12,000, the contractor should prepare and submit a bid for the project.

Example 5.3 An actuary, a statistician employed by an insurance company, determines the premiums the company should charge for its insurance. Consider the problem of determining the yearly premium for a $100,000 automobile liability insurance policy. The policy covers the type of events that, over a long period of time, evidence shows have occurred at the rate of 3 times per 5000 drivers each year.

Let y be the yearly financial gain to the insurance company resulting from the sale of the policy and let C be the unknown yearly premium. We will calculate the value of C such that the expected gain $E(y)$ is 0. Then C is the premium required to break even. To this value the company would add administrative costs and a margin of profit.

Solution We assume that the expected gain $E(y)$ depends on C. Using the requirement that the expected gain must equal 0, we have

$$E(y) = \sum_y yp(y) = 0$$

We must solve this equation for C.

The first step in the solution is to determine the values that the gain y may take. Then we determine $p(y)$. If the event does not occur during the year, the insurance company will gain the premium, or $y = C$ dollars. If the event does occur, the gain will be negative—will be a loss—amounting to $y = -(100,000 - C)$ dollars. The probabilities associated with these two values of y are $4997/5000$ and $3/5000$, respectively. The probability distribution for the gain is shown in the table.

y, *Gain*	C	$-(100,000 - C)$
$p(y)$	$4997/5000$	$3/5000$

Setting the expected value of y equal to 0 and solving for C, we have

$$E(y) = \sum_y yp(y) = C\left(\frac{4997}{5000}\right) + [-(100,000 - C)]\left(\frac{3}{5000}\right) = 0$$

or

$$\left(\frac{4997}{5000}\right)C + \left(\frac{3}{5000}\right)C - 60 = 0$$

$$C = \$60$$

Thus if the insurance company were to charge a yearly premium of $60, the average gain calculated for a large number of similar policies would be 0. The actual premium the company would charge would be $60 plus administrative costs and profit.

The insurance premium problem can be extended to include any number of gains to the insurance company by solving for the premium C in the expression

$$E(y) = \sum_y yp(y) = 0$$

where y is the random variable representing the insurance company's gains. The difficulty in a practical problem of this type is to identify each gain and associate with that gain a meaningful probability. This is the responsibility of the actuary.

Exercises

5.10. Let y represent the number of machine failures for any given day by two automatic canning machines in a food-processing plant. Assume that the probability distribution of y (which is suggested by an analysis of historical data) is as follows:

y	0	1	2	3
p(y)	.35	.35	.25	.05

Find the expected value of y. Construct a graph of the probability distribution. Locate μ on the graph.

5.11.   Let y represent the number of times a housewife visits a grocery store in a one-week period. Assume that the following table is the probability distribution of y.

y	0	1	2	3
p(y)	.1	.2	.3	.4

Find the expected value of y. Construct a graph of the probability distribution. Locate μ on the graph.

5.12.   Risky investment opportunities are often classified according to their expected profitability. Those with a positive expected profitability are subject to further analysis, while those with negative expected profitability are rejected outright. As an example, consider an investment of $1,000 in an oil drilling venture that is believed to return either a $10,000 profit or nothing. If the probability of success of the oil drilling venture is .1, find the expected profit for investing in the venture.

5.13.   Since farming is a very risky business due to the uncertainties of the weather, many farmers now insure their crops each year against possible losses due to rain, hail, flood, wind, or freezing temperatures. Follow the guidelines of example 5.3 to determine the annual premium for an insurance policy to cover a farmer's $200,000 wheat crop. Insurance actuaries suggest that chances are 1 in 50 that the crop will be destroyed by adverse weather conditions.

5.6 The Variance of a Random Variable

The expected value of a random variable is a measure of central tendency for its probability distribution in the same way that $\bar{y}$ is a measure of central tendency for a relative frequency histogram. The term "expected value" possesses an intuitive meaning in the sense that we are more likely to draw a value of y near the center of the probability distribution.

However, just as $\bar{y}$ does not provide a complete description of a sample relative frequency histogram, the expected value μ does not provide a complete description of the behavior of a random variable. For a better description we need a measure of the spread of the probability distribution p(y). To illustrate, suppose that you are managing a hospital and that you must decide on the number of shots of antitetanus vaccine that must be maintained in inventory. Furthermore, assume that you must always have an adequate supply

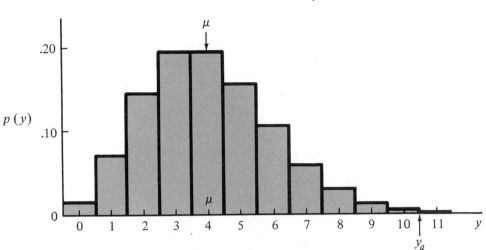

Figure 5.7 *Probability distribution for the number y of tetanus shots used in a three-day period; μ = 4*

on hand to meet a three-day demand and that the probability distribution for y, the number of shots required for a three-day period, appears as shown in figure 5.7.

You can see from figure 5.7 that the expected demand μ (located on the figure) does not tell us the largest number of tetanus shots that might be required for a three-day period. Thus we need to know the value of y, say y_a, in the right tail of $p(y)$ such that the probability of demand exceeding y_a is very small (see figure 5.7).

Finding the number y_a of shots that should be maintained in inventory would be relatively easy if we knew $p(y)$, because we could move to the right along the y-axis in figure 5.7 until we found y_a such that $P(y \geq y_a)$ is small. If $p(y)$ is not known, we would be satisfied with a measure of spread for $p(y)$. This measure of variability would enable us to locate an approximate value for y_a.

Because a probability distribution is (in a sense) a theoretical relative frequency histogram for the random variable y, it is natural for us to describe the variability of y by finding its variance and its standard deviation. Then we can interpret these descriptive measures using Tchebysheff's theorem and the Empirical Rule. For the tetanus shot supply problem, we would be able to find a value for y_a that would be in the upper tail of $p(y)$.

You will recall (from chapter 3) that the variance of a set of measurements is the average of the square of the deviations of the measurements from their mean. Because taking an expectation is equivalent to "averaging," we define the *variance* and the *standard deviation* as given in the boxes.

Definition

Let y be a discrete random variable with probability distribution $p(y)$ and expected value $E(y) = \mu$. The **variance** of the probability distribution $p(y)$ is

$$\sigma^2 = E(y - \mu)^2 = \sum_y (y - \mu)^2 \, p(y)$$

where the elements are summed over all values of the random variable y.

Definition

The **standard deviation** σ of a random variable y is equal to the square root of its variance.

Example 5.4 Find the variance σ^2 for the population associated with example 5.1, the coin-tossing problem. The expected value of y was shown to equal 1.

Solution The variance is equal to the expected value of $(y - \mu)^2$, or

$$\sigma^2 = E(y - \mu)^2 = \sum_y (y - \mu)^2 \, p(y)$$
$$= (0 - 1)^2 \, p(0) + (1 - 1)^2 \, p(1) + (2 - 1)^2 \, p(2)$$
$$= (1)(\tfrac{1}{4}) + (0)(\tfrac{1}{2}) + (1)(\tfrac{1}{4}) = \tfrac{1}{2}$$

Then $\sigma = \sqrt{\tfrac{1}{2}} = .707$. Note how $\mu = 1$ and $\sigma = .707$ describe the probability distribution shown in figure 5.1.

Example 5.5 In a study of the mobility of corporate purchasing executives, it was found that the following distribution provides an adequate approximation to the probability distribution of y, the number of firms in which purchasing positions have been held by purchasing executives currently employed by industrial organizations:*

y	1	2	3	4	5
$p(y)$	.52	.22	.19	.04	.03

Find the mean and standard deviation of y.

*I. V. Fine and J. H. Westling, "Organizational Characteristics of Purchasing Personnel in Public and Private Hierarchies," *Journal of Purchasing*, August 1973.

Solution The mean is equal to $E(y)$, or

$$\mu = E(y) = \sum_y yp(y)$$

$$= 1(.52) + 2(.22) + 3(.19) + 4(.04) + 5(.03) = 1.84$$

The variance is equal to the expected value of $(y - \mu)^2$, or

$$\sigma^2 = E(y - \mu)^2 = \sum_y (y - \mu)^2 p(y)$$

$$= (1 - 1.84)^2(.52) + (2 - 1.84)^2(.22) + (3 - 1.84)^2(.19)$$
$$+ (4 - 1.84)^2(.04) + (5 - 1.84)^2(.03)$$

$$= (.7056)(.52) + (.0256)(.22) + (1.3456)(.19)$$
$$+ (4.6656)(.04) + (9.9856)(.03)$$

$$= 1.1144$$

The standard deviation of the distribution is

$$\sigma = \sqrt{\sigma^2} = \sqrt{1.1144} = 1.055$$

The mean $\mu = 1.84$ and the standard deviation $\sigma = 1.055$ can be used together with Tchebysheff's theorem to describe the probability distribution for the number of firms in which purchasing positions have been held by purchasing executives currently employed by industrial organizations. Even though the distribution is not mound-shaped, you can see from the table that the probability that y will fall outside the interval $(\mu \pm 2\sigma)$, or $(-.27, 3.95)$, is very small (only .07).

Example 5.6 As you will subsequently learn (in chapter 6), it is quite likely that the number y of people requiring a tetanus shot in a three-day period will follow a Poisson probability distribution. You will also learn that for a Poisson random variable, the variance of y equals its mean. Suppose that past experience shows that the average number of shots required for a three-day period is 4. Use this information to find an approximate value for y_a, the number of tetanus shots to stock for a three-day demand.

Solution We are given the information that the expected demand is 4. Therefore, if the demand y follows a Poisson probability distribution (and we will assume that it does), we have

$$\sigma^2 = \mu = 4$$
$$\sigma = \sqrt{4} = 2$$

From our knowledge of Tchebysheff's theorem and the Empirical Rule, we would not expect y to fall more than three standard deviations from μ. Or, in other words, we would not expect y to exceed

$$\mu + 3\sigma = 4 + 3(2) = 10$$

If we choose $y_a = 10$ as the number of shots to stock, we can be reasonably certain that the demand y will not exceed this value.

Exercises

5.14.   Now that you know how to find the standard deviation of a random variable, let us reconsider exercise 5.10, which is restated here for your convenience.

Let y represent the number of machine failures for any given day by two automatic canning machines in a food processing plant. Assume that the following table is the probability distribution of y, obtained from historical data.

y	0	1	2	3
$p(y)$	.35	.35	.25	.05

Find the expected value and variance of y. Construct a graph of the probability distribution. What is the probability that y will fall more than three standard deviations from its mean? Two standard deviations?

5.15.   The extent of a firm's involvement in new product research depends on the industries in which the firm competes. A survey was made of firms in high-technology fields to find the number of new products each firm introduced during 1975. Based on the survey, the probability distribution shown in the table was determined for y, the number of new products introduced by each high-technology firm during 1975.

Number of New Products, y	3	4	5	6	7	8	9
$p(y)$	.08	.14	.22	.30	.14	.08	.04

a. Find the expected value and variance of y.
b. Give the proportion of high-technology firms for which y exceeds $(\mu + 2\sigma)$.

5.16.   Past experience has shown that the number of industrial accidents per month in a manufacturing plant has a Poisson probability distribution, with the expected value of y equal to 6. Use the information provided in example 5.6 to give an upper limit to the number of accidents that we might expect in any month.

5.17.   Refer to exercise 5.7. The probability distribution for y, the number of submissions until a contract is won, is known as a *geometric probability distribution*. It is shown in the accompanying table, with calculations rounded to the nearest

y	$p(y)$
1	.6
2	$(.4)\,(.6) = .24$
3	$(.4)^2\,(.6) = .096$
4	$(.4)^3\,(.6) = .038$
5	$(.4)^4\,(.6) = .015$
6	$(.4)^5\,(.6) = .006$
7	$(.4)^6\,(.6) = .002$
8	$(.4)^7\,(.6) = .001$
> 8	.005

thousandth. Individual values of $p(y)$ for y larger than 8 are omitted, but their total probability is .005.

 a. It can be shown that for the above distribution $\mu = 2$ and $\sigma = \sqrt{10/9}$. Construct a graph of $p(y)$ using these values. Locate μ on the graph.

 b. Give the probability that y will be more than two standard deviations from its mean. Confirm that this result agrees with Tchebysheff's theorem.

 c. Give the probability that y will be more than three standard deviations from its mean. Confirm that this result agrees with Tchebysheff's theorem.

 d. Compare your answers to parts b and c with the approximate values given by the Empirical Rule. Note that despite the fact that $p(y)$ is highly skewed to the right, the Empirical Rule provides a satisfactory description of the variability of the random variable y.

5.7 Summary

 Random variables, representing numerical events defined over a sample space, may be classified as **discrete** or **continuous** random variables, depending on whether the number of sample points in the sample space is or is not countable. The theoretical population frequency distribution for a discrete random variable is called a **probability distribution** and may be derived by using the techniques of chapter 4. The model for the frequency distribution for a continuous random variable is a mathematical function $f(y)$, called a **probability distribution** or **probability density function**. This function, usually a smooth curve, is defined over a line interval and is chosen so that the total area under the curve is equal to 1. The probabilities associated with a continuous random variable are given as areas under the probability distribution $f(y)$.

 The expected value of a random variable is the average of that variable calculated for the theoretical population that is defined by its probability distribution. The standard deviation for a random variable can be interpreted in the same way as a sample standard deviation. Using Tchebysheff's theorem and the Empirical Rule, the standard deviation provides a measure of variability for a random variable.

Supplementary Exercises

 5.18. An investor has decided to invest his money in two different stocks. He has his choice narrowed down to three common stocks and two preferred stocks. Unable to choose between these, he randomly chooses three stocks from the five. Find the probability distribution for y, the number of preferred stocks he chooses. Construct a graph of the probability distribution for y.

 5.19. The American economy is characterized by its mass purchasing power. To a great degree this is a result of the rapid increase in family income we have enjoyed. In 1950, 40% of American families had annual incomes exceeding $10,000, while records show that 60% currently have incomes in excess of $10,000.* Suppose that 60% of all families in a particular city have incomes in excess of $10,000. In

---

*Department of Commerce, Bureau of the Census.

a consumer expenditure study, a sample of three families is randomly selected from a city directory in order to study the purchasing habits of the families. Find the probability distribution for y, the number of families in the sample whose current annual income exceeds $10,000. Construct a probability histogram for y.

5.20.   Simulate the experiment described in exercise 5.18 by marking five pieces of cardboard, or coins, so that three represent common stocks and two represent preferred stocks. Place these in a hat, mix, draw three, and record y, the number of preferred stocks observed. Replace the three you have drawn back in the hat and repeat the process until a total of $n = 100$ observations on y have been recorded. Construct a relative frequency histogram for this sample and compare it with the population probability distribution you constructed in exercise 5.18. If the sampling process were repeated $n = 100,000$ times (instead of $n = 100$) and you constructed a relative frequency histogram for the data, how would your histogram compare with the population probability distribution of exercise 5.18?

5.21.   Identify the following as discrete or continuous random variables:
a. The number of defective light bulbs in a package containing four bulbs.
b. The time required to transport a shipment of supplies to a buyer.
c. The weights of blue fin tuna sold each day to a commercial seafood cannery.
d. The number of incoming telephone calls received during a day by an information operator for a large office building.
e. The transaction price for single-family residences sold during a given week in Utica, New York.

5.22.   One researcher suggests that fashion periods vary widely depending on the product class.* On the basis of past experience, a clothing manufacturer has developed the following probability distribution for y, the length (in years) of the fashion period for women's dresses:

y	1	2	3	4
$p(y)$	1/12	3/12	4/12	4/12

Find the expected value and variance of y.

5.23.   Refer to exercise 5.22. Construct a probability histogram for y. Calculate σ and find the probability that y will lie within two standard deviations (2σ) of μ. Does this agree with Tchebysheff's theorem? Do μ and σ characterize $p(y)$?

5.24.   Refer to exercise 5.18. Find the expected value and variance of y. If you performed the simulation in exercise 5.20, calculate $\bar{y}$ and s^2 for the 100 measurements. Compare these values with the expected value and variance of y.

5.25.   Refer to exercise 5.24. Calculate σ and find the probability that y will lie within 2σ of μ. Does this agree with Tchebysheff's theorem? Do μ and σ characterize $p(y)$? What proportion of the 100 measurements lie within $2s$ of $\bar{y}$?

5.26.   Suppose a corporate personnel director wishes to select three of five candidates for three managerial trainee positions and that all candidates have the same chance of being selected. Three of the candidates are members of racial minority groups and two are not. Let y be the number of minority candidates appearing in the personnel director's selection; that is, $y = 1, 2,$ or 3. Find the probability distribution for y. Construct a probability histogram for y. What is the probability that he selects at least two of the three minority group members for the managerial trainee positions?

*D. E. Robinson, "Style Changes: Cyclical, Inexorable, and Foreseeable," *Harvard Business Review*, December 1975.

5.27.   A mail-order magazine subscription service receives orders for one-, two-, and three-year subscriptions, with probabilities $1/2$, $1/4$, and $1/4$, respectively. For each subscription year it receives a \$1 commission. What is the subscription service's expected commission for each order received?

5.28.   A potential customer for a \$20,000 fire insurance policy has a home in an area that, according to experience, may sustain a total loss in a given year with a probability of .001 and a 50% loss with a probability of .01. Ignoring all other partial losses, what premium should the insurance company charge for a yearly policy in order to break even?

5.29.   Suppose the owner of a delicatessen knows from past experience that the daily demand for fresh pastrami is uniformly distributed over the range from 20 pounds to 40 pounds. That is, all demand levels, from 20 to 40 pounds per day, are equally likely to occur. The owner realizes that if demand falls below her inventory, she must sell leftover pastrami at a loss in the delicatessen's prepared sandwich section. How many pounds of pastrami should the owner prepare (or purchase from a supplier) each morning if she wishes the chances to be no more than 25% that she is left with excess inventory at the day's end? (Note: This problem illustrates an elementary form of the demand-inventory model that we will explore in chapter 10. This model has many valuable applications to retail inventory management.)

5.30.   Refer to exercise 5.29. Find the probability that on a given day the demand for pastrami by the customers of the delicatessen:
a. does not exceed 30 pounds
b. is at least 25 pounds
c. is between 22 and 30 pounds

5.31.   In a county containing a large number of rural homes, 60% are thought to be insured against fire. Three rural homeowners are chosen at random from the entire population and a number y are found to be insured against a fire. Find the probability distribution for y. What is the probability that at least two of the three will be insured?

5.32.   A manufacturing representative is considering the option of taking out an insurance policy to cover possible losses incurred by marketing a new product. If the product is a complete failure, the representative feels that a loss of \$80,000 would be incurred; if it is only moderately successful, a loss of \$25,000 would be incurred. Insurance actuaries have determined from market surveys and other available information that the probabilities that the product will be a failure or only moderately successful are .01 and .05, respectively. Assuming that the manufacturing representative would be willing to ignore all other possible losses, what premium should the insurance company charge for the policy in order to break even?

Experiences with Real Data

Title VII of the Civil Rights Act of 1964, as amended by the Equal Employment Opportunity Act of 1972, forbids discrimination on the basis of race, color, national origin, religion, or sex by almost all private and public employers of more than 15 persons. These laws provide that recruitment for employment vacancies in most organizations must be undertaken without designation or identification by race, sex, or ethnicity.

The law specifies that organizations must establish hiring goals to correct current deficiencies caused by past hiring practices. These goals must be stated in terms of the percentage of qualified women and minorities that exist in the pool of applicants.

To illustrate, suppose a large insurance company has advertised openings for 5 experienced claims adjustors. An advertisement of the position generated applications from 10 qualified applicants, 4 of whom are women. Since 40% (4 of 10) of the qualified applicants are women, the insurance company should then establish a goal of filling at least 40% of its openings (at least 2 of the 5 openings) with women. If the insurance company randomly selects 5 people from the applicant pool of 10 qualified claims adjustors, find the probability distribution for y, the number of women in the selected list of 5. What is the probability that the insurance company has met their required hiring goal?

Simulate this experiment by placing $N = 10$ poker chips in a bowl, such that 4 chips are of one color—say, blue—and 6 are of another color—say, white. Blue chips thus represent female applicants and white chips represent male applicants. Select $n = 5$ chips without replacement from the bowl and record y, the number of blue chips (female applicants) in your selection. Repeat this process 100 times, so that you have 100 observations for the random variable y. Then construct a table of relative frequencies for $y = 0$, 1, 2, 3, 4. This table is an approximation of the *actual* probability distribution for y, which is provided in the table. You will learn in chapter 6 that this distribution is called the hypergeometric probability distribution. Compare your derived distribution with the actual distribution. How do you account for any differences?

y	0	1	2	3	4
$p(y)$	.0238	.2381	.4762	.2381	.0238

Returning to the example, find the probability that the insurance company will meet its affirmative action hiring goal (at least 2 of the 5 chosen candidates are women) by using:

i. your empirical results
ii. the actual hypergeometric probabilities.

Note the empirical results obtained by other students in your class. Derive the probability distribution for y by using the collective results of all students who performed the simulation experiment. Does this derived distribution provide a closer approximation to the actual hypergeometric probabilities than the distribution based only on your own empirical results?

References

FRASER, D. A. S. *Statistics: An Introduction.* New York: Wiley, 1958. Chapters 3 and 4.

HADLEY, G. *Introduction to Business Statistics.* San Francisco: Holden-Day, 1968. Chapter 2.

MOSTELLER, F. R. E.; K. ROURKE; and G. B. THOMAS, JR. *Probability with Statistical Applications.* 2d ed. Reading, Mass.: Addison-Wesley, 1970. Chapter 5.

SPURR, W. A., and C. P. BONINI. *Statistical Analysis for Business Decisions.* Rev. ed. Homewood, Ill.: Richard D. Irwin, 1973. Chapters 7 and 8.

TSOKOS, C. P. *Probability Distributions: An Introduction to Probability Theory with Applications.* Belmont, Calif.: Duxbury Press, 1972.

chapter objectives

GENERAL OBJECTIVES A method for finding the probabilities associated with specific values of random variables was presented in chapter 5. Specifically, we learned that discrete and continuous random variables required different probability distributions and that these distributions were subject to different interpretations. Now we turn to some specific applications of these notions and present three useful discrete random variables and their probability distributions. In concluding the chapter we will show you how these probability distributions can be used in making business decisions.

SPECIFIC OBJECTIVES

1. To describe the characteristics of a binomial experiment, to indicate types of business data that represent measurements on binomial random variables, and to present the formula for the binomial probability distribution and to show how it is used to calculate probabilities for a binomial random variable.
Sections 6.2, 6.3

2. To present the formulas for the expected value and variance of a binomial random variable. These parameters will be used to describe a binomial probability distribution and will be used in chapter 7 for a simple procedure for approximating binomial probabilities.
Section 6.4

3. To identify business data for which a Poisson probability distribution would be appropriate and to demonstrate the use of the Poisson probability distribution in calculating the probabilities of specific numerical observations (optional).
Section 6.5

4. To describe the sampling situations for which a hypergeometric probability distribution would be appropriate and to show how this probability distribution can be used to calculate the probabilities of specific numerical observations (optional).
Section 6.6

5. To show how a discrete probability distribution, the binomial, can be used to make inferences about a population parameter, namely, the proportion of defectives in a lot of incoming (or outgoing) products in a manufacturing plant.
Section 6.7

6. To show how sample data can be used to make a decision about the parameter p of a binomial population. This example of decision making introduces the reasoning employed in a statistical test of an hypothesis.
Sections 6.8, 6.9

chapter six

Three Useful Discrete Probability Distributions

6.1 Introduction

In chapter 5 we found that random variables defined over a finite or countably infinite number of points are called discrete random variables. Examples of discrete random variables abound in business and economics, but three discrete probability distributions serve as models for a large number of these applications. These three distributions are the binomial, the Poisson, and the hypergeometric probability distributions. In this chapter we will study these distributions, noting their development as logical models for discrete processes observed in different business settings.

Throughout the study of chapter 6, it will be necessary to refer back to chapter 5 and the definition of a probability distribution—a formula or model that assigns a probability to each possible numerical outcome of an experiment. Thus the nature of the experiment itself and the numerical outcomes of the experiment must be considered before selecting the appropriate probability distribution that will serve as a model for the process.

6.2 The Binomial Experiment

One of the most elementary, useful, and interesting discrete random variables—the binomial—is associated with the coin-tossing experiment. In this experiment either one coin is tossed n times or n coins are each tossed once. One observation, "head" or "tail," is then recorded for each toss. In an abstract sense, numerous coin-tossing experiments of practical importance

145

are conducted daily in the social sciences, physical sciences, and industry.

As an illustration, consider a sample survey conducted to predict voter preference in a political election. Interviewing a single voter bears a similarity, in many respects, to tossing a single coin, because the voter's response may be in favor of our candidate—a "head"—or it may be against (or indicate indecision)—a "tail." In most cases the fraction of voters favoring a particular candidate does not equal one-half, but even this similarity to the coin-tossing experiment is satisfied in national presidential elections. History demonstrates that the fraction of the total vote favoring the winning presidential candidate in most national elections is very near one-half.

Similar polls are conducted in the social sciences, in industry, and in education. The sociologist is interested in the fraction of rural homes that have been electrified; the cigarette manufacturer wishes to know the fraction of smokers who prefer his brand; the teacher is interested in the fraction of students who pass the course. Each person sampled is analogous to the toss of an unbalanced coin (since the probability of a head is usually not $\frac{1}{2}$ for an unbalanced coin).

Firing a projectile at a target is similar to a coin-tossing experiment if the outcome "hit the target" and the outcome "miss the target" are regarded as a head and a tail, respectively. A single missile results in a successful or an unsuccessful launching. A new drug is effective or ineffective when administered to a single patient. A manufactured item selected from a production line is defective or nondefective. With each contact, either a salesperson will consummate a sale or no sale will result. Although dissimilar in some respects, the experiments described above will often exhibit, to a reasonable degree of approximation, the characteristics of a *binomial experiment.*

Definition

A binomial experiment is an experiment that possesses the following properties:

1. The experiment consists of n identical trials.
2. Each trial results in one of two outcomes. For lack of a better nomenclature, we will call one outcome a success, S, and the other a failure, F.
3. The probability of success on a single trial is equal to p and remains the same from trial to trial. The probability of a failure is equal to $(1 - p) = q$.
4. The trials are independent.
5. The experimenter is interested in y, the number of successes observed during the n trials.

Example 6.1 Suppose that there are approximately 1,000,000 potential buyers for a manufacturer's product and that an unknown proportion p favor the product over all its competitors. A sample of 1,000 is selected in such a way that every one of the 1,000,000 buyers has an equal chance of being selected and each potential buyer is asked whether he or she prefers the manufacturer's product over all its competitors. (The ultimate objective of this market survey is to estimate the unknown proportion p, a problem that we will learn how to solve in chapter 8.) Is this a binomial experiment?

Solution To decide whether this is a binomial experiment, we must see if the sampling satisfies the five characteristics described in the box.

1. The sampling consists of $n = 1,000$ identical trials. One trial represents the selection of a single person from the 1,000,000 potential buyers.
2. Each trial results in one of two outcomes: a person prefers the product (call this a success) or does not (a failure).
3. The probability of a success is equal to the proportion of potential buyers. For example, if 400,000 of the 1,000,000 potential buyers favor the product, then the probability of selecting a person favoring the product out of the 1,000,000 potential buyers is $p = .4$. For all practical purposes this probability will remain the same from trial to trial, even though persons selected in the earlier trials are not replaced as the sampling continues.
4. For all practical purposes the probability of a success on any one trial is unaffected by the outcome on any of the others (it will remain very close to p).
5. We are interested in the number y of people in the sample of 1,000 who favor the manufacturer's product.

Because the survey satisfies the five characteristics reasonably well, for all practical purposes it can be viewed as a binomial experiment.

Example 6.2 A purchaser who has received a boxcar containing 20 large electronic computers wishes to sample 3 of the computers to see if they are in working order before he unloads the shipment. The 3 nearest the door of the boxcar are removed for testing and, afterward, are declared either defective or nondefective. Unknown to the purchaser 2 of the 20 computers are defective. Is this a binomial experiment?

Solution As in example 6.1 we check the sampling procedure against the characteristics of a binomial experiment.

1. The experiment consists of $n = 3$ identical trials. Each trial represents the selection and testing of one computer from the total of 20.
2. Each trial results in one of two outcomes: Either a computer is defective (call this a success) or it is not (a failure).
3. Suppose that the computers were randomly loaded into the boxcar so that any one of the 20 computers could have been placed near the boxcar door. Then the unconditional probability of drawing a defective computer on a given trial is $\frac{2}{20}$.

4. The condition of independence between trials *is not* satisfied because the probability of drawing a defective computer on the second and third trials is dependent on the outcome of the first trial. For example, if the first trial results in a defective computer, then there is only one defective left in the remaining 19 computers in the boxcar. Therefore, the conditional probability of success on trial two, given a success on trial one, is $\frac{1}{19}$. This differs from the unconditional probability of a success on the second trial (which is $\frac{2}{20}$). Therefore, the trials are dependent and the sampling does not represent a binomial experiment.

Example 6.1 illustrates an important point. Very few real life situations will completely satisfy the requirements for a binomial experiment, but this is of little consequence as long as the lack of agreement is moderate and does not affect the end result. For instance, the probability of drawing a buyer favoring a particular product in a marketing research survey remains approximately constant from trial to trial as long as the population of buyers is relatively large in comparison with the sample. If 50% of a population of 1000 buyers prefer Product A, then the probability of drawing an A on the first interview is $\frac{1}{2}$. The probability of an A on the second draw is 499/999 or 500/999, depending on whether the first draw was favorable or unfavorable to A. Both are near $\frac{1}{2}$ and would continue to be for the third, fourth, and *n*th trial as long as *n* is not too large. Hence $P(A)$ remains approximately $\frac{1}{2}$ from trial to trial and, for all practical purposes, we could regard the trials as independent. Thus when the sample is small in comparison to the number of elements in the population from which it is selected, the market survey will represent a binomial experiment. On the other hand, if the number of buyers in the population is 10, and 5 favor Product A, then the probability of A on the first trial is $\frac{1}{2}$; the probability of A on the second trial is $\frac{4}{9}$ or $\frac{5}{9}$, depending on whether A was or was not drawn on the first trial. For small populations the probability of A will vary appreciably from trial to trial, independence will not exist, and the resulting experiment will not be a binomial experiment.

Exercises

6.1.   Which of the following business problems can be modeled by the binomial distribution? For those that cannot be modeled by the binomial distribution, explain why not.

a. Determination of the probability that a sales agent will consummate two sales in five contacts if the probability is .25 that the agent consummates a sale on any given contact.

b. Determination of the probability that a manufacturer receives no defective assembly machines in a shipment of three machines from a firm that has three defective and seven nondefective assembly machines in its warehouse.

c. Determination of the probability that the gasoline mileage on a new car will exceed 25 mpg if EPA reports suggest the average mileage rating for such a car is 28 mpg.

d. Determination of the probability that at least 20 people respond to the mailing of 100 advertising circulars when it is known that the usual response rate is 15%.

e. Determination of the probability that no more than 1 of 10 items of machine output is defective when the items are selected over time and it is known that the defective rate increases with excessive machine wear over time.

6.2. Why do we sometimes use the binomial experiment for computing probabilities when sampling from a finite set of objects? What are the conditions under which we are justified in using the binomial distribution in such a case?

6.3 The Binomial Probability Distribution

Having defined the binomial experiment and suggested several practical applications, we now turn to a derivation of the probability distribution for the random variable y, the number of successes observed in n trials. Rather than attempt a direct derivation, we will obtain $p(y)$ for experiments containing $n = 1, 2,$ and 3 trials and leave the general formula to your intuition.

For $n = 1$ trial, we have two sample points, E_1 representing a success S and E_2 representing a failure F, with probabilities p and $q = (1 - p)$, respectively. Since y is the number of successes for the one ($n = 1$) trial and since E_1 implies a success, we assign $y = 1$ to E_1. Similarly, since E_2 represents a failure for the single trial, we assign $y = 0$ to this sample point. The resulting probability distribution for y is given in table 6.1.

Table 6.1 $p(y)$ *for a binomial experiment when* $n = 1$

SAMPLE POINTS	OUTCOMES*	$P(E_i)$	y		y	$p(y)$
E_1	S	p	1		0	q
E_2	F	q	0		1	p

$$\sum_{y=0}^{1} p(y) = q + p = 1$$

*S represents a success on a single trial; F denotes a failure.

The probability distribution for an experiment consisting of $n = 2$ trials is derived in a similar manner and is presented in table 6.2. The four sample points associated with the experiment are presented in the first column; the notation SF denotes a success on the first trial and a failure on the second.

The probabilities of the sample points are easily calculated because each point is an intersection of two independent events, the outcomes of the first and second trials. Thus $P(E_i)$ can be obtained by applying the multiplicative law of probability:

$$P(E_1) = P(SS) = P(S)P(S) = p^2$$

$$P(E_2) = P(SF) = P(S)P(F) = pq$$

$$P(E_3) = P(FS) = P(F)P(S) = qp$$

$$P(E_4) = P(FF) = P(F)P(F) = q^2$$

Table 6.2 $p(y)$ *for a binomial experiment when* $n = 2$

SAMPLE POINTS	OUTCOMES	$P(E_i)$	y	y	$p(y)$
E_1	SS	p^2	2	0	q^2
E_2	SF	pq	1	1	$2pq$
E_3	FS	qp	1	2	p^2
E_4	FF	q^2	0		

$$\sum_{y=0}^{2} p(y) = (q + p)^2 = (1)^2 = 1$$

The value of y assigned to each sample point is given in the fourth column. Note that the numerical event $y = 0$ contains sample point E_4, the event $y = 1$ contains sample points E_2 and E_3, and the event $y = 2$ contains sample point E_1. The probability distribution $p(y)$, presented to the right in table 6.2, reveals a most interesting consequence: the probabilities $p(y)$ are terms of the expansion of $(q + p)^2$.

Summing, we obtain

$$\sum_{y=0}^{2} p(y) = (q + p)^2 = q^2 + 2pq + p^2 = 1$$

The point we wish to make is now quite clear: the probability distribution for a binomial experiment consisting of n trials is obtained by expanding

Table 6.3 $p(y)$ *for a binomial experiment when* $n = 3$

SAMPLE POINTS	OUTCOMES	$P(E_i)$	y	y	$p(y)$
E_1	SSS	p^3	3	0	q^3
E_2	SSF	$p^2 q$	2	1	$3pq^2$
E_3	SFS	$p^2 q$	2	2	$3p^2 q$
E_4	SFF	pq^2	1	3	p^3
E_5	FSS	$p^2 q$	2		
E_6	FSF	pq^2	1		
E_7	FFS	pq^2	1		
E_8	FFF	q^3	0		

$$\sum_{y=0}^{3} p(y) = (q + p)^3 = 1$$

$(q + p)^n$. The proof of this statement is omitted but we will reassure ourselves that it is true by deriving the probability distribution for a binomial experiment consisting of $n = 3$ trials. The computations for the derivation are presented in table 6.3.

Once again you will note that $p(0)$, $p(1)$, $p(2)$, and $p(3)$ are terms of the expansion of $(q + p)^n$, or in this case $(q + p)^3$.

Since the probability associated with a particular value of y is the term involving p to the power y in the expansion of $(q + p)^n$, we write the probability distribution for the binomial experiment as given in the box.

Binomial Probability Distribution

$$p(y) = C_y^n p^y q^{n-y} = \frac{n!}{y!(n-y)!} p^y q^{n-y}$$

where y may take values 0, 1, 2, 3, 4, ... , n and C_y^n is a symbol used to represent the expression

$$\frac{n!}{y!(n-y)!}$$

You may recall from section 4.7 that the factorial notation $n!$ is a short way of writing $n(n - 1)(n - 2) \cdots (3)(2)(1)$. Thus $3! = 3 \cdot 2 \cdot 1$, and $0!$ is defined to be equal to 1. The notation C_y^n is a short way of writing

$$\frac{n!}{y!(n-y)!}$$

This notation is useful because it occurs so often when working with the binomial probability distribution.

Graphs of three binomial probability distributions are shown in figure 6.1, the first for $n = 10$, $p = .1$, the second for $n = 10$, $p = .5$, and the third for $n = 20$, $p = .5$. The binomial distribution for $n = 20$, $p = .5$ was also shown for the discussion of the consumer preference poll in section 5.3.

Let us now consider examples that illustrate applications of the binomial distribution.

Example 6.3 A study on the relative influence of husbands and wives in consumer purchasing, which appeared in a recent issue of *Time* magazine, reported that the husband exerts the primary influence in selecting the make of a new automobile in about 70% of all new car purchases by families. Suppose 4 families have decided to buy a new car.

a. What is the probability that in *exactly* 2 of the 4 families the husband will exert the primary influence in choosing the make of a car?

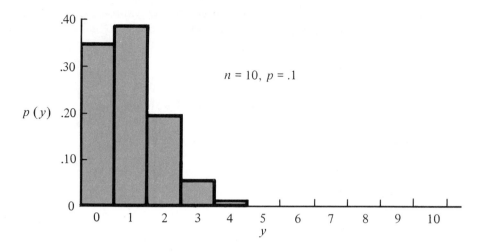

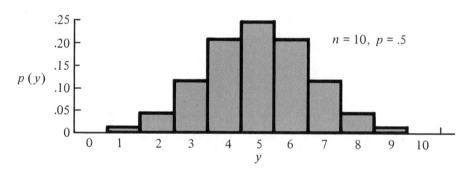

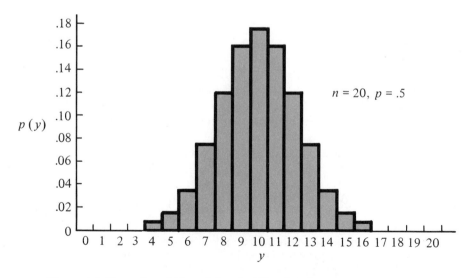

Figure 6.1 *Examples of binomial probability distributions*

b. What is the probability that the husband will exert the primary influence in choosing the make of a car in *at least* 2 of the 4 families?

c. What is the probability that the husband will select the make of car in all 4 families?

Solution Assuming that the family purchase decisions are independent and that *p* remains constant from one family to another, then $n = 4$ and $p = .7$. Let *y* denote the number of families in which the husband exerts the primary influence in selecting a new automobile. Then for $y = 0, 1, 2, 3, 4$, we have

$$p(y) = C_y^4 (.7)^y (.3)^{4-y}.$$

a.

$$p(2) = C_2^4 (.7)^2 (.3)^2 = \frac{4!}{2!\,2!} (.49)(.09) = .2646$$

The probability is .2646 that in exactly 2 of the 4 families the husband will exert the primary influence in choosing the make of a car.

b.

$$P(\text{at least two}) = P(y \geq 2) = p(2) + p(3) + p(4)$$
$$= 1 - p(0) - p(1) = 1 - C_0^4 (.7)^0 (.3)^4 - C_1^4 (.7)^1 (.3)^3$$
$$= 1 - .0081 - .0756 = .9163$$

The probability is .9163 that the husband selects the make of car in at least 2 of the families.

c.

$$p(4) = C_4^4 (.7)^4 (.3)^0 = \frac{4!}{4!\,0!} (.7)^4 (1) = .2401$$

The probability is .2401 that the husband selects the make of car in all 4 families.

Note that these probabilities would be incorrect if the members of one family were in any way influenced by the purchase decision of one of the other 3 families. In that case the purchase decisions (trials) would be dependent upon one another and *p* would very likely be different from family to family.

Example 6.4 Large lots of an incoming product at a manufacturing plant are inspected for defectives by means of a sampling scheme. Ten items are examined and the lot rejected if 2 or more defectives are observed. If a lot contains exactly 5% defectives, what is the probability that the lot is accepted? Rejected? Assume independence between successive draws from the lot.

Solution Let *y* be the number of defectives observed. Then $n = 10$, and the probability of observing a defective on a single trial is $p = .05$. So

$$p(y) = C_y^{10}(.05)^y(.95)^{10-y}$$

and

$$P(\text{accept}) = p(0) + p(1) = C_0^{10}(.05)^0(.95)^{10} + C_1^{10}(.05)^1(.95)^9$$

$$= .599 + .315 = .914$$

$$P(\text{reject}) = 1 - P(\text{accept}) = 1 - .914 = .086$$

Example 6.5 A new serum was tested to determine its effectiveness in preventing the common cold. Ten people were injected with the serum and observed for a period of one year. Eight survived the winter without a cold. Suppose it is known that when a serum is not used, the probability of surviving a winter without a cold is .5 and that whether an individual survives a winter without a cold is independent of the state of health of any other person. What is the probability of observing 8 or more survivors, given that the serum is ineffective in increasing bodily resistance to colds?

Solution Assuming that the vaccine is ineffective, the probability of surviving the winter without a cold is $p = .5$. The probability distribution for y, the number of survivors, is

$$p(y) = C_y^{10}(.5)^y(.5)^{10-y} = C_y^{10}(.5)^{10}$$

Then we have

$$P(8 \text{ or more}) = p(8) + p(9) + p(10)$$

$$= C_8^{10}(.5)^{10} + C_9^{10}(.5)^{10} + C_{10}^{10}(.5)^{10}$$

$$= .0439 + .0098 + .0010 = .055$$

Calculating binomial probabilities is a tedious task when n is large, as you perhaps noticed in examples 6.4 and 6.5. To simplify our calculations the sum of the binomial probabilities from $y = 0$ to $y = a$ is presented in the appendix in table 1, for $n = 5, 10, 15, 20,$ and 25.

To illustrate how we use table 1, suppose we wish to find the sum of the binomial probabilities from $y = 0$ to $y = 3$ for $n = 5$ trials and $p = .6$. That is, we wish to find

$$P(y \leq 3) = \sum_{y=0}^{3} p(y) = p(0) + p(1) + p(2) + p(3)$$

where

$$p(y) = C_y^5(.6)^y(.4)^{5-y}$$

Since the table values give

$$P(y \leq a) = \sum_{y=0}^{a} p(y)$$

we seek the table value in the row corresponding to $a = 3$ and the column for $p = .6$. The table value, .663, is shown in the table here as it appears in table 1(a). Therefore, the sum of the binomial probabilities from $y = 0$ to $y = a = 3$ (for $n = 5$, $p = .6$) is .663.

$n = 5$ p

a	0.01	0.05	0.10	0.20	0.30	0.40	0.50	0.60	0.70	0.80	0.90	0.95	0.99	a
0	—	—	—	—	—	—	—	—	—	—	—	—	—	0
1	—	—	—	—	—	—	—	—	—	—	—	—	—	1
2	—	—	—	—	—	—	—	—	—	—	—	—	—	2
3	—	—	—	—	—	—	—	.663	—	—	—	—	—	3
4	—	—	—	—	—	—	—	—	—	—	—	—	—	4

To find an individual binomial probability, say $p(3)$ for $n = 5$, $p = .6$, we calculate

$$p(3) = \sum_{y=0}^{3} p(y) - \sum_{y=0}^{2} p(y) = .663 - .317 = .346$$

We use table 1 in the following example.

Example 6.6 Refer to example 6.5 and use table 1 in the appendix to calculate the probability of 8 or more survivors, given that the probability of a single person surviving the winter without a cold is $p = .5$.

Solution For this example $n = 10$ and $p = .5$. Consequently, we consult table 1(b) in the appendix. Since we know that the sum of the binomial probabilities from $y = 0$ to $y = 10$ is 1, we have that

$$P(y \geq 8) = \sum_{y=8}^{10} p(y) = 1 - \sum_{y=0}^{7} p(y)$$

The quantity

$$\sum_{y=0}^{7} p(y)$$

can be found by moving across the top of table 1(b) to the column for $p = .5$ and down that column to the row corresponding to $a = 7$. We then read

$$\sum_{y=0}^{7} p(y) = .945$$

Then

$$P(y \geq 8) = \sum_{y=8}^{10} p(y) = 1 - \sum_{y=0}^{7} p(y) = 1 - .945 = .055$$

Examples 6.3, 6.4, and 6.5 illustrate the use of the binomial probability distribution in calculating the probability associated with values of y, the number of successes in n trials defined for the binomial experiment. Thus we note that the probability distribution, $p(y) = C_y^n p^y q^{n-y}$, provides a simple formula for calculating the probabilities of numerical events y applicable to a broad class of experiments that occur in everyday life. This statement must be accompanied by a word of caution: Each physical application must be carefully checked against the defining characteristics of the binomial experiment (presented in section 6.2) to determine whether the binomial experiment is a valid model for the given application.

Exercises

6.3.   Imported cars comprise 20% of all new car sales in the United States.* Suppose we randomly select $n = 4$ people who have purchased a new car during the past week.
  a. Find the probability that all of them purchased an imported car.
  b. Find the probability that one of them purchased an imported car.
  c. Find the probability that none of them purchased an imported car.

6.4.   Refer to exercise 6.3 and let y be the number of buyers in the sample of $n = 4$ new car buyers who purchase an imported car. Construct a probability histogram for $p(y)$.

6.5.   Many utility companies have begun to promote energy conservation by offering discounted rates to consumers who keep their energy usage below certain established subsidy standards. A recent EPA report notes that 70% of the island residents of Puerto Rico have reduced their electricity usage sufficiently to qualify for discounted rates. Suppose five residential subscribers are randomly selected from San Juan, Puerto Rico.
  a. Find the probability that all five qualify for the favorable rates.
  b. Find the probability that at least four qualify for the favorable rates.

6.6.   A number of airlines run a commuter service between Washington, D.C., and New York City. Due to the traffic congestion at the airports of both cities, commuter flights are often delayed by as much as 2 hours. Recent evidence shows that 25% of the Washington–New York commuter flights run more than 30 minutes late. Suppose five commuters each take a different Washington–New York commuter flight on five different days. (Assume that the flights are independent of each other.)
  a. Find the probability that all five arrive in New York within 30 minutes of their expected time of arrival.
  b. Find the probability that no more than two of the five commuters are more than 30 minutes late in arriving in New York.

6.7.   Maintenance records suggest that only 1 of every 100 electric typewriters of a certain brand require major service and repair during the first year of use. An office manager has purchased 10 typewriters of this brand.
  a. Find the probability that none of the typewriters require major service and repair during the first year of use.

Time, 15 September 1975.

b. Find the probability that two of the typewriters require major service and repair during the first year of use.

6.8.   Refer to exercise 6.7. If you are the office manager and find that 2 of the 10 typewriters require major service and repair during the first year of use, how would you feel about the manufacturer's claim that only 1 in 100 require such repair? (We are asking here for an inference, a topic that will be discussed in section 6.6.)

6.4 The Mean and Variance for the Binomial Random Variable

As we saw, the calculation of $p(y)$ becomes very tedious for large values of n. Thus it is convenient to describe the binomial probability distribution by using its mean and standard deviation. This will enable us to identify values of y that are highly improbable simply by using our knowledge of Tchebysheff's theorem and the Empirical Rule. A more precise method for approximating binomial probabilities will be presented in chapter 7 and this method will rely on knowledge of the mean and standard deviation of y, namely μ and σ. Consequently, we need to know the expected value and variance of the binomial random variable y.

The formulas for the mean, the variance, and the standard deviation of the binomial random variable are given in the box.

Mean, Variance, and Standard Deviation for a Binomial Random Variable

$$mean \quad \mu = E(y) = np$$

$$variance \; \sigma^2 = npq$$

$$standard \; \sigma = \sqrt{npq}$$
$$dev.$$

The following example illustrates how to use these quantities to describe a binomial probability distribution.

Example 6.7 A manufacturer believes that 30% of all consumers favor her product. To check her belief she randomly samples 800 consumers and counts the number y favoring her product. If 30% of all consumers favor the manufacturer's product, within what limits would you expect y to fall?

Solution It would be difficult to calculate the probabilities for y because n is so large. Consequently, we will describe the probability distribution by using μ and σ.

Since $n = 800$ and $p = .3$, we have that

$$\mu = np = (800)(.3) = 240$$

$$\sigma = \sqrt{npq} = \sqrt{(800)(.3)(.7)} = \sqrt{168} = 12.96$$

Based on Tchebysheff's theorem and the Empirical Rule, we would expect y to fall within the interval $(\mu \pm 2\sigma)$ with a high probability and almost certainly within the interval $(\mu \pm 3\sigma)$. The intervals are

$$(\mu \pm 2\sigma) = (240 \pm 25.92) \qquad \text{or} \qquad 214.08 \text{ to } 265.92$$

$$(\mu \pm 3\sigma) = (240 \pm 38.88) \qquad \text{or} \qquad 201.12 \text{ to } 278.88$$

(Comment: A histogram of the binomial probability distribution will be very mound shaped for $n = 800$ and $p = .3$. Hence we would expect the Empirical Rule to work very well. Why this is true will be explained in chapter 7.)

Exercises

6.9.   The Energy Policy Center of the Environmental Protection Agency reports that 75% of the homes in New England are heated by oil-burning furnaces. If a certain New England community is known to have 2500 homes, find the expected number of homes in the community that are heated by oil furnaces. If y is the number of homes in the community that are heated by oil, find the variance and standard deviation of y. Use Tchebysheff's theorem to describe limits within which you could expect y to fall.

6.10.   The annual report is one of the most important documents produced by publicly owned companies and a document that incurs considerable expense in its production. However, a recent study suggests that 40% of stockholders spend 5 minutes or less reading their company's annual report.* Suppose 100 stockholders of a publicly owned company are randomly selected from the firm's registry of stockholders.

a. Find the expected value of y, the number of stockholders who spend no more than 5 minutes reading their company's annual report.

b. Determine the standard deviation of y.

c. If the controller observed $y = 25$ of the 100 selected stockholders who spend no more than 5 minutes reading the annual report, does it appear that the proportion of all stockholders spending 5 minutes or less reading his company's annual report is really 40%? Explain.

6.11.   Much debate exists in modern day corporate boardrooms about the propriety of "social auditing," the monitoring of the firm's impact on society and its attainment of preestablished social goals and responsibilities. The president of a large publicly owned firm claims that 50% of the firm's stockholders favor a periodic social audit of the firm. To test his claim 1000 stockholders are randomly selected from the firm's registry of stockholders and asked their opinion about a periodic social audit.

a. What is the expected value of y, the number of stockholders favoring a social audit?

b. What is the standard deviation of y?

c. Suppose that $y = 450$. Is this a value that might be expected with reasonable probability if, in fact, $p = .5$? How might you explain this observed result?

*F. C. Coy, "Annual Reports Don't Have to Be Dull," *Harvard Business Review,* January–February 1973.

6.12.   It is known that 10% of a certain brand of self-winding wristwatches will have to be replaced under the warranty. If 2500 watches are sold, find the expected value and variance of y, the number of watches that must be replaced under the warranty. Within what limits would y be expected to fall? (Hint: Use Tchebysheff's theorem.)

6.5 The Poisson Probability Distribution (Optional)

In this section and the next, we present two additional discrete random variables and their probability distributions, namely, the Poisson and the hypergeometric random variables. The probability distributions for both these random variables are related to the binomial probability distribution but both have important applications in their own right. The Poisson random variable is discussed in this section; the hypergeometric is presented in section 6.6.

If, in a binomial experiment, the sample size n is large and the probability of success p is small, the binomial probabilities are often approximated by the corresponding probabilities given by the Poisson probability distribution. The formula for the Poisson probability distribution is shown in the box.

Poisson Probability Distribution

$$p(y) = \frac{\mu^y e^{-\mu}}{y!} \qquad y = 0, 1, 2, 3, \ldots$$

where μ is the mean of the probability distribution and $e = 2.71828\ldots$ (e is the base of natural logarithms). To approximate the binomial probability distribution, let $\mu = np$.

For your convenience, values of $e^{-\mu}$ are presented in table 2 of the appendix for values of μ from 0 through 10, in increments of .1.

Graphs of the Poisson probability distribution for $\mu = .5$, 1, and 4 are shown in figure 6.2.

Binomial tables are seldom available for n greater than 100. However, many applications of a binomial experiment with $n = 100$ or more arise in business and economics. Consequently, we need simple, easy-to-compute approximation procedures for calculating binomial probabilities. The Poisson probability distribution provides good approximations to binomial probabilities when n is large and $\mu = np$ is small, preferably with $np \leq 7$. An approximation procedure suitable for larger values of $\mu = np$ will be presented in chapter 7.

As an illustration of the Poisson approximation procedure, consider the

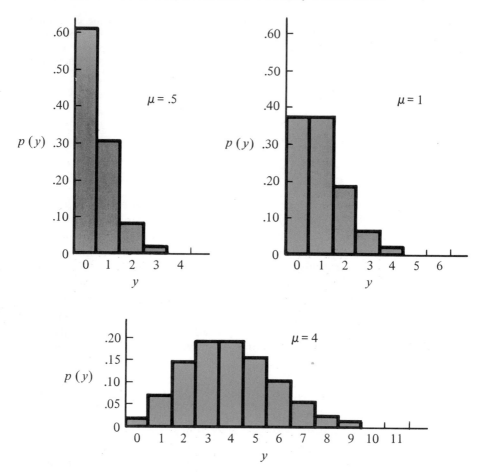

Figure 6.2 *Poisson probability distributions for* μ = .5, 1, and 4

following application. Suppose that a life insurance company insures the lives of 5000 men of age 42. If actuarial studies show the probability of any 42-year-old man dying in a given year to be .001, the exact probability that the company will have to pay $y = 4$ claims during a given year is given by the binomial distribution as

$$P(y = 4) = p(4) = \frac{5000!}{4!4996!}(.001)^4(.999)^{4996}$$

for which binomial tables are not available. To compute $P(y = 4)$ by hand is out of the question, but the Poisson distribution can be used to provide a good approximation to $P(y = 4)$. Computing $\mu = np = (5000)(.001) = 5$ and substituting into the formula for the Poisson probability distribution, we have

$$P(4) \approx \frac{\mu^4 e^{-\mu}}{4!} = \frac{5^4 e^{-5}}{4!} = \frac{(625)(.0067)}{24} = .1745$$

Example 6.8 Suppose that a large food-processing and canning plant has 20 automatic canning machines in operation at all times. If the probability that an individual canning machine breaks down during a given day is .05, find the probability that during a given day 2 canning machines fail. Use the binomial distribution to compute the exact probability and then compute the Poisson approximation.

Solution This is a binomial experiment with $n = 20$ and $p = .05$. The expected number of machine breakdowns in a given day is $\mu = np = 20(.05) = 1.0$. Thus, using the binomial distribution and the binomial tables (table 1, appendix), we have

$$P(y = 2) = p(2) = \sum_{y=0}^{2} p(y) - \sum_{y=0}^{1} p(y) = .925 - .736 = .189$$

Using the Poisson distribution to approximate this binomial experimental outcome, we find from table 2 in the appendix that

$$P(y = 2) \approx p(2) = \frac{1^2 e^{-1}}{2!} = \frac{.367879}{2}$$

Thus, rounded to three decimal places, we have

$$P(y = 2) \approx p(2) = .184$$

You can see that the Poisson approximation, .184, is quite close to the exact value of the binomial probability, .189. The larger the value of n (for a fixed value of $\mu = np$), the better will be the Poisson approximation.* In particular, we recommend that n be large and that $\mu = np$ be less than or equal to 7.

Example 6.9 A manufacturer of power lawn mowers buys 1-horsepower two-cycle engines in lots of 1000 from a supplier. He then equips each of the mowers produced by his plant with one of the engines. Past history shows that the probability of any one engine purchased from the supplier proving unsatisfactory is .001. In a shipment of 1000 engines, what is the probability that none are defective? One is defective? Two are? Three are? Four are?

Solution This is a binomial experiment with $n = 1000$ and $p = .001$. The expected number of defectives in a shipment of $n = 1000$ engines is $\mu = np = (1000)(.001) = 1$. Since this is a binomial experiment, the probability of y defective engines in the shipment may be approximated by

$$p(y) = \frac{\mu^y e^{-\mu}}{y!} = \frac{1^y e^{-1}}{y!} = \frac{e^{-1}}{y!}$$

*See Feller (1968).

[since $1^y = 1$ for any value of y]. Therefore, we have

$$p(0) \approx \frac{e^{-1}}{0!} = \frac{.368}{1} = .368$$

$$p(1) \approx \frac{e^{-1}}{1!} = \frac{.368}{1} = .368$$

$$p(2) \approx \frac{e^{-1}}{2!} = \frac{.368}{2} = .184$$

$$p(3) \approx \frac{e^{-1}}{3!} = \frac{.368}{6} = .061$$

$$p(4) \approx \frac{e^{-1}}{4!} = \frac{.368}{24} = .015$$

Notice that $p(y)$ exists for any integer value of y greater than or equal to 0 and that its value decreases as y becomes large. The tailing off of $p(y)$ for increasing y is consistent with the intent of the Poisson distribution—to provide a probability model for count data where the counts represent numbers of "rare events."

The Poisson probability distribution has numerous applications other than approximating certain binomial probabilities. It provides a good model for the count data resulting from any experiment where the count y represents the number of rare events observed in a given unit of time or space. Some examples of experiments for which the random variable y can be considered a Poisson random variable are these:

1. The number of calls received by a switchboard during a given small period of time.
2. The number of claims against an insurance company during a given week.
3. The number of arrivals at a checkout stand during a given minute.
4. The number of machine breakdowns during a given day.
5. The number of sales by a real estate agent in a given day.

In each example y represents the **number of rare events during a period of time over which an average of** μ **such events can be expected to occur.** The only assumptions needed when using the Poisson distribution to model experiments such as those just described are that the counts or events occur **randomly** and **independently** of one another.

Example 6.10 Bank closures in the United States due to financial difficulties have occurred at the average rate of 5.7 closures per year since

1960.* Assume that the number of closures y in a given period possesses a Poisson probability distribution.

   a. Find the probability of no bank closures during a given four-month period.

   b. Find the probability of at least 3 bank closures during a given year.

   Solution

   a. If closures occur at the rate of 5.7 per year, then for four months ($\frac{1}{3}$ of a year) we could expect

$$\mu = 5.7(\tfrac{1}{3}) = 1.9$$

closures during any four-month period. Therefore, the probability of no closures in a given four-month period is

$$p(0) = \frac{1.9^0 e^{-1.9}}{0!} = \frac{e^{-1.9}}{1} = .150$$

(Note: $\mu^0 = 1$.)

   b. During a given year we could expect $\mu = 5.7$ bank closures. Since

$$P(y \geq 3) = \sum_{y=3}^{\infty} p(y) = 1 - p(0) - p(1) - p(2)$$

where

$$p(0) = \frac{5.7^0 e^{-5.7}}{0!} = .0033$$

$$p(1) = \frac{5.7^1 e^{-5.7}}{1!} = (5.7)(.0033) = .0188$$

$$p(2) = \frac{5.7^2 e^{-5.7}}{2!} = \frac{(32.5)(.0033)}{2} = .0536$$

then

$$P(y \geq 3) = 1 - .0033 - .0188 - .0536 = 1 - .0757 = .9243$$

Exercises

6.13.   Perceived quality is at least as important, if not more so, than the actual quality of goods as determined by impartial testing laboratories. Researcher Nagashima reports that 40% of Japanese businessmen believe Japanese electrical appliances are a greater value than U.S. manufactured appliances, but only 5% of U.S. businessmen favor the Japanese appliances.[†] Perceived quality could be responsible for the difference. Use the Poisson approximation to the binomial to estimate the probability of exactly 5 in a sample of 100 U.S. businessmen favoring Japanese

FDIC Annual Report (1973), p. 227.

[†] A. Nagashima, "A Comparison of Japanese and U.S. Attitudes Toward Foreign Products," *Journal of Marketing*, January 1970.

electrical appliances. Estimate the probability that no more than 5 of 100 favor the Japanese appliances.

6.14. To illustrate how well the Poisson probability distribution approximates the binomial probability distribution, calculate the Poisson approximate values for $p(0)$ and $p(1)$ for a binomial probability distribution with $n = 25$ and $p = .05$. Compare the answers with the exact values obtained from table 1 in the appendix.

6.15. The Labor-Management Reporting and Disclosure Act of 1959 prescribes fiduciary responsibilities for union officials and makes embezzlement of union funds a federal offense. Since 1961 civil suits have been filed under this law randomly and independently of one another at the average rate of 2.9 suits per month.*

a. Find the probability of no suits being filed under the LMRDA during a given month.

b. Find the probability of no more than 4 suits being filed under the law during a given two-month span. (Hint: See example 6.10.)

6.16. The local manager of a rental car organization buys tires in lots of 500 in order to take advantage of volume price discounts from the supplier. From experience the manager knows that 1% of all new tires purchased from the supplier are defective and must be replaced within the first week of usage. Find the probability that a shipment of 500 tires from the supplier will contain no defective tires. One defective tire. No more than three defective tires.

6.17. A telephone switchboard for a medical office building can handle at most 5 incoming calls a minute. If past experience suggests that an average of 120 incoming calls per hour are received by the switchboard, find the probability that the switchboard is overloaded during any given minute.

6.6 The Hypergeometric Probability Distribution (Optional)

Suppose you are selecting a sample of elements from a population and you record whether each element does or does not possess a certain characteristic. Consequently, you are dealing with the "success" or "failure" type of data encountered in the binomial experiment. The consumer preference survey of example 6.1 (section 6.2) or the sampling for defectives of example 6.2 (section 6.2) are practical illustrations of these sampling situations.

If the number of elements in the population is large in relation to the number in the sample (as for example 6.1), the probability of selecting a success on a single trial is equal to the proportion p of successes in the population. Because the population is large in relation to the sample size, this probability will remain constant (for all practical purposes) from trial to trial and the number y of successes in the sample will follow a binomial probability distribution. However, if the number of elements in the population is small in relation to the sample size, the probability of a success for a given trial is dependent on the outcomes of preceding trials. Then the number y of successes follows what is known as a *hypergeometric probability distribution*.

*U.S. Department of Labor, *Compliance Enforcement and Reporting in 1973* (1974).

Before giving the formula for a hypergeometric probability distribution, we must define the following notation:

N = the number of elements in the population

k = the number of elements in the population that are successes (that is, the number possessing one of the two characteristics)

$N - k$ = the number of elements in the population that are not successes

n = the number of elements in the sample, selected from the N elements in the population

y = the number of successes in the sample

Then the hypergeometric probability distribution for the random variable y is as given in the box.

Hypergeometric Probability Distribution

$$p(y) = \frac{C_y^k \, C_{n-y}^{N-k}}{C_n^N} \quad \text{for} \quad \begin{array}{l} y = 0, 1, 2, \dots, n \quad \text{if} \quad n < k \\ y = 0, 1, 2, \dots, k \quad \text{if} \quad n \geq k \end{array}$$

where C_y^k, C_{n-y}^{N-k}, and C_n^N are combination symbols as defined in section 4.7. For example,

$$C_y^k = \frac{k!}{y! \, (k - y)!}$$

You may recall that several exercises in chapter 4 portrayed situations for which the hypergeometric probability distribution would have been applicable. For those exercises N and n were kept small and our intention was to use the exercises to develop your ability to solve probability problems. We can now solve similar but more complex probability problems using the hypergeometric probability distribution.

Example 6.11 An important problem encountered by personnel directors and others faced with the selection of the "best" in a finite set of elements is indicated by the following situation. From a group of 20 PhD engineers, 10 are randomly selected for employment. What is the probability that the 10 selected include all the 5 best engineers in the group of 20?

Solution For this example $N = 20$, $n = 10$, $k = 5$, and $N - k = 15$. That is, there are only 5 in the set of 5 best engineers, and we seek the probability that $y = 5$, where y denotes the number of best engineers among the 10 selected. Then

$$p(y) = \frac{C_y^k \, C_{n-y}^{N-k}}{C_n^N}$$

$$p(5) = \frac{C_5^5 \, C_5^{15}}{C_{10}^{20}} = \frac{\left(\dfrac{5!}{5! \, 0!}\right)\left(\dfrac{15!}{5! \, 10!}\right)}{\dfrac{20!}{10! \, 10!}}$$

$$= \left(\frac{15!}{5! \, 10!}\right)\left(\frac{10! \, 10!}{20!}\right) = \frac{21}{1292} = .0163$$

(Remember that 0! = 1.)

Example 6.12 A particular industrial product is shipped in lots of 20. Testing to determine whether an item is defective is costly, and hence the manufacturer samples production rather than using a 100% inspection plan. A sampling plan constructed to minimize the number of defectives shipped to customers calls for sampling 5 items from each lot and rejecting the lot if more than 1 defective is observed. (If rejected, each item in the lot is then tested.) If a lot contains 4 defectives, what is the probability that it will be accepted?

Solution Let y be the number of defectives in the sample. Then $N = 20$, $k = 4$, $N - k = 16$, and $n = 5$. The lot will be rejected if $y = 2$, 3, or 4. Then

$$P(\text{accept the lot}) = P(y \le 1) = p(0) + p(1)$$

$$= \frac{C_0^4 \, C_5^{16}}{C_5^{20}} + \frac{C_1^4 \, C_4^{16}}{C_5^{20}}$$

$$= \frac{\left(\dfrac{4!}{0! \, 4!}\right)\left(\dfrac{16!}{5! \, 11!}\right)}{\dfrac{20!}{5! \, 15!}} + \frac{\left(\dfrac{4!}{1! \, 3!}\right)\left(\dfrac{16!}{4! \, 12!}\right)}{\dfrac{20!}{5! \, 15!}}$$

$$= \frac{91}{323} + \frac{455}{969} = .2817 + .4696 = .7513$$

You will note that example 6.12 is similar to example 6.4. The only difference is that the number y of defectives possesses a hypergeometric probability distribution when the number of elements N in the population is small in relation to the sample size n.

Being familiar with discrete probability distributions and the properties of the experiments that generate them is extremely helpful. Rather than solving the same probability problem over and over again from first principles (as was done in chapter 4), all you need to do is to recognize the type of random variable involved and then substitute into the formula for its probability distribution. The first three exercises in the exercise set that follows have been selected from those given in chapter 4.

Exercises

6.18. (exercise 4.3) A local school board consists of five residents of the community, two of whom are lawyers. The mayor plans to select two board members at random from the school board to form a subcommittee to meet with the local teachers' union on a collective bargaining issue. We are interested in the composition of the subcommittee.

a. Let y be the number of lawyers in the sample of $n = 2$. Explain why y possesses a hypergeometric probability distribution. Give the formula for $p(y)$.

b. What is the probability that the subcommittee will consist of at least one lawyer? [Hint: $p(1) + p(2) = 1 - p(0)$.]

6.19. (exercise 4.43) A large municipal bank chain has ordered five mini-computers to distribute among their five branches in a certain city. Unknown to the purchaser three of the five computers are defective. Before installing the computers an agent of the bank selects two of the five computers from the shipment, thoroughly tests them, and then classifies each as either defective or nondefective.

a. Let y be the number of defective computers in the sample of $n = 2$. Explain why y possesses a hypergeometric probability distribution. Give the formula for $p(y)$.

b. What is the probability that both computers in the sample will be defective?

c. Calculate the values of $p(y)$ for $y = 0, 1, 2$. Graph $p(y)$.

6.20. (exercise 4.47) One researcher noted that affirmative action commitments by industrial organizations have led to an increase in the number of women in executive positions.* Suppose a company has vacancies for two positions of vice-president. The vacancies are to be filled by randomly selecting two people from a list of four candidates. On the list are two women and two men, all of whom have worked for the company for a long period of time.

a. Let y be the number of women selected to fill the two vice-presidential positions. Explain why y possesses a hypergeometric probability distribution. Give the formula for $p(y)$.

b. What is the probability that at least one of the women will be selected?

c. What is the probability that neither woman will be selected?

6.21. A particular antibiotic is shipped to drug stores in cases, each of which contains 24 bottles. Having doubts about the potency of the drug, the druggist decides to have 5 bottles of the drug tested. Suppose that actually 10 of the 24 bottles are understrength.

a. Let y be the number of understrength bottles in the sample of 5. Explain why y is or is not a hypergeometric random variable.

b. Find the probability that none of the bottles sent to the testing company is understrength.

c. Find the probability that exactly 1 of the 5 is understrength.

*S. Reynolds, "Women on the Line," *MBA*, February 1975.

6.7 Making Decisions: Lot Acceptance Sampling

You may have noted that examples 6.3 through 6.5, examples of applications of the binomial probability distribution, were problems in probability rather than statistics. In those examples the composition of the binomial population, characterized by p, the probability of a success on a single trial, was assumed known. We were interested in calculating the probability of certain numerical events.

Let us now reverse the procedure; that is, let us assume that we possess a sample from the population and wish to make inferences concerning p. The physical settings for examples 6.4 and 6.5 supply excellent practical situations in which the ultimate objective is statistical inference. We will consider these two problems in greater detail in succeeding sections.

Example 6.4 describes a plan for accepting or rejecting large lots of incoming items in a manufacturing plant. The objective is to screen the lots to eliminate those containing a high fraction of defective items. For example, suppose that a manufacturer purchases large lots, each containing 1000 ball bearings of a specific type. He is unwilling to pay for defective bearings and hence will accept lots containing no more than $p = .05$ defective. Thus a good lot is defined as one for which $p \leq .05$ and, correspondingly, an unacceptable lot is defined as one for which $p > .05$. The manufacturer wishes to identify unacceptable lots so that they can be returned to the supplier.

One way to detect unacceptable lots is to inspect every item in the lot. This procedure is not only costly but is also imperfect. That is, inspector fatigue and other human factors permit defective items to pass through the screen undetected. A less costly and more effective means of screening is to reject or accept lots on the basis of a sample. Thus a sample of n items is drawn from the lot, inspected, and then the number y of defectives is recorded. The lot is accepted if the number of defectives is less than or equal to a preselected number a, called the **acceptance number**. If y exceeds a, the lot is rejected and returned to the supplier.

Every sampling plan is defined by a sample size n and an acceptance number a. Changing one or both of these numbers changes the sampling plan and the characteristics of the screen. These characteristics are measured by the probability of accepting a lot containing a specified fraction of defectives, as indicated in example 6.4. Quality control engineers characterize the goodness of a sampling plan by calculating the probability of lot acceptance for various fractions of defectives in a lot. The result is presented in a graphic form and is called the **operating characteristic curve** for the sampling plan. A typical operating characteristic curve is shown in figure 6.3.

For the screen to operate satisfactorily, we would like the probability of accepting lots with a low lot defective fraction to be high and the probability of accepting lots with a high lot defective fraction to be low. From figure

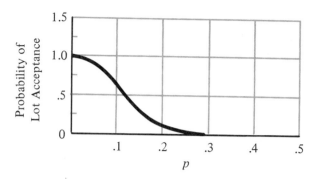

Figure 6.3 *Typical operating characteristic curve for a sampling plan*

6.3 you can see that the probability of acceptance always drops as the fraction of defectives increases—a result that is in agreement with our intuition.

We illustrate these ideas with some examples.

Example 6.13 Calculate the probability of lot acceptance for a sampling plan with sample size $n = 6$ and acceptance number $a = 0$ for lot defective fractions of $p = .1, .3,$ and $.5$. Sketch the operating characteristic curve for the plan.

Solution For us to accept the lot, the sample must contain no defectives. Therefore, we have

$$P(\text{accept}) = p(0) = C_0^6 p^0 q^6 = q^6$$

$$P(\text{accept when } p = .1) = (.9)^6 = .531$$

$$P(\text{accept when } p = .3) = (.7)^6 = .118$$

$$P(\text{accept when } p = .5) = (.5)^6 = .016$$

A sketch of the operating characteristic curve can be obtained by plotting the three points determined in the calculations above. In addition, we know that the probability of acceptance must equal 1 when $p = 0$ and must equal 0 when $p = 1$. The operating characteristic curve is given in figure 6.4.

Example 6.14 Construct the operating characteristic curve for a sampling plan with $n = 15$ and $a = 1$.

Solution The probability of lot acceptance will be calculated for $p = .1,$ .2, .3, .5.

$$P(\text{accept}) = p(0) + p(1) = \sum_{y=0}^{a=1} p(y)$$

$$P(\text{accept when } p = .1) = .549$$

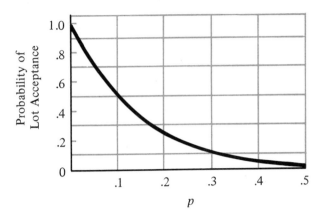

Figure 6.4 *Operating characteristic curve, for n = 6, and a = 0; example 6.13*

$$P(\text{accept when } p = .2) = .167$$

$$P(\text{accept when } p = .3) = .035$$

$$P(\text{accept when } p = .5) = .000$$

(Note that these probabilities can be found by using table 1(c) in the appendix.) The operating characteristic curve for the sampling plan is given in figure 6.5.

Sampling inspection plans are widely used in industry. Each sampling plan possesses its own unique operating characteristic curve that characterizes the plan and, in a sense, describes the size of the holes in the screen. The quality control engineer will choose the plan that satisfies the requirements of the situation. Increasing the acceptance number increases the probability

Figure 6.5 *Operating characteristic curve for n = 15, and a = 1; example 6.14*

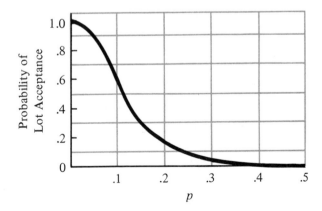

of acceptance and hence increases the size of the holes in the screen. Increasing the sample size provides more information upon which to base the decision and hence improves the discriminatory power of the decision procedure. Thus when n is large, the operating characteristic curve will drop rapidly as p increases. You may verify these remarks by working the exercises at the end of the chapter.

Note that lot acceptance sampling is an example of statistical inference, because the procedure implies a decision concerning the lot defective fraction p. If you accept a lot, you infer that the true lot defective fraction p is some relatively small acceptable value. If you reject a lot, it is clear that you think p is too large. Consequently, lot acceptance sampling is a procedure that implies inference making about the lot defective fraction. The operating characteristic curve for the sampling plan provides a measure of the goodness of this inferential procedure.

Exercises

6.22. A buyer and seller agree to use a sampling plan with sample size $n = 5$ and acceptance number $a = 0$. What is the probability that the buyer would accept a lot having the following fractions of defectives?
a. $p = .1$ b. $p = .3$ c. $p = .5$ d. $p = 0$ e. $p = 1$
Construct the operating characteristic curve for this plan.

6.23. Repeat exercise 6.22 for $n = 5$ and $a = 1$.

6.24. Repeat exercise 6.22 for $n = 10$ and $a = 0$.

6.25. Repeat exercise 6.22 for $n = 10$ and $a = 1$.

6.26. Graph the operating characteristic curves for the four plans given in exercises 6.22 through 6.25 on the same sheet of graph paper. What is the effect of increasing the acceptance number a when n is held constant? What is the effect of increasing the sample size n when a is held constant?

6.27. A radio and television manufacturer who buys large lots of transistors from an electronics supplier selects $n = 25$ transistors from each lot shipped by the supplier and notes the number of defectives.
a. On the same sheet of graph paper, construct the operating characteristic curves for the sampling plans $n = 25$ with $a = 1, 2,$ and 3.
b. Which sampling plan best protects the supplier from having acceptable lots rejected and returned by the manufacturer?
c. Which sampling plan best protects the manufacturer from accepting lots for which the fraction of defectives is exceedingly large?
d. How might the sampling inspector arrive at an acceptance level that compromises between the risk to the producer and the risk to the consumer?

6.28. A buyer and a seller agree to use sampling plan ($n = 15, a = 0$) or sampling plan ($n = 25, a = 1$). Sketch the operating characteristic curves for the two sampling plans. If you were a buyer, which of the two sampling plans would you prefer? Why?

6.29. Refer to exercise 6.27 and assume that the manufacturer wishes the probability to be at least .90 of his accepting lots containing 1% defective and the probability to be about .90 of rejecting any lot with 10% or more defective. If the manufacturer's sampling inspector samples $n = 25$ items from the supplier's incoming shipments, what is the acceptance level a that meets the above requirements?

6.8 Making Decisions: A Test of an Hypothesis

The cold vaccine problem of example 6.5 illustrates a **statistical test of an hypothesis.** The practical question to be answered concerns the effectiveness of the vaccine. Do the data contained in the sample present sufficient evidence to indicate that the vaccine is effective?

The reasoning employed in testing an hypothesis bears a striking resemblance to the procedure used in a court trial. In trying a man for theft, the court assumes the accused innocent until proved guilty. The prosecution collects and presents all available evidence in an attempt to contradict the "not guilty" hypothesis and hence to obtain a conviction.

The statistical problem portrays the vaccine as the accused. The hypothesis to be tested, called the **null hypothesis,** is that the vaccine is ineffective. The evidence in the case is contained in the sample drawn from the population of potential vaccine customers. The experimenter, playing the role of the prosecutor, believes that an **alternative hypothesis** is true, namely, that the vaccine is really effective. Hence the experimenter attempts to use the evidence contained in a sample to reject the null hypothesis (vaccine ineffective) and thereby to support the alternative hypothesis, the contention that the vaccine is, in fact, a very successful cold vaccine. You will recognize this procedure as an essential feature of the scientific method, where all proposed theories must be compared with reality.

Intuitively we would select the number y of survivors as a measure of the quantity of evidence in the sample. If y is very large, we would be inclined to reject the null hypothesis and conclude that the vaccine is effective. On the other hand, a small value of y would provide little evidence to support the rejection of the null hypothesis. As a matter of fact, if the null hypothesis were true and the vaccine were ineffective, the probability of surviving a winter without a cold would be $p = \frac{1}{2}$ and the average value of y would be

$$E(y) = np = 10(\tfrac{1}{2}) = 5$$

Most individuals utilizing their own built-in decision makers would have little difficulty arriving at a decision for the cases $y = 10$ or $y = 5, 4, 3, 2,$ or 1, which, on the surface, appear to provide substantial evidence to support rejection or acceptance, respectively. But what can be said concerning less obvious results, say $y = 7, 8,$ or 9? Clearly, whether we employ a subjective or an objective decision-making procedure, we would choose the procedure that gives the smallest probability of making an incorrect decision.

As statisticians, we test the null hypothesis in an objective manner similar to our intuitive procedure. A decision maker, commonly called a **test statistic,**

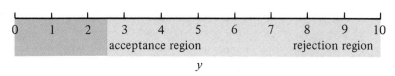

Figure 6.6 *Possible values for the test statistic y*

is calculated from information contained in the sample. In our example the number y of survivors would suffice for a test statistic. We then consider all possible values the test statistic may assume, for example, $y = 0, 1, 2, \ldots, 9, 10$. These values are divided into two groups, as shown in figure 6.6—one called the **rejection region** and the other the **acceptance region**. An experiment is then conducted and the decision maker y observed. If y takes a value in the rejection region, the hypothesis is rejected. Otherwise the hypothesis is accepted. (Caution: As you will subsequently learn, you will reject or accept the null hypothesis only if the risks of a wrong decision are small for these two actions.)

For example, in our experiment we might choose $y = 8, 9,$ or 10 as the rejection region and assign the remaining values of y to the acceptance region. Since we observed $y = 8$ survivors in the experiment, we reject the null hypothesis that the vaccine is ineffective and conclude that the probability of surviving the winter without a cold is greater than $p = \frac{1}{2}$ when the vaccine is used.

What is the probability that we will reject the null hypothesis when, in fact, it is true? The probability of falsely rejecting the null hypothesis is the probability that y will equal 8, 9, or 10 given that $p = \frac{1}{2}$, and this is the probability computed in example 6.5 and found to equal .055. Since we have decided to reject the null hypothesis and note that this probability is small, we are reasonably confident that we have made the correct decision.

Upon reflection, you will observe that the cold vaccine manufacturer is faced with two possible types of error. On the one hand, he might reject the null hypothesis and falsely conclude that the vaccine was effective. Proceeding with a more thorough and expensive testing program or a pilot plant production of the vaccine would result in a financial loss. On the other hand, he might decide not to reject the null hypothesis and falsely conclude that the vaccine was ineffective. This error would result in the loss of potential profits that could be derived through the sale of a successful vaccine.

Definition

Rejecting the null hypothesis when it is true is called a **type I error** for a statistical test. The probability of making a type I error is denoted by the symbol α (Greek letter alpha).

The probability α will increase or decrease as we increase or decrease the size of the rejection region. Then why not decrease the size of the rejection region and make α as small as possible? For example, why not choose $y = 10$ as the rejection region? Unfortunately, decreasing α increases the probability of not rejecting the null hypothesis when it is false and some alternative hypothesis is true. This second type of error is called the type II error for a statistical test and its probability is denoted by the symbol β.

Definition

Accepting the null hypothesis when it is false is called a **type II error** for a statistical test. The probability of making a type II error when some specific alternative is true is denoted by the symbol β (Greek letter beta).

For a fixed sample size, α and β are inversely related; as one increases the other decreases. Increasing the sample size provides more information upon which to base the decision and hence reduces both α and β. In an experimental situation the probabilities of the type I and type II errors for a test measure the risk of making an incorrect decision. The experimenter selects values for these probabilities, and the rejection region and sample size are chosen accordingly.

Example 6.15 Refer to the cold vaccine study and the statistical test based on the rejection region shown in figure 6.6 (i.e., reject the null hypothesis if $y = 8, 9, 10$).
    a. State the null hypothesis and the alternative hypothesis for the test.
    b. Find α for the test.
    c. Find β, the probability of accepting the null hypothesis when the probability of survival for a vaccinated person is $p = .9$.

Solution
    a. The null hypothesis is that $p = .5$ or, equivalently, that the vaccine is ineffective. The alternative hypothesis is that $p > .5$, that is, that the vaccine is effective.
    b. The probability of rejecting the null hypothesis when it is true $(p = .5)$ is

$$\alpha = P(y = 8, 9, 10 \text{ given } p = .5) = \sum_{y=8}^{10} p(y)$$

where $p(y)$ is a binomial probability distribution with $p = .5$. Then

$$\alpha = \sum_{y=8}^{10} C_y^{10} (.5)^y (.5)^{10-y}$$

is found by using table 1 in the appendix; $\alpha = .055$. The probability dis-

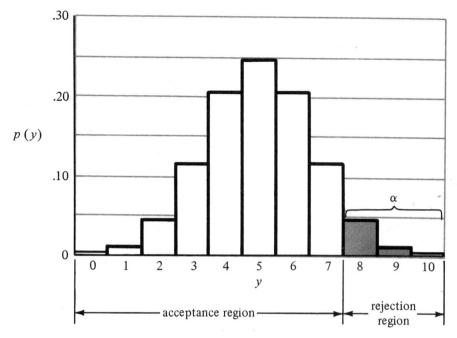

Figure 6.7 *Binomial probability distribution, for n = 10 and p = .5; example 6.15*

tribution for $n = 10$ and $p = .5$, is shown in figure 6.7; α is represented by the shaded portion of the probability distribution.

c.

$$\beta = P(\text{accepting the null hypothesis when } p = .9)$$

$$= P(y = 0, 1, 2, \ldots, 6, 7 \text{ given } p = .9)$$

$$= \sum_{y=0}^{7} p(y)$$

where $p(y)$ is a binomial probability distribution with $p = .9$. Thus

$$\beta = \sum_{y=0}^{7} C_y^{10} (.9)^y (.1)^{10-y} = .07$$

which can be obtained directly from table 1 in the appendix. The probability distribution for $n = 10$ and $p = .9$ is shown in figure 6.8; β is represented by the shaded portion of the probability distribution.

To summarize the implications of parts a and b, $\alpha = .055$ and $\beta = .07$ give measures of the risks of making the two (and only two) types of errors for this statistical test. The probability that the test statistic will, by chance, fall in the rejection region when the null hypothesis is true is only .055. That is, the probability of concluding that the vaccine is somewhat effective, when in fact it is worthless, is only .055. But suppose that the

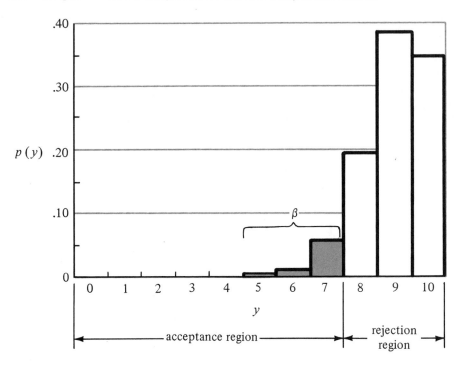

Figure 6.8 *Binomial probability distribution, n = 10 and p = .9; example 6.15*

vaccine is really effective and that the probability of surviving a winter without a cold, when the vaccine has been used, is .9. What is the probability of accepting the null hypothesis that the vaccine is ineffective? We have shown that the risk of making this type II error is only $\beta = .07$.

Exercises

6.30.   Brand preference studies are often conducted by providing a complementary supply of two competing brands in similar unmarked packages to a selected group of consumers. After a trial period each consumer then states his or her brand preference. Under the hypothesis of no preference between brands, it is assumed that $p = 1/2$, where p is the probability a selected consumer favors Brand A. Suppose a brand preference study of two brands is conducted among five consumers, with the agreement to reject the hypothesis of equal preference if none or all five consumers favor Brand A.

a. What is the probability of a type I error for this test?

b. If the actual proportion of consumers favoring Brand A is .7, what is the probability of a type II error for this test?

6.31.   Refer to exercise 6.30 and answer parts a and b when the agreement is to reject the hypothesis of equal preference if 0, 1, 4, or 5 consumers favor Brand A.

6.32.   A packaging experiment was conducted by placing two different package designs for a breakfast food side by side on a supermarket shelf. The objective of the experiment was to see if buyers indicated a preference for one of the two package designs. In a given day six customers purchased a package from the supermarket, with one choosing package design no. 1 and five choosing design no. 2.

   a. State the hypothesis to be tested.

   b. Let y be the number of buyers who choose the second package design. What is the value of α for the test if the rejection region includes $y = 0$ and $y = 6$?

   c. What is the value of β for the alternative $p = .8$ (that is, 80% of the buyers actually favor the second package design)?

   d. In context of the problem, give a practical interpretation of the type I error and the type II error.

   6.33.   Some researchers have suggested that public and private organizations should attempt to dissuade consumers in times of excess demand and supply shortages.* Their comments have been accepted by the public sector as an effective means of dealing with energy supply shortages, but business has been more reluctant to accept the notion of demarketing. Consider a manufacturer of small pocket-sized cameras who has found herself facing runaway demand for her product. Two options are available to the company: intentionally demarket the product or ration supplies to better dealers. To determine whether the corporate executives favor one of these two alternatives over the other, 10 executives are chosen from the organization, with $y = 8$ stating a preference for rationing supplies to better dealers. Do these data present sufficient evidence to indicate that the executives favor rationing to demarketing? State the null hypothesis to be tested and use y as a test statistic.

   6.34.   A certain machine is said to be in control if the proportion of defective items manufactured by the machine is not greater than 10%. To check whether the machine is in control, 10 finished items are randomly selected from its output. The implicit hypothesis that the machine is in control will be rejected if 3 or more defectives are found.

   a. What is the probability of the type I error for this test?

   b. If the machine is really out of control and the probability of a defective is .3, what is the probability of the type II error for the test?

6.9 Choosing the Null Hypothesis

The reasoning employed in a statistical test of an hypothesis runs counter to our everyday method of thinking. That is, it is similar to the mathematical method of proof by contradiction. The hypothesis that scientists wish to "prove" (that is, support) is the alternative hypothesis (often called the "research hypothesis" by social scientists). To do this they test the converse (opposite) of the research hypothesis, the null hypothesis. They hope that the data will support its rejection because this implies support for the alternative or research hypothesis, which was the research objective. You can see that this is exactly what we did with the cold vaccine experiment. We showed support for the effectiveness of the cold vaccine by rejecting the null hypothesis that the vaccine was not effective.

*P. Kotler and S. J. Levy, "Demarketing, Yes, Demarketing," *Harvard Business Review*, November–December 1971.

Why employ this reverse type of thinking, gaining support for a theory by showing that there is little evidence to support its converse? Why not test the alternative or research hypothesis? The answer lies in the problem of evaluating the probabilities of incorrect decisions.

If the research hypothesis is true, testing the null hypothesis (the converse of the research hypothesis) should lead to its rejection. Then the probability of making an incorrect decision is readily at hand. It is α, a probability that was specified in setting up the rejection region. Thus if we reject the null hypothesis (which is what we hope will occur), we immediately know the probability of making an incorrect decision. This gives us a measure of confidence in our conclusion.

Suppose that we had taken the opposite tack, testing the alternative (research) hypothesis that the vaccine is effective. If the research hypothesis is true, the test statistic will most probably fall in the acceptance region (instead of the rejection region). Now to find the probability of an incorrect decision, we must evaluate β, the probability of accepting the null hypothesis when it is false. Although this is not an insurmountable task for the cold vaccine problem, it is extra work. And for many statistical tests it is very difficult to calculate β.

So, to summarize, it is a lot easier to follow the route of "proof by contradiction." Thus the statistician will select the converse of the research hypothesis as the null hypothesis and hope the test leads to its rejection. If it does, the statistician knows α and immediately has a measure of the confidence that he or she can place in this conclusion.

6.10 A General Comment

A discussion of the theory of tests of hypotheses may seem a bit premature at this point, but it provides an introduction to a line of reasoning that is sometimes difficult to grasp and that is best understood when it is allowed to incubate in your mind over a period of time. Thus some of the exercises at the end of chapter 6 involve the use of the binomial probability distribution and, at the same time, lead you to utilize the reasoning involved in statistical tests of hypotheses. We will expand on these ideas through examples and exercises in chapter 7, and we will discuss in detail the topic of statistical tests of hypotheses in chapter 8 and succeeding chapters.

In closing, we direct your attention to the similarity of the lot acceptance sampling problem and the statistical test of an hypothesis. Theoretically they are equivalent because each involves an inference, formulated as a decision, concerning the value of p, the unknown parameter of a binomial population.

6.11 Summary

Three useful discrete probability distributions were presented in chapter 6—the binomial, the Poisson, and the hypergeometric. These probability distributions, discussed in sections 6.1 through 6.6, enabled us to calculate the probabilities associated with several events that are of interest in business. But more importantly, they provided the necessary mechanism to illustrate how statistical inferences are made. Thus in section 6.7 we used the binomial probability distribution to make inferences concerning the proportion of defectives for a lot acceptance sampling plan and then employed it again in section 6.8 to make an inference about the effectiveness of a cold vaccine. Both examples of statistical inference resulted in a decision concerning the parameter p of a binomial population.

The binomial probability distribution gives the probability of y successes in a series of n identical independent trials, where the probability of a success in a single trial is equal to p. The binomial experiment is an excellent model for many sampling situations, particularly market surveys that result in "yes" or "no" type of data.

The Poisson probability distribution is important because it can be used to approximate corresponding binomial probabilities when n is large and p is small. Consequently, it can greatly reduce the computations involved in calculating binomial probabilities. In addition, the Poisson probability distribution is important in its own right. It provides an excellent probabilistic model for the number of occurrences of rare events in time or space.

The hypergeometric probability distribution is also related to the binomial probability distribution. It gives the probability of drawing y elements of a particular type from a population where the number N of elements in the population is small in relation to the sample size n. The binomial probability distribution applies to the same situation except that it is appropriate only when N is large in relation to n.

Perhaps the most important part of chapter 6 is the introduction to statistical inference—making an inference about a population based on information contained in a sample. For illustrative purposes we chose to make inferences about the binomial parameter p, the proportion of elements in a large population that possesses a specified characteristic. For the lot acceptance sampling of section 6.7, we made a decision concerning the fraction p of defectives in a lot of incoming (or departing) goods. For the cold vaccine we made a decision about whether the probability of surviving the winter without a cold was increased by taking a cold vaccine. Both examples show how probability is used to make inferences (for these examples the inferences were decisions) and, more importantly, they show how to evaluate the "goodness" of the decision (the probability of making erroneous decisions). We will expand on these ideas in the following chapters.

Supplementary Exercises

6.35.   Which of the following business problems can be modeled by the binomial distribution? For those that cannot be modeled by the binomial distribution, explain why not.

a. Determination of the probability that 5 of 10 assembly machines break down in a given day when the probability any one will break down in a day is .15.

b. Determination of the probability that at least one assembly machine will break down in a given day when the probability that any one will break down in a given day is .15.

c. Determination of the probability of selecting 2 defective transistors in a sample of 5 transistors drawn from a bin containing 100 transistors of which 10 are defective.

d. Determination of the probability of selecting 2 defective transistors in a sample of 5 transistors drawn from an assembly line in which each transistor produced is defective with a probability of .1.

e. Determination of the probability of selecting only 2 men in a group of 6 selected from a committee consisting of 10 men and 4 women.

6.36.   The proportion of residential households in Burlington, Vermont, that are heated by natural gas is approximately .2. A random sample of 20 residences is selected from the city of Burlington. Assume that the properties of a binomial experiment are satisfied.

a. Find the probability that none of the households is heated by natural gas.

b. Find the probability that no more than 4 of the 20 are heated by natural gas.

6.37.   A salesman figures that each contact results in a sale with a probability of 1/2. On a given day he contacts three prospective clients. Let y be the number of contacts who actually sign a sales contract with the salesman.

a. Use the formula for the binomial probability distribution to calculate the probabilities associated with $y = 0, 1, 2$, and 3.

b. Construct a probability distribution similar to figure 5.1.

c. Find the expected value and standard deviation of y, using the formulas $E(y) = np$ and $\sigma = \sqrt{npq}$.

d. Using the probability distribution in part b, find the fraction of the population measurements lying within one standard deviation of the mean. Repeat for two standard deviations. How do your results agree with Tchebysheff's theorem and the Empirical Rule? Of what practical use are these results?

6.38.   Suppose that a saleswoman figures that each of her contacts results in a sale with a probability of 1/10. Follow the instructions of parts a, b, and c in exercise 6.37. Note that this exercise illustrates the fact that the probability distribution loses its symmetry and becomes skewed when p is not equal to 1/2.

6.39.   It is known that 10% of a brand of television tubes will burn out before their guarantee has expired. If 1000 tubes are sold, find the expected value and variance of y, the number of original tubes that must be replaced. Within what limits would y be expected to fall? (Hint: Use Tchebysheff's theorem.)

6.40.   Because of increased technical complexity in product lines, there has been a trend toward product line specialization and away from traditional general line activities among independent wholesalers. *Industrial Distribution*, which conducts an annual census of industrial distributors, reports that currently only 10% of industrial distributors categorize themselves as "general line" wholesalers. A manufacturer has randomly selected five wholesalers from a list of distributors in order to discuss with them the possibility of serving as the distribution agent for the firm's products.

a. What is the probability that none of the wholesalers is a general line wholesaler?

b. What is the probability that no more than one is a general line wholesaler?

6.41. An old established family-owned company has decided to "go public." It has been reported that 90% of the brokerage houses are recommending the issue to their clients.

a. Assuming this to be true, find the probability of contacting two brokerage houses at random and finding that neither of them recommends purchase of this new stock issue.

b. Find the probability of contacting four brokerage houses at random and finding no more than two recommending purchase of the stock issue.

6.42. Calculate the probability of lot acceptance for a sampling plan with sample size $n = 20$ and acceptance number $a = 0$ for lot defective fractions of $p = .1, .3,$ and $.5$. Sketch the operating characteristic curve for the plan. (Assume that the lot size is large.)

6.43. Repeat exercise 6.42 for $n = 20$ and $a = 1$.

6.44. Repeat exercise 6.42 for $n = 20$ and $a = 2$.

6.45. Repeat exercise 6.42 for $n = 20$ and $a = 3$.

6.46. A gasoline service station offers gasoline for sale at a 2-cents-per-gallon discount if the customer pays in cash and does not use a credit card. Past evidence has shown that 40% of all customers choose to pay in cash. During a given day 25 customers buy gasoline at the service station. (Note: Use table 1 in the appendix.)

a. Find the probability that at least 10 pay in cash.

b. Find the probability that no more than 20 pay in cash.

c. Find the probability that more than 10 but less than 15 pay in cash.

6.47. It is known that 90% of those who purchase a color television will not have claims against the guarantee during the duration of the guarantee. Suppose that each of 20 customers buys a color television set from a certain appliance dealer. What is the probability that at least 2 of these 20 customers will have claims against the guarantee? (Use table 1 in the appendix.)

6.48. Referring to exercise 6.47, what is the expected value and standard deviation of y, the number of claims from 20 buyers? Within what limits would y be expected to fall?

6.49. Suppose that it is known that 1 of 10 undergraduate college textbooks is an outstanding financial success. A publisher has selected 10 new textbooks for publication.

a. What is the probability that exactly 1 will be an outstanding financial success?

b. What is the probability that at least 1 will be an outstanding financial success?

c. What is the probability that at least 2 will be outstanding financial successes?

6.50. A manufacturing process that produces electron tubes is known to have a 5% defective rate. Suppose a sample of $n = 25$ is selected from the manufacturing process.

a. Find the probability that no more than 2 defectives are found.

b. Find the probability that exactly 4 defectives are found.

c. Find the probability that at least 3 defectives are found.

6.51. Since commission costs paid by pension fund owners to fund managers are now negotiated, pension fund officers have noticed that these costs vary widely among the different pension fund management groups. However, it has been reported that only 70% of corporate fund representatives have undertaken a discussion with their managers to see if their commissions are consistent with going rates.* Suppose 25 pension fund groups are contacted to investigate the issue of commission costs paid to fund managers.

*"Monitoring Commission Costs," *Institutional Investor*, September 1975.

a. Find the probability that no more than 15 of the fund groups have discussed commissions relative to the going rate with their fund managers.

b. Find the probability that at least 20 of the fund groups discussed this issue with their fund managers.

6.52. A mail-order magazine subscription service considers any mail advertisement successful if at least 20% of those receiving an advertisement respond favorably by ordering a subscription. Let p be the probability that a single recipient of advertising will order a subscription. What is the smallest value for p in order to have a probability of 90% that at least 20% of 25 recipients (i.e., at least 5) of an advertisement respond favorably?

6.53. A retail variety store that advertises extensively by mail circulars expects a sale from 1 of every 20 mailings. Suppose 25 prospects are randomly selected from a city-wide mailing.

a. How many sales can the store expect to result from this sample of 25?

b. What is the probability that no sales will result from mailings to this group of 25 prospects?

c. What is the probability that at least three sales will result from mailings to the 25 prospects?

d. Suppose that the 25 prospects had been selected from a single city neighborhood. Would this satisfy the properties of a binomial experiment?

6.54. A manufacturer of floor wax has developed two new brands, A and B, that he wishes to subject to a housewife evaluation to determine which of the two is superior. Both waxes A and B are applied to floor surfaces in each of 15 homes. If there is actually no difference in the quality of the brands, what is the probability that 10 or more housewives would state a preference for Brand A?

6.55. Continuing exercise 6.54, let p be the probability that a housewife will choose Brand A in preference to B, and suppose that we wish to test the hypothesis that there is no observable difference between the brands—in other words, that $p = 1/2$. Let y, the number of times that A is preferred to B, be the test statistic.

a. Calculate the value of α for the test if the rejection region is chosen to include $y = 0$, 1, 14, and 15.

b. If p is really equal to .8, what is the value of β for the test defined in part a? (Note that this is the probability that $y = 2$, 3, ... , 12, 13 given that $p = .8$.)

6.56. Continuing exercise 6.54, suppose that the rejection region is enlarged to include $y = 0$, 1, 2, 13, 14, 15.

a. What is the value of α for the test? Should this probability be larger or smaller than the answer found in exercise 6.55a?

b. If p is really equal to .8, what is the value of β for the test? Compare your answer here to your answer in part b of exercise 6.55.

6.57. Refer to exercise 6.51. Suppose that 21 of the 25 pension fund groups contacted indicate that they have discussed commission charges relative to the going rates with their fund managers. Would this provide sufficient evidence to reject the claim reported in exercise 6.51 and to assume that, in fact, more than 70% of the fund groups are discussing the commission issue with their fund managers?

6.58. When evaluating competing capital items, a manufacturer must consider the cost and performance characteristics of each item. Suppose a manufacturer of electrical fuses must decide between two assembly machines A and B. Since the machines have identical costs, the manufacturer decides to base his choice on the performance characteristics of the machines. The number of defective electrical fuses proceeding from each of the two machines A and B is recorded daily for a period of 10 days, with the following results:

Day	1	2	3	4	5	6	7	8	9	10
A	172	165	206	184	174	142	190	169	161	200
B	201	179	159	192	177	170	182	179	169	210

Assume that both assembly machines produce the same daily output. Compare the number of defectives produced by A and B each day and let y be the number of days when B exceeds A. Do the data present sufficient evidence to indicate that the number of defectives per day from Machine B exceeds the number from Machine A on more than half of all days? State the null hypothesis to be tested and use y as a test statistic.

6.59.   A shipment of 200 portable television sets is received by a retailer. To protect herself against a bad shipment, she will inspect 5 sets and accept the entire lot if she observes 0 or 1 defectives. Suppose that there are actually 20 defective sets in the shipment.

a. What is the probability that she accepts the entire shipment?

b. Given that the retailer accepts the entire lot, what is the probability that she observed exactly 1 defective set?

6.60.   Under what conditions can the Poisson distribution be used to approximate certain binomial probabilities? What other application does the Poisson distribution have other than to estimate certain binomial probabilities?

6.61.   Which of the following business problems can be modeled by the Poisson distribution? For those that cannot be modeled by the Poisson distribution, explain why not.

a. Determination of the probability that 2 of 10 city buses will break down during a given day when the probability that any one will break down is .01.

b. Determination of the probability that an insurance company will not have to pay out on any fire damage claims during a year given that the company has insured 1000 firms against fire damage and the probability that any one of the firms incurs a fire during a given year is .002.

c. Determination of the probability that a telephone switchboard receives at least five incoming calls during a given hour, when incoming calls normally arrive randomly and independently of one another at an average rate of one every 15 minutes.

d. Determination of the probability that a saleswoman consummates at least 25 sales in 100 contacts knowing that the probability that she consummates a sale on any contact is .4.

6.62.   Highway engineers and patrol officers naturally tend to focus attention on roadways that have higher-than-average accident rates. In doing so they consider policies such as reduced speed limits or the placement of caution or stop signals that may alleviate the problem. Assume that accidents occur randomly and independently of one another over a specified section of the highway at the average rate of two per week. If officials devote particular attention to the specified section of the highway during a given week, find the probability that they observe no accidents during that period.

6.63.   Logging trucks have a special problem with tire failure because of the rough terrain they are often required to traverse. Suppose that a logging company with 100 trucks has reason to believe that the average number of trucks with at least one tire failure in a given day is 5.

a. Find the probability that during a given day none of the trucks has tire failure.

b. Find the probability that during a given day 5 have tire failure.

c. Find the probability that during a given day not more than 3 have tire failure.

6.64.   Refer to exercise 6.63. Suppose that a truck has a mechanical breakdown during any given day with a probability of .01.

a. Find the probability that during a given day none of the trucks has a mechanical breakdown.

b. Find the probability that during a given day at least 2 of the trucks has a mechanical breakdown.

c. Assuming that tire failures and mechanical breakdowns are independent occurrences in the 100 logging trucks, find the probability that during a given day none of the 100 logging trucks has either a tire or a mechanical problem.

6.65.   In an attempt to minimize the chances of loan defaults, banks, credit unions, and other loan-granting institutions employ a set of rigid criteria to evaluate loan applications. Nevertheless, loan defaults still occur. A full-service commercial bank in Portland, Oregon, reports that defaults on personal loans of less than $2500 have occurred randomly and independently of one another since January 1971 at the average rate of 1.5 defaults per month.

a. Find the probability of no defaults on personal loans of less than $2500 during a given month.

b. Find the probability of no more than 1 loan default during a given two-month span.

6.66.   In a certain large manufacturing plant, serious industrial accidents occur randomly and independently of one another at the rate of 1 every 10 working days. Find the probability of no more than 1 serious accident in the plant over the next month (30 working days).

6.67.   A manufacturer of small electronic desk calculators knows from experience that 1% of all the calculators manufactured and sold by his firm are defective and will have to be replaced under the warranty. A large accounting firm purchases 500 calculators from the manufacturer for use by its employees.

a. Find the probability that none of the calculators will have to be replaced.

b. Find the probability that no more than 4 will have to be replaced.

c. Find the probability that at least 2 will have to be replaced.

d. What is the expected number of desk calculators, purchased by the accounting firm, which can be expected to fail and must be replaced under the warranty?

Experiences with Real Data

1. The results of public opinion surveys published by such organizations as Louis Harris and Associates, the A. C. Nielsen Company, and George Gallup, Inc., are often criticized by readers because they believe the number of sample respondents is too small. A common reply is, "How can the responses of 2000 individuals accurately reflect the opinions of over 200 million Americans?" Let us consider, for purposes of illustration, an excerpt from an article that appeared in a number of newspapers across the nation in late 1975.

Majority Favor Gun Control

In the past 12 years we have witnessed the assassination of three of our leaders and attempts on the lives of at least two others. During this period we have also seen a soaring crime rate in America despite numerous social programs designed to alleviate the sources of crime. Nevertheless, Congress appears firm in its resolve not to restrict the sale of handguns. They continue to hold this view even though a majority of Americans favor gun control. In a survey of approximately 2200 adults representing a cross section of America, 57% indicated that they supported restricting the sale of handguns to all but law enforcement agencies.

If the respondents truly represent a cross section of American adults, do you have faith in the published results? [Hint: To answer this question let y denote the number of respondents in the sample of $n = 2200$ that indicate they favor control of handguns and assume that y has a binomial probability distribution with $p = .5$. Find μ and σ and use these parameters to characterize $p(y)$, assuming that Americans are equally divided on this issue.]

How would your acceptance of the published results have varied had the 57% favoring handgun control been based on a sample of $n = 22$ or $n = 220$ instead of $n = 2200$? Note that if $p = .5$, the population is truly divided equally on this issue—and thus they do not favor handgun control—and it is far less likely that we could find 57% favoring control when $n = 2200$ than when $n = 220$. Likewise, it is less likely that we could find 57% favoring control in a sample of $n = 220$ than in a sample of $n = 22$.

2. The Poisson probability distribution has been offered as a model for estimating binomial probabilities in certain instances and for modeling the probability of the occurrences of rare events that occur randomly and independently of one another in a given unit of time. In section 6.5 we offered some examples of experiments for which the observed outcome y can be considered a Poisson random variable. Select one of these experiments—or design your own experiment such that the observed outcome y represents the number of rather uncommon events that have occurred randomly and independently of one another in a given unit of time or space.

For the experiment you have selected, collect some data on y and compare the relative frequency distribution for these data with the theoretical frequencies provided by the Poisson probability distribution. First select $n = 200$ values of y. Then construct the relative frequencies for $y = 0, 1, 2, \ldots$ and calculate the sample mean $\bar{y}$. Let $\bar{y}$ be an estimate for μ in the Poisson distribution and calculate the Poisson probabilities $p(0), p(1), p(2), \ldots$. Superimpose the probability histogram defined by the Poisson probabilities $p(0), p(1), p(2), \ldots$ over the relative frequency histogram for these data. Does it appear that the Poisson probability distribution is a good model for the population frequency distribution for your random variable?

References

FELLER, W. *An Introduction to Probability Theory and Its Applications.* Vol. 1. 3d ed. New York: Wiley, 1968. Chapter 6.

HILLIER, F. S., and G. J. LIEBERMAN. *Introduction to Operations Research.* San Francisco: Holden-Day, 1967. Chapter 10.

MOSTELLER, F.; R. E. K. ROURKE; and G. B. THOMAS, JR. *Probability with Statistical Applications.* 2d ed. Reading, Mass.: Addison-Wesley, 1970. Chapter 7.

NATIONAL BUREAU OF STANDARDS. *Tables of the Binomial Probability Distribution.* Washington, D.C.: U.S. Government Printing Office, 1949.

SAATY, T. L. *Elements of Queueing Theory: With Applications.* New York: McGraw-Hill, 1961.

SPURR, W., and C. P. BONINI. *Statistical Analysis for Business Decisions.* Revised. Homewood, Ill.: Richard D. Irwin, 1973. Chapter 8.

chapter objectives

GENERAL OBJECTIVES Three important discrete random variables and their probability distributions were presented in chapter 6. This chapter presents the normal random variable, one of the most important and most common continuous random variables. We explain why the normal random variable occurs so frequently, give its probability distribution, and show how the probability distribution can be used. We will take advantage of the normal distribution to reinforce your understanding of the basic concepts involved in a statistical test of an hypothesis.

SPECIFIC OBJECTIVES

1. To explain why normally distributed random variables occur so frequently in nature. The Central Limit Theorem is one of the reasons presented for this situation.
                            Sections 7.1, 7.2

2. To present the Central Limit Theorem as a reason for studying the normal probability distribution and to explain why it plays an important role in statistics.
                            Section 7.2

3. To give a precise definition of the term "random sampling." The Central Limit Theorem is based on random sampling, as are many important estimation and decision-making procedures that will be discussed in subsequent chapters.
                            Section 7.3

4. To present the normal probability distribution and to explain how to find the probability that a random variable falls in a particular interval.
                            Section 7.4

5. To demonstrate the applicability of the Central Limit Theorem by showing how the normal probability distribution can be used to approximate binomial probabilities when the number n of trials is large.
                            Section 7.5

6. To use examples to reinforce the inferential ideas introduced in chapter 6.
                            Section 7.5

chapter seven

The Normal Probability Distribution

7.1 Introduction

Continuous random variables, as noted in section 5.4, are associated with sample spaces representing the infinitely many sample points contained on a line interval. These are some common examples of continuous random variables:

1. The heights or weights of a group of people.
2. The length of life of a perishable product, such as a light bulb, a machine part, or a food product.
3. The time it takes for an individual to perform a task.
4. The measurement errors resulting from laboratory experiments.

In short, any random variable whose values are measurements, as opposed to counts, is a continuous random variable. Reviewing section 5.4 we note that the probabilistic model for the frequency distribution of a continuous random variable is represented by a curve, usually a smooth curve, called the probability distribution, or the probability density function. While these distributions may assume a variety of shapes, it is interesting to note that many random variables observed in nature possess a frequency distribution that is approximately bell-shaped or, as the statistician would say, is approximately a normal probability distribution.

Mathematically speaking, the normal probability density function is given by the equation of the bell-shaped curve shown in figure 7.1. The equation for the density function is constructed so that the area under the curve represents probability. Hence the total area is equal to 1.

In practice we seldom encounter variables that range in value from "minus infinity" to "plus infinity," whatever meaning we may wish to attach

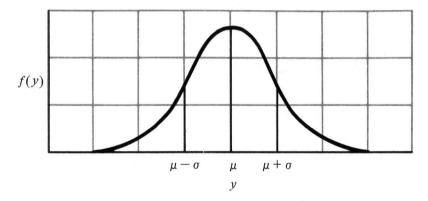

Figure 7.1 *Normal probability density function*

Normal Probability Density Function

$$f(y) = \frac{e^{-(y-\mu)^2/2\sigma^2}}{\sigma \sqrt{2\pi}} \qquad -\infty < y < +\infty$$

The symbols e and π represent irrational numbers whose values are approximately 2.7183 and 3.1416, respectively, while μ and σ are the population mean and standard deviation.

to these phrases. Certainly the height of humans, the weight of a species of beetle, or the length of life of a light bulb do not satisfy this requirement. Nevertheless, a relative frequency histogram plotted for many types of measurements will generate a bell-shaped figure, which may be approximated by the function shown in figure 7.1.

In addition, the continuous normal distribution can often be used to approximate the probability distributions for discrete random variables. Why so many random variables possess probability distributions that are closely approximated by the normal probability distribution is a matter for conjecture. However, one explanation is provided by the *Central Limit Theorem*, a theorem that may be regarded as the most important in statistics.

7.2 The Central Limit Theorem

The Central Limit Theorem states that under rather general conditions, sums and means of samples of random measurements drawn from a population tend to possess, approximately, a bell-shaped distribution in repeated sampling. The significance of this statement is perhaps best illustrated by an example.

Consider a population of die throws generated by tossing a die infinitely

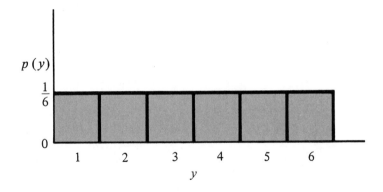

Figure 7.2 *Probability distribution for y, the number appearing on a single toss of a die*

many times, with a resulting probability distribution as given by figure 7.2. Draw a sample of $n = 5$ measurements from the population by tossing a die five times and record each of the five observations, as indicated in table 7.1. Note that the numbers observed in the first sample are $y = 3, 5, 1, 3, 2$. Calculate the sum of the five measurements as well as the sample mean $\bar{y}$.

For experimental purposes repeat the sampling procedure 100 times or preferably an even larger number of times. The results for 100 samples are given in table 7.1 along with the corresponding values of

$$\sum_{i=1}^{5} y_i \quad \text{and} \quad \bar{y}$$

Construct a frequency histogram for

$$\bar{y} \quad \left(\text{or for } \sum_{i=1}^{5} y_i\right)$$

for the 100 samples and observe the resulting distribution, shown in figure 7.3. You will observe an interesting result: although the values of y in the population ($y = 1, 2, 3, 4, 5, 6$) are equiprobable and hence possess a probability distribution that is perfectly horizontal, the distribution of the sample means (or sums) chosen from the population forms a bell-shaped distribution. We will add one additional comment without proof. If we should repeat the study outlined here by combining two samples of $n = 5$ to obtain larger samples of size $n = 10$, we would find that the distribution of the sample means tends to become more nearly bell-shaped.

Note that a proper evaluation of the form of the probability distribution of the sample means would require an infinitely large number of samples, or, at the very least, far more than the 100 samples contained in our experiment. Nevertheless, the 100 samples illustrate the basic idea involved in the Central Limit Theorem.

Table 7.1 *Sampling from the population of die throws*

SAMPLE NUMBER	y, SAMPLE MEASUREMENTS	$\sum y_i$	$\bar{y}$	SAMPLE NUMBER	y, SAMPLE MEASUREMENTS	$\sum y_i$	$\bar{y}$
1	3,5,1,3,2	14	2.8	51	2,3,5,3,2	15	3.0
2	3,1,1,4,6	15	3.0	52	1,1,1,2,4	9	1.8
3	1,3,1,6,1	12	2.4	53	2,6,3,4,5	20	4.0
4	4,5,3,3,2	17	3.4	54	1,2,2,1,1	7	1.4
5	3,1,3,5,2	14	2.8	55	2,4,4,6,2	18	3.6
6	2,4,4,2,4	16	3.2	56	3,2,5,4,5	19	3.8
7	4,2,5,5,3	19	3.8	57	2,4,2,4,5	17	3.4
8	3,5,5,5,5	23	4.6	58	5,5,4,3,2	19	3.8
9	6,5,5,1,6	23	4.6	59	5,4,4,6,3	22	4.4
10	5,1,6,1,6	19	3.8	60	3,2,5,3,1	14	2.8
11	1,1,1,5,3	11	2.2	61	2,1,4,1,3	11	2.2
12	3,4,2,4,4	17	3.4	62	4,1,1,5,2	13	2.6
13	2,6,1,5,4	18	3.6	63	2,3,1,2,3	11	2.2
14	6,3,4,2,5	20	4.0	64	2,3,3,2,6	16	3.2
15	2,6,2,1,5	16	3.2	65	4,3,5,2,6	20	4.0
16	1,5,1,2,5	14	2.8	66	3,1,3,3,4	14	2.8
17	3,5,1,1,2	12	2.4	67	4,6,1,3,6	20	4.0
18	3,2,4,3,5	17	3.4	68	2,4,6,6,3	21	4.2
19	5,1,6,3,1	16	3.2	69	4,1,6,5,5	21	4.2
20	1,6,4,4,1	16	3.2	70	6,6,6,4,5	27	5.4
21	6,4,2,3,5	20	4.0	71	2,2,5,6,3	18	3.6
22	1,3,5,4,1	14	2.8	72	6,6,6,1,6	25	5.0
23	2,6,5,2,6	21	4.2	73	4,4,4,3,1	16	3.2
24	3,5,1,3,5	17	3.4	74	4,4,5,4,2	19	3.8
25	5,2,4,4,3	18	3.6	75	4,5,4,1,4	18	3.6
26	6,1,1,1,6	15	3.0	76	5,3,2,3,4	17	3.4
27	1,4,1,2,6	14	2.8	77	1,3,3,1,5	13	2.6
28	3,1,2,1,5	12	2.4	78	4,1,5,5,3	18	3.6
29	1,5,5,4,5	20	4.0	79	4,5,6,5,4	24	4.8
30	4,5,3,5,2	19	3.8	80	1,5,3,4,2	15	3.0
31	4,1,6,1,1	13	2.6	81	4,3,4,6,3	20	4.0
32	3,6,4,1,2	16	3.2	82	5,4,2,1,6	18	3.6
33	3,5,5,2,2	17	3.4	83	1,3,2,2,5	13	2.6
34	1,1,5,6,3	16	3.2	84	5,4,1,4,6	20	4 0
35	2,6,1,6,2	17	3.4	85	2,4,2,5,5	18	3.6
36	2,4,3,1,3	13	2.6	86	1,6,3,1,6	17	3.4
37	1,5,1,5,2	14	2.8	87	2,2,4,3,2	13	2.6
38	6,6,5,3,3	23	4.6	88	4,4,5,4,4	21	4.2
39	3,3,5,2,1	14	2.8	89	2,5,4,3,4	18	3.6
40	2,6,6,6,5	25	5.0	90	5,1,6,4,3	19	3.8
41	5,5,2,3,4	19	3.8	91	5,2,5,6,3	21	4.2
42	6,4,1,6,2	19	3.8	92	6,4,1,2,1	14	2.8
43	2,5,3,1,4	15	3.0	93	6,3,1,5,2	17	3.4
44	4,2,3,2,1	12	2.4	94	1,3,6,4,2	16	3.2
45	4,4,5,4,4	21	4.2	95	6,1,4,2,2	15	3.0
46	5,4,5,5,4	23	4.6	96	1,1,2,3,1	8	1.6
47	6,6,6,2,1	21	4.2	97	6,2,5,1,6	20	4.0
48	2,1,5,5,4	17	3.4	98	3,1,1,4,1	10	2.0
49	6,4,3,1,5	19	3.8	99	5,2,1,6,1	15	3.0
50	4,4,4,4,4	20	4.0	100	2,4,3,4,6	19	3.8

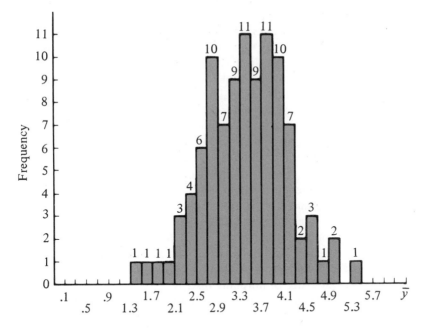

Figure 7.3 *Histogram of sample means for the die-tossing experiment*

The Central Limit Theorem

If random samples of n observations are drawn from a population with finite mean μ and standard deviation σ, then, when n is large, the sample mean $\bar{y}$ is approximately normally distributed, with mean μ and standard deviation $\sigma/\sqrt{n}$. The approximation will become more and more accurate as n becomes large.

The Central Limit Theorem can be restated to apply to the sum of the sample measurements

$$\sum_{i=1}^{n} y_i$$

which, as n becomes large, would also tend to possess a normal distribution, in repeated sampling, with mean $n\mu$ and standard deviation $\sigma\sqrt{n}$.

It can be shown (proof omitted) that the mean and standard deviation of the distribution of sample means are always related to the mean and standard deviation of the sampled population as well as to the sample size n. The two distributions have the same mean μ, and the standard deviation of the distribution of sample means is equal to the population standard deviation σ divided by $\sqrt{n}$ (it can be shown that this relationship is true regardless of the sample size n). Consequently, the spread of the distribution of sample means is considerably less ($1/\sqrt{n}$ as large) as the spread of the population distribution. We will forego discussion of this point for the moment and consider the relevance of the Central Limit Theorem to our previous work.

The significance of the Central Limit Theorem is twofold. First, it explains why some measurements tend to possess, approximately, a normal distribution. We might imagine the height of a person as being composed of a number of elements, each random, associated with such things as the height of the mother, the height of the father, the activity of a particular gland, the environment, and diet. If each of these effects tends to add to the others to yield the measurement of height, then height is the sum of a number of random variables, and the Central Limit Theorem may become effective and yield a distribution of heights that is approximately normal. All of this is conjecture, of course, because we really do not know the true situation. Nevertheless, the Central Limit Theorem, along with other theorems dealing with normally distributed random variables, provides an explanation of the rather common occurrence of normally distributed random variables in nature.

The second and most important contribution of the Central Limit Theorem is in statistical inference. Many estimators that are used to make inferences about population parameters are sums or averages of the sample measurements. When this is true and when the sample size n is sufficiently large, we expect the estimator to possess (approximately) a normal probability distribution in repeated sampling according to the Central Limit Theorem. We can then use the Empirical Rule discussed in chapter 3 to describe the behavior of the inference maker. This aspect of the Central Limit Theorem will be utilized in section 7.4 and in later chapters dealing with statistical inference.

One disturbing feature of the Central Limit Theorem, and of most approximation procedures, is that we must have some idea of how large the sample size n must be for the approximation to give useful results.

Figure 7.4 *Frequency distributions for $\bar{y}$ for two different probability distributions; $n = 2, 5, 10, 25$*

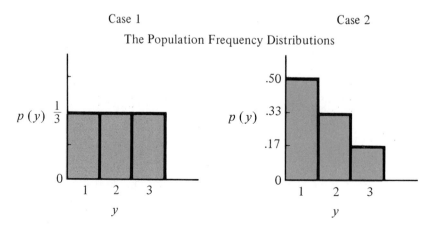

Figure 7.4 *Continued*

The Sampling Distributions

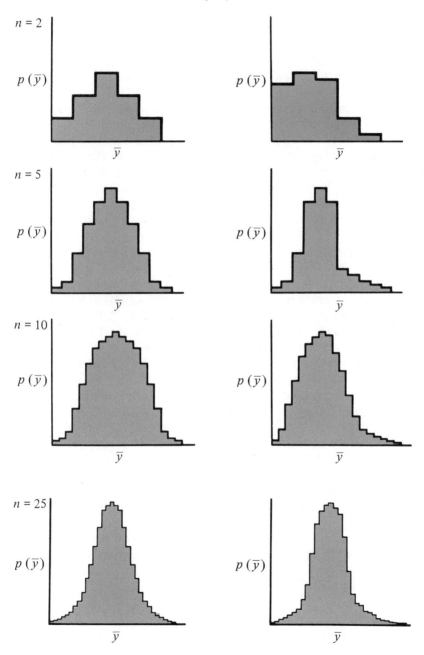

Unfortunately, there is no clear-cut solution to this problem, as the appropriate value for *n* will depend on the population probability distribution as well as the use we will make of the approximation. Although the preceding comment sidesteps the difficulty and suggests that we must rely soley on experience, we may take comfort in the results of the die-tossing experiment discussed earlier in this section. Note that in repeated sampling, the distribution of $\bar{y}$, based on a sample of only *n* = 5 measurements, tends to be approximately bell-shaped. Generally speaking, the Central Limit Theorem functions very well, even for small samples, but this is not always true.

In figure 7.4 frequency distributions for $\bar{y}$ are shown for two different probability distributions, the uniform distribution on the left and a nonsymmetric distribution on the right, for sample sizes ranging (down the page) from *n* = 2 to *n* = 25. The sample means for each frequency histogram were obtained by drawing a large number of samples from the population, each of fixed sample size, and then computing $\bar{y}$ for each. Figure 7.4 illustrates two significant characteristics of the Central Limit Theorem. The first is the somewhat remarkable fact that the Central Limit Theorem applies regardless of the shape of the underlying population frequency distribution. Clearly, neither of the population frequency distributions depicted in the first row of figure 7.4 is bell-shaped. But in both case 1 and case 2, the frequency distributions of the sample mean $\bar{y}$ are effectively approximated by the normal distribution, when the sample size is sufficiently large. Note how the frequency distribution for $\bar{y}$ approaches a bell shape as the sample size increases from *n* = 2 to *n* = 25 (moving from top to bottom of figure 7.4).

A second point to note is that the distribution of $\bar{y}$ for case 1 is closely approximated by a smooth bell-shaped curve for samples as small as *n* = 10, but a much larger sample size is required to gain an effective normal approximation to the distribution of $\bar{y}$ for case 2. This is explained by the fact that when the probability distribution of *y* is symmetric about its mean μ, the Central Limit Theorem will apply very well to small sample sizes, often *n* = 10 or less. However, when the population frequency distribution is skewed (nonsymmetric), larger sample sizes are required to yield an effective approximation to the distribution of $\bar{y}$ by the normal probability distribution.

Some authors offer as a rule of thumb a required minimal sample size of *n* = 30 to guarantee an effective normal approximation, regardless of the shape of the population frequency distribution. However, there are cases when *n* = 30 may be much too small [the binomial experiment when either *p* or (1 − *p*) is small is such a case]. Consequently, we will not rely on such a rule. The appropriate sample size *n* will be given for specific applications of the Central Limit Theorem as they are encountered in section 7.5 and later in the text.

Exercises

7.1.   Let *y* be the number of dots appearing on the up face when a single six-sided die is tossed.

a. Verify that the mean and the standard deviation of *y* are μ = 3.5 and σ = 1.71, respectively.

b. Repeat the sampling experiment of table 7.1 by tossing a die $n = 5$ times and recording the five observations. Repeat this procedure 100 times so that you have 100 samples, each consisting of $n = 5$ observations. Compute the sample mean for the $n = 5$ observations in each of the 100 samples.

c. Construct a frequency histogram to represent your 100 sample means and compare it with figure 7.2. What are the mean and standard deviation of the probability distribution for the sample mean? (Hint: See the Central Limit Theorem.)

d. Find the mean and standard deviation of your 100 sample means and compare them with the mean and standard deviation of the probability distribution for $\bar{y}$ determined in part c.

7.2. Repeat the sampling experiment of exercise 7.1 by selecting 25 samples, each consisting of $n = 20$ observations of y.

a. Compute the sample mean for the $n = 20$ observations in each of the 25 samples.

b. Construct a frequency histogram to represent your 25 sample means and compare it with figure 7.2. What are the mean and standard deviation of the probability distribution for the sample mean? (Hint: See the Central Limit Theorem.)

c. Find the mean and standard deviation of your 25 sample means and compare them with the mean and standard deviation of the probability distribution for $\bar{y}$ given in exercise 7.1d.

d. Compare the results of exercise 7.2 with those observed in exercise 7.1.

7.3 Random Samples

In previous sections we referred to representative samples, sampling in a "random manner," and "random samples" without attempting an explicit definition of these phrases. But perhaps you noted that the Central Limit Theorem applies only when the sampling is conducted in a random manner. What is a random sample and why, in general, is the method of sampling important to our objective of statistical inference? The latter question is more basic and will be answered first.

Consider the decision-making procedures discussed in chapter 6 in connection with lot acceptance sampling and the test of an hypothesis concerning the effectiveness of the cold vaccine. In each case a sample was drawn from the population of interest in order to make an inference (a decision in each of these examples) concerning a parameter of the population. If, after sampling, we observe what we consider to be a highly improbable result (that is, an improbable sample, assuming the null hypothesis to be true), we reject the null hypothesis. If the sample is quite probable, assuming the null hypothesis to be true, we do not reject. In other words, we must know the probability of obtaining the observed sample in order to arrive at a statistical inference. And the way the sample is selected affects its probability. For this reason we will consider a variety of sampling procedures, the simplest of which is random sampling. Others will be presented in chapters 13 and 16.

Random sampling is defined as given in the box.

Definition

Suppose that a sample of *n* measurements is selected from a finite population of *N* measurements. If the sampling is conducted in such a way that every different sample of *n* measurements has an equal probability of being selected, the sampling is said to be **random** and the result is said to be a **random sample**.

Perfect random sampling is difficult to achieve in practice. If the population is not too large, we might place each of the *N* numbers on a poker chip, mix, and select a sample of *n* chips. The numbers on the poker chips would specify the measurements to appear in the sample. Another technique, which uses a table of random numbers, is presented in chapter 16.

In many situations the population is conceptual, as in an observation made during a laboratory experiment. Here the population is considered to be the infinitely many measurements obtained when the experiment is repeated over and over again. If we wish a sample of $n = 10$ measurements from this population, we repeat the experiment 10 times and hope that the results represent a random sample, to a reasonable degree of approximation.

As noted in the preceding chapters, how the sample is selected greatly affects the quantity of information in the sample data. Random sampling and other strategies for the selection of a sample will be discussed in chapter 16.

7.4 Tabulated Areas of the Normal Probability Distribution

You will note that the equation for the normal probability distribution, given in section 7.1, depends on the numerical values of μ and σ and that by supplying various values for these parameters, we could generate infinitely many bell-shaped normal distributions. A separate table of areas for each of these curves is obviously impractical; rather, we would like one table of areas to be applicable to all. The easiest way to do this is to work with areas lying within a specified number of standard deviations of the mean, as was done in the case of the Empirical Rule. For instance, we know that approximately .68 of the area lies within one standard deviation of the mean, .95 within two, and almost all within three. But what fraction of the total area lies within .7 standard deviations, for instance? This question, as well as others, is answered by table 3 in the appendix.

Since the normal curve is symmetrical about the mean, half of the area

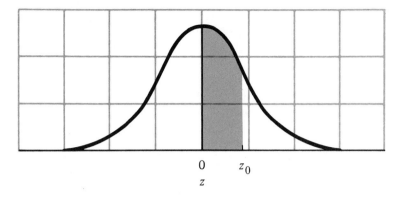

Figure 7.5 *Standardized normal distribution*

under the curve lies to the left of the mean and half to the right (see figure 7.5). Also, because of the symmetry we can simplify our table of areas by listing the areas between the mean and a specified number z of standard deviations to the right of μ. An area to the left of the mean can be calculated by using the corresponding and equal area to the right of the mean. The distance from the mean to a given value of y is $(y - \mu)$. Expressing this distance in units of the standard deviation σ, we obtain

$$z = \frac{y - \mu}{\sigma}$$

Note that there is a one-to-one correspondence between z and y and, particularly, that $z = 0$ when $y = \mu$. The probability distribution for z is often called the **standardized normal distribution,** because its mean is 0 and its standard deviation is 1. It is shown in figure 7.5. The area under the normal curve between the mean $z = 0$ and a specified value of z, say z_0, is the probability $P(0 \le z \le z_0)$. This area is recorded in table 3 of the appendix and is shown as the shaded area in figure 7.5. An abbreviated version of table 3 in the appendix is shown in table 7.2.

Table 7.2 *Abbreviated version of table 3 in the appendix*

z	.00	.01	.02	.03	.04	.05	.06	.07	.08	.09
0.0	.0000	.0040	.0080	.0120	.0160	.0199	.0239	.0279	.0319	.0359
0.1	.0398	.0438	.0478	.0517	.0557	.0596	.0636	.0675	.0714	.0753
0.2	.0793	.0832	.0871	.0910	.0948	.0987	.1026	.1064	.1103	.1141
0.3	.1179	.1217	.1255	.1293	.1331	.1368	.1406	.1443	.1480	.1517
0.4	.1554	.1591	.1628	.1664	.1700	.1736	.1772	.1808	.1844	.1879
0.5	.1915	.1950	.1985	.2019	.2054	.2088	.2123	.2157	.2190	.2224
0.6	.2257	⋮	⋮	⋮	⋮	⋮	⋮	⋮	⋮	⋮
0.7	.2580									
⋮	⋮									
1.0	.3413									
⋮	⋮									
2.0	.4772									

Note that z, correct to the nearest tenth, is recorded in the left-hand column. The second decimal place for z, corresponding to hundredths, is given across the top row. Thus the area between the mean and $z = .7$ standard deviations to the right, read in the second column of the table opposite $z = .7$, is found to be .2580. Similarly, the area between the mean and $z = 1.0$ is .3413. The area lying within one standard deviation on either side of the mean would be two times .3413, or .6826. The area lying within two standard deviations of the mean, correct to four decimal places, is 2(.4772) = .9544. These numbers provide the approximate values, 68% and 95%, used in the Empirical Rule in chapter 3.

To find the area between the mean and a point $z = .57$ standard deviations to the right of the mean, proceed down the left-hand column to the 0.5 row. Then move across the top row of the table to the .07 column. The intersection of this row-column combination gives the approximate area, .2157.

As you begin to find probabilities associated with normal random variables, you will encounter one result that differs greatly from your experiences with discrete random variables. The probabilities for continuous random variables are defined as areas under a probability density function. Consequently, if y is a normal random variable, the probability that y takes some specific value, say 10, is 0. This is because there is no area under the probability density function over the point $y = 10$. Therefore, the probability $P(y \leq 10)$ is the same as $P(y < 10)$, because the probability that $y = 10$ is 0.

Keep in mind that this phenomenon applies only to continuous random variables. Later in this chapter we will use the normal probability distribution to approximate the binomial probability distribution. The binomial random variable y is a discrete random variable. Hence, as you know, the probability that y takes some specific value, say $y = 10$, may not equal 0. Consequently, for discrete random variables $P(y \leq y_0)$ is not the same as $P(y < y_0)$.

Let us now consider some examples.

Example 7.1 Find $P(0 \leq z \leq 1.63)$. This probability corresponds to the area between the mean ($z = 0$) and a point $z = 1.63$ standard deviations above the mean (see figure 7.6).

Solution The area is shaded and indicated by the symbol A in figure 7.6. Since table 3 in the appendix gives areas under the normal curve to the right of the mean, we need only find the table value corresponding to $z = 1.63$. Go down the left-hand column of the table to the row corresponding to $z = 1.6$ and across the top of the table to the column marked .03. The intersection of this row and column combination gives the area, $A = .4484$. Therefore, $P(0 \leq z \leq 1.63) = .4484$.

Example 7.2 Find $P(-.5 \leq z \leq 1.0)$. This probability corresponds to the area between $z = -.5$ and $z = 1.0$, as shown in figure 7.7.

Solution The area required is equal to the sum of A_1 and A_2 shown in

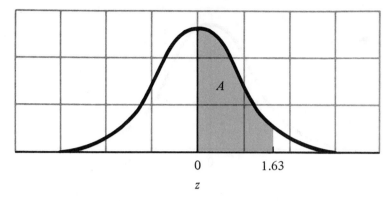

Figure 7.6 *Probability required for example 7.1*

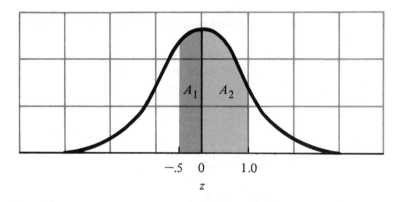

Figure 7.7 *Area under the normal curve in example 7.2*

figure 7.7. From table 3 in the appendix, we read $A_2 = .3413$. The area A_1 equals the corresponding area between $z = 0$ and $z = .5$, or $A_1 = .1915$. Thus the total area is

$$A = A_1 + A_2 = .1915 + .3413 = .5328$$

Hence $P(-.5 \leq z \leq 1.0) = .5328$.

Example 7.3 Find the value of z, say z_0, such that (to four decimal places) .95 of the area is within $\pm z_0$ standard deviations of the mean.

Solution Half of the .95 area will lie to the left of the mean and half to the right, because the normal distribution is symmetrical. Thus we seek the value z_0 corresponding to an area equal to .475. The area .475 falls in the row corresponding to $z = 1.9$ and the .06 column. Hence $z_0 = 1.96$. Note that this is very close to the approximate value, $z = 2$, used in the Empirical Rule.

Example 7.4 Let y be a normally distributed random variable, with a

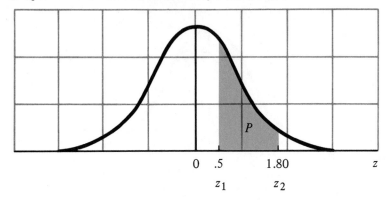

Figure 7.8 *Area under the normal curve in example 7.4*

mean of 10 and a standard deviation of 2. Find the probability that y lies between 11 and 13.6.

Solution As a first step we must calculate the values of z corresponding to $y_1 = 11$ and $y_2 = 13.6$. Thus we have

$$z_1 = \frac{y_1 - \mu}{\sigma} = \frac{11 - 10}{2} = .5$$

$$z_2 = \frac{y_2 - \mu}{\sigma} = \frac{13.6 - 10}{2} = 1.80$$

The desired probability is therefore $P(.5 \le z \le 1.80)$ and is the area lying between z_1 and z_2, as shown in figure 7.8. The area between $z = 0$ and z_1 is $A_1 = .1915$, and the area between $z = 0$ and z_2 is $A_2 = .4641$, which are obtained from table 3. The desired probability is equal to the difference between A_1 and A_2; that is,

$$P(.5 \le z \le 1.80) = A_2 - A_1 = .4641 - .1915 = .2726$$

Example 7.5 Studies show that gasoline usage for compact cars sold in the United States is normally distributed, with a mean usage of 25.5 miles per gallon (mpg) and a standard deviation of 4.5 mpg. What percentage of compacts obtain 30 or more miles per gallon?

Solution The proportion P of compacts obtaining 30 or more miles per gallon is given by the shaded area in figure 7.9.

First we must find the z-value corresponding to $y = 30$. Substituting into the formula for z, we obtain

$$z = \frac{y - \mu}{\sigma} = \frac{30 - 25.5}{4.5} = 1.0$$

The area A to the right of the mean, corresponding to $z = 1.0$, is .3413 (from table 3). Then the proportion of compacts having a miles-per-gallon

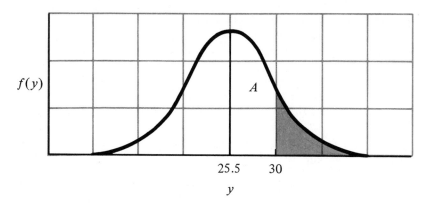

Figure 7.9 *Area under the normal curve for example 7.5*

ratio equal to or greater than 30 is equal to the entire area to the right of the mean, .5, minus the area A.

$$P(y \geq 30) = .5 - P(0 \leq z \leq 1) = .5 - .3413 = .1587$$

The percentage exceeding 30 mpg is

$$100(.1587) = 15.87\%$$

Example 7.6 Refer to example 7.5. In times of scarce energy resources, a competitive advantage is given to an automobile manufacturer who can produce a car obtaining substantially better fuel economy than the competitors' cars. If a manufacturer wishes to develop a compact car that outperforms 95% of the current compacts in fuel economy, what must be the gasoline usage rate for the new car?

Solution Let y be a normally distributed random variable, with a mean of 25.5 and a standard deviation of 4.5. We want to find the value y_0 such that

$$P(y \leq y_0) = .95$$

As a first step we find

$$z_0 = \frac{y_0 - \mu}{\sigma} = \frac{y_0 - 25.5}{4.5}$$

and note that our required probability is the same as the area to the left of z_0 for the standardized normal distribution. Therefore,

$$P(z \leq z_0) = .95$$

The area to the left of the mean is .5. The area to the right of the mean between z_0 and the mean is $.95 - .5 = .45$. Thus from table 3 we find that z_0 is between 1.64 and 1.65. Notice that the area .45 is exactly halfway between the areas for $z = 1.64$ and $z = 1.65$. Thus z_0 is exactly halfway between 1.64 and 1.65; that is, $z_0 = 1.645$.

Substituting $z_0 = 1.645$ into the equation for z_0, we have

$$1.645 = \frac{y_0 - 25.5}{4.5}$$

and solving for y_0 we obtain

$$y_0 = 1.645(4.5) + 25.5 = 32.9$$

The manufacturer's new compact car must therefore obtain a fuel economy of 32.9 mpg to outperform 95% of the compact cars currently available on the U.S. market.

Exercises

7.3.   Using table 3 in the appendix calculate the area under the normal curve between these values:
a. $z = 0$ and $z = 1.4$ b. $z = 0$ and $z = .74$
c. $z = 0$ and $z = -.26$ d. $z = -.62$ and $z = 1.25$
e. $z = .73$ and $z = 1.62$ f. $z = -.33$ and $z = -1.57$
g. $z = -.29$ and $z = .84$ h. $z = -1.71$ and $z = 2.26$

7.4.   Use table 3 to calculate these areas under the normal curve:
a. to the right of $z = .49$ b. to the left of $z = -1.27$
c. to the right of $z = -.66$ d. to the left of $z = 1.28$
e. between $z = -.60$ and $z = 1.96$ f. between $z = 1.24$ and $z = 1.70$

7.5.   Find z_0 for these probabilities:
a. $P(z > z_0) = .05$ b. $P(z < z_0) = .05$
c. $P(z > z_0) = .8051$ d. $P(z < z_0) = .1314$
e. $P(-z_0 < z < z_0) = .733$ f. $P(-z_0 < z < z_0) = .90$
g. $P(-z_0 < z < z_0) = .95$

7.6.   Suppose a normal random variable y possesses a mean of 10 and a standard deviation of 3. Find these probabilities:
a. that y falls between 10 and 11.8   b. that y falls between 6 and 12
c. that y exceeds 14.2   d. that y is less than 12
e. that y falls between 8 and 10.5   f. that y exceeds 15.4

7.7.   The length of life of a certain type of automatic washer is approximately normally distributed, with a mean of 3.1 years and a standard deviation of 1.2 years. If this type of washer is guaranteed for 1 year, what fraction of original sales will require replacement?

7.8.   The average yield to maturity of industrial bonds issued during the quarter ending 31 March 1975 was 8.55, with a standard deviation of yield of .70. Suppose the bond yields were approximately normally distributed and that the bond yield for a certain firm during that quarter was 7.10. Give the percentile corresponding to a yield of 7.10. Recalling that the yield to maturity of an industrial bond is partly dependent on the issuing firm's bond rating, what can you infer about the financial state of the issuing firm during the first quarter of 1975?

7.9.   In anticipation of federal legislation regarding aerosal can propellents, manufacturers of hairsprays and deodorants have been introducing liquid sprays. A new machine used for filling cans of liquid hairspray can be set for any average fill. If the amount of fill is normally distributed around a setting with a standard deviation of .05 ounces, what setting will cause 95% of the cans to contain 12.00 or more ounces of liquid?

7.5 The Normal Approximation to the Binomial Distribution

In chapter 6 we considered several applications of the binomial probability distribution, all of which required that we calculate the probability that y, the number of successes in n trials, falls in a given region. For the most part we restricted our attention to examples where n was small because of the tedious calculations necessary in the computations of $p(y)$. Let us now consider the problem of calculating $p(y)$, or the probability that y falls in a given region, when n is large, say $n = 1000$.

A direct calculation of $p(y)$ for large values of n is not an impossibility, but it does provide a formidable task which we would prefer to avoid. Fortunately, the Central Limit Theorem provides a solution to this dilemma since we may view y, the number of successes in n trials, as a sum that satisfies the conditions of the Central Limit Theorem. Each trial results in either 0 or 1 success with probability q and p, respectively. Therefore, each of the n trials may be regarded as an independent observation drawn from a simpler binomial experiment consisting of one trial, and y, the total number of successes in n trials, is the sum of these n independent observations. Then if n is sufficiently large, the binomial variable y will be approximately normally distributed, with mean and variance (obtained in chapter 6) of np and npq, respectively. We can then use areas under a fitted normal curve to approximate the binomial probabilities.

For instance, consider a binomial probability distribution for y when $n = 10$ and $p = \frac{1}{2}$. Then $\mu = np = 10(\frac{1}{2}) = 5$ and $\sigma = \sqrt{npq} = \sqrt{2.5} = 1.58$. Figure 7.10 shows the corresponding binomial probability distribution and the approximating normal curve on the same graph. A visual comparison of the figures suggests that the approximation is reasonably good, even though

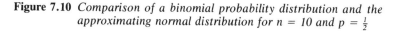

Figure 7.10 *Comparison of a binomial probability distribution and the approximating normal distribution for $n = 10$ and $p = \frac{1}{2}$*

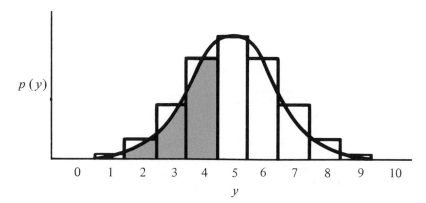

a small sample, $n = 10$, was necessary for this graphic illustration.

The probability that $y = 2, 3,$ or 4 is exactly equal to the area of the three rectangles lying over $y = 2, 3,$ and 4. We may approximate this probability with the area under the normal curve from $y = 1.5$ to $y = 4.5$, which is shaded in figure 7.10. Note that the area under the normal curve between 2.0 and 4.0 would not provide a good approximation to the probability that $y = 2, 3,$ or 4 because it excludes half of the probability rectangles corresponding to $y = 2$ and $y = 4$. Thus it is important to remember to approximate the entire areas for $y = 2$ and $y = 4$ by including the area over the interval from 1.5 to 4.5.

Although the normal probability distribution provides a reasonably good approximation to the binomial probability distribution (figure 7.10), this will not always be the case. When n is small and p is near 0 or 1, the binomial probability distribution is nonsymmetrical; that is, its mean is located near 0 or n. For example, when p is near 0, most values of y are small, producing a distribution that is concentrated near $y = 0$ and that tails gradually toward n (see figure 7.11). Certainly when this is true, the normal distribution, symmetrical and bell-shaped, provides a poor approximation to the binomial probability distribution. How, then, can we tell whether n and p are such that the binomial distribution will be symmetrical?

Recalling the Empirical Rule from chapter 3, approximately 95% of the measurements associated with a normal distribution lie within two standard deviations of the mean and almost all lie within three. We suspect that the binomial distribution would be nearly symmetrical if the distribution were

Figure 7.11 *Comparison of a binomial probability distribution (shaded) and the approximating normal distribution; $n = 10$ and $p = .1$ ($\mu = np = 1$; $\sigma = \sqrt{npq} = .95$)*

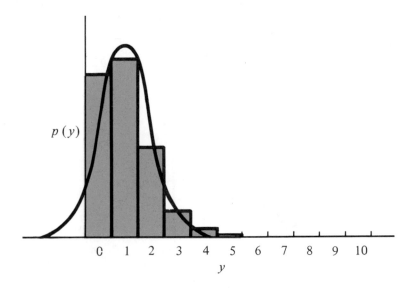

able to spread out a distance equal to two standard deviations on either side of the mean and this is, in fact, the case. Hence to determine when the normal approximation is adequate, calculate $\mu = np$ and $\sigma = \sqrt{npq}$. If the interval $(\mu \pm 2\sigma)$ lies within the binomial bounds 0 and n, the approximation is reasonably good. Note that this criterion is satisfied for the example of figure 7.10 but is not satisfied for figure 7.11.

Example 7.7 To see how well the normal curve can be used to approximate binomial probabilities, refer to the binomial experiment illustrated in figure 7.10, where $n = 10$ and $p = .5$. Calculate the probability that $y = 2$, 3, or 4, correct to three places, using table 1 in the appendix. Then calculate the corresponding normal approximation to this probability and compare.

Solution The exact probability P_1 can be calculated using table 1(b). Thus we have

$$P_1 = \sum_{y=2}^{4} p(y) = \sum_{y=0}^{4} p(y) - \sum_{y=0}^{1} p(y) = .377 - .011 = .366$$

As noted earlier in this section, the normal approximation requires the area lying between $y_1 = 1.5$ and $y_2 = 4.5$, where $\mu = 5$ and $\sigma = 1.58$. Thus we have

$$\sum_{y=2}^{4} p(y) \approx P(z_1 \leq z \leq z_2) = P_2$$

where

$$z_1 = \frac{y_1 - \mu}{\sigma} = \frac{1.5 - 5}{1.58} = -2.22$$

$$z_2 = \frac{y_2 - \mu}{\sigma} = \frac{4.5 - 5}{1.58} = -.32$$

Figure 7.12 *Area under the normal curve in example 7.7*

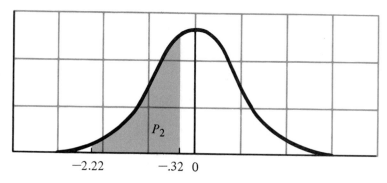

-2.22 $-.32$ 0

z

(Note: The symbol $\approx$ means "approximately equal to.") The probability P_2 is shown in figure 7.12. The area between $z = 0$ and $z = 2.22$ is $A_1 = .4868$. Likewise, the area between $z = 0$ and $z = .32$ is $A_2 = .1255$. You can see from figure 7.12 that

$$P_2 = P(z_1 \le z \le z_2) = P(-2.22 \le z \le -.32)$$
$$= A_1 - A_2 = .4868 - .1255 = .3613$$

Note that the normal approximation is quite close to the binomial probability obtained from table 1(b).

Example 7.8 According to *Finance Facts Yearbook*, 1973, 50% of all loans extended by consumer finance associations are taken for the purpose of consolidating existing bills. Find the probability that exactly 45 of 100 loans randomly selected from the files of a consumer loan agency were extended for the purpose of debt consolidation.

Solution The exact probability of observing $y = 45$ successes in $n = 100$ independent trials of a binomial experiment, with a probability of success of $p = .5$ at each trial, is found by evaluating

$$p(45) = C_{45}^{100}(.5)^{45}(.5)^{55}$$

Since this form is very tedious to evaluate, we will employ the normal approximation method. The mean and standard deviation of the binomial distribution are

$$\mu = np = 100(.5) = 50$$
$$\sigma = \sqrt{npq} = \sqrt{100(.5)(.5)} = 5$$

The probability of exactly 45 successes is approximated by the area under the normal curve between 44.5 and 45.5 (see figure 7.13). The z-values corresponding to 44.5 and 45.5 are

Figure 7.13 *Area under the normal curve in example 7.8*

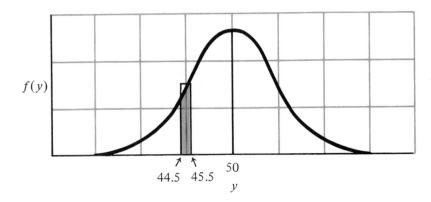

$$z_1 = \frac{44.5 - 50}{5} = -1.1$$

$$z_2 = \frac{45.5 - 50}{5} = -.9$$

The area between $z_1 = -1.1$ and $z_2 = -.9$, and hence the probability of exactly 45 loans for debt consolidation, is

$$p(45) \approx P(z_1 \le z \le z_2) = P(-1.1 \le z \le -.9)$$

This area under the normal curve is equal to the difference between $P(0 \le z \le 1.1)$ and $P(0 \le z \le .9)$. Or,

$$p(45) = P(0 \le z \le 1.1) - P(0 \le z \le .9) = .3643 - .3159 = .0484$$

Thus the probability is .0484 that exactly 45 of 100 loans were extended for the purpose of debt consolidation.

Example 7.9 The reliability of an electrical fuse can be stated as the probability that a fuse, chosen at random from production, will function under the conditions for which it has been designed. A random sample of 1000 fuses was tested and $y = 27$ defectives were observed. Calculate the probability of observing 27 or more defectives, assuming that fuse reliability is .98.

Solution The probability of observing a defective when a single fuse is tested is $p = .02$, given that fuse reliability is .98. Then

$$\mu = np = 1000(.02) = 20$$

$$\sigma = \sqrt{npq} = \sqrt{1000(.02)(.98)} = 4.43$$

The probability of 27 or more defective fuses, given $n = 1000$, is

$$P = P(y \ge 27)$$

$$= p(27) + p(28) + p(29) + \cdots + p(999) + p(1000)$$

Figure 7.14 *Normal approximation to the binomial in example 7.9*

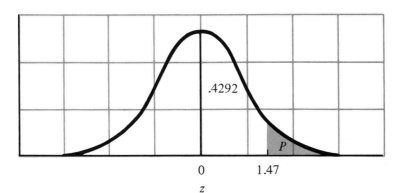

The normal approximation to P is the area under the normal curve to the right of $y = 26.5$. (Note that we must use $y = 26.5$ rather than $y = 27$ so as to include the entire probability rectangle associated with $y = 27$.) The z-value corresponding to $y = 26.5$ is

$$z = \frac{y - \mu}{\sigma} = \frac{26.5 - 20}{4.43} = \frac{6.5}{4.43} = 1.47$$

and the area between $z = 0$ and $z = 1.47$ is equal to .4292, as shown in figure 7.14. Since the total area to the right of the mean is equal to .5, then

$$P = P(y \geq 27) \approx P(z \geq 1.47) = .5 - .4292 = .0708$$

Thus the probability of observing 27 or more defectives is .0708.

Example 7.10 A popular television advertisement claims that "taste experts" prefer the taste of a certain brand of vegetable oil spread to butter. To test the truth of this claim, 100 chefs were given two pieces of bread to taste, one with the vegetable oil spread, the other with butter. Sixty-eight of the chefs claimed that the vegetable oil spread had a richer, more satisfying taste. If the difference in taste between the spread and butter is undetectable, it is assumed that the probability is $\frac{1}{2}$ that a chef will choose either piece as having superior taste. On the basis of the results of the above experiment, what conclusions would you make about the truth of the advertisement?

Solution Translating the question into an hypothesis concerning the parameter of a binomial population, we wish to test the null hypothesis that p, the probability of a chef choosing the piece of bread with the vegetable oil spread as the piece with superior taste, is equal to .5. Or, equivalently, the content of the spread is such that it is undetectable in taste from butter. The alternative hypothesis that the spread is preferred over butter would imply rejection of the null hypothesis when y, the number of chefs preferring the spread, is large.

Since the normal approximation to the binomial is adequate for this example, we interpret a large and improbable value of y to be one that lies several standard deviations away from the hypothesized mean, $\mu = np = 100(.5) = 50$. Noting that

$$\sigma = \sqrt{npq} = \sqrt{(100)(.5)(.5)} = 5$$

we may arrive at a conclusion without bothering to locate a specific rejection region. The observed value of y, 68, lies more than 3σ away from the hypothesized mean $\mu = 50$. Specifically, y lies

$$z = \frac{y - \mu}{\sigma} = \frac{68 - 50}{5} = 3.6$$

standard deviations away from the hypothesized mean. This result is so improbable, assuming that the spread and butter are undetectable in taste, that we reject the null hypothesis and conclude that the probability of choosing the spread as having a taste superior to butter is greater than $p = .5$. (You will observe that the area above $z = 3.6$ is so small that it is not included in table 3.)

Rejecting the null hypothesis raises some questions in the minds of prospective wholesalers of margarine and butter. What proportion of consumers will prefer the taste of the vegetable oil spread, and is its taste sufficiently superior to that of butter to warrant a change in inventory policy? The former question leads to an estimation problem, a topic discussed in chapter 8, while the latter, involving an inventory decision, would utilize the results of a consumer preference experiment as well as a study of unit costs.

Example 7.11 The probability α of a type I error and the location of the rejection region for a statistical test of an hypothesis are usually specified before the data are collected. In the case of the previous example, a type I error is the error of assuming the vegetable oil spread is superior in taste to butter when, in fact, the taste difference between the two is undetectable (i.e., the probability of rejecting $p = .5$ when it is true). For the taste-testing problem of example 7.10, find the appropriate rejection region for the test of the null hypothesis $p = .5$ if we wish α, the probability of falsely concluding the vegetable oil spread has a superior taste to butter, to be approximately equal to .05. (See figure 7.15.)

Solution We stated in example 7.10 that y, the number indicating a preference for the vegetable oil spread, would be used as a test statistic and that the rejection region would be located in the upper tail of the probability distribution for y. Desiring α approximately equal to .05, we seek a value of y, say y_α, such that

$$P(y \geq y_\alpha) \approx .05$$

This can be determined by first finding the corresponding z_α that gives the number of standard deviations between the mean $\mu = 50$ and y_α. Since the total area to the right of $z = 0$ is .5, the area between $z = 0$ and

Figure 7.15 *Location of the rejection region in example 7.11*

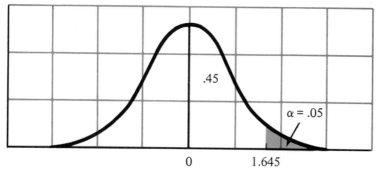

z_α equals .45 (see figure 7.15). Checking table 3 we find that $z = 1.64$ corresponds to an area equal to .4495 and $z = 1.65$ to an area of .4505. Thus, as we determined in example 7.6,

$$z_\alpha = 1.645$$

Recall the relation between z and y,

$$z_\alpha = \frac{y_\alpha - \mu}{\sigma}$$

and substitute $z_\alpha = 1.645$ into the expression. Then

$$1.645 = \frac{y_\alpha - 50}{5}$$

Solving for y_α we obtain

$$y_\alpha = 58.225$$

Since we cannot observe $y = 58.225$ chefs who indicate a preference for the vegetable oil spread, we must choose 58 or 59 as the point were the rejection region begins.

Suppose that we decide to reject when y is greater than or equal to 59. Then the actual probability α of the type I error for the test is

$$P(y \geq 59) = \alpha$$

which can be approximated by using the area under the normal curve above $y = 58.5$, a problem similar to that encountered in example 7.6. The z-value corresponding to $y = 58.5$ is

$$z = \frac{y - \mu}{\sigma} = \frac{58.5 - 50}{5} = 1.7$$

and the tabulated area between $z = 0$ and $z = 1.7$ is .4554. Then

$$\alpha = .5 - .4554 = .0446$$

While the method described above provides a more accurate value for α, there is very little practical difference between an α of .0446 and one of .05. When n is large, time and effort may be saved by using z as a test statistic rather than y. This method was employed in example 7.8. We would then reject the null hypothesis when z is greater than or equal to 1.645.

Example 7.12 One of the major determinants of a company's sales is the total amount of business available to it in the market areas where it competes. An individual company's share of the total market is called its *market share*. Sales goals and market objectives are usually stated in terms of target market share percentages. Consider, for instance, a cigarette manufacturer who states that the firm controls about 10% of the market of cigarette sales through the sales of his product, Brand A. To test this belief 2500 smokers are selected at random from the population of cigarette smokers and questioned concerning their cigarette brand pref-

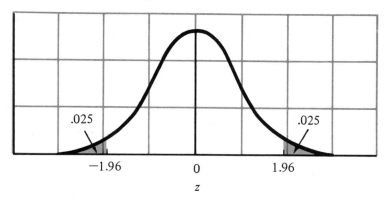

Figure 7.16 *Location of the rejection region in example 7.12*

erence. A total of $y = 218$ express a preference for Brand A. Do these data provide sufficient evidence to contradict the hypothesis that 10% of all smokers favor Brand A and, hence, that the company's share of the cigarette market is 10%? Conduct a statistical test by using α equal to .05.

Solution We wish to test the null hypothesis that p, the probability that a single smoker prefers Brand A, is equal to .1 against the alternative that p is greater than or less than .1. The rejection region corresponding to $\alpha = .05$ is located as shown in figure 7.16. We will reject the null hypothesis when $z > 1.96$ or $z < -1.96$. In other words, we will reject when y lies more than approximately two standard deviations away from its hypothesized mean. Note that half of α is placed in one tail of the distribution and half in the other because we wish to reject the null hypothesis when p is either larger or smaller than $p = .1$. This is called a *two-tailed* statistical test, in contrast to the one-tailed test discussed in examples 7.10 and 7.11, where the alternative to the null hypothesis was that p was only larger than the hypothesized value.

Assuming the null hypothesis to be true, the mean and standard deviation for y are

$$\mu = np = (2500)(.1) = 250$$

$$\sigma = \sqrt{npq} = \sqrt{(2500)(.1)(.9)} = 15$$

The z-value corresponding to the observed $y = 218$ is

$$z = \frac{y - \mu}{\sigma} = \frac{218 - 250}{15} = \frac{-32}{15} = -2.13$$

Noting that z falls in the rejection region, we reject the null hypothesis and conclude that less than 10% of all smokers prefer Brand A.*

What is the probability that we have made an incorrect decision? The

*To be exact we should make the half-unit correction (i.e., use $y = 217.5$) when calculating z. However, for large n the correction will have little effect on either the computed value of z or on the test conclusion.

answer, of course, is either 1 or 0, depending on whether our decision is correct or incorrect in this specific case. However, we know that if this statistical test were employed over and over again, the probability of rejecting the null hypothesis when it is true is only $\alpha = .05$. Hence we are reasonably certain that we have made the correct decision.

Exercises

7.10.   Let y be a binomial random variable obtained from a binomial experiment with $n = 25$ and $p = .4$.

a. Use table 1 in the appendix, to calculate $P(8 \leq y \leq 11)$.

b. Find μ and σ for y and use the normal approximation to the binomial distribution to approximate the probability $P(8 \leq y \leq 11)$. Note that this value is a good approximation to the exact value of $P(8 \leq y \leq 11)$.

7.11.   Consider a binomial experiment with $n = 20$ and $p = .2$.

a. Calculate $P(y \geq 5)$ by use of table 1 in the appendix.

b. Calculate $P(y \geq 5)$ by use of the normal approximation to the binomial probability distribution. Compare the approximation with the exact value found in part a.

7.12.   Sensitivity training is designed to remove barriers of communication among employees in a firm and to create an atmosphere where everyone feels free to speak openly about any issue, personal or otherwise, that affects the firm. A research note in the December 1974 issue of the *Academy of Management Journal* reported that 50% of those managers who have participated in sensitivity training would personally recommend that other firms emphasize sensitivity training in their management development programs. Suppose a random sample of 100 managers who have participated in sensitivity training are selected.

a. What is the probability that 60 or more managers would not recommend sensitivity training to other firms?

b. What is the probability that more than 47 but less than 52 managers would recommend sensitivity training to other firms?

7.13.   In an attempt to check the truth of the claim reported in exercise 7.12, suppose a sample of 100 managers who have participated in sensitivity training is selected, with $y = 38$ indicating they would recommend that other firms emphasize sensitivity training in their management development programs. Do these data provide sufficient evidence to reject the claim reported in the *Academy of Management Journal?*

7.14.   The survival of many of the programs presented by the major television networks depends on the weekly Nielsen ratings. Last week's Nielsen ratings showed that 20% of all television viewers watch a particular program. In a random sample of $n = 1000$ viewers, $y = 184$ viewers watch the program. Do these data present sufficient evidence to contradict last week's Nielsen ratings?

7.6 Some Hints for Problem Solving

Two suggestions that will aid you in solving exercises in this chapter are given in the box.

Hints for Problem Solving

1. Always sketch a normal curve and locate the probability areas pertinent to the exercise. If you are approximating a binomial probability distribution, sketch in the probability rectangles as well as the normal curve.
2. Read each exercise carefully to see whether the data come from a binomial experiment or whether they possess a distribution that, by its very nature, is approximately normal. If you are approximating a binomial probability distribution, do not forget to make a half-unit correction so that you will include the half rectangles at the ends of the interval. If the distribution is not binomial, *do not* make the half-unit corrections. If you make a sketch (as suggested in 1), you will see why the half-unit correction is or is not needed.

7.7 Summary

Many continuous random variables observed in nature possess a probability distribution that is bell-shaped and may be approximated by the normal probability distribution discussed in section 7.1. The common occurrence of normally distributed random variables may be partly explained by the **Central Limit Theorem,** which states that under rather general conditions, the sum or the mean of a random sample of n measurements drawn from a population is approximately normally distributed in repeated sampling when n is large.

As a case in point, the number y of successes associated with a binomial experiment may be regarded as a sum of n sample measurements that possess, approximately, a normal probability distribution when n, the total number of trials, is large. This application of the Central Limit Theorem provides a method for calculating, with reasonable accuracy, the probabilities of the binomial probability distribution by using corresponding areas under the normal probability distribution. Since the opportunities for the application of the binomial probability distribution within business are numerous, the normal approximation to the binomial is an essential topic in our study of business statistics. While other applications of the Central Limit Theorem and the normal distribution will be encountered in succeeding chapters, we particularly note that the Central Limit Theorem provides justification for the use of the Empirical Rule (chapter 3). Furthermore, we observe that the contents of this chapter provide an extension and refinement of the thought embodied in the Empirical Rule.

Supplementary Exercises

7.15.   Using table 3 in the appendix, calculate these areas under the normal curve:
a. between $z = 0$ and $z = 1.2$ b. between $z = 0$ and $z = -.9$
c. between $z = 0$ and $z = 1.6$ d. between $z = 0$ and $z = .75$
e. between $z = 0$ and $z = 1.45$ f. between $z = 0$ and $z = -.42$
g. between $z = .3$ and $z = 1.56$ h. between $z = .2$ and $z = -.2$
i. between $z = 1.21$ and $z = 1.75$ j. between $z = -1.96$ and $z = 1.96$

7.16.   Find z_0 for these probabilities:
a. $P(z > z_0) = .5$ b. $P(z < z_0) = .8643$
c. $P(-z_0 < z < z_0) = .5$ d. $P(-z_0 < z < z_0) = .99$

7.17.   A normally distributed random variable y possesses a mean of 300, and a standard deviation of 20. Find these probabilities:
a. that y exceeds 315   b. that y falls between 290 and 310
c. that y is less than 275   d. that y falls between 305 and 318

Give the percentile values corresponding to these values:
e. $y = 330$ f. $y = 303$ g. $y = 287$ h. $y = 270$

7.18.   An auditor has found that the credit records of a large mail-order house are approximately normally distributed and show an average account billing error of $0 and a standard deviation of $1. (Billing errors may be positive or negative according to whether the purchaser was overcharged or undercharged.) Suppose one credit account is randomly selected from the files of the mail-order house.
   a. Find the probability that it contains a billing error between $0 and $1.50.
   b. Find the probability that it contains a billing error between −$2.00 and $0.
   c. Find the probability that it contains a billing error of at least $1.75.
   d. Find the probability that it contains a billing error that is not an overcharge.
   e. Find the probability that it contains a billing error between −$1.50 and $1.25.
   f. Find the probability that it contains a billing error between −$2.00 and −$1.00.

7.19.   The loan departments of a large system of state banks found that home loans they issued over the past year were approximately normally distributed, with a mean of $43,000 and a standard deviation of $8,500. Suppose loan conditions at the banks remain essentially the same next year.
   a. What proportion of new loans would you expect to be $35,000 or less?
   b. What proportion of new loans would you expect to be greater than $50,000?
   c. What proportion of new loans would you expect to be between $30,000 and $45,000?
   d. What proportion of new loans would you expect to be between $35,000 and $40,000?
   e. What proportion of new loans would you expect to be between $41,000 and $45,000?
   f. What proportion of new loans would you expect to be less than $30,000?

7.20.   The 13 October 1975 issue of *U.S. News and World Report* lists the current pay scales for federal workers in a variety of job types. Grade 9 experienced accountants were reported to earn salaries that are approximately normally distributed, with a mean of $15,089 and a standard deviation of $1,035.
   a. What proportion of government accountants earn more than $17,000 in annual salary?
   b. What is the fraction of government accountants who earn more than $15,000 in annual salary?

7.21.   The average length of time required to complete the civil service examination for prospective U.S. Treasury Department employees is found to equal

70 minutes, with a standard deviation of 12 minutes. When should the examination be terminated if the examination supervisor wishes to allow sufficient time for 90% of the applicants to complete the test? (Assume that the time required to complete the examination is normally distributed.)

7.22. All prospective employees of the U.S. Treasury Department are required to pass a special civil service examination before they are hired. Over a period of time, the examination scores have been normally distributed, with a mean of $\mu = 85$ and a standard deviation of $\sigma = 4$. Suppose that a large number of applicants for jobs with the Treasury Department sit for the test.

a. What proportion can be expected to obtain an examination score of 90 or greater?

b. An examination score of less than 80 is considered as failing. What is the probability that an individual will fail the examination?

c. Of the large number sitting for the examination, what proportion would you expect to pass?

7.23. The owner of a fast-food restaurant has found the daily demand for ground beef at her restaurant to be normally distributed, with a mean of 240 lb and a standard deviation of 23 lb. Since the fast-food business is very competitive, the manager would like to insure that sufficient ground beef is available in her restaurant each day so that the probability is no greater than 1% that the day's supply is exhausted and a stockout occurs. How many pounds of ground beef should the manager have available for use each day?

7.24. Consider a binomial experiment with $n = 25$ and $p = .3$.

a. Calculate $P(6 \le y \le 9)$ by using the binomial probabilities in table 1 of the appendix.

b. Calculate $P(6 \le y \le 9)$ by using the normal approximation to the binomial.

7.25. Consider a binomial experiment with $n = 25$ and $p = .2$.

a. Calculate $P(y \le 4)$ by using table 1 of the appendix.

b. Calculate $P(y \le 4)$ by using the normal approximation to the binomial.

7.26. Current value accounting is not too popular among financial executives, but it is realistic. The current value approach records what an asset is worth today—as opposed to the historical cost approach, which records what was paid for the asset at the date of acquisition. Current value reporting, although highly dependent on judgmental evidence, better reflects today's fast-changing and uncertain economic environment.* Recent studies suggest that 30% of the financial executives associated with the *Fortune* 500 companies favor current value accounting over a historical cost approach. If $n = 25$ financial executives are randomly selected from this group, find the probability that at least 8 but not more than 11 of them favor current value accounting, using, successively:

a. the binomial probabilities, table 1 of the appendix
b. the normal approximation to the binomial

7.27. An assembly machine is known to produce items of which 10% are defective. Suppose that a sample of $n = 100$ is selected from the output of this machine. Use the normal approximation to the binomial.

a. Calculate the probability of at least 10 but not more than 15 defectives in the sample.

b. Calculate the probability of no more than 5 defectives in the sample.

7.28. A rather large and well-known company has been giving serious consideration to adopting flexitime, a system in which employees can work any hours during the week as long as they work 40 hours from Monday through Friday. The

---

*See K. D. Bowes, "The Current Value of Current Value Accounting," *Financial Executive*, November 1975.

results of a preliminary survey show that 30% of the employees are not in favor of adopting flexitime procedures. To check these results a sample of $n = 20$ employees is selected. Use the normal approximation to the binomial to calculate:

     a. the probability that at least 2 but no more than 4 of the 20 employees oppose flexitime

     b. the probability that no more than 4 of the employees are opposed to flexitime.

     7.29.   A salesperson for a pollution abatement equipment manufacturer has found that, on the average, the probability of a sale of a certain piece of equipment on any given contact is equal to .7. If the salesperson contacts 50 customers, what is the probability that at least 10 will buy? (Assume that y, the number of sales, follows a binomial probability distribution.)

     7.30.   Voters in a certain city were sampled concerning their voting preference in a primary election. Suppose that Candidate A could win if he could poll 40% of the vote. If 920 of a sample of 2500 voters favored A, does this contradict the hypothesis that A will win?

     7.31.   In light of the diminishing supply and increasing costs of heating fuels, an eastern manufacturer has developed a new line of Plexiglass storm doors and windows to better protect homes against heat loss during the winter. To determine market acceptability of the product, advertising brochures describing the products and their costs were mailed to 1000 New England homeowners. Sixty-three homeowners responded to the brochure indicating a positive interest in the products. Does this sample present sufficient evidence to indicate that the new product line meets the firm's criteria for success—namely, that more than 5% of the homeowners show a positive interest in the product line?

     7.32.   Market segmentation is concerned with the recognition of significant differences in buyer characteristics. Generally the market is partitioned into distinct groups for the purpose of designing products and marketing programs to satisfy the distinctive consumer characteristics of each group. Recognizing that the market is often segmented according to the sex of the consumer, a newly designed portable radio was styled by an electronics manufacturer on the assumption that 50% of all purchasers are female. If the manufacturer's assumption is correct and if a random sample of 400 purchasers is selected, what is the probability that the number of female purchasers in the sample will be greater than 175?

     7.33.   It is known from experience that 10% of those who register their intent to attend a certain convention do not show up. Suppose that 100 people indicate an intent to attend.

     a. Find the probability that 10 or more fail to show.

     b. Find the probability that no more than 5 fail to show.

     7.34.   Airlines and hotels often grant reservations in excess of capacity to minimize losses due to no-shows. Reservation agents must use discretion, however, as the loss potential may be great when they cannot accommodate an arriving customer with a valid reservation. (Consumer advocate Ralph Nader recently won a substantial sum in a class action suit after he was bumped from a flight although he had a valid reservation.) Suppose an airline reservation office has found that 5% of the passengers reserving a place on a particular flight will not show for the flight. If the reservation office accepts 160 reservations for the flight and there are only 155 available seats on the plane, what is the probability that a seat will be available for every passenger who arrives with a valid reservation?

     7.35.   Refer to exercise 7.34. If the airline wishes to grant sufficient reservations so that the probability is .99 that a seat is available for each arriving passenger with a valid reservation, how many reservations should they grant if the airplane capacity is 200? If capacity is 100?

7.36.   It is known that 30% of all calls coming into a telephone exchange are long-distance calls. If 200 calls come into the exchange, what is the probability that at least 50 will be long-distance calls?

7.37.   A machine operation produces bearings whose diameters are normally distributed, with a mean of .498 and a standard deviation of .002. If specifications require that the bearing diameter equal .500 inches plus or minus .004 inches, what fraction of the production will be unacceptable?

7.38.   Generally speaking, preventive maintenance is less expensive to perform than maintenance after failure or breakdown, because preventive maintenance can be scheduled at noncritical, or nonpeak, time intervals. A manufacturing plant utilizes 3000 electric light bulbs that have a length of life which is normally distributed, with a mean of 500 hours and a standard deviation of 50 hours. To minimize the number of bulbs that burn out during operating hours, all the bulbs are replaced after a given period of operation. How often should the bulbs be replaced if the plant wishes no more than 1% of the bulbs to burn out between replacement periods?

7.39.   Hair dryers and blowers have recently become very popular as a result of longer hairstyles. To ascertain the quality of various hair blowers on the market, a well-known testing service conducted an intensive study of these products. As part of their study it was reported that the length of life of a certain product tested was normally distributed, with a mean life of 1200 hours and a standard deviation of 100 hours. How long should the guarantee time be if the manufacturer of the product wants to replace only 10% of the hair blowers sold?

7.40.   The probability that a certain kind of component will fail in 1000 hours or less is .20. Let y be the random number of components that fail in 1000 hours or less in a sample of size 100.
   a. What is the expected value of y? (Hint: See chapter 6.)
   b. What is the standard deviation of y? (Hint: See chapter 6.)
   c. Using the normal approximation, find $P(y \leq 30)$.

7.41.   A national poll claims that 60% of all American voters favor a popular vote for presidential nominees. Investigation indicates that only 100 voters were sampled and, of those, $y = 54$ favored the popular vote. Suppose that the fraction of voters p favoring a popular vote is actually .6. Find the probability that $y \leq 54$ given that $p = .60$.

7.42.   The average personal yearly income in a given state is $6200, with a standard deviation of $400.
   a. If a sample of 64 people is randomly chosen from this state, find the probability that the mean income for the sample exceeds $6300.
   b. If a second independent sample of 64 people is randomly chosen from this state, find the probability that both sample means exceed $6300. (Hint: See the Central Limit Theorem.)

7.43.   Many state legislatures have been closely monitoring the effects of the Oregon bottle bill, an enactment that, in an attempt to reduce litter, prohibits the use of nonreturnable containers for beer or soft drinks. A prominent official in another state claims that at least 50% of the residents of her state favor such legislation. To check her claim 64 residents are randomly selected from the state, with $y = 27$ indicating they favor the enactment of a bottle bill in their state. Do these data present sufficient evidence to reject the official's claim?

7.44.   When a doctor writes a prescription, often the patient takes it to the nearest drugstore and learns what the pills cost only when they are handed to him. This happens because pharmacies are prohibited from advertising prices for prescription drugs. A study of pharmacies in the San Francisco area found the prices charged for 100 tablets of ampicillan to be approximately normally distributed, with a mean price of $8.50 and a standard deviation of $2.00. Find the probability that a particular San Francisco pharmacy charges between $10.00 and $12.00 for 100 tablets of ampicillan.

7.45. A common theme of advertising is to state that a certain product is superior to its competitors in some superficial way (e.g., shinier floors, brighter teeth, richer flavor). A similar theme offers a product as a panacea for certain real problems without reference to a competitor (e.g., using Brand X reduces cavities). Experimental evidence is needed but seldom offered to check the veracity of such advertised claims. Consider, for example, a new serum that a commercial laboratory claims is effective in preventing the common cold. Suppose 40 people were injected with the serum and observed for a period of one year. Twenty-eight survived the winter without a cold. According to prior information it is known that the probability of surviving the winter without a cold is .5 when the serum is not used. On the basis of the results of this experiment, what conclusions would you make regarding the effectiveness of the serum?

7.46. The probability α of a type I error and location of the rejection region for a statistical test of an hypothesis are usually specified before the data are collected. In the case of exercise 7.45, a type I error is the error of assuming that the cold serum is effective as a cold deterrent when, in fact, the serum is ineffective (i.e., rejecting $p = .5$ when it is true). For the cold serum problem, find the appropriate rejection region for the test of the null hypothesis $p = .5$ if we wish α, the probability of falsely concluding that the serum is effective, to be approximately equal to .05. (See figure 7.15.)

7.47. College bookstores face a delicate inventory management problem when ordering from a publisher. If the order is too small, they incur costly emergency procurement charges; if they overorder, they may be left with useless inventory if the text is never used again. Suppose records show that attendance in Accounting 1 at a particular university possesses a probability distribution that is approximately normal, with a mean of 150 students a semester and a standard deviation of 20 students. How many Accounting 1 textbooks should be ordered if the university bookstore would like to allow for no more than a 5% chance of a stockout?

7.48. Advertising agencies compete vigorously with one another to sponsor network television programs that attract a significant portion of the television viewing audience. The producer of a certain television program has stated that 20% of all television viewers watch his program. An advertising agency, with interest in sponsoring the program, would like to check the accuracy of the producer's claim. In a random sample of 1600 viewers contacted during a showing of the program, $y = 284$ were watching the program. Do these data present sufficient evidence to contradict the producer's claim? Conduct a statistical test, using α equal to .05.

Experiences with Real Data

1. One business process that is subject to a variety of random disturbances is the securities market. Uncertain political, economic, social, and market factors act collectively to cause security prices to move unpredictably over time. A considerable volume of research effort has led to the development of a "random walk" hypothesis to define the movement of security prices over time. One version of this theory suggests that the price changes of a security from one period to the next are approximately normally distributed.

From past issues of the *Wall Street Journal* or some other source, record the daily price changes of some listed security over 100 consecutive market days. Construct a relative frequency histogram for these data. Compare all the histograms in the class and note the number that show a bell-shaped distribution. Do your findings and those of other members of your class appear to support the random walk hypothesis that security price changes are approximately normally distributed?

2. The Central Limit Theorem states that if measurements are selected from a population with a finite mean and variance, then sums and means computed from the measurements tend to be approximately normally distributed. What this means in the context of our security price data is that even if the daily price changes are not approximately normally distributed, weekly changes, which are the sum of daily price changes, are likely to be approximately normally distributed.

Using the daily price changes you collected for the security (exercise 1) plus additional price data on this security, compute the *weekly* price changes of the security over 50 consecutive weeks. (Use only those weeks with five market trading days.) Construct a relative frequency histogram for these data. Does your histogram appear to be approximately normal?

References

HOEL, P. G. *Elementary Statistics*. 3d ed. New York: Wiley, 1971. Chapter 4.

JOHNSON, R. R. *Elementary Statistics*. N. Scituate, Mass.: Duxbury Press, 1973. Chapter 6.

SUMMERS, G. W., and W. S. PETERS. *Basic Statistics in Business and Economics*. Belmont, Calif.: Wadsworth, 1973. Chapter 8.

chapter objectives

GENERAL OBJECTIVES In the preceding chapters we have established the groundwork for the study of statistical inference making—our main objective—by introducing and discussing the fundamental ideas of statistics. In this chapter we will apply these ideas by considering the basic concepts of statistical inference. We will use four very practical inference-making situations to illustrate the ideas involved, and we will show how the Central Limit Theorem, presented in chapter 7, justifies each of these inference-making procedures.

SPECIFIC OBJECTIVES

1. To review the objective of statistics—statistical inference—and to explain how inferences about population parameters can be made.
Sections 8.2, 8.3

2. To explain how we can measure the goodness of statistical estimators.
Sections 8.4, 8.5

3. To give examples of some useful statistical estimators.
Sections 8.6 through 8.11

4. To explain the role of the Central Limit Theorem in large-sample estimation.
Section 8.8

5. To stress, once again, the importance of planning an experiment—the most elementary step being the choice of the sample size.
Section 8.12

6. To review the concepts of statistical tests of hypotheses and to illustrate their application to tests of hypotheses about population means.
Sections 8.13, 8.14, 8.15

chapter eight

Large-Sample Statistical Inference

8.1 A Brief Summary

The preceding seven chapters set the stage for the objective of this text, developing an understanding of statistical inference and the role it plays in business decision making. In chapter 1 we stated that statisticians are concerned with making inferences about populations of measurements based on information contained in samples. We showed you how you phrase an inference—that is, how you describe a set of measurements—in chapter 3. We discussed probability, the mechanism for making inferences, in chapter 4, and we followed that with three chapters about probability distributions—a general presentation in chapter 5, useful discrete probability distributions in chapter 6, and the normal distribution in chapter 7.

To get you thinking about statistical inference, we introduced to you in chapter 6 the useful application of the binomial probability distribution to lot acceptance sampling and the test of an hypothesis concerning the effectiveness of a cold vaccine. We touched lightly on these topics again in the examples and exercises of chapter 7. Now we are ready to utilize the foundation we have laid—to study the basic concepts involved in statistical inference.

Perhaps the most important contribution to our preparation for a study of statistical inference is the Central Limit Theorem of chapter 7. This theorem, which justifies the approximate normality of the probability distribution of sample means for large samples, was used to justify the normal approximation to the binomial probability distribution in chapter 7. But more important, it will be used to justify the approximate normality of the probability distributions of estimators and decision makers encountered in this chapter. Let us turn now to the objective of our study, statistical inference.

8.2 Inference: The Objective of Statistics

Inference, specifically decision making and prediction, is centuries old and plays an important role in our individual lives. Each of us is faced with daily personal decisions and situations that require predictions about the future. The government is concerned with predicting the flow of gold to Europe. The broker seeks knowledge concerning the behavior of the stock market. The metallurgist wishes to use the results of an experiment to infer whether a new type of steel is more resistant to temperature changes than another. The housewife wishes to know whether Detergent A is more effective than Detergent B in her washing machine. These inferences are supposedly based on relevant bits of available information, which we call observations or data.

In many practical situations the relevant information is abundant, seemingly inconsistent, and, in many respects, overwhelming. As a result, our carefully considered decision or prediction is often little better than an outright guess. You need only refer to the "Market Views" section of the *Wall Street Journal* to observe the diversity of expert opinion concerning future stock market behavior. Similarly, a visual analysis of data by scientists and engineers will often yield conflicting opinions regarding conclusions to be drawn from an experiment. While many individuals tend to feel that their own built-in inference-making equipment is quite good, experience suggests that most people are incapable of utilizing large amounts of data, mentally weighing each bit of relevant information, and arriving at a good inference. (You may test your individual inference-making equipment by using the exercises in this chapter and the next. Scan the data and make an inference before using the appropriate statistical procedure. Compare the results.) Certainly a study of inference-making systems is desirable, and this is the objective of the mathematical statistician.

Although we have purposely touched upon some of the notions involved in statistical inference in preceding chapters, it is beneficial to organize our knowledge at this point as we attempt an elementary presentation of some of the basic ideas involved in statistical inference.

> The objective of statistics is to make inferences about a population based on information contained in a sample.

Since populations are characterized by numerical descriptive measures called parameters, statistical inference is concerned with making inferences about population parameters. Typical population parameters are the mean, the standard deviation, the area under the probability distribution, above or below some value of the random variable, or the area between two values of the variable. Indeed, all the practical problems mentioned in the first paragraph of this section can be restated in the framework of a population with a specified parameter of interest.

Methods for making inferences about parameters fall into one of two categories. We may make decisions concerning the value of the parameter, as exemplified by the lot acceptance sampling and test of an hypothesis described in chapter 6. Or we may estimate or predict the value of the parameter. While some statisticians view estimation as a decision-making problem, it is convenient for us to retain the two categories and, particularly, to concentrate separately on estimation and on tests of hypotheses.

A statement of the objective and the types of statistical inference would be incomplete without reference to a measure of the goodness of inferential procedures. We may define numerous objective methods for making inferences, in addition to our own individual procedures based on intuition. Therefore, a measure of goodness must be defined so that one procedure may be compared with another. More than that, we wish to state the goodness of a particular inference in a given physical situation. Thus to predict that the price of a stock will be $80 next Monday would be insufficient and would stimulate few of us to take action to buy or sell. We also wish to know whether the estimate is correct to within plus or minus $1, $2, or $10.

Statistical inference in a practical situation contains two elements: (1) the inference and (2) a measure of its goodness.

Which method of inference should be used; that is, should the parameter be estimated or should we test an hypothesis concerning its value? The answer to this question is dictated by the practical question posed and is often determined by personal preference. Some people like to test theories concerning parameters, while others prefer to express their inference as an estimate. Inasmuch as both estimation and tests of hypotheses are frequently used in scientific literature, we would be remiss in excluding one or the other from our discussion.

Types of Estimators

Estimation procedures may be divided into two types, point estimation and interval estimation. Suppose as a merchandising executive you are concerned with the "shrinkage rate" (disappearance of merchandise due to internal and external theft) of items in your store. To estimate the rate you sample *n* items from inventory and record the ratio

$$\text{shrinkage rate} = \frac{(\text{book inventory value}) - (\text{value on hand})}{\text{book inventory value}}$$

$$= \frac{\text{loss}}{\text{book inventory value}}$$

From these sample data you could form two types of estimates of the mean shrinkage rate μ for all items in the store. For example, if the sample mean is .20, you use this single number as an estimate of μ. This would imply that you estimate the mean loss to be 20% of inventory value. Or you might estimate the mean shrinkage loss to lie in the interval from .15 to .25; or, equivalently, you estimate that the mean loss is between 15% and 25% of inventory value. The first type of estimate is called a **point estimate** because a single number, representing the estimate, may be associated with a point on a line. The second type, involving two points and defining an interval on a line, is called an **interval estimate.** We will consider each of these methods of estimation in turn.

A point estimation procedure utilizes information in a sample to arrive at a single number or point that estimates the population parameter of interest. The actual estimation is accomplished by an **estimator.** An estimator is a rule that tells us how to calculate the estimate based on information in the sample; it is generally expressed as a formula. For example, the sample mean

$$\bar{y} = \frac{\sum\limits_{i=1}^{n} y_i}{n}$$

is an estimator of the population mean μ and explains exactly how the actual numerical value of the estimate may be obtained once the sample values $y_1, y_2, \ldots, y_n$ are known. On the other hand, an interval estimator uses the data in the sample to calculate two points that are intended to enclose the value of the population parameter estimated.

Definition

A **point estimator** of a population parameter is a rule that tells you how to calculate a single number based on sample data. The resulting number is called a **point estimate** of the parameter.

> **Definition**
>
> An **interval estimator** of a population parameter is a rule that tells you how to calculate two numbers based on sample data.

> **Definition**
>
> When an interval estimator is employed to estimate a population parameter, the pair of numbers obtained from the estimator is called an **interval estimate** or **confidence interval** for the parameter. The larger number, which locates the upper end of the interval, is called the **upper confidence limit** and is denoted by the symbol UCL. Similarly, the number that locates the lower extreme of the interval is called the **lower confidence limit** and is denoted by the symbol LCL.

Both point estimates and interval estimates are employed in surveys of consumer preference but point estimates seem to be the most common. When estimation arises in connection with industrial experimentation, confidence intervals seem to be most commonly used.

8.4   Evaluating the Goodness of a Point Estimator

Returning to the estimation of the shrinkage rate, suppose your sample mean is .20 and you report this value to the store manager. Would your manager be satisfied with this estimate or would he or she ask you, "Plus or minus what?" In other words, how much faith can the manager place in your estimate? Is it an accurate estimate of μ? How far away from μ might you expect this estimate to be?

To answer this question let us consider an analogy that may help to explain the reasoning employed in evaluating the goodness of a point estimator. Point estimation is similar, in many respects, to firing a revolver at a target. The estimator, which generates estimates, is analogous to the revolver; a particular estimate is analogous to the bullet; and the parameter of interest is analogous to the bull's-eye. Drawing a sample from the population and estimating the value of the parameter is equivalent to firing a single shot at the target.

Suppose that a man fires a single shot at a target and the shot pierces the bull's-eye. Do we conclude that he is an excellent shot? The answer

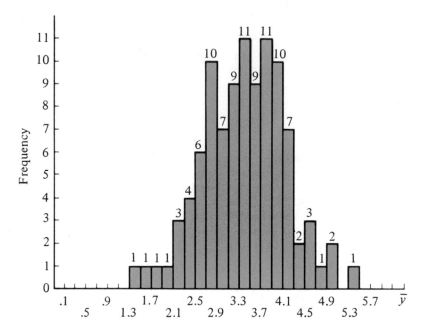

Figure 8.1 *Histogram of sample means for the die-tossing experiments in section 7.2*

is no, because not one of us would consent to hold the target while a second shot is fired. On the other hand, if 1 million shots in succession hit the bull's-eye, we might acquire sufficient confidence in the marksman to hold the target for the next shot, if the compensation were adequate. The point we wish to make is that we cannot evaluate the goodness of an estimation procedure on the basis of a single estimate. Rather, we must observe the results when the estimation procedure is used over and over again, many, many times—we then observe how closely the shots are distributed about the bull's-eye. In fact, since the estimates are numbers, we would evaluate the goodness of the estimator by constructing a frequency distribution of the estimates obtained in repeated sampling and note how closely the distribution centers about the parameter of interest.

To illustrate, recall the die-tossing experiment used in chapter 7 where we generated 100 samples of $n = 5$ measurements each and calculated the mean for each sample. Since we know the mean value μ of the number showing on a die toss ($\mu = 3.5$), we can use the results of the die-tossing experiment to see how well the mean of a sample of $n = 5$ measurements estimates μ.

The frequency histogram of the 100 sample means is shown in figure 8.1. Notice how the estimates group about the population mean $\mu = 3.5$, and that they range from 1.3 to 5.5. Surely this distribution of estimates tells us something about how good a new estimate of μ would be if we were to draw one more sample of $n = 5$ measurements and compute the sample mean $\bar{y}$.

We would know the characteristics of every point estimator if for each estimator we had a relative frequency histogram (similar to figure 8.1) that depicted its behavior in repeated sampling for a given sample size. Thus instead of drawing 100 samples and computing 100 values of $\bar{y}$, as was done in the study illustrated in figure 8.1, think of drawing thousands or even infinitely many samples. If we were able to draw infinitely many samples of a fixed size, the resulting relative frequency distribution of the $\bar{y}$'s would be the probability distribution or, as it is frequently called, the *sampling distribution* for the estimator $\bar{y}$.

Definition

The probability distribution for an estimator is called its **sampling distribution**.

The acquisition of sampling distributions for various estimators is the work of research statisticians. Fortunately, the sampling distributions of most point estimators are known.

Now that we see that the properties of a point estimator are revealed by its probability distribution, we might ask what properties are most desirable. Essentially, there are two, and these can be seen by viewing figure 8.2. First, we want the sampling distribution of estimates to center about the parameter of interest. For example, if we are estimating μ, we would like the sampling distribution of the estimator to center on μ, as shown in figure 8.2(a). Such an estimator is said to be *unbiased*. The sampling distribution for a *biased* estimator is shown in figure 8.2(b). (Note: A hat (^) over a parameter is a symbol used to denote an estimator of a parameter.)

Figure 8.2 *Distributions for unbiased and biased estimators*

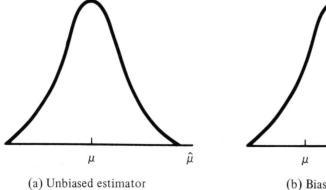

(a) Unbiased estimator

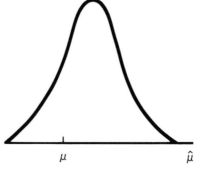

(b) Biased estimator

Definition

An estimator of a population parameter is said to be **unbiased** if the mean of its sampling distribution is equal to the parameter. Otherwise the estimator is said to be **biased**.

The second desirable property of a point estimator is that the standard deviation of its sampling distribution be small. Thus we would like the spread of the distribution of estimates (see figure 8.2) to be as small as possible. For most estimators the standard deviation of the sampling distribution is controllable. That is, we can make the standard deviation (which measures the spread of the distribution) as small as we wish by increasing the sample size.

For example, in chapter 7 we learned from the Central Limit Theorem that the sampling distribution of the sample mean $\bar{y}$ was approximately normally distributed, with mean μ and standard deviation $\sigma/\sqrt{n}$ (μ and σ are the mean and standard deviation of the population from which the sample was selected and n is the sample size). Consequently, we can make the standard deviation of the sampling distribution of $\bar{y}$, denoted by the symbol $\sigma_{\bar{y}}$, as small as we choose by increasing the value of n. The way the sampling distribution for $\bar{y}$ depends on the sample size n is shown in figure 8.3, where we show three sampling distributions for $\bar{y}$ sketched on the same graph. The sampling distributions, corresponding to $n = 5$, $n = 20$, and $n = 80$,

Figure 8.3 *Sampling distributions for $\bar{y}$; n = 5, 20, and 80*

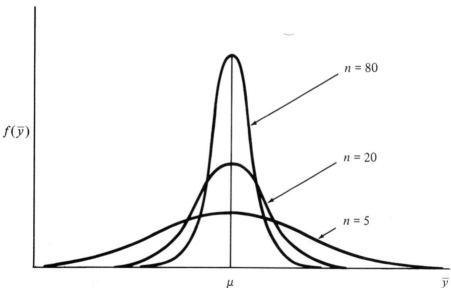

show how the spread of the sampling distributions decreases as n increases. The sampling distributions also show why an estimate based on $n = 80$ measurements will, with a high probability, lie much closer to μ than one based on a sample of $n = 5$ measurements.

In a real life sampling situation, you may know that the sampling distribution of an estimator centers over the parameter that you are attempting to estimate, but you do not know the value of the parameter. All that you have is the estimate computed from the n measurements contained in the sample. How far will your particular estimate be from the estimated parameter? Since the parameter usually lies in the center of the sampling distribution (it is usually the mean of the distribution), the distance between the estimate and the parameter, called the *error of estimation*, is less than or equal to the distance between the center and the tails of the distribution.

Definition

The distance between an estimate and the estimated parameter is called the **error of estimation**.

In fact, the error of estimation should be less than two standard deviations of the sampling distribution, with a probability equal to at least .75 (Tchebysheff's theorem) and most likely very close to .95 (the Empirical Rule). For example, the sampling distribution of the sample mean is approximately normal, with mean μ and standard deviation $\sigma / \sqrt{n}$ (from the Central Limit Theorem), and appears as shown in figure 8.4. You can see that the probability that an estimate falls within $2\sigma / \sqrt{n}$, either above or below μ, is approximately equal to .95. Or equivalently, the probability that the error of estimation is less than $2\sigma / \sqrt{n}$ is approximately equal to .95.

So to summarize, the properties of a point estimator are characterized

Figure 8.4 *The sampling distribution of $\bar{y}$*

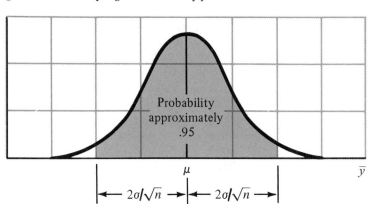

by its sampling distribution. We prefer estimators that are unbiased and have small standard deviations. And if we are dissatisfied with the standard deviation of an estimator, we can reduce it by increasing the sample size n. The error of estimation—the distance between the estimate and the estimated parameter—should be less than two standard deviations of the sampling distribution, with probability at least equal to .75 and very likely near .95.

8.5 Evaluating the Goodness of an Interval Estimator

Constructing an interval estimate is like attempting to rope a dead steer. In this case the parameter that you wish to estimate corresponds to the steer and the interval corresponds to the loop formed by the cowboy's lariat. Each time you draw a sample, you construct a confidence interval for a parameter and you hope to "rope it," that is, include it in the interval. You will not be successful for every sample. The probability that an interval will enclose the estimated parameter is called the *confidence coefficient*.

Definition

The probability that a confidence interval will enclose the estimated parameter is called the **confidence coefficient**.

To consider a practical example, suppose that you wish to estimate the mean profit per week of a small company. If we were to draw 10 samples, each containing $n = 20$ weekly profit observations, and construct a confidence interval for the population mean μ for each sample, the intervals might appear as shown in figure 8.5. The horizontal line segments represent the 10 intervals and the vertical line represents the location of the true mean weekly profit. Note that all but one of the intervals enclose μ for these particular samples.

So, what is a good confidence interval? The answer is, one that is as narrow as possible and has a large confidence coefficient, near 1. The narrower the interval, the more exactly we have located the estimated parameter. The larger the confidence coefficient, the more confidence we have that a particular interval encloses the estimated parameter. Remember, the confidence coefficient gives the probability that the interval estimator will produce confidence limits that enclose the estimated parameter. It gives you a measure of the confidence you can place in the confidence limits constructed from the data contained in a sample. In that sense, the width of an interval and its associated confidence coefficient measure the goodness of the confidence interval.

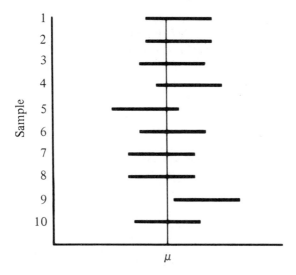

Figure 8.5 *Ten confidence intervals for mean weekly profit (each based on a sample of n = 20 observations)*

What is the effect of larger samples on the width of a confidence interval? Larger samples provide more information to use in forming the interval estimate. Therefore, for a given confidence coefficient, the larger the sample, the narrower will be the resulting confidence interval.

8.6 Point Estimation of a Population Mean

Business decision problems often require the estimation of a population mean μ. For example, we are concerned with the average daily production of workers on an assembly line, with the average strength of a new type of steel, with the average number of accidents per month in a manufacturing plant, or with the average demand for a new product. Conveniently, the estimation of μ serves as a very practical application of statistical inference as well as an excellent illustration of the principles of point estimation discussed in section 8.4.

Many estimators are available for estimating the population mean μ, including the sample median, the average between the largest and smallest measurements in the sample, and the sample mean $\bar{y}$. Each generates a sampling distribution in repeated sampling and, depending on the population and practical problem involved, each possesses certain advantages and disadvantages. The sample median and the average of the sample extremes are usually easy to calculate. The sample mean $\bar{y}$ is usually superior to them, however, since for some populations its standard deviation is a minimum and, furthermore, it is always unbiased, regardless of the population.

Three facts emerge from a study of the probability distribution of $\bar{y}$ in repeated random sampling of n measurements from a population with mean μ and variance σ^2. Regardless of the probability distribution of the population, these three features are always true.

Characteristics of the Probability Distribution of $\bar{y}$

1. The expected value of $\bar{y}$ is equal to the population mean μ.
2. The standard deviation of $\bar{y}$ is

$$\sigma_{\bar{y}} = \frac{\sigma}{\sqrt{n}} \sqrt{\frac{N-n}{N-1}}$$

where N is the number of measurements in the population. In the following discussion we will assume that N is large relative to the sample size n and hence that $\sqrt{(N-n)/(N-1)}$ is approximately equal to 1. Then

$$\sigma_{\bar{y}} = \frac{\sigma}{\sqrt{n}}$$

3. When n is large, $\bar{y}$ is approximately normally distributed, according to the Central Limit Theorem (assuming that μ and σ are finite numbers).

Thus $\bar{y}$ is an unbiased estimator of μ with a standard deviation that is proportional to the population standard deviation σ and inversely proportional to the square root of the sample size n. While we give no proof of these results, we suggest that they are intuitively reasonable. Certainly the more variable the population data, as measured by σ, the more variable will be $\bar{y}$. On the other hand, more information is available for estimating as n becomes large. Hence the estimates should fall closer to μ and $\sigma_{\bar{y}}$ should decrease.

In addition to knowledge of the mean and standard deviation of the probability distribution for $\bar{y}$, the Central Limit Theorem provides information of its form. That is, when the sample size n is large, the distribution of $\bar{y}$ is approximately normal. This distribution is shown in figure 8.6.

To construct a point estimate for a population mean μ, you calculate the value of the sample mean. Then, as explained in section 8.4, you calculate the two-standard-deviation bound on the error of estimation. For the sample mean the bound on the error of estimation is

$$\text{bound on error} = \frac{2\sigma}{\sqrt{n}}$$

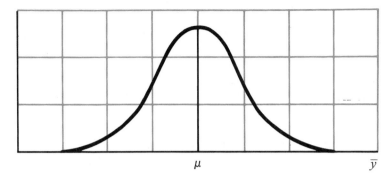

Figure 8.6 *Distribution of ȳ for large n*

The probability that the error of estimation is less than the bound on error is approximately .95.

Most often we do not know the value of the population standard deviation σ that appears in the formula for the bound on error. Thus we must approximate it. Most often we will substitute the sample standard deviation s for σ. This approximation is quite adequate if the sample size is 30 or larger. When σ is not known and the sample size is small (say less than 30), we employ small-sample estimation procedures; these procedures will be presented in chapter 9.

Point Estimator of a Population Mean

Estimator: $\bar{y}$.

Bound on error: $2\sigma_{\bar{y}} = \dfrac{2\sigma}{\sqrt{n}}$.

Note: If σ is unknown and n is 30 or more, you can use the sample standard deviation s to approximate the value of σ.

We illustrate the estimation procedure with an example.

Example 8.1 Suppose that we wish to estimate the average daily yield of a chemical manufactured in a chemical plant. The daily yield, recorded for $n = 50$ days, produced a mean and a standard deviation of

$$\bar{y} = 871 \text{ tons} \qquad \text{and} \qquad s = 21 \text{ tons}$$

Estimate the average daily yield μ.

Solution The estimate of the daily yield is $\bar{y} = 871$ tons. The bound on the error of estimation is

$$2\sigma_{\bar{y}} = \frac{2\sigma}{\sqrt{n}} = \frac{2\sigma}{\sqrt{50}}$$

Although σ is unknown, we can approximate its value by using s, the estimator of σ. Thus the bound on the error of estimation is approximately

$$\frac{2s}{\sqrt{n}} = \frac{2(21)}{\sqrt{50}} = \frac{42}{7.07} = 5.94$$

We feel fairly confident that our estimate of 871 tons is within 5.94 tons of the true average yield.

Example 8.1 deserves further comment in regard to two points. You should be careful not to use the erroneous 2σ as a bound on the error of estimation rather than the correct value $2\sigma_{\bar{y}}$. Certainly if we wish to discuss the distribution of $\bar{y}$, we must use its standard deviation $\sigma_{\bar{y}}$ to describe its variability. Care must be taken not to confuse the descriptive measures of one distribution with another.

The second point concerns the use of s to approximate σ. This approximation is reasonably good when n is large, say 30 or greater. If the sample size is small, two techniques are available. Sometimes experience or data obtained from previous experiments will provide a good estimate of σ. You can then substitute this value for σ in the formula for $\sigma_{\bar{y}}$. When an approximate value for σ is not available and when the sample has been selected from a population that has an approximately normal (or at least mound-shaped) relative frequency distribution, we may resort to a small-sample procedure to be described in chapter 9. The choice of $n = 30$ as the division between "large" and "small" samples is arbitrary. The reasoning for its selection will be discussed in chapter 9.

Exercises

8.1.   In its publication *Environmental News*, the Environmental Protection Agency has noted that for the past several years a large East Coast community has been plagued with severe summer water shortages. To monitor water usage city officials randomly selected and monitored $n = 100$ residential water meters to estimate the average daily water consumption per household over a specified dry spell. The sample mean and standard deviation were found to be 117.5 gallons and 16.8 gallons, respectively. Estimate μ, the average daily household consumption for the community, and place a bound on the error of estimation.

8.2.   According to one researcher, when attempting to evaluate the internal control procedures of an organization by studying its accounting records, businesses are too large to allow for an examination of all, or even a quarter, of the year's transactions.* An audit was conducted by selecting $n = 50$ bills of lading transacted during the past year by a trucking firm. The mean and standard deviation of the bills were found to be \$2160 and \$575, respectively. Estimate μ, the average bill of lading transacted by the firm during the past year. Place a bound on the error of estimation.

*T. G. Haworth, "Statistical Sampling: A Practical Approach," *Journal of Accountancy*, February 1969.

8.3.   The mean and standard deviation of the after-tax profits of 100 manufacturers for a certain period were found to be 4.5 cents per dollar revenue and 1.3 cents per dollar revenue, respectively. Estimate the mean after-tax profits for this period for the population of manufacturers from which this sample was drawn. Place a bound on the error of estimation.

8.4.   Suppose the population mean after-tax profits for all manufacturers during a certain period is really 4.8 cents per dollar revenue, with $\sigma = 1.5$ cents. What is the probability that the mean after-tax profits of a random sample of $n = 100$ manufacturers would exceed 5 cents?

8.7 Interval Estimation of a Population Mean

The interval estimator, or **confidence interval,** for a population mean may be easily obtained from the results of section 8.6. It is possible that $\bar{y}$ might lie either above or below the population mean, although we would not expect it to deviate more than approximately $2\sigma_{\bar{y}}$ from μ. Hence if we choose $(\bar{y} - 2\sigma_{\bar{y}})$ as the lower point of the interval, called the **lower confidence limit,** or LCL, and $(\bar{y} + 2\sigma_{\bar{y}})$ as the upper point, or **upper confidence limit,** UCL, the interval most probably will enclose the true population mean μ. In fact, if n is large and the distribution of $\bar{y}$ is approximately normal, we would expect approximately 95% of the intervals obtained in repeated sampling to enclose the population mean μ.

The confidence interval described above is called a **large-sample** confidence interval (or confidence limits) because n must be large enough for the Central Limit Theorem to be effective and hence for the distribution of $\bar{y}$ to be approximately normal. Since σ is usually unknown, the sample standard deviation s must be used to estimate σ. As a rule of thumb, this **confidence interval is appropriate when $n = 30$ or more.**

The confidence coefficient .95 corresponds to $\pm 2\sigma_{\bar{y}}$, or, more exactly, $1.96\sigma_{\bar{y}}$. Recalling that .90 of the measurements in a normal distribution fall within $z = 1.645$ standard deviations of the mean (table 3 in the appendix), we could construct 90% confidence intervals by using

$$\text{LCL} = \bar{y} - 1.645\sigma_{\bar{y}} = \bar{y} - 1.645\frac{\sigma}{\sqrt{n}}$$

$$\text{UCL} = \bar{y} + 1.645\sigma_{\bar{y}} = \bar{y} + 1.645\frac{\sigma}{\sqrt{n}}$$

In general, we may construct confidence intervals corresponding to any desired confidence coefficient, say $(1 - \alpha)$, by using the formula given in the box.

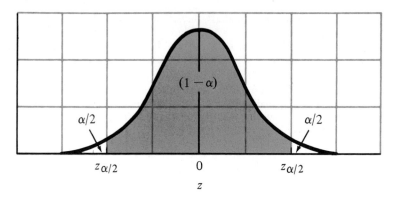

Figure 8.7 *Location of* $z_{\alpha/2}$

A (1 − α)100% Large-Sample Confidence Interval for μ

$$\bar{y} \pm z_{\alpha/2}\frac{\sigma}{\sqrt{n}}$$

The normal curve z-value, $z_{\alpha/2}$, that appears in the formula for the confidence interval is located as shown in figure 8.7. For example, if you want a confidence coefficient $(1 - \alpha)$ equal to .95, then the tail-end area is $\alpha = .05$, and half of α (.025) is placed in each tail of the distribution. Then $z_{.025}$ is the table z-value corresponding to an area of .475 to the right of the mean, or

$$z_{.025} = 1.96$$

Confidence limits corresponding to some of the commonly used confidence coefficients are shown in table 8.1.

Table 8.1 *Confidence limits for* μ

CONFIDENCE COEFFICIENT, $(1 - \alpha)$	α	$z_{\alpha/2}$	LCL	UCL
.90	.10	1.645	$\bar{y} - 1.645\dfrac{\sigma}{\sqrt{n}}$	$\bar{y} + 1.645\dfrac{\sigma}{\sqrt{n}}$
.95	.05	1.96	$\bar{y} - 1.96\dfrac{\sigma}{\sqrt{n}}$	$\bar{y} + 1.96\dfrac{\sigma}{\sqrt{n}}$
.99	.01	2.58	$\bar{y} - 2.58\dfrac{\sigma}{\sqrt{n}}$	$\bar{y} + 2.58\dfrac{\sigma}{\sqrt{n}}$

Example 8.2 Find a 90% confidence interval for the population mean of example 8.1. Recall that $\bar{y} = 871$ tons and $s = 21$ tons.

Solution The 90% confidence limits are

$$\bar{y} \pm 1.645\, \frac{\sigma}{\sqrt{n}}$$

Using s to estimate σ, we obtain

$$871 \pm (1.645)\, \frac{21}{\sqrt{50}} = 871 \pm 4.89$$

Therefore, we estimate that the average daily yield μ lies in the interval from 866.11 to 875.89 tons. The confidence coefficient .90 implies that in repeated sampling 90% of the confidence intervals similarly formed would enclose μ.

Example 8.3 A sample of $n = 100$ employees from a company was selected and the annual salary for each was recorded. The mean and standard deviation of their salaries were found to be

$$\bar{y} = \$7750 \qquad \text{and} \qquad s = \$900$$

Construct the 95% confidence interval for the population average salary μ.

Solution The 95% confidence limits are

$$\bar{y} \pm 1.96\, \frac{\sigma}{\sqrt{n}}$$

Using s to estimate σ, we obtain

$$\$7750 \pm (1.96)\, \frac{\$900}{\sqrt{100}} = \$7750 \pm \$176.40$$

Thus the average salary for the employees of the company is contained within the interval from $7573.60 to $7926.40, with confidence coefficient .95.

The confidence interval of example 8.3 is approximate because we substituted s as an approximation for σ. That is, instead of the confidence coefficient being .95, the value specified in the example, the true value of the coefficient may be .92, .94, or .97. But this is of little concern from a practical point of view because as far as our "confidence" is concerned, there is little difference among these confidence coefficients. Most interval estimators employed in statistics yield approximate confidence intervals, because the assumptions upon which they are based are not satisfied exactly. Having made this point, we will not continue to refer to confidence intervals as "approximate." It is of little practical concern as long as the actual confidence coefficient is near the value specified.

Note in table 8.1 that for a fixed sample size, the width of the confidence interval increases as the confidence coefficient increases, a result that is in agreement with our intuition. Certainly if we wish to be more confident that the interval will enclose μ, we would increase the width of the interval. Since we prefer narrow confidence intervals and large confidence coefficients, it is clear that we must reach a compromise in choosing the confidence coefficient.

The choice of the confidence coefficient to be used in a given situation is made by the experimenter and depends on the degree of confidence the experimenter wishes to place in the estimate. Most confidence intervals are constructed using one of the three confidence coefficients shown in table 8.1. The most popular seems to be 95% confidence intervals. Use of 99% confidence intervals is less common because of the wider interval width. Of course, you can always decrease the width by increasing the sample size n.

The frequent use of the .95 confidence coefficient raises a question asked by many beginners. Should you use $z = 1.96$ or $z = 2$ in the confidence interval? The answer is that it does not really make much difference which value is used. The value $z = 1.96$ is more exact for a .95 confidence coefficient, but the error introduced by using $z = 2$ will be very small. The use of $z = 2$ simplifies the calculations, particularly when the computing is done manually. We will agree to use two standard deviations when placing bounds on the error of a point estimator, but we will use $z = 1.96$ when constructing a confidence interval, simply to remind you that this is the z-value obtained from the table of areas under the normal curve.

Note the fine distinction between point estimators and interval estimators. Note also that in placing bounds on the error of a point estimate, for all practical purposes we are constructing an interval estimate when a population mean is being estimated. While this close relationship exists for most of the parameters estimated in this text, the two methods of estimation are not equivalent. For instance, it is not obvious that the best point estimator falls in the middle of the best interval estimator—in many cases it does not. Furthermore, it is not necessarily true that the best interval estimator is even a function of the best point estimator. Although these problems are of a theoretical nature, they are important and worth mentioning. From a practical point of view, the two methods are closely related and the choice between the point and the interval estimator in an actual problem depends on the preference of the experimenter.

Exercises

8.5.   Forecasting sales revenue in a given geographical region is a difficult task and often requires an estimate of consumer expenditures for a particular product class. A market research study was undertaken by a major oil company to determine the amount spent on gasoline and heating oil during a particular year by the residential households of a certain city. The city directory was used to select a random sample of n = 64 households from the city. The sample mean and standard deviation of their expenditures on gasoline and heating oil during the year were $836 and $178, respectively. Find a 90% confidence interval for the average annual expenditure on gasoline and heating oil by the residential households of the city.

8.6.   Investors who purchase income-producing stocks are generally interested only in the annual yield produced by this class of common stocks. To assist investors of this type, a leading stock brokerage firm randomly sampled 50 income-producing common stocks. The average annual yield and standard deviation of yield per security over the past five years was observed to be $\bar{y}$ = 8.71% and s = 2.1%. Estimate the true average annual yield μ for this class of common stocks, using a 90% confidence interval.

8.7.   Meaningful union-management wage negotiations require a precise estimate of current union member wages. A random sample of 60 unionized elevator operators in New York City was selected by a labor arbitrator. The mean and standard deviation of the weekly wage rates for the unionized elevator operators were found to be $147.45 and $11.60, respectively. Find a 95% confidence interval for the average weekly wage for all unionized New York City elevator operators.

8.8.   An audit of the inventory of a retailer was conducted by randomly selecting the purchase invoices from n = 100 unsold units in stock. The average purchase price per unit was found to be'$17.50, with a standard deviation of $6.75.

a. Find a 95% confidence interval for the mean purchase price of all units of inventory held by the retailer.

b. Audit results are interpreted in terms of the stated precision and the materiality of the derived estimate. The precision of the estimate is the level of confidence in the confidence interval, while materiality is synonymous with accuracy—the absolute difference between μ and its estimate $\bar{y}$. Holding the sample size constant, how can the estimate derived in part a be made more precise? With n fixed, how can greater materiality be achieved?

8.8 Estimation from Large Samples

Estimation of a population mean, presented in sections 8.6 and 8.7, sets the stage for the other estimation problems to be discussed in this chapter. Like the sample mean, every estimator presented in this chapter possesses a sampling distribution that is approximately normal, due all, or in part, to the Central Limit Theorem. All are unbiased estimators and the standard deviations of their sampling distributions are known. Consequently, the point estimators and confidence intervals will be constructed in exactly the same way as for the population mean μ.

We will encounter other estimators in later chapters that will possess sampling distributions that are approximately normal but here we will concentrate on four situations that occur frequently in practice. Particularly, we consider inferences concerning population means (sections 8.6 and 8.7), a comparison of two means, a binomial parameter p, and a comparison of two binomial parameters.

Business surveys frequently have as their objective the estimation of a population mean μ or a binomial parameter p. For example, estimating the mean shrinkage of items of merchandise in a department store is an example of estimating a population mean μ. The random selection of a sample of people to determine the proportion who favor a particular household product has as its objective the estimation of a binomial parameter p.

Other business surveys may seek a comparison of these parameters for two or more populations. For example, suppose you plan to purchase one of two automobiles and wish to choose the one that will give the more economical fuel consumption (the greater miles per gallon, mpg). Since you know that the EPA ratings are far from realistic, and that the mpg ratings vary considerably from one automobile to another, you plan to base your decision on a comparison of the mpg measurements for two randomly selected samples of automobiles, n_1 of the first type and n_2 of the second. Your decision would be based on the estimated difference in the mean mpg ratings for the two automobiles. Similarly, you might wish to estimate the difference in the proportions of defectives for two large lots of industrial products based on random samples selected from each.

The four situations described above identify the four estimation problems covered in this chapter. We will give estimates for these situations:

1. A single population mean μ.
2. The difference between two population means.
3. A single population proportion p.
4. The difference between two population proportions.

Since we have already stated that the sampling distributions for these estimators possess the same properties, we can give the general form of the point and interval estimators for all four estimation problems. Although the formulas for the point estimators differ, the bound on the error for each is as given in the box.

Bound on Error of Estimation for a Large-Sample

Point Estimator

bound on error = 2 standard deviations of the sampling
distribution of the point estimator

Similarly, the large-sample confidence intervals for each parameter are as given here.

A $(1 - \alpha)100\%$ Large-Sample Confidence Interval

point estimate $\pm z_{\alpha/2} \times$ (standard deviation of point estimator)

where $z_{\alpha/2}$ is obtained from table 8.1.

The specific formulas for estimating the difference between two means, for estimating a single binomial proportion, and for estimating the difference between two proportions will be presented in sections 8.9, 8.10, and 8.11.

8.9 Estimating the Difference Between Two Means

A problem of equal importance to the estimation of a population mean is the estimation of the difference between two population means. For instance, we might wish to estimate the difference in mean lengths of time required to assemble an industrial product using two different methods of assembly. Assemblers could be randomly divided into two groups, the first using assembly method 1 and the second using method 2. We could then make inferences concerning the difference in mean time to assemble the device for the two assembly methods. Or we might wish to compare the average yield in a chemical plant using raw materials furnished by two suppliers, A and B. Samples of daily yield, one for each of the two raw materials, could be recorded and used to make inferences concerning the difference in mean yields.

For each of these examples, there are two populations, the first with mean and variance μ_1 and σ_1^2, and the second with mean and variance μ_2 and σ_2^2. A random sample of n_1 measurements is drawn from population 1 and n_2 from population 2, where the samples are assumed to have been drawn independently of one another. Finally, the estimates of the population parameters $\bar{y}_1$, s_1^2, $\bar{y}_2$, and s_2^2 are calculated from the sample data.

The point estimator of the difference $(\mu_1 - \mu_2)$ between the population means is $(\bar{y}_1 - \bar{y}_2)$, the difference between the sample means. If repeated pairs of samples of n_1 and n_2 measurements are drawn from the two populations and the estimate $(\bar{y}_1 - \bar{y}_2)$ calculated for each pair, a distribution of estimates will result. What are the properties of this sampling distribution? It can be shown that the sampling distribution of the sum or the difference between two normally distributed sample means is normally distributed, with a mean equal to $(\mu_1 - \mu_2)$, the difference in the means for the two populations.

The standard deviation of the sampling distribution is

$$\sigma_{(\bar{y}_1 - \bar{y}_2)} = \sqrt{\frac{\sigma_1^2}{n_1} + \frac{\sigma_2^2}{n_2}}$$

where σ_1^2 and σ_2^2 are the two population variances and n_1 and n_2 are the corresponding sample sizes. The sampling distribution is shown in figure 8.8 on the next page.

To summarize, the point estimator of the difference between two population means is as given in the box.

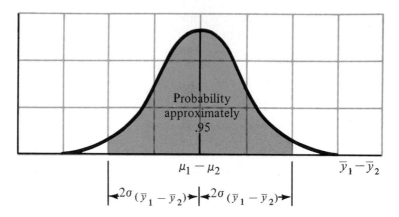

Figure 8.8 *The distribution of $(\bar{y}_1 - \bar{y}_2)$ for large samples*

Point Estimation of $(\mu_1 - \mu_2)$

Estimator: $(\bar{y}_1 - \bar{y}_2)$.

Bound on error: $2\sigma_{(\bar{y}_1 - \bar{y}_2)} = 2\sqrt{\dfrac{\sigma_1^2}{n_1} + \dfrac{\sigma_2^2}{n_2}}$.

Note: If σ_1^2 and σ_2^2 are unknown, but both n_1 and n_2 are 30 or more, you can use the sample variances s_1^2 and s_2^2 to estimate σ_1^2 and σ_2^2.

We illustrate this point estimation procedure with an example.

Example 8.4 A comparison of the wearing quality of two types of automobile tires was obtained by road-testing samples of $n_1 = n_2 = 100$ tires for each type. The number of miles until wear-out was recorded, where wear-out was defined as a specific amount of tire wear. The test results were as follows:

TIRE 1	TIRE 2
$\bar{y}_1 = 26{,}400$ miles	$\bar{y}_2 = 25{,}100$ miles
$s_1^2 = 1{,}440{,}000$	$s_2^2 = 1{,}960{,}000$

Estimate $(\mu_1 - \mu_2)$, the difference in mean miles to wear-out, and place bounds on the error of estimation.

Solution The point estimate of $(\mu_1 - \mu_2)$ is

$$(\bar{y}_1 - \bar{y}_2) = 26{,}400 - 25{,}100 = 1{,}300 \text{ miles}$$

Thus we have

$$\sigma_{(\bar{y}_1 - \bar{y}_2)} = \sqrt{\frac{\sigma_1^2}{n_1} + \frac{\sigma_2^2}{n_2}} \approx \sqrt{\frac{s_1^2}{n_1} + \frac{s_2^2}{n_2}}$$

$$= \sqrt{\frac{1,440,000}{100} + \frac{1,960,000}{100}} = \sqrt{34,000} = 184 \text{ miles}$$

We would expect the error of estimation to be less than 2(184) = 368 miles. If the point estimate of the difference in mean miles to wear-out is 1,300 miles and the error of estimation is less than 368 miles (with a high probability), it seems fairly conclusive that there is a substantial difference in the mean miles to wear-out for the two types of tires. In fact, it would appear that Tire 1 is superior to Tire 2 in wearing quality when subjected to the road test.

Following the procedure of section 8.7, we can construct a confidence interval for $(\mu_1 - \mu_2)$ with confidence coefficient $(1 - \alpha)$.

A $(1 - \alpha)$100% Confidence Interval for $(\mu_1 - \mu_2)$

$$(\bar{y}_1 - \bar{y}_2) \pm z_{\alpha/2} \sqrt{\frac{\sigma_1^2}{n_1} + \frac{\sigma_2^2}{n_2}}$$

As a rule of thumb we will require both n_1 and n_2 to be equal to 30 or more in order that s_1^2 and s_2^2 provide good estimates of their respective population variances.

Example 8.5 Place a confidence interval on the difference in the mean miles to wear-out for the problem described in example 8.4. Use a confidence coefficient of .99.

Solution The confidence interval is

$$(\bar{y}_1 - \bar{y}_2) \pm 2.58 \sqrt{\frac{\sigma_1^2}{n_1} + \frac{\sigma_2^2}{n_2}}$$

Using the results of example 8.4, we find that the confidence interval is

$$1,300 \pm 2.58(184)$$

Therefore, LCL = 825, UCL = 1,775, and the difference in the mean miles to wear-out is estimated to lie between these two points. Note that the confidence interval is wider than the $\pm 2\sigma_{(\bar{y}_1 - \bar{y}_2)}$ used in example 8.4, because we have chosen a larger confidence coefficient.

Exercises

8.9. New EPA noise standards, which apply to all trucks and buses over 10,000 pounds engaged in interstate commerce, limit the amount of noise from the vehicles to 90 decibels at 50 feet in zones with speed limits above 35 miles per hour. Lesser amounts of noise are allowed in slower speed zones.* To determine whether to place initial major emphasis on enforcement of the noise regulation with the trucking industry or with buses, an EPA official conducted an experiment to estimate the difference between trucks and buses in mean noise level measured at 50 feet in zones with speed limits above 35 miles per hour. Fifty trucks and 50 buses of various makes and satisfying the EPA weight condition were subjected to a test. The trucks showed an average noise level of 91.5 decibels, with a standard deviation of 5.2 decibels, while the buses had an average and standard deviation of 88.6 decibels and 3.8 decibels, respectively. Find a 95% confidence interval for the difference in the mean noise recorded for trucks and buses at 50 feet in zones with speed limits above 35 miles per hour.

8.10.   Almost all graduate business schools require a score on the Graduate Management Admission Test (GMAT) from each applicant for admission. The GMAT, like the Graduate Record Examination, measures a student's verbal ability and quantitative ability on separate 800-point scales. Two graduate schools decided to compare the quantitative scores for students admitted to their graduate programs. Random samples of the GMAT scores of 75 students were selected for each school. The average and standard deviations were

$$\text{school 1:} \quad \bar{y}_1 = 525 \qquad s_1 = 52$$

$$\text{school 2:} \quad \bar{y}_2 = 564 \qquad s_2 = 70$$

Estimate the difference in the mean GMAT quantitative scores for the two graduate business schools, and place bounds on the error of estimation.

8.11.   Consumer activism, legislation, and heightened consumer expectations have placed more responsibility on the manufacturer for the performance of his goods. Because of this, a manufacturer of two distinct products is interested in estimating the difference in mean monthly complaints received concerning the two products. Over a period of four years (48 months) the number of complaints concerning each product produced the results shown in the table. Find a 90% confidence interval

PRODUCT 1	PRODUCT 2
$\bar{y}_1 = 17.2$	$\bar{y}_2 = 25.1$
$s_1 = 4.6$	$s_2 = 5.3$

for the difference in the mean monthly complaints received concerning the two products. (Assume that the numbers of complaints for the two products can be regarded as independent random samples.)

8.10 Estimating the Parameter of a Binomial Population

As we have noted, many surveys have as their objective the estimation of the proportion of people or objects in a large group that possess a particular attribute. Such a survey is a practical example of the binomial experiment

*Environmental Protection Agency, *Environmental News*, August 1975.

discussed in chapter 6. Estimating the proportion of sales that can be expected in a large number of customer contacts is a practical problem requiring the estimation of a binomial parameter p.

The best point estimator of the binomial parameter p is also the estimator that would be chosen intuitively. That is, the estimator of p, denoted by the symbol, $\hat{p}$, is the total number y of successes divided by the total number n of trials. That is,

$$\hat{p} = \frac{y}{n}$$

where y is the number of successes in n trials. (Note: A "hat" (ˆ) over a parameter is the symbol used to denote the estimator of the parameter.) By "best" we mean that $\hat{p}$ is unbiased and possesses a smaller variance than other possible estimators.

As noted in section 8.8, the estimator $\hat{p}$ possesses a sampling distribution that is normally distributed because of the Central Limit Theorem. It is an unbiased estimator of the population proportion p, with a standard deviation of

$$\sigma_{\hat{p}} = \sqrt{\frac{pq}{n}}$$

where p is the unknown population proportion, $q = (1 - p)$, and n is the sample size. Then from section 8.8 the point estimation procedure for p is as given in the box.

Point Estimator for p

Estimator: $\hat{p} = \dfrac{y}{n}$.

Bound on error: $2\sigma_{\hat{p}} = 2\sqrt{\dfrac{pq}{n}}$.

The corresponding large-sample confidence interval with confidence coefficient $(1 - \alpha)$ is shown in the box.

A $(1 - \alpha)100\%$ Confidence Interval for p

$$\hat{p} \pm z_{\alpha/2}\sqrt{\frac{\hat{p}\hat{q}}{n}}$$

The sample size will be considered large when we can assume that the distribution of $\hat{p}$ is approximately normal. These conditions were discussed in section 7.5.

The only difficulty encountered in our procedure will be in calculating $\sigma_{\hat{p}}$, which involves p (and $q = 1 - p$), an unknown. You will note that we have substituted $\hat{p}$ for the parameter p in the standard deviation $\sqrt{pq/n}$. When n is large, little error will be introduced by this substitution. As a matter of fact, the standard deviation changes only slightly as p changes. This can be observed in table 8.2, where $\sqrt{pq}$ is recorded for several values of p. Note that $\sqrt{pq}$ changes very little as p changes, especially when p is near .5.

Table 8.2 *Some calculated values of $\sqrt{pq}$*

p	$\sqrt{pq}$
.5	.50
.4	.49
.3	.46
.2	.40
.1	.30

Example 8.6 A random sample of $n = 100$ voters in a community produced $y = 59$ voters in favor of Candidate A. Estimate the fraction of the voting population favoring A, and place a bound on the error of estimation.

Solution The point estimate of p is

$$\hat{p} = \frac{y}{n} = \frac{59}{100} = .59$$

and the bound on the error of estimation is

$$2\sigma_{\hat{p}} = 2\sqrt{\frac{\hat{p}\hat{q}}{n}} = 2\sqrt{\frac{(.59)(.41)}{100}} = .098$$

A 95% confidence interval for p is

$$\hat{p} \pm 1.96\sqrt{\frac{\hat{p}\hat{q}}{n}} = .59 \pm 1.96(.049)$$

Thus we estimate that the interval from .494 to .686 includes p, with confidence coefficient .95.

Exercises

8.12.   The National Labor Relations Board requires that a certain percentage of workers form union affiliation before workers can vote on which union they wish to represent their concerns. Sixty out of 200 workers polled in a nonunion shop favored union affiliation. Estimate the proportion of workers favoring union affiliation, and place a bound on the error of estimation.

8.13.   Generally, feasibility studies include some measure of demand to ascertain the potential profitability of the given product or service. In studying the feasibility of expanding public television programming, an investigator found that 76 of 180 randomly chosen households with television sets watch at least two hours of public television programming per week. Find a 90% confidence interval for p, the proportion of households in the population that watch at least two hours of public television programming per week.

8.14.   In late 1975 many U.S. newspapers contained the following article:*

Nearly 30% Can't Pinpoint Importance of 1776

With the nation about to celebrate its 200th anniversary, nearly 3 Americans in 10 are unable to say what important event occurred in the year 1776. In a random sample of 550 Americans, only 396 respondents were able to correctly identify 1776 as the year our country declared its independence from Great Britain by the signing of the Declaration of Independence.

Based on these findings, estimate the proportion of all Americans who can identify the importance of 1776, and place a bound on the error of estimation.

8.15.   A recent Gallup poll reported that 82% of 1200 residents of greater São Paulo, Brazil, polled in a recent survey,[†] consider air pollution in their city to be "very serious." Find a 95% confidence interval for the proportion of the population of São Paulo who consider air pollution in their city to be very serious.

8.16.   A buyer intention survey conducted by a representative of the New Car Dealers Association of a certain city showed that 20% of a random sample of $n = 1500$ consumers were planning to buy new cars during the coming year. Find a 90% confidence interval for the fraction of consumers in the city planning to buy new cars during the coming year.

8.11 Estimating the Difference Between Two Binomial Parameters

The fourth and final estimation problem considered in this chapter is the estimation of the difference between the parameters of two binomial populations. Assume that the two populations 1 and 2 possess parameters p_1 and p_2, respectively. Independent random samples consisting of n_1 and n_2 trials are drawn from the population and the estimates $\hat{p}_1$ and $\hat{p}_2$ are calculated.

We have already noted in section 8.10 that the sample proportion

---

Oregonian, 30 November 1975.
[†]Environmental Protection Agency, *Environmental News*, July, 1975.

$$\hat{p} = \frac{y}{n}$$

of a random sample of n items from a binomial population possesses a sampling distribution that is approximately normal when n is large because of the Central Limit Theorem. Then, like the difference between two sample means, the difference in the sample proportions

$$(\hat{p}_1 - \hat{p}_2)$$

is approximately normally distributed when both n_1 and n_2 are large. Further, we state (without proof) that the estimator $(\hat{p}_1 - \hat{p}_2)$ is an unbiased estimator of the difference $(p_1 - p_2)$ in the population proportions, and that it possesses a standard deviation of

$$\sigma_{(\hat{p}_1 - \hat{p}_2)} = \sqrt{\frac{p_1 q_1}{n_1} + \frac{p_2 q_2}{n_2}}$$

where p_1 and p_2 are the two population proportions, $q_1 = (1 - p_1)$, $q_2 = (1 - p_2)$, and n_1 and n_2 are the sizes of the samples selected from the two populations.

Using the procedure of section 8.8, the point estimator for $(p_1 - p_2)$ is as given in the box.

Point Estimator of $(p_1 - p_2)$

Estimator: $(\hat{p}_1 - \hat{p}_2)$.

Bound on error: $2\sigma_{(\hat{p}_1 - \hat{p}_2)} = 2\sqrt{\dfrac{p_1 q_1}{n_1} + \dfrac{p_2 q_2}{n_2}}$.

Note: The estimates $\hat{p}_1$ and $\hat{p}_2$ must be substituted for p_1 and p_2 to obtain the bound on the error of estimation.

The $(1 - \alpha)$ confidence interval, appropriate when n_1 and n_2 are large, is shown in the box.

A $(1 - \alpha)100\%$ Confidence Interval for $(p_1 - p_2)$

$$(\hat{p}_1 - \hat{p}_2) \pm z_{\alpha/2} \sqrt{\frac{\hat{p}_1 \hat{q}_1}{n_1} + \frac{\hat{p}_2 \hat{q}_2}{n_2}}$$

Example 8.7 A manufacturer of fly sprays wishes to compare two new concoctions, 1 and 2. Two rooms of equal size, each containing 1000 flies, are employed in the experiment, one treated with Fly Spray 1 and the other treated with an equal amount of Fly Spray 2. Totals of 825 and 760 flies succumb to sprays 1 and 2, respectively. Estimate the difference in the rate of kill for the two sprays when used in the test environment.

Solution The point estimate of $(p_1 - p_2)$ is

$$(\hat{p}_1 - \hat{p}_2) = .825 - .760 = .065$$

The bound on the error of estimation is

$$2 \sqrt{\frac{p_1 q_1}{n_1} + \frac{p_2 q_2}{n_2}} \approx 2 \sqrt{\frac{(.825)(.175)}{1000} + \frac{(.76)(.24)}{1000}} = .036$$

The corresponding confidence interval, using confidence coefficient .95, is

$$(\hat{p}_1 - \hat{p}_2) \pm 1.96 \sqrt{\frac{\hat{p}_1 \hat{q}_1}{n_1} + \frac{\hat{p}_2 \hat{q}_2}{n_2}}$$

The resulting confidence interval is

$$.065 \pm .035$$

Hence we estimate that the difference $(p_1 - p_2)$ between the rates of kill falls in the interval from .030 to .100. That is, we estimate that p_1 exceeds p_2 by as little as .030 or as much as .100. We are fairly confident of this estimate, because we know that if our sampling procedure were repeated over and over again, each time generating an interval estimate, approximately 95% of the estimates would enclose the quantity $(p_1 - p_2)$.

Exercises

8.17. In a poll taken among the stockholders of a company, 300 of 500 men favored the decision to adopt a new product line, whereas 64 of 100 women stockholders favored the new product. Estimate the difference in the fractions favoring the decision, and place a bound on the error of estimation.

8.18. Participative decision making as a managerial strategy has been advocated as a means of improving both the performance and participation of individuals in organizations. Two groups of employees, which differed substantially in the opportunity for employee participation in decision making, were asked whether or not they were satisfied with their present jobs. Seventy-seven of 110 employees from a group where employee participation was encouraged indicated they were satisfied with their jobs, whereas 52 of 125 employees from a group where employee participation was discouraged indicated they were satisfied with their jobs.

a. Estimate the difference in the fraction of employees satisfied with their jobs, and place a bound on the error of estimation.

b. Find a 90% confidence interval for the difference in the proportion of employees who are satisfied with their jobs.

8.19. Tomeski and Lazarus have reported on the use of computerized information systems for the management of personnel records by state and local governments.* Suppose a similar study shows that 33 of 40 city governments selected use computerized information systems for the management of personnel records, while

*E. A. Tomeski and H. Lazarus, "Computerized Information Systems in Personnel," *Academy of Management Journal*, March 1974.

25 of 32 state governments are found to use such a system. Estimate the difference in the fraction of city and state governments in the country that use computerized information systems for the management of personnel records; use a 95% confidence interval.

8.12 Choosing the Sample Size

The design of an experiment is essentially a plan for purchasing a quantity of information. This information, like any other commodity, may be acquired at varying prices, depending on the manner in which the data are obtained. Some measurements contain a large amount of information about the parameter of interest; others may contain little or none. Since the sole product of research is information, we should try to purchase it at minimum cost.

The sampling procedure, or experimental design as it is usually called, affects the quantity of information per measurement. This, along with the sample size n, controls the total amount of relevant information in a sample. With few exceptions we will be concerned with the simplest sampling situation—random sampling from a relatively large population—and will devote our attention to the selection of the sample size n.

The researcher makes little progress in planning an experiment before encountering the problem of selecting the sample size. Indeed, perhaps one of the most frequent questions asked of the statistician is, how many measurements should be included in the sample? Unfortunately, the statistician cannot answer this question without knowing how much information the experimenter wishes to buy. Certainly the total amount of information in the sample will affect the measure of goodness of the method of inference and must be specified by the experimenter. Referring specifically to estimation, we would like to know how accurate the experimenter wishes his estimate to be. This may be stated by specifying a bound on the error of estimation.

For instance, suppose that we wish to estimate the average daily yield μ of a chemical (see example 8.1), and we wish the error of estimation to be less than 4 tons with a probability of .95. Since approximately 95% of the sample means will lie within $2\sigma_{\bar{y}}$ of μ in repeated sampling, we are asking that $2\sigma_{\bar{y}}$ equal 4 tons (see figure 8.9). Then

$$2\sigma_{\bar{y}} = 4 \qquad \text{or} \qquad \frac{2\sigma}{\sqrt{n}} = 4$$

Solving for n we obtain

$$n = \frac{\sigma^2}{4}$$

This gives the minimum sample size such that the error of estimation will be less than $2\sigma_{\bar{y}}$ with probability near .95.

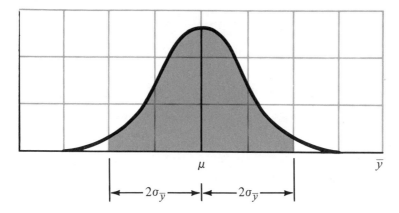

Figure 8.9 *Approximate distribution of $\bar{y}$ for large samples*

You will quickly note that we cannot obtain a numerical value for n unless the population standard deviation σ is known. And certainly this is exactly what we should expect, because the variability of the sample mean $\bar{y}$ depends on the variability of the population from which the sample was drawn. Lacking an exact value for σ, we would use the best approximation available, such as an estimate s obtained from a previous sample or knowledge of the range in which the measurements fall. Since the range is approximately equal to 4σ (the Empirical Rule), one-fourth of the range provides an approximate value for σ. For our example we would use the results of example 8.1, which provided a reasonably accurate estimate of σ equal to $s = 21$. Then

$$n = \frac{\sigma^2}{4} \approx \frac{(21)^2}{4} = 110.25 \approx 111$$

Using a sample size $n = 111$, we would be reasonably certain (with probability approximately equal to .95) that our estimate lies within $2\sigma_{\bar{y}} = 4$ tons of the true average daily yield.

Actually we should expect the error of estimation to be much less than 4 tons. According to the Empirical Rule, the probability is approximately equal to .68 that the error of estimation is less than $\sigma_{\bar{y}} = 2$ tons. You will note that the probabilities .95 and .68 used in these statements are inexact, since s was substituted for σ. Although this method of choosing the sample size is only approximate for a specified desired accuracy of estimation, it is the best available and is certainly better than selecting the sample size on the basis of our intuition.

The method of choosing the sample size for all the large-sample estimation procedures discussed in preceding sections is identical to that described above. The experimenter must specify a desired bound on the error of estimation and an associated confidence level $(1 - \alpha)$.

Procedure for Choosing the Sample Size

Let θ (Greek letter theta) be the parameter to be estimated and let $\sigma_{\hat{\theta}}$ be the standard deviation of the point estimator. Then proceed as follows:

1. Choose *B*, the bound on the error of estimation, and a confidence coefficient $(1 - \alpha)$.
2. Solve the following equation for the sample size *n*:

$$z_{\alpha/2}/\sigma_{\hat{\theta}} = B$$

Note: For most estimators (all presented in this text), σ_{θ} is a function of the sample size *n*.

We illustrate these ideas with two examples.

Example 8.8 One of the more baffling problems of organizations that depend on consumer information to direct organizational policy is the failure of selected consumers to respond to a survey questionnaire. Chervany and Heinlen have reported their experience with pilot surveys to estimate the response rate for different versions of a questionnaire.* Suppose an organizational executive would like to estimate the response rate for a newly designed survey questionnaire by observing the response rate in a pilot survey. How many consumers should be selected for the pilot survey if the executive wishes to estimate *p*, the response rate, with a probability of .90 that the error of estimation is less than .06? Assume that he expects *p* to lie somewhere in the neighborhood of .6.

Solution Since the confidence coefficient is $(1 - \alpha) = .90$, α must equal .10, and $\alpha/2 = .05$. The *z*-value corresponding to an area of .05 in the upper tail of the *z* distribution is $z_{\alpha/2} = 1.645$. We then require

$$1.645\sigma_{\hat{p}} = .06 \qquad \text{or} \qquad 1.645\sqrt{\frac{pq}{n}} = .06$$

Since the variability of $\hat{p}$ is dependent on *p*, which is unknown, we must use the guessed value of $p = .6$, which is provided by the executive as an approximation. Then

$$1.645\sqrt{\frac{(.6)(.4)}{n}} = .06$$

Solving for *n* we obtain

$$\sqrt{n} = \frac{1.645}{.06}\sqrt{(.6)(.4)} = 13.43$$

$$n = 180.4 \approx 181$$

*N. L. Chervany and J. S. Heinlen, "The Structure of a Student Project Course," *Decision Sciences*, January 1975.

That is, the pilot survey should include 181 consumers.

Example 8.9 An experimenter wishes to compare the effectiveness of two methods of training industrial employees to perform a certain assembly operation. Selected employees are to be divided into two equal groups, the first receiving training method 1 and the second, training method 2. Each will perform the assembly operation, and the length of assembly time will be recorded. It is expected that the measurements for both groups will have a range of approximately 8 minutes. If the estimate of the difference in mean time to assemble is desired correct to within 1 minute, with probability approximately equal to .95, how many workers must be included in each training group?

Solution Equating $2\sigma_{(\bar{y}_1 - \bar{y}_2)}$ to 1 minute, we obtain

$$2\sqrt{\frac{\sigma_1^2}{n_1} + \frac{\sigma_2^2}{n_2}} = 1$$

Since we desire n_1 to equal n_2, we may let $n_1 = n_2 = n$ and obtain the equation

$$2\sqrt{\frac{\sigma_1^2}{n} + \frac{\sigma_2^2}{n}} = 1$$

As noted above, the variability of each method of assembly is approximately the same, and hence $\sigma_1^2 = \sigma_2^2 = \sigma^2$. Since the range, equal to 8 minutes, is approximately equal to 4σ, then we have

$$4\sigma \approx 8$$

$$\sigma \approx 2$$

Substituting this value for σ_1 and σ_2 in the above equation, we obtain

$$2\sqrt{\frac{(2)^2}{n} + \frac{(2)^2}{n}} = 1$$

$$\sqrt{n} = 2\sqrt{8}$$

or $n = 32$. Thus each group should contain $n = 32$ members.

Exercises

8.20. Faulty products are costly to a manufacturer in terms of replacement costs and loss of product image among the consuming public. A manufacturer of portable tape recorders believes that no more than 10% of the products produced by the firm are faulty. If the manufacturer wishes to estimate the actual fraction of faulty recorders to within .03, how large a sample of finished recorders should be selected?

8.21. The maintenance of charge accounts may become too costly if the average account purchase falls below a certain level. A department store manager would like to estimate the average amount of account purchases per month by its customers with charge accounts to within $2.50, with a probability of approximately

.95. How many accounts should be selected from the store's records if the standard deviation of monthly account balances is known to be $7.50?

8.22.   To bid competitively for the lumbering rights on a certain tract of land, a wood products firm needs to know the mean diameter of the trees on the tract to within 2 inches, with a probability of .90. How large a sample of trees should be selected from the tract to estimate the mean diameter if it is known that the diameters of usable timber on the tract range from 10 inches to 38 inches?

8.23.   Are subcompacts noticeably less expensive to operate than compact cars? According to an article in *U.S. News and World Report*, the answer is, not necessarily.* To explore this issue a private testing agency wishes to estimate the difference in the average operating cost per mile between subcompacts and compacts, accurate to within 1 cent per mile, with a probability of approximately .95. If it is known from past experience that the standard deviation of total operating costs per mile (including depreciation, maintenance, gas and oil, insurance, and taxes) is about 3 cents, how many cars should the testing agency include in each group? (Assume the same number of cars will be used in each group.)

8.13 A Statistical Test of an Hypothesis

The basic reasoning employed in a statistical test of an hypothesis was outlined in section 6.8 in connection with the test of the effectiveness of a cold vaccine. In this section we will attempt to condense the basic points involved and we refer you to section 6.8 for an intuitive presentation of the subject.

Parts of a Statistical Test

The objective of a statistical test is to test an hypothesis concerning the values of one or more population parameters. A statistical test involves four elements:

1. Null hypothesis.
2. Alternative hypothesis.
3. Test statistic.
4. Rejection region.

The specification of these four elements defines a particular test; changing one or more of the parts creates a new test.

The relationship between the null and the alternative hypotheses was discussed in section 6.9. The **alternative (or research) hypothesis** is the hypothesis

---

*"How Much Does Car Ownership Really Cost?" *U.S. News and World Report*, November 1975.

that the researcher wishes to support. The **null hypothesis** is a contradiction of the alternative hypothesis; that is, if the null hypothesis is false, the research hypothesis must be true. For reasons you will subsequently see, it is easier to show support for the research hypothesis by presenting evidence (sample data) that indicates that the null hypothesis is false. Thus we are building a case in support of the research hypothesis using a method that is analogous to proof by contradiction.

Even though we wish to gain evidence in support of the alternative hypothesis (denoted by the symbol H_a), the null hypothesis (indicated by the symbol H_0) is the hypothesis to be tested. Thus H_0 will specify hypothesized values for one or more population parameters. For example, we might wish to test the null hypothesis that a population mean is equal to 50, hoping to show, in fact, that the mean exceeds 50. Or we might wish to test the null hypothesis that two population means, say μ_1 and μ_2, are equal, hoping to show, perhaps, that μ_1 is larger that μ_2.

The decision to reject or accept the null hypothesis is based on information contained in a sample drawn from the population of interest. The sample values are used to compute a single number, corresponding to a point on a line, which operates as a decision maker. This decision maker is called the **test statistic**. The entire set of values that the test statistic may assume is divided into two sets, or regions, one corresponding to the **rejection region** and the other to the **acceptance region**. If the test statistic computed from a particular sample assumes a value in the rejection region, the null hypothesis is rejected and the alternative hypothesis H_a (the research hypothesis) is accepted. If the test statistic falls in the acceptance region, either the null hypothesis is accepted or the test is judged to be inconclusive. The circumstances leading to this latter decision are explained subsequently.

The decision procedure described above is subject to two types of errors, which are prevalent in a two-choice decision problem. We may reject the null hypothesis H_0 when, in fact, it is true, or we may accept H_0 when it is false and some alternative hypothesis is true. These errors are called the type I and type II errors, respectively, for the statistical test.

Definition

A **type I error** for a statistical test is the error made by rejecting the null hypothesis when it is true. The probability of making a type I error is denoted by the symbol α.

A **type II error** for a statistical test is the error made by accepting (not rejecting) the null hypothesis when it is false and some alternative hypothesis is true. The probability of making a type II error is denoted by the symbol β.

The two possibilities for the null hypothesis, that is, true or false, along

Table 8.3 *Decision table*

	NULL HYPOTHESIS	
DECISION	True	False
reject H_0	type I error	correct decision
accept H_0	correct decision	type II error

with the two decisions the experimenter can make, are indicated in the two-way table, table 8.3. The occurrences of the type I and type II errors are indicated in the appropriate cells.

The goodness of a statistical test of an hypothesis is measured by the probabilities of making a type I or a type II error, denoted by the symbols α and β, respectively. These probabilities, calculated for the elementary statistical tests presented in the exercises for chapter 6, illustrate the basic relationship among α, β, and the sample size n. Since α is the probability that the test statistic falls in the rejection region, assuming H_0 to be true, an increase in the size of the rejection region increases α and, at the same time, decreases β for a fixed sample size. Reducing the size of the rejection region decreases α and increases β. If the sample size n is increased, more information is available upon which to base the decision and both α and β will decrease.

The probability β of making a type II error varies depending upon the true value of the population parameter. For instance, suppose that we wish to test the null hypothesis that the binomial parameter p is equal to p_0 = .4. (We will use a subscript 0 to indicate the parameter value specified in the null hypothesis H_0.) Furthermore, suppose that H_0 is false and that p is really equal to an alternative value, say p_a. Which will be more easily detected, a p_a = .4001 or a p_a = 1.0? Certainly, if p is really equal to 1.0, every single trial will result in a success and the sample results will produce strong evidence to support a rejection of H_0: p_0 = .4. On the other hand, p_a = .4001 lies so close to p_0 = .4 that it would be extremely difficult to detect p_a without a very large sample. In other words, the probability β of accepting H_0 will vary depending on the difference between the true value of p and the hypothesized value p_0. A graph of the probability β of a type II error as a function of the true value of the parameter is called the **operating characteristic curve** for the statistical test. Note that the operating characteristic curves for the lot acceptance sampling plans in chapter 6 were really graphs expressing β as a function of p.

Since the rejection region is specified and remains constant for a given test, α also remains constant and, as in a lot acceptance sampling, the operating characteristic curve describes the characteristics of the statistical test. An increase in the sample size n will decrease β and reduce its value for all alternative values of the parameter tested. Thus we possess an operating characteristic curve corresponding to each sample size. This property of the operating characteristic curve was illustrated in the exercises for chapter 6.

Ideally, you will have in mind some values α and β which measure the risks of the respective errors you are willing to tolerate. You will also have in mind some deviation from the hypothesized value of the parameter, which you consider of practical importance and wish to detect. The rejection region for the test will then be located in accordance with the specified value of α. Finally, you choose the sample size necessary to achieve an acceptable value of β for the specified deviation that you wish to detect. This could be done by consulting the operating characteristic curves, corresponding to various sample sizes, for the chosen test.

In practice, β is often unknown, either because it was never computed before the test was conducted or because it may be extremely difficult to compute for the test. Then, rather than accepting the null hypothesis when the test statistic falls in the acceptance region, you should withhold judgment. That is, you should not accept the null hypothesis unless you know the risk (measured by β) of making an incorrect decision. Notice that you will never be faced with this "no conclusion" situation when the test statistic falls in the rejection region. Then you can reject the null hypothesis (and accept the alternative hypothesis) because you always know the value of α, the probability of rejecting the null hypothesis when it is true. The fact that β is often unknown explains why we attempt to support the alternative hypothesis by rejecting the null hypothesis. When we reach this decision, the probability α that such a decision is incorrect is known.

We will observe in the next section that the alternative hypothesis H_a assists in the location of the rejection region.

8.14 A Large-Sample Statistical Test

Large-sample tests of hypotheses concerning the population parameters μ, p, $(\mu_1 - \mu_2)$, and $(p_1 - p_2)$ are based on a normally distributed test statistic and for that reason may be regarded as one and the same test. We will present the reasoning involved in the test in a very general manner, referring to the parameter of interest as θ. Thus we could imagine θ as representing μ, $(\mu_1 - \mu_2)$, p, or $(p_1 - p_2)$. The specific test for each parameter will be illustrated by examples.

Suppose that we wish to test an hypothesis concerning a parameter θ and that an unbiased point estimator $\hat{\theta}$ is available and known to be normally distributed, with standard deviation $\sigma_{\hat{\theta}}$. If the null hypothesis

$$H_0: \theta = \theta_0$$

is true, then the sampling distribution for $\hat{\theta}$ is normally distributed, with a mean equal to θ_0, as shown in figure 8.10.

Suppose that the objective of our experiment is to gain evidence to show that θ is greater than θ_0. The alternative hypothesis is $H_a: \theta > \theta_0$, and we would reject the null hypothesis H_0 when $\hat{\theta}$ is too large. "Too

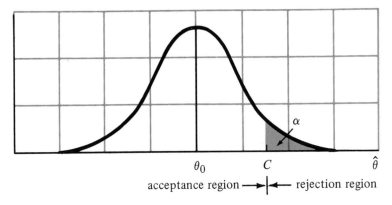

Figure 8.10 *Distribution of $\hat{\theta}$ when H_0 is true*

large," of course, means too many standard deviations, $\sigma_{\hat{\theta}}$, away from θ_0. The rejection region for the test is shown in figure 8.10. The value of $\hat{\theta}$ that separates the rejection and acceptance regions, denoted by the symbol C, is called the **critical value** of the test statistic. The probability of rejecting, assuming the null hypothesis to be true, is equal to the area under the normal curve lying above the rejection region. This area α is shaded in figure 8.10. Thus if we desire $\alpha = .05$, we would reject the null hypothesis when $\hat{\theta}$ is more than $1.645\sigma_{\hat{\theta}}$ to the right of θ_0.

Definition

A **One-tailed statistical test** is one that locates the rejection region in only one tail of the sampling distribution of the test statistic. To detect $\theta > \theta_0$ place the rejection region in the upper tail of the distribution of $\hat{\theta}$. To detect $\theta < \theta_0$ place the rejection region in the lower tail of the distribution of $\hat{\theta}$.

Figure 8.11 *Distribution of $\hat{\theta}$ when H_0 is false and $\theta = \theta_a$*

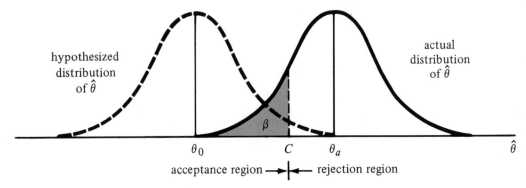

If we wish to detect departures either greater than or less than θ_0, the alternative hypothesis is

$$H_a: \theta \neq \theta_0$$

that is,

$$\theta > \theta_0 \quad \text{or} \quad \theta < \theta_0$$

In this case the probability α of a type I error is equally divided between the two tails of the normal distribution, resulting in a *two-tailed statistical test*.

Definition

A **two-tailed statistical test** is one that locates the rejection region in both tails of the sampling distribution of the test statistic. Two-tailed tests are used to detect either $\theta > \theta_0$ or $\theta < \theta_0$.

The calculation of β for the one-tailed statistical test described previously can be facilitated by considering figure 8.11. When H_0 is false and $\theta = \theta_a$, the test statistic $\hat{\theta}$ will be normally distributed about a mean θ_a rather than θ_0. The distribution of $\hat{\theta}$, assuming $\theta = \theta_a$, is shown by the solid normal curve. The hypothesized distribution of $\hat{\theta}$ (that is, the distribution of $\hat{\theta}$ under the null hypothesis), shown by a dashed normal curve, is used to locate the rejection region and the critical value of $\hat{\theta}$, C. Since β is the probability of accepting H_0, given $\theta = \theta_a$, β is equal to the area under the solid curve located above the acceptance region. This area, which is shaded, could be calculated by using the methods described in chapter 7.

You will note that all the point estimators discussed in the preceding section satisfy the requirements of the test described above when the sample size n is large. That is, the sample size must be large enough so that the point estimator is approximately normally distributed, according to the Central Limit Theorem, and permits a reasonably good estimate of its standard deviation. We may therefore test hypotheses concerning μ, p, $(\mu_1 - \mu_2)$, and $(p_1 - p_2)$.

The mechanics of testing are simplified by using

$$z = \frac{\hat{\theta} - \theta_0}{\sigma_{\hat{\theta}}}$$

as a test statistic, as noted in example 7.11. Note that z is simply the deviation of a normally distributed random variable $\hat{\theta}$ from θ_0 expressed in units of $\sigma_{\hat{\theta}}$. Thus for a two-tailed test with $\alpha = .05$, we would reject H_0 when $z > 1.96$ or $z < -1.96$, since $P(z < -1.96 \text{ or } z > 1.96) = .05$ when H_0 is true.

As we previously stated, the method of inference used in a given situation often depends on the preference of the experimenter. Some people wish to express an inference as an estimate; others prefer to test an hypothesis concerning the parameter of interest.

The following example demonstrates the use of the z test in testing an hypothesis concerning a population mean and, at the same time, it illustrates the close relationship between the statistical test and the large-sample confidence intervals discussed in the preceding sections.

Example 8.10   Refer to example 8.1. Test the hypothesis that the average daily yield of the chemical is $\mu = 880$ tons per day against the alternative that μ is either greater or less than 880 tons per day. The sample (example 8.1), based on $n = 50$ measurements, yielded $\bar{y} = 871$ and $s = 21$ tons.

Solution   The null hypothesis and alternative hypothesis are

$$H_0: \mu = 880 \quad \text{and} \quad H_a: \mu \neq 880$$

The point estimate for μ is $\bar{y}$. Therefore, the test statistic is

$$z = \frac{\bar{y} - \mu_0}{\sigma_{\bar{y}}} = \frac{\bar{y} - \mu_0}{\sigma/\sqrt{n}}$$

Using s to approximate σ, we obtain

$$z = \frac{871 - 880}{21/\sqrt{50}} = -3.03$$

For $\alpha = .05$ the rejection region is $z > 1.96$ or $z < -1.96$. Since the calculated value -3.03 of z falls in the rejection region, we reject the hypothesis that $\mu = 880$ tons and conclude that it is less. The probability of rejecting, assuming H_0 to be true, is $\alpha = .05$. Hence we are reasonably confident that our decision is correct.

The statistical test based on a normally distributed test statistic, with a given α, and the $(1 - \alpha)100\%$ confidence interval of section 8.7 are related. The interval $(\bar{y} \pm 1.96\sigma/\sqrt{n})$ or approximately (871 ± 5.82) for example 8.10, is constructed so that in repeated sampling $(1 - \alpha)$ of the intervals will enclose μ. Noting that $\mu = 880$ does not fall in the interval, we would be inclined to reject $\mu = 880$ as a likely value and conclude that the mean daily yield was, indeed, less.

There is another similarity between this test and the confidence interval of section 8.7. The test is "approximate" because we substituted s, an approximate value, for σ. That is, the probability α of a type I error selected for the test is not .05. It is close to .05, close enough for all practical purposes, but it is not exact. This will be true for most statistical tests because one or more assumptions will not be satisfied exactly.

The following example demonstrates the calculation of β for the statistical test of example 8.10.

Example 8.11 Referring to example 8.10, calculate the probability β of accepting H_0 when μ is actually equal to 870 tons.

Solution The acceptance region for the test of example 8.10 is located in the interval $(\mu_0 \pm 1.96\sigma_{\bar{y}})$. Substituting numerical values, we obtain

$$880 \pm 1.96 \frac{21}{\sqrt{50}} \quad \text{or} \quad 874.18 \text{ to } 885.82$$

The probability of accepting H_0, given $\mu = 870$, is equal to the area under the frequency distribution for the test statistic $\bar{y}$ in the interval from 874.18 to 885.82. Since $\bar{y}$ is normally distributed with a mean of 870 and $\sigma_{\bar{y}} = 21/\sqrt{50} = 2.97$, β is equal to the area under the normal curve located between 874.18 and 885.82 (see figure 8.12). Calculating the z-values corresponding to 874.18 and 885.82, we obtain

$$z_1 = \frac{\bar{y} - \mu}{\sigma/\sqrt{n}} = \frac{874.18 - 870}{21/\sqrt{50}} = 1.41$$

$$z_2 = \frac{\bar{y} - \mu}{\sigma/\sqrt{n}} = \frac{885.82 - 870}{21/\sqrt{50}} = 5.33$$

Then

$$\beta = P(\text{accept } H_0 \text{ when } \mu = 870) = P(874.18 < \bar{y} < 885.82 \text{ when } \mu = 870)$$

$$= P(1.41 < z < 5.33)$$

You can see from figure 8.12 that the area under the normal curve above $\bar{y} = 885.82$ (or $z = 5.33$) is negligible. Therefore,

$$\beta = P(z > 1.41)$$

From table 3 in the appendix, we find that the area between $z = 0$ and $z = 1.41$ is .4207. Therefore, $\beta = .5 - .4207 = .0793$.

Figure 8.12 *Calculating β in example 8.11*

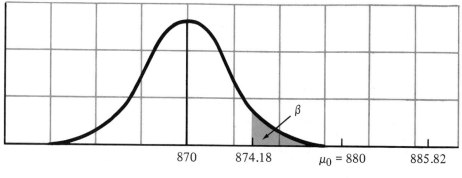

Thus the probability of accepting H_0, given that μ is really equal to 870, is .0793 or, approximately, 8 chances in 100.

You can see that if you choose to calculate β for a value of μ_a closer to $\mu_0 = 880$, the distribution shown in figure 8.12 will shift to the right and β will increase. Conversely, the greater the distance between μ_a and μ_0, the smaller will be the value of β. That is, you make fewer mistakes in deciding that the mean differs from μ_0 (H_0 is false) when the difference between μ_a and μ_0 is large.

Example 8.12 It is known that approximately 1 in 10 smokers favors cigarette Brand A. After a promotional campaign in a given sales region, 200 cigarette smokers were interviewed to determine the effectiveness of the campaign. The result of this sample survey showed that a total of 26 people expressed a preference for Brand A. Do these data present sufficient evidence to indicate an increase in the acceptance of Brand A in the region? (Note that for all practical purposes, this problem is identical to the vegetable oil problem in example 7.10.)

Solution It is assumed that the sample satisfies the requirements of a binomial experiment. The question posed may be answered by testing the hypothesis

$$H_0: p = .10 \text{ (the program is ineffective)}$$

against the alternative

$$H_a: p > .10 \text{ (the program is effective)}$$

A one-tailed statistical test will be utilized, because we are primarily concerned with detecting a value of p greater than .10. For this situation it can be shown that the probability β of a type II error is minimized by placing the entire rejection region in the upper tail of the distribution of the test statistic.

The point estimator of p is $\hat{p} = y/n$ and the test statistic is

$$z = \frac{\hat{p} - p_0}{\sigma_{\hat{p}}} = \frac{\hat{p} - p_0}{\sqrt{p_0 q_0 / n}}$$

(Note: This test statistic is algebraically equivalent to

$$z = \frac{y - np_0}{\sqrt{np_0 q_0}}$$

the test statistic used in example 7.10.)

Once again we require a value of p so that $\sigma_{\hat{p}} = \sqrt{pq/n}$, which appears in the denominator of z, may be calculated. Since we have hypothesized that $p = p_0$, it seems reasonable to use p_0 as an approximation for p. Note that this differs from the estimation procedure where, lacking knowledge of p, we chose $\hat{p}$ as the best approximation. This apparent inconsistency will have a negligible effect on the inference, whether it is the result of a test or an estimation, when n is large.

Choosing $\alpha = .05$, we will reject H_0 when $z > 1.645$. Substituting the numerical values into the test statistic, we obtain

$$z = \frac{\hat{p} - p_0}{\sqrt{\dfrac{p_0 q_0}{n}}} = \frac{.13 - .10}{\sqrt{\dfrac{(.10)(.90)}{200}}} = 1.41$$

The calculated value $z = 1.41$ does not fall in the rejection region and hence we do not reject H_0, the hypothesis that the proportion of smokers of cigarette Brand A is .10. There is not sufficient evidence to indicate that the advertising campaign has been effective and that the percentage of smokers has increased as a result of the advertising effort.

Do we accept H_0? No, not until we have stated some alternative value of p that is larger than $p_0 = .10$ and is considered to be of practical significance. The probability β of a type II error should be calculated for this alternative. If β is sufficiently small, we would accept H_0 and would do so with the risk of an erroneous decision fully known.

Examples 8.10 and 8.12 illustrate an important point. If the data present sufficient evidence to reject H_0, the probability α of an erroneous conclusion is known in advance because α is used in locating the rejection region. Since α is usually small, we are fairly certain that we have made a correct decision. On the other hand, if the data present insufficient evidence to reject H_0, the conclusions are not so obvious. It does not necessarily follow that p is equal to p_0 when we fail to reject the null hypothesis. Rather, we conclude that there is insufficient evidence in the sample to indicate that p is greater than p_0. In example 8.12, at the $\alpha = .05$ significance level, we would have failed to reject any hypothesized value for p greater than $p_0 = .095$. We could have assumed that p is not significantly greater than .095, .097, .10, .13, or any other value greater than .095. However, only one of these values, $p = .10$, has a special importance in the context of the problem. Ideally, following the statistical test procedures outlined in section 8.13, we would have specified a practically significant alternative p_a in advance and chosen n such that β would be small. Unfortunately, many experiments are not conducted in this ideal manner. Someone chooses a sample size and the experimenter or statistician is left to evaluate the evidence.

The calculation of β is not too difficult for the statistical test procedures outlined in this section but may be extremely difficult, if not beyond the capability of the beginner, in other test situations. A much simpler procedure is **to not reject** H_0 rather than to accept it; then **estimate** p using a confidence interval. The interval will give a range of possible values for p.

Example 8.13 With hopes of reducing its massive 350-million-gallons-a-year gasoline bill, the U.S. Postal Service has ordered a number of electric mail trucks on an experimental basis. Officials estimate that savings will result if they use electric trucks on flat routes and use a special off-peak-hours rate for recharging. In an experiment to compare operating costs, $n_1 = 100$ conventional gasoline mail trucks and $n_2 = 100$ electric mail trucks were placed in service for a period of time. The cost per mile for operating the gasoline trucks averaged $\bar{y}_1 = 6.70$ cents per mile with variance $s_1^2 = .36$, while for the electric trucks the mean and variance of the cost per mile was $\bar{y}_2 = 6.54$ and $s_2^2 = .40$, respectively. Do these data present sufficient evidence to indicate a difference in the mean operating cost between conventional gasoline and electric-powered mail trucks?

Solution We wish to test the null hypothesis that the difference $(\mu_1 - \mu_2)$ between two population means equals some specified value, say D_0. (For our example we would hypothesize that $D_0 = 0$, that is, there is no difference in mean operating costs for the two types of trucks.)

Recall that $(\bar{y}_1 - \bar{y}_2)$ is an unbiased point estimator of $(\mu_1 - \mu_2)$, which will be approximately normally distributed in repeated sampling when n_1 and n_2 are large. Furthermore, the standard deviation of $(\bar{y}_1 - \bar{y}_2)$ is

$$\sigma_{(\bar{y}_1 - \bar{y}_2)} = \sqrt{\frac{\sigma_1^2}{n_1} + \frac{\sigma_2^2}{n_2}}$$

Then

$$z = \frac{(\bar{y}_1 - \bar{y}_2) - D_0}{\sqrt{\dfrac{\sigma_1^2}{n_1} + \dfrac{\sigma_2^2}{n_2}}}$$

will serve as a test statistic when σ_1^2 and σ_2^2 are known or when s_1^2 and s_2^2 provide a good approximation for σ_1^2 and σ_2^2 (that is, when n_1 and n_2 are larger than 30).

For our example, we have

$$H_0: \mu_1 - \mu_2 = D_0 = 0$$
$$H_a: \mu_1 - \mu_2 \neq 0$$

Substituting the numerical values into the formula for the test statistic, we obtain

$$z = \frac{6.70 - 6.54}{\sqrt{\dfrac{.36}{100} + \dfrac{.40}{100}}} = 1.83$$

Using a two-tailed test with $\alpha = .05$, we will reject H_0 when $z > 1.96$ or $z < -1.96$. Since z does not fall in the rejection region, we do not reject the null hypothesis.

Note, however, that if we choose $\alpha = .10$, the rejection region will be $z > 1.645$ or $z < -1.645$, and the null hypothesis will be rejected.

The decision to reject or accept depends on the risk that we are willing to tolerate. In example 8.13, if we choose $\alpha = .05$, the null hypothesis is not rejected, but we could not accept H_0 (that is, $\mu_1 = \mu_2$) without investigating the probability of a type II error.

On the other hand, if α were chosen equal to .10, the null hypothesis would be rejected. With no other information given, we would be inclined to reject the null hypothesis that there is no difference in the mean operating cost for conventional gasoline and electric-powered mail trucks. The chance of rejecting H_0, assuming H_0 true, is only α = .10, and hence we would be inclined to think that we had made a reasonably good decision.

Example 8.14 Hospital administrators are often responsible for gathering and computing certain medical statistics that are vital to doctors and hospital policymakers. The records of a particular hospital show that 52 men in a sample of 1000 men versus 23 women in a sample of 1000 women were admitted because of heart disease. Do these data present sufficient evidence to indicate a higher rate of heart disease among men admitted to the hospital?

Solution We will assume that the number of patients admitted for heart disease follows approximately a binomial probability distribution for both men and women, with parameters p_1 and p_2, respectively. Stated generally, we wish to test the hypothesis that a difference exists between p_1 and p_2, say $(p_1 - p_2) = D_0$. (For our example we wish to test the hypothesis that $D_0 = 0$.) Recall that for large samples the point estimator of $(p_1 - p_2)$, namely $(\hat{p}_1 - \hat{p}_2)$, is approximately normally distributed in repeated sampling, with a mean of $(p_1 - p_2)$ and a standard deviation of

$$\sigma_{(\hat{p}_1 - \hat{p}_2)} = \sqrt{\frac{p_1 q_1}{n_1} + \frac{p_2 q_2}{n_2}}$$

Then

$$z = \frac{(\hat{p}_1 - \hat{p}_2) - (p_1 - p_2)}{\sigma_{(\hat{p}_1 - \hat{p}_2)}}$$

possesses a standardized normal distribution in repeated sampling. Hence z can be employed as a test statistic to test

$$H_0 : (p_1 - p_2) = D_0$$

when suitable approximations are used for p_1 and p_2, which appear in $\sigma_{(\hat{p}_1 - \hat{p}_2)}$. Approximations are available for two cases, as discussed below.

Case 1. If we hypothesize that p_1 equals p_2, that is,

$$H_0 : p_1 = p_2 \qquad \text{or} \qquad (p_1 - p_2) = 0$$

then $p_1 = p_2 = p$ and the best estimate of p is obtained by pooling the data from both samples. Thus if y_1 and y_2 are the numbers of successes obtained from the two samples, then

$$\hat{p} = \frac{y_1 + y_2}{n_1 + n_2}$$

The test statistic is

$$z = \frac{(\hat{p}_1 - \hat{p}_2) - 0}{\sqrt{\dfrac{\hat{p}\hat{q}}{n_1} + \dfrac{\hat{p}\hat{q}}{n_2}}} = \frac{\hat{p}_1 - \hat{p}_2}{\sqrt{\hat{p}\hat{q}\left(\dfrac{1}{n_1} + \dfrac{1}{n_2}\right)}}$$

Case 2. On the other hand, if we hypothesize that D_0 is not equal to zero, that is,

$$H_0: (p_1 - p_2) = D_0$$

where $D_0 \neq 0$, then the best estimates of p_1 and p_2 are $\hat{p}_1$ and $\hat{p}_2$, respectively. The test statistic is

$$z = \frac{(\hat{p}_1 - \hat{p}_2) - D_0}{\sqrt{\dfrac{\hat{p}_1\hat{q}_1}{n_1} + \dfrac{\hat{p}_2\hat{q}_2}{n_2}}}$$

For most practical problems involving the comparison of two binomial populations, the experimenter will wish to test the null hypothesis that $(p_1 - p_2) = D_0 = 0$. For our example we test

$$H_0: (p_1 - p_2) = 0$$

against the alternative

$$H_a: (p_1 - p_2) > 0$$

Note that a one-tailed statistical test will be employed, because if a difference exists, we wish to detect $p_1 > p_2$. Therefore, we will reject the null hypothesis in favor of the alternative hypothesis if $(\hat{p}_1 - \hat{p}_2)$, or, equivalently, if the calculated value of z, is large. In fact, if we choose $\alpha = .05$, we will reject H_0 when $z > 1.645$.

The pooled estimate of p required for $\sigma_{(\hat{p}_1 - \hat{p}_2)}$ is

$$\hat{p} = \frac{y_1 + y_2}{n_1 + n_2} = \frac{52 + 23}{1000 + 1000} = .0375$$

The test statistic is

$$z = \frac{\hat{p}_1 - \hat{p}_2}{\sqrt{\hat{p}\hat{q}\left(\dfrac{1}{n_1} + \dfrac{1}{n_2}\right)}} = \frac{.052 - .023}{\sqrt{(.0375)(.9625)\left(\dfrac{1}{1000} + \dfrac{1}{1000}\right)}} = 3.41$$

Since the computed value of z falls in the rejection region, we reject the hypothesis that $p_1 = p_2$ and conclude that the data present sufficient evidence to indicate that the percentage of men entering the hospital because of heart disease is higher than that of women. Note that this does not imply that the incidence of heart disease is higher in men. Perhaps fewer women enter the hospital when afflicted with the disease.

Exercises

8.24. A manufacturer claims that at least 95% of the equipment that he supplied to a factory conformed to specifications. An examination of 700 pieces of equipment reveals that 53 are faulty. Do these results provide sufficient evidence to reject the manufacturer's claim? Use $\alpha = .05$.
  a. State the null hypothesis to be tested.
  b. State the alternative hypothesis.
  c. Conduct a statistical test of the null hypothesis and state your conclusions.

8.25. In the face of diminishing energy resources, the National Aeronautics and Space Administration (NASA) is currently working with utilities throughout the nation to find sites for large wind machines for generating electric power. D. J. Vargo, the NASA official in charge of this project, has said that wind speeds must average at least 15 miles per hour for a site to be acceptable.* Thirty-six wind speed recordings were taken at random intervals on a site under consideration for a wind machine; the wind speeds averaged 14.2 mph, with a standard deviation of 3 mph. Do these data indicate that the site fails to meet NASA requirements for acceptability as a site for the location of an electric power-generating wind machine? Use $\alpha = .10$.

8.26. Continuing exercise 8.25, discuss the meaning of type I and type II errors in the site selection problem. If you were a NASA official, would you recommend the selection of an extremely small (.01) value for α, a moderate value (.05), or a reasonably large value (.10) when testing for the acceptability of a site on the basis of its average wind speed? Explain.

8.27. With yields near historic highs, many investors are turning away from the securities market in favor of U.S. Treasury notes and bonds. But since the price is not set until after the sale, note and bond yields are not known in advance. The yields to maturity on 30 recently issued Treasury notes and 36 recently issued Treasury

TREASURY NOTES	TREASURY BONDS
$\bar{y}_1 = 7.91\%$	$\bar{y}_2 = 8.12\%$
$s_1^2 = 1.85$	$s_2^2 = 1.98$

bonds were recorded by an investor. The statistics shown in the table were obtained from the data. Do these data present sufficient evidence to indicate a difference in the average yield to maturity between U.S. Treasury notes and bonds? (Test with $\alpha = .10$.)

8.28. In a study to assess various effects of using a female model in automobile advertising, each of 100 male subjects was shown photographs of two automobiles matched for price, color, and size but of different makes. One automobile was shown with a female model and one without a model to 50 of the subjects (Group A), and both automobiles were shown without a model to the other 50 subjects (Group B). In Group A the automobile shown with the model was judged as more expensive by 37 subjects, while in Group B the same automobile was judged as the more expensive by 23 subjects. Do these results indicate that using a female model influences the perceived expensiveness of an automobile? Use a one-tailed test with $\alpha = .05$.

*"Wind Power: Still Becalmed," *Environmental News*, July 1975.

8.15 Some Comments on the Theory of Tests of Hypotheses

As outlined in section 8.13, the theory of a statistical test of an hypothesis is indeed a very clear-cut procedure, enabling the experimenter to either reject or accept the null hypothesis with measured risks α and β. Unfortunately, as we noted, the theoretical framework does not suffice for all practical situations.

The crux of the theory requires that we be able to specify a meaningful alternative hypothesis that permits the calculation of the probability β of a type II error for all alternative values of the parameter(s). This indeed can be done for many statistical tests, including the test discussed in section 8.13, although the calculation of β for various alternatives and sample sizes may, in some cases, be a formidable task. On the other hand, in some test situations it is extremely difficult to clearly specify alternatives to H_0 that have *practical* significance. This may occur when we wish to test an hypothesis concerning the values of a set of parameters, a situation that we will encounter in chapter 17 in analyzing enumerative data.

The obstacle that we mention does not invalidate the use of statistical tests. Rather, it urges caution in drawing conclusions when insufficient evidence is available to reject the null hypothesis. The difficulty of specifying meaningful alternatives to the null hypothesis, together with the difficulty encountered in the calculation and tabulation for β for other than the simplest statistical tests, justifies skirting this issue in an introductory text. Hence, we will agree to adopt the procedure described in example 8.11 when tabulated values of β (the operating characteristic curve) are unavailable for the test. When the test statistic falls in the acceptance region, we will "not reject" rather than "accept" the null hypothesis. Further conclusions may be made by calculating an interval estimate for the parameter or by consulting one of several published statistical handbooks for tabulated values of β. We will not be too surprised to learn that these tabulations are inaccessible, if not completely unavailable, for some of the more complicated statistical tests.

The probability α of making a type I error is often called the **significance level** of the statistical test, a term that originated in the following way. The probability of the observed value of the test statistic, or some value even more contradictory to the null hypothesis, measures, in a sense, the weight of evidence favoring rejection. Some experimenters report test results as being significant (we would reject) at the 5% significance level but not at the 1% level. This means that we would reject H_0 if α were .05 but not if α were .01. This line of thought does not conflict with the procedure of choosing the test in advance of the data collection. Rather, it presents a convenient way of publishing the statistical results of a scientific investigation, permitting the reader to choose his own α and β as he pleases.

Finally, we might comment on the choice between a one- or a two-tailed test for a given situation. We emphasize that this choice is dictated by the practical aspects of the problem and will depend on the alternative value of the parameter, say θ, the experimenter is trying to detect. If we were to sustain a large financial loss if θ were greater than θ_0, but not if it were less, we would concentrate our attention on the detection of values of θ greater than θ_0. Hence we would reject in the upper tail of the distribution for the test statistics previously discussed. On the other hand, if we are equally interested in detecting values of θ that are either less than or greater than θ_0, we would employ a two-tailed test.

8.16 Summary

The material presented in chapter 8 has been directed toward two objectives. First, we wanted to discuss the various methods of inference along with procedures for evaluating their goodness. Second, we wished to present a number of estimation procedures and statistical tests of hypotheses which, owing to the Central Limit Theorem, make use of the results of chapter 7. The resulting techniques possess practical value and, at the same time, illustrate the principles involved in statistical inference.

Inferences concerning the parameter(s) of a population may be made by estimating or by testing hypotheses about their value. A parameter may be estimated by using either a point or an interval estimator. The measure of goodness, or confidence, that we can place in a point estimator is given by the bound on the error of estimation. Similarly, the confidence coefficient and the width of the interval measure the goodness of an interval estimator.

A statistical test of an hypothesis or theory concerning the population parameter(s) ideally will result in its rejection or acceptance. Practically, we may be forced to view this decision in terms of rejection or nonrejection. The probabilities of making the two possible incorrect decisions, resulting in type I and type II errors, measure the goodness of the decision procedure. While a test of an hypothesis may be best suited for some physical situations (for example, lot acceptance sampling), it seems that estimation would be the eventual goal of many experimental investigations and hence would be desirable if we were permitted an option in our choice of a method of inference.

All the confidence intervals and statistical tests described in this chapter are based on the Central Limit Theorem and hence apply to large samples. When n is large, each of the respective estimators and test statistics possesses, for all practical purposes, a normal distribution in repeated sampling. This result, along with the properties of the normal distribution studied in chapter 7, permits the construction of the confidence intervals and the calculation of α and β for the statistical tests.

Supplementary Exercises

8.29. Define and discuss the following concepts with regard to their role in statistical inference:

a. type I and type II errors

b. the probabilities α and β
c. the level of significance
d. a confidence interval
e. the z test

8.30.   Explain the role of the Central Limit Theorem in the topics presented in this chapter.

8.31.   Distinguish between a one-tailed test and a two-tailed test of an hypothesis.

8.32.   In each of the following hypothesis-testing situations, suggest whether a one-tailed test or a two-tailed test of an hypothesis is appropriate. In each case justify your answer and then write the null and alternative hypotheses.

a. A study by an economist to determine whether the unemployment level has changed significantly during the past year.

b. A study to check the assumption of a credit manager who claims that the average account balance of credit customers for a certain store is at least $50.

c. A marketing research study to determine whether the proportion of buyers who favor your product is greater than the proportion who favor the product of your competitor.

d. A study to determine whether the average salary of employees from the Ford Motor Company differs from the average salary of a General Motors employee.

e. A study to determine whether pension benefits at a certain large firm have increased during the past year.

8.33.   The mean and standard deviation for the life of a random sample of 100 light bulbs were calculated to be 1280 hours and 142 hours, respectively. Estimate the mean life of the population of light bulbs from which the sample was drawn, and place bounds on the error of estimation.

8.34.   Suppose that the population mean in exercise 8.33 was really 1285 hours, with $\sigma = 150$ hours. What is the probability that the mean of a random sample of $n = 100$ measurements would exceed 1300 hours?

8.35.   Using a confidence coefficient of .90, place a confidence interval on the mean life of the light bulbs of exercise 8.33.

8.36.   In industrial selling, negotiation costs represent a significant percentage of a firm's sales since the seller must formulate a sales strategy before negotiating with a business or governmental client. Several researchers interviewed 500 firms involved in industrial selling and found their average negotiation cost per transaction to be $11,336, with a standard deviation of $1,951.* Estimate the average negotiation cost per transaction for all industrial sellers, and place a bound on the error of estimation.

8.37.   Using a confidence coefficient of .99, place a confidence interval on the mean negotiation cost per transaction for the industrial sellers of exercise 8.36.

8.38.   Manufacturers of photographic equipment have introduced many new easy-to-use cameras, film types, and flash equipment in recent years. A new type of flashbulb was tested to estimate the probability p that the new bulb would produce the required light output at the appropriate time. A sample of 1000 bulbs was tested and 920 were observed to function according to specifications. Estimate p and place bounds on the error of estimation.

8.39.   Place a confidence interval on p of exercise 8.38 by using a confidence coefficient of .99. Interpret this interval.

8.40.   The 10 November 1975 issue of *Business Week* reports that in an attempt

---

*M. Bird, E. Clayton, and L. Moore, "Sales Negotiation Cost Planning for Corporate Level Sales," *Journal of Marketing*, April 1973.

to draw more tourists and conventioneers into its stores, many department store chains are now accepting Bank Americard and Master Charge cards. In a random sample of 60 department store chains, 10 are now accepting the above charge cards. Find a 98% confidence interval on the true proportion of department stores now accepting Bank Americard and Master Charge cards.

8.41.   The mean and standard deviation of 49 randomly selected purchases on credit by the credit customers of a large department store were found to be $6.50 and $2.10, respectively. Find a 98% confidence interval for the mean of the population of all credit purchases of the store.

8.42.   The percentage of pages containing at least one typing mistake was recorded for a secretary who used first a standard typewriter and then an electric typewriter. For the standard typewriter, 64 of 200 typewritten pages contained at least one typing mistake, while for the electric typewriter, 36 of 180 pages contained at least one mistake. Estimate the difference in the percentage of typewritten pages containing at least one mistake when using the standard and the electric typewriter. Place bounds on the error of estimation.

8.43.   The marketing research department of a soap and detergent manufacturing company conducted a survey of housewives to determine the fraction who prefer the company's brand, Detergent A. Sixty out of 87 housewives prefer Detergent A. If the 87 housewives represent a random sample from the population of all potential purchasers, estimate the fraction of total housewives favoring Detergent A. Use a 90% confidence interval.

8.44.   In examining the credit accounts of a department store, an auditor selected a random sample of 64 accounts and found the average account error to be $-\$42$, with a standard deviation of $16. Construct a confidence interval for the true average account error, using a confidence coefficient of .95.

8.45.   A manufacturing firm recently instituted an occupational safety program designed to reduce lost time due to on-the-job accidents. In the 48 months since the institution of the program, employee time lost due to on-the-job accidents has averaged 91 hours per month, with a standard deviation of 14 hours per month. In the 50 months prior to the safety program, lost time due to accidents averaged 108 hours per month, with a standard deviation of 12 hours. Estimate the difference in mean employee time lost per month due to on-the-job accidents before and after the occupational safety program. Place bounds on the error of estimation.

8.46.   Refer to exercise 8.45. Estimate the difference in the mean employee time lost per month by using a confidence coefficient of .90.

8.47.   Manufacturers of golf balls are concerned with a scientific concept called the coefficient of restitution, defined as the ratio of the relative velocity of the ball and club after impact to the relative velocity before impact. A manufacturer has developed a new solid-core golf ball which he wishes to test and compare with his firm's standard brand. Fifty of the new golf balls and 50 of the standard brand are subjected to a test. The results are given in the table. Estimate the difference

SOLID-CORE BALL	STANDARD BRAND
$\bar{y}_1 = .65$	$\bar{y}_2 = .59$
$s_1 = .16$	$s_2 = .20$

in the mean coefficient of restitution for the new solid-core ball and the firm's standard brand of golf ball. Place bounds on the error of estimation.

8.48.   Estimate the difference in the mean coefficient of restitution for the two types of golf balls of exercise 8.47 by using a confidence coefficient of .99.

8.49.   Ideally, inventory control avoids both excessive carrying costs and stockout costs. A large trade association wishes to estimate the difference in the proportion of retail firms that utilize formal inventory control techniques and the proportion of wholesalers that utilize such techniques. In a random sample of 91 retail firms, it was found that 62 employ formal inventory control techniques, while 37 of 65 firms primarily involved in wholesale trade were found to use such procedures. Estimate the difference in the proportions using formal inventory control procedures, and place bounds on the error of estimation.

8.50.   Refer to exercise 8.49. Estimate the difference in the proportions using formal inventory control procedures by using a confidence coefficient of .99.

8.51.   Past experience shows that the standard deviation of the yearly income of textile workers in a certain state is $400. How large a sample of textile workers would we need to take if we wished to estimate the population mean to within $50.00, with a probability of .95 of being correct? Given that the mean of the sample in this problem is $4800, determine a 95% confidence interval for the population mean.

8.52.   How many voters must be included in a sample collected to estimate the fraction of the popular vote favorable to a presidential candidate in a national election if the estimate is desired correct to within .005? Assume that the true fraction will lie somewhere in the neighborhood of .5.

8.53.   A quality control engineer wishes to estimate the fraction of defectives in a large lot of light bulbs. From previous experience she feels that the actual fraction of defectives should be somewhere around .2. How large a sample should she take if she wants to estimate the true fraction to within .01 by using a 95% confidence interval?

8.54.   Bird notes that opinion is quite divided among corporate executives about the most important training topic for purchasing managers.* However, traditional theory has held that the most important training topic for purchasing managers is the development of negotiation skills. In light of Bird's article, a researcher would like to see if the traditional theory still holds true. How large a sample of corporate executives should the researcher select if he wishes to estimate the fraction who hold to the traditional theory correct to within 5%? Assume that past evidence suggests that the proportion who believe the development of negotiation skills to be the most important training topic is about .60.

8.55.   The manager of the chamber of commerce of a convention city wishes to estimate the average amount spent by each convention attendee correct to within $20.00, with a probability of .95. If a random sample of convention attendees is to be selected and asked to keep financial expenditure data, how many must be included in the sample? Assume the manager knows only that the range of expenditures will vary, approximately, from $200.00 to $500.00.

8.56.   It is desired to estimate the difference in the average time to assemble an electronic component for two factory workers to within 2 minutes, with a probability of .95. If the standard deviation of the assembly times is approximately equal to 6 minutes for each worker, how many component assembly times must be observed for each worker? (Assume that the number of component assembly times observed for each worker will be the same.)

8.57.   The Florida Highway Commission has recently instituted an inspection

---

*M. M. Bird, "Research and Training Needs of Industrial Purchasing Managers," *Journal of Purchasing*, February 1973.

campaign in an attempt to reduce highway accidents resulting from mechanical defects. A random sample of 100 automobiles is selected for inspection. Of these, 20 are found to have defective brakes. Construct a 95% confidence interval for the fraction of automobiles on Florida highways having defective brakes. Suppose that a more accurate estimate is needed than the one just obtained. How large a sample would be necessary to reduce the bound on the error to .04?

8.58.   In the past a chemical plant has produced an average of 1100 pounds of chemical per day. A random sample of 260 operating days from the past year shows that $\bar{y}$ = 1060 lb/day and s = 340 lb/day. It is desired to test whether the average daily production $\bar{y}$ has dropped significantly over the past year.

a. Give the appropriate null and alternative hypotheses.

b. If z is used as a test statistic, determine the rejection region corresponding to a level of significance of α = .05.

c. Do the data provide sufficient evidence to indicate a drop in average daily production?

✓ 8.59.   The daily wages in a particular industry are normally distributed, with a mean of $43.20 and a standard deviation of $2.50. If a company in this industry randomly samples 40 of its workers and finds the average wage to be $42.20, can this company be accused of paying inferior wages, at the 1% level of significance? Interpret your results.

8.60.   A manufacturer of automatic washers provides a particular model in one of three colors, A, B, or C. Of the first 1000 washers sold, it is noticed that 400 of the washers were of color A. Would you conclude that more than $\frac{1}{3}$ of all customers have a preference for color A? Use α = .01.

8.61.   A manufacturer claims that at least 20% of the public prefers her product. A sample of 100 persons is taken to check her claim. With α = .05, how small would the sample percentage need to be before the claim could be rightfully refuted? (Note: This problem requires a one-tailed test of an hypothesis.)

8.62.   Refer to exercise 8.61. Sixteen people in the sample of 100 consumers expressed a preference for the manufacturer's product. Does this present sufficient evidence to reject the manufacturer's claim? Test at the 10% level of significance.

8.63.   Weekly ratings published by the A. C. Nielsen Company have made television producers and advertisers very sensitive to claims regarding the proportion of the total viewing audience reached by a particular program. A television station claims that its six o'clock evening news reaches 50% of the viewing audience in the area. A firm considering the purchase of advertising time during the six o'clock time slot wishes to test the validity of the station's claim. How large a sample should the firm select if it wants the bound on the error of estimation to be 5%?

8.64.   Refer to exercise 8.63. Suppose that a sample of 100 television viewers is selected from the potential viewing audience of the television station and 38 indicate that they watch the station's six o'clock evening news. Is this sufficient evidence to refute the claim of the television station that its six o'clock evening news reaches 50% of the viewing audience? (Test at the 1% level of significance.)

8.65.   A company marketing representative wishes to determine the acceptability of a new product in a particular community. If the company could feel confident that about 50% of the community's residents would buy its product, the marketing representative would suggest that it be marketed in that community. A random sample of n = 64 is selected from the target community, and 24 of those sampled state that they would buy the product. What conclusions can you draw concerning the marketability of the product in the target community?

8.66.   A test of the breaking strengths of two different types of cables was conducted, using samples of $n_1 = n_2 = 100$ pieces of each type of cable. The data are shown in the table. Do the data provide sufficient evidence to indicate a difference in the mean breaking strengths of the two cables? Use $\alpha = .10$.

CABLE 1	CABLE 2
$\bar{y}_1 = 1925$	$\bar{y}_2 = 1905$
$s_1 = 40$	$s_2 = 30$

8.67.   Refer to Exercise 8.45. Do the data indicate that the firm's occupational safety program has been effective in reducing on-the-job accidents? Use a significance level of .05.

8.68.   Refer to exercise 8.47. Do the test results indicate a difference in the mean coefficient of restitution for the new solid-core ball and the firm's standard brand of golf ball? Use $\alpha = .01$.

8.69.   A survey of buying habits was conducted in Boston and Seattle. In Boston 200 housewives were interviewed, and it was found that they spend an average of $190 on food per month, with a standard deviation of $25. In Seattle 175 housewives reported an average monthly food expenditure of $180, with a standard deviation of $35. Using a significance level of .05, test the hypothesis that there is no difference in the average amount spent on food per month by housewives in Boston and by housewives in Seattle. Interpret your results.

8.70.   Random samples of 200 bolts manufactured by Machine A and 200 bolts manufactured by Machine B showed 16 and 8 defective bolts, respectively. Do these data present sufficient evidence to suggest a difference in the performances of the machines? Use a .05 level of significance.

8.71.   The mean lifetime of a sample of 100 fluorescent bulbs produced by a company is computed to be 1570 hours, with a standard deviation of 120 hours. If μ is the mean lifetime of all the bulbs produced by the company, test the hypothesis $\mu = 1600$ hours against the alternative hypothesis $\mu < 1600$, using a level of significance of .05.

8.72.   All corporate directors need timely and accurate information about the company each serves. One researcher suggests that the amount of information needed by an executive is dependent on the industry within which the firm operates.* Suppose a random sample of 36 corporate directors of manufacturing firms shows that they receive an average of 6 reports during the course of a month from top management, with a standard deviation of 1.8, whereas a random sample of 30 corporate directors from the transportation industry was found to receive, on the average, 3.1 reports per month, with a standard deviation of 1.4. Do these data suggest that the mean number of reports received by corporate directors from these two industries differ? Use $\alpha = .05$.

8.73.   Refer to exercise 8.49. Do these sample data indicate that a greater fraction of retail firms than wholesaling firms use formal inventory control techniques? Test at a level of significance such that, if the trade association finds retailing firms do make greater use of formal inventory control techniques, they are incurring a very small risk of an erroneous conclusion.

8.74.   Presently, 20% of potential customers buy a certain brand of soap,

*M. Mace, "Management Information Systems for Directors," *Harvard Business Review*, December 1975.

say Brand A. To increase sales an extensive advertising campaign is conducted. At the end of the campaign, a sample of 400 potential customers is interviewed to determine if the campaign was successful.

a. State H_0 and H_a in terms of p, the probability that a customer prefers Brand A.

b. It is decided to conclude that the advertising campaign was a success if at least 92 of the 400 customers interviewed prefer Brand A. Find α. (Use the normal approximation to the binomial distribution to evaluate the desired probability α.)

8.75. Suppose that the true fraction p in favor of wage and price controls is the same for labor as it is for management. Independent random samples are selected, one consisting of 800 laborers and the other of 800 managers. Find the probability that the sample fraction of laborers favoring controls exceeds that of the managers by more than .03 if the true proportion who favor controls is p = .5. What is this probability if p = .1?

8.76. A union official believes that the fraction p_1 of laborers in favor of wage and price controls is greater than the fraction p_2 of managers in favor of wage and price controls. The official acquired independent random samples of 200 laborers and 200 managers and found 46 laborers and 34 managers favoring controls. Does this evidence provide statistical support at the .05 level of significance for the union official's belief?

8.77. Educational psychologists are often interested in factors that affect academic performance. One such psychologist conducted an investigation to determine whether car ownership was detrimental to academic achievement. Two random samples of 100 students were drawn from the student population on the basis of car ownership. The grade point average for the n_1 = 100 non-car owners possessed an average and variance of $\bar{y}_1$ = 2.70 and s_1^2 = .36, as opposed to $\bar{y}_2$ = 2.54 and s_2^2 = .40 for the n_2 = 100 car owners. Do these data present sufficient evidence to indicate a difference in the mean achievement between car owners and non-car owners?

8.78. When evaluating a loan applicant, a financial officer is faced with the problem of granting loans to those who are good risks and denying loans to those who appear to be poor risks. In a sense, one could say that the financial officer is testing the statistical hypothesis

$$H_0: \text{The applicant is a good risk}$$

against the alternative

$$H_a: \text{The applicant is a poor risk}$$

for each loan applicant. A type I error is then committed by the loan officer if he rejects an applicant who is actually a good risk; a type II error is committed if he grants a loan to an applicant who is a poor risk. Discuss the selection of a significance level α in the following instances:

a. Lending money is tight; interest rates are high and loan applicants are numerous.

b. Lending money is plentiful; interest rates are moderate and there is competition for loan applicants.

8.79. Refer to exercise 8.25. Use the procedure described in example 8.11 to calculate β for several alternative values of μ. For example, you might calculate β for μ = 13, 13.5, 14, and 14.5. Use these computed values of β to construct an operating characteristic curve for the statistical test.

8.80. A magazine subscription service in a city containing 25,000 households is interested in estimating the average number of magazine subscriptions by the residents of each household. Suppose that we let y equal the number of magazine subscriptions held by the residents of a sample household. In a random sample of 401 households selected from the city, it was found that

$$\sum_{i=1}^{401} y_i = 785 \quad \text{and} \quad \sum_{i=1}^{401} y_i^2 = 2{,}015$$

a. Find a point estimate for the average number of magazine subscriptions per resident household in the city.

b. Does the point estimate computed in part a supply the reader with any measure of confidence in its closeness to μ, the unknown mean number of magazine subscriptions per resident household in the entire city? Explain.

c. Place bounds on the error of estimation of μ.

d. Suppose that the magazine subscription service wishes to be 99% confident that they obtain an interval estimate that contains μ. Find an interval estimate that bounds μ with 99% confidence. What is the meaning of this interval estimate?

e. Without gathering additional sample information, how can the magazine subscription service reduce the width of the confidence interval for μ? What price must be paid for the reduction?

f. If a separate sample of $n = 401$ households was collected, would you expect the 99% confidence interval for μ based on the new sample to be the same as the interval estimate computed in part d for the original sample? Explain.

g. If the two sets of sample observations are combined, what will be the effect on the width of the resulting confidence interval (for a fixed confidence level)?

h. Based on the results obtained in part d and from the known number of households in the city, find a 99% confidence interval for the *total* number of magazine subscriptions by all the residents of the city.

i. The director of the magazine subscription service has issued the following statement: "We shall undertake an extensive sales campaign in the city unless there is sufficient evidence to indicate that the average number of subscriptions per household is 1.8 or greater. We are further willing to assume no more than a 5% chance of failing to undertake the sales campaign when we should have." Based on the sample information, should the subscription service undertake a sales campaign in the city?

8.81.   It was mentioned in the Experiences with Real Data section of chapter 4 that if a customer is more likely to purchase an item from a department store on noninstallment credit than to pay cash for the item, sufficient demand exists to justify the use of an in-house credit card for the department store. Use the collective data gathered by the students in your class for the Experiences with Real Data of chapter 4 to see if sufficient demand exists to justify the use of an in-house credit card for the department store involved in your class study. (Hint: Test the hypothesis that the fraction p of buyers demanding noninstallment credit is equal to .5 against the alternative that p is greater than .5.)

Experiences with Real Data

Advertisers are interested in the extent to which media sources carrying their messages reach the ultimate consumer. Television and radio station managers as well as magazine and newspaper editors must be able to demonstrate the degree to which their sources penetrate the market in order to attract advertisers. Most colleges and universities have only one campus newspaper but they, too, must demonstrate their degree of market penetration to advertisers since all students do not regularly read the campus newspaper.

Conduct a sample survey to determine the proportion of students on your campus who read your student newspaper at least *three* days each week. Randomly select a large sample of n students (say $n = 400$ or more) from your student directory. Use the random number tables described in section 16.3 to assist in the sample selection process in order to provide maximal assurance that sampling is conducted in a random manner. Since each of the *n* students in the sample must be contacted to determine whether he or she is a regular reader of the student newspaper (i.e., they read it at least three days a week), this chore should be divided so that a small group of students is responsible for contacting a fixed number, say 25, of the total sample. Telephone interviews can be made when possible; when not possible, those chosen in the sample should be contacted personally.

When the data have been collected, pool the *n* student responses and obtain an estimate of the proportion of students who read your campus newspaper at least three days a week. Place a bound on your error of estimation.

To observe sampling variation let each team use the *n* = 25 student responses they gathered to estimate the proportion of regular readers. Collect the estimates from each team and construct a histogram of the estimates. Locate on the graph the estimate based on all *n* students. Notice how the estimates based on the samples of 25 students tend to vary and that the large-sample estimate (based on all *n* students) falls near the center of the histogram.

The calculations for this experiment can easily be accomplished with the aid of a desk calculator. However, for other analyses considered in this chapter, you may wish to use an available library program and an electronic computer to perform the required calculations. Useful library programs, which should be available in your campus computing center, are the Biomed series, produced by the UCLA Health Sciences Computing Facility, and SPSS, the Statistical Package for the Social Sciences.

References

DIXON, W. J., and F. J. MASSEY, JR. *Introduction to Statistical Analysis*. 3d ed. New York: McGraw-Hill, 1969. Chapters 6 and 7.

FREUND, J. E., and F. J. WILLIAMS; revised by B. Perles and C. Sullivan. *Modern Business Statistics*. Englewood Cliffs, N.J.: Prentice-Hall, 1969. Chapters 9, 10, and 11.

HARNETT, D. L., and J. L. MURPHY. *Introductory Statistical Analysis*. Reading, Mass.: Addison-Wesley, 1975. Chapters 6 and 7.

HOEL, P. G., and R. J. JESSEN. *Basic Statistics for Business and Economics*. New York: Wiley, 1971. Chapters 6 and 7.

SUMMERS, G. W., and W. S. PETERS. *Basic Statistics in Business and Economics*. Belmont, Calif.: Wadsworth, 1973. Chapters 9 and 10.

chapter objectives

GENERAL OBJECTIVE The basic concepts of statistical estimation were presented in chapter 8, along with a summary of the concepts involved in a statistical test of an hypothesis. Large-sample estimation and test procedures for population means and proportions were used to illustrate concepts as well as to give you some useful tools for solving business problems. Because all these techniques rely on the Central Limit Theorem to justify the normality of the estimators and test statistics, they apply only when the sample sizes are large. Consequently, the objective of chapter 9 is to generalize the results of chapter 8. We will present small-sample statistical test and estimation procedures for population means and variances and show how these techniques can be applied in business decision making. These techniques differ substantially from those of chapter 8 because they require that the relative frequency distributions of the sampled populations be approximately normal.

SPECIFIC OBJECTIVES

1. To present the Student's t statistic and identify the relationship between it and the z statistic of chapter 8.
                    Section 9.2

2. To point out the relationship between the normal distribution and a Student's t distribution and to explain how to use the table values for the Student's t statistic.
                    Section 9.2

3. To show how the t statistic is used to construct small-sample confidence intervals and tests of hypotheses for population means and the difference between two population means.
                    Sections 9.3, 9.4

4. To present an example of an experiment designed to reduce the cost of experimentation, the paired-difference experiment. We will show how the techniques of section 9.2 can be used to analyze these data.
                    Section 9.5

5. To show how to estimate and test hypotheses about population variances and to give examples and exercises that indicate the practical applications of these techniques to the solution of certain business problems.
                    Sections 9.6, 9.7

chapter nine

Inferences from Small Samples

9.1 Introduction

Large-sample methods for making inferences about population means and the difference between two means were discussed, with examples, in chapter 8. Frequently, however, cost and available time limit the size of the sample that can be acquired. When this occurs, the large-sample procedures of chapter 8 are inadequate and other tests and estimation procedures must be used. In this chapter we will study several small-sample inferential procedures, which are closely related to the large-sample methods presented in chapter 8. Specifically, we will consider methods for estimating and testing hypotheses about population means, the difference between two means, a population variance, and a comparison of two population variances. (Small-sample tests about the binomial parameter p were discussed in chapter 6.)

9.2 Student's t Distribution

We introduce our topic by considering a problem. A very costly experiment has been conducted to evaluate a new process for producing synthetic diamonds. Six diamonds have been generated by the new process, with recorded weights of .46, .61, .52, .48, .57, and .54 carats.

A study of the process costs indicates that the average weight of the diamonds must be greater than .5 carats if the process is to be operated at a profitable level. Do the six diamond weight measurements present sufficient evidence to indicate that the average weight of the diamonds produced by the process is in excess of .5 carats? That is, we wish to test the null hypothesis that $\mu = .5$ against the alternative hypothesis that $\mu > .5$.

Recall that, according to the Central Limit Theorem, the test statistic

$$z = \frac{\bar{y} - \mu}{\sigma / \sqrt{n}}$$

possesses, approximately, a normal distribution in repeated sampling when n is large. For $\alpha = .05$ we could employ a one-tailed statistical test and reject the null hypothesis if $z > 1.645$. This, of course, assumes that σ is known or that a good estimate s is available and is based on a reasonably large sample (we have suggested $n \geq 30$). Unfortunately, the latter requirement is not satisfied for the $n = 6$ diamond weight measurements. How, then, may we test the hypothesis that $\mu = .5$ against the alternative that $\mu > .5$ when we have a small sample?

The problem we pose is not new; it received serious attention from statisticians and experimenters at the turn of the century. If a sample standard deviation s is substituted for σ in z, does the resulting test statistic possess, approximately, a standardized normal distribution in repeated sampling? More specifically, is the rejection region $z > 1.645$ appropriate—that is, do approximately 5% of the values of the test statistic, computed in repeated sampling, exceed 1.645 when H_0 is true?

The answer to these questions, not unlike many of the problems encountered in the sciences, may be resolved by experimentation. In other words, we could draw a small sample, say $n = 6$ measurements, and compute the value of the test statistic. Then we would repeat this process many, many times and construct a frequency distribution for the computed values of the test statistic. The general shape of the distribution and the location of the rejection region would then be evident.

The distribution of the test statistic

$$t = \frac{\bar{y} - \mu}{s / \sqrt{n}}$$

for samples drawn from a normally distributed population was discovered by W. S. Gosset and published (in 1908) under the pen name "Student." He referred to the quantity under study as t and it has ever since been known as Student's t. We omit the complicated mathematical expression for the sampling distribution for t but we describe some of its characteristics.

In repeated sampling the distribution of the test statistic

$$t = \frac{\bar{y} - \mu}{s / \sqrt{n}}$$

is, like z, mound-shaped and perfectly symmetrical about $t = 0$. Unlike z, it is much more variable, tailing rapidly out to the right and left, a phenomenon that may be readily explained. The variability of z in repeated sampling is

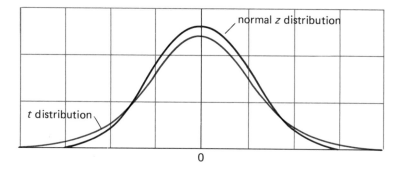

Figure 9.1 *Standard normal z and the t distribution based on n = 6 measurements (5 d.f.)*

due solely to $\bar{y}$; the other quantities appearing in z (n and σ) are nonrandom. On the other hand, the variability of t is contributed by two random quantities, $\bar{y}$ and s, which can be shown to be independent of one another. When $\bar{y}$ is very large, s may be very small, and vice versa. As a result, t is more variable than z in repeated sampling (see figure 9.1). Finally, as we might surmise, the variability of t decreases as n increases because the estimate s of σ is based on more and more information. When n is infinitely large, the t and z distributions are identical. Thus Gosset discovered that the distribution of t depended on the sample size n.

The divisor $(n - 1)$ of the sum of squares of deviations that appears in the formula for s^2 is called the number of **degrees of freedom** associated with s^2. The origin of the term "degrees of freedom" is linked to the statistical theory underlying the probability distribution of s^2. We will not pursue this point further except to note that we may say that the test statistic t is based

Table 9.1 *Format of the Student's t table, table 4 of the appendix*

d.f.	$t_{.100}$	$t_{.050}$	$t_{.025}$	$t_{.010}$	$t_{.005}$	d.f.
1	3.078	6.314	12.706	31.821	63.657	1
2	1.886	2.920	4.303	6.965	9.925	2
3	1.638	2.353	3.182	4.541	5.841	3
4	1.533	2.132	2.776	3.747	4.604	4
5	1.476	2.015	2.571	3.365	4.032	5
6	1.440	1.943	2.447	3.143	3.707	6
7	1.415	1.895	2.365	2.998	3.499	7
8	1.397	1.860	2.306	2.896	3.355	8
⋮	⋮	⋮	⋮	⋮	⋮	⋮
27	1.314	1.703	2.052	2.473	2.771	27
28	1.313	1.701	2.048	2.467	2.763	28
29	1.311	1.699	2.045	2.462	2.756	29
inf.	1.282	1.645	1.960	2.326	2.576	inf.

on a sample of n measurements or we may say that it possesses $(n - 1)$ degrees of freedom.

The critical values of t that separate the rejection and acceptance regions for the statistical test are presented in table 4 of the appendix. Table 9.1 gives the format and shows some of the entries for table 4. The tabulated value t_α is the value of t such that an area α lies to its right, as shown in figure 9.2. The degrees of freedom d.f. associated with s^2 are shown in the first and last columns of the table, and the t_α's corresponding to various values of α appear in the top row. Thus if we wish to find the value of t such that 5% of the area lies to its right, we would use the column marked $t_{.05}$. The critical value of t for our example, found in the $t_{.05}$ column opposite the row corresponding to d.f. $= n - 1 = 6 - 1 = 5$ is $t_{.05} = 2.015$. This entry is shaded in table 9.1. Thus we would reject H_0: $\mu = .5$ when $t > 2.015$.

Note that the critical value of t is always larger than the corresponding critical value of z for a specified α. For example, when $\alpha = .05$, the critical value of t for $n = 2$ is $t = 6.314$, which is very large when compared with the corresponding $z = 1.645$. Proceeding down the $t_{.05}$ column, we note that the critical value of t decreases, reflecting the effect of a larger sample size on the estimation of σ. Finally, when n is infinitely large, the critical value of t equals 1.645.

The reason for choosing $n = 30$ (an arbitrary choice) as the dividing line between large and small samples becomes apparent when you examine table 9.1. For $n = 30$ the critical value of $t_{.05} = 1.699$ is numerically quite close to $z_{.05} = 1.645$. For a two-tailed test based on $n = 30$ measurements and $\alpha = .05$, we would place .025 in each tail of the t distribution and reject H_0: $\mu = \mu_0$ when $t > 2.045$ or $t < -2.045$. Note that this is very close to the value $z_{.025} = 1.96$ employed in the z test.

It is important to note that the Student's t and the corresponding tabulated critical values are based on the assumption that the sampled population possesses a normal probability distribution. This indeed is a very restrictive assumption because in many sampling situations, the properties of the population will be completely unknown and may be nonnormal. If this were to affect seriously

Figure 9.2 *Tabulated values of Students t*

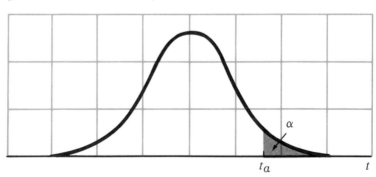

the distribution of the t statistic, the application of the t test would be very limited. Fortunately, this point is of little consequence, as it can be shown that the distribution of the t statistic possesses nearly the same shape as the theoretical t distribution for populations that are nonnormal but possess a mound-shaped probability distribution. This property of the t statistic and the common occurrence of mound-shaped distributions of data in nature enhance the value of the Student's t for use in statistical inference.

We should also note that $\bar{y}$ and s^2 must be independent (in a probabilistic sense) in order that the quantity

$$\frac{\bar{y} - \mu}{s/\sqrt{n}} \quad .$$

possess a t distribution in repeated sampling. As mentioned previously, this requirement will automatically be satisfied when the sample has been randomly drawn from a normal population.

Having discussed the origin of the Student's t and the tabulated critical values (table 4 of the appendix), we now return to the problem of making an inference about the mean diamond weight based on our sample of $n = 6$ measurements. Prior to considering the solution, you may wish to test your built-in inference-making equipment by glancing at the six measurements and arriving at a conclusion about the significance of the data.

9.3 Small-Sample Inferences About a Population Mean

The statistical test of an hypothesis about a population mean may be stated as follows:

Small-Sample Test for a Population Mean

Null hypothesis H_0: $\mu = \mu_0$.

Test statistic: $t = \dfrac{\bar{y} - \mu_0}{s/\sqrt{n}}$.

Alternative hypothesis H_0: Specified by the experimenter, depending on the alternative values of μ which he or she wishes to detect.

Rejection region: See the critical values of t in table 4 of the appendix.

The mean and standard deviation for the six diamond weights are .53

and .0559, respectively, and the elements of the test as defined above are as follows:

Null hypothesis H_0: $\mu = .5$.

Test statistic: $t = \dfrac{\bar{y} - \mu_0}{s/\sqrt{n}} = \dfrac{.53 - .5}{.0559/\sqrt{6}} = 1.31$.

Alternative hypothesis H_0: $\mu > .5$.

Rejection region: The rejection region for $\alpha = .05$ is $t > 2.015$.

Noting that the calculated value of the test statistic does not fall in the rejection region, we do not reject H_0. This implies that the data do not present sufficient evidence to indicate that the mean diamond weight exceeds .5 carats.

The calculation of the probability β of a type II error for the t test is difficult and is beyond the scope of this text. Therefore, we will avoid this problem and obtain an interval estimate for μ, as noted in section 8.15.

We recall that the large-sample confidence interval for μ is

$$\bar{y} \pm z_{\alpha/2} \frac{\sigma}{\sqrt{n}}$$

where $z_{\alpha/2} = 1.96$ for a confidence coefficient of .95. This result assumes that σ is known and simply involves a measurement of $1.96\,\sigma_{\bar{y}}$ (or approximately $2\sigma_{\bar{y}}$) on either side of $\bar{y}$, in conformity with the Empirical Rule. When σ is unknown and must be estimated by a small-sample standard deviation s, the large-sample confidence interval based on z will not enclose μ 95% of the time in repeated sampling. Although we omit the derivation, it seems fairly clear that the corresponding small-sample confidence interval for μ will be as given in the box.

Small-Sample $(1 - \alpha)$ 100% Confidence Interval for μ

$$\bar{y} \pm t_{\alpha/2} \frac{s}{\sqrt{n}}$$

where $s/\sqrt{n}$ is the estimated standard deviation of $\bar{y}$.

For our example a 95% confidence interval for μ is

$$\bar{y} \pm t_{\alpha/2} \frac{s}{\sqrt{n}} = .53 \pm 2.571 \frac{.0559}{\sqrt{6}} = .53 \pm .059$$

Therefore, the interval estimate for μ is .471 to .589, with confidence coefficient of .95. If the experimenter wishes to detect a small increase in mean diamond weight in excess of .5 carats, the width of the interval must be reduced by obtaining more diamond weight measurements. This will decrease both $1/\sqrt{n}$ and $t_{\alpha/2}$ and thereby decrease the width of the interval. Or, looking at it from the standpoint of a statistical test of an hypothesis, more information will be available upon which to base a decision, and the probability of making a type II error will decrease.

Example 9.1 A manufacturer of gunpowder has developed a new powder that is designed to produce a muzzle velocity of 3000 feet per second. Eight shells are loaded with the charge and the muzzle velocities measured. The resulting velocities are shown in the table. Do the data

MUZZLE VELOCITY (FT/SEC)	
3005	2995
2925	3005
2935	2935
2965	2905

present sufficient evidence to indicate that the average velocity differs from 3000 feet per second?

Solution Testing the null hypothesis that $\mu = 3000$ feet per second against the alternative that μ is either greater than or less than 3000 feet per second results in a two-tailed statistical test. Thus

$$H_0: \mu = 3000$$
$$H_a: \mu \neq 3000$$

Using $\alpha = .05$ and placing .025 in each tail of the t distribution, we find that the critical value of t for $n = 8$ measurements [or $(n - 1) = 7$ d.f.] is $t = 2.365$. Hence we will reject H_0 if $t > 2.365$ or $t < -2.365$. (Recall that the t distribution is symmetrical about $t = 0$.)
The sample mean and standard deviation for the recorded data are

$$\bar{y} = 2958.75 \quad \text{and} \quad s = 39.26$$

Then

$$t = \frac{\bar{y} - \mu_0}{s/\sqrt{n}} = \frac{2958.75 - 3000}{39.26/\sqrt{8}} = -2.97$$

Since the observed value of t falls in the rejection region, we reject H_0 and conclude that the average velocity is less than 3000 feet per second. Furthermore, we are reasonably confident that we have made the correct decision. Using our procedure we should erroneously reject H_0 only $\alpha = .05$ of the time in repeated applications of the statistical test.
   A 95% confidence interval provides additional information about μ. Calculating

$$\bar{y} \pm t_{\alpha/2} \frac{s}{\sqrt{n}}$$

we obtain

$$2958.75 \pm (2.365) \frac{39.26}{\sqrt{8}} = 2958.75 \pm 32.83$$

Thus we estimate that the average muzzle velocity lies in the interval from 2925.92 to 2991.58 feet per second. A more accurate estimate can be obtained by increasing the sample size.

Exercises

9.1.   When is the z test inappropriate as a test statistic when the sample size is small? Why is it inappropriate?

9.2.   The specifications for a new heat-resistant alloy state that the amount of copper in the alloy must be less than 23.2%. A sample of 10 analyses of a current batch of the product showed a mean copper content of 23.0% and a standard deviation of .24%. Do the analyses provide sufficient evidence to indicate that the amount of copper in the batch is less than the specified limit? Use $\alpha = .10$.

9.3.   Refer to exercise 9.2. Estimate the mean copper content of the current batch of the alloy using a 90% confidence interval.

9.4.   A random sampling of $n = 24$ items in a supermarket showed a difference between the actual versus the recorded values of the items. The mean and standard deviation of the differences between actual versus recorded values for the 24 items were $-\$37.14$ and $\$6.42$, respectively. Find a 95% confidence interval for the mean difference between actual versus recorded values per item in the supermarket.

9.5.   A building contractor has built a large number of houses of about the same size and value. The contractor claims that the average value of these houses (or similar houses that he might have built) does not exceed $35,000. A real estate appraiser randomly selected five of the new homes built by the contractor and assessed their values as $34,500, $37,000, $36,000, $35,000, and $35,500. Do the five appraisals contradict the building contractor's claim regarding the mean value of his houses? Test at the $\alpha = .05$ level of significance.

9.6.   Fast-food service companies try to devise wage plans that provide incentive and produce salaries for their managers that are competitive with corresponding positions in competing companies. A random sampling of 12 unit managers for one company shows that they earn an average salary of $16,750, with a standard deviation of $3,100. Do these data suggest that the mean salary earned by the company's unit managers differs from $18,500, the mean annual salary paid by a competitive firm to their managers? Test the null hypothesis that $\mu = \$18,500$ against the alternative that $\mu \neq \$18,500$ at the 5% level of significance.

9.7.   Owing to the variability of trade-in allowance, the profit per new car sold by an automobile dealer varies from car to car. The profit per sale, tabulated for the past week, was (in hundreds of dollars)

$$2.1, \quad 3.0, \quad 1.2, \quad 6.2, \quad 4.5, \quad 5.1$$

Find a 90% confidence interval for the average profit per sale.

9.8.   Continuing exercise 9.7, suppose that the profit goal of the dealership is to average at least $480 on the sale of each new car. On the basis of the available data, is there evidence to indicate that the dealership has not met its profit goal? Test at the 5% level of significance.

9.9.   A new gasoline additive has been developed by a West Coast firm, and the developer claims the additive will result in fuel savings of at least 15%.* A city police agency conducted an experiment by using the additive in eight of its cars for a one-week period. The recorded gasoline savings, as measured by the percentage decrease in fuel usage per mile after using the additive, are

$$
\begin{array}{cccc}
15.2 & 14.1 & 13.7 & 15.2 \\
18.6 & 15.0 & 14.5 & 13.8
\end{array}
$$

Do these data contradict the developer's claim that the additive will reduce fuel usage by at least 15%? Use $\alpha = .05$.

9.10.   Continuing exercise 9.9, estimate the mean gasoline savings per car to the police agency if they use the additive regularly, using a 95% confidence interval.

9.4 Small-Sample Inferences About the Difference Between Two Means

The physical setting for the problem we consider here is identical to that discussed in section 8.9. Independent random samples of n_1 and n_2 measurements, respectively, are drawn from two populations, which possess means and variances μ_1, σ_1^2 and μ_2, σ_2^2. Our objective is to make inferences concerning the difference $(\mu_1 - \mu_2)$ between the two population means.

The following small-sample methods for testing hypotheses and placing a confidence interval on the difference between two means are, like the case for a single mean, founded on assumptions regarding the probability distributions of the sampled populations. Specifically, **we will assume that both populations possess a normal probability distribution and also that the population variances σ_1^2 and σ_2^2 are equal.** In other words, we assume that the variability of the measurements in the two populations is the same and can be measured by a common variance, which we will designate as σ^2; that is, $\sigma_1^2 = \sigma_2^2 = \sigma^2$. Although the assumption of equal population variances may be surprising, it is quite reasonable for many sampling situations.

The point estimator of $(\mu_1 - \mu_2)$, $(\bar{y}_1 - \bar{y}_2)$, the difference between the sample means, was discussed in section 8.9, where it was observed to be unbiased and to possess a standard deviation of

$$
\sigma_{(\bar{y}_1 - \bar{y}_2)} = \sqrt{\frac{\sigma_1^2}{n_1} + \frac{\sigma_2^2}{n_2}}
$$

in repeated sampling. This result was used in placing bounds on the error

Oregonian, 30 October 1975.

of estimation, for the construction of a large-sample confidence interval, and for the z test statistic

$$z = \frac{(\bar{y}_1 - \bar{y}_2) - D_0}{\sqrt{\dfrac{\sigma_1^2}{n_1} + \dfrac{\sigma_2^2}{n_2}}}$$

for testing an hypothesis $H_0: \mu_1 - \mu_2 = D_0$, where D_0 is the hypothesized difference between the means. Utilizing the assumption that $\sigma_1^2 = \sigma_2^2 = \sigma^2$, the z test statistic can be simplified as follows:

$$z = \frac{(\bar{y}_1 - \bar{y}_2) - D_0}{\sqrt{\dfrac{\sigma^2}{n_1} + \dfrac{\sigma^2}{n_2}}} = \frac{(\bar{y}_1 - \bar{y}_2) - D_0}{\sigma \sqrt{\dfrac{1}{n_1} + \dfrac{1}{n_2}}}$$

For small-sample tests of the hypothesis $H_0: \mu_1 - \mu_2 = D_0$, it seems reasonable to use the test statistic given in the box.

Small-Sample Test Statistic t

$$t = \frac{(\bar{y}_1 - \bar{y}_2) - D_0}{s \sqrt{\dfrac{1}{n_1} + \dfrac{1}{n_2}}}$$

That is, we substitute a sample standard deviation s for σ. Surprisingly enough, this test statistic possesses a Student's t distribution in repeated sampling when the stated assumptions are satisfied, a fact that can be proved mathematically or verified by experimental sampling from two normal populations.

The estimate s used in the t statistic can be either s_1 or s_2, the standard deviations for the two samples, although the use of either would be wasteful since both estimate σ. Since we wish to obtain the best estimate available, it seems reasonable to use an estimator that pools the information from both samples. This estimator, utilizing the sums of squares of the deviations about the mean for both samples, is given in the next box. Note that the pooled estimator may also be written as

$$s^2 = \frac{(n_1 - 1)s_1^2 + (n_2 - 1)s_2^2}{n_1 + n_2 - 2}$$

As in the case for the single sample, the denominator $(n_1 + n_2 - 2)$ in the formula for s^2 is called the "number of degrees of freedom" associated with s^2. It can be proved mathematically that the expected value of the pooled estimator s^2 is equal to σ^2 and hence that s^2 is an unbiased estimator

---

Pooled Estimator of σ^2

$$s^2 = \frac{\sum\limits_{i=1}^{n_1} (y_i - \bar{y}_1)^2 + \sum\limits_{i=1}^{n_2} (y_i - \bar{y}_2)^2}{n_1 + n_2 - 2}$$

where

$$s_1^2 = \frac{\sum\limits_{i=1}^{n_1} (y_i - \bar{y}_1)^2}{n_1 - 1} \quad \text{and} \quad s_2^2 = \frac{\sum\limits_{i=1}^{n_2} (y_i - \bar{y}_2)^2}{n_2 - 1}$$

---

of the common population variance. Finally, we note that the divisors of the sums of squares of deviations in s_1^2 and s_2^2, $(n_1 - 1)$ and $(n_2 - 1)$, respectively, are the numbers of degrees of freedom associated with these two independent estimators of σ^2. It is interesting to note that an estimator using the pooled information from both samples possesses $(n_1 - 1) + (n_2 - 1)$, or $(n_1 + n_2 - 2)$, degrees of freedom.

Small-Sample Statistical Test for the Difference Between Two Means

$H_0: \mu_1 - \mu_2 = D_0$.

Test statistic: $t = \dfrac{(\bar{y}_1 - \bar{y}_2) - D_0}{s \sqrt{\dfrac{1}{n_1} + \dfrac{1}{n_2}}}$.

The alternative hypothesis H_a and α are specified by the experimenter and are used to locate the critical value of t for the rejection region. s is the square root of s^2, the pooled estimator of σ^2.

The critical value of t can be obtained from table 4 in the appendix. Thus if $n_1 = 10$ and $n_2 = 12$, we would use the t-value corresponding to d.f. $= (n_1 + n_2 - 2) = 20$ degrees of freedom. The following example illustrates the test procedure.

Example 9.2 Manufacturing organizations incur considerable cost in the training of new employees. Not only is there a direct cost involved in the training program, there is also an indirect cost to the firm since employees in training do not contribute directly to the firm's manufacturing process. Hence such organizations seek training programs that can bring new employees to maximum efficiency in the shortest possible time.

Table 9.2 *Length of time (seconds) to assemble device, example 9.2*

STANDARD PROCEDURE		NEW PROCEDURE	
32	44	35	40
37	35	31	27
35	31	29	32
28	34	25	31
41		34	

   An assembly operation in a manufacturing plant requires approximately a one-month training period for a new employee to reach maximum efficiency. A new method of training was suggested and a test conducted to compare the new method with the standard procedure. Two groups of nine new employees were trained for a period of three weeks, one group using the new method and the other following the standard training procedure. The length of time (in minutes) required for each employee to assemble the device was recorded at the end of the three-week period. These measurements appear in table 9.2. Do the data present sufficient evidence to indicate that the mean time to assemble at the end of the three-week training period is less for the new training procedure?

   Solution Let μ_1 and μ_2 be the mean time to assemble for the standard and the new assembly procedures, respectively. Then since we seek evidence to support the theory that $\mu_1 > \mu_2$, we will test the null hypothesis H_0: $\mu_1 = \mu_2$ (i.e., $\mu_1 - \mu_2 = 0$) against the alternative hypothesis H_a: $\mu_1 > \mu_2$ (i.e., $\mu_1 - \mu_2 > 0$). To conduct this test assume that the variability in mean time to assemble is essentially a function of individual differences, that the population distributions of measurements are approximately normal, and that the variability for the two populations of measurements are approximately equal.
   The sample means and sums of squares of deviations are

$$\bar{y}_1 = 35.22 \quad \text{and} \quad \sum_{i=1}^{9} (y_i - \bar{y}_1)^2 = 195.56$$

$$\bar{y}_2 = 31.56 \quad \text{and} \quad \sum_{i=1}^{9} (y_i - \bar{y}_2)^2 = 160.22$$

Then the pooled estimate of the common variance is

$$s^2 = \frac{\sum_{i=1}^{9} (y_i - \bar{y}_1)^2 + \sum_{i=1}^{9} (y_i - \bar{y}_2)^2}{n_1 + n_2 - 2} = \frac{195.56 + 160.22}{9 + 9 - 2} = 22.24$$

and the standard deviation is $s = 4.72$.
   The alternative hypothesis H_a: $\mu_1 > \mu_2$, or, equivalently, $\mu_1 - \mu_2 > 0$, implies that we should use a one-tailed statistical test and that the rejection region for the test is located in the upper tail of the t distribution. Referring to table 4 in the appendix, we note that the critical value

of t for $\alpha = .05$ and $(n_1 + n_2 - 2) = 16$ degrees of freedom is 1.746. Therefore, we will reject the null hypothesis when the calculated value of t is greater than 1.746.

The calculated value of the test statistic is

$$t = \frac{(\bar{y}_1 - \bar{y}_2)}{s \sqrt{\dfrac{1}{n_1} + \dfrac{1}{n_2}}} = \frac{35.22 - 31.56}{4.72 \sqrt{\dfrac{1}{9} + \dfrac{1}{9}}} = 1.64$$

Comparing this with the critical value, $t_{.05} = 1.746$, we note that the calculated value does not fall in the rejection region. Therefore, we must conclude that there is insufficient evidence to indicate that the new method of training is superior, at the .05 level of significance.

The small-sample confidence interval for $(\mu_1 - \mu_2)$ is based on the same assumptions as the statistical test procedure. This confidence interval, with confidence coefficient $(1 - \alpha)$, is given by the formula in the box.

Small-Sample $(1 - \alpha)$ 100% Confidence Interval

for $(\mu_1 - \mu_2)$

$$(\bar{y}_1 - \bar{y}_2) \pm t_{\alpha/2} s \sqrt{\frac{1}{n_1} + \frac{1}{n_2}}$$

Note the similarity in the procedures for constructing the confidence intervals for a single mean, section 9.3, and for the difference between two means. In both cases the interval is constructed by using the appropriate point estimator and then adding and subtracting an amount equal to $t_{\alpha/2}$ times the estimated standard deviation of the point estimator.

Example 9.3 Refer to example 9.2. Since there is expense as well as risk associated with abandoning a working procedure, management would like to estimate the difference $(\mu_1 - \mu_2)$ in learning time by using a confidence coefficient of .95. Find an interval estimate for $(\mu_1 - \mu_2)$ by using a confidence coefficient of .95.

Solution Substituting into the formula

$$(\bar{y}_1 - \bar{y}_2) \pm t_{\alpha/2} s \sqrt{\frac{1}{n_1} + \frac{1}{n_2}}$$

we find that the interval estimate (or 95% confidence interval) is

$$(35.22 - 31.56) \pm (2.120)(4.72) \sqrt{\frac{1}{9} + \frac{1}{9}} = 3.66 \pm 4.71$$

Thus we estimate that the difference $(\mu_1 - \mu_2)$ in mean time to assemble falls in the interval from -1.05 to 8.37. Note that the interval width is considerable and that it seems advisable to increase the size of the samples and reestimate the interval.

Before concluding our discussion it is necessary to comment on the two assumptions upon which our inferential procedures are based. **Moderate departures from the assumption that the populations possess a normal probability distribution do not seriously affect the distribution of the test statistic and the confidence coefficient for the corresponding confidence interval. On the other hand, the population variances should be nearly equal in order to ensure that the procedures given above are valid.**

If there is reason to believe that the population variances are unequal, an adjustment must be made in the test procedures and the corresponding confidence interval. We omit a discussion of these techniques and refer you to the text by Li (1961) listed in the references.

A procedure will be presented in section 9.7 for testing an hypothesis concerning the equality of two population variances.

Exercises

9.11. A manufacturing plant has two assembly machines that perform identical operations on different assembly lines. As a result of their constant use, machine breakdowns occur quite frequently. The time between 10 consecutive breakdowns was recorded for each machine. Suppose that the time between breakdowns for each machine is normally distributed, with a common variance σ^2. The sample means and variances of recorded times between breakdowns are given in the table. Do

MACHINE 1	MACHINE 2
$\bar{y}_1 = 60.4$ min	$\bar{y}_2 = 65.3$ min
$s_1^2 = 31.40$ min^2	$s_2^2 = 44.82$ min^2

these data present sufficient evidence to indicate a difference in the population mean machine breakdown times? Test at the $\alpha = .10$ level of significance.

9.12. A sample of five vice-presidents and four market research analysts of a large industry were asked to estimate what they consider to be the optimal market share for their company. Their responses were as follows:

Vice-Presidents	22.5	25.0	30.0	27.5	20.0
Market Analysts	21.0	17.5	17.0	20.0	

Do these data suggest that corporate vice-presidents and market research analysts tend to disagree when estimating their firm's optimal market share? Test at a 5% level of significance.

9.13. Worker productivity is highly dependent on a number of factors, such as wage rates, task complexity, and the work environment. But oftentimes the design of the job (i.e., the ordered sequencing of worker movements and material inputs) is the most crucial factor influencing worker productivity. Two work designs are under consideration for adoption in a plant. A time and motion study shows that 12 workers using Design A have a mean assembly time of 304 seconds, with a standard deviation of 18 seconds, and that 15 workers using Design B have a mean assembly time of 335 seconds, with a standard deviation of 24 seconds. Do these data present sufficient evidence to indicate a difference in the rate of worker productivity for the two job designs? Test by using $\alpha = .01$.

9.14.   Refer to exercise 9.13. Suppose the manufacturing plant currently uses work Design B in its assembly operations. Design A is an experimental work design and one which should be considered as a possible replacement to Design B if shown to be effective in reducing assembly time. Does this additional information change the statistical test you would use? How?

9.15.   The development of synthetic materials such as nylon, polyester, and latex and their introduction into the marketplace has stirred much debate about the comparative wearing quality and strength of human-made materials versus natural materials. A manufacturer of a new synthetic fiber claims that her product has greater tensile strength than natural fibers. Samples of 10 synthetic fibers and 10 natural fibers were randomly selected and tested for breaking strength. The means and variances

NATURAL FIBER	SYNTHETIC FIBER
$\bar{y}_1 = 272$ lb	$\bar{y}_2 = 335$ lb
$s_1^2 = 1636$ lb^2	$s_2^2 = 1892$ lb^2

for the two samples are given in the table. Do these data support the manufacturer's claim? Test by using $\alpha = .10$.

9.16.   Refer to exercise 9.15. Find a 95% confidence interval for the difference in mean breaking strength between the natural fiber and the synthetic fiber.

9.17.   Continuing exercises 9.15 and 9.16, suppose you wish to estimate the difference in mean breaking strength correct to within 10 pounds, with a probability of approximately .95.

a. Assuming that the sample sizes for natural and synthetic fibers are equal, approximately how large a sample would be required for each fiber?

b. To conduct the experiment using the sample sizes in part a will require additional expenditures of time and money. Can the sample sizes be reduced from those specified in part a and still allow us to achieve the 10-pound bound on the error of estimation?

9.5 A Paired-Difference Test

A manufacturer wishes to compare the wearing qualities of two different types of automobile tires, A and B. To make the comparison a tire of type A and one of type B are randomly assigned and mounted on the rear wheels of each of five automobiles. The automobiles are then operated for a specified number of miles and the amount of wear is recorded for each tire. These measurements appear in table 9.3. Do the data present sufficient evidence to indicate a difference in average wear for the two tire types?

Table 9.3 *Tire wear for two types of tires*

| | TIRE TYPE | |
AUTOMOBILE	A	B
1	10.6	10.2
2	9.8	9.4
3	12.3	11.8
4	9.7	9.1
5	8.8	8.3
	$\bar{y}_1 = 10.24$	$\bar{y}_2 = 9.76$

Analyzing the data we note that the difference between the two sample means is $(\bar{y}_1 - \bar{y}_2) = .48$, a rather small quantity considering the variability of the data and the small number of measurements involved. At first glance it would seem that there is little evidence to indicate a difference between the population means, a conjecture that we may check by the method outlined in section 9.4.

The pooled estimate of the common variance σ^2 is

$$s^2 = \frac{\sum_{i=1}^{n_1} (y_i - \bar{y}_1) + \sum_{i=1}^{n_2} (y_i - \bar{y}_2)}{n_1 + n_2 - 2} = \frac{6.932 + 7.052}{5 + 5 - 2} = 1.748$$

and thus

$$s = 1.32$$

The calculated value of t used to test the hypothesis that $\mu_1 = \mu_2$ is

$$t = \frac{\bar{y}_1 - \bar{y}_2}{s \sqrt{\dfrac{1}{n} + \dfrac{1}{n}}} = \frac{10.24 - 9.76}{1.32 \sqrt{\dfrac{1}{5} + \dfrac{1}{5}}} = .58$$

a value that is not nearly large enough to reject the hypothesis that $\mu_1 = \mu_2$.

The corresponding 95% confidence interval is

$$(\bar{y}_1 - \bar{y}_2) \pm t_{\alpha/2} s \sqrt{\frac{1}{n_1} + \frac{1}{n_2}} = (10.24 - 9.76) \pm (2.306)(1.32) \sqrt{\frac{1}{5} + \frac{1}{5}}$$

or -1.45 to 2.41. Note that the interval is quite wide, considering the small difference between the sample means.

A second glance at the data reveals a marked inconsistency with the conclusion. We note that the wear measurement for type A is larger than the corresponding value for type B for each of the five automobiles. These

Table 9.4 *Differences in tire wear, using the data of table 9.3*

AUTOMOBILE	$d = A - B$
1	.4
2	.4
3	.5
4	.6
5	.5
	$\bar{d} = .48$

differences, recorded as $d = A - B$, are shown in table 9.4.

than B, as a test statistic, as was done in exercise 6.30 in chapter 6. Then the probability that A is larger than B for a given automobile, assuming no difference between the wearing quality of the tires, is $p = \frac{1}{2}$, and so y would be a binomial random variable.

If we choose $y = 0$ and $y = 5$ as the rejection region for a two-tailed test, then $\alpha = p(0) + p(5) = 2(\frac{1}{2})^5 = \frac{1}{16}$. We would then reject H_0: $\mu_1 = \mu_2$ with a probability of a type I error equal to $\alpha = \frac{1}{16}$. Certainly this is evidence to indicate that a difference exists in the mean wear of the two tire types.

You will note that we have employed two different statistical tests to test the same hypothesis. Is it not peculiar that the t test, which utilizes more information (the actual sample measurements) than the binomial test, fails to supply sufficient evidence for rejection of the hypothesis $\mu_1 = \mu_2$?

The explanation of this seeming inconsistency is quite simple. The t test described in section 9.4 is not the proper statistical test to be used for our example. The statistical test procedure of section 9.4 requires that the two samples be **independent and random**. Certainly the independence requirement was violated by the manner in which the experiment was conducted. The (pair of) measurements, an A and a B tire, for a particular automobile are definitely related. A glance at the data shows that the readings have approximately the same magnitude for a particular automobile but vary from one automobile to another. This, of course, is exactly what we might expect. Tire wear, in a large part, is determined by driver habits, the balance of the wheels, and the road surface. Since each automobile has a different driver, we would expect a large amount of variability in the data from one automobile to another.

The familiarity we have gained with interval estimation has shown us that the width of the large-sample and small-sample confidence intervals depends on the magnitude of the standard deviation of the point estimator of the parameter. The smaller its value, the better the estimate and the more likely that the test statistic will provide evidence to reject the null hypothesis if it is, in fact, false. Knowledge of this phenomenon was utilized in designing the tire wear experiment. The experimenter realized that the wear measurements would vary greatly from auto to auto and that this variability could not be separated from the data if the tires were assigned to the 10 wheels in a random manner. (A random assignment of the tires would have implied that the data be analyzed according to the procedure of section 9.4.) Instead, a comparison of the wear between the tire types A and B made on each automobile resulted in the five difference measurements. This design eliminates the effect of the car-to-car variability and yields more information on the mean difference in the wearing quality for the two tire types.

A proper analysis of the data would utilize the five difference measurements to test the hypothesis that the average difference is equal to zero, a statement that is equivalent to the null hypothesis $\mu_1 = \mu_2$.

You may verify that the average and standard deviation of the five difference measurements are

$$\bar{d} = .48 \quad \text{and} \quad s_d = .0837$$

Then

$$H_0: \mu_d = 0$$

and

$$t = \frac{\bar{d} - 0}{s_d/\sqrt{n}} = \frac{.48}{.0837/\sqrt{5}} = 12.8$$

The critical value of t for a two-tailed statistical test with $\alpha = .05$ and 4 degrees of freedom is 2.776. Certainly the observed value of $t = 12.8$ is extremely large and highly significant. Hence we conclude that the average amount of wear for tire type B is less than that for type A.

A 95% confidence interval for the difference between the mean wear is

$$\bar{d} \pm t_{\alpha/2} \frac{s_d}{\sqrt{n}} = .48 \pm (2.776) \frac{.0837}{\sqrt{5}} = .48 \pm .10$$

The statistical design of the tire experiment is a simple example of a **randomized block design** and the resulting statistical test is often called a **paired-difference test**. You will note that the **pairing occurred when the experiment was planned and not after the data were collected.** Comparisons of tire wear were made within relatively homogeneous blocks (automobiles), with the tire types randomly assigned to the two automobile wheels.

An indication of the gain in the amount of information obtained by blocking the tire experiment may be observed by comparing the calculated confidence interval for the unpaired (and incorrect) analysis with the interval obtained for the paired-difference analysis. The confidence interval for $(\mu_1 - \mu_2)$ that might have been calculated, had the tires been randomly assigned to the 10 wheels (unpaired), is unknown but probably would have been of the same magnitude as the interval -1.45 to 2.41, which was calculated by analyzing the observed data in an unpaired manner. Pairing the tire types on the automobiles (blocking) and the resulting analysis of the differences produced the interval estimate .38 to .58. Note the difference in the width of the intervals, which indicates the very sizable increase in information obtained by blocking in this experiment.

Although blocking proved to be very beneficial in the tire experiment, this may not always be the case. We observe that the degrees of freedom available for estimating σ^2 are less for the paired than for the corresponding unpaired experiment. If there were actually no difference between the blocks, the reduction in the degrees of freedom would produce a moderate increase in the $t_{\alpha/2}$ employed in the confidence interval and hence would increase the width of the interval. This, of course, did not occur in the tire experiment because the large reduction in the standard deviation of $\bar{d}$ more than compensated for the loss in degrees of freedom.

Before concluding we want to reemphasize a point. Once you have used a paired design for an experiment, you no longer have the option of using the unpaired analysis of section 9.4. The assumptions upon which that test is based have been violated. Your only alternative is to use the correct method of analysis, the paired-difference test (and associated confidence interval) of this section.

Exercises

9.18. Although not very popular with management because of the cost of implementing its requirements, the Occupational Safety and Health Act (OSHA) has been effective in reducing industrial accidents.* The following data were collected on lost-time accidents (the figures given are mean man-hours lost per month over a period of one year), both before and after OSHA came into effect. Data were

	PLANT NUMBER					
	1	2	3	4	5	6
Before OSHA	38	64	42	70	58	30
After OSHA	31	58	43	65	52	29

recorded for six industrial plants. Do these data provide sufficient evidence to indicate that OSHA has been effective in reducing lost-time accidents? Test at the $\alpha = .10$ level of significance.

9.19. Consider the test of hypothesis in exercise 9.18 from the point of view of (i) management, who wishes to be absolutely certain of the positive effects of OSHA in order to justify its cost of implementation, and (ii) government, who wishes to provide workers with all available health and personal injury safeguards at any reasonable cost. Recommend ‚the selection of a significance level α for the test of hypothesis concerning the effectiveness of OSHA from these points of view:
a. management   b. government

9.20. The Food and Drug Administration requires precise measurements of

*F. Foulkes, "Learning to Live with OSHA," *Harvard Business Review*, December 1973.

meat fat content for labeling and pricing procedures. A meat packing company is considering the use of two different methods to determine the percentage of fat content in samples of meat. Both methods were used to evaluate the fat content in eight different meat samples. The results are given in the table. Do these data

PERCENTAGE FAT	MEAT SAMPLE							
CONTENT, USING	1	2	3	4	5	6	7	8
method 1	23.1	27.1	25.0	27.6	22.2	27.1	23.2	24.7
method 2	22.7	27.4	24.9	27.2	22.5	27.4	23.6	24.4

suggest that the two methods differ in their measurements of meat fat content? Test at the $\alpha = .05$ level of significance.

9.21. Continuing exercise 9.20, use a 90% confidence interval to estimate the difference in mean fat content measurements obtained by using the two measurement methods.

9.22. Five secretaries were selected at random from among the secretaries of a large insurance company. The typing speed (words per minute) was recorded for each secretary on an electric typewriter and on a standard typewriter. The results are given in the accompaning table. Do these data justify the conclusion that the

SECRETARY	ELECTRIC TYPEWRITER	STANDARD TYPEWRITER
1	82	73
2	77	69
3	79	75
4	68	62
5	84	71

secretaries' typing speeds are greater with an electric typewriter than with a standard typewriter? Use $\alpha = .10$.

9.23. Why would you use paired observations to estimate the difference between two population means in preference to estimation based on independent random samples selected from the two populations? For example, in exercise 9.22 what advantage is gained by using the same five secretaries in *both* experimental settings as opposed to using five different secretaries in *each* experiment? Is a paired experiment always preferable? Explain.

9.6 Inferences About a Population Variance

We have seen in the preceding sections that an estimate of the population variance σ^2 is fundamental to procedures for making inferences about the population means. Moreover, there are many practical situations where σ^2

is the primary objective of an experimental investigation; thus σ^2 assumes a position of far greater importance than that of the population mean.

Scientific measuring instruments must provide unbiased readings with a very small error of measurement. An aircraft altimeter that measures the correct altitude on the *average* would be of little value if the standard deviation of the error of measurement were 5000 feet. Indeed, bias in a measuring instrument can often be corrected, but the precision of the instrument, measured by the standard deviation of the error of measurement, is usually a function of the design of the instrument itself and cannot be controlled.

Machined parts in a manufacturing process must be produced with minimum variability in order to reduce out-of-size, and hence defective, products. In general, it is desirable to maintain a minimum variance in the measurements of the quality characteristics of an industrial product in order to achieve process control and therefore minimize the percentage of poor-quality product.

The sample variance

$$s^2 = \frac{\sum_{i=1}^{n} (y_i - \bar{y})^2}{n - 1}$$

is an unbiased estimator of the population variance σ^2. The distribution of sample variances generated by repeated sampling will have a probability distribution that begins at $s^2 = 0$ (since s^2 cannot be negative), with a mean equal to σ^2. Unlike the distribution of $\bar{y}$, the distribution of s^2 is nonsymmetrical, the exact form being dependent on the probability distribution of the population from which the sample measurements were drawn.

For the methodology that follows, we will assume that the sample is drawn from a normal population and that s^2 is based on a random sample of n measurements. Or, using the terminology of section 9.2, we would say that s^2 possesses $(n - 1)$ degrees of freedom.

The next step is to consider the distribution of s^2 in repeated sampling from a specified normal distribution—one with a specific mean and variance—and to tabulate the critical values of s^2 for some of the commonly used tail areas. If this is done, we will find that the distribution of s^2 is independent of the population mean μ but possesses a distribution for each sample size and each value of σ^2. This task is quite laborious, but fortunately it may be simplified by **standardizing,** as was done by using z in the normal tables. The quantity we use for standardizing is shown in the box.

A Chi-square Random Variable

$$\chi^2 = \frac{(n - 1)s^2}{\sigma^2}$$

Table 9.5 *Format of the chi-square table, table 5 in the appendix*

d.f.	$\chi^2_{0.995}$	$\cdots$	$\chi^2_{0.950}$	$\chi^2_{0.900}$	$\chi^2_{0.100}$	$\chi^2_{0.050}$	$\cdots$	$\chi^2_{0.005}$	d.f.
1	0.0000393		0.0039321	0.0157908	2.70554	3.84146		7.87944	1
2	0.0100251		0.102587	0.210720	4.60517	5.99147		10.5966	2
3	0.0717212		0.351846	0.584375	6.25139	7.81473		12.8381	3
4	0.206990		0.710721	1.063623	7.77944	9.48773		14.8602	4
5	0.411740		1.145476	1.61031	9.23635	11.0705		16.7496	5
6	0.675727		1.63539	2.20413	10.6446	12.5916		18.5476	6
$\vdots$	$\vdots$		$\vdots$	$\vdots$	$\vdots$	$\vdots$		$\vdots$	$\vdots$
15	4.60094		7.26094	8.54675	22.3072	24.9958		32.8013	15
16	5.14224		7.96164	9.31223	23.5418	26.2962		34.2672	16
17	5.69724		8.67176	10.0852	24.7690	27.5871		35.7185	17
18	6.26481		9.39046	10.8649	25.9894	28.8693		37.1564	18
19	6.84398		10.1170	11.6509	27.2036	30.1435		38.5822	19

The quantity χ^2, called a chi-square variable by statisticians (χ is the Greek letter chi), admirably suits our purposes. Its distribution in repeated sampling is called, as we might suspect, a chi-square sampling distribution. The equation of the sampling distribution for the chi-square variable is well known to statisticians, who have tabulated critical values corresponding to various tail areas of the distribution. These values are presented in table 5 of the appendix. The format of table 5 in the appendix is shown in table 9.5.

The shape of the chi-square distribution, like that of the *t* distribution, varies with the sample size or, equivalently, with the degrees of freedom associated with s^2. Thus table 5 is constructed in exactly the same manner as the *t* table, with the degrees of freedom shown in the first and last columns. The symbol χ^2_α indicates that the tabulated χ^2-value is such that an area α lies to its right (see figure 9.3). Stated in probabilistic terms,

Figure 9.3 *A chi-square distribution*

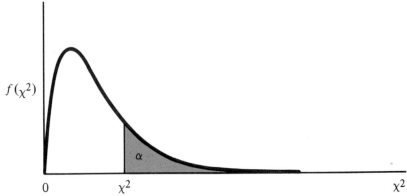

$$P(\chi^2 > \chi_\alpha^2) = \alpha$$

Thus 99% of the area under the χ^2 distribution lies to the right of $\chi^2_{.99}$. We note that the extreme values of χ^2 must be tabulated for both the lower and upper tails of the distribution because it is nonsymmetrical.

You can check your ability to use the table by verifying the following statements. The probability that χ^2, based on $n = 16$ measurements (d.f. $= 15$), will exceed 24.9958 is .05. For a sample of $n = 6$ measurements (d.f. $= 5$), 95% of the area under the χ^2 distribution lies to the right of $\chi^2 = 1.145476$. These entries are shaded in table 9.5.

The statistical test of a null hypothesis concerning a population variance,

$$H_0: \sigma^2 = \sigma_0^2$$

employs the test statistic

$$\chi^2 = \frac{(n - 1)s^2}{\sigma_0^2}$$

If σ^2 is really greater than the hypothesized value σ_0^2, the test statistic will tend to be large and will probably fall toward the upper tail of the distribution. If $\sigma^2 < \sigma_0^2$, the test statistic will tend to be small and will probably fall toward the lower tail of the χ^2 distribution. As in other statistical tests, we may use either a one- or a two-tailed statistical test, depending on the alternative hypothesis that we choose.

Test of an Hypothesis About a Population Variance

Null hypothesis $H_0: \sigma^2 = \sigma_0^2$.

Alternative hypothesis $H_a: \sigma^2 \neq \sigma_0^2$, for a two-tailed statistical test.

Test statistic: $\chi^2 = \dfrac{(n - 1)s^2}{\sigma_0^2}$.

Rejection region: For a two-tailed test, reject H_0 if $\chi^2 < \chi^2_{1-\alpha/2}$ or $\chi^2 > \chi^2_{\alpha/2}$, where $\chi^2_{1-\alpha/2}$ is that value of χ^2 such that $100(1 - \alpha/2)\%$ of the area under the χ^2 distribution, with $(n - 1)$ degrees of freedom, lies to the right.

For alternatives $H_a: \sigma^2 > \sigma_0^2$ or $H_a: \sigma^2 < \sigma_0^2$, use a one-tailed test and place all of α in the appropriate tail (upper or lower) of the χ^2 distribution.

We illustrate a test of an hypothesis about a variance with an example.

Example 9.4 A cement manufacturer claims that concrete prepared from its product possesses a relatively stable compressive strength and that the strength, measured in kilograms per square centimeter, lies within a range of 40 kilograms per square centimeter. A sample of $n = 10$ measurements produced a mean and variance of

$$\bar{y} = 312 \quad \text{and} \quad s^2 = 195$$

Do these data present sufficient evidence to reject the manufacturer's claim?

Solution As stated, the manufacturer claims that the range of the strength measurements equals 40 kilograms per square centimeter. We will suppose that the manufacturer meant that the measurements lie within this range 95% of the time and therefore that the range equals approximately 4σ and that $\sigma = 10$. Then we wish to test the null hypothesis

$$H_0: \sigma^2 = (10)^2 = 100$$

against the alternative

$$H_a: \sigma^2 > 100$$

The alternative hypothesis requires a one-tailed statistical test, with the entire rejection region located in the upper tail of the χ^2 distribution. The critical value of χ^2 for $\alpha = .05$ and $n = 10$ is $\chi^2 = 16.9190$, which implies that we will reject H_0 if the test statistic exceeds this value.

Calculating, we obtain

$$\chi^2 = \frac{(n - 1)s^2}{\sigma_0^2} = \frac{1755}{100} = 17.55$$

Since the value of the test statistic falls in the rejection region, we conclude that the null hypothesis is false and that the range of concrete strength measurements actually exceeds the manufacturer's claim.

A confidence interval for σ^2 with a $(1 - \alpha)$ confidence coefficient is shown in the box.

A $(1 - \alpha)$ 100% Confidence Interval for σ^2

$$\frac{(n-1)s^2}{\chi_U^2} < \sigma^2 < \frac{(n-1)s^2}{\chi_L^2}$$

where χ_L^2 and χ_U^2 are the lower and upper χ^2-values that locate one-half of α in each tail of the chi-square distribution.

For example, a 90% confidence interval for σ^2 in example 9.4 would use

$$\chi_L^2 = \chi_{.95}^2 = 3.32511$$

$$\chi_U^2 = \chi_{.05}^2 = 16.9190$$

Then the interval estimate for σ^2 would be

$$\frac{(9)(195)}{16.9190} < \sigma^2 < \frac{(9)(195)}{3.32511}$$

$$103.73 < \sigma^2 < 527.80$$

Example 9.5 An experimenter is convinced that his measuring equipment possesses a variability measured by a standard deviation $\sigma = 2$. During an experiment he recorded the measurements 4.1, 5.2, and 10.2. Do these data disagree with his assumption? Test the hypothesis H_0: $\sigma = 2$, or $\sigma^2 = 4$, and place a 90% confidence interval on σ^2.

Solution The calculated sample variance is $s^2 = 10.57$. Since we wish to detect $\sigma^2 > 4$ as well as $\sigma^2 < 4$, we should employ a two-tailed test. Using $\alpha = .10$ and placing .05 in each tail, we will reject H_0 when $\chi^2 > 5.99147$ or $\chi^2 < .102587$.
The calculated value of the test statistic is

$$\chi^2 = \frac{(n-1)s^2}{\sigma_0^2} = \frac{(2)(10.57)}{4} = 5.29$$

Since the test statistic does not fall in the rejection region, the data do not provide sufficient evidence to reject the null hypothesis H_0: $\sigma^2 = 4$.
The corresponding 90% confidence interval is

$$\frac{(n-1)s^2}{\chi_U^2} < \sigma^2 < \frac{(n-1)s^2}{\chi_L^2}$$

The values of χ_L^2 and χ_U^2 are

$$\chi_L^2 = \chi_{.95}^2 = .102587$$

$$\chi_U^2 = \chi_{.05}^2 = 5.99147$$

Substituting these values into the formula for the interval estimate, we obtain

$$\frac{2(10.57)}{5.99147} < \sigma^2 < \frac{2(10.57)}{.102587}$$

$$3.53 < \sigma^2 < 206.1$$

Thus we estimate that the population variance falls in the interval from 3.53 to 206.1. This very wide confidence interval indicates how little information about the population variance is obtained in a sample of only three measurements. Consequently, it is not surprising that there was insufficient evidence to reject the null hypothesis $\sigma^2 = 4$. To obtain more information on σ^2, the experimenter needs to increase the sample size.

$$\frac{ns^2}{\sigma^2}$$

$n = 25$
$\bar{y} = 438$
$s = 29$

Exercises

9.24.   Many manufacturers of large household appliances have recently changed from riveting procedures to spot welding in an attempt to cut costs. However, it is important that shear strength and strength uniformity be maintained with spot welding. A manufacturer of an arc welder claims his product can produce spot welds on household appliances that range in shear strength from 400 to 500 pounds. A sample of $n = 25$ spot welds produced by the manufacturer's arc welder were subjected to a shear strength test. The mean and standard deviation of the recorded shear strengths were $\bar{y} = 438$ pounds and $s = 29$ pounds. Do these data present sufficient evidence to reject the arc welder manufacturer's claim? Test by using $\alpha = .05$.

9.25.   Refer to exercise 9.24.

a. Estimate the mean shear strength of spot welds produced by the manufacturer's arc welder, using a 90% confidence interval.

b. Find a 90% confidence interval for the variance of the shear strengths of spot welds produced by the arc welder.

9.26.   Precision instruments, such as those designed to measure volume, temperature, pressure, meat fat content, or content mixture, must be designed to provide a measure that not only is correct on the average but also possesses very little variation around the true value. As an elementary example, consider the volume meter on a gasoline station fuel pump. The station operator insists that the meter not underregister the dispensed volume, while the buyer demands that it not register in excess of the true dispensed amount. Suppose a manufacturer claims that his volume meters register accurately to within .1 gallon of the actual amount of gasoline dispensed through the meter. The manufacturer's meter is installed in a gasoline pump and five different samples of exactly 10 gallons of gasoline are dispersed through the meter and the pump. The recorded volumes on the meter for the five samples are as follows: 10.05, 10.00, 9.90, 9.95, 10.15. Do these data support or refute the meter manufacturer's claim? Conduct the test at a 5% level of significance. (Hint: If the manufacturer claims his product is accurate to within .1 gallon of the actual amount, he is specifying that the recorded amount will vary by no more than .1 gallon *in either direction* from the actual amount. Therefore, he is specifying that the *range* of readings for a particular amount dispensed will be no more than .2 gallons.)

9.7 Comparing Two Population Variances

The need for statistical methods to compare two population variances is readily apparent from the discussion in section 9.6. We may frequently wish to compare the precision of one measuring device with that of another, the stability of one manufacturing process with that of another, or the variability in the grading procedure of one college professor with that of another.

Intuitively, we might compare two population variances σ_1^2 and σ_2^2 by using the ratio of the sample variances s_1^2/s_2^2. If s_1^2/s_2^2 is nearly equal to 1, we would find little evidence to indicate that σ_1^2 and σ_2^2 are unequal. On the other hand, a very large or small value for s_1^2/s_2^2 would provide evidence of a difference in the population variances.

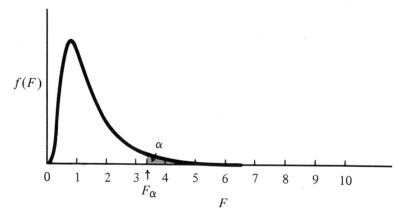

Figure 9.4 *An F distribution with* $v_1 = 10$ *and* $v_2 = 10$

How large or small must s_1^2 / s_2^2 be to provide sufficient evidence for rejecting the null hypothesis $H_0: \sigma_1^2 = \sigma_2^2$? The answer to this question may be obtained by studying the distribution of s_1^2 / s_2^2 in repeated sampling.

When independent random samples are drawn from two normal populations with equal variances, that is, $\sigma_1^2 = \sigma_2^2$, then s_1^2 / s_2^2 possesses a sampling distribution that is known to statisticians as an **F distribution**. We need not concern ourselves with the equation for the sampling distribution for F except to state that, as we might surmise, it is reasonably complex. For our purposes it will suffice to accept the fact that the distribution is well known and that critical values have been tabulated. These appear in tables 6 and 7 of the appendix.

The shape of the F distribution is nonsymmetrical and depends on the number of degrees of freedom associated with s_1^2 and s_2^2. We represent these quantities as v_1 and v_2, respectively. An F distribution with $v_1 = 10$ and $v_2 = 10$ is shown in figure 9.4. This fact complicates the tabulation of critical values of the F distribution and necessitates the construction of a table for each value that we may choose for a tail area α. Tables 6 and 7 of the appendix present critical values corresponding to $\alpha = .05$ and $\alpha = .01$, respectively.

For example, table 6 in the appendix records the value F such that the probability that F exceeds $F_{.05}$ is .05. Another way of saying this is that 5% of the area under the F distribution lies to the right of $F_{.05}$ (see figure 9.4). The degrees of freedom v_1 for s_1^2 are indicated across the top of the table, while the degrees of freedom v_2 for s_2^2 appear in the first column on the left (see table 9.6).

Referring to table 9.6, we note that $F_{.05}$ for sample sizes $n_1 = 7$ and $n_2 = 10$ (that is, $v_1 = 6$, $v_2 = 9$) is 3.37. Likewise, the critical value $F_{.05}$ for sample sizes $n_1 = 9$ and $n_2 = 16$ ($v_1 = 8$, $v_2 = 15$) is 2.64. These entries are shaded in table 9.6.

In a similar manner, the critical values for a tail area of $\alpha = .01$ are

Table 9.6 *Format of the F table, table 6 in the appendix, α = .05*

					NUMERATOR DEGREES OF FREEDOM									
ν_1	1	2	3	4	5	6	7	8	9	$\cdots$	60	120	∞	ν_1
ν_2														ν_2
1	161.4	199.5	215.7	224.6	230.2	234.0	236.8	238.9	240.5	$\cdots$	252.2	253.3	254.3	1
2	18.51	19.00	19.16	19.25	19.30	19.33	19.35	19.37	19.38	$\cdots$	19.48	19.49	19.50	2
3	10.13	9.55	9.28	9.12	9.01	8.94	8.89	8.85	8.81	$\cdots$	8.57	8.55	8.53	3
4	7.71	6.94	6.59	6.39	6.26	6.16	6.09	6.04	6.00	$\cdots$	5.69	5.66	5.63	4
5	6.61	5.79	5.41	5.19	5.05	4.95	4.88	4.82	4.77	$\cdots$	4.43	4.40	4.36	5
6	5.99	5.14	4.76	4.53	4.39	4.28	4.21	4.15	4.10	$\cdots$	3.74	3.70	3.67	6
7	5.59	4.74	4.35	4.12	3.97	3.87	3.79	3.73	3.68	$\cdots$	3.30	3.27	3.23	7
8	5.32	4.46	4.07	3.84	3.69	3.58	3.50	3.44	3.39	$\cdots$	3.01	2.97	2.93	8
9	5.12	4.26	3.86	3.63	3.48	3.37	3.29	3.23	3.18	$\cdots$	2.79	2.75	2.71	9
$\vdots$	$\vdots$	$\vdots$	$\vdots$	$\vdots$	$\vdots$	$\vdots$	$\vdots$	$\vdots$	$\vdots$		$\vdots$	$\vdots$	$\vdots$	$\vdots$
15	4.54	3.68	3.29	3.06	2.90	2.79	2.71	2.64	2.59	$\cdots$	2.16	2.11	2.07	15
16	4.49	3.63	3.24	3.01	2.85	2.74	2.66	2.59	2.54	$\cdots$	2.11	2.06	2.01	16
17	4.45	3.59	3.20	2.96	2.81	2.70	2.61	2.55	2.49	$\cdots$	2.06	2.01	1.96	17
$\vdots$														$\vdots$

Denominator Degrees of Freedom (row label, left side)

presented in table 7 of the appendix. Thus

$$P(F > F_{.01}) = .01$$

The statistical test of the null hypothesis

$$H_0: \sigma_1^2 = \sigma_2^2$$

utilizes the test statistic

$$F = \frac{s_1^2}{s_2^2}$$

When the alternative hypothesis implies a one-tailed test, that is,

$$H_a: \sigma_1^2 > \sigma_2^2$$

we may use the tables directly. However, when the alternative hypothesis requires a two-tailed test, that is,

$$H_a: \sigma_1^2 \neq \sigma_2^2$$

we note that the rejection region will be divided between the lower and upper tails of the F distribution and that tables of critical values for the lower tail are conspicuously missing. The reason for their absence is not difficult to explain.

We are at liberty to identify either of the two populations as population 1. If the population with the larger sample variance is designated as population 2, then $s_2^2 > s_1^2$ and we will be concerned with rejection in the lower tail of the F distribution. **Since the identification of the population is arbitrary, we may avoid this difficulty by designating the population with the larger variance as population 1.** In other words, **always place the larger sample variance in the numerator of**

$$F = \frac{s_1^2}{s_2^2}$$

and designate that population as 1. Then since the area in the right-hand tail represents only $\alpha/2$, we double this value to obtain the correct value for the probability α of a type I error. Hence if we use table 6 for a two-tailed test, the probability of a type I error is $\alpha = .10$.

Test of an Hypothesis About the Equality of Two

Population Variances

Null hypothesis H_0: $\sigma_1^2 = \sigma_2^2$.
Alternative hypothesis H_a: $\sigma_1^2 \neq \sigma_2^2$, for a two-tailed statistical test.
Test statistic: $F = s_1^2/s_2^2$, where s_1^2 is the larger sample variance.
Rejection region: For a two-tailed test, reject H_0 if $F > F_{\alpha/2}$, where $F_{\alpha/2}$
                       is based on $(n_1 - 1)$ and $n_2 - 1)$ degrees of freedom
                       and the value $\alpha/2$ is doubled to obtain the correct
                       value for the probability α of a type I error. (If we
                       use table 6 for a two-tailed test, the probability of a
                       type I error is $\alpha = .10$.)

We illustrate these ideas with some examples.

Example 9.6 The risk of alternative investments is generally evaluated by the variance of returns associated with each investment.* The distribution of returns for two alternative investments A and B is shown in figure 9.5. The expected rate of return on each investment is 17.8%, but on the basis of returns over the past 10 years for investment A and 8 years

Figure 9.5 *Distribution of rates of return for investments A and B*

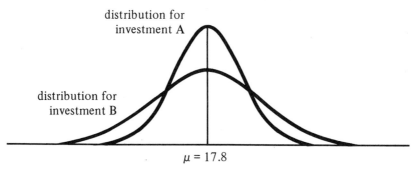

$\mu = 17.8$

Rate of Return

*W. F. Sharpe, *Portfolio Theory and Capital Markets* (New York: McGraw-Hill, 1970).

for investment B, the variance of returns for the two investments are 3.21 and 7.14, respectively. Do these variances present sufficient evidence to indicate that the risks of investments A and B are unequal? (That is, is there a difference in the population variances?)

Solution Assume that the populations possess probability distributions that are reasonably mound-shaped and hence will satisfy, for all practical purposes, the assumption that the populations are normal.

We wish to test the null hypothesis

$$H_0: \sigma_A^2 = \sigma_B^2$$

against the alternative

$$H_a: \sigma_A^2 \neq \sigma_B^2$$

Using table 6 and doubling the tail area, we will reject H_0 when $F > 3.29$ with $\alpha = .10$, since $n_A = 10$ and $n_B = 8$. (Note that $F_{.05} = 3.29$ is based on $v_1 = 7$ and $v_2 = 9$ degrees of freedom.)

The calculated value of the test statistic is

$$F = \frac{s_B^2}{s_A^2} = \frac{7.14}{3.21} = 2.22$$

Noting that the test statistic does not fall in the rejection region, we do not reject $H_0: \sigma_A^2 = \sigma_B^2$. Thus there is insufficient evidence to indicate a difference between the population variances.

Example 9.7 Consistency in the taste of beer is an important quality in retaining customer loyalties. The variability in the taste of a given beer can be affected by the length of brewing, variations in ingredients, and differences in the equipment used in the brewing process. A brewery with two production lines, 1 and 2, has made a slight adjustment to line 2, hoping to reduce the variability as well as the average taste index. Random selections of $n_1 = 25$ and $n_2 = 25$ eight-ounce glasses of beer were selected from the two production lines and were measured using an instrument designed to index the beer taste. The two samples produced means and variances as follows:

$$\bar{y}_1 = 3.2 \qquad \bar{y}_2 = 3.0$$
$$s_1^2 = 1.04 \qquad s_2^2 = .51$$

Do the data present sufficient evidence to indicate that the process variability is less for process 2? (That is, test the null hypothesis H_0: $\sigma_1^2 = \sigma_2^2$ against the alternative $H_a: \sigma_1^2 > \sigma_2^2$.)

Solution Testing the null hypothesis

$$H_0: \sigma_1^2 = \sigma_2^2$$

against the alternative

$$H_a: \sigma_1^2 > \sigma_2^2$$

at an $\alpha = .05$ significance level, we will reject H_0 when F is greater than $F_{.05} = 1.98$; that is we will employ a one-tailed statistical test.

We observe that the calculated value of the test statistic,

$$F = \frac{s_1^2}{s_2^2} = \frac{1.04}{.51} = 2.04$$

falls in the rejection region, and hence we conclude that the variability of process 2 is less than that for process 1.

The test for the equality of two population variances does not assume that the means are equal, but we might notice that the process averages in both examples 9.6 and 9.7 are nearly equal. What the test does examine is the **comparative uniformity of individual values within each population.** For instance, in example 9.6 if we had rejected our hypothesis, we would have concluded that the closing prices of one stock are significantly more volatile than the closing prices of the other. Thus we would conclude that even though the average prices of the two stocks are nearly equal, one stock involves more risk than the other, as indicated by its larger variance. Since the hypothesis of equal variances was not rejected for that example, there is not sufficient evidence to indicate a difference in risks for the two stocks. Note: A difference in risks may exist but it was not evident based on samples of sizes $n_A = 10$ and $n_B = 8$.

Exercises

9.27. A production plant has two extremely complex fabricating systems, with one being twice the age of the other. Both systems are checked, lubricated, and maintained once every two weeks. The number of finished products fabricated daily by each of the systems is recorded for 30 working days. The results are given

NEW SYSTEM	OLD SYSTEM
$\bar{y}_1 = 246$	$\bar{y}_2 = 240$
$s_1 = 15.6$	$s_2 = 28.2$

in the table. Do these data present sufficient evidence to conclude that the variability in daily production warrants increased maintenance of the older fabricating system? Use a 5% level of significance.

9.28. The stability of measurements of the characteristics of a manufactured product is important in maintaining product quality. In fact, it is sometimes better to have a small variation in the measured value of some important characteristic of a product and have the process mean slightly off target than to suffer wide variation with a mean value that perfectly fits requirements. The latter situation may produce a higher percentage of defective products than the former. A manufacturer of light bulbs suspects that one of her production lines is producing bulbs with a higher variation in length of life. To test this theory she compares the lengths of life of $n = 50$ bulbs randomly sampled from the suspect line and $n = 50$ from a line that seems to be in control. The sample means and variances for the two samples are given in the table. Do these data present sufficient evidence to indicate that bulbs

SUSPECT LINE	LINE IN CONTROL
$\bar{y}_1 = 1,520$	$\bar{y}_2 = 1,476$
$s_1^2 = 92,000$	$s_2^2 = 37,000$

produced by the suspect line possess a larger variance in length of life than those produced by the line that is assumed to be in control? Use $\alpha = .05$.

9.29.   In the metal-fabricating industry, productivity and subsequent profitability are highly dependent on the quality and uniformity of needed raw materials. Suppose there are two principal sources of raw materials under consideration for use by a metal fabricator in a certain process. Both sources appear to have similar quality characteristics, but the manufacturer is not certain about their respective uniformity of impurities content. Ten 100-pound samples of each source are selected and the amount, in pounds, of impurities is measured for each sample. The results

RAW MATERIAL A	RAW MATERIAL B
$\bar{y}_1 = 41.3$	$\bar{y}_2 = 39.6$
$s_1^2 = 18.75$	$s_2^2 = 7.85$

are given in the table. Do these data suggest that a difference exists in the uniformity of impurities content in the two raw materials? Test with a 10% level of significance.

9.30.   Continuing exercise 9.29, test to see if a difference exists in the mean impurities content within the 100-pound samples of the two raw materials. Use $\alpha = .05$.

9.8 Summary

It is important to note that the t, χ^2, and F statistics employed in the small-sample statistical methods discussed in the preceding sections are based on the assumption that the sampled **populations possess a normal probability distribution.** This requirement will be satisfied for many types of experimental measurements.

You will observe the close relationship connecting the Student's t and the z statistic and therefore the similarity of the methods for testing hypotheses and the construction of confidence intervals. The χ^2 and F statistics employed in making inferences about population variances do not, of course, follow this pattern, but the reasoning employed in the construction of the statistical tests and confidence intervals is identical for all the methods we have presented.

Supplementary Exercises

9.31.   What assumptions are made when a Student's t test is employed to test an hypothesis about a population mean? What is the result if these assumptions are not valid but a Student's t test is used anyway?

9.32.   A chemical process has produced, on the average, 800 tons of chemical per day. The daily yields for the past week are 785, 805, 790, 793, and 802 tons. Do these data indicate that the average yield is less than 800 tons and hence that something is wrong with the process? Test at the 5% level of significance.

9.33.   Find a 90% confidence interval for the mean yield in exercise 9.32.

9.34.   Refer to exercises 9.32 and 9.33. How large should the sample be in order that the width of the confidence interval be reduced to approximately 5 tons?

9.35.   A random sample of $n = 20$ secretaries employed by public agencies in a certain city was randomly chosen and found to earn an average hourly wage of $y = \$3.95$, with a standard deviation of $s = \$.58$. Find a 95% confidence interval for the mean hourly wage rate for secretaries employed by public agencies in the city.

9.36.   A coin-operated soft drink machine was designed to discharge, on the average, 7 ounces of beverage per cup. To test the machine 10 cups of beverage were drawn from the machine and measured. The mean and standard deviation of the 10 measurements were 7.1 ounces and .12 ounces, respectively. Do these data present sufficient evidence to indicate that the mean discharge differs from 7 ounces? Test at the 10% level of significance.

9.37.   Find a 90% confidence interval for the mean discharge in exercise 9.36.

9.38.   A manufacturer can tolerate a small amount (.05 milligrams per liter) of impurities in a raw material needed for manufacturing its product. Because the laboratory test for the impurities is subject to experimental error, the manufacturer tests each batch 10 times. Assume that the mean value of the experimental error is 0 and hence that the mean value of the 10 test readings is an unbiased estimate of the true amount of the impurities in the batch. For a particular batch of the raw material, the mean of the 10 test readings is .058 milligrams per liter (mg/l), with a standard deviation of .012 mg/l. Do the data provide sufficient evidence to indicate that the amount of impurities in the batch exceeds .05 mg/l?

9.39.   Continuing exercise 9.38, find a 90% confidence interval for the mean milligrams/liter of impurities in the batch of the raw material.

9.40.   A cigarette manufacturer claims that the mean amount of nicotine in his cigarettes does not exceed 25 milligrams of nicotine. A sample of 16 cigarettes yielded a mean and standard deviation of 26.4 and 2, respectively. Do the data provide sufficient evidence to refute the manufacturer's claim? Use $\alpha = .10$.

9.41.   The vice-president of a large commercial banking chain has recorded the percentage gain in demand deposit accounts for 17 of the branch banks during the past year. The average gain computed from the sample was 11.3% and the standard deviation was 3.4%.

a. Find a 90% confidence interval for the mean growth rate of demand deposits for the banking chain during the past year.

b. Give a practical interpretation of the confidence interval calculated in part a.

9.42.   A manufacturer of television sets claims that her product possesses an average defect-free life of 3 years. Three households in a community have purchased the sets and all three sets are observed to fail before 3 years, with failure times of 2.5, 1.9, and 2.9 years. Do these data present sufficient evidence to contradict the manufacturer's claim? Test at the $\alpha = .05$ level of significance.

9.43.   Calculate a 90% confidence interval for the mean life of the television sets in exercise 9.42.

9.44. Refer to exercises 9.42 and 9.43. Approximately how many observations would be required to estimate the mean life of the television sets correct to within .2 years with a probability of .90?

9.45. A cannery prints "weight 16 ounces" on its label. The quality control supervisor selects nine cans at random and weighs them. He finds $\bar{y} = 15.7$ and $s = .5$. Do the data present sufficient evidence to indicate that the mean weight of the cans is less than the amount claimed on the label? Use $\alpha = .05$.

9.46. A magazine subscription service conducts two sales training programs for prospective salespeople. Each salesperson participates in one training program or the other, but not both. To measure the efficiency of each training program, a company representative randomly selects a group of salespeople and records the number of sales they consummated in the past month. The sample means and variances computed from the sales data are given in the table. Do the data present sufficient

	TRAINING PROGRAM 1	TRAINING PROGRAM 2
Number of Salespeople	11	14
$\bar{y}$	64	69
s^2	52	71

evidence to indicate a difference in the mean number of monthly sales for the populations associated with the two sales training programs? Test at the $\alpha = .05$ level of significance.

9.47. A comparison of the time before recognition of the product for two different-colored advertising layouts of a common product produced the following results (in seconds) when applied to a random sample of 16 people:

Layout 1	1	3	2	1	2	1	3	2
Layout 2	4	2	3	3	1	2	3	3

a. Do the data present sufficient evidence to indicate a difference in mean recognition time for the two layouts? Test at the $\alpha = .05$ level of significance.
b. Obtain a 90% confidence interval for $(\mu_1 - \mu_2)$.
c. Give a practical interpretation for the confidence interval in part b.

9.48. Refer to exercise 9.47. Suppose that the product recognition experiment had been conducted by using people as blocks and by making a comparison of recognition time within each person; that is, each of the eight persons would be subjected to both layouts in a random order. The data for this experiment (in seconds) are given in the table.

Person	1	2	3	4	5	6	7	8
Layout 1	3	1	1	2	1	2	3	1
Layout 2	4	2	3	1	2	3	3	3

a. Do the data present sufficient evidence to indicate a difference in mean recognition time for the two layouts? Test at the $\alpha = .05$ level of significance.
b. Obtain a 95% confidence interval for $(\mu_1 - \mu_2)$.

c. Give a practical interpretation for the confidence interval obtained in part b.

9.49.   Analyze the data in exercise 9.48 as though the experiment had been conducted in an unpaired manner. Calculate a 95% confidence interval for $(\mu_1 - \mu_2)$ and compare it with the interval obtained in part b of exercise 9.48. Does it appear that blocking increased the amount of information available in the experiment?

9.50.   There are numerous methods available for evaluating inventories. Two of the more common methods are LIFO and FIFO, with inherent advantages and disadvantages for each. However, in times of inflation, LIFO tends to reduce taxes and improve cash flow.* A multiproduct firm is contemplating changing from FIFO to LIFO as a method for evaluating inventories. Five different finished goods inventories are evaluated at the end of the year by using both FIFO and LIFO. The results

| | INVENTORY VALUE ($\times$ 1000) | |
PRODUCT	FIFO	LIFO
1	121	117
2	217	198
3	92	105
4	98	86
5	52	49

are given in the table. Do these data suggest that LIFO is effective in reducing the inventory value of the firm's finished goods inventories? Test with an $\alpha = .05$ level of significance.

9.51.   An auditor for a hardware store chain wished to compare the efficiency of two different auditing techniques. To do so he selected a sample of nine store accounts, applied auditing technique A, then selected another nine store accounts

TECHNIQUE A		TECHNIQUE B
	125	89
	116	101
	133	97
	115	95
	123	94
	120	102
	132	98
	128	106
	121	98
Σy	1,113	880
$\bar{y}$	123.7	97.8
Σy^2	137,973	86,240

and applied auditing technique B. The number of material (accounting) errors found in each store's account is shown in the table. Determine whether there is evidence

*See O. R. Keister, "LIFO and Inflation," *Management Accounting*, May 1975.

of a difference in the mean number of accounting errors detected by the two auditing techniques. Test at the $\alpha = .10$ level of significance.

9.52. Would the amount of information extracted from the data in exercise 9.51 be increased by pairing successive observations and analyzing the differences? Suppose we assume that only nine store accounts were selected and that the data from exercise 9.51 represent a paired experiment that gives the number of accounting errors found in each of the nine store accounts using each of the two auditing techniques.

a. Calculate a 90% confidence interval for $(\mu_1 - \mu_2)$ assuming the data of exercise 9.51 are from a paired experiment.

b. Calculate a 90% confidence interval for $(\mu_1 - \mu_2)$ assuming the data of exercise 9.51 are from an unpaired experiment.

c. Compare the widths of the intervals in parts a and b.

9.53. Since shelf space is a limited resource of a retail store, product selection, shelf space allocation, and shelf space placement decisions must be made according to a careful analysis of profitability and inventory turnover. The manager of a chain of variety stores wishes to see whether shelf location affects the sales of a certain product. She believes that placing the product at eye level will result in greater sales than will placing the product on a lower shelf. The data shown in the table represent

| | NUMBER SOLD | |
STORE	Lower Shelf	Eye-Level Shelf
1	27	33
2	22	23
3	32	38
4	32	33

the number of sales of the product in four different stores. Sales were observed over two weeks, with product placement at eye level one week and on a lower shelf the other week. Test, at the 5% level of significance, to determine whether placement of the product at eye level significantly increases sales.

9.54. An investor wishes to determine if growth stocks associated with utility companies tend to outperform industrial high-yield stocks. From a list of high-yield stocks recommended by several leading brokerage houses, the investor randomly selects six industrial stocks and five utilities. Using Standard and Poor's value line estimate of yield (dividend as a percentage of price) for each chosen stock, the investor obtained

Industrials	7.8	10.3	7.9	8.7	9.2	8.9
Utilities	9.2	9.1	11.1	8.8	9.6	

the data shown in the table. Do these data suggest that growth stocks of utilities outperform industrial growth stocks? Test at the 5% level of significance.

9.55. Refer to exercise 9.54.

a. Estimate the difference in mean yield between industrial growth stocks and utility growth stocks, using a 95% confidence interval.

b. Find a 95% confidence interval for the average yield of utility growth stocks.

9.56. A manufacturer of a machine that packages soap powder claims that his machine can fill cartons at a given weight with a range of no more than $\frac{2}{3}$ oz.

The mean and variance of a sample of eight 3-lb boxes were found to be 3.1 lb and .018 oz^2, respectively. Do these results tend to refute the manufacturer's claim? Test by using a 5% level of significance. (Hint: If the manufacturer's claim is that the range is less than or equal to .4 oz, then, since the range is approximately 4σ, an hypothesis consistent with the manufacturer's claim is $\sigma^2 = .01$ against the alternative $\sigma^2 > .01$.)

9.57.   Continuing exercise 9.56, find a 90% confidence interval for σ^2, the variance of fill using the manufacturer's fill machine. Use these results to find a 90% confidence interval for the *range* of fill using the manufacturer's fill machine.

9.58.   A dairy is in the market for a new bottle-filling machine and is considering models A and B manufactured by Company A and Company B, respectively. If ruggedness, cost, and convenience are comparable in the two models, the deciding factor is the variability of fills (the model producing fills with the smaller variance being preferred.) Wishing to demonstrate that the variability of fills is less for Model A than for Model B, a salesman for Company A acquired a sample of 30 fills from a machine of Model A and a sample of 10 fills from a machine of Model B. The sample variances were $s_A^2 = .027$ and $s_B^2 = .065$. Do these sample variances provide statistical support at the .05 level of significance for the salesman's belief?

9.59.   The temperature of operation of two paint-drying ovens associated with two manufacturing production lines was recorded for 20 days. (Pairing was ignored.) The means and variances of the two samples are

$$\bar{y}_1 = 164 \qquad \bar{y}_2 = 168$$
$$s_1^2 = 81 \qquad s_2^2 = 172$$

Do the data present sufficient evidence to indicate a difference in temperature variability for the two ovens? Test the hypothesis that $\sigma_1^2 = \sigma_2^2$ at the $\alpha = .10$ level of significance.

9.60.   Refer to exercise 9.59. Do these data suggest that there is a difference in mean drying temperature for the two paint-drying ovens? Use $\alpha = .10$.

9.61.   A chemical manufacturer claims that the purity of his product never varies more than 2%. Five batches were tested and gave purity readings of 98.2%, 97.1%, 98.9%, 97.7%, 97.9%. Do these data provide sufficient evidence to contradict the manufacturer's claim? Test by using a 5% level of significance.

9.62.   The office of the Lane County Assessor is considering the use of a computer valuation model for obtaining the assessed valuations for each of the residential dwellings of Lane County. The assessor is interested in seeing how assessed valuations obtained from the model compare with assessments made by an "expert"

| | ASSESSED VALUATIONS | |
DWELLING	Computer Model	Expert Valuer
1	$21,000	$20,000
2	37,500	36,000
3	42,000	40,000
4	28,000	28,500
5	30,000	31,000
6	36,500	35,000
7	44,500	44,000
8	23,000	24,500
9	46,000	44,000
10	30,000	29,500

valuer. To note the comparison the assessor randomly selects 10 residential dwellings from Lane County, computes their assessed valuation using the computer model, and obtains an assessed valuation for each from an expert valuer. The results are given in the table.

a. What is the advantage of conducting the experiment as a paired experiment by blocking on the houses?

b. What would be the effect of testing the hypothesis H_0: $\mu_1 = \mu_2$ against the alternative H_a: $\mu_1 \neq \mu_2$ by using the methods of section 9.4 and assuming that the two sets of values constitute two independent samples? Would such an analysis be (theoretically) incorrect?

c. Analyzing the problem as a paired-difference experiment, do the sample data present sufficient evidence to indicate a difference in mean assessed valuation for the two assessment procedures? Use $\alpha = .05$.

d. Obtain a 95% confidence interval for $(\mu_1 - \mu_2)$. Interpret this interval.

e. Suppose that the above sample information is truly representative of all 40,000 residential dwelling units in Lane County. Furthermore, assume that the expert valuer is giving as true a measure as is possible for each residential dwelling unit he is asked to value. What is the estimated gain (loss) in *total* Lane County property valuation by using the computer model for valuation instead of hiring the more reliable, but also slower and more expensive, expert valuer?

f. Obtain a 95% confidence interval for the gain (loss) in total Lane County property valuation by using the computer valuation model.

g. Property taxes are assessed in Lane County at the annual rate of 3% of the assessed valuation of the residential dwelling. What estimated additional (lesser) amount of property tax revenue can Lane County expect to incur by using the computer valuation model?

h. Obtain a 95% confidence interval for the additional (lesser) amount of property tax revenue Lane County is expected to receive if they decide to use the computer valuation model instead of hiring an expert valuer.

Experiences with Real Data

The purpose of the following two experiments is to compare the means of two populations by using random samples selected from the two populations. The first experiment is a paired comparison test (section 9.5), while the second involves the analysis of independent samples drawn from the two populations (section 9.4). The focus of the study in our two experiments is on the comparison of the performance and price characteristics of products sold by discount retail establishments with those sold by ordinary (nondiscount) retail establishments. Conduct a statistical test to determine whether your data provide sufficient evidence to indicate a difference in the population means in each exercise.

1. Is there a difference in the performance characteristics of gasoline purchased from discount service stations and gasoline purchased from a station of one of the major national oil companies? Have a group of $n = 10$ or more of the students in your class who own cars perform a mileage experiment to provide the experimental data for your study. At the start of the experiment, carefully record the mileage of each car and make sure the gasoline tank is almost empty. Have some of the car owners participating in the experiment purchase a tank of gasoline at a discount station and the rest start out with a tank purchased at a station affiliated with one of the major national oil companies. Have each individual operate his car as he would normally until the fuel tank is almost empty. Then reverse the process. Those who began with discount gasoline should fill their tanks with gasoline purchased from a major national oil company station, while the others should obtain a tank of gasoline from a discount station. Then compute the mileage, in terms of the average

miles per gallon, obtained by each car when using each type of gasoline. You should then have two figures associated with each car: one measuring its mileage using the discount gasoline, the other its mileage using the gasoline purchased from an outlet of one of the major national oil companies. Do your data provide sufficient evidence to assume that the average gasoline mileage differs for cars fueled by gasoline purchased from these two sources?

2. Do discount department stores and variety stores really provide a cost savings to the consumer, compared to what they would pay for similar items in a regular (nondiscount) store? Select a number of typical consumer items such as, say, a man's dress shirt, a child's sweater, a record album, a pair of women's shoes, a set of bed sheets, a package of tennis balls, or any other items you care to list. Price these items in each of n_1 different discount stores. Then price items of *similar quality* to those priced in the discount stores in n_2 different nondiscount retail stores. Find the total cost for purchasing your list of consumer items in each store where prices were observed. Using the methods of section 9.4, do your data suggest that the mean price of your list of consumer items purchased in discount stores is less than the mean price of a list of items of similar quality purchased from nondiscount stores?

Calculation of the required statistics in these two exercises can easily be accomplished on a calculator, but you can use a library program and an electronic computer to perform the calculations. Useful library programs are available in the Biomed series and the SPSS package; but more likely than not, the computer on your campus has a built-in subroutine to compute means and variances of data sets. You should check with someone in your computer center to find out what programs are available and which statistical subroutines are built into the computing system of your campus.

References

FREUND, J. E., and F. J. WILLIAMS; revised by B. Perles and C. Sullivan. *Modern Business Statistics.* Englewood Cliffs, N.J.: Prentice-Hall, 1969. Chapters 9, 10, and 11.

HOEL, P. G., and R. J. JESSEN. *Basic Statistics for Business and Economics.* New York: Wiley, 1971. Chapter 7.

LI, J. C. R. *Introduction to Statistical Inference.* Ann Arbor, Mich.: J. W. Edwards, Publisher, 1961. Chapter 10.

WONNACOTT, T. H., and R. J. WONNACOTT. *Introductory Statistics.* New York: Wiley, 1969. Chapters 7, 8, and 9.

chapter objectives

GENERAL OBJECTIVE In this chapter we show how decision analysis extends the classical approach to decision making (chapters 8 and 9) by formally integrating the decision maker's personal preferences and perceptions regarding uncertainty and value into the decision framework. Decision analysis requires that the decision maker identify available alternatives and payoffs associated with each alternative and assess probabilities to the uncertain states of nature. Therefore, decision analysis forces the decision maker to assume a more active role in the decision process than that which is required by the classical approach.

SPECIFIC OBJECTIVES

1. To show how decision analysis extends the classical approach to decision making.
> **Section 10.1**

2. To illustrate the various components of the decision analysis model and illustrate its use in practice.
> **Sections 10.2, 10.3, 10.4**

3. To justify the expected monetary (value) objective of the decision analysis model.
> **Sections 10.5, 10.6**

4. To show how the decision analysis model can simultaneously accommodate both judgmental and empirical information.
> **Section 10.7**

5. To identify other topics in decision analysis.
> **Section 10.8**

318

chapter ten

Decision Analysis

10.1 Introduction

The statistical tests of hypotheses and inferential methods presented in chapters 8 and 9 are all based on sample information. As we recall, the objective of this classical, or empirical, approach to statistics is to make inferences about certain characteristics of a population—usually its mean or proportion of successes—based on information contained in a sample drawn from the population. Certain assumptions are made about the distribution of the population as suggested by the nature of the population or as implied by the Central Limit Theorem. On the basis of these assumptions, we can determine a sampling distribution for our statistic of interest and hence construct an appropriate confidence interval or test of our hypothesis.

The classical approach was extended by Robert Schlaifer and others in the early 1960s to enable the decision maker to formally integrate his or her personal preferences and perceptions regarding uncertainty and value into the decision framework. This modern approach to decision making, which has come to be known as decision analysis, can be defined as the logical and quantitative analysis of all the factors that influence a decision. Decision makers are thus forced to assume more active roles in the decision-making process. In the end, they rely more on rules consistent with their logic and personal behavior and less on the mechanical use of a set of formulas and tabulated probabilities.

One particular area of difference between the classical approach to statistics and decision analysis concerns the treatment of errors that may result from the application of each procedure. The classical approach employs the probabilities α and β of the two types of errors to evaluate the goodness of a test. In application, the classical approach often causes an arbitrary reliance on rules and formulas—as evidenced by the almost religious zeal attached to the significance level $\alpha = .05$. We know that great care should be taken when selecting the error probabilities α and β in a statistical test of an hypothesis. For example, these probabilities should consider losses associated with the errors they define as well as any prior information that may not be formally involved with the sample data used to calculate the test statistic. But quite often an experimenter *arbitrarily* selects an α-value (say $\alpha = .05$). Since the choice of α is often arbitrary, the decision analysis approach suggests that the *expected loss* is a more practical and realistic criterion for comparing two statistical test procedures. As we will see, decision analysis utilizes the concept of gain (or loss) associated with every possible decision available to the decision maker and selects the decision that maximizes the expected gain.

The primary purpose of decision analysis is to increase the likelihood of good outcomes by making good decisions. A good outcome is one that we would like to occur, while a good decision is one that is consistent with the information and preferences of the decision maker. Thus decision analysis attempts to provide a decision-making framework based on any and all information, whether it be sample information, judgmental information, or a combination of both.

In summary, decision analysis is not really in conflict with the classical approach to statistical inference. The difference is simply in the degree of formality of the decision-making procedures employed and the extent to which the decision maker himself plays an active or a passive role in the formulation of the decision.

10.2 Certainty Versus Uncertainty

In most decision-making situations the decision maker must choose one of several possible actions or alternatives. Usually the alternatives and their associated payoffs are known in advance to the decision maker. An investor choosing one investment from several investment opportunities, a store owner determining how many of a certain type of commodity to stock, and a company executive making capital-budgeting decisions are all examples of a business decision maker selecting from a multitude of alternatives. However, the decision maker does not know which alternative will be best in each case unless he or she also knows with certainty the values of the economic variables that affect profit. We call these variables states of nature since they represent different events that may occur or states over which the decision maker has no control.

Generally, the states of nature in a decision problem are denoted by the symbols $s_1, s_2, \ldots, s_k$. That is, the first state of nature is denoted by s_1, the second by the symbol s_2, and so forth. It is assumed that the states of nature are **mutually exclusive** (no two states can be in effect at the same time) and **collectively exhaustive** (all possible states are included within the analysis).

If n actions or alternatives are available to the decision maker, they are generally labeled a_1, a_2, ... , a_n. It is also assumed that the alternatives constitute a mutually exclusive, collectively exhaustive set.

The exact identity of the alternatives and states of nature associated with a decision-making situation is unique to that problem. An investor may wish to invest a certain sum of money in a savings account, a treasury bond, or a number of shares of a mutual fund. His available alternatives are the investment of his money in one of the three mentioned investment opportunities, while the states of nature that may influence his economic payoffs are the possible states of the securities market in the near future. Another investor may wish to invest none, part, or all of the same or a different sum of money among the same three investment opportunities. His alternatives include all possible splits of his investment sum among the three investment opportunities and may be practically endless. In such cases alternatives may have to be grouped and others eliminated in order to arrive at an analytical solution to the decision-making problem. In any decision-making problem, ingenuity and perceptiveness are required of the decision maker since he or she must identify the available actions or alternatives and their associated payoffs and must identify the states of nature that may affect the outcome.

When the state of nature s_j is known, or, if unknown, has no influence on the outcome of the alternatives, we say that the decision maker is operating under **certainty**. Otherwise, we say the decision is made under

Decision making under certainty is usually the simpler of the two techniques. The decision maker simply evaluates the outcome of each alternative and selects the one that best meets his objective. However, if the number of alternatives is very large, even in the absence of uncertainty, the best decision may be difficult to identify. For example, consider the problem of a delivery agent who must make 100 or so deliveries to different residences scattered over a broad urban area. There may be literally thousands of different alternate routes the agent could select, yet only one supplies a least cost alternative. However, if he had only 3 stops to make, the agent could easily find the least cost route.

Under uncertainty the decision-making task is always complicated. Probability theory and mathematical expectation offer tools for establishing the logical procedures for selecting the best alternative. Statistics provides the structure to reach the decision. But the decision maker must inject his intuition and knowledge of the problem into the decision-making framework to arrive at the decision that is both theoretically justifiable and intuitively appealing. A good theoretical framework and a commonsense approach are both essential ingredients for decision making under uncertainty. One cannot function properly without the other.

To understand these ideas, consider an investor who wishes to invest

$1000 in one of three possible investment opportunities. Investment A is a savings plan with a commercial bank returning 6% annual interest, and investment B is a government bond returning the equivalent of $4\frac{1}{2}\%$ annual interest. These two investments involve no risk and hence no uncertainty to the investor.

Suppose that investment C consists of shares of a mutual fund with a wide diversity of available holdings from the securities market. The annual return from an investment in C depends on the *uncertain* behavior of the mutual fund under varying economic conditions.

Assuming that the investor either will invest his $1000 in one of the aforementioned investment opportunities or will not invest at all, his available actions are as follows:

a_1: Do not invest.
a_2: Select investment A, the 6% commercial bank savings plan.
a_3: Select investment B, the $4\frac{1}{2}\%$ government band.
a_4: Select investment C, the uncertain mutual fund.

Actions a_1, a_2, and a_3 do not involve uncertainty as the outcomes associated with the selection of any of these actions do not depend on the uncertain market conditions. We could therefore say it is clear that action a_2 dominates actions a_1 and a_3. In general, one action is said to dominate another when the second action is never preferred to the first, regardless of the state of nature. Unless the investor attaches patriotic considerations to the selection of the government bond, investment B would never be selected when investment A is available. Similarly, action a_1, which provides for no growth of the principal amount, is clearly inferior to the risk-free positive-growth investment alternatives a_2 and a_3.

Investment C, the mutual fund investment, is another matter. This alternative (a_4) is associated with an uncertain outcome that, depending on the state of the economy, may produce either a negative return or a positive return. Thus there exists no immediately apparent dominance relationship between action a_4 and action a_2 (the best among the actions involving no uncertainty, a_1, a_2, and a_3).

Suppose that the investor believes that if the market is down in the next year, an investment in the mutual fund would lose 10%; if the market stays the same, the investment would stay the same; and if the market is up, the investment would gain 20%. The investor has thus defined the states of nature for his investment decision-making problem as follows:

s_1: The market is down.
s_2: The market remains unchanged.
s_3: The market is up.

A study of the market combined with economic expectations for the coming year may lead the investor to attach subjective probabilities of $\frac{1}{4}$, $\frac{1}{4}$, and

$\frac{1}{2}$, respectively, to the states of nature s_1, s_2, and s_3. The question of interest is, then, how can the investor use the foregoing information regarding investments A, B, and C and the expected market behavior as an aid in selecting the investment that best satisfies his objectives? We will consider this question in sections 10.3 and 10.4.

Exercises

10.1.   For each of the following business decision-making problems, list the actions available to the businessperson and the states of nature that might result to affect the payoff.

a. The marketing of a new brand of soft drink in a midwestern state.

b. The decision of a delicatessen manager about the number of loaves of French bread he should order when he knows the daily demand ranges from 10 to 15 loaves a day.

c. The decision of a building contractor about the number of speculative single-family homes he should build when he has the initial resources to build up to seven homes.

d. Deciding whether or not to bid on a construction contract to build an office building.

e. The pricing of a newly developed airless spray dispenser for paint.

f. The decision of an owner of a city taxi service about the automobile manufacturer from which to buy a new fleet of cabs.

10.2.   Is decision making under certainty always a simpler task than decision making under uncertainty? Explain.

10.3.   Select a major decision which you have had to make recently, such as the choice of where to go to college, the choice of which job offer to accept upon graduation, the choice of where to invest a certain part of your income, or some other appropriate decision problem. Identify the alternatives available to you and the states of nature associated with your decision problem. Did your decision problem involve decision making under certainty or under uncertainty?

10.3 Analysis of the Decision Problem

In any problem involving the choice of one from many alternatives, we first must identify all the actions we may take and all the states of nature whose occurrence may influence our decision. The action to take none of the listed alternatives whose outcome is known with certainty may also be included within this list. Associated with each action is a list of payoffs to the decision maker. These are the value consequences to him if he takes a specific action, assuming that each of the states of nature occurs. Of course, if an action does not involve risk, the payoff will be the same no matter which state of nature occurs.

The payoffs associated with each possible outcome in a decision problem are listed in a payoff table, which is defined as follows:

Definition

A **payoff table** is a listing in tabular form of the value payoffs associated with all possible actions under every state of nature in a decision problem.

The payoff table is usually displayed in grid form, with the states of nature indicated in the columns and the actions in the rows. If we label the actions $a_1, a_2, \ldots, a_n$ and the states of nature $s_1, s_2, \ldots, s_k$, a payoff table for a decision problem would appear as shown in table 10.1. A payoff is entered in each of the nk cells of the payoff table, one for the payoff associated with each action under every possible state of nature.

It is important at this point to digress for a moment and examine what is meant by the concept of a "payoff." What is the decision maker attempting to accomplish? Assuming that he is a rational decision maker, he will study the list of alternatives and select the action that best satisfies his objectives. The payoffs he assigns to each possible outcome must be assessed in value units consistent with his objectives.

These value units may, in fact, not be measurable in monetary units at all. Or the value consequences may be represented in a payoff table as some measure combining profit and a subjective measure such as "possible environmental impact."

The following problem settings illustrate the importance of satisfying a goal by other than short-term profit maximization.

A doctor is faced with the problem of treating a group of factory employees, all of whom exhibit a rash on their upper bodies. She does not know the source of the rash, but for similar rashes she knows that an inexpensive medicated lotion works quite well. She also has access to a new experimental, but costly, drug that may prove effective. Ten patients are administered the lotion and 10 are given the experimental drug. After three days those who received the new drug are all back at work and completely free of the effects of the rash. Nine of the 10 patients who received the lotion are cured within

Table 10.1 *Payoff table*

ACTION	STATE OF NATURE				
	s_1	s_2	s_3	$\cdots$	s_k
a_1					
a_2					
a_3					
$\vdots$					
a_n					

six days, but the lotion does not effectively cure the tenth patient at all. The new experimental drug is then administered to all future employees who exhibit the rash. Clearly the cost of administering the new drug is incidental in this case. Cure rate and time lost from the job were the measures of payoff considered.

A company is considering building a subsidiary plant in one of two locations. The most costly location for building is in an area with chronic unemployment and low per capita income. The more expensive site location is chosen. Thus the company executives have sought to maximize some social goal, perhaps identified as the firm's public image, instead of minimizing building costs. Payoff has been measured in social value units or public image units and is represented accordingly in the payoff table.

Defense planning is made using a probability of damage payoff measure to rank alternatives. Timber companies adopt a payoff model combining timber harvesting costs and environmental considerations, such as clear-cutting versus selective harvesting. Scientific expeditions measure payoff on an arbitrary scale in terms of the comparative worth of expected findings to the scientific community and the public.

An example of the construction of a **value model** follows. This example involves the selection of one from among three alternatives under certainty since the effect of each attribute of value is known in advance to the decision maker.

Example 10.1 The president of a large manufacturing company is considering three potential locations as sites at which to build a subsidiary plant. To decide which location to select for the subsidiary plant (only one location will be selected), the president will determine the degree to which each location satisfies the company's objective of minimizing transportation costs, minimizing the effect of local taxation, and having access to an ample pool of available semiskilled workers. Construct a payoff model and select payoff measures that effectively rank each potential location according to the degree to which each satisfies the company's objectives for transportation costs, taxation costs, and the available work force.

Solution Let us call the three potential locations site A, site B, and site C. To determine a payoff measure to associate with each attribute (company objective) under each alternative, the president subjectively assigns a rating on a 0-to-10 scale to measure the degree to which each location satisfies the company's objectives for transportation costs, taxation costs, and the available work force. (For each objective, a rating of 0 indicates complete dissatisfaction while a rating of 10 indicates complete satisfaction.) The results are shown in table 10.2.

To combine the components of payoff, the president asks himself, what are the relative measures of importance of the three company objectives I have considered as components of payoff? Suppose he decides that minimizing transportation costs is most important and twice as important as either the minimization of state and local taxation or the size

Table 10.2 *Ratings for the three alternative plant sites, example 10.1*

| | ALTERNATIVE | | |
COMPANY OBJECTIVE	Site A	Site B	Site C
transportation costs	6	4	10
taxation costs	6	9	5
work force pool	7	6	4

of the work force available. He thus assigns a weight of 2 to the transportation costs and weights of 1 to taxation costs and work force. The following payoff measures result:

$$\text{payoff (site A)} = 6(2) + 6(1) + 7(1) = 25$$

$$\text{payoff (site B)} = 4(2) + 9(1) + 6(1) = 23$$

$$\text{payoff (site C)} = 10(2) + 5(1) + 4(1) = 29$$

It is clear that before a decision maker can rationally evaluate a set of alternatives, he must **clearly identify his goals and objectives.** Having done this he must then **define a payoff measure that can effectively rank the outcomes according to the amount by which they satisfy his goals and objectives.** The payoff measure is thus unique to the decision maker and his preferences and certainly unique to each new problem setting.

Advanced methods can and do handle most classes of nonmonetary outcomes in a decision analysis. However, these methods are reserved for a later course. For economy of time and space, the examples in this chapter will assume an analysis involving **monetary payoff,** measurable by the **profit** or **opportunity loss** associated with each.

Definition

The **opportunity loss** $L_{i,j}$ for selecting action a_i, given a state of nature s_j, is the difference between the maximum profit that could be realized if s_j occurs and the profit obtained by selecting action a_i. The opportunity loss is never negative.

The opportunity loss is zero at the optimal action for each state of nature. A nonzero opportunity loss does not necessarily imply an accounting loss but only that the decision maker had the opportunity to realize $L_{i,j}$ more in profit had he selected the best action instead of action a_i. The concept of an opportunity loss (sometimes called the **regret**) need not be concerned with lost profits. In a general sense, an opportunity loss is a measure of the amount of payoff that has been lost by taking a particular action a_i under state of nature s_j.

A payoff table displaying the opportunity losses in a decision analysis is called an **opportunity loss table,** and one showing profits is called a **profit table.**

The decision process consists of selecting from the list of possible alternatives the action that best satisfies the decision maker's objectives. This decision is called the **optimal decision.** Various economic objectives and the decision processes necessary to accomplish these objectives will be examined in sections 10.4, 10.7, and 10.8.

Example 10.2 For the investment problem discussed in section 10.2, construct the opportunity loss table.

Solution Define the actions "select no investment," "select investment A," "select investment B," and "select investment C" as a_1, a_2, a_3, and a_4, respectively. Also, let s_1, s_2, and s_3 represent the states of the securities market "down," "about the same," and "up," respectively.

Actions a_1, a_2, and a_3 are not affected by the states of nature, but action a_4 is. The resultant profits associated with this problem are listed in table 10.3.

Table 10.3 *Profit table for the investment problem (dollar units)*

ACTION	STATE OF NATURE		
	s_1, Down	s_2, The Same	s_3, Up
a_1 (none)	0	0	0
a_2 (A)	60	60	60
a_3 (B)	45	45	45
a_4 (C)	−100	0	200

We can see that if state of nature s_1 or s_2 is in effect, the best action is to select a_2, the 6% savings plan. If the securities market is up, the $200 profit for investing in investment C far exceeds what we could realize by selecting any other action.

From the definition of opportunity loss given earlier and from the information in table 10.3, we can now construct the opportunity loss table. It is given in table 10.4.

Table 10.4 *Opportunity loss table for the investment problem*

ACTION	STATE OF NATURE		
	s_1, Down	s_2, The Same	s_3, Up
a_1 (none)	60	60	200
a_2 (A)	0	0	140
a_3 (B)	15	15	155
a_4 (C)	160	60	0

Example 10.3 A retailer must decide each morning how many of a perishable commodity to include in her inventory. The commodities cost her

$2 per unit and are sold for $5 per unit, realizing a profit of $3 per unit. Any unsold items at the day's end are a total loss to the retailer. Construct the opportunity loss table.

Solution In this problem the states of nature are the demand levels, and the actions are the possible inventory levels.

The profits associated with each action are given in table 10.5, where inventory levels 1, 2, and 3 are denoted by a_1, a_2, and a_3, respectively.

Table 10.5 *Profit table for the demand-inventory problem*

	STATE OF NATURE, DEMAND		
ACTION, INVENTORY	$s_1(1)$	$s_2(2)$	$s_3(3)$
$a_1(1)$	3	3	3
$a_2(2)$	1	6	6
$a_3(3)$	−1	4	9

Similarly, the demand levels are labeled s_1, s_2, and s_3. It then follows directly that the opportunity losses would be as shown in table 10.6.

Table 10.6 *Opportunity loss table for the demand-inventory problem*

	STATE OF NATURE, DEMAND		
ACTION, INVENTORY	$s_1(1)$	$s_2(2)$	$s_3(3)$
$a_1(1)$	0	3	6
$a_2(2)$	2	0	3
$a_3(3)$	4	2	0

In the investment problem of section 10.2, the states of nature that will be in effect have no bearing on the outcome of investments A and B. But with investment C, the unknown state of nature that will be in effect does influence the outcome. We hope that this illustrates the importance of an earlier remark: Personal judgment and subjective reasoning are required of the decision maker to identify the relevant states of nature and to place precise probabilities on their occurrence. The intuition, perception, and judgment of the decision maker are thus an integral part of the analysis in modern decision making.

Definition

The probabilities representing the chances of occurrence of the identifiable states of nature in a decision problem before gathering any sample information are called **prior probabilities**.

How should the prior probabilities be determined? As indicated in example 10.2, we may sometimes be able to estimate the probabilities of different states of nature from previous experience. Our probability estimate may then just be the relative frequency of occurrence of that particular state of nature in the past. However, in other decision problems our feeling about which states of nature are more likely to occur cannot be summarized solely by such historical relative frequencies. In those cases some other method of assessing prior probabilities is required. [See Howard (1966) or Winkler (1972) listed in the references for some techniques available in eliciting personal opinion concerning the prior probabilities.]

In short, there is no simple answer to the question of how to select the prior probabilities. Realizing that the optimal action is often quite sensitive to only a slight change in the set of prior probabilities, the decision maker seeks a precise and meaningful set of prior probabilities. To accomplish this end he uses all available information at his disposal, and he relies on his judgment and experience to process that information and to identify, as well as possible, his prior probabilities.

We will not expand on the necessary aspects of the subjective elements of our decision framework and will assume now that they have been formulated. The discussion in the remainder of this chapter presents some logical procedures for selecting the best action after all states of nature have been identified and their associated payoffs and probabilities have been computed.

Exercises

10.4.   Reconstruct your decision to attend the college at which you are now currently enrolled. Pick one or two other colleges that you might have chosen and construct a payoff model for your selection decision like the one discussed in example 10.1. As the value measures, choose educational quality and social environment of the college and location, then rate each college on a 0-to-10 scale according to each dimension of value. Combine the value measures according to their comparative measures of importance. Was your decision to attend the college at which you are currently enrolled consistent with the outcome of your analysis?

10.5.   An owner of a newsstand buys a certain weekly newsmagazine for $.60 a copy and sells it for $1.00 a copy. Past evidence suggests the weekly demand for the newsmagazine ranges from 9 to (but not including) 13 copies. If unsold copies cannot be returned for refund at the week's end, construct the owner's weekly profit table for the newsmagazine. Construct an opportunity loss table.

10.6.   One researcher notes that floating currency exchange rates have created a great amount of uncertainty for those dealing in import and export trade.* For example, consider the case of an importer of goods who is considering an offer from a New Zealand manufacturer of sheepskin rugs to distribute the manufacturer's finished goods in the United States. Recent New Zealand currency devaluations have created a favorable pricing situation for New Zealand products in foreign markets, but an upward float of the New Zealand dollar against the U.S. dollar would, of course, have an opposite effect. If the importer accepts the manufacturer's offer, he envisions the revenues shown in the table (net cost of goods sold) under each

*W. D. Serfass, Jr., "You Can't Outguess the Foreign Exchange Market," *Harvard Business Review*, March–April 1976.

EXCHANGE RELATIONSHIP	NET REVENUES
downward float of N.Z. dollar	$30,000
no change	15,000
upward float of N.Z. dollar	7,500

of three possible exchange relationships that might occur between U.S. and New Zealand currencies over the next fiscal year. If the import agreement is accepted, the rugs would be transported to the U.S. market from New Zealand by air freight at an annual cost of $9,000.

a. Identify the alternatives and the states of nature for the importer's decision problem.

b. Construct the importer's profit table.

c. Construct the importer's opportunity loss table.

10.4 Expected Monetary Value Decisions

One decision-making procedure, which employs both the payoff table and the prior probabilities associated with the states of nature to arrive at a decision, is called the expected monetary value decision procedure.

Definition

An **expected monetary value decision** is the selection of an available action based on either the expected opportunity loss or the expected profit of the action.

The optimal expected monetary value decision is the selection of the action associated with the minimum expected opportunity loss or the action associated with the maximum expected profit, depending on the decision maker's objective. We will note later that expected opportunity loss decisions and expected profit decisions are always associated with the same optimal decision.

The concept of expected monetary value is an application of mathematical expectation (section 5.5), where opportunity loss or profit is the random variable and the prior probabilities represent the probability distribution associated with the random variable. The expected opportunity loss for each action a_i is found by evaluating the formula given in the box.

If we are interested in computing the expected profits for each action, the analysis would be the same except that we would substitute the profits, denoted by the symbol $G_{i,j}$ for the opportunity losses $L_{i,j}$ in the formula given in the box. (The symbol G, for gain, is used to denote profits in order

---

Computing the Expected Opportunity Loss by Using
Prior Probabilities

$$E(L_i) = \sum_{\text{all } j} L_{i,j} P(s_j) \qquad i = 1, 2, \ldots$$

where $L_{i,j}$ is the opportunity loss for selecting action a_i given that state
of nature s_j occurs and $P(s_j)$ is the prior probability assigned to state of
nature s_j.

---

to avoid the confusion which might result by using P, which throughout
this text is used to represent the probability of an event.)

The following examples should clarify the computational procedures
involved with the expected monetary value decision procedure.

Example 10.4 Refer to example 10.2. Find the optimal decision that
minimizes the investor's expected opportunity loss.

Solution The prior probabilities associated with the states of nature s_1,
s_2, and s_3 are $P(s_1) = \frac{1}{4}$, $P(s_2) = \frac{1}{4}$, and $P(s_3) = \frac{1}{2}$. Thus the expected
opportunity losses are

$$E(L_1) = \$60(\tfrac{1}{4}) + \$60(\tfrac{1}{4}) + \$200(\tfrac{1}{2}) = \$130$$

$$E(L_2) = \$0(\tfrac{1}{4}) + \$0(\tfrac{1}{4}) + \$140(\tfrac{1}{2}) = \$70$$

$$E(L_3) = \$15(\tfrac{1}{4}) + \$15(\tfrac{1}{4}) + \$155(\tfrac{1}{2}) = \$85$$

$$E(L_4) = \$160(\tfrac{1}{4}) + \$60(\tfrac{1}{4}) + \$0(\tfrac{1}{2}) = \$55$$

Action a_4, the mutual fund investment, is the optimal expected monetary
value decision, since its expected opportunity loss is less than those for
other actions.

Actions a_1 and a_3 could have been eliminated from the analysis at the
beginning, since their payoffs are never better than those for a_2. They
were retained in our example for illustrative purposes.

As an aside, suppose that we compute the expected profits for each
action of example 10.2. If G_i represents the profit realized by selecting action
a_i, then we find

$$E(G_1) = \$0(\tfrac{1}{4}) + \$0(\tfrac{1}{4}) + \$0(\tfrac{1}{2}) = \$0$$

$$E(G_2) = \$60(\tfrac{1}{4}) + \$60(\tfrac{1}{4}) + \$60(\tfrac{1}{2}) = \$60$$

$$E(G_3) = \$45(\tfrac{1}{4}) + \$45(\tfrac{1}{4}) + \$45(\tfrac{1}{2}) = \$45$$

$$E(G_4) = -\$100(\tfrac{1}{4}) + \$0(\tfrac{1}{4}) + \$200(\tfrac{1}{2}) = \$75$$

Consistent with our analysis based on opportunity losses, we see that action a_4 is best, since it is associated with the maximum expected profit. This is no accident but will always happen. A proof of this assertion follows from the definition of opportunity loss. **The difference between the expected opportunity losses of any two actions is equal in magnitude to, but opposite in sign from, the difference between their expected profits.** For instance, in example 10.4 if we look at actions a_2 and a_4, we find that

$$E(L_4) - E(L_2) = \$55 - \$70 = -\$15$$

$$E(G_4) - E(G_2) = \$75 - \$60 = \$15$$

Under either method action a_4 is \$15 better than a_2.

Example 10.5 By recording the daily demand over a period of time for the perishable commodity described in example 10.3, the retailer was able to construct the following probability distribution for the daily demand level(s):

s_j	$P(s_j)$
1	.5
2	.3
3	.2
4 or more	0

Find the inventory level that minimizes the expected opportunity loss for the demand-inventory problem described in example 10.3.

Solution The historical frequencies of demand—the prior probabilities for this case—are given in the table. The expected opportunity losses are found by evaluating

$$E(L_i) = \sum_{j=1}^{3} L_{i,j} P(s_j)$$

for each inventory level, $i = 1, 2, 3$.
    The expected opportunity losses at each inventory level are

$$E(L_1) = \$0(.5) + \$3(.3) + \$6(.2) = \$2.10$$

$$E(L_2) = \$2(.5) + \$0(.3) + \$3(.2) = \$1.60$$

$$E(L_3) = \$4(.5) + \$2(.3) + \$0(.2) = \$2.60$$

Thus the retailer's optimal decision is to stock 2 units of the commodity.

Note that in example 10.5 the prior probabilities assigned to the states of nature were constructed empirically from historical data. However, subjectivity and intuition are also implicit in this case, as evidenced by the fact that past historical frequencies are believed to adequately depict the future. In any decision problem the prior probabilities reflect all available information, both empirical and subjective, related to the likelihoods of occurrence of the states of nature.

Exercises

10.7.   Refer to exercise 10.5. On the basis of historical evidence, the owner of the newsstand believes the weekly demand for the newsmagazine is as follows:

Demand	9	10	11	12
Probability of Demand	.20	.15	.35	.30

Find the number of issues of the weekly newsmagazine the newsstand owner should buy each week in order to maximize his expected weekly profits.

10.8.   Refer to exercise 10.6. Suppose that after consultation with an expert on international monetary issues, the importer decides that the prior probabilities are as follows: the probability that the New Zealand dollar will float downward from the U.S. dollar within the next year is .10, the probability that the exchange rate will stay about the same is .50, and the probability that an upward float of the New Zealand dollar will occur is .40. If the importer wishes to maximize his expected profits, should he accept the importing agreement with the New Zealand manufacturer? What is the expected profit associated with the decision to accept the importing agreement?

10.9.   An investor is considering one of two alternatives for the investment of $25,000. On the one hand, he can accept a risk-free investment by investing the sum at 7% annual interest with a local savings and loan institution. On the other hand, he can invest the sum with a speculative developer who is building an elegant restaurant in a reasonably undeveloped area of the city. If successful, the restaurant investment is seen to return a net present value of $100,000 at the end of one year on the investor's $25,000 investment, with a probability of .25. If unsuccessful, it could be run as a fast-food restaurant with a one-year net present value return of $15,000, or, if that proves unsuccessful, it could be turned into a storage warehouse at a return of $7,500 on the original $25,000 investment. The probabilities of these latter two events occurring are seen to be .50 and .25, respectively. If the investor wishes to maximize his expected return over the next year, should he invest in the risk-free savings account or invest in the speculative restaurant venture? What is the investor's expected capital gain for a one-year investment in each of the two investment opportunities?

10.10.   Decision analysis is often used in new product introduction decisions.* For instance, suppose an executive must decide whether or not her company should introduce onto the market a new product that would appeal to a limited class of consumers. The executive has determined that the revenues returned to the firm are directly related to the proportion of consumers who buy the product by the equation $revenue = \$50,000 + \$100,000(p)$, where p is the proportion of consumers who buy. Based on a careful analysis of the market of potential consumers, the executive has developed the following probability distribution for p:

Proportion of Consumers, p	.10	.20	.30	.40	.50	.60
Probability, $P(p)$	.10	.15	.30	.25	.15	.05

*See F. J. Anscombe, "Bayesian Statistics," *American Statistician*, February 1961.

If development, advertising, and promotional costs would run to as much as $75,000 for the product, should the executive recommend that her company introduce the new product? (Hint: Do expected revenues cover costs?)

10.5 Justification of Expected Monetary Value Decisions

Expected profit or opportunity loss decisions are most easily interpreted if we look at the decision in the long run. If the optimal action is repeated many times, the average of the opportunity losses incurred using one procedure would be less than if the decision maker employed some other decision-making scheme. This idea is easy to see in example 10.5. If the retailer stocks 2 units every day, her average of all daily opportunity losses will be a minimum and will be about $1.60, which is less than the average loss incurred by stocking 1 or 3 units each day.

In an experiment that cannot be repeated, such as the investment problem of section 10.2 or a capital-budgeting problem, expected monetary value decisions are not so easily justified. For the investment problem we would expect the average profit for investment C to be $75, or a return of $7\frac{1}{2}\%$ on the investment. Thus the expected monetary value decision maker selects investment C, since its *expected* return of $7\frac{1}{2}\%$ is greater than the *sure* returns of 6% for investment A and $4\frac{1}{2}\%$ for investment B.

The best way of interpreting nonrepetitive expected monetary value decisions is to examine the expected opportunity losses. The expected opportunity loss associated with the optimal decision represents the loss to be expected if we select the optimal decision. This does not imply that the opportunity loss we will actually realize by selecting the optimal decision will equal this value. Often it is impossible for this value to be realized. Recalling example 10.4, the optimal decision was to select action a_4, with an associated expected opportunity loss of $55. Note, however, that if action a_4 is selected, the only possible opportunity losses that can be realized are $160, $60, and $0, none of which equals the expected opportunity loss of $55.

What is implied is that the minimum expected opportunity loss measures the expected **cost of uncertainty** due to our uncertain knowledge about which state of nature will be in effect. If the opportunity presented itself, we should be willing to pay up to the amount of the cost of uncertainty to obtain information identifying the state of nature that will be in effect. Minimizing our expected cost of uncertainty seems to be a logical objective, and this is what we are doing by selecting the action associated with the maximum expected profit and the minimum expected opportunity loss.

Another way to justify the use of this decision method is to examine the components of the expected monetary value. The profit or loss values are economic outcomes, while their relative frequencies of occurrence are the prior probabilities assigned to the states of nature. Thus an expected monetary analysis provides a model that combines both real economic data and qualitative or subjective information available to the decision maker related to the outcome of the economic data. It is the only decision-making model available to the decision maker at the time that incorporates all available information into the decision. Also, it makes the decision maker more than just an impartial observer by forcing him to construct meaningful prior probabilities to associate with the states of nature. This latter concept is the most important and often the most difficult task in decision making under uncertainty.

A reassessment of the values assigned to the prior probabilities often changes the optimum decision. Suppose that, after reconsidering his prior beliefs, the investor confronted with the investment problem of section 10.2 decides to assign these prior probabilities:

$$P(s_1) = \tfrac{1}{4} \qquad P(s_2) = \tfrac{1}{2} \qquad P(s_3) = \tfrac{1}{4}$$

His expected profit for investment C is now

$$E(G_4) = -\$100(\tfrac{1}{4}) + \$0(\tfrac{1}{2}) + \$200(\tfrac{1}{4}) = \$25$$

But the expected profits for investments A and B remain at $60 and $45, respectively. Hence under the new set of prior probabilities, investment C is only third best behind the "new" optimum decision to select investment A.

Expected monetary value decisions, since they employ probabilities in the decision analysis, assume that the prior probabilities assigned to the states of nature are the true probabilities for the problem. The optimal expected monetary value decision is meaningful only in terms of its associated prior probabilities. If the decision maker does not have complete confidence in the prior probabilities that he has assigned to the states of nature, he should try other likely sets of prior probabilities in order to examine the sensitivity of the decision to the selection of the prior probabilities. The conduct of a sensitivity analysis is demonstrated in the following example.

Example 10.6 Brown offers an elementary example of a sensitivity analysis,* which we will revise in our discussion here. A New York metal broker has the opportunity to buy 100,000 tons of iron ore from a foreign government, at the advantageous price of $5 a ton. The broker is certain that he can sell the ore in the United States for $8 a ton, but there is a chance that the U.S. government will refuse to grant him an import license. In such an event the import contract would be annulled, at a cost of $1 a ton.

a. If the broker seeks to maximize his expected profit, should he buy

---

*R. V. Brown, "Do Managers Find Decision Theory Useful?" *Harvard Business Review*, May–June 1970.

the ore from the foreign government if he believes the chances are 50-50 that the U.S. government will grant him an import license?

b. Because of conflicting reports from a number of advisors, suppose the broker is very uncertain about the chances of the U.S. government granting him an import license. How great does the probability that the U.S. government will grant him a license have to be in order for the broker to accept the advantageous offer from the foreign government?

Solution The alternatives available to the broker are that he accept the offer from the foreign government and buy the ore (a_1) or that he not buy the ore (a_2). The states of nature concern the actions of the U.S. government regarding the granting of an import license; either the license is granted (s_1) or it is not granted (s_2). The payoff for each possible outcome is listed in the payoff table. The payoffs listed in the payoff table are profits, or the difference between revenues and costs for each possible outcome.

	STATE OF NATURE	
ACTION	License Granted, s_1	Not Granted, s_2
buy ore, a_1	\$300,000	-\$100,000
do not buy ore, a_2	0	0

a. If the chances of a license being granted are 50-50, then $P(s_1) = \frac{1}{2}$ and $P(s_2) = \frac{1}{2}$. Then we find that the expected profit associated with each alternative is

$$E(G_1) = \$300,000(\tfrac{1}{2}) - \$100,000(\tfrac{1}{2}) = \$100,000$$

$$E(G_2) = \$0(\tfrac{1}{2}) + \$0(\tfrac{1}{2}) = \$0$$

Therefore, the broker is maximizing his expected profits by buying the ore.

b. If the broker does not buy the ore, his expected profit is \$0 (nothing ventured, nothing gained or lost). Therefore, he should buy the ore as long as $E(G_1) > E(G_2)$, or as long as $E(G_1) > 0$. Suppose we let p be the probability that the U.S. government will grant the broker a license. Then we find

$$E(G_1) = \$300,000(p) - \$100,000(1 - p)$$

Then $E(G_1) > 0$ if and only if $\$300,000(p) - \$100,000(1 - p) > 0$. Solving for p we find that this implies that $E(G_1) > 0$ if and only if $p > \frac{1}{4}$. Therefore, the probability that the U.S. government grants him a license has to be at least $\frac{1}{4}$ before the broker decides to buy the ore. If the probability of a license being granted is less than $\frac{1}{4}$, he will not buy the ore.

Exercises

10.11.   The president of a manufacturing plant is faced with the problem of accommodating a possible drastic increase in demand for his firm's product. His alternatives are to ignore the increase in demand, since his firm is currently operating at production capacity, or to buy out a small competing firm at a cost of $100,000 and use its resources to generate additional production. The firm stands to gain an additional $600,000 in revenues if the additional demand materializes, but it will not gain in additional revenues if demand continues to be stable. The chance that demand for the firm's product will increase is believed to be .10.

a. Construct an opportunity loss table for the decision problem.

b. If the president wishes to minimize his firm's expected opportunity loss, should he buy the competing firm to allow for additional production capacity?

c. Suppose the president is very undecided about the chance that the demand for his firm's product will show a sizable increase. How great must the chance of a sizable increase in demand be in order for the president to be justified in buying the competing firm?

10.12.   A building contractor is considering whether to submit a bid on a contract to build a large modern apartment complex. Analysis of the proposed plan and other preliminary planning required by the contractor and his firm before submitting a bid on this contract would cost the building firm $50,000. If they bid, they will submit a bid large enough to realize a $250,000 profit for themselves (including the cost of preliminary planning). If the contractor's objective is to maximize his expected profit, how large must the probability of his winning the contract be before he will decide to submit a bid on the apartment contract?

10.6 The Economic Impact of Uncertainty

Ideally we would like to know in advance which state of nature will be in effect. In the investment problem, knowing that the market will be down or about the same, the decision maker would select investment A, the 6% commercial bank savings account; knowing that the market will be up, he would select investment C, the mutual fund investment. In the demand-inventory example, if the retailer knows which demand level will be in effect, she selects an inventory level to meet demand exactly.

Under normal conditions uncertainty prevails, and the decision maker must act according to his best information and interests. Not knowing the state of nature that will be in effect, he cannot expect a return as great as the return he would expect if the true state of nature were known.

Using an expected monetary value approach, the decision maker, in the face of uncertainty, selects the alternative that maximizes his expected profit and minimizes his expected opportunity loss. We will now define a concept that was mentioned earlier only in passing.

Definition

The expected opportunity loss associated with the optimal decision is called the **cost of uncertainty.**

The cost of uncertainty is the maximum amount the decision maker would pay to know precisely the state of nature that will be in effect. Therefore, it is the value the decision maker would place on perfect information (information removing all uncertainty) offered to him. The cost of uncertainty is sometimes called the **expected value of perfect information,** or simply EVPI.

In example 10.4 the optimal decision is to select action a_4, with expected opportunity loss $E(L_4) = \$55$. We saw in example 10.5 that action a_2, with expected opportunity loss $E(L_2) = \$1.60$, is best. These two values are the costs of uncertainty associated with, respectively, the investment problem and the demand-inventory problem. In each case these values represent the maximum amount the decision maker would be willing to pay to know precisely the state of nature that will be in effect.

To better understand the concept of the cost of uncertainty or EVPI, let us focus on example 10.4, the investment problem. Suppose that the decision maker knows with *certainty* which state of nature will be in effect. If he knows that the state will be s_1 or s_2, he selects action a_2, since the opportunity losses for selecting a_2 under states s_1 and s_2 are the least. Similarly, if he knows that state of nature s_3 will be in effect, he selects action a_4. In each case the opportunity loss is zero, so the decision maker's expected opportunity loss *under certainty* is

$$E(L_0) = \$0(\tfrac{1}{4}) + \$0(\tfrac{1}{4}) + \$0(\tfrac{1}{2}) = \$0$$

The best the decision maker can expect *under uncertainty* is the expected opportunity loss of $55 by selecting the optimal action a_4. Therefore, the effect of the uncertainty of market conditions costs the investor, on the average, $55 each time he faces an investment decision such as the one outlined in section 10.2.

The cost of uncertainty of EVPI may be more easily interpreted if defined in terms of expected profits. Under certainty of the state of nature, the decision maker can expect a profit of

$$E(G_0) = \$60(\tfrac{1}{4}) + \$60(\tfrac{1}{4}) + \$200(\tfrac{1}{2}) = \$130$$

by selecting action a_2 under states s_1 and s_2 and by selecting action a_4 under state of nature s_3. This represents his expected profit under perfect information.

The best the investor can expect under *uncertainty* is $75, the expected profit associated with the optimal decision, action a_4. The difference between expected profits under certainty and uncertainty is $130 $-$ \$75 = \$55$, the EVPI. **The cost of uncertainty, or, equivalently, the EVPI, is then the amount of profits foregone or additional losses incurred due to uncertain conditions**

affecting the outcome of a decision problem.

Suppose that the investor has the opportunity to hire the services of the Securities Consulting Agency (SCA). This group would survey many market experts to determine whether they believe the securities market will fall, stay about the same, or rise within the next year. The SCA would then sell the consensus of the survey to the investor.

How much should the investor pay for the SCA report? If the SCA provides *perfect* market information, the investor would pay up to $55 for the report, but no more. That is, the value of the perfect information supplied by the SCA does not exceed $55. The reasoning behind this limiting value is easy to see. For instance, if the decision maker did pay $60 to the SCA for perfect market information, his expected profit would be $130 (the expected profit for perfect information) minus $60 (the cost of buying the perfect information), or $70. But an expected profit of $70 is $5 less than the $75 expected profit by taking action a_4 under *uncertainty* and without auxiliary information.

Naturally the investor does not expect perfect market information from the SCA. Thus the amount he should pay for the SCA report is some amount less than $55, an amount that is determined from the measures of **reliability** of the SCA market analysis techniques. The concept of placing a value on *imperfect* sample information will be discussed in the next section.

Exercises

10.13. Inflation tends to boost interest rates on loans of all types. As a partial hedge against inflation, some lending institutions offer variable-rate loans to borrowers with high credit ratings. The borrower can then accept either a fixed-rate loan with a certain interest obligation or a variable-rate loan that may result in either interest savings or additional interest costs to the borrower, depending on the future state of the economy. Suppose an individual wishes to borrow $10,000 for one year when the prevailing interest rate is 8%. After consulting with an economist, the individual establishes the following probability distribution for the prevailing lending rate over the next year:

Interest Rate, i	7%	7.5%	8%	8.5%	9%
Probability, $P(i)$	.10	.25	.50	.10	.05

The borrower must decide whether to accept the prevailing interest rate (8%) or to undertake a variable-rate loan and pay interest according to the prevailing rate at the end of the borrowing period.

a. Construct an opportunity loss table for the borrower.

b. If the borrower wishes to minimize his expected opportunity loss, which type of loan should he undertake?

c. How much would the borrower be willing to pay to know exactly the interest rate that will be in effect at the end of the year?

10.14. Refer to exercise 10.5. How much would the newsstand owner be willing to pay to know exactly the demand for the weekly newsmagazine during a particular week?

10.15.   Refer to exercise 10.10. Suppose a market research firm has approached the executive of the firm considering the introduction of the new product with the offer that for $2000 they can conduct a market survey and determine with virtual certainty the proportion of consumers who will buy the firm's product. Should the executive accept the offer of the market research firm? Explain.

10.16.   Find the cost of uncertainty associated with the investor's decision in exercise 10.9. Give a practical interpretation for the cost of uncertainty from the investor's point of view.

10.7 Decision Making That Involves Sample Information

In the definition and discussion of the prior probabilities in sections 10.3 and 10.5, we said that they are acquired either by subjective selection or by computation from historical data. No current information describing the chance of occurrence of the states of nature was assumed to be available.

Perhaps observational information or other evidence may be available to the decision maker either for purchase or at the cost of experimentation. For instance, a retailer whose business depends on the weather may consult a meteorologist before making his decision, or an investor may hire a market consultant before investing. Market research surveys before the release of a new product represent another area in which the decision maker seeks additional information. In each instance the decision maker is acquiring information relative to the occurrence of the states of nature from a source other than that from which the prior probabilities were computed.

If such information is available, Bayes's law can be employed to revise the prior probabilities to reflect the new information. These revised probabilities are called *posterior probabilities*.

Definition

The **posterior probability** $P(s_k|x)$ represents the chance of occurrence of the state of nature s_k given the experimental information x. This probability is

$$P(s_k|x) = \frac{P(x|s_k)P(s_k)}{\displaystyle\sum_{\text{all } i} P(x|s_i)P(s_i)}$$

The probabilities $P(x|s_i)$ are the conditional probabilities of observing the observational information x under the states of nature s_i, and the probabilities $P(s_i)$ are the prior probabilities. The notation used here for

Bayes's law is different from what was presented in chapter 4, but the forms are equivalent.

The expected monetary value decisions are formulated in the same way as before except that we use the posterior probabilities in place of the prior probabilities. If our objective is to minimize the expected opportunity loss, the quantity is computed for each action a_i. As before, the optimal decision is to select the action associated with the smallest expected opportunity loss.

Computing the Expected Opportunity Loss by Using Posterior Probabilities

$$E(L_i) = \sum_{all\ j} L_{i,j} P(s_j|x) \qquad i = 1, 2, ...$$

Example 10.7 It is known that an assembly machine operates at a 5% or a 10% defective rate. When believed to be running at the 10% defective rate, the machine is judged out of control, shut down, and readjusted. From past history the machine is known to run at the 5% defective rate 90% of the time. A sample of $n = 20$ has been selected from the output of the machine and $y = 2$ defectives have been observed. Based on both the prior and sample information, what is the probability that the assembly machine is in control (5% defective rate)?

Solution The states of nature in this example relate to the possible assembly machine defective rates. Thus the states of nature are

$$s_1 = .05 \qquad \text{and} \qquad s_2 = .10$$

with assumed prior probabilities of occurrence of .90 and .10, respectively. We want to use these prior probabilities, in light of the observed sample information, to find the posterior probability associated with state s_1.

In this example the "experimental information x" is the observation of $y = 2$ defectives from a sample of $n = 20$ items selected from the output of the assembly machine. It is necessary to find the probability that the experimental information x could arise under each state of nature s_j. This can be accomplished by referring to the tables for the binomial probability distribution (table 1 of the appendix). Under state of nature $s_1 = .05$, we find

$$P(x|.05) = P(n = 20, y = 2|.05) = .925 - .736 = .189$$

Similarly, under state of nature $s_2 = .10$, we find

$$P(x|.10) = .285$$

Now we are ready to employ Bayes's law to find the posterior probability that the machine is in control (s_1) based on both the prior and experimental information.

Table 10.7 *Columnar approach to use of Bayes's law, example 10.7*

(1) STATE OF NATURE, s_j	(2) PRIOR, $P(s_j)$	(3) EXPERIMENTAL INFORMATION, $P(x\|s_j)$	(4) PRODUCT, $P(s_j)P(x\|s_j)$	(5) POSTERIOR, $P(s_j\|x)$
s_1	.05	.9	.189	.1701	.86
s_2	.10	.1	.285	.0285	.14
		1.0		.1986	1.00

The application of Bayes's law is most clearly illustrated by using a columnar approach, as shown in table 10.7. Using a columnar approach, the states of nature, their associated prior probabilities, and the probabilities that the experimental information x could arise under each state of nature are listed in the first three columns. In column (4) we compute the product of the corresponding entries from columns (2) and (3). These values measure the **joint probabilities** $P(xs_j) = P(s_j)P(x|s_j)$. We then sum the entries in column (4). This sum is the term in the denominator of the formula for Bayes's law and measures the **marginal probability** of observing the experimental information x. The posterior probabilities, column (5), are thus obtained by taking each entry in column (4) and dividing it by the sum of the entries in column (4). This simply rescales the joint probabilities of column (4), causing them to sum to 1.00. For instance,

$$P(s_1|x) = \frac{.1701}{.1986} \qquad P(s_2|x) = \frac{.0285}{.1986}$$

Even though we found 10% of the sample items to be defective (2 of 20), the posterior probability that the machine is running at the 10% defective rate (out of control) is only .14, very little greater than the prior probability that the machine is out of control. Therefore, the probability that the machine is not running out of control is .86.

In reviewing the solution to example 10.7, you may have noticed that the columnar approach is nothing but a restructuring of Bayes's law. It is bound to give the same numerical answers as would be obtained by substituting the appropriate numbers into the formula for Bayes's law. For those who are not convinced of this fact, example 10.7 should be repeated by using the formula substitution approach instead of the columnar approach.

The experimental information x often presents itself other than as sample information. It may be in the form of an expert opinion, the result of an engineering survey or a structural test, or the observance of a sudden, unpredictable economic trend. How can such information best be used to help us in the task of decision making under conditions involving uncertainty? We must first evaluate the reliability of the information and then incorporate these measures of reliability into the decision analysis with the other problem data.

The discussion involving the hypothetical Securities Consulting Agency presented in the preceding section offers a case in point. Notice that the "experimental information" in this example is in the form of an opinion; based on their studies, the SCA will indicate the market condition (state of nature) which it believes will occur. In this example the experimental information x may then present itself at any of three levels. That is, the SCA may report one of these conditions:

$x_1 =$ the market will be down

$x_2 =$ the market will remain at the same level

$x_3 =$ the market will be up

The probabilities $P(x|s_j)$ should be rewritten as $P(x_i|s_j)$ to denote the different possible responses x_i, $i = 1, 2, 3$, which may be given by the SCA. These probabilities measure the reliability of the experimental information x_i provided by the SCA. If the information provides complete certainty, then

$$P(x_i|s_j) = \begin{cases} 1 & \text{if } i = j \\ 0 & \text{if } i \neq j \end{cases}$$

The further removed the $P(x_i|s_j)$ values are from these limits, the less reliable, and hence the less valuable, is the experimental information that is available.

Example 10.8 Reconsider the investment problem of example 10.2. Suppose that the investor is considering hiring the services of the Securities Consulting Agency, the group discussed in section 10.6. In the past when the SCA has conducted a market survey for a client, the results shown in table 10.8 have been noted. For example, $P(x_1|s_2) = .15$ is the probability that the SCA concludes the market will drop when it actually stays the same. Find the investor's best decision based on the expected opportunity losses associated with each possible response from the SCA.

Table 10.8 *Conditional probabilities indicating the reliability of SCA market surveys,* $P(x_i|s_j)$

| RESPONSE, x | STATE OF NATURE | | |
	s_1, Down	s_2, The Same	s_3, Up
x_1, down	.60	.15	.05
x_2, the same	.30	.50	.25
x_3, up	.10	.35	.70

Solution Recall from table 10.3 that regardless of which state of nature is in effect, action a_2 guarantees a greater profit than either action a_1 or action a_3. Thus we could ignore actions a_1 and a_3, which are dominated by a_2, reducing the investor's decision to one of choosing either action a_2 or action a_4, depending on the information received from the SCA. An analysis of the investor's problem involves three decisions, one for each

Table 10.9 *Columnar approach for using Bayes's law, example 10.8*

(1) s_j	(2) $P(s_j)$	(3) $P(x_1 \| s_j)$	(4) $P(s_j)P(x_1 \| s_j)$	(5) $P(s_j \| x_1)$
s_1, down	1/4	.60	.1500	.706
s_2, the same	1/4	.15	.0375	.176
s_3, up	1/2	.05	.0250	.118
			.2125	1.000

response from the SCA: "down," "stay the same," or "up." We first need to compute the revised probabilities—the posterior probabilities—for each response.

If the SCA suggests that the market will drop, we employ Bayes's law, using the columnar approach, and find (see table 10.9)

$$P(s_1|x_1) = .706 \qquad P(s_2|x_1) = .176 \qquad P(s_3|x_1) = .118$$

Similarly, under the response from the SCA that the market will stay the same, we find

$$P(s_1|x_2) = .230 \qquad P(s_2|x_2) = .385 \qquad P(s_3|x_2) = .385$$

If they say it will rise, we find

$$P(s_1|x_3) = .054 \qquad P(s_2|x_3) = .189 \qquad P(s_3|x_3) = .757$$

Now suppose that the SCA reports that the market will drop. Then using the opportunity loss values from table 10.4, the expected opportunity losses associated with actions a_2 and a_4 are

$$E(L_2) = \$0(.706) + \$0(.176) + \$140(.118) = \$16.47$$
$$E(L_4) = \$160(.706) + \$60(.176) + \$0(.118) = \$123.53$$

Thus if the SCA suggests the market will drop, the investor should choose the 6% savings plan, action a_2. The expected opportunity losses for the other two SCA responses have been computed and are listed in table 10.10. The investor's optimal decisions are easily read from this table. He should select action a_2 if "down" or "the same" is reported, but he should select a_4 if the SCA reports that the market will be up in the coming year.

The value of revising our prior probabilities and arriving at this more complex solution is twofold. First, the independent source of information

Table 10.10 *Expected opportunity losses for example 10.8*

	RESPONSE FROM THE SCA		
ACTION	Down	The Same	Up
a_2	$ 16.47	$53.85	$105.95
a_4	123.53	60.00	20.00

we are using to compute the posterior probabilities gives us a check that the values we assign to the prior probabilities of the states of nature are near what the true values should be. That is, we are updating our guesses with someone else's guesses or with experimental evidence. This procedure allows us to incorporate all available information into our decision analysis. As a second point, related to the first, the additional information allows us to tailor our decision to the specific outcome of the experimental evidence. Thus instead of having a single decision, we have a whole array of decisions corresponding to the nature of the experimental evidence. In the investment problem each response from the SCA initiates a decision by the investor. Thus he is able to "partition" his decision and obtain the most use possible from the additional information.

What is the value of the sample information? Certainly the investor would not pay as much as $55 for the sample information, but how much should he be willing to pay? We will answer that question in the paragraphs that follow.

Although it was not mentioned earlier, we can find the marginal probabilities by using the formula given in the box.

Computing the Marginal Probability $P(x_i)$

$$P(x_i) = \sum_{\text{all } j} P(x_i|s_j)P(s_j) \qquad i = 1, 2, \ldots$$

This is the sum of the entries in column (4) when using the columnar approach. Therefore, the probability that the SCA report will indicate that the market will be down (x_1) is

$$P(x_1) = (.6)(\tfrac{1}{4}) + (.15)(\tfrac{1}{4}) + (.05)(\tfrac{1}{2}) = .2125$$

Similarly, the probabilities that their report will indicate that the market will be about the same (x_2) and up (x_3) can be found by using the same formula and are, respectively,

$$P(x_2) = .3250 \qquad \text{and} \qquad P(x_3) = .4625$$

Returning to example 10.8, we recall that when the SCA reports $x_1 =$ down or $x_2 =$ the same, the investor should select action a_2. However, when the report suggests $x_3 =$ up, action a_4 is best. Referring back to table 10.10, we can see that the investor's expected opportunity loss using the SCA service is

$$E(L) = \$16.47P(x_1) + \$53.85P(x_2) + \$20.00P(x_3)$$

$$= \$16.47(.2125) + \$53.85(.3250) + \$20.00(.4625) = \$30.25$$

Thus the experimental information is worth, at most,

$$\$55.00 - \$30.25 = \$24.75$$

since the experimental information has effectively reduced the investor's expected cost of uncertainty by that amount. This value, called the **expected value of sample information** (EVSI), represents the maximum amount the investor would pay for the SCA market service. Under no conditions would the investor pay more than $24.75. For any amount less than $24.75 which he pays for the SCA report, the investor incurs a net savings by buying the additional information.

Exercises

10.17. A newspaper publisher with headquarters in Portland, Oregon, is considering the distribution of her newspaper in Eugene, 100 miles to the south of Portland. Her potential profits will depend on which of two methods of transportation she uses for the distribution of the newspaper from Portland to Eugene and the proportion of the 50,000 Eugene households that subscribe to her newspaper. The two options for transporting the newspaper are to contract for daily shipment with an interstate bus service (method A) or to sublet the transportation between the two cities to an independent trucker (method B). The daily profits associated with each method of distribution under every assumed likely proportion of subscribing households are given in the table.

NUMBER OF HOUSEHOLDS SUBSCRIBING	PROPORTION SUBSCRIBING, p	$P(p)$	DAILY PROFITS	
			Method A	Method B
20,000	.4	.3	$1000	$1200
25,000	.5	.5	1400	1500
30,000	.6	.2	2100	1800

a. Based only on the above information, what is the best method of distribution for the publisher?

b. How much would the publisher be willing to pay to know the exact proportion of households who would subscribe to her newspaper?

10.18. Refer to exercise 10.17. To reduce her uncertainty suppose the publisher commissions a group to conduct a door-to-door survey of the Eugene households to estimate the proportion who would subscribe to her newspaper. The results of the survey show that 25 households were randomly selected from among the 50,000 households of Eugene, with 14 indicating an interest in subscribing to the newspaper. What is the best method of distribution in light of this sample information?

10.19. Manufacturers seek to design product packages that fit the retailers' available shelf space, are easily stored, and appeal to the consumer. The marketing manager for a food products company is considering a new package design for his company's primary product. Before deciding whether to adopt the new design, the manager must decide whether to hire a market survey group, at a cost of $2,500, to obtain buyer reaction to the new design. It has been determined that if the new design is introduced and sales are "good," $20,000 will be returned monthly to the company. If sales are "fair," $10,000 will be returned, and if sales are "poor," $3,000 will be returned. Past experience with the market survey group leads the marketing manager to develop the probabilities shown in the table regarding the reliability of the market survey group's information, $P(x_i|s_j)$.

SURVEY CONCLUSION	STATE OF NATURE, ACTUAL SALES		
	Good, s_1	Fair, s_2	Poor, s_3
favorable, x_1	.75	.40	.10
unfavorable, x_2	.25	.60	.90

If a decision is made to not introduce the new design, the old design will be continued, with a monthly return of $8,000 to the company. The monthly advertising cost is $5,000 for the new design and $2,500 for the old design. It is assumed that the probabilities of good, fair, or poor sales resulting after switching to the new design are .425, .25, and .325, respectively.

a. Construct a payoff table for the decision problem. (Use either profit or opportunity loss as the payoff measure.)

b. If the marketing manager seeks to maximize his firm's expected profits, should he recommend that his firm adopt the new package design?

c. What is the cost of uncertainty to the firm associated with the decision problem?

d. If the market survey group is employed to conduct a survey of consumer preferences, should the new design be employed if the survey conclusion is favorable? If it is unfavorable?

e. What is the expected value of the sample information provided by the market survey group? Would the manager then recommend hiring their services for a $2,500 fee?

10.8   Other Topics in Decision Analysis

Over the past twenty years decision analysis has evolved from a controversial set of rules of thumb into a rigorous discipline involving techniques now generally accepted by both the theoretician and the practitioner. Only the most important topics involved with decision analysis have been included in the chapter up to this point. Other topics and techniques of lesser importance will be introduced in this section. For a complete discussion of the modern concepts of decision analysis, see the texts by Schlaifer (1969), Winkler (1972), or Brown, Kahr, and Peterson (1974), listed in the references.

Decisions Ignoring Prior Information

Occasionally the decision maker's attitude toward risk suggests that a limit exists on the amount of money he can afford to lose. Such an individual may be typified by an investor with a very small bankroll or a retailer on the verge of bankruptcy. The prior probabilities attached to the states of nature are ignored by such an individual, even if the probabilities associated with unfavorable states are extremely small. He looks only at the magnitudes of the losses for each action, shows virtually no interest in the magnitude of potential profits, and makes his decision without computing an expected monetary value. A decision criterion that satisfies a conservative economic objective like that described above is provided by the *minimax* decision.

Definition

The **minimax** decision is the selection of the action whose maximum possible opportunity loss is a minimum.

The minimax decision maker says that in order to remain competitive in his business, he must, at all costs, avoid the large opportunity losses. He is minimizing his maximum opportunity loss and, by so doing, he is usually rejecting those alternatives associated with the greatest possible return. He thus assures himself of, at best, a stable financial position.

Example 10.9 Find the minimax decision for the investment problem of example 10.2.

Solution The maximum opportunity losses associated with each action are given in the table.

Action	a_1	a_2	a_3	a_4
Maximum Opportunity Loss	$200	$140	$155	$160

Thus the minimax decision is to select action a_2, the 6% savings plan.

Another decision criterion is the maximax decision method, which selects the action that maximizes the maximum possible profit. The individual who employs this technique has a very small aversion toward risk and might be categorized as a gambler. He stands a great chance of losing everything but is concerned only with the possibility of making a quick fortune.

Maximin, maximizing the minimum profit, is similar in nature but somewhat more conservative than maximax.

All these techniques have two common traits. First, the decision maker employing minimax, maximax, or maximin strategies tends to totally ignore prior information relating to the probabilities of occurrence of the states of nature. His attitudes toward risk make some economic payoffs appear attractive and others unattractive, regardless of the occurrence of the states of nature. The second similarity is that minimax, maximax, or maximin decisions are usually one-shot decisions. If they were not, the "averaging effect" of employing minimax, maximax, or maximin continuously could tend to deplete one's profits, since whether accepted or not, the states of nature do occur according to some probability distribution.

The expected monetary value decision and the minimax, maximax, or maximin decisions are often the same. This only occurs by chance and does not imply any similarity between objectives. The basic difference is that expected monetary value decisions incorporate prior information (probabilities) into the decision analysis, while the others ignore this information.

Decision Trees

A tree diagram is a useful device to illustrate a multistage probability problem. When used in a decision analysis application, the tree diagram is called a **decision tree.**

Figure 10.1 *Decision tree for the investment problem of example 10.8*

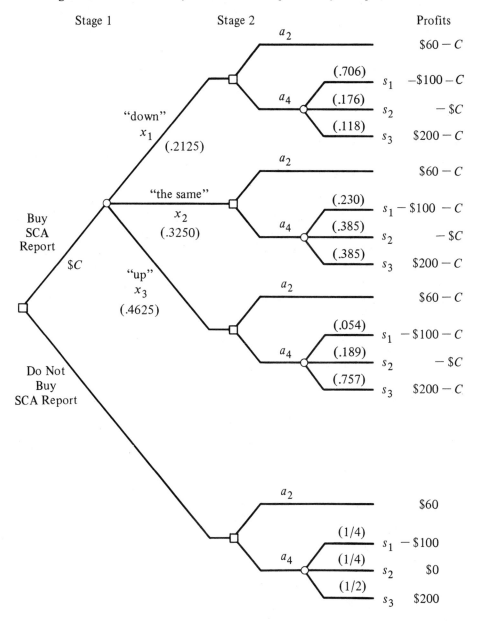

Decision trees are most useful for multistage decision problems, especially decision problems sequenced over time. The decision tree graphically X-rays the decision problem and displays the anatomy of the decision. The decision tree becomes more useful as the decision problem becomes more complex, because it focuses attention on the interrelationship of activities and events over time.

To properly clarify interpretation of the tree, decision points are represented by squares while chance points, points over which the decision maker has no control, are represented by circles. At the base of the tree are the available first-stage alternatives. From each alternative the decision tree constructs the chronological path, leading through chance points and other decision points, to each assumed possible terminal outcome. Payoffs associated with each path and probabilities of occurrence associated with the chance events are shown on the tree.

In figure 10.1 a decision tree is used to illustrate the investment problem. Notice that the anatomy of the problem appears on the tree according to how its activities relate over time. Before the investor decides between actions a_2 and a_4 (a_1 and a_3 have been eliminated because of their dominance by a_2), he must first decide whether or not to buy the SCA report at a cost of $C. His stage 1 activity, and hence the base of the decision tree, is the decision problem involving the experimental information offered by the SCA report. Each branch of the tree illustrates a possible sequence of activities from that initial decision to each possible terminal outcome.

There is no rigid set of rules for constructing a decision tree. The only requirements are that the decision tree illustrate the anatomy of the decision problem while properly ordering the problem activities over time. The decision maker then has a clear picture of the decision problem and can proceed, working from the terminal outcomes backward toward the base of the tree, to find the set of decisions that best satisfies his objectives.

The Utility for Money

Suppose the owner of a $60,000 home has been informed by insurance actuaries that his home stands a 1:200 chance of being destroyed by fire during a given year. If the homeowner can purchase a fire insurance policy to cover the possible loss of his home due to fire for an annual premium of $150, should he make the purchase?

The payoffs associated with each action available to the homeowner are as follows:

	FIRE, S_1	NO FIRE, S_2
Buy Insurance, a_1	−$350	−$350
No Insurance, a_2	−60,000	0

His expected gains (payoffs) for selecting each available action are, respectively,

$$E(G_1) = -\$350\left(\frac{1}{200}\right) - \$350\left(\frac{199}{200}\right) = -\$350$$

$$E(G_2) = -\$60,000\left(\frac{1}{200}\right) + \$0\left(\frac{199}{200}\right) = -\$300$$

Based solely on expected monetary value criteria, the homeowner *would not buy* the insurance. However, we can be certain that almost any rational individual *would buy* such an insurance policy.

The problem is that it is often inappropriate to perform a decision analysis based solely on expected monetary value considerations. **An expected monetary value decision carries the implicit assumptions that the value of a dollar does not differ from one person to the next and that the value of D dollars is equal to D times the value of a single dollar.** These assumptions are invalid when we consider that we each have different measures of personal wealth and thus have financial limits beyond which we are not willing to venture.

The theory of utility prescribes a method whereby the outcomes of a decision problem can be scaled according to their relative value to the decision maker. These measures of relative value collectively describe the decision maker's **utility curve.** The decision analysis is then performed by substituting utility values for monetary values and computing the expected utility associated with each action. Maximization of utility then ensures maximization of *value* to the decision maker in terms of how he perceives the comparative personal measures of value of the possible outcomes associated with his decision problem. The actual assessment of utility for an individual is a complex activity involving the quantification of his subjective feelings toward risk, and the topic will not be discussed in this text. For detailed discussions of some modern developments in utility assessment, refer to the texts by Winkler (1972) or Brown, Kahr, and Peterson (1974) listed in the references.

Utility measures are needed primarily when the decision problem contains some possible outcomes that, if incurred, may place the decision maker in financial or personal jeopardy. As opposed to adopting a minimax decision, the decision maker may choose to rescale the monetary outcomes according to his preferences and perform an expected utility value analysis. This allows him full utilization of all available resources, including probability measures associated with the uncertain events which the minimax decision ignores, and provides a decision that is guaranteed to maximize his personal interpretation of value.

Exercises

10.20.   Construct an opportunity loss table and then find the minimax decision for the decision problem associated with each of the following exercises:
a. exercise 10.5   b. exercise 10.6   c. exercise 10.9
d. exercise 10.11   e. exercise 10.13

10.21.   Refer to exercise 10.19. Construct a decision tree to graphically portray the decision problem faced by the marketing manager. As shown in figure 10.1, indicate the chance points by circles and the decision points by squares.

10.22. Shown in the accompanying table is a set of utility values that have been assessed for the associated dollar-valued outcomes by a decision maker. If the decision maker wishes to maximize his expected utility, how should he act on each of the following investment problems?

Dollar-valued Outcome	−$10,000	−$5,000	−$1,000	$0	$5,000	$10,000	$25,000
Utility	0	.45	.50	.55	.70	.80	1.0

a. The investment of $1,000 in an oil drilling venture returning either a $10,000 profit or nothing. The probability of success in the oil drilling venture is estimated to be .1.

b. The investment of $10,000 in a new motel-restaurant facility. Depending on the success of the project, the investment is estimated to return a $25,000 profit with a probability of .2, a $5,000 profit with a probability of .3, a $5,000 loss with a probability of .4, or a loss of the entire $10,000 investment with a probability of .1.

c. In both of the preceding investment problems, compare the optimal decision using a maximum expected utility objective with the optimal decision using a maximum expected payoff objective. How do you account for any differences in the selection of an optimal decision between these two objectives?

10.9 Summary

Decision analysis is characterized by a decision maker selecting from among many alternatives, some of which have uncertain outcomes. The decision maker must determine which among the alternatives available to him best satisfies his objective.

Usually the objective in a decision analysis is involved with an expected monetary value decision. If so, the decision maker proceeds as follows:

1. List all possible actions that may be taken and all the states of nature that may affect the outcomes of those actions.
2. List the payoffs associated with each action under every state of nature.
3. Determine meaningful probabilities to assign to the states of nature to represent their likelihoods of occurrence.
4. Compute the expected profit or expected opportunity loss for each action.
5. Select the action associated with the maximum expected profit or the minimum expected opportunity loss, both of which arrive at the same decision.

If additional information is available to the decision maker, Bayes's law can be employed to revise the prior probabilities based on experimental information. The posterior probabilities then reflect all available information, both subjective and experimental, relevant to the likelihood of the occurrence of the states of nature.

Supplementary Exercises

10.23. How are the decision-making methods of chapter 10 different from the decision-making methods of chapters 8 and 9?

10.24. What is the justification for using an expected monetary value objective in decision problems involving uncertainty?

10.25. For each of the following business decision-making problems, list the actions available to the decision maker and the states of nature that might result to affect the payoff.

a. The replacement of manually operated packaging machines by a fully automated machine.

b. The leasing of a computer by a commercial bank to process checks and handle internal accounting.

c. The expansion of the market of a regional brewery from a two-state market to either a four-state market or a seven-state market.

d. The daily demand for pastrami by the customers of a delicatessen is known to range from 13 pounds to 41 pounds of pastrami. The owner of the delicatessen must decide how much pastrami to order the morning of each business day.

e. The assignment of seven secretaries to seven executives.

f. The investment of a company pension fund.

10.26. Refer to exercise 10.25 and indicate whether the outcomes associated with the decision maker's available actions in parts a through f are known with certainty or involve uncertainty. Do any of the six decision-making problems involve an exercise of decision making under certainty?

10.27. Refer to exercise 10.25 and identify an appropriate payoff measure for each of the decision-making problems in parts a through f.

10.28. A special commission appointed by the governor of the state of Oregon is currently studying the issue of site location for a nuclear power–generating plant. The alternatives they are currently considering are (a_1) to locate the plant at the mouth of the Columbia River, 85 miles west of Portland, (a_2) to locate the plant inland on the Willamette River near Salem, and (a_3) not to build the plant and depend on hydroelectric power. Because of the heat generated in cooling operations, a_1 may prove damaging to the commercial salmon fishing industry, vital to the economy of Oregon. However, there may be insufficient water flow available at site a_2 for adequate cooling. If alternative a_3 is chosen, sufficient energy resources may not be available to meet future needs.

a. If you were an advisor to the commission, what would you recommend be used as a measure of payoff for evaluating the three alternatives?

b. Suppose the commission has decided that the important attributes of value in the decision model are (i) the ability of each alternative to meet future commercial power needs, (ii) the ability of each alternative to meet future residential power needs, (iii) the effect on the state's salmon industry, and (iv) other environmental effects. How would you recommend the commission incorporate these four attributes into a model to derive some reasonable measure of payoff associated with each alternative?

c. Using your own judgment, follow the format prescribed in example 10.1 to assign a payoff measure to each of the three alternatives while incorporating the four attributes of value agreed upon in part b.

10.29. Gregson owns and operates a small magazine and book store in a town where no other magazine and book store exists; thus competition is not a problem. However, Gregson must still attempt to gauge the demand for each different magazine and book before ordering.

a. Is the problem of estimating the demand level for a magazine or book an exercise in decision making under certainty or uncertainty?

b. What information does Gregson consider before determining the number of copies of a magazine or book that she should have in stock?

c. If Gregson's objective is to select the inventory level that maximizes her expected profit, how would she use the information obtained from part b to place meaningful prior probabilities on the possible demand levels?

10.30.   Conservation is usually more visible when there are immediate, direct dollar savings to the consumer. But since energy efficient items usually cost more and their cost-saving benefits are usually not recovered in the short run, many consumers are not willing to undertake the initial purchase of such items.* A utility company is considering offering clock thermostats for nightly cutbacks in room temperature to their customers. The utility would pay $30 for each thermostat and would offer them to customers for a price of $45. A careful survey of the utility's customers suggests the following probability distribution for the number who would buy a thermostat:

Demand	50	60	70	80	90
Probability of Demand	.15	.25	.35	.20	.05

Assume that unsold thermostats cannot be returned to the manufacturer.

a. Construct the profit table.

b. Construct the opportunity loss table.

c. Find the minimax decision.

d. Find the utility company's optimal order quantity if it wishes to maximize its expected profits on the sale of thermostats.

e. If unsold thermostats can be returned to the manufacturer at full refund of cost, how many should the utility company order? Explain your answer.

10.31.   A businessman is trying to decide whether to take one of two contracts or neither one. He has simplified the situation somewhat and feels it is sufficient to imagine that the contracts provide the alternatives shown in the table.

CONTRACT A		CONTRACT B	
Profit	Probability	Profit	Probability
$100,000	.2		
50,000	.4	$40,000	.3
0	.3	10,000	.4
−30,000	.1	−10,000	.3

a. Which contract should the businessman select if he wishes to maximize his expected profit?

b. What is the expected profit associated with the optimal decision?

*See Environmental Protection Agency, "How to Save Energy—and Money—at the Same Time," *Environmental News*, July 1975.

c. Give a practical interpretation for the expected profit from part b. Is it actually the profit the businessman can expect if he selects the optimal contract?

10.32. A retail trade association is considering publishing a quarterly trade journal describing the activities of the firms in their association. The journal is a feasible undertaking only if a sufficient number of members agree to subscribe. The association believes that the following probabilities adequately represent the chance that a given percentage of the trade association membership will subscribe to the journal:

Percentage, p	.30	.40	.50	.60
Probability, $P(p)$	.1	.3	.4	.2

Suppose there are 1000 members in the trade association and that fixed costs of printing the journal will amount to $160 per quarter, with a variable cost of $.25 per journal. How much would the trade association have to charge per journal in order to break even?

10.33. The Chemical Corporation is interested in determining the potential loss on a binding purchase contract that will be in effect at the end of the fiscal year. The corporation produces a chemical compound that deteriorates and must be discarded if it is not sold by the end of the month during which it was produced. The total variable cost of the manufactured compound is $25 per unit and it is sold for $40 per unit. The compound can be purchased from a competitor for $40 per unit plus $5 freight per unit. It is estimated that failure to fill orders would result in the complete loss of customers' orders for the compound. The corporation has sold the compound for the past 30 months. Demand has been irregular and sales per month during this time have been as shown in the table.

UNITS SOLD PER MONTH	NUMBER OF MONTHS
4000	6
5000	15
6000	9

a. Find the probability of sales of 4000, 5000, or 6000 units in any month.
b. Assuming that all orders will be filled, construct the profit table and the opportunity loss table.
c. What is the expected monthly income of the corporation if 5000 units are manufactured every month and all orders are filled?
d. What is the inventory level that maximizes the corporation's expected monthly income?

10.34. George Smith will buy a one-year term life insurance policy from either Firm A or Firm B. Firm A offers a $10,000 policy for $200 while Firm B offers a $25,000 policy for $250. Actuarial tables suggest that men of Smith's age have a probability of .99 of living through the next year.
a. Construct Smith's payoff (profit) table, listing the actions that he may take and the states of nature that may result.
b. Which policy maximizes Smith's expected gain?
c. Recalling our discussion of similar problems from section 5.5, criticize the premiums placed on each of the policies.

10.35.   Factors such as demand elasticity, competitive retaliation, threat of future price weakness, and potential entry of new competitors influence the pricing policy of manufacturers.* The marketing manager for a nationwide brewery has suggested that his firm reduce the price of its product to 20% in the hopes of inducing greater market penetration. Studies indicate that if the firm's primary competitor does not also reduce its price, the brewery would incur an annual gain in profits of $500,000, but if the competitor does retaliate with a price reduction, annual profits would drop by $200,000. If the objective of the brewery is to maximize expected profits, what is the limiting value of the probability that the competitor will retaliate with a price reduction in order to justify the suggestion of the marketing manager?

10.36.   A manufacturer of electric power tools is contemplating the purchase of $\frac{1}{4}$-horsepower electric motors from a Japanese manufacturer. The motors are currently produced internally and, from experience, it is known that about 97% of all internally produced motors are nondefective. The product from the Japanese manufacturer, although available at a unit cost of $.50 less than the cost to produce a motor internally, has an uncertain quality. Based on the best available information, the following probability distribution describes the quality (fraction defective) of motors available from the Japanese manufacturer:

FRACTION DEFECTIVE	PROBABILITY
.01	.3
.05	.5
.10	.2

The cost to the power tool manufacturer of having to replace or repair a power tool with a defective motor is estimated to be $10 per tool. Assuming that the manufacturer will need 50,000 electric motors during the next year, should he purchase the motors from the Japanese supplier or manufacture them internally?

10.37.   The Electronics Corporation produces electron tubes for use in transmitting equipment. Currently the principal material used for the tube casings is made of di-essolan. The management is considering the use of tetra-essolan for the casings to be used on production runs where a relatively low proportion of the tubes are defective. Tetra-essolan is more expensive than di-essolan but is also more attractive to the consumer. Since there is great difficulty determining in advance of a production run what the proportion of defective tubes will be, it is not certain which would be the best material to use for the casings. After considerable research the corporation has determined that the information shown in the table best describes the relationship

PROPORTION OF TUBES DEFECTIVE PER PRODUCTION RUN	PRIOR PROBABILITY OF PROPORTION DEFECTIVE	CASING MATERIAL	
		Di-essolan	Tetra-essolan
.01	.5	$ 50	$180
.05	.2	80	100
.10	.2	120	70
.20	.1	160	40

*P. E. Green, "Bayesian Decision Theory in Pricing Strategy," *Journal of Marketing,* January 1963.

among proportion defective, their associated probabilities of occurrence (prior probabilities), and the dollar profits per production run by using each of the casing materials.

a. If management's objective is to maximize the expected profits, which casing material should be used?

b. What is the expected profit associated with the optimal decision?

c. How much would the corporation be willing to pay to know in advance of a production run the exact proportion of tubes that will be defective?

10.38. A building contractor must decide how many speculative homes to build. Each home will be of the same design and will cost the contractor $40,000 to build. He plans to sell each for $45,000 but intends to auction off unsold homes at the end of six months to recover construction costs. Auction prices on homes of the design proposed by the contractor have been bringing $37,500. Suppose that the contractor's prior beliefs regarding the likelihood of selling any number of the speculative homes within six months are as shown in the table. If the contractor's

NUMBER OF SPECULATIVE HOMES SOLD WITHIN SIX MONTHS	PROBABILITY
0	.05
1	.15
2	.45
3	.25
4	.10
5 or more	0

objective is to maximize his expected return, how many of the speculative homes should he build?

10.39. The manager of a manufacturing plant that produces an item for which the demand has been quite variable must decide whether to buy a new assembly machine or to have the current assembly machine repaired, at a cost of $500. The new machine would cost $7000 and would assemble finished items at a cost of $.60 per unit, while the current machine incurs a variable cost of $1.10 per unit. Units are produced as demand occurs so that management will not be left with unsold merchandise. The selling price of each unit of production is $2.00, and the manager believes that demand will occur according to the following probability distribution:

Remaining Demand	5,000	10,000	25,000	50,000
Probability	.2	.4	.3	.1

a. Construct the profit table, listing the actions and possible states of nature.

b. If the manager's objective is to maximize the expected profits to the manufacturing plant, what is the optimal decision?

c. What is the expected profit associated with the optimal decision?

d. Give a practical interpretation to the expected profit calculated in part c.

10.40. Refer to exercise 10.39 and suppose that another new machine is available; it would cost $2000 and would incur a variable cost of $1 per unit in the process of assembling a finished unit. Thus the manager can choose from among three machines, two new ones and the old one. If his objective is to maximize the manufacturing plant's expected profit, what is the manager's optimal decision?

10.41.   Refer to exercise 10.36. Before deciding whether or not to purchase the 50,000 motors from the Japanese supplier, suppose that the power tool manufacturer obtains a sample of 25 motors from the Japanese supplier, subjects the motors to a battery of rigorous tests, and finds three of them to be defective. Based on the information contained in exercise 10.36 and this sample information, should the manufacturer buy from the Japanese supplier?

10.42.   Refer to exercise 10.37. Suppose that the Electronic Corporation has decided to select a sample of 25 finished tubes from the production process before deciding whether or not to switch to tetra-essolan for the remainder of a day's production run. Three defective tubes are found in the sample. From the tables for the binomial distribution (table 1 of the appendix), the probability of observing three defectives in a sample of size $n = 25$, given that the fraction defective within the population is p_D, has been determined and is given in the accompanying table for each possible value of p_D.

p_D	$P(n = 25, 3 \text{ defectives} | p_D)$
.01	.002
.05	.093
.10	.227
.20	.142

a. Using Bayes's law, construct the posterior probabilities associated with the fraction defective (p_D) levels.
b. Incorporating both the prior and sample information, should the Electronics Corporation use di-essolan or tetra-essolan for the casings during that day?
c. How much would the Electronics Corporation have been willing to pay to obtain the given sample information? Explain.

10.43.   Refer to exercises 10.37 and 10.42. Suppose that during another day a sample of $n = 25$ finished tubes is selected from the process and that two defectives are found. Use the binomial probability tables (table 1 of the appendix) to compute the probability of observing two defectives in a sample of size $n = 25$, given that the fraction defective within the population is .01, .05, .10, and .20. With regard to the sample information of observing two defectives in a sample of size $n = 25$ and their associated probabilities for each p_D level, answer the questions posed in exercise 10.42.

10.44.   A sampling plan is defined as a decision rule by which the decision maker selects a sample of n items and makes one decision if not more than a defectives are observed but makes a complementary decision if more than a defectives are observed in the sample. With regard to the results of exercises 10.42 and 10.43, suggest a sampling plan for the Electronics Corporation, using a sample size of $n = 25$.

10.45.   A publisher is considering marketing a new monthly magazine BLASH (Buy Low and Sell High) with articles and other information of special interest to investors. Based on past experience and her perceptions of potential demand for a monthly such as BLASH, the publisher has established the profit table shown here to represent her annual profits under three assumed possible levels of buyer response. The publisher estimates the probabilities of occurrence of the three buyer responses to be $P(R_1) = .5$, $P(R_2) = .2$, and $P(R_3) = .3$. Should the publisher go ahead with the project and publish BLASH?

	ANNUAL PROFITS	
BUYER RESPONSE	Do Not Publish	Publish
poor (R_1)	$0	-$2,500,000
fair (R_2)	0	500,000
good (R_3)	0	3,000,000

10.46.   Refer to exercise 10.45. Suppose the publisher recognizes that competition would be very fierce owing to the presence of many current and successful publications dealing with the same subject as BLASH. She thus wishes to test-market BLASH and base her decision on information other than her own experience and perceptions. A test market will result in either an unfavorable reaction (O_1) or a favorable reaction (O_2). Based on historical experience from market tests on other publications, the publisher has established the following conditional probabilities, given buyer response:

$$P(O_1|R_1) = .9 \qquad P(O_1|R_2) = .4 \qquad P(O_1|R_3) = .3$$
$$P(O_2|R_1) = .1 \qquad P(O_2|R_2) = .6 \qquad P(O_2|R_3) = .7$$

a. What is the publisher's best decision if the market test is unfavorable?
b. What is the publisher's best decision if the market test is favorable?
c. Construct a decision tree to illustrate the publisher's decision problem.

10.47.   Mr. Owner, a property owner, has offered to list Angelacres with Mr. Realtor, a real estate broker, on a one-month guaranteed sale basis for a price of $25,000. If Angelacres is sold within the stipulated time, Mr. Owner will offer either Blackacre or Chancyhill to Mr. Realtor on a 90-day guaranteed sale basis at minimum prices of $50,000 and $100,000, respectively. Mr. Owner will not offer both properties to the broker, however. Mr. Realtor estimates that broker-absorbed advertising and administrative costs and the chances of selling each property under the stipulated conditions in the open market are as listed in the table. Assume that the 4% standard sales commission is appropriate for all transactions.*

PROPERTY NAME	GUARANTEED SALE PRICE	ESTIMATED SALES COST	PROBABILITY OF SALE IN STIPULATED TIME
Angleacres	$ 25,000	$800	.7
Blackacre	50,000	200	.6
Chancyhill	100,000	400	.5

a. Construct a decision tree to illustrate Mr. Realtor's decision problem.
b. If Mr. Realtor's objective is to maximize his expected profit, should he accept Mr. Owner's offer to list Angelacres?
c. If Mr. Realtor is able to sell Angelacres within the stipulated time, should he then list Blackacre, Chancyhill, or neither?

10.48.   Refer to exercise 10.47. An econometrician who is an acquaintance of Mr. Realtor has assisted him in developing a personal utility function to measure

*This exercise has been extracted in part from J. E. Reinmuth and J. B. Weidler, "To List or Not to List, That Is the Question," *Real Estate Appraiser*, March 1970, Society of Real Estate Appraisers, Chicago.

his risk preferences toward various possible dollar-valued outcomes. Utility values that have been assessed for certain dollar-valued outcomes are listed in the table.

DOLLAR-VALUED OUTCOME	UTILITY
$5000	.80
3800	.75
2000	.65
1000	.55
700	.50
200	.40
0	.35
−200	.30
−500	.20
−800	.00

a. If Mr. Realtor's objective is to maximize his expected utility, should he accept Mr. Owner's offer to list Angelacres?

b. If Mr. Realtor is able to sell Angelacres within the stipulated time, should he then list Blackacre, Chanceyhill, or neither, assuming that he wishes to maximize his expected utility?

c. Explain any difference in the optimal decisions for exercises 10.47 and 10.48.

Experiences with Real Data

Select a journal appropriate to your area of study. (See Experiences with Real Data, chapter 1, for a partial list of appropriate journals.) Scan past issues of the journal you have chosen and find an article dealing with an application of decision analysis (modern decision making, Bayesian statistics, decision theory) to a real business problem. After carefully reading the article, identify the states of nature used in the decision analysis and the actions available to the decision maker. Examine the use of probabilistic information in the decision analysis and then answer the following questions:

1. On what basis was a prior probability distribution assigned to the states of nature? That is, were the probabilities based on historical frequencies, on some assumed theoretical probability distribution, or on the basis of the decision maker's judgment?

2. What numerical values were assigned to the states of nature as prior probabilities?

3. Did the decision analysis involve the use of indirectly relevant or sample information? If so, how was it incorporated into the analysis and what are the numerical values for the derived posterior probabilities?

Explain the objective of the decision analysis application you have chosen. That is, did the analysis seek to find an action that minimizes the expected opportunity loss, maximizes the expected profit, maximizes the expected utility, or maximizes some other measure of payoff? What was the final outcome of the study and what effect did the use of decision analysis have in arriving at the final outcome? That is, in your opinion, was the final outcome self-evident? Could it have been found by the use of an alternative statistical decision-making procedure? Was it a direct result of the application of decision analysis?

Listed below are some articles appearing in the literature, which demonstrate an application of decision analysis to real world business problems. If you have trouble finding an appropriate article, you may choose one of those listed below.

Accounting: J. A. Knoblett, "The Applicability of Bayesian Statistics in Auditing," *Decision Sciences* 1 (July–October 1970): 423-440.

Economics: J. C. Wiginton, "A Bayesian Approach to Discrimination Among Economic Models," *Decision Sciences* 5 (April 1974): 182-194.

Finance: D. B. Hertz, "Risk Analysis in Capital Investment," *Harvard Business Review* 42 January–February 1964: 95-107.

Management: S. H. Archer, "The Structure of Management Decision Theory," *Academy of Management Journal* 7 December 1964: 269-287.

Marketing: R. D. Buzzell and C. C. Slater, "Decision Theory and Marketing Management," *Journal of Marketing* 26 (July 1962): 7-17.

Real Estate: R. U. Ratcliff and B. Schwab, "Contemporary Decision Theory and Real Estate Investment," *Appraisal Journal* 38 (January 1970): 165-188.

References

BROWN, R. V.; A. S. KAHR; and C. PETERSON. *Decision Analysis for the Manager.* New York: Holt, Rinehart and Winston, 1974.

HADLEY, G. *Introduction to Probability and Statistical Decision Theory.* San Francisco: Holden-Day, 1967. Chapters 3, 4, 8, and 9.

HOWARD, R. "Decision Analysis: Applied Decision Theory." *Proceedings, Fourth International Conference on Operational Research*, 1966.

JEDAMUS, P., and R. J. FRAME. *Business Decision Theory.* New York: McGraw-Hill, 1969.

RAIFFA, H. *Decision Analysis.* Reading, Mass.: Addison-Wesley, 1968.

SCHLAIFER, R. *Analysis of Decisions Under Uncertainty.* New York: McGraw-Hill, 1969.

WINKLER, R. L. *An Introduction to Bayesian Inference and Decision.* New York: Holt, Rinehart and Winston, 1972.

chapter objectives

GENERAL OBJECTIVES Chapters 8 and 9 presented methods for making inferences about population means based on large random samples (chapter 8) and small random samples (chapter 9). The object of this chapter is to extend this methodology to consider the case in which the mean value of y is related to another variable, call it x. By making simultaneous observations on y and the x variable, we can use information contained in the x measurements to estimate the mean value of y and to predict particular values of y for preassigned values of x. This chapter will be devoted to the case where y is a linear function of one predictor variable x. The general case, where y is related to one or more predictor variables, say x_1, x_2, ..., x_k, will be discussed in chapter 12.

SPECIFIC OBJECTIVES

1. To give practical illustrations of the types of business problems that can be solved using the techniques of linear regression and correlation.
Sections 11.1 through 11.9

2. To distinguish between deterministic and probabilistic models and to identify their advantages and limitations, and to present a linear probabilistic model for relating a response y to a single independent variable x.

3. To explain how the method of least squares can be used to fit a linear probabilistic model to data.
Section 11.3

4. To provide a method for determining whether x contributes information for the prediction of y.
Sections 11.4, 11.5

5. To present a confidence interval for estimating the mean value of y for a given value of x.
Section 11.6

6. To give a prediction interval for predicting a particular value of y for a given value of x.
Section 11.7

7. To present measures of the strength of the linear relationship between y and x—the simple linear coefficient of correlation and the coefficient of determination.
Sections 11.8, 11.9

chapter eleven

Linear Regression and Correlation

11.1 Introduction

An estimation problem of particular importance in almost every field of study is the problem of **forecasting**, or **predicting**, the value of a process variable from known, related variables. Practical examples of prediction problems are numerous in business, industry, and the sciences. The stockbroker wishes to predict stock market behavior as a function of a number of observable "key indices." The sales manager of a chain of retail stores wishes to predict the monthly sales volume of each store from the number of credit customers and the amount spent on advertising. The manager of a manufacturing plant would like to relate the yield of a chemical to a number of process variables. He will then use the prediction equation to find the settings for the controllable process variables that will provide the maximum yield of the chemical. The personnel director of a corporation, like the admissions director of a university, wishes to test and measure individual characteristics so that she may hire the person best suited for a job. The political scientist may wish to relate success in a political campaign to the characteristics of a candidate, the opposition, and various campaign issues and promotional techniques. Certainly, all the prediction problems are identical in many respects.

In a sense, the statistical approach to each problem is a formalization of the procedure we might follow intuitively. Suppose that a security analyst is attempting to predict the price of a firm's securities on the securities market from the Dow-Jones industrial average, the prime interest rate, and the sales volume for the firm over the past month. He could expect the security price to rise as the Dow-Jones average and the sales volume rise. However, a rise in the prime interest rate would tend to drive money out of the securities market into more risk-free investments and hence would be accompanied by a decrease in security prices. Each of the variables above seems to individually influence the security price in some way. But when observed together, these variables may have an interactive effect that influences the security price differently from the way either variable does alone. For instance, how would a firm's security price be affected if a rise in sales volume were accompanied by an increase in the prime interest rate? The true relationship in this case could probably best be seen by introducing another variable into the model, say the amount of new debt undertaken by the firm in the past month, which would be related to both the prime interest rate and the sales volume. Carrying this line of thought to the ultimate and idealistic extreme, we would expect the price of a firm's securities to be a mathematical **function** of the preceding variables plus any others that may influence price and be easily measurable. Ideally we would like to possess a mathematical equation that relates the price of a particular firm's securities to all relevant variables, so that the equation could be used for prediction.

Observe that the problem we have defined is of a very general nature. We are interested in a random variable y that is related to a number of **independent predictor variables** x_1, x_2, x_3, ... The variable y for our example is the price of a particular firm's securities at a certain time, and the independent predictor variables might be

$$x_1 = \text{Dow-Jones industrial average}$$

$$x_2 = \text{prime interest rate}$$

$$x_3 = \text{last month's retail sales volume}$$

and so on. The ultimate objective is to measure x_1, x_2, x_3, ... for a particular firm, substitute these values into the prediction equation, and thereby predict the price of the firm's securities. To accomplish this we must first locate the related variables x_1, x_2, x_3, ... and obtain a measure of the strength of their relationship to y. Then we must construct a good prediction equation that expresses y as a function of the selected independent predictor variables.

In this chapter we will restrict our attention to the simple problem of predicting y as a linear function of a single variable. The solution for the multivariable problem—for example, predicting the price of a firm's securities as a function of more than one predictor variable—will be the subject of chapter 12.

11.2 A Simple Linear Probabilistic Model

For purposes of illustration we introduce our topic by considering the problem of predicting the gross monthly sales volume y for a corporation that is not subject to substantial seasonal variation in its sales volume. As

Table 11.1 *Advertising expenditures and sales volumes for a corporation during 10 randomly selected months*

MONTH	ADVERTISING EXPENDITURES, x (× $10,000)	SALES VOLUME, y (× $10,000)
1	1.2	101
2	.8	92
3	1.0	110
4	1.3	120
5	.7	90
6	.8	82
7	1.0	93
8	.6	75
9	.9	91
10	1.1	105

the predictor variable x we will use the amount spent by the company on advertising during the month of interest. As noted in section 11.1, we wish to determine whether advertising is actually worthwhile—that is, whether advertising is actually related to the firm's sales volume—and, in addition, we wish to obtain an equation that will be useful for predicting the monthly sales y as a function of advertising expenditures x. The evidence, presented in table 11.1, represents a sample of advertising expenditures and the associated sales volumes for the company during 10 randomly selected months. We will assume that the advertising expenditures and sales volumes for these 10 months constitute a random sample of measurements for all past and current months' operations for the company.

Our initial approach to the analysis of the data of table 11.1 is to plot the data as points on a graph, representing a month's sales volume as y and the corresponding advertising expenditure as x. The graph, shown in figure 11.1, is called a **scatter diagram.** You will probably observe that y appears to increase as x increases. (Could this arrangement of the points occur by chance even if x and y were unrelated?)

Figure 11.1 *Plot of the data of table 11.1*

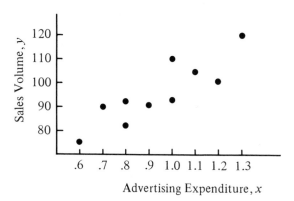

Advertising Expenditure, x

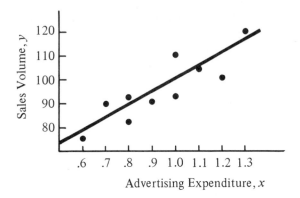

Figure 11.2 *Fitting a line by eye*

One method of obtaining a prediction equation relating y to x is to place a ruler on the graph and move it about until it seems to pass through the points, thus providing what we might regard as the "best fit" to the data. Indeed, if we draw a line through the points, it would appear that our prediction problem was solved (see figure 11.2). Certainly we can now use the graph to predict the company's monthly sales volume y as a function of the amount x budgeted for advertising during that month.

Let us review several facts concerning the graphing of mathematical functions. First, the mathematical equation of a straight line is

$$y = \beta_0 + \beta_1 x$$

where β_0 is the y-**intercept** and β_1 is the **slope** of the line. Second, the line that we may graph for each linear equation is unique. Each equation corresponds to only one line, and vice versa. Thus when we draw a line through the points, we automatically choose a mathematical model for the response y:

$$y = \beta_0 + \beta_1 x$$

where β_0 and β_1 have unique numerical values.

The linear model $y = \beta_0 + \beta_1 x$ is said to be a **deterministic** mathematical model, because when a value of x is substituted into the equation, the value of y is determined and no allowance is made for error. Fitting a straight line through a set of points by eye produces a deterministic model. Many other examples of deterministic mathematical models may be found by leafing through the pages of elementary chemistry, physics, economics, or engineering textbooks.

Deterministic models are quite suitable for explaining and predicting phenomena when the error of prediction is negligible for practical purposes. Thus Newton's law, which expresses the relation between the force F imparted by a moving body with mass m and acceleration a, given by the deterministic model

$$F = ma$$

predicts force with very little error for most practical applications. "Very little" is, of course, a relative concept. An error of .1 inch in forming a beam for a bridge is extremely small, but the same error is impossibly large in the manufacture of parts for a wristwatch. Thus in many physical situations the error of prediction cannot be ignored. Indeed, consistent with our stated philosophy, we would hesitate to place much confidence in a prediction unaccompanied by a measure of its goodness. For this reason a visual choice of a line to relate the advertising expenditures to the sales volume would have limited value.

Now let us return to our example. What is wrong with using a deterministic model to relate sales volume to advertising expenditures? The answer is that although deterministic models permit us to predict y for various values of x (by substituting values of x into the prediction equation), they do not provide us with a way to evaluate the error of prediction. For example, if we use the line in figure 11.2 to predict sales volume y when advertising expenditure is $x = 1.2$, we obtain a predicted value $y = 111$. But plus or minus what? You can see that the plotted points deviate substantially from the line and do so in a seemingly random pattern. What is the bound on the error of prediction? A businessperson would need this information to be able to decide whether it would be profitable to make a particular advertising expenditure. Consequently, we need a method for fitting a model to data, a method that will enable us to place bounds on the errors of estimating β_0 and β_1 and bounds on the error of predicting y for given values of x. More important, we want a method that can be used to fit models when more than one predictor variable x is involved.

The probabilistic model we use to relate sales volume y and advertising expenditures x is a simple modification of the deterministic model. Rather than saying that y and x are related by the deterministic model

$$y = \beta_0 + \beta_1 x$$

we say that the mean (or expected) value of y for a given value of x, denoted by the symbol $E(y|x)$, has a graph that is a straight line. That is, we let

$$E(y|x) = \beta_0 + \beta_1 x$$

For any given values of x, y-values will vary in a random manner about the mean $E(y|x)$. For example, if the company spends \$10,000 ($x = 1.0$) per month on advertising, the gross monthly sales y could assume some value from a population of possible values. But the *mean* gross monthly sales is precisely determined by substituting $x = 1.0$ into the equation

$$E(y|x) = \beta_0 + \beta_1 x$$

So to summarize, we write the probabilistic model for any particular observed value of y as

$$y = \text{(mean value of } y \text{ for given value of } x\text{)} + \text{(random error)}$$

$$= \overbrace{\beta_0 + \beta_1 x}^{E(y|x)} + \epsilon$$

where ϵ is a random error, the difference between an observed value of y and the mean value of y for a given value of x. Thus we assume that for any given value of x, the observed value of y varies in a random manner and possesses a probability distribution with a mean value $E(y|x)$. To assist in conveying this idea, the probability distributions of y about the line of means $E(y|x)$ are shown in figure 11.3 for three hypothetical values of x, x_1, x_2, and x_3, although, as noted, we imagine a distribution of y-values for every value of x.

What properties will we assume for the probability distribution of y for a given value of x? Assumptions that seem to fit many practical situations, and upon which the methodology of sections 11.4, 11.5, 11.6, and 11.7 are based, are given in the box.

Assumptions for the Probabilistic Model

For any given value of x, y possesses a normal distribution, with a mean value given by the equation

$$E(y|x) = \beta_0 + \beta_1 x$$

and with a variance of σ^2. Furthermore, any one value of y is independent of every other value.

These assumptions permit us to construct tests of hypotheses and confidence intervals for β_0, β_1, and $E(y|x)$. In addition, we are able to

Figure 11.3 *Linear probabilistic model*

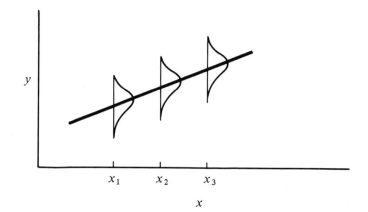

give a prediction interval for y when we use the fitted model (the line that we find to fit a particular set of data) to predict some new value of y for a particular set of x. The practical problems we can solve by using these procedures will become apparent when we have worked some examples.

In the next section we will consider the problem of finding the "best fitting" line for a given set of data. This will give us a prediction equation, or *regression line* as it is known in statistics.

Exercises

11.1.   For the linear equation $y = 5 - 4x$:
a. Give the y-intercept and the slope of the line.
b. Graph the line corresponding to the equation.

11.2.   Graph the line corresponding to the equation $2y = 3x + 1$.

11.3.   What is the difference between deterministic and probabilistic mathematical models?

11.3 The Method of Least Squares

The statistical procedure for finding the "best-fitting" straight line for a set of points is, in many respects, a formalization of the procedure employed when we fit a line by eye. For instance, when we visually fit a line to a set of data, we move the ruler until we think that we have minimized the deviations of the points from the prospective line. If we denote the predicted value of y obtained from the fitted line as $\hat{y}$, then the prediction equation is

$$\hat{y} = \hat{\beta}_0 + \hat{\beta}_1 x$$

where $\hat{\beta}_0$ and $\hat{\beta}_1$ represent estimates of the true β_0 and β_1. This line for the data of table 11.1 is shown in figure 11.4. The vertical lines drawn from

Figure 11.4 *Linear prediction equation*

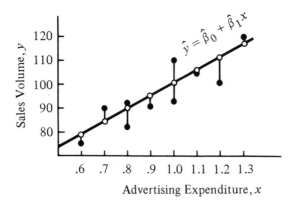

the prediction line to each point represent the deviations of the points from the predicted value of y. Thus the deviation of the ith point is

$$y_i - \hat{y}_i$$

where

$$\hat{y}_i = \hat{\beta}_0 + \hat{\beta}_1 x_i$$

Having decided that in some manner or other we will attempt to minimize the deviations of the points in choosing the best-fitting line, we must now define what we mean by "best." That is, we wish to define a criterion for "best fit" that seems intuitively reasonable, is objective, and under certain conditions gives the best prediction of y for a given value of x.

We will employ a criterion of goodness that is known as the principle of least squares, which may be stated as follows. Choose as the best-fitting line that line that minimizes the sum of squares of the deviations of the observed values of y from those predicted. Expressed mathematically, we wish to minimize

$$SSE = \sum_{i=1}^{n} (y_i - \hat{y}_i)^2$$

The term SSE represents the sum of squares of deviations or, as it is commonly called, the sum of squares for error.

Since the predicted value of y_i corresponding to $x = x_i$ is $\hat{\beta}_0 + \hat{\beta}_1 x_i$, we can substitute this quantity for $\hat{y}_i$ in SSE and obtain the following expression:

$$SSE = \sum_{i=1}^{n} [y_i - (\hat{\beta}_0 + \hat{\beta}_1 x_i)]^2$$

The method for finding the numerical values of $\hat{\beta}_0$ and $\hat{\beta}_1$ that minimize SSE utilizes differential calculus and hence is beyond the scope of this text. We simply state that it can be shown that $\hat{\beta}_0$ and $\hat{\beta}_1$ are given by the following formulas:

Least Squares Estimators of β_0 and β_1

$$\hat{\beta}_1 = \frac{SS_{xy}}{SS_x} \quad \text{and} \quad \hat{\beta}_0 = \bar{y} - \hat{\beta}_1 \bar{x}$$

where

$$SS_x = \sum_{i=1}^{n} (x_i - \bar{x})^2 = \sum_{i=1}^{n} x_i^2 - \frac{\left(\sum_{i=1}^{n} x_i\right)^2}{n}$$

$$SS_{xy} = \sum_{i=1}^{n} (x_i - \bar{x})(y_i - \bar{y}) = \sum_{i=1}^{n} x_i y_i - \frac{\left(\sum_{i=1}^{n} x_i\right)\left(\sum_{i=1}^{n} y_i\right)}{n}$$

Note that SS_x (sum of squares for x) is computed using the familiar shortcut formula for calculating sums of squares of deviations that was used in calculating s^2 in chapter 3. SS_{xy} is computed using a very similar formula and hence should be easy to remember. Once $\hat{\beta}_0$ and $\hat{\beta}_1$ have been computed, we substitute their values into the equation of a line to obtain the least squares prediction equation

$$\hat{y} = \hat{\beta}_0 + \hat{\beta}_1 x$$

There is one important point to note here. Rounding errors can greatly affect the answer you obtain in calculating SS_x and SS_{xy}. If you must round a number, it is recommended that you carry six or more significant figures in the calculations. (Note also that in working exercises, rounding errors might cause some slight discrepancies between your answers and the answers given in the text.)

The use of these formulas for finding $\hat{\beta}_0$, $\hat{\beta}_1$, and the least squares line is illustrated by an example.

Example 11.1 Obtain the least squares prediction line for the data of table 11.1.

Solution The calculation of $\hat{\beta}_0$ and $\hat{\beta}_1$ for the data of table 11.1 is simplified by the use of table 11.2.

Substituting the appropriate sums from table 11.2 into the least squares equations, we obtain the calculations shown next.

Table 11.2 *Calculations for the data of table 11.1*

y_i	x_i	x_i^2	$x_i y_i$	y_i^2
101	1.2	1.44	121.2	10,201
92	.8	.64	73.6	8,464
110	1.0	1.00	110.0	12,100
120	1.3	1.69	156.0	14,400
90	.7	.49	63.0	8,100
82	.8	.64	65.6	6,724
93	1.0	1.00	93.0	8,649
75	.6	.36	45.0	5,625
91	.9	.81	81.9	8,281
105	1.1	1.21	115.5	11,025
Sums 959	9.4	9.28	924.8	93,569

$$SS_x = \sum_{i=1}^{n} x_i^2 - \frac{\left(\sum_{i=1}^{n} x_i\right)^2}{n} = 9.28 - \frac{(9.4)^2}{10} = .444$$

$$SS_{xy} = \sum_{i=1}^{n} x_i y_i - \frac{\left(\sum_{i=1}^{n} x_i\right)\left(\sum_{i=1}^{n} y_i\right)}{n} = 924.8 - \frac{(9.4)(959)}{10} = 23.34$$

$$\bar{y} = \frac{\sum_{i=1}^{n} y_i}{n} = \frac{959}{10} = 95.9 \qquad \bar{x} = \frac{\sum_{i=1}^{n} x_i}{n} = \frac{9.4}{10} = .94$$

Hence

$$\hat{\beta}_1 = \frac{SS_{xy}}{SS_x} = \frac{23.34}{.444} = 52.5676 \approx 52.57$$

$$\hat{\beta}_0 = \bar{y} - \hat{\beta}_1 \bar{x} = 95.9 - (52.5676)(.94) \approx 46.49$$

Then according to the principle of least squares, the best-fitting straight line (often called the *regression line*) relating the advertising expenditures to the sales volume is

$$\hat{y} = \hat{\beta}_0 + \hat{\beta}_1 x \qquad \text{or} \qquad \hat{y} = 46.49 + 52.57x$$

The graph of this prediction equation is shown in figure 11.4.

We may now predict y for a given value of x by referring to figure 11.4 or by substituting into the prediction equation. For example, if the corporation has budgeted \$10,000 for advertising in a month, their predicted sales volume is

$$\hat{y} = \hat{\beta}_0 + \hat{\beta}_1 x = 46.49 + (52.57)(1.0) = 99.06$$

or, expressing the sales volume in dollars, \$990,600.

Keep in mind that the best-fitting straight line, the prediction equation

$$\hat{y} = \hat{\beta}_0 + \hat{\beta}_1 x$$

also estimates the line of means

$$E(y|x) = \beta_0 + \beta_1 x$$

That is, in example 11.1 the best estimate of the *mean* gross monthly sales for an $x = 1.0$ (\$10,000) advertising expenditure is also equal to \$990,600. Thus the prediction equation can be used either to predict some future value of y or to estimate the mean value of y for a given advertising expenditure x.

The next step in our procedure is to place a bound on our error of estimation. We will consider this and related problems in succeeding sections.

Exercises

11.4. Suppose you are given the five points whose coordinates are as shown in the table.

x	-3	-1	1	1	2
y	6	4	3	1	1

a. Find the least squares line for the data.
b. As a check on the calculations in part a, plot the five points and graph the least squares line.

11.5. The flexible budget is an expression of management's expectations concerning revenues and costs for some future period and serves to communicate top management goals to the various managers of the organization. Suppose that the management of an organization is interested in establishing a flexible budget for purposes of estimating overhead costs over a certain range of production. Historical cost and production data are given in the table.

Production ($\times$ 10,000)	3	4	5	6	7	8	9
Overhead Costs ($\times$ \$1,000)	12	10.5	13	12	13	13.3	16.5

a. Find the least squares line to allow for the estimation of overhead costs from production (that is, the least squares line relating costs to production).
b. As a check on your calculations, plot the seven points and graph the least squares line.

11.4 Calculating s^2, an Estimator of σ^2

Recall that the probabilistic model for y in section 11.2 assumes that y is related to x by the equation

$$y = \beta_0 + \beta_1 x + \epsilon$$

For a given value of x, the y-values are independent and possess a normal probability distribution with mean $E(y|x)$ and variance σ^2. Thus each observed value of y is subject to a random error ϵ that enters into the computations of $\hat{\beta}_0$ and $\hat{\beta}_1$ and introduces error into these estimates. Furthermore, if we use the least squares line

$$\hat{y} = \hat{\beta}_0 + \hat{\beta}_1 x$$

to predict some future value of y, the random errors affect the error of prediction. Consequently, the variability of the random errors, measured by

σ^2, plays an important role when estimating or predicting using the least squares line.

The first step toward acquiring a bound on a prediction error requires that we estimate σ^2, the variance of y for a given value of x. For this purpose it seems reasonable to use SSE, the sum of squares of deviations (sum of squares for error) about the predicted line. Indeed, it can be shown that the formula given in the box provides a good estimator for σ^2, which will be unbiased and will be based on $(n - 2)$ degrees of freedom.

An Estimator for σ^2

$$\hat{\sigma}^2 = s^2 = \frac{\text{SSE}}{n - 2}$$

The sum of squares of deviations SSE may be calculated directly by using the prediction equation to calculate $\hat{y}$ for each point, then calculating the deviations $(y_i - \hat{y}_i)$, and finally calculating

$$\text{SSE} = \sum_{i=1}^{n} (y_i - \hat{y}_i)^2$$

This tends to be a tedious procedure and is rather poor from a computational point of view because the numerous subtractions tend to introduce computational rounding errors. An easier and computationally better procedure is to use the formula given in the box.

Formula for Calculating SSE

$$\text{SSE} = \text{SS}_y - \hat{\beta}_1 \text{SS}_{xy}$$

where

$$\text{SS}_y = \sum_{i=1}^{n} (y_i - \bar{y})^2 = \sum_{i=1}^{n} y_i^2 - \frac{\left(\sum_{i=1}^{n} y_i\right)^2}{n}$$

$$\text{SS}_{xy} = \sum_{i=1}^{n} x_i y_i - \frac{\left(\sum_{i=1}^{n} x_i\right)\left(\sum_{i=1}^{n} y_i\right)}{n}$$

Observe that SS_{xy} was used in the calculation of $\hat{\beta}_1$ and hence has already been computed. Furthermore, note that if you must round a number, it is desirable to retain six or more significant figures in the calculations to avoid serious rounding errors in the final answer.

Example 11.2 Calculate an estimate of σ^2 for the data of table 11.1.

Solution First we calculate

$$SS_y = \sum_{i=1}^{n} (y_i - \bar{y})^2 = \sum_{i=1}^{n} y_i^2 - \frac{\left(\sum_{i=1}^{n} y_i\right)^2}{n}$$

$$= 93,569 - \frac{(959)^2}{10} = 1,600.9$$

Substituting this value of SS_y and the values of $\hat{\beta}_1$ and SS_{xy} calculated in example 11.1 into the formula for SSE, we obtain

$$SSE = SS_y - \hat{\beta}_1 SS_{xy} = 1,600.9 - (52.5676)(23.34) = 373.97$$

Then

$$s^2 = \frac{SSE}{n-2} = \frac{373.97}{8} = 46.75$$

Exercises

11.6.   Calculate SSE and s^2 for the data of exercise 11.4.

11.7.   Calculate SSE and s^2 for the data of exercise 11.5.

11.5 Inferences Concerning the Slope β_1 of a Line

The initial inference desired in studying the relationship between y and x concerns the **existence** of the relationship. That is, do the data present sufficient evidence to indicate that x contributes information for the prediction of y over the region of observation? Or is it quite probable that when y and x are completely unrelated, the points would fall on the graph in a manner similar to that observed in figure 11.1?

The practical question we pose concerns the value of β_1, which is the average change in y for a one-unit change in x. Stating that x does not contribute information for the prediction of y is equivalent to saying that $\beta_1 = 0$. (If $\beta_1 = 0$, we always predict the same value of y regardless of the value of x.) We should first test the hypothesis that $\beta_1 = 0$ against the alternative that $\beta_1 \neq 0$. As we might suspect, the estimator $\hat{\beta}_1$ is extremely useful in constructing a test statistic for this hypothesis. Therefore, we wish to examine the distribution of the estimates $\hat{\beta}_1$ that would be obtained when samples, each containing n points, are repeatedly drawn from the population of interest.

Based on our earlier assumptions concerning the probability distribution of y for a given value of x, it can be shown that both $\hat{\beta}_0$ and $\hat{\beta}_1$ are normally distributed in repeated sampling and that the expected value and variance of $\hat{\beta}_1$ are

$$E(\hat{\beta}_1) = \beta_1 \quad \text{and} \quad \sigma_{\hat{\beta}_1}^2 = \frac{\sigma^2}{\text{SS}_x}$$

Thus $\hat{\beta}_1$ is an unbiased estimator of β_1; we know its standard deviation, and hence we can construct a z statistic in the manner described in section 8.14. Then

$$z = \frac{\hat{\beta}_1 - \beta_1}{\sigma_{\hat{\beta}_1}} = \frac{\hat{\beta}_1 - \beta_1}{\sigma / \sqrt{\text{SS}_x}}$$

possesses a standardized normal distribution in repeated sampling. Since the actual value of σ^2 is unknown, we should obtain the estimated standard deviation of $\hat{\beta}_1$, which is

$$\frac{s}{\sqrt{\text{SS}_x}}$$

Substituting s for σ in z, we obtain, as in chapter 9, a test statistic

$$t = \frac{\hat{\beta}_1 - \beta_1}{s / \sqrt{\text{SS}_x}} = \frac{\hat{\beta}_1 - \beta_1}{s} \sqrt{\text{SS}_x}$$

which can be shown to follow a Student's t distribution in repeated sampling, with $(n - 2)$ degrees of freedom. Note that the number of degrees of freedom associated with s^2 determines the number of degrees of freedom associated with t.

We observe that the test of an hypothesis that β_1 equals some particular numerical value, say $\beta_1 = 0$, is the familiar t test encountered in chapter 9. Let β_{10} be the hypothesized value of β_1. Then the test is as given in the box.

Test of an Hypothesis Concerning the Slope of a Line

Null hypothesis H_0: $\beta_1 = \beta_{10}$.

Alternative hypothesis: Specified by the experimenter, depending on the values of β_1 which he or she wishes to detect.

Test statistic: $t = \dfrac{\hat{\beta}_1 - \beta_{10}}{s} \sqrt{\text{SS}_x}$.

Rejection region: See the critical values of t, table 4 of the appendix, for $(n - 2)$ degrees of freedom.

Example 11.3 Use the data of table 11.2 to determine whether there is evidence to indicate that β_1 differs from 0 by using a linear relationship between advertising expenditure x and monthly sales volume y.

Solution We wish to test the null hypothesis

$$H_0: \beta_1 = 0$$

against the alternative hypothesis

$$H_a: \beta_1 \neq 0$$

for the sales volume and advertising expenditure data in table 11.2. The test statistic is

$$t = \frac{\hat{\beta}_1 - 0}{s} \sqrt{SS_x}$$

and, if we choose $\alpha = .05$, we will reject H_0 when $t > 2.306$ or $t < -2.306$. The critical value of t is obtained from the t table using $(n - 2) = 8$ degrees of freedom. Substituting values determined in examples 11.1 and 11.2 into the test statistic, we obtain

$$t = \frac{\hat{\beta}_1}{s} \sqrt{SS_x} = \frac{52.57}{6.84} \sqrt{.444} = 5.12$$

Observing that the test statistic exceeds the critical value of t, 2.306, we reject the null hypothesis $\beta_1 = 0$ and conclude that there is evidence to indicate that advertising expenditures provide information for the prediction of gross monthly sales volume.

Once we have decided that β_1 differs from 0, we are interested in examining this relationship in detail. If x increases by one unit, what is the estimated change in y and how much confidence can be placed in the estimate? In other words, we require an estimate of the slope β_1. You probably will not be surprised to observe a continuity in the procedures of chapters 9 and 11. That is, the $(1 - \alpha)100\%$ confidence interval for β_1 can be shown to be as given in the box.

A Confidence Interval for β_1

$$\hat{\beta}_1 \pm \frac{t_{\alpha/2} s}{\sqrt{SS_x}}$$

Example 11.4 Find a 95% confidence interval for β_1 based on the data of table 11.1.

Solution The 95% confidence interval for β_1, based on the data of table 11.1, is

$$\hat{\beta}_1 \pm \frac{t_{.025}\, s}{\sqrt{SS_x}}$$

Substituting we obtain

$$52.57 \pm \frac{(2.306)(6.84)}{\sqrt{.444}} = 52.57 \pm 23.67$$

This means that we estimate the increase in monthly sales volume for a one-unit ($10,000) increase in advertising expenditure to fall in the interval from 28.90 to 76.24, or, in the original units for y, from $289,000 to $762,400.

Several points concerning the interpretation of our results deserve particular attention. As we have noted, β_1 is the slope of the assumed line over the region of observation and indicates the linear change in $E(y|x)$ for a one-unit change in x. If we do not reject the null hypothesis $\beta_1 = 0$, it does not mean that x and y are unrelated. In the first place, we must be concerned with the probability of committing a type II error, that is, of accepting the null hypothesis that the slope equals 0 when this hypothesis is false. Second, it is possible that x and y might be perfectly related in a curvilinear, but not linear, manner. For example, figure 11.5 depicts a curvilinear relationship between y and x over the domain of x: $a \le x \le f$. We note that a straight line would provide a good predictor of y if fitted over a small interval in the x domain, say $b \le x \le c$. The resulting line is line 1. On the other hand, if we attempt to fit a line over the region $c \le x \le d$, then β_1 equals zero and the best fit to the data is the horizontal line 2. This would occur even though all the points fell perfectly on the curve and y and x possessed a functional relation as defined in section 11.2. We must take care in drawing conclusions if we do not find evidence to indicate that β_1 differs from zero. Perhaps we have chosen the wrong type of probabilistic model for the physical situation.

Figure 11.5 *Curvilinear relation*

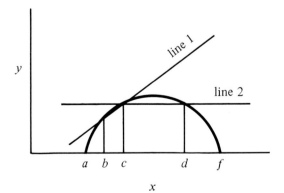

Note that the comments contain a second implication. **If the data provide values of** x **in an interval** $b \le x \le c$**, then the calculated prediction equation is appropriate only over this region.** Extrapolation in predicting y for values of x outside the region $b \le x \le c$ for the situation indicated in figure 11.5 would result in a serious prediction error.

If the data present sufficient evidence to indicate that β_1 differs from zero, we do not conclude that the true relationship between y and x is linear. Undoubtedly y is a function of a number of variables, which demonstrate their existence to a greater or lesser degree in terms of the random error ϵ that appears in the model. This, of course, is why we have been obliged to use a probabilistic model in the first place. Large errors of prediction imply curvatures in the true relation between y and x, the presence of other important variables that do not appear in the model, or both, as most often is the case. All we can say is that we have evidence to indicate that y changes as x changes and that we may obtain a better prediction of y by using x and the linear predictor than simply using $\bar{y}$ and ignoring x. Note that this **does not imply a causal relationship between** x **and** y**.** Some third variable may have caused the change in both x and y, producing the relationship that we have observed.

The standard deviation of the estimator of β_1,

$$\sigma_{\hat{\beta}_1} = \frac{\sigma}{\sqrt{SS_x}} = \frac{\sigma}{\sqrt{\sum_{i=1}^{n}(x_i - \bar{x})^2}}$$

sheds information on the way to select the x-values, that is, on the way to design an experiment to obtain the best estimate of the slope β_1 and of the mean value of y for a given value of x, $E(y|x)$. To illustrate, note that

$$\sum_{i=1}^{n}(x_i - \bar{x})^2$$

appears in the formula for $\sigma_{\hat{\beta}_1}$. This quantity measures the spread (variation) of the x-values. The greater the spread of the x-values, the larger will be

$$\sum_{i=1}^{n}(x_i - \bar{x})^2$$

Now examine the formula for $\sigma_{\hat{\beta}_1}$. You will see that the larger the value of

$$\sum_{i=1}^{n}(x_i - \bar{x})^2$$

the smaller will be $\sigma_{\hat{\beta}_1}$. This provides the clue to finding a good design of an experiment to find the best-fitting straight line to fit a set of data. Locate the majority of the x-values at the extremities of the experimental region, half at each end. Locate a few *x*-values near the middle of the region to detect curvature (similar to that shown in figure 11.5) if it exists.

Exercises

11.8.   Do the data of exercise 11.4 present sufficient evidence to indicate that *y* and *x* are linearly related? (Test the hypothesis that $\beta_1 = 0$; use $\alpha = .05$.)

11.9.   Do the data of exercise 11.5 present sufficient evidence to indicate that the organization's overhead costs and production are linearly related? Use $\alpha = .05$.

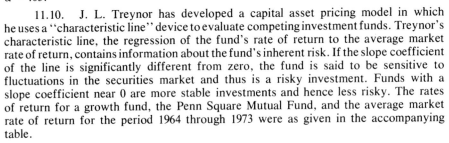

11.10.   J. L. Treynor has developed a capital asset pricing model in which he uses a "characteristic line" device to evaluate competing investment funds. Treynor's characteristic line, the regression of the fund's rate of return to the average market rate of return, contains information about the fund's inherent risk. If the slope coefficient of the line is significantly different from zero, the fund is said to be sensitive to fluctuations in the securities market and thus is a risky investment. Funds with a slope coefficient near 0 are more stable investments and hence less risky. The rates of return for a growth fund, the Penn Square Mutual Fund, and the average market rate of return for the period 1964 through 1973 were as given in the accompanying table.

Year	1964	1965	1966	1967	1968	1969	1970	1971	1972	1973
Penn Square	18.4	29.7	−12.3	10.8	23.6	−16.2	5.8	7.2	7.7	−8.8
Average Market Return	12.9	9.1	−13.1	20.1	7.7	−11.4	.1	10.8	15.6	−17.4

Source: Wiesenberger Financial Services; *Investment Companies, Mutual Funds and Other Types,* 1974.

a. Find the characteristic line for the Penn Square Mutual Fund (that is, the least squares line relating the return on Penn Square to the market rate of return).

b. Plot the points and graph the least squares line as a check on your calculations.

c. Describe the risk characteristics of the Penn Square Mutual Fund (that is, test the hypothesis that $\beta_1 = 0$; use $\alpha = .05$).

11.11.   Find a 95% confidence interval for the slope of the characteristic line for the Penn Square Mutual Fund of exercise 11.10.

11.12.   Find a 90% confidence interval for the slope of the line relating overhead costs to production in exercise 11.5.

11.13.   A marketing research experiment was conducted to study the relationship between the length of time necessary for a buyer to reach a decision and the number of alternative package designs of a product presented. Brand names were eliminated from the packages to reduce the effects of brand preferences. The buyers made their selections by using the manufacturer's product descriptions on the packages as the only buying guide. The length of time necessary to reach a decision was recorded for 15 participants in the marketing research study.

Length of Decision Time (sec)	5,8,8,7,9	7,9,8,9,10	10,11,10,12,9
Number of Alternatives	2	3	4

a. Find the least squares line appropriate for these data.

b. Plot the points and graph the line as a check on your calculations.

c. Calculate s^2.

d. Do the data present sufficient evidence to indicate that the length of decision time is linearly related to the number of alternative package designs? (Test at the $\alpha = .05$ level of significance.)

11.6 Estimating $E(y|x)$, the Expected Value of y for a Given Value of x

In chapters 8 and 9 we studied methods for estimating a population mean μ and encountered numerous practical applications of these methods in the examples and exercises. Now let us consider a generalization of this problem.

Estimating the mean value of y for a given value of x [that is, estimating $E(y|x)$] can be a very important practical problem. A corporate safety director might wish to estimate the mean number of accidents (of a particular type) given the number of hours of safety education each employee receives. Or a company personnel director might wish to estimate the mean number of years a new employee will stay with the company given the score on a test designed to test the employee's job compatibility. If a corporation's profit y is linearly related to advertising expenditures x, the marketing director may wish to estimate the mean profit for a given expenditure x. For example, if the corporation invests \$10,000 in advertising, what can it expect the mean sales volume to be? Finding a confidence interval for $E(y|x)$ will be the topic of this section.

Let us assume that x and y are linearly related according to the probabilistic model defined in section 11.2 and therefore that $E(y|x) = \beta_0 + \beta_1 x$ represents the expected value of y for a given value of x. The fitted line

$$\hat{y} = \hat{\beta}_0 + \hat{\beta}_1 x$$

attempts to estimate the line of means $E(y|x)$ (that is, to estimate β_0 and β_1). Thus $\hat{y}$ can be used to estimate the expected value of $\bar{y}$ as well as to predict some value of y that might be observed in the future. It seems quite reasonable to assume that the errors of estimation and prediction differ for these two cases. Consequently, the two estimation procedures differ. In this section we consider the estimation of the expected value of y for a given value of x.

Observe the two lines in figure 11.6. The first line represents the line of means

$$E(y|x) = \beta_0 + \beta_1 x$$

and the second is the fitted prediction equation

$$\hat{y} = \hat{\beta}_0 + \hat{\beta}_1 x$$

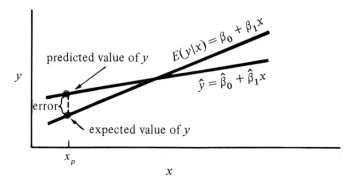

Figure 11.6 *Expected and predicted values for y*

We observe from the figure that the error in estimating the expected value of y when $x = x_p$ is the deviation between the two lines above the point x_p. Also, this error increases as we move to the endpoints of the interval over which x has been measured.

It can be shown that the predicted value

$$\hat{y} = \hat{\beta}_0 + \hat{\beta}_1 x$$

is an unbiased estimator of $E(y|x)$, that is, $E(\hat{y}) = \beta_0 + \beta_1 x$, and that $\hat{y}$ is normally distributed, with variance

$$\sigma_{\hat{y}}^2 = \sigma^2 \left[\frac{1}{n} + \frac{(x_p - \bar{x})^2}{SS_x} \right]$$

The corresponding estimated variance of $\hat{y}$ uses s^2 in place of σ^2 in the expression above.

The results outlined above may be used to test an hypothesis about the average or expected value of y for a given value of x, say x_p.* (This, of course, would also enable us to test an hypothesis concerning the y-intercept β_0, which is the special case where $x_p = 0$.) The null hypothesis is

$$H_0 : E(y|x = x_p) = E_0$$

where E_0 is the hypothesized numerical value of $E(y)$ when $x = x_p$. Once again, it can be shown that the quantity

$$t = \frac{\hat{y} - E_0}{\text{estimated } \sigma_{\hat{y}}} = \frac{\hat{y} - E_0}{s\sqrt{\dfrac{1}{n} + \dfrac{(x_p - \bar{x})^2}{SS_x}}}$$

*See cautionary comments in section 11.5. It is best if x_p lies within the range of the observed values of x.

follows a Student's *t* distribution in repeated sampling with $(n - 2)$ degrees of freedom. The statistical test is conducted in exactly the same manner as the other *t* test discussed previously.

A Test Concerning the Expected Value of y

Null hypothesis H_0: $E(y|x = x_p) = E_0$.

Alternative hypothesis: Specified by the experimenter, depending on the values of $E(y|x)$ that he or she wishes to detect.

Test statistic: $t = \dfrac{\hat{y} - E_0}{s\sqrt{\dfrac{1}{n} + \dfrac{(x_p - \bar{x})^2}{SS_x}}}$.

Rejection region: See the critical values of *t*, table 4 in the appendix, for $(n - 2)$ degrees of freedom.

The corresponding confidence interval, with confidence coefficient $(1 - \alpha)$, for the expected value of *y* given $x = x_p$ is given in the box.

A Confidence Interval for E(y|x)

$$\hat{y} \pm t_{\alpha/2}s\sqrt{\frac{1}{n} + \frac{(x_p - \bar{x})^2}{SS_x}}$$

Example 11.5 Using the data of table 11.1, find a 95% confidence interval for the expected monthly sales volume for an advertising expenditure of $x = 1.0$ ($10,000).

Solution To estimate the mean monthly sales volume for an advertising expenditure $x_p = 1.0$, we use

$$\hat{y} = \hat{\beta}_0 + \hat{\beta}_1 x_p$$

to calculate $\hat{y}$, the estimate of $E(y|x = 1.0)$. Then using values calculated in previous examples, we find

$$\hat{y} = 46.49 + (52.57)(1.0) = 99.06 \qquad \text{or} \qquad \$990,600$$

The formula for the 95% confidence interval is

$$\hat{y} \pm t_{.025}s\sqrt{\frac{1}{n} + \frac{(x_p - \bar{x})^2}{SS_x}}$$

Substituting into this expression, we find that the 95% confidence interval for the expected (mean) monthly sales volume, given an advertising expenditure of 1.0, is

$$99.06 \pm (2.306)(6.84) \sqrt{\frac{1}{10} + \frac{(1.0 - .94)^2}{.444}}$$

Performing these calculations, we have

$$99.06 \pm 5.19 \quad \text{or} \quad 93.87 \text{ to } 104.25$$

Recall that each unit of sales volume represents $10,000. Therefore, we estimate that the mean monthly sales volume for the population of months for which the corporation invests $10,000 in advertising falls in the interval from $938,700 to $1,042,500.

Exercises

11.14. Refer to exercise 11.4. Estimate the expected value of y given $x = 1$, using a 90% confidence interval.

11.15. Refer to exercise 11.5. Estimate the mean overhead cost associated with the production of 55,000 units ($x = 5.5$). Use a 95% confidence interval.

11.16. The rising price of petroleum products over the past few years has led to continually increasing costs to the manufacturer for shipping his goods to the market. These costs have led manufacturers to seek cheaper, but oftentimes slower, means of goods shipment, such as substituting rail freight for air freight services. In a study of shipping costs incurred by his firm, a company controller has randomly selected $n = 9$ air freight invoices from current shippers in order to estimate the relationship between shipping costs and distance for a given volume of goods. The results of his sample are given in the table.

Distance (× 100 miles)	6	13	27	15	9	11	21	14	12
Invoice Charge	$49	$93	$159	$115	$66	$90	$139	$98	$88

a. Find the least squares line for estimating invoice charges (y) from distance (x) when using current air freight shippers.

b. Logically, if the distance traveled is 0 miles, the invoice charge should be $0, since no service has been rendered. Explain why the least squares line does not go through the origin. Should it?

c. Estimate the mean invoice charge for air freight shipped a distance of 1700 miles ($x = 17$), using a 90% confidence interval.

11.7 Predicting a Particular Value of y for a Given Value of x

Suppose that the prediction equation obtained for the 10 measurements in table 11.1 were used to predict the corporation's sales volume for a month selected at random. Although the expected value of y for a particular value

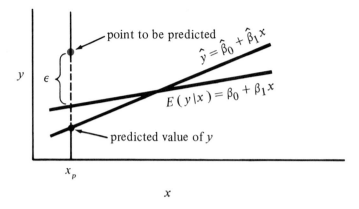

point to be predicted

$\hat{y} = \hat{\beta}_0 + \hat{\beta}_1 x$

$E(y|x) = \beta_0 + \beta_1 x$

predicted value of y

Figure 11.7 *Error in predicting a particular value of y*

of x is of interest for our example (table 11.1), we are primarily interested in *using* the prediction equation $\hat{y} = \hat{\beta}_0 + \hat{\beta}_1 x$, based on our observed data, to predict the sales volume for a month during which the corporation is or has been in operation. If the corporation's advertising expenditures during the month of interest are x_p, we intuitively see that the error of prediction (the deviation between $\hat{y}$ and the actual sales volume y that will occur during that month) is composed of two elements. $(y - \hat{y})$ equals the deviation between $\hat{y}$ and the expected value of y, described in section 11.6 (and shown in figure 11.6) plus a random error ϵ that represents the difference between the actual value of y and its expected value (see figure 11.7). Thus the variability in the error for predicting a single value of y exceeds the variability for estimating the expected value of y.

It can be shown that the variance of the error $(y - \hat{y})$ of predicting a particular value of y when $x = x_p$ is

$$\sigma^2_{\text{error}} = \sigma^2 \left[1 + \frac{1}{n} + \frac{(x_p - \bar{x})^2}{SS_x} \right]$$

When n is very large, the second and third terms in the brackets become small and the variance of the prediction error approaches σ^2. These results may be used to construct the prediction interval for y, given $x = x_p$. The confidence coefficient for the prediction interval is $(1 - \alpha)$.

A Prediction Interval for y

$$\hat{y} \pm t_{\alpha/2} s \sqrt{1 + \frac{1}{n} + \frac{(x_p - \bar{x})^2}{SS_x}}$$

Example 11.6 Find a 95% prediction interval for the next month's sales for the corporation if the advertising expenditure is \$10,000, assuming that other economic conditions remain approximately the same as during the months included in table 11.1.

Solution If in a particular month the advertising expenditures were \$10,000, then $x_p = 1.0$, and we would predict that the sales volume would be

$$99.06 \pm (2.306)(6.84) \sqrt{1 + \frac{1}{10} + \frac{(1.0 - .94)^2}{.444}} = 99.06 \pm 16.60$$

or 82.46 to 115.66. Keep in mind that each unit of sales volume represents \$10,000. Then the 95% prediction interval for the next month's sales volume is \$990,600 $\pm$ \$166,000 or

$$\$824,600 \text{ to } \$1,156,600$$

Note that in a practical situation we would probably have data on the sales volume and advertising expenditures from more than the $n = 10$ months indicated in table 11.1. More data would reduce somewhat the width of the prediction interval by decreasing the quantity under the square root sign in the expression above.

Again note the distinction between the confidence interval for $E(y|x)$

Figure 11.8 *Confidence intervals for E(y|x) and prediction intervals for y based on data of table 11.1*

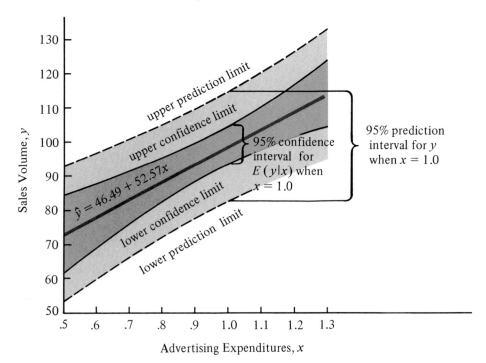

discussed in section 11.6 and the prediction interval presented in this section. $E(y|x)$ is a mean, a parameter of a population of y-values, and y is a random variable that varies in a random manner about $E(y|x)$. The mean value of y when $x = 1.0$ is vastly different from some value of y chosen at random from the set of all y-values for which $x = 1.0$. To make this distinction when making inferences, we always *estimate* the value of a parameter and *predict* the value of a random variable. As noted in our earlier discussion and as shown in figures 11.7 and 11.8, the error of predicting y is different from the error of estimating $E(y|x)$. This is evident in the difference in widths of the prediction and confidence intervals.

A graph of the confidence interval for $E(y|x)$ and the prediction interval for a particular value of y for the data of table 11.1 is shown in figure 11.8. The plot of the confidence interval is shown by solid lines, the prediction interval is identified by dashed lines. Note how the widths of the intervals increase as you move to the right or left of $\bar{x} = .94$. Particularly, note the confidence interval and prediction interval for $x = 1.0$, which were calculated in examples 11.5 and 11.6.

Exercises

11.17. Refer to exercise 11.5. Suppose that the production during a certain period is 55,000 units. Find a 95% prediction interval for y, the overhead cost incurred during the period. Compare this prediction interval with the confidence interval for $E(y|x = 5.5)$ computed in exercise 11.15.

11.18. Refer to exercise 11.13. Suppose a particular buyer is shown three alternative package designs. Find a 90% prediction interval for y, the length of time for the buyer to reach a decision on choice of design.

11.19. An experiment was conducted in a supermarket to observe the relation between the amount of display space allotted to a brand of coffee (Brand A) and its weekly sales. The amount of space allotted to Brand A was varied over 3-, 6-, and 9-square-foot displays in a random manner over 12 weeks, while the space allotted to competing brands was maintained at a constant 3 square feet for each. The data are given in the table.

Weekly Sales, y (dollars)	526	421	581	630	412	560	434	443	590	570	346	672
Space Allotted, x (ft²)	6	3	6	9	3	9	6	3	9	6	3	9

a. Find the least squares line appropriate for these data.
b. Plot the points and graph the least squares line as a check on your calculations.
c. Calculate s^2.
d. Find a 95% confidence interval for the mean weekly sales if the space allotted is 6 square feet.
e. Suppose you intend to allot 6 square feet next week. Find a 95% prediction interval for the weekly sales. Explain the difference between this interval and the interval obtained in part d.

11.8 A Coefficient of Correlation

It is sometimes desirable to obtain an indicator of the strength of the linear relationship between two variables y and x that is independent of their respective scales of measurement. We call this a measure of the **linear correlation** between y and x.

The measure of linear correlation commonly used in statistics is called the Pearson product-moment coefficient of correlation between y and x. This quantity, denoted by the symbol r, is computed as follows:

Pearson Product-Moment Coefficient of Correlation

$$r = \frac{SS_{xy}}{\sqrt{SS_x SS_y}}$$

Example 11.7 Calculate the coefficient of correlation for the advertising expenditure and sales volume data of table 11.1.

Solution The coefficient of correlation for the advertising expenditure and sales volume data of table 11.1 may be obtained by using the formula for r and the quantities

$$SS_{xy} = 23.34 \qquad SS_x = .444 \qquad SS_y = 1600.9$$

which were computed previously. Then

$$r = \frac{SS_{xy}}{\sqrt{SS_x SS_y}} = \frac{23.34}{\sqrt{.444(1600.9)}} = .88$$

A study of the coefficient of correlation r yields rather interesting results and explains the reason for its selection as a measure of linear correlation. We note that the denominators used in calculating r and $\hat{\beta}_1$ will always be positive since they both involve sums of squares of numbers. We also note that the numerator used in calculating r is identical to the numerator of the formula for the slope $\hat{\beta}_1$. Therefore, the coefficient of correlation r will assume exactly the same sign as $\hat{\beta}_1$ and will equal zero when $\hat{\beta}_1 = 0$. Thus $r = 0$ implies no linear correlation between y and x. A positive value for r implies that the line slopes upward to the right; a negative value indicates that it slopes downward to the right.

Figure 11.9 shows six typical scatter diagrams and their associated correlation coefficients. Note that $r = 0$ implies no linear correlation, not simply "no correlation." A pronounced curvilinear pattern may exist, as in figure 11.9(b), but its linear correlation coefficient may equal 0. In general,

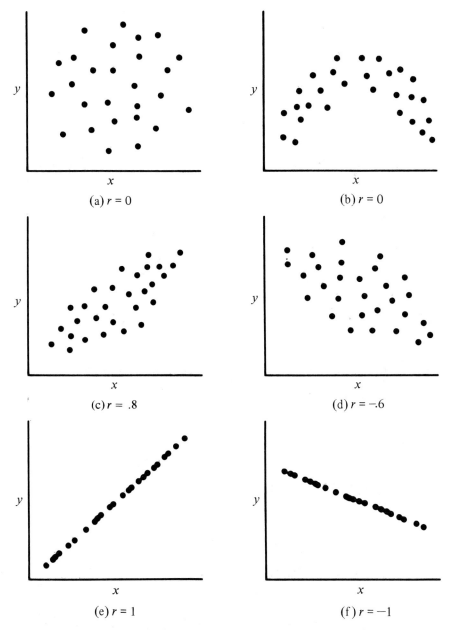

Figure 11.9 *Some typical scatter diagrams and their associated correlation coefficients*

we can say that *r* measures the linear association of the two variables *y* and *x*. When *r* = 1 or −1, all the points fall on a straight line; when *r* = 0, they are scattered and give no evidence of a *linear* relation. Any other value of *r* suggests the degree to which the points tend to be linearly related.

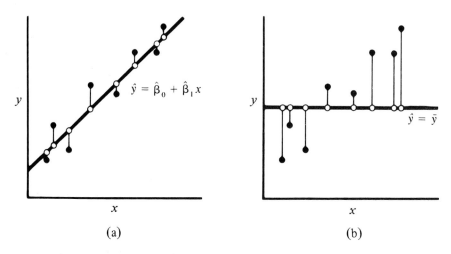

Figure 11.10 *Two models fit to the same data*

The interpretation of nonzero values of r may be obtained by comparing the errors of prediction for the prediction equation

$$\hat{y} = \hat{\beta}_0 + \hat{\beta}_1 x$$

with the predictor of y, $\bar{y}$, that would be employed if x were ignored. Figures 11.10(a) and (b) show the lines $\hat{y} = \hat{\beta}_0 + \hat{\beta}_1 x$ and $\hat{y} = \bar{y}$ fit to the same set of data. Certainly if x is of any value in predicting y, then SSE, the sum of squares of deviations of y about the linear model, should be less than the sum of squares of deviations about the predictor $\bar{y}$, which is

$$SS_y = \sum_{i=1}^{n} (y_i - \bar{y})^2$$

Indeed, we see that SSE can *never* be larger than

$$SS_y = \sum_{i=1}^{n} (y_i - \bar{y})^2$$

because

$$SSE = SS_y - \hat{\beta}_1 SS_{xy} = SS_y - \left(\frac{SS_{xy}}{SS_x}\right) SS_{xy} = SS_y - \frac{(SS_{xy})^2}{SS_x}$$

Therefore, SSE is equal to SS_y minus a positive quantity. Consequently, SSE must always be less than SS_y.

Furthermore, with the aid of a bit of algebraic manipulation, we can show that

$$r^2 = 1 - \frac{SSE}{SS_y} = \frac{SS_y - SSE}{SS_y}$$

In other words, r^2 lies in the interval

$$0 \le r^2 \le 1$$

and r will equal $+1$ or -1 only when all the points fall exactly on the fitted line, that is, when SSE equals zero.

Actually, we see that r^2 is equal to the ratio of the reduction in the sum of squares of deviations obtained by using the linear model to the total sum of squares of deviations about the sample mean $\bar{y}$, which would be the predictor of y if x were ignored. Thus r^2, called the *coefficient of determination,* would seem to give a more meaningful interpretation of the strength of the relation between y and x than would the correlation coefficient r.

The Coefficient of Determination

$$r^2 = \frac{SS_y - SSE}{SS_y} = \frac{\sum\limits_{i=1}^{n} (y_i - \bar{y})^2 - SSE}{\sum\limits_{i=1}^{n} (y_i - \bar{y})^2}$$

You will observe that the sample correlation coefficient r is an estimator of a population correlation coefficient ρ (Greek letter rho), which would be obtained if the coefficient of correlation were calculated using all the points in the population. A discussion of a test of an hypothesis concerning the value of ρ is omitted here, as well as a discussion on the bound on the error of estimation. Ordinarily we would be interested in testing the null hypothesis that $\rho = 0$. In fact, a common test statistic for testing the null hypothesis $\rho = 0$ is the Student's t statistic:

$$t = \frac{r\sqrt{n-2}}{\sqrt{1-r^2}}$$

where t is based on $(n - 2)$ degrees of freedom. Using ordinary algebra we can show that

$$t = \frac{r\sqrt{n-2}}{\sqrt{1-r^2}} = \frac{\hat{\beta}_1}{s}\sqrt{SS_x}$$

and hence that the t test of the null hypothesis $\rho = 0$ is equivalent to the t test of section 11.5, the test of the null hypothesis $\beta_1 = 0$. While this test is sometimes of interest, most investigations will have as their objective the estimation of the mean of y or the prediction of a particular value of y for a given value of x.

Although r gives a rather nice measure of the goodness of fit of the least squares line to the fitted data, its use in making inferences concerning ρ appears to be of dubious practical value in many situations. It seems unlikely that a phenomenon *y* observed in the physical sciences and especially in economics would be a function of a single variable. Thus the correlation coefficient between the monthly sales volume of a firm and any one variable probably would be quite small and of questionable value. A larger reduction in SSE could possibly be obtained by constructing a predictor of *y* based on a set of variables x_1, x_2, ...

One further reminder is worthwhile concerning the interpretation of *r*. It is not uncommon for researchers in some fields to speak proudly of sample correlation coefficients *r* in the neighborhood of .5 (and, in some cases, as low as .1) as being indicative of a "relation" between *y* and *x*. Certainly, even if these values were accurate estimates of ρ, only a very weak relation would be indicated. A value *r* = .5 implies that the use of *x* in predicting *y* reduces the sum of squares of deviations about the prediction line by only r^2 = .25, or 25%. A correlation coefficient of *r* = .1 implies only an r^2 = .01, or a 1% reduction in the total sum of squares of deviations that could be explained by *x*.

If the linear coefficient of correlation between *y* and each of two variables x_1 and x_2 were calculated to be .4 and .5, respectively, it does not follow that a predictor using both variables would account for a $(.4)^2 + (.5)^2$ = .41, or a 41%, reduction in the sum of squares of deviations. Actually, x_1 and x_2 might be highly correlated and therefore contribute virtually the same information for the prediction of *y*.

Finally, we remind you that *r* is a measure of **linear correlation** and that *x* and *y* could be perfectly related by a *curvilinear* function even when the observed value of *r* is equal to zero.

Exercises

11.20. Describe the significance of the algebraic sign and the magnitude of *r*.

11.21. What is implied when the value of *r* is very close to 0?

11.22. What value does *r* assume if all the sample points fall on the same straight line and if the line has a positive slope? If the line has a negative slope?

11.23. Calculate the coefficient of correlation *r* between the annual rates of return for the Penn Square Mutual Fund and the average market rate of return given in exercise 11.10.

11.24. Is there a relationship between a country's energy usage and its gross national product? One would tend to think that the countries whose residents have greater personal wealth would require more energy per capita than those with lesser personal wealth. To examine this notion we have randomly selected 12 countries and recorded the per capita energy usage (in pounds) and the per capita gross national product (in U.S. dollars) for each selected country. The results are given in the table.

a. Compute the correlation *r* between per capita energy usage and per capita gross national product.

COUNTRY	PER CAPITA ENERGY USAGE	PER CAPITA GNP	COUNTRY	PER CAPITA ENERGY USAGE	PER CAPITA GNP
United States	25,598	5,515	Canada	23,715	4,704
Australia	12,568	3,370	Denmark	12,273	3,978
Norway	10,227	3,779	France	9,156	3,810
Japan	7,167	2,757	Italy	6,164	2,170
Venezuela	5,452	1,291	Greece	3,543	1,382
Brazil	1,173	513	India	410	98

Source: *U.N. Statistical Yearbook*, 1973, and *U.S. Statistical Abstract*, 1974.

b. Do the data present sufficient evidence to indicate that a linear relationship exists between per capita energy usage and per capita GNP? Test the hypothesis that $\rho = 0$; use $\alpha = .05$. (Note: A test of the hypothesis $\rho = 0$ is equivalent to testing the hypothesis $\beta_1 = 0$. See section 11.5.)

11.25. Refer to exercise 11.19. Compute the coefficient of determination r^2 and discuss the meaning of this term.

11.26. Refer to exercise 11.5. By what percentage is the sum of squares of deviations in operating cost reduced by using the prediction equation $\hat{y}$ rather than using $\bar{y}$ as a predictor of y?

11.27. An independent variable that shows a strong negative relationship with y is as useful as one that exhibits a positive relationship. The important feature is the absolute magnitude of the correlation between y and x, not the direction of relationship. Consider, for instance, interest rates and housing starts. Interest rates (x) provide an excellent leading indicator for predicting housing starts (y). As interest rates decline, housing starts increase, and vice versa. Suppose the data given in the accompanying table represent the prevailing interest rates on first mortgages and the recorded building permits in a certain region over an eight-year span.

Year	1969	1970	1971	1972	1973	1974	1975	1976
Interest Rates (%)	6.5	6.0	6.5	7.5	8.5	9.5	10.0	9.0
Building Permits	2165	2984	2780	1940	1750	1535	962	1310

a. Find the least squares line to allow for the estimation of building permits from interest rates. Plot the data points and graph the least squares line as a check on your calculations.

b. Calculate the correlation coefficient r for these data. Is the correlation between building permits and interest rates significantly different from zero? Use $\alpha = .05$.

c. By what percentage is the sum of squares of deviations of building permits reduced by using interest rates as a predictor rather than using the average annual building permits $\bar{y}$ as a predictor of y for these data?

d. If economic indicators suggest that the prevailing interest rate on first mortgages will be 8.5% in the region next year, predict the number of building permits to be issued for the region during the year, using a 95% prediction interval.

11.9 The Additivity of Sums of Squares

An important property of a regression analysis is that it tends to partition the total sum of squares of deviations

$$SS_y = \sum_{i=1}^{n} (y_i - \bar{y})^2$$

into two parts. One part is attributable to the sum of squares of deviations of the y-values about the fitted line SSE. The other part represents the reduction in the sum of squares of deviations that results from information contributed by the auxiliary variable x.

To understand the partitioning of sums of squares, note that a regression analysis tends to partition the deviation of each measurement from its mean, $(y_i - \bar{y})$, into two parts. Thus

$$(y_i - \bar{y}) = (y_i - \hat{y}_i) + (\hat{y}_i - \bar{y})$$

The partitioning of $(y_i - \bar{y})$ can be seen in figure 11.11.

Taking the sum of squared deviations over all observations for each expression within the partitioning of $(y_i - \bar{y})$, it can be shown that

$$SS_y = \sum_{i=1}^{n} (y_i - \bar{y})^2 = \sum_{i=1}^{n} (\hat{y}_i - \bar{y})^2 + \sum_{i=1}^{n} (y_i - \hat{y}_i)^2$$

Thus the total sum of squares of y, denoted by the symbol SS_y (called total SS), can be partitioned into two components.

Partitioning the Total Sum of Squares of y

Total SS = SSR + SSE

where

$$SSR = \sum_{i=1}^{n} (\hat{y}_i - \bar{y})^2 = \text{sum of squares due to regression}$$

which is the amount of total variation explained by the auxiliary variable x, and

$$SSE = \sum_{i=1}^{n} (y_i - \hat{y}_i)^2 = \text{sum of squares for error}$$

which is the amount of total variation unexplained by the auxiliary variable x.

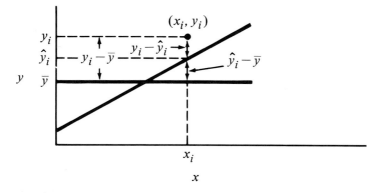

Figure 11.11 *Partitioning of* $(y_i - \bar{y})$ *into* $(y_i - \hat{y}_i)$ *and* $(\hat{y}_i - \bar{y})$

The partitioning of the sums of squares is important for two reasons. It gives an important additive relationship for the sums of squares,

$$\text{Total SS} = \text{SSR} + \text{SSE}$$

that can often reduce computational effort in a regression analysis. (We can compute two of the quantities and obtain the third by subtraction.) Second, it helps to explain the contribution of the auxiliary variable x in providing information for the prediction of y. Finally, note that we can write r^2 as

$$r^2 = \frac{\text{SS}_y - \text{SSE}}{\text{SS}_y} = \frac{\text{Total SS} - \text{SSE}}{\text{Total SS}} = \frac{\text{SSR}}{\text{Total SS}}$$

11.10   Summary

Although it was not stressed, you will observe that the prediction of a particular value of a random variable y was considered for the most elementary situation in chapters 8 and 9. Thus if we possessed no information concerning variables related to y, the sole information available for predicting y would be provided by its probability distribution. As we noted in chapter 5, the probability that y falls between two specific values, say y_1 and y_2, equals the area under the probability distribution curve over the interval $y_1 \leq y \leq y_2$. And if we were to select randomly one member of the population, we would most likely choose μ, or some other measure of central tendency, as the most likely values of y to be observed. Thus we would wish to estimate μ, and this of course was considered in chapters 8 and 9.

Chapter 11 was concerned with the problem of predicting y when auxiliary information is available on other variables, say x_1, x_2, x_3, ..., which are related to y and hence assist in its prediction. We have concentrated primarily on the problem of predicting y as a linear function of a single variable x, which provides the simplest extension of the prediction problem beyond that considered in chapters 8 and 9. The more interesting case, where y is a linear function of a set of independent variables, is the subject of chapter 12.

Supplementary Exercises

11.28.   What assumptions are necessary to use the prediction equation $\hat{y} = \hat{\beta}_0 + \hat{\beta}_1 x$ to predict the value of a dependent variable y from a predictor variable x?

11.29.   For what configurations of sample points will s^2 be zero?

11.30.   For what parameter is s^2 an unbiased estimator? Explain how this parameter enters into the description of the probabilistic model.

11.31.   For the linear equation $y = 6 + 3x$:
a. Give the y-intercept and the slope of the line.
b. Graph the line corresponding to the equation.

11.32.   Follow the instructions given in exercise 11.31 for the linear equation $2x - 3y - 5 = 0$.

11.33.   Suppose you are given five points whose coordinates are as follows:

y	0	0	1	1	3
x	-2	-1	0	1	2

a. Find the least squares line for the data.
b. As a check on the calculations in a, plot the five points and graph the line.
c. Calculate s^2.
d. Do the data present sufficient evidence to indicate that y and x are linearly related? (Test the hypothesis that $\beta_1 = 0$, using $\alpha = .05$.)

11.34.   The following data represent the number y of workdays absent during the past year and the number x of years employed by the company for seven employees randomly selected from a large company.

y	2	0	5	6	4	9	2
x	7	8	2	3	5	3	7

a. Find the least squares line for the data.
b. As a check on the calculations in a, plot the seven points and graph the line.
c. Calculate s^2.
d. Do the data present sufficient evidence to indicate that y and x are linearly related? (Test the hypothesis that $\beta_1 = 0$, using $\alpha = .05$.)

11.35.   Find a 90% confidence interval for the slope of the line in exercise 11.33.

11.36.   Find a 95% confidence interval for the slope of the line in exercise 11.34.

11.37. Refer to exercise 11.34. Obtain a 95% confidence interval for the mean number of workdays absent during the year per employee for employees who have worked for the firm for the past five years. (That is, find a 95% confidence interval for the expected value of y when $x = 5$.)

11.38. Refer to exercise 11.34. Find a 95% prediction interval for the number of workdays absent during the year by a particular employee who has been employed by the firm for five years. Compare this prediction interval with the interval computed in exercise 11.37. Explain the difference in the inferential objectives of exercises 11.37 and 11.38.

11.39. Calculate the coefficient of correlation r for the data in exercise 11.34. Do these data suggest that the underlying population coefficient of correlation ρ is significantly different from zero? Use $\alpha = .05$.

11.40. By what percentage is the sum of squared deviations of the number of workdays absent during the year reduced by using the auxiliary information provided by the number of years employed rather than using $\bar{y}$ as a prediction of y in exercise 11.37?

11.41. Since long-run forecasts of temperature are better indicators of fuel usage than are direct forecasts of heating oil sales and demand for other heating fuels, most distributors of fuels use the relationship between temperature and fuel sales to determine their proper levels of inventory. A regional distributor of heating oils has recorded the monthly sales volume and the average daily high temperature for nine randomly selected months. The results are given in the table.

Sales Volume, y ($\times$ 100 gal)	26.2	17.4	7.8	12.3	35.9	42.1	26.4	19.0	10.1
Average Daily High Temperature, x(°F)	46.5	54.6	65.2	62.3	41.9	38.6	43.7	52.0	59.8

a. Find the least squares line appropriate for these data.
b. Plot the points and graph the least squares line as a check on your calculations.
c. Calculate s^2.
d. Do these data present sufficient evidence to indicate that fuel oil sales are linearly related to temperature? Test by using $\alpha = .05$.

11.42. Continuing exercise 11.41, find a 90% confidence interval for the expected sales volume for those months with an average daily high temperature of 45°F.

11.43. Refer to exercise 11.41. Partition the total sum of squares into its two components, regression and the sum of squares for error. What is the percentage of variation in monthly sales volume that is explained by the average daily high temperature?

11.44. In exercise 11.10 we discussed Treynor's characteristic line device for evaluating investment funds. The funds studied in exercise 11.10 were growth funds, developed under the objective of capital value appreciation. Another common class of funds are balanced funds which, through greater diversification than growth funds, usually are less risky and place more emphasis on dividend accrual. One of the larger balanced funds is the George Putnam Fund of Boston. Listed in the table are its annual returns and the average market rate of return for the years 1964 through 1973.

Year	1964	1965	1966	1967	1968	1969	1970	1971	1972	1973
Putnam Fund	11.4	13.0	−3.8	18.7	10.6	−6.6	2.6	17.9	19.9	−10.8
Average Market Return	12.9	9.1	−13.1	20.1	7.7	−11.4	.1	10.8	15.6	−17.4

Source: Wiesenburger Financial Services; *Investment Companies, Mutual Funds and Other Types,* 1974.

a. Find the characteristic line for the Putnam Fund (that is, the least squares line relating the return on the Putnam Fund to the market rate of return).

b. Plot the points and graph the least squares line as a check on your calculations.

c. Describe the risk characteristics of the Putnam Fund (that is, test the hypothesis that $\beta_1 = 0$; use $\alpha = .05$).

11.45.   Calculate the coefficient of correlation r between the annual rates of return on the George Putnam Fund of Boston and the average market rate of return in exercise 11.44.

11.46.   In the accompanying table, x is the tensile force applied to a steel specimen in thousands of pounds and y is the resulting elongation in thousandths of an inch.

x	1	2	3	4	5
y	2	4	5	6	8

a. Assuming the regression of y on x to be linear, find the least squares line for the data.

b. Plot the points and graph the line found in part a as a check on your calculations.

c. Calculate SSE and s for the data.

11.47.   Refer to exercise 11.46. Find a 90% confidence interval for the mean change in elongation of the specimen per thousand pounds of tensile stress.

11.48.   Refer to exercise 11.46. Use a 95% prediction interval to predict the elongation if the experiment is to be run one more time at a tensile force of 4000 pounds ($x = 4$).

11.49.   Refer to exercise 11.46. If a force of 0 pounds is applied, the resulting elongation should be 0 units. Perform a test of an hypothesis to see if the line is consistent with the preceding statement. That is, test the hypothesis that $E(y|x)$ is related to x by a straight line that passes through the point (0, 0). Use $\alpha = .05$. [Note: The point $x = 0$ lies outside the region of experimentation. Rejection of the hypothesis that the line passes through the point (0, 0) does not imply that your model is unsatisfactory over the interval $1 \le x \le 5$. A straight line might provide a very good model for the relationship between $E(y|x)$ and x over the interval $1 \le x \le 5$.]

11.50.   An economist wished to develop a model enabling her to examine the relationship between the Federal Reserve discount rate and bank debits. In the process of her study, the data given in the accompanying table were obtained for the years 1962 through 1968.

Federal Reserve Discount Rate, y	3.00	3.38	3.88	4.38	4.50	4.25	5.25
Bank Debits, x (trillions of dollars)	3.43	3.75	4.52	5.13	5.94	6.35	7.99

  a. Find the least squares prediction equation appropriate for the data.
  b. Graph the points and the least squares line as a check on your calculations.
  c. Calculate s^2.

  11.51.   Do the data in exercise 11.50 present sufficient evidence to indicate that bank debits are useful in predicting the Federal Reserve discount rate? Test by using $\alpha = .01$.

  11.52.   Calculate r^2, the proportion of variation in Federal Reserve discount rates explained by bank debits, for the data in exercise 11.50.

  11.53.   Refer to exercise 11.50. Obtain a 90% confidence interval for the expected Federal Reserve discount rate knowing that the bank debit level is $7 trillion.

  11.54.   An experiment was conducted by a pharmaceutical manufacturer to observe the effect of an increase in temperature on the potency of an antibiotic. Three 1-ounce portions of the antibiotic were stored for equal lengths of time at each of the following temperatures: 30°, 50°, 70°, and 90°. The potency readings observed at the temperature of the experimental period are given in the table.

Potency Readings	38,43,29	32,26,33	19,27,23	14,19,21
Temperature	30	50	70	90

  a. Find the least squares equation appropriate for these data.
  b. Plot the points and graph the line as a check on your calculations.
  c. Calculate s^2.

  11.55.   Refer to exercise 11.54. Estimate the change in potency for a one-unit change in temperature, using a 90% confidence interval.

  11.56.   Refer to exercise 11.54. Estimate the mean potency corresponding to a temperature of 50°, using a 90% confidence interval.

  11.57.   Refer to exercise 11.54. Suppose a batch of the antibiotic was stored at 50° for the same length of time as the experimental period. Predict the potency of the batch at the end of the storage period, using a 90% prediction interval.

  11.58.   Calculate the coefficient of correlation for the data in exercise 11.54. How much reduction in SSE is obtained by using the least squares predictor rather than using $\bar{y}$ in predicting y for the data given in exercise 11.54?

  11.59.   A comparison of the undergraduate grade point averages of 12 corporate employees with their scores on a managerial trainee examination produced the results shown in the table.

Exam Score, y	76	89	83	79	91	95	82	69	66	75	80	88
GPA, x	2.2	2.4	3.1	2.5	3.5	3.6	2.5	2.0	2.2	2.6	2.7	3.3

  a. Find the least squares prediction equation appropriate for the data.
  b. Graph the points and the least squares line as a check on your calculations.
  c. Calculate s^2.
  d. Do the data present sufficient evidence to indicate that x (undergraduate GPA) is useful in predicting y (managerial trainee exam score)? Test by using $\alpha = .05$.

  11.60.   Calculate the coefficient of correlation for the data in exercise 11.59. By what percentage is the sum of squares of deviations reduced by using the least squares predictor $\hat{y} = \hat{\beta}_0 + \hat{\beta}_1 x$ rather than $\bar{y}$ as a predictor of y for the data in exercise 11.59?

11.61.   Refer to exercise 11.59. Obtain a 90% confidence interval for the expected exam score for all managerial trainees whose undergraduate GPA was 2.6.

11.62.   Use the least squares equation derived in exercise 11.59 to predict the exam score for a particular managerial trainee whose undergraduate GPA was 2.6. Obtain a 90% prediction interval for this individual's true exam score.

11.63.   What is the difference in the inferential objectives of exercises 11.61 and 11.62?

Experiences with Real Data

In the text of section 11.2, we discussed the relationship of advertising to sales and expressed the relationship for a particular company through an illustrative example. Even though most of us may assume that a distinct relationship exists between advertising and sales, some authors dispute this claim. Mason suggests that there is, in fact, no consistent correlation between the level of spending on advertising and sales.* He notes that the correlation which does exist between advertising and sales is probably inflated by the fact that companies continue to advertise products that sell well while ceasing to advertise unsuccessful products.

Conduct a study to determine the correlation that exists between advertising expenditures and sales for a certain selected company. You may obtain data for this study from company annual reports (which may be available in your college or university library). Gather "net sales" and "advertising expenditures" data for the company of your choice for each of the past 15 years. (Some companies do not separately list advertising expenditures in annual reports but include these costs within a "selling and administrative costs" category. Use these costs in place of advertising expenditures if the latter are not available.)

1. Fit a least squares line to the data relating net sales y to advertising expenditures (or selling and administrative costs) x.

2. Calculate r^2 and the linear coefficient of correlation r. Describe the strength of the relationship between net sales y and advertising expenditures x.

3. Do these data provide evidence to indicate a linear correlation between net sales and advertising expenditures for the company you have chosen? (Hint: Recall that a test of the hypothesis $\rho = 0$ is equivalent to testing the hypothesis that $\beta_1 = 0$.)

4. Compare the results of your study with those of other students in your class who have performed similar analyses for different companies.

What conclusions have you drawn about the relationship between net sales and advertising expenditures?

Computations associated with this study are acquired most easily using either a calculator or an electronic computer. Your college or university computing center should have an available regression analysis program allowing for the rapid computation of the statistics requested in this exercise. Particularly good programs are those in the Biomed series (Biomedical Programs of the UCLA Health Sciences Computing Facility), SAS (Statistical Analysis System), and SPSS (Statistical Package for the Social Sciences) program libraries. Ask a computer consultant about the availability of these or other regression analysis computer programs.

*K. Mason, "How Much Do You Spend on Advertising? Product Is Key," *Advertising Age*, 12 June 1972.

References

DRAPER, N., and H. SMITH. *Applied Regression Analysis.* New York: Wiley, 1966.

EZEKIEL, M., and K. A. FOX. *Methods of Correlation and Regression Analysis.* 3d ed. New York: Wiley, 1959.

LI, J. C. R. *Introduction to Statistical Inference.* Ann Arbor, Mich.: J. W. Edwards, Publisher, 1961. Chapter 16.

MENDENHALL, W. *An Introduction to Linear Models and the Design and Analysis of Experiments.* Belmont, Calif.: Wadsworth, 1967. Chapters 6 and 7.

RAO, P., and R. L. MILLER. *Applied Econometrics.* Belmont, Calif.: Wadsworth, 1971.

WONNACOTT, R. J., and T. H. WONNACOTT. *Econometrics.* New York: Wiley, 1970. Chapters 1, 2, and 5.

chapter objectives

GENERAL OBJECTIVES In chapter 11 we considered the problem of estimating the mean value of y and predicting some future value of y based on a simple linear model. The objective of chapter 12 is to extend those inferential techniques to the case in which a response y and the mean value of y are a linear function of a set of independent parameters and to show how to use this multivariable model for solving business problems.

SPECIFIC OBJECTIVES

1. To present a multivariable prediction model and to explain how it can be used for making inferences.
                    Sections 12.1, 12.2

2. To explain how to fit a multivariable prediction model to a set of data by using the method of least squares.
                    Sections 12.3, 12.4

3. To give a confidence interval and a test of an hypothesis for an individual β parameter.
                    Section 12.5

4. To explain the difficulties encountered in interpreting the estimates of the β parameters—the problem of multicollinearity.
                    Section 12.6

5. To explain how to measure how well a model fits a given set of data.
                    Section 12.7

6. To explain how to test the utility of a regression model—stepwise regression.
                    Section 12.8

7. To explain how the prediction equation can be used for estimation and prediction.
                    Section 12.9

8. To show how to decide whether sets of terms should be included in a model (an important problem in model building).
                    Section 12.11

9. To summarize the tests and estimation and prediction procedures that can be used in a multiple regression analysis.
                    Section 12.12

10. To present applications of multiple regression to the solution of business problems.
                    Sections 12.3 through 12.13

chapter twelve

Multiple Regression

12.1 Introduction

Multiple regression, the extension of the methodology of chapter 11 to more than one independent variable, has some exciting applications. Several years ago one of the major newsmagazines told the story of a computer programmer working for a company in a midwestern state. Although he undoubtedly performed his company's chores with zeal, he possessed a curious nature that motivated him to speculate on the relationship between certain commodity prices and various variables to which he thought they might be related. The article went on to state that with the aid of the company's computer (and, we would add, some knowledge of multiple regression analysis), the programmer developed very accurate multivariable prediction equations for the prices of certain commodities. How accurate? Well, the article concluded by stating that he made a million dollars in trading within a few years and that he had left his company to practice investment counseling on the West Coast.

Not all attempts to construct multivariable prediction equations for response measurements are as successful or spectacular as that described in the newsmagazine. But there are a sufficient number of successful applications to establish multiple regression analysis as a very powerful statistical tool that can be applied in many diverse areas of business.

To illustrate, think of your special area of business (or your anticipated special area) and then think of some criterion variable y that measures success in the performance of that specialty. For example, if you are majoring in marketing, you might think of sales volume as a measure of success in that field. A small businessperson would likely use profit, and the director of

403

security for a large department store might measure performance by the value of merchandise lost by theft.

Now suppose that you possessed a multivariable prediction equation that gave accurate predictions of values of y for given values of the x's. Think of the benefits to be derived from this tool. You would be able to predict values for your criterion variable for various values of the x's and just noting which x variables enter into the equation, you would likely develop a better understanding of how to control the criterion variable y and make it take values advantageous to you.

Finding a multivariable prediction equation is the subject of this chapter. The employment of this methodology—the statistical tests and estimation procedures—to a set of data is often called **multiple regression analysis.**

12.2 Linear Statistical Models

A multivariable prediction equation (or prediction model) is an extension of the simple linear model of chapter 11. Typical models for response variables are

(1) $y = \beta_0 + \beta_1 x_1 + \beta_2 x_2 + \beta_3 x_3 + \epsilon$

(2) $y = \beta_0 + \beta_1 x_1 + \beta_2 x_1^2 + \epsilon$

(3) $y = \beta_0 + \beta_1 x_1 + \beta_2 x_2 + \beta_3 x_1 x_2 + \beta_4 x_1^2 + \beta_5 x_2^2 + \epsilon$

where x_1, x_2, and x_3 are predictor variables and ϵ is a random error. For all the models discussed in this chapter, we will assume that ϵ is normally distributed, with a mean of 0 and a variance of σ^2. Furthermore, we will assume that the random errors associated with any pair of y-values are independent in a probabilistic sense (we made the same assumptions about the properties of ϵ for the simple linear model in chapter 11).

Since the expected value of ϵ is zero, it follows that the mean value of y for specified values of the predictor variables is given by the deterministic portion of the prediction model. Thus the expected (mean) values of y for the three models given above are*

(1) $E(y) = \beta_0 + \beta_1 x_1 + \beta_2 x_2 + \beta_3 x_3$

(2) $E(y) = \beta_0 + \beta_1 x_1 + \beta_2 x_1^2$

(3) $E(y) = \beta_0 + \beta_1 x_1 + \beta_2 x_2 + \beta_3 x_1 x_2 + \beta_4 x_1^2 + \beta_5 x_2^2$

---

*If the expected value of y is a function of the predictor variables x_1, x_2, and x_3, it is usually denoted by the symbol $E(y|x_1, x_2, x_3)$. That is, $E(y|x_1, x_2, x_3)$ represents the expected value of y for given values of x_1, x_2, and x_3. Since we may have many different predictor variables appearing in a model, this expectation symbol can become quite cumbersome. Consequently, to simplify the notation in this chapter, we will represent the mean value of y for given values of the predictor variables as $E(y)$.

Models (1), (2), and (3) are called **linear statistical models** because the right sides of the expressions for $E(y)$ are linear functions of the β parameters. In contrast,

$$(4) \quad y = \beta_0 x^{\beta_1} + \epsilon$$

is not a linear model because the right side of the prediction equation is not a linear function of the unknown β parameters β_0 and β_1. A multiple regression analysis is based on the assumption that y is represented by a linear statistical model. Consequently, only linear models are employed in this chapter.

12.3 The Least Squares Equations for a Multivariable Prediction Model

A prediction equation based on a number of variables x_1, x_2, ..., x_k can be obtained by the method of least squares in exactly the same manner as that employed for the simple linear model. For example, we might wish to fit the model

$$y = \beta_0 + \beta_1 x_1 + \beta_2 x_2 + \beta_3 x_3 + \epsilon$$

where y is the price of a firm's securities at the end of some month, x_1 is earnings per share during the past fiscal year, x_2 is the gross sales volume for the firm during the preceding month, x_3 is the firm's profit margin during the past month, and ϵ is a random error. (Note that we could add other variables, as well as the squares, cubes, and cross products of x_1, x_2, and x_3.)

We would need a random sample of the recorded values for y, x_1, x_2, and x_3 for n randomly selected months during which the firm was in operation. The set of measurements y, x_1, x_2, x_3 for each of the n months could be regarded as the coordinates of a point in four-dimensional space. Then, ideally, we would like to possess a multidimensional "ruler" (in our case, a plane) that we could visually move about among the n points until the deviations of the observed values of y from the predicted values would in some sense be a minimum. Although we cannot graph points in four dimensions, you can readily see that this device is provided by the method of least squares, which, mathematically, performs this task for us.

The sum of squares of deviations of the observed value of y from the fitted model is

$$\text{SSE} = \sum_{i=1}^{n} (y_i - \hat{y}_i)^2 = \sum_{i=1}^{n} [y_i - (\hat{\beta}_0 + \hat{\beta}_1 x_{1i} + \hat{\beta}_2 x_{2i} + \hat{\beta}_3 x_{3i})]^2$$

where $\hat{y} = \hat{\beta}_0 + \hat{\beta}_1 x_1 + \hat{\beta}_2 x_2 + \hat{\beta}_3 x_3$ is the fitted model and $\hat{\beta}_0$, $\hat{\beta}_1$, $\hat{\beta}_2$, and $\hat{\beta}_3$ are estimates of the model parameters. We would then use the calculus to find the estimates $\hat{\beta}_0$, $\hat{\beta}_1$, $\hat{\beta}_2$, and $\hat{\beta}_3$ that make SSE a minimum. The estimates, as for the simple linear model, would be obtained as the solution of a set of simultaneous linear equations known as the least squares equations.

In the case above, with **three** independent variables x_1, x_2, and x_3, the least squares equations (sometimes called normal equations) give **four** linear equations in the four unknowns $\hat{\beta}_0$, $\hat{\beta}_1$, $\hat{\beta}_2$, and $\hat{\beta}_3$. The four least squares equations, which we do not derive but simply state, are

$$\hat{\beta}_0 n \quad + \hat{\beta}_1 \sum x_1 \quad + \hat{\beta}_2 \sum x_2 \quad + \hat{\beta}_3 \sum x_3 \quad = \sum y$$
$$\hat{\beta}_0 \sum x_1 + \hat{\beta}_1 \sum x_1^2 \quad + \hat{\beta}_2 \sum x_1 x_2 + \hat{\beta}_3 \sum x_1 x_3 = \sum x_1 y$$
$$\hat{\beta}_0 \sum x_2 + \hat{\beta}_1 \sum x_1 x_2 + \hat{\beta}_2 \sum x_2^2 \quad + \hat{\beta}_3 \sum x_2 x_3 = \sum x_2 y$$
$$\hat{\beta}_0 \sum x_3 + \hat{\beta}_1 \sum x_1 x_3 + \hat{\beta}_2 \sum x_2 x_3 + \hat{\beta}_3 \sum x_3^2 \quad = \sum x_3 y$$

where each summation sign indicates that the quantity within the summation sign is to be summed over all data points $i = 1, 2, ..., n$.

Note the pattern formed by the terms in the least squares equations above and you will surmise a truth—that this pattern holds for any number of independent variables. For the regression model

$$y = \beta_0 + \beta_1 x_1 + \beta_2 x_2 + \epsilon$$

with **two** independent variables, the **three** least squares equations in the three unknowns $\hat{\beta}_0$, $\hat{\beta}_1$, and $\hat{\beta}_2$ are

$$\hat{\beta}_0 n \quad + \hat{\beta}_1 \sum x_1 \quad + \hat{\beta}_2 \sum x_2 \quad = \sum y$$
$$\hat{\beta}_0 \sum x_1 + \hat{\beta}_1 \sum x_1^2 \quad + \hat{\beta}_2 \sum x_1 x_2 = \sum x_1 y$$
$$\hat{\beta}_0 \sum x_2 + \hat{\beta}_1 \sum x_1 x_2 + \hat{\beta}_2 \sum x_2^2 \quad = \sum x_2 y$$

Note the location of these terms in the equations for the three-variable model.

Now block out the appropriate terms for the simple model

$$y = \beta_0 + \beta_1 x_1 + \epsilon$$

with **one** independent variable x. The **two** least squares equations in the two unknowns $\hat{\beta}_0$ and $\hat{\beta}_1$ are

$$\hat{\beta}_0 n \quad + \hat{\beta}_1 \sum x \ = \sum y$$
$$\hat{\beta}_0 \sum x + \hat{\beta}_1 \sum x^2 = \sum xy$$

Solving these two equations for the unknowns gives *exactly* the same values for $\hat{\beta}_0$ and $\hat{\beta}_1$ as would be obtained by using the formulas for $\hat{\beta}_0$ and $\hat{\beta}_1$ on page 370.

By an extension of the preceding discussion, for a regression model with k independent variables, there are $(k + 1)$ least squares equations in the $(k + 1)$ unknowns $\hat{\beta}_0, \hat{\beta}_1, \hat{\beta}_2, ..., \hat{\beta}_k$. The form of the $(k + 1)$ least squares equations can be established from the pattern set by the least squares equations for the models with one, two, and three independent variables.

12.4 Solving the Least Squares Equations

Solution of the least squares equations for $\hat{\beta}_0, \hat{\beta}_1, \hat{\beta}_2, ..., \hat{\beta}_k$ *guarantees* that the resulting estimates, substituted into the prediction equation

$$\hat{y}_i = \hat{\beta}_0 + \hat{\beta}_1 x_{1i} + \hat{\beta}_2 x_{2i} + \cdots + \hat{\beta}_k x_{ki}$$

minimize the sum of square deviations

$$\text{SSE} = \sum_{i=1}^{n} (y_i - \hat{y}_i)^2$$

That is, no other set of estimates for the betas provides a smaller SSE.

A set of *m* equations in *m* unknowns is commonly solved in the following ways:

1. By expressing the simultaneous equations in matrix form and solving by means of matrix algebra (employing matrix inverse and matrix multiplication operations).
2. Or by a process of elimination, individually solving for each unknown.

Both approaches are quite tedious when the number of equations, and hence the number of unknowns, exceeds three. Preprogrammed routines for use on a computer are available and should be employed in these cases.

We illustrate the solution of the least squares equations with three unknowns in the following example.

Example 12.1 The owner of an automobile dealership believes the relationship between the number *y* of new cars sold by his agency in a given month is related to the number *x* of his agency's full-page newspaper advertisements during the month by the model

$$y = \beta_0 + \beta_1 x_1 + \beta_2 x_2 + \epsilon$$

where $x_1 = x$ and $x_2 = x^2$. Over a period of six months, he noticed the results given in the table. Fit the model $y = \beta_0 + \beta_1 x_1 + \beta_2 x_2 + \epsilon$ to the

	MONTH					
	1	2	3	4	5	6
y	10	10	15	20	30	40
x	0	1	2	2	3	4

Table 12.1 *Calculations for the data of example 12.1*

MONTH, i	y_i	x_{1i}	x_{2i}	x_{1i}^2	x_{2i}^2	$x_{1i}x_{2i}$	$x_{1i}y_i$	$x_{2i}y_i$
1	10	0	0	0	0	0	0	0
2	10	1	1	1	1	1	10	10
3	15	2	4	4	16	8	30	60
4	20	2	4	4	16	8	40	80
5	30	3	9	9	81	27	90	270
6	40	4	16	16	256	64	160	640
Sums	125	12	34	34	370	108	330	1060

data by solving the least squares equations to obtain estimates of the unknown parameters β_0, β_1, and β_2.

Solution The model $y = \beta_0 + \beta_1 x_1 + \beta_2 x_2 + \epsilon$, or, equivalently, the model $y = \beta_0 + \beta_1 x + \beta_2 x^2 + \epsilon$, is an example of a second-order polynomial. It requires the solution of three least squares equations in the three unknowns $\hat{\beta}_0$, $\hat{\beta}_1$, and $\hat{\beta}_2$. In table 12.1 the necessary elements are calculated from the data so that the least squares equations for this problem can be identified. Remember, for the calculations within table 12.1, $x_{1i} = x_i$ and $x_{2i} = x_i^2$. Therefore, $x_{1i}^2 = x_i^2$, $x_{2i}^2 = x_i^4$, $x_{1i}x_{2i} = x_i^3$, $x_{1i}y_i = x_iy_i$, and $x_{2i}y_i = x_i^2y_i$.

Substituting the appropriate sums from table 12.1 into the least squares equations, we find

(I) $$6\hat{\beta}_0 + 12\hat{\beta}_1 + 34\hat{\beta}_2 = 125$$

(II) $$12\hat{\beta}_0 + 34\hat{\beta}_1 + 108\hat{\beta}_2 = 330$$

(III) $$34\hat{\beta}_0 + 108\hat{\beta}_1 + 370\hat{\beta}_2 = 1060$$

For convenience we have labeled the three least squares equations as (I), (II), and (III). We will now employ the process of elimination to solve for the unknowns $\hat{\beta}_0$, $\hat{\beta}_1$, and $\hat{\beta}_2$.

Suppose that we subtract 2(I) from equation (II). We then have

$$10\hat{\beta}_1 + 40\hat{\beta}_2 = 80$$

or

$$\hat{\beta}_1 = 8 - 4\hat{\beta}_2$$

Now we subtract $(\frac{34}{6})$(I) from equation (III) and find that

$$40\hat{\beta}_1 + 177.33\hat{\beta}_2 = 351.67$$

$$\hat{\beta}_1 = 8.79 - 4.43\hat{\beta}_2$$

We now have eliminated $\hat{\beta}_0$ and have two equations in two unknowns. Setting these two equations equal to each other, we can solve for $\hat{\beta}_2$:

$$8 - 4\hat{\beta}_2 = 8.79 - 4.43\hat{\beta}_2$$

$$.43\hat{\beta}_2 = .79$$

Thus $\hat{\beta}_2 = 1.84$.

From an earlier equation, we solve for $\hat{\beta}_1$:

$$\hat{\beta}_1 = 8 - 4\hat{\beta}_2 = 8 - 4(1.84) = .64$$

Returning to equation (I), we find that it can be rewritten as

$$\hat{\beta}_0 = \tfrac{1}{6}(125 - 12\hat{\beta}_1 - 34\hat{\beta}_2)$$

Therefore,

$$\hat{\beta}_0 = \tfrac{1}{6}[125 - 12(.64) - 34(1.84)] = \tfrac{1}{6}(54.76) = 9.13$$

The equation for predicting monthly sales y from the number x of full-page newspaper advertisements during the month and the quadratic term $x_2 = x^2$ is

$$\hat{y} = 9.13 + .64x_1 + 1.84x_2$$

The errors of prediction (often called **residuals**) can be found by substituting the predictor variables x_1 and x_2 into the prediction equation and computing the estimated sales $\hat{y}$. For month 1 we have

$$\hat{y}_1 = 9.13 + .64(0) + 1.84(0) = 9.13$$

Therefore, the error of prediction for month 1 is

$$e_1 = y_1 - \hat{y}_1 = 10 - 9.13 = .87$$

The estimated sales $\hat{y}_i$ and errors of prediction are given in the accompanying table.

y_i	10	10	15	20	30	40
$\hat{y}_i$	9.13	11.61	17.77	17.77	27.61	41.13
e_i	.87	-1.61	-2.77	2.23	2.39	-1.13

The sum of squared errors is

$$\text{SSE} = \sum_{i=1}^{6} (y_i - \hat{y}_i)^2 = \sum_{i=1}^{6} e_i^2 = 22.9838$$

We are guaranteed that no other estimates of the unknown parameters β_0, β_1, and β_2 would result in an SSE smaller than 22.9838.
    The graph for the equation

$$\hat{y} = 9.13 + .64x_1 + 1.84x_2 \qquad (x_1 = x, \ x_2 = x^2)$$

is shown in figure 12.1. Note that the straight-line model

$$y = \beta_0 + \beta_1 x + \epsilon$$

would have provided a much poorer fit to the data points than the polynomial model that was used.

    For a prediction model with three or more independent variables, it is practically imperative that we employ an electronic computer to estimate the

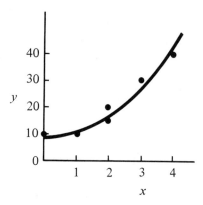

Figure 12.1 *Plot of the data and the prediction model for example 12.1*

unknown regression parameters, β_0, β_1, β_2, ..., β_k. Almost every computing facility has access to at least one "canned" regression analysis program, which requires only that the user execute the proper commands to activate the program and then to submit the problem data. We illustrate a computer solution in the following example.

Example 12.2 Consider a study designed to examine the role of television viewing in the lives of a selected group of people over 65 years of age. The purpose of the study was to provide guidelines for developing television programming that would adequately meet the special needs of this audience. A sample of $n = 25$ senior citizens was selected and from each senior citizen the following data were obtained: y = the average number of hours per day an interviewee spends watching television, x_1 = the marital status of the interviewee ($x_1 = 1$ if the interviewee is living with his or her spouse, $x_1 = 0$ if not), x_2 = the age of the interviewee, and x_3 = the number of years of education of the interviewee.* The data are listed in table 12.2.

The objective of this study is to relate y, the average daily hours an interviewee spends watching television, to the descriptive variables x_1, x_2, and x_3. For purposes of simplicity we select the prediction model

$$y = \beta_0 + \beta_1 x_1 + \beta_2 x_2 + \beta_3 x_3 + \epsilon$$

Find the least squares prediction equation for the data of table 12.2.

Solution Fitting the prediction equation to the data of table 12.2 by computer eliminates the need to find and to solve the least squares equations. To convince you of the labor-saving contribution of the computer,

---

*Variable x_1 is an example of a *dummy variable*, a frequently employed independent variable designed to include the effect of a *qualitative* factor into a regression model. Dummy variables serve to partition the regression model into several (two in this case) separate components,

$$\hat{y} = (\hat{\beta}_0 + \hat{\beta}_1) + \hat{\beta}_2 x_2 + \hat{\beta}_3 x_3$$
$$\hat{y} = \hat{\beta}_0 + \hat{\beta}_2 x_2 + \hat{\beta}_3 x_3$$

the former modeling response in the presence of the qualitative factor, the other, in its absence.

Table 12.2 *Daily hours spent watching television, marital status, age, and years of education of 25 randomly selected senior citizens; example 12.2*

INDIVIDUAL	HOURS, y	MARITAL STATUS, x_1	AGE, x_2	EDUCATION, x_3
1	.5	1	73	14
2	.5	1	66	16
3	.7	0	65	15
4	.8	0	65	16
5	.8	1	68	9
6	.9	1	69	10
7	1.1	1	82	12
8	1.6	1	83	12
9	1.6	1	81	12
10	2.0	0	72	10
11	2.5	1	69	8
12	2.8	0	71	16
13	2.8	0	71	12
14	3.0	0	80	9
15	3.0	0	73	6
16	3.0	0	75	6
17	3.2	0	76	10
18	3.2	0	78	6
19	3.3	1	79	6
20	3.3	0	79	4
21	3.4	1	78	6
22	3.5	0	76	9
23	3.6	0	65	12
24	3.7	0	72	12
25	3.7	0	80	6

the least squares equations are shown below.

$$25\hat{\beta}_0 + 10\hat{\beta}_1 + 1{,}846\hat{\beta}_2 + 254\hat{\beta}_3 = 58.5$$
$$10\hat{\beta}_0 + 10\hat{\beta}_1 + 748\hat{\beta}_2 + 105\hat{\beta}_3 = 16.2$$
$$1{,}846\hat{\beta}_0 + 748\hat{\beta}_1 + 137{,}086\hat{\beta}_2 + 18{,}509\hat{\beta}_3 = 4{,}376.0$$
$$254\hat{\beta}_0 + 105\hat{\beta}_1 + 18{,}509\hat{\beta}_2 + 2{,}892\hat{\beta}_3 = 533.4$$

(You can verify that these coefficients are correct by calculating the sums, sum of squares, and sums of cross products given in the pattern for the least squares equations in section 12.3.) As you can see, performing the necessary calculations to find the equations and to solve them is a tedious and time-consuming task. But this work is quickly accomplished by a computer.

Table 12.3 reproduces the computer output obtained by applying a commonly used regression analysis program to the data of table 12.2. In this example we are only concerned with the estimates of β_0, β_1, β_2, and β_3, which are given in the shaded portion of table 12.3. The other portions of the output will be explained in sections 12.5 and 12.7.

The shaded portion of table 12.3, titled INDIVIDUAL ANALYSIS OF VARI-

Table 12.3 *Computer output for senior citizen television-viewing data of table 12.2*

MULTIPLE R	.7918			
R SQUARE	.6269			
STD. ERROR OF EST.	.7524			

ANALYSIS OF VARIANCE

	DF	SUM OF SQUARES	MEAN SQUARE	F RATIO
REGRESSION	3	19.972	6.657	11.760
RESIDUAL	21	11.888	.566	

INDIVIDUAL ANALYSIS OF VARIABLES

VARIABLE	COEFFICIENT	STD. ERROR	F VALUE
(CONSTANT	1.41411)		
MARITAL STATUS	−1.17396	.31445	13.9380
AGE	.03971	.03191	1.5480
YEARS EDUCATION	−.15106	.05023	9.0456

ABLES, contains four columns. The second column, under the heading COEF-
FICIENT, contains the estimates of β_0, β_1, β_2, and β_3, in order from top
to bottom. Column 1, headed VARIABLE, gives the programmer's identifica-
tion of the parameter. Thus

$$\hat{\beta}_0 = 1.41411 \qquad \hat{\beta}_1 = -1.17396 \qquad \hat{\beta}_2 = .03971 \qquad \hat{\beta}_3 = -.15106$$

and it follows that the prediction equation is

$$\hat{y} = 1.41411 - 1.17396\,x_1 + .03971\,x_2 - .15106\,x_3$$

For this particular model β_1, β_2, and β_3 represent the change in the
mean value of y, $E(y)$, for a one-unit change in x_1, x_2, and x_3, respectively. For example, $\hat{\beta}_2 = .03971$ is the estimated mean change in television viewer time for a one-year increase in x_2 (age of the interviewee).
The coefficient $\hat{\beta}_1$ of the dummy variable x_1 represents the difference in
mean viewing time between interviewees who are living with spouses and
those who are not. The estimate of β_1 is −1.17396 hours. Thus we estimate that unmarried interviewees spend, on the average, approximately
1.17 more hours per day watching television than the interviewees who
live with their spouses.

12.5 Confidence Intervals and Tests of Hypotheses for the β Parameters

The shaded portion of table 12.3 also provides the information needed
to form confidence intervals for the β parameters in the linear model for
the television study of example 12.2 and for tests of hypotheses concerning

these parameters. The estimated standard deviations, $s_{\hat{\beta}_1}$, $s_{\hat{\beta}_2}$, and $s_{\hat{\beta}_3}$, respectively, of the regression coefficient estimates $\hat{\beta}_1$, $\hat{\beta}_2$, and $\hat{\beta}_3$ are given under the column headed STD. ERROR. These statistics enable us to construct confidence intervals for the regression coefficients, the parameters β_1, β_2, and β_3.

The procedure for forming confidence intervals for the β parameters is identical to the procedure employed in section 11.5 for the simple linear model except that the computing formulas for $s_{\hat{\beta}_1}$, $s_{\hat{\beta}_2}$, and $s_{\hat{\beta}_3}$ are much more complex than the corresponding quantity calculated in section 11.5. Thus the formula for a $(1 - \alpha)100\%$ confidence interval for a parameter, say β_i, is as given in the box.

$(1 - \alpha)100\%$ Confidence Interval for β_i

$$\hat{\beta}_i \pm t_{\alpha/2} s_{\hat{\beta}_i}$$

The number of degrees of freedom for $t_{\alpha/2}$ equals the number n of data points less one degree of freedom for every β parameter appearing in the model.

We illustrate the procedure for forming a confidence interval with an example.

Example 12.3 Refer to example 12.2. Find a 95% confidence interval for β_1, the mean difference in daily television-viewing time between interviewees who are living with their spouses and those who are not.

Solution Checking under the STD. ERROR column, we find that $s_{\hat{\beta}_1} = .31445$. The table t-value, $t_{.025}$, is based on (n − number of β parameters in the model) $= 25 - 4 = 21$ degrees of freedom. From table 4 of the appendix, we find this value to be $t_{.025} = 2.080$. Then the 95% confidence interval for β_1 is

$$\hat{\beta}_1 \pm t_{\alpha/2} s_{\hat{\beta}_1} = -1.17396 \pm (2.080)(.31445)$$

or from -1.82802 to $-.5199$. Since $x_1 = 1$ for interviewees living with their spouses, 0 if not, we estimate that the "loners" watch television, on the average, between .52 and 1.83 more hours per day than those living with their spouses.

A test of an hypothesis that a particular parameter, say β_i, equals 0 can be conducted by using a t statistic:

$$t = \frac{\hat{\beta}_i - 0}{s_{\hat{\beta}_i}} = \frac{\hat{\beta}_i}{s_{\hat{\beta}_i}}$$

The procedure is identical to the procedure employed in testing an hypothesis about the slope β_1 in the simple linear model (section 11.5) except that the formula for computing $s_{\hat{\beta}_i}$ in a multiple regression analysis is much more complex.

The test can also be conducted by using the F statistic (presented in section 9.7). We can use the F statistic here because the square of a t statistic (with v degrees of freedom) is equal to an F statistic with 1 degree of freedom in the numerator and v degrees of freedom in the denominator. That is,

$$t^2_{v\text{d.f.}} = F_{1,v}$$

Some multiple regression computer programs utilize a t test to test the null hypothesis

$$H_0: \beta_i = 0$$

against the two-sided alternative hypothesis

$$H_a: \beta_i \neq 0$$

However, the program shown in table 12.3 utilizes the F statistic. The computed F-values for testing the null hypotheses that β_1, β_2, and β_3 equal 0 are shown in table 12.3 under the heading F VALUE.

When the null hypothesis is true, the F statistic is the ratio of two unbiased estimates (called *mean squares*) of σ^2, the variance of the random error ϵ that appears in the linear model. Thus

$$F = \frac{\text{numerator mean square}}{\text{denominator mean square}}$$

The denominator mean square is always s^2, the estimator of σ^2. The numerator mean square depends solely on the parameter, say β_i, being tested. If $\beta_i = 0$, the expected value of the numerator mean square is σ^2. If β_i differs from 0, the expected value of the numerator mean square is larger than σ^2 and F is larger than would have been expected if H_0 were true. Therefore,

Figure 12.2 *Location of the rejection region for the F test*

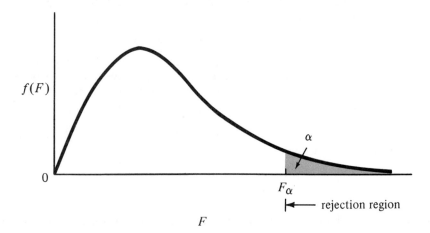

we reject the null hypothesis $H_0: \beta_i = 0$ only for large values of F. The rejection region for the F test is located in the upper tail of the F distribution, as shown in figure 12.2.

The critical value of the F statistic, F_α, depends on two quantities, ν_1 and ν_2, where

ν_1 = number of degrees of freedom associated with numerator mean square

ν_2 = number of degrees of freedom associated with s^2

To test the null hypothesis $\beta_i = 0$, the degrees of freedom ν_1 and ν_2 are as follows:

$\nu_1 = 1$

ν_2 = number n of data points less one degree of freedom for every parameter appearing in the model

Critical values of F for various combinations of ν_1 and ν_2 are tabulated in tables 6 and 7 in the appendix. Table 6 gives values of $F_{.05}$; table 7 gives values of $F_{.01}$. (For more information on how to read these tables, see section 9.7.)

We illustrate the use of the F test with an example.

Example 12.4 Refer to example 12.2. Test the null hypothesis that β_1 (the mean difference in average daily television-viewing time between interviewees who live with their spouses and those who do not) is equal to 0. Test at the $\alpha = .05$ level of significance.

Solution The critical value for this F test is based on

$\nu_1 = 1$

$\nu_2 = n$ − number of β parameters in the model
$= 25 - 4 = 21$

degrees of freedom. Entering table 6 in the appendix with $\nu_1 = 1$ and $\nu_2 = 21$, we find $F_{.05} = 4.32$.

The computed value of F in table 12.3 corresponding to β_1 is 13.9380. Since this F-value exceeds the critical value $F_{.05} = 4.32$, the null hypothesis is rejected. Therefore, the data provide sufficient evidence to indicate a difference in mean daily television-viewing time between interviewees who live with their spouses and those who do not.

Example 12.5 To show the equivalence of the F and t tests for a test of the null hypothesis $H_0: \beta_i = 0$, conduct the test of example 12.4 by using the t statistic.

Solution As noted earlier, we can test the null hypothesis

$$H_0: \beta_i = 0$$

against the alternative hypothesis

$$H_a: \beta_i \neq 0$$

by using

$$t = \frac{\hat{\beta}_i}{s_{\hat{\beta}_i}}$$

Then because the alternative hypothesis implies a two-tailed test, we will reject the null hypothesis if $t > t_{.025}$ or $t < -t_{.025}$, where $t_{.025}$ (like the denominator of F) is based on ($n -$ number of β parameters in the model) degrees of freedom. The number of degrees of freedom for the t statistic is 21, and from table 4 of the appendix, $t_{.025} = 2.080$.

To find the value of the test statistic, we first need to obtain $\hat{\beta}_1$ and $s_{\hat{\beta}_1}$ from the computer output of table 12.3. These values are

$$\hat{\beta}_1 = -1.17396 \qquad s_{\hat{\beta}_1} = .31445$$

Then the observed value of the test statistic is

$$t = \frac{\hat{\beta}_1}{s_{\hat{\beta}_1}} = \frac{-1.17396}{.31445} = -3.733$$

Since this computed value of t is less than the critical value $-t_{.025} = -2.080$, we reject the null hypothesis that $\beta_1 = 0$ (the same conclusion as the conclusion in example 12.4).

Now note the equivalence of the t and F tests. The critical values for the two tests are

$$F_{.05} = 4.32 \qquad \text{and} \qquad t_{.025} = 2.080$$

Squaring $t_{.025}$ we see that $t^2 = (2.080)^2 = 4.326 \approx F_{.05}$. (The slight discrepancy is due to the fact that the table F- and t-values are rounded to two and three decimal places.) As you would suspect, the same equivalence holds true for the computed values of F and t. The values of F (shown in table 12.3) and t calculated from the data are

$$F = 13.9380 \qquad \text{and} \qquad t = 3.733$$

Then $t^2 = (3.733)^2 = 13.935$. Again note that $F \approx t^2$.

Notice that the F test can only be used to test $H_0: \beta_i = 0$ against the two-sided alternative $H_a: \beta_i \neq 0$. In contrast, the t statistic can also be used to test H_0 against the one-sided alternative hypothesis $H_a: \beta_i > 0$ (or the alternative hypothesis $H_a: \beta_i < 0$). For example, if you wished only to support the hypothesis that single interviewees watched television more than those living with spouses, you would utilize the one-sided alternative $H_a: \beta_1 < 0$. Then you would reject H_0 only if $t < -t_\alpha$. The F statistic would not be appropriate for this one-tailed test. For this reason computer packages that print the values of the t statistics are slightly more versatile than those that print the F statistics.

Example 12.6 Refer to the television viewer printout of table 12.3. Test the null hypothesis $H_0: \beta_1 = 0$ against the alternative hypothesis H_a:

$\beta_1 < 0$. That is, we wish to determine if the data provide sufficient evidence to indicate that, on the average, "loner" interviewees watch television more than those who live with their spouses. Test at the $\alpha = .05$ level of significance.

Solution Since this is a one-tailed test, we locate the entire $\alpha = .05$ in the lower tail of the t distribution. Then we will reject $H_0: \beta_1 = 0$ if $t < -t_{.05}$. The critical value of t, based on 21 degrees of freedom, is $-t_{.05} = -1.721$.

Now we compare the computed value of t, $t = -3.733$ (which was calculated in example 12.5) with the critical value of t. Since the computed value is less than $-t_{.05} = -1.721$, we reject the null hypothesis. It appears that "loners" spend more time watching television (on the average) than interviewees who live with their spouses.

Before concluding this section, we note a very important point. When fitting a model to a set of data, the number n of data points must be larger than the number of parameters in the model in order that a sufficient number of degrees of freedom be associated with s^2 (and consequently with the t and F statistics). How many? The more the better. You can see from the t and F tables that the tabulated t- and F-values are large for a small number of degrees of freedom. Certainly we would want at least 5 degrees of freedom available for estimating σ^2, and preferably many more (10, 20, or more).

If your original data was used to build a model (finding a model that provides a good fit to the data), you should have some new data to test the model you have selected. This procedure follows the logic of the scientific method. First you observe nature and formulate a theory (collect data and find a model that best fits the data). Then you test the theory against observation (test the fitted model against a new set of data). To accomplish this many researchers split their data set into two parts, one part to use in model building and the other to test the model derived from the first part.

12.6 The Problem of Correlated Estimates: Multicollinearity

If you fit a multiple regression model

$$y = \beta_0 + \beta_1 x_1 + \beta_2 x_2 + \cdots + \beta_k x_k + \epsilon$$

you must be very careful in interpreting the results of t tests on the β parameters. One reason for this caution is that the estimators may be correlated. For example, if one β parameter is overestimated, another might tend to be underestimated. We call this phenomenon **multicollinearity**. That is, some of the information contributed by two or more of the independent variables for predicting y may be different, but some information may be identical.

To illustrate, suppose you wish to construct a model to predict the price of an automobile as a function of a set of independent variables x_1, x_2, ..., x_k and that two of the variables are

$$x_1 = \text{weight of the automobile}$$

$$x_2 = \text{horsepower of the engine}$$

In general, you would expect heavier automobiles and those with larger engines (greater horsepower) to cost more. That is, both of these variables contribute information for the prediction of price but some of the information (although not all) is the same. This is because weight and horsepower are correlated. Heavy cars require larger engines.

When two or more of the independent variables are correlated, you cannot determine their respective individual contributions to the reduction in SSE, the sum of squares of the deviations between the observed and the predicted values of y. Therefore, the contribution of information by a particular independent variable to the prediction of y depends on the other independent variables included in the model. If two variables contribute overlapping information, a test of the β parameter (β_1) for x_1 might indicate statistical significance (rejection of the hypothesis that the β parameter equals 0), while a test of the other parameter (β_2) might indicate nonsignificance. Actually, the second variable x_2 might be causally related even though the t test (or F test) did not lead to rejection of the hypothesis that $\beta_2 = 0$. Thus when multicollinearity exists, it is the complete model that is important, not the individual β parameters.

You can test to determine if the model is contributing information for the prediction of y by testing the hypothesis

$$H_0: \beta_1 = \beta_2 = \cdots = \beta_k = 0$$

We will show you how to conduct this test in section 12.8. We will also show you how to measure the adequacy of the model (how well it fits the data) in section 12.7.

An additional problem sometimes results when the multiple regression model is sequenced over time and involves **time series data**, as in the use of the model for applications to sales forecasting, demand analysis, and econometric studies. When one or more key variables have been omitted from the multiple regression model for time series data, the residuals are often dependent upon one another (said to be **autocorrelated**, or **serially correlated**). For example, suppose that the number of new housing starts for each of the past 60 months is regressed against the prime lending rate. The omission of population size as a separate independent variable may lead to serially correlated residuals if population size is correlated with the prime lending rate over this 60-month span.

Serial correlation affects the precision but not the accuracy of the estimation of the β parameters in a multiple regression model. The estimates are unbiased but their true variances are underestimated when serial correlation is present. As a result, the SSE may seriously underestimate the true unexplained variation, causing the t-values (or F-values) to be larger than they should be. This could lead to the conclusion that certain β parameters are statistically significant when in fact they are not. Thus the effect of serial correlation is opposite to that of multicollinearity.

The most common test for the presence of serial correlation is the Durbin-Watson test, which is described in Neter and Wasserman on pages 358–360 (see the references). An easier test to apply, and one which overcomes the limitations of the Durbin-Watson test, is a runs test applied to the residuals. This test will be discussed in section 18.6 of this text.

To summarize, exercise caution when interpreting the t tests (or F tests) concerning the individual β parameters that appear in the model.

Exercises

12.1. The owner of an automobile dealership undertook a study to determine the relationship among

y = number of new cars sold per month by his dealership

x_1 = number of 10-minute local TV spot advertisements during the month

x_2 = number of full-page newspaper advertisements during the month

			MONTH			
	1	2	3	4	5	6
y	10	10	20	30	40	40
x_1	0	1	2	2	3	4
x_2	1	0	2	3	3	3

Over a period of six months, the owner noticed the results shown in the table. Use the process of elimination to fit the model $y = \beta_0 + \beta_1 x_1 + \beta_2 x_2 + \epsilon$ to the data by solving the least squares equations for the unknown parameters $\hat{\beta}_0$, $\hat{\beta}_1$, and $\hat{\beta}_2$.

Use the computer output of table 12.3 to solve exercises 12.2 through 12.6.

12.2. Test the null hypothesis that the mean increase (or decrease) in daily television-viewing time for a one-year increase in the age of interviewees is zero. That is, test H_0: $\beta_2 = 0$ against the alternative hypothesis $\beta_2 \neq 0$. Test at the $\alpha = .05$ level of significance, using the F test.

12.3. Repeat exercise 12.2 but use the t test. Show that the square of the calculated value of t equals the value of F obtained in exercise 12.2.

12.4. Find a 95% confidence interval for the mean increase in daily television viewer time for a one-year increase in the age of interviewees.

12.5. Find a 95% confidence interval for the mean increase in daily television viewer time for a one-year increase in the number of years of education of the interviewee (i.e., find a 95% confidence interval for β_3).

12.6. Suppose you have a theory that the mean daily television-viewing time decreases as the age of the viewer increases. Test the null hypothesis H_0: $\beta_3 = 0$

against the one-sided alternative $H_a: \beta_3 < 0$. Test at the $\alpha = .05$ level of significance.

12.7.   A land developer was interested in creating a model to use for estimating the selling price of beach lots on the Oregon coast. To do so he recorded the following items for each of 20 beach lots recently sold:

$$y = \text{sale price of the beach lot (in \$1000 units)}$$

$$x_1 = \text{area of the lot (in hundreds of square feet)}$$

$$x_2 = \text{elevation of the lot}$$

$$x_3 = \text{slope of the lot}$$

The land developer then employed a regression analysis computer program and obtained the output that follows.

MULTIPLE R	.8854
R SQUARE	.7838
STD. ERROR OF EST.	.6075

ANALYSIS OF VARIANCE

	DF	SUM OF SQUARES	MEAN SQUARE	F RATIO
REGRESSION	3	21.409	7.136	19.345
RESIDUAL	16	5.903	.369	

INDIVIDUAL ANALYSIS OF VARIABLES

VARIABLE	COEFFICIENT	STD. ERROR	F VALUE
(CONSTANT	−2.491)		
AREA	.099	.058	2.935
ELEVATION	.029	.006	23.327
SLOPE	.086	.031	7.841

a. Give the prediction equation for the linear model relating selling price to the area, elevation, and slope of a beach lot.

b. Which of the predictor variables contributes information for the prediction of y? Determine this by using the appropriate statistical test. Use $\alpha = .05$.

12.8.   Suppose that before you collected the data for the analysis of exercise 12.7, you had a theory that sloping beach lots were preferred over those with lesser slope. Do the data provide sufficient evidence to indicate that sales price increases as the slope increases? (Hint: Test $H_0: \beta_3 = 0$ against the one-sided alternative $H_a: \beta_3 > 0$. Use $\alpha = .05$.)

12.9.   Refer to exercise 12.7. Find a 90% confidence interval for the regression parameter relating area to selling price in the presence of elevation and slope. For given values of elevation and slope, provide an interpretation of this confidence interval.

12.7 Measuring the Goodness of Fit of a Model

We noted in section 11.9 that an important property of a regression analysis is that the total sum of squares of deviations of the y-values about their mean partitions into two quantities,

$$\text{SSE} = \sum_{i=1}^{n} (y_i - \hat{y})^2 = \text{sum of squares for error}$$

$$\text{SSR} = \sum_{i=1}^{n} (\hat{y}_i - \bar{y})^2 = \text{sum of squares due to regression}$$

Also,

$$\text{Total SS} = \sum_{i=1}^{n} (y_i - \bar{y})^2 = \text{SSR} + \text{SSE}$$

SSE, the sum of squares of deviations of the y-values about their predicted values (those values calculated from the prediction equation), divided by the appropriate number of degrees of freedom is equal to s^2, an estimate of σ^2.

In addition, we showed that r^2, the coefficient of determination, is equal to

$$r^2 = \frac{\text{Total SS} - \text{SSE}}{\text{Total SS}} = \frac{\text{SSR}}{\text{Total SS}}$$

You will recall that r^2 measures the proportion of the total SS that can be explained by the single predictor variable x. Consequently, r^2, which assumes values in the interval $0 \le r^2 \le 1$, measures the goodness of fit of the simple linear regression model.

The Total SS,

$$\sum_{i=1}^{n} (y_i - \bar{y})^2$$

is partitioned in exactly the same way for a multiple regression analysis. Thus

$$\text{Total SS} = \text{SSR} + \text{SSE}$$

and SSR and SSE are defined in exactly the same way as they were for a simple linear regression analysis. The only difference here is that y is a function of more than one predictor variable.

Suppose you fit the multiple regression model

$$y = \beta_0 + \beta_1 x_1 + \beta_2 x_2 + \cdots + \beta_k x_k + \epsilon$$

to a set of data. Then the quantity

$$R^2 = \frac{\text{Total SS} - \text{SSE}}{\text{Total SS}} = \frac{\text{SSR}}{\text{Total SS}}$$

gives the proportion of the Total SS that is explained by the predictor variables x_1, x_2, ..., x_k. The remainder is explained by the omission of important information-contributing variables from the model, an incorrect formulation of the model, and experimental error. Just like r^2, the simple linear coefficient of determination, the multiple coefficient of determination R^2 takes values in the interval

$$0 \le R^2 \le 1$$

A small value of R^2 means that $x_1, x_2, \ldots, x_k$ contribute very little information for the prediction of y; a value of R^2 near 1 means that $x_1, x_2, \ldots, x_k$ provide almost all the information necessary for the prediction of y. Thus, just as r^2 provides a measure of the fit of the simple linear model, R^2 provides a measure of the fit of a more complex regression model.

To illustrate, let us return to the computer output (table 12.3) for example 12.2, the television viewer data analysis. We repeat the computer output here in table 12.4, shading the information that is pertinent to the preceding discussion.

The first line of computer output in table 12.4, labeled MULTIPLE R, gives the value of the multiple correlation coefficient R. R measures the correlation of y with the portion of the model involving $x_1, x_2, \ldots, x_k$. Thus R is the multivariable counterpart to the simple correlation coefficient r. You will note that $R = .7918$ for the television viewer example.

The second line of the computer output, labeled R SQUARE, gives the value of the multiple coefficient of determination R^2. This value, $R^2 = .6269$, is much more meaningful than R as a measure of the goodness of fit of the model. It tells us that only 62.69% of the total variation of the y-values about their mean can be explained by the terms in the model. The remainder, a rather large 37.31%, is unexplained. The relatively poor fit of this model could be due to the fact that x_1, x_2, and x_3 are not entered properly into

Table 12.4 *Computer output for senior citizen television-viewing data of table 12.2*

MULTIPLE R	.7918
R SQUARE	.6269
STD. ERROR OF EST.	.7524

ANALYSIS OF VARIANCE

	DF	SUM OF SQUARES	MEAN SQUARE	F RATIO
REGRESSION	3	19.972	6.657	11.760
RESIDUAL	21	11.888	.566	

INDIVIDUAL ANALYSIS OF VARIABLES

VARIABLE	COEFFICIENT	STD. ERROR	F VALUE
(CONSTANT	1.41411)		
MARITAL STATUS	−1.17396	.31445	13.9380
AGE	.03971	.03191	1.5480
YEARS EDUCATION	−.15106	.05023	9.0456

the model (perhaps we should include terms involving x_2^2, x_3^2, $x_1 x_2$, $x_1 x_3$, $x_2 x_3$, etc.). Or perhaps y, the average daily viewing time of interviewees, is a function of many other variables besides x_1, x_2, and x_3. For example, you might wish to include a variable x_4 that measures an interviewee's propensity to read and a dummy variable x_5 that takes a value 1 if the interviewee is employed and a value 0 if not employed. You may think of many other variables that might affect the length of television viewer time. Actually, the fact that R^2 is as low as .6269 is probably due to both reasons. It is likely that x_1, x_2, and x_3 are not entered into the model in the best way (some comments on model formulation are given in section 12.10) and that the model does not include an adequate number of predictor variables related to y.

The third line of the computer output of table 12.4, titled STD. ERROR OF EST., is the value of s for the regression analysis. Thus s is the square root of s^2, where s^2, the estimate of σ^2 (the variance of the y-values for a given value of x_1, x_2, ..., x_k), is equal to SSE divided by the appropriate number of degrees of freedom. You will recall that for the simple linear model (which contains two β parameters), you divided SSE by $(n - 2)$. In the general case s^2 equals SSE divided by n, the number of data points, less one degree of freedom for every β parameter that appears in the model. You can see from table 12.4 that $s = .7524$ for the television viewer data.

Estimate of Variance for Multiple Regression

$$s^2 = \frac{\text{SSE}}{n - (\text{number of } \beta \text{ parameters in the model})}$$

Some computer outputs print SSE, some print s^2, and some print s. Of course, once you have one of the quantities, you can find any other. Of what value are they? The answer is that, as in simple linear regression, s appears in all the formulas for confidence intervals and tests of hypotheses. Not all the confidence intervals or test statistics appear in the printouts. Hence s is included in case you need it for some special confidence interval or test of an hypothesis.

But there is a much more direct application of s. It can be used as a check to detect errors in the computation of the prediction equation. Perhaps some data points were incorrectly entered into the computer or perhaps the computer program is sensitive to rounding errors and gives incorrect answers. To detect errors of this type, calculate the deviation between the observed and the predicted values of y. Most of these errors of prediction should be less than $2s$ and almost all should be less than $3s$. If the y-values do not conform to these rule-of-thumb limits, you might wish to check your calculations.

12.8 Testing the Utility of the Regression Model

The partitioning of the total SS into SSR and SSE is called an "analysis of variance." This name is used because when $x_1, x_2, ..., x_k$ contribute no information for the prediction of y (i.e., the model is worthless), each of the quantities SSR and SSE provides an independent (in a probabilistic sense) estimate of σ^2, the variance of y for given values of $x_1, x_2, ..., x_k$. These estimates are called **mean squares**. Thus

$$\text{MSR} = \text{mean square for regression} = \frac{\text{SSR}}{\nu_1}$$

$$\text{MSE} = s^2 = \text{mean square for error} = \frac{\text{SSE}}{\nu_2}$$

where

ν_1 = one less than the number of β parameters in the model = k

$\nu_2 = n -$ (number of β parameters in the model) $= n - (k + 1)$

Then for any multiple regression model, say one containing k predictor variables,

$$y = \beta_0 + \beta_1 x_1 + \beta_2 x_2 + \cdot \cdot \cdot + \beta_k x_k + \epsilon$$

we use MSR and MSE to test the null hypothesis that $x_1, x_2, ..., x_k$ contribute no information for the prediction of y. This is equivalent to hypothesizing that

$$\beta_1 = \beta_2 = \cdot \cdot \cdot = \beta_k = 0$$

If the data provide sufficient information to reject this hypothesis, it means that at least one of the predictor variables $x_1, x_2, ..., x_k$ contributes information for the prediction of y.

The test statistic for this test is

$$F = \frac{\text{MSR}}{\text{MSE}} = \frac{\text{MSR}}{s^2}$$

This statistic possesses an F distribution with ν_1 and ν_2 degrees of freedom, where, as previously explained,

ν_1 = numerator degrees of freedom

= one less than the number of β parameters in the model = k

ν_2 = denominator degrees of freedom

$= n -$ (number of β parameters in the model) $= n - (k + 1)$

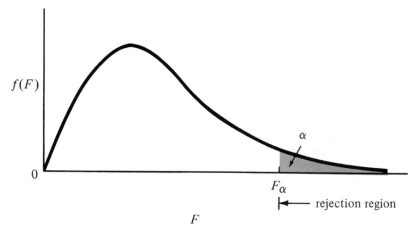

Figure 12.3 *Rejection region for the F test of H_0: $\beta_1 = \beta_2 = \cdots = \beta_k$* = 0

When the null hypothesis is false (the model *is* useful in predicting y), SSR will tend to be larger than expected and hence F will be large. Thus we will reject H_0: $\beta_1 = \beta_2 = \cdots = \beta_k = 0$ for values of F that exceed $F_{.05}$, an upper-tail value in the F distribution (see figure 12.3). The degrees of freedom associated with $F_{.05}$ are those for MSR and s^2.

We illustrate the test procedure with an example.

Example 12.7 Test the adequacy of the regression model for predicting daily television viewing (example 12.2). Use the computer output of table 12.4 to conduct the test.

Solution The computer output for the television viewer data of table 12.2 is shown here in table 12.5. The portion of the output devoted to an analysis of variance is shaded. The third column of the table, under the heading SUM OF SQUARES, gives SSR and SSE. Thus

$$\text{SSR} = 19.972 \qquad \text{SSE} = 11.888$$

The second column of the table, under the heading DF, gives the number of degrees of freedom associated with each sum of squares. Thus v_1, the number of degrees of freedom associated with SSR, is 3. Similarly, v_2, the number of degrees of freedom associated with SSE, is 21.

The fourth column of the table, under the heading MEAN SQUARE, gives MSR and MSE. Thus

$$\text{MSR} = \frac{\text{SSR}}{3} = \frac{19.972}{3} = 6.657$$

$$\text{MSE} = s^2 = \frac{\text{SSE}}{21} = \frac{11.888}{21} = .566$$

The fifth column of the table, under the heading F RATIO, gives the

Table 12.5 *Computer output for senior citizen television-viewing data of table 12.2*

MULTIPLE R	.7918
R SQUARE	.6269
STD. ERROR OF EST.	.7524

ANALYSIS OF VARIANCE

	DF	SUM OF SQUARES	MEAN SQUARE	F RATIO
REGRESSION	3	19.972	6.657	11.760
RESIDUAL	21	11.888	.566	

INDIVIDUAL ANALYSIS OF VARIABLES

VARIABLE	COEFFICIENT	STD. ERROR	F VALUE
(CONSTANT	1.41411)		
MARITAL STATUS	−1.17369	.31445	13.9380
AGE	.03971	.03191	1.5480
YEARS EDUCATION	−.15106	.05023	9.0456

computed value of the test statistic. Thus

$$F = \frac{MSR}{MSE} = \frac{MSR}{s^2} = \frac{6.657}{.566} = 11.760$$

We compare this calculated value of F with the critical value of $F_{.05}$ based on $v_1 = 3$ and $v_2 = 21$ degrees of freedom and we reject H_0 if $F > F_{.05}$. Checking table 6 of the appendix, we find $F_{.05} = 3.07$. Since the value of F calculated from the data, $F = 11.760$, exceeds $F_{.05} = 3.07$, we reject the null hypothesis

$$H_0: \beta_1 = \beta_2 = \beta_3 = 0$$

It appears that at least one of the predictor variables contributes information for the prediction of y.

Exercises

12.10. A marketing representative for a company that sells soybeans as a meat supplement in the United States is interested in constructing a model to predict the sales of soybeans in different market areas. Sales data (in thousands of dollars) were obtained from 25 different market areas and were related to each area's measurement of the following:

x_1 = the coefficient of cross-elasticity between soybeans and beef

x_2 = per capita income (in thousands of dollars)

x_3 = average utilities consumption index

x_4 = unit price of one pound of packaged soybeans

x_5 = proportion of total expenses devoted to advertising

x_6 = 1 if the area is a major beef-producing area, 0 if it is not

The marketing representative then employed a regression analysis computer program, which listed the results that follow.

```
MULTIPLE R            .9825
R SQUARE              .9654
STD. ERROR OF EST.    1.7098
```

ANALYSIS OF VARIANCE

	DF	SUM OF SQUARES	MEAN SQUARE	F RATIO
REGRESSION	6	1468.034	244.672	
RESIDUAL	18	52.614	2.923	

INDIVIDUAL ANALYSIS OF VARIABLES

VARIABLE	COEFFICIENT	STD. ERROR	F VALUE
(CONSTANT	−51.034)		
CROSS-ELAS.	57.600	13.813	17.356
PER CAP. INC.	2.956	2.548	1.347
UTIL. IDX.	−.934	1.129	.682
PRICE	−46.542	31.661	2.160
ADV. EXP.	46.355	9.499	23.843
BEEF AREA	−1.805	.775	5.431

a. Calculate the missing F RATIO and test the hypothesis associated with this missing value, using $\alpha = .05$.

b. What proportion of the variation of soybean sales is explained by the six predictor variables in the model?

c. Give the regression equation for predicting the sales of soybeans in a particular market area.

d. In the presence of the other variables in the model, which is the most significant predictor of sales? Which is the least significant?

e. Suppose it is very difficult to obtain a precise measure of the per capita income of a market area. Does the prediction model lose much information if we eliminate per capita income as a predictor of sales? If we do so, how would the revised model be established?

f. Find a 90% confidence interval for the difference in predicted sales between areas which are major beef-producing regions and those which are not (i.e., for β_6).

12.11.   Table 12.6 lists the sales price y as a substitute for value and seven assumed related predictor variables $x_1, x_2, \ldots, x_7$ for each of 50 single-family residences. These are actual data recorded from a sampling of residential sales in Eugene, Oregon, during 1974 by the Lane County Assessor's Office for the purpose of developing a residential value appraisal model. The linear model $y = \beta_0 + \beta_1 x_1 \beta_2 x_2 + \cdots + \beta_7 x_7 + \epsilon$ was fitted to the data set using a standard regression analysis program. The results of the analysis follow.

```
MULTIPLE R            .9490
R SQUARE              .9006
STD. ERROR OF EST.    3.0051
```

ANALYSIS OF VARIANCE

	DF	SUM OF SQUARES	MEAN SQUARE	F RATIO
REGRESSION	7	3438.183	491.169	54.389
RESIDUAL	42	379.291	9.031	

INDIVIDUAL ANALYSIS OF VARIABLES

VARIABLE	COEFFICIENT	STD. ERROR	F VALUE
(CONSTANT	−13.4617)		
SQ. FT.	1.0207	.1110	84.5510
NO. BDRMS.	1.6270	1.4813	1.2065
NO. BTHRMS.	−3.8510	1.3634	7.9778
TOT. ROOMS	2.8936	.9192	9.9095
AGE	−.0242	.1031	.0550
AT. GAR.	1.2497	1.1239	1.2363
VIEW	−1.2897	1.4965	.7428

Table 12.6 *Measurements taken on 50 single-family residences; exercise 12.11*

RESIDENCE, i	SALES PRICE, y (× $1000)	SQUARE FEET, x_1 (× 100)	BEDROOMS, x_2	BATHROOMS, x_3	TOTAL ROOMS, x_4	AGE, x_5	ATTACHED GARAGE, x_6	VIEW, x_7
1	10.2	8.0	2	1	5	5	0	0
2	10.5	9.5	2	1	5	8	0	0
3	11.1	9.1	3	1	6	2	0	0
4	15.3	9.5	3	1	6	6	0	0
5	15.8	12.0	3	2	7	5	0	0
6	16.3	10.0	3	1	6	11	0	0
7	17.2	11.8	3	2	7	8	0	0
8	17.7	10.0	2	1	7	15	1	0
9	18.0	13.8	3	2	7	10	0	0
10	18.1	12.5	3	2	7	11	0	0
11	18.4	15.0	3	2	7	12	0	0
12	18.4	12.0	3	2	7	8	0	0
13	18.9	16.0	3	2	7	9	1	1
14	19.3	16.5	3	2	7	15	0	0
15	19.5	16.0	3	2	7	11	1	0
16	19.9	16.8	2	2	7	12	0	0
17	20.3	15.0	3	1	7	8	1	0
18	20.3	17.8	3	2	8	13	1	0
19	20.8	17.9	3	2	7	18	1	0
20	21.0	19.0	2	2	7	22	0	0
21	21.5	17.6	3	1	6	17	0	0
22	22.0	18.5	3	2	8	11	1	0
23	22.1	18.0	3	2	7	5	0	0
24	22.5	17.0	2	3	8	2	1	0
25	22.8	18.7	3	1	6	6	0	0
26	22.8	20.0	3	2	7	16	0	0
27	22.9	20.0	3	2	7	12	0	0
28	23.2	21.0	3	2	7	10	1	0
29	23.5	20.5	2	2	7	11	1	0
30	24.9	19.9	3	1	7	13	1	1
31	25.0	21.5	2	2	7	8	0	0
32	25.1	20.5	3	1	7	9	1	0
33	26.6	22.0	3	2	7	10	0	0
34	26.9	22.0	3	2	7	6	1	1
35	26.9	21.8	2	1	6	15	1	0
36	27.8	22.5	3	2	7	11	1	0
37	28.0	24.0	3	2	7	17	0	0
38	28.7	23.5	3	2	8	12	0	0
39	29.0	25.0	3	2	7	11	1	0
40	30.1	25.6	3	2	7	15	1	0
41	32.0	25.0	4	2	8	12	1	0
42	33.8	25.0	2	2	8	8	0	1
43	35.3	26.8	3	2	7	6	1	0
44	37.1	22.1	3	2	8	18	1	0
45	37.5	27.5	3	2	8	12	1	0
46	38.0	25.0	4	2	8	10	1	0
47	38.4	24.0	3	2	8	13	1	1
48	39.0	31.0	4	3	9	25	1	0
49	43.0	21.0	4	2	9	18	1	0
50	55.0	40.0	5	3	12	22	1	0

Discuss the computer output for this analysis by discussing the same topics that were addressed in the television-viewing example.

12.12. Refer to exercise 12.11. Use the model derived in that exercise to establish assessed valuations for each of five Eugene residences. The data are given in the accompanying table.

RESIDENCE, i	SQUARE FEET, x_1	BEDROOMS, x_2	BATHROOMS, x_3	TOTAL ROOMS, x_4	AGE, x_5	ATTACHED GARAGE, x_6	VIEW, x_7
1	22.4	4	2	7	18	1	1
2	15.3	3	2	7	6	0	0
3	17.2	4	1	7	4	1	0
4	31.7	5	3	9	24	0	0
5	20.0	4	2	8	11	1	1

12.13. We are all aware of the effect of inflation on property values; that is, they tend to appreciate at least as rapidly as the local rate of inflation. This requires local assessment officers to periodically update the assessed valuations of properties in their jurisdiction. There are three approaches that might be adopted by the assessor to update property valuations:

i. Each listed valuation could be increased by the percentage increase in inflation since the previous assessment.
ii. New sales data could be obtained and pooled with past available sales data to develop a new regression assessment model.
iii. A new regression assessment model could be developed based only on current sales data, ignoring all past sales data.

Which approach would you suggest? Explain.

12.14. Is there a consistent relationship between capital-budgeting practices and earnings performance? If so, this evidence would support the practice of many firms that practice capital-budgeting techniques in their investment programs. To study this issue Kim and Kwak* regressed y, the estimated value of the average earnings per share, on the following:

x_1 = degree of sophistication of the capital-budgeting system (0–100)

x_2 = size of firm (average annual total assets)

x_3 = capital intensity (depreciation / total assets)

x_4 = risk (standard deviation of annual earnings per share)

x_5 = capitalization (debt / total assets)

x_6 = price-earnings ratio

Data were taken from each of $n = 114$ machinery firms with revenue in excess of $50 million in 1974. Data were used for the period from 1969 through 1974. The results of their analysis follow:

*S. H. Kim and N. K. Kwak, "Capital Budgeting Practices and Their Impact on Earnings Performance," *Proceedings of the American Institute for Decision Sciences,* November 1976.

$$R^2 = .776$$
$$F = 61.991$$

VARIABLE	CONSTANT	X1	X2	X3	X4	X5	X6
COEFFICIENT	−1.613	.040	.001	.090	−.072	.018	.010
T VALUE		10.282	.464	2.232	−.559	4.143	.806

a. Does the Kim-Kwak study support the notion that there is a significant relationship between capital-budgeting practices and earnings performance for the class of firms studied in their analysis? Explain.

b. Which of the capital-budgeting variables used in the analysis contribute information for the prediction of earnings performance? Determine this by using appropriate tests of hypotheses. Use $\alpha = .05$.

c. Explain and interpret the $R^2 = .776$ value listed for this analysis.

12.9 Using the Prediction Equation for Estimation and Prediction

Suppose you are a research analyst for a brokerage firm and you wish to investigate the relationship between the price y of an electric utility stock and a set of independent variables x_1, x_2, ..., x_k, where x_1, x_2, x_3, and x_4 might be

$$x_1 = \text{prime interest rate}$$

$$x_2 = (\text{prime interest rate})^2 = x_1^2$$

$$x_3 = \text{earnings per share}$$

$$x_4 = \text{dividend rate}$$

The remaining predictor variables, $x_5, x_6, ..., x_k$, are other variables that you suspect would contribute information for the prediction of electric utility stock price y, or the squares, cross products, cubes, and so on of variables already appearing in the model. For example, to adjust for the effect of time, you might include the GNP as one of the predictor variables. Or if you thought that the prime interest rate x_1 tended to have a greater (or lesser) effect on the stock price y as x_1 increased, as shown in figure 12.4, you would include a term in the model containing $x_2 = x_1^2$. Then the model for the price y of an electric utility stock might be

$$y = \beta_0 + \beta_1 x_1 + \beta_2 x_1^2 + \beta_3 x_3 + \beta_4 x_4 + \cdots + \beta_k x_k + \epsilon$$

Suppose you select a random sample of electric utility stocks at certain points within a two-year period, fit the model to the data, and obtain the prediction equation

$$\hat{y} = 18.7 - .6x_1 - .2x_1^2 + 1.1x_3 + 8.3x_4 + \cdots$$

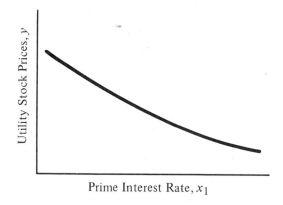

Figure 12.4 *A graph of the hypothetical effect of prime interest rate x_1 on the price y of an electric utility stock*

What can you learn from the estimates $\hat{\beta}_0 = 18.7$, $\hat{\beta}_1 = -.6$, $\hat{\beta}_2 = -.2$, $\hat{\beta}_3 = 1.1$, $\hat{\beta}_4 = 8.3$, and so on? The answer is, very little. As noted in sections 12.7 and 12.8, you can test the null hypothesis that a particular β parameter equals 0 (i.e., that the corresponding x variable contributes no information for the prediction of y in the presence of the other x variables contained in the model), but what does this really tell you?

If you reject the null hypothesis, it appears that this x variable contributes some information for the prediction of y. It does not tell you that the x variable caused y to change. And it does not tell you anything about the amount of information contributed in comparison to the other x variables in the model (because of the multicollinearity problem described in section 12.6).

If you do not reject the null hypothesis that the β parameter equals 0, it does not mean that "the x variable contributes no information for the prediction of y." Perhaps you do not have sufficient data to detect information contributed by the x variable. Or the x variable may contain a substantial amount of information on y but this same information may already have been contributed by the other x variables in the model.

In either case, it seems that for most business investigations, tests of hypotheses or estimation of the individual model parameters are of much less value than the use of the complete prediction equation. The prediction equation can be of value in three ways:

1. It can be used to estimate the mean value of y for given values of the predictor variables.
2. It can be used to predict some future value of y for given values of $x_1, x_2, ..., x_k$.
3. If a prediction equation provides a good fit to a set of data (R^2 is large) and the number of predictor variables is not too large, then the equation may help you to understand the process that you are investigating.

Estimates of the mean value of y for given values of $x_1, x_2, ..., x_k$ or predictions of specific values of y (to be observed in the future) for given values of $x_1, x_2, ..., x_k$ can be obtained by substituting numerical values of $x_1, x_2, ..., x_k$ into the prediction equation.

For example, suppose that a mail-order firm wished to relate the Christmas holiday sales y to two predictor variables (we will keep the number of predictor variables small for purposes of illustration), the number of mailings x_1 and the number of months x_2 that the mailing preceded Christmas. Furthermore, suppose the company varied x_1 and x_2 over 20 somewhat similar sales regions and measured the amount of sales y for each. Then they fitted a prediction equation to the data they collected and obtained the following result:

$$\hat{y} = 1.3 + 1.7x_1 - .1x_1^2 + 1.0x_1x_2 - .3x_1^2x_2$$

where

x_1, expressed in units of 100,000, is measured over the interval $.5 \le x_1 \le 2.5$

x_2, expressed in months, is measured over the interval $1 \le x_2 \le 3$

y is measured in units of $100,000

The best estimate of the mean sales $E(y)$ for a given combination of values of x_1 and x_2, say $x_1 = 1$ (100,000 mailings) and $x_2 = 2$ (mailings sent two months before Christmas), is obtained by substituting $x_1 = 1$ and $x_2 = 2$ into the prediction equation. Then

$$\hat{y} = 1.3 + 1.7x_1 - .1x_1^2 + 1.0x_1x_2 - .3x_1^2x_2$$
$$= 1.3 + 1.7(1) - .1(1)^2 + 1.0(1)(2) - .3(1)^2(2) = 4.3$$

or $\hat{y} = \$430,000$.

As noted in sections 11.6 and 11.7, not only is $\hat{y}$ the best estimate of the mean value of y for given values of x_1 and x_2, but it also gives the best prediction of some value of y to be observed in the future. That is, if we were to select a region, send out 100,000 mailings ($x_1 = 1$) two months ($x_2 = 2$) prior to Christmas, the predicted holiday sales for the region would be $430,000.

We can construct a confidence interval for $E(y)$ and a prediction interval for y using a procedure similar to that employed for the simple linear model in sections 11.6 and 11.7. However, the formulas for these intervals are much too complex to present in a text at this level. Fortunately, the computed intervals are included as an option in some computer program packages (see section 12.12). If you do not have such a computer program available, you will need to become familiar with the actual formulas, which, because of their complexity, are always expressed in matrix notation. This notation and the formulas for the confidence interval for $E(y)$ and the prediction interval for some future value of y are presented and explained in *An Introduction to Linear Models and the Design and Analysis of Experiments*, by W. Mendenhall (see the references).

Example 12.8 Although we would expect demand to decline as price increases when competing products are available at a lower price, that is not always the case. In fact, the relationship between demand and price can often be approximated by a second-order model; slight price increases result in decreased demand while significant price increases result in perception of increased quality and demand. In an attempt to study the relationship between price and demand and to develop an appropriate pricing policy, a distributor of a common brand of whiskey, which ordinarily sells for $5.00 a fifth, conducted a pricing experiment in 15 different market areas over a 12-month period using five different price levels. The results of the experiment are shown in the table.

NO. OF CASES SOLD PER MONTH PER 10,000 POPULATION, y	PRICE PER FIFTH, x
23, 20, 21	$5.00
19, 21, 18	5.50
18, 17, 20	6.00
21, 19, 20	6.50
25, 24, 22	7.00

a. Fit the second-order model $y = \beta_0 + \beta_1 x + \beta_2 x^2 + \epsilon$ to these data.

b. Predict y, the number of cases likely to be sold per month per 10,000 population in a sales area in which the price per fifth is $5.00. Predict y when $x = \$6.00$; when $x = \$7.00$.

c. Construct a 95% prediction interval for y when $x = \$5.00$; when $x = \$6.00$; when $x = \$7.00$.

d. Construct a 95% confidence interval for $E(y)$, the average number of cases sold per month per 10,000 population, when $x = \$5.00$; when $x = \$6.00$; when $x = \$7.00$.

Solution Using a standard regression program with an option providing the computations of prediction intervals for y and confidence intervals for $E(y)$, the results that follow were obtained.

```
MULTIPLE R           .8393
R SQUARE             .7045
STD. ERROR OF EST.   1.3292
```

ANALYSIS OF VARIANCE

	DF	SUM OF SQUARES	MEAN SQUARE	F RATIO
REGRESSION	2	50.533	25.266	14.301
RESIDUAL	12	21.200	1.767	

INDIVIDUAL ANALYSIS OF VARIABLES

VARIABLE	COEFFICIENT	STD. ERROR	F VALUE
(CONSTANT	156.1306)		
PRICE (X)	−46.9325	9.8564	22.6729
PRICE SQ. (X2)	3.9999	.8204	23.7729

OBSERVATION	OBS. X VALUE	OBS. Y VALUE	PREDICTED Y VALUE	LOWER 95% CL FOR E(Y)	UPPER 95% CL FOR E(Y)
1	5.0000	23.0000	21.4681	18.9971	23.9391
2	5.0000	20.0000	21.4681	18.9971	23.9391
3	5.0000	21.0000	21.4681	18.9971	23.9391
4	5.5000	19.0000	19.0018	16.8414	21.1622
5	5.5000	21.0018	19.0018	16.8414	21.1622
6	5.5000	18.0000	19.0018	16.8414	21.1622
7	6.0000	18.0000	18.5356	16.3039	20.7673
8	6.0000	17.0000	18.5356	16.3039	20.7673
9	6.0000	20.0000	18.5356	16.3039	20.7673
10	6.5000	21.0000	20.0693	17.9089	22.2297
11	6.5000	19.0000	20.0693	17.9089	22.2297
12	6.5000	20.0000	20.0693	17.9089	22.2297
13	7.0000	25.0000	23.6031	21.1343	26.0719
14	7.0000	24.0000	23.6031	21.1343	26.0719
15	7.0000	22.0000	23.6031	21.1343	26.0719

OBSERVATION	OBS. X VALUE	OBS. Y VALUE	PREDICTED Y VALUE	LOWER 95% CL FOR Y	UPPER 95% CL FOR Y
1	5.0000	23.0000	21.4681	18.1736	24.7626
2	5.0000	20.0000	21.4681	18.1736	24.7626
3	5.0000	21.0000	21.4681	18.1736	24.7626
4	5.5000	19.0000	19.0018	15.9333	22.0703
5	5.5000	21.0000	19.0018	15.9333	22.0703
6	5.5000	18.0000	19.0018	15.9333	22.0703
7	6.0000	18.0000	18.5356	15.4166	21.6546
8	6.0000	17.0000	18.5356	15.4166	21.6546
9	6.0000	20.0000	18.5356	15.4166	21.6546
10	6.5000	21.0000	20.0693	17.0008	23.1378
11	6.5000	19.0000	20.0693	17.0008	23.1378
12	6.5000	20.0000	20.0693	17.0008	23.1378
13	7.0000	25.0000	23.6031	20.3102	26.8960
14	7.0000	24.0000	23.6031	20.3102	26.8960
15	7.0000	22.0000	23.6031	20.3102	26.8960

a. The prediction equation for estimating y, the predicted number of cases which will be sold per month per 10,000 population in a sales area in which the price per unit is given in terms of x, is

$$\hat{y} = 156.1306 - 46.9325x + 3.9999x^2$$

b. At $x = \$5.00$, we have

$$\hat{y} = 156.1306 - 46.9325(5.0) + 3.9999(25.0) = 21.4681$$

or more than 21 cases. Similarly, at other unit prices, we find the following as estimates of y:

$$\hat{y} = 18.5356 \quad \text{when} \quad x = \$6.00$$

$$\hat{y} = 23.6031 \quad \text{when} \quad x = \$7.00$$

The model for estimating a particular outcome y or the mean response $E(y)$ is the same. For example, the estimated mean number of cases sold per month per 10,000 population when the price per fifth is \$5.00 is $\hat{E}(y) = 21.4681$. Similarly, when $x = \$6.00$ and $x = \$7.00$, the mean estimates are, respectively, $\hat{E}(y) = 18.5356$ and $\hat{E}(y) = 23.6031$.

c., d. The 95% prediction intervals for y and the 95% confidence intervals for $E(y)$ can be read directly from the computer printout. They are given in the table.

PRICE PER FIFTH	95% PREDICTION INTERVAL FOR y	95% CONFIDENCE INTERVAL FOR $E(y)$
\$5.00	18.1736 to 24.7626	18.9971 to 23.9391
6.00	15.4166 to 21.6546	16.3039 to 20.7673
7.00	20.3102 to 26.8960	21.1343 to 26.0719

Notice that in each case the 95% prediction interval for y is wider than the 95% confidence interval for $E(y)$. As we noted in sections 11.6 and 11.7, this is a reflection of the fact that the variance of the error of predicting a particular value y exceeds the variance of the error of estimating the mean value $E(y)$. In addition, since each of these variances depends on the particular values selected for the independent variables in computing y, the three prediction intervals and the three confidence intervals vary in width.

As we have seen, the prediction equation is of value for estimating $E(y)$ and for predicting some future value of y. But it may also help us to understand the process under study. For example, consider the prediction equation for the sales of the mail-order firm. An easy way to view this relationship is to graph the holiday sales y as a function of the number of mailings x_1 for various dates of mailings, say $x_2 = 1$, 2, or 3 months prior to Christmas. For example, substituting $x_2 = 1$ month into the prediction equation, we obtain

$$\hat{y} = 1.3 + 1.7x_1 - .1x_1^2 + 1.0x_1x_2 - .3x_1^2x_2$$
$$= 1.3 + 1.7x_1 - .1x_1^2 + 1.0x_1(1) - .3x_1^2(1)$$
$$= 1.3 + 2.7x_1 - .4x_1^2$$

This gives the predicted sales if the mailings are made $x_2 = 1$ month before Christmas. Similarly, for $x_2 = 2$ months and $x_2 = 3$ months, we obtain the following two prediction equations:

$$\text{for} \quad x_2 = 2: \quad \hat{y} = 1.3 + 3.7x_1 - .7x_1^2$$
$$\text{for} \quad x_2 = 3: \quad \hat{y} = 1.3 + 4.7x_1 - 1.0x_1^2$$

These three equations give the predicted holiday sales as a function of the

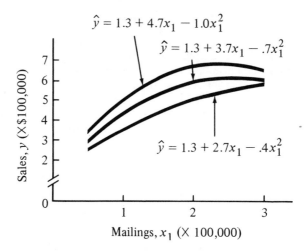

$$\hat{y} = 1.3 + 4.7x_1 - 1.0x_1^2$$

$$\hat{y} = 1.3 + 3.7x_1 - .7x_1^2$$

$$\hat{y} = 1.3 + 2.7x_1 - .4x_1^2$$

Sales, y (×$100,000)

Mailings, x_1 (× 100,000)

Figure 12.5 *A graph of holiday sales y as a function of the number of mailings x_1*

number of mailings x_1. Graphs of these three sales curves are shown in figure 12.5.

Note how the shapes of the sales curves change for the three values of x_2. This tells us that the relationship between predicted holiday sales y and the number of mailings x_1 is dependent on the time of mailing x_2. When this situation occurs, we say that x_1 and x_2 **interact**. Or putting it another way, the effect of x_1 on the predicted value of y is dependent on the value of x_2 (or vice versa). This example illustrates how graphs of $\hat{y}$ help us to understand the relationship between the predicted value of y and a set of predictor variables.

Example 12.9 A study was undertaken to examine the profit y per sales dollar earned by a construction company and its relationship to the size x_1 of the construction contract and the number x_2 of years of experience of the construction superintendent. An additional purpose of the study was to investigate the interaction effect of the size of the contract and the experience of the construction superintendent on profit. Data were obtained from a sample of $n = 18$ construction projects undertaken by the construction company over the past two years. These data are shown in the accompanying table.

Fit the model

$$y = \beta_0 + \beta_1 x_1 + \beta_2 x_2 + \beta_3 x_2^2 + \beta_4 x_1 x_2 + \epsilon$$

to the construction contract data. Carefully interpret your analysis. Graph the profit y per sales dollar as a function of contract size x_1 for the three levels of experience x_2. What does this graph suggest about the effect of the interaction between contract size and superintendent experience?

Solution Using a standard regression analysis program, the results that follow were obtained:

PROFIT, y	CONTRACT SIZE, x_1	SUPERINTENDENT EXPERIENCE, x_2 (years)	PROFIT, y	CONTRACT SIZE, x_1	SUPERINTENDENT EXPERIENCE, x_2 (years)
2.0	5.1	4	5.0	4.3	6
3.5	3.5	4	6.0	2.9	2
8.5	2.4	2	7.5	1.1	2
4.5	4.0	6	4.0	2.6	4
7.0	1.7	2	4.0	4.0	6
7.0	2.0	2	1.0	5.3	4
2.0	5.0	4	5.0	4.9	6
5.0	3.2	2	6.5	5.0	6
8.0	5.2	6	1.5	3.9	4

MULTIPLE R   .9289
R SQUARE   .8628
STD. ERROR OF EST.   .9708

ANALYSIS OF VARIANCE

	DF	SUM OF SQUARES	MEAN SQUARE	F RATIO
REGRESSION	4	77.026	19.256	20.432
RESIDUAL	13	12.252	.942	

INDIVIDUAL ANALYSIS OF VARIABLES

VARIABLE	COEFFICIENT	STD. ERROR	F VALUE
(CONSTANT	19.3048)		
CONT. SIZE (X1)	−1.4874	1.1779	1.5944
EXPER. (X2)	−6.3707	1.0424	37.3506
SIZE SQ. (X1 SQ.)	−.7521	.2253	11.1460
SIZE × EXP. (X1X2)	1.7169	.2538	45.7593

An analysis of the most important quantities appearing in the computer output follows.

R^2 *(coefficient of determination).* The proportion of variation in profit per sales dollar explained by the model involving contract size and superintendent experience is

$$R^2 = .8628$$

Thus approximately 14% of the variation in profit per sales dollar is left unexplained, an amount that might be reduced by the inclusion of additional predictive variables not included in our analysis.

F ratio. The F ratio provides a test of the hypothesis

$$H_0: \beta_1 = \beta_2 = \beta_3 = \beta_4 = 0$$

The calculated F ratio from the data, $F = 20.432$, exceeds the critical value of $F_{.05}$ based on $v_1 = 4$ and $v_2 = 13$ degrees of freedom, which is $F_{.05} = 3.18$. So we reject the null hypothesis and conclude that our chosen model contributes information for the prediction of y.

Variable coefficients. The prediction equation relating profit per sales dollar to the size x_1 of the contract and the experience x_2 of the construction superintendent is

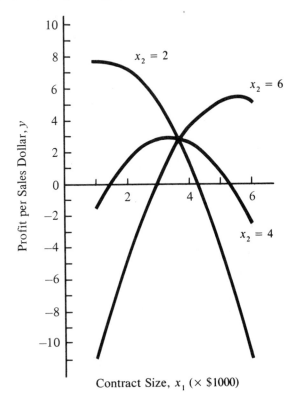

Figure 12.6 *A graph of the profit y per sales dollar as a function of the size of the contract x_1*

$$\hat{y} = 19.3048 - 1.4874x_1 - 6.3707x_2 - .7521x_1^2 + 1.7169x_1x_2$$

F-value. Each computed *F*-value provides a separate test of the hypothesis

$$H_0: \beta_j = 0 \qquad j = 1, 2, 3, 4$$

for its associated regression parameter β_j. The tabulated $F_{.05}$ with $v_1 = 1$ and $v_2 = 13$ degrees of freedom is $F_{.05} = 4.67$. Thus the parameters β_2, β_3, and β_4 are assumed to be significantly different from 0, but β_1 is not. Recall that this does *not* imply that the term x_1 contributes no information for the prediction of *y*. It could mean that the contribution of x_1 is small when in the presence of the other predictors x_1^2, x_2, and x_1x_2.

The *F*-value associated with the contract size and construction superintendent term (x_1x_2) is $F = 45.7593$, which greatly exceeds the critical value $F_{.05} = 4.67$. There appears to be a strong interaction relationship between these two factors and profit per sales dollar. This is especially evident when we view figure 12.6, which provides a graph of the separate functions relating profits *y* to the size x_1 of the contract for the three different levels of superintendent experience x_2.

The three separate functions appearing in figure 12.6 are as follows:

for $x_2 = 2$: $\hat{y} = 6.5634 + 1.9464x_1 - .7521x_1^2$

for $x_2 = 4$: $\hat{y} = -6.1780 + 5.3802x_1 - .7521x_1^2$

for $x_2 = 6$: $\hat{y} = -18.9194 + 8.8140x_1 - .7521x_1^2$

Notice that the profit per sales dollar declines rapidly as the size of the contract increases for superintendents with $x_2 = 2$ years of experience. The opposite effect occurs for superintendents with $x_2 = 6$ years of experience. On the basis of these sample contract data, an appropriate company policy is to assign construction superintendents with a few years of experience only to the smaller jobs while assigning those with the most experience to construction jobs with larger-sized contracts.

12.10 Some Comments on Formulating a Model

The independent variables that contribute information for the prediction of y can be of two types, quantitative and qualitative. As you will subsequently see, the way you enter an independent variable into a prediction equation depends on its type.

Definition

A **quantitative independent variable** is a variable that can take values corresponding to the points on a real line. Independent variables that are not quantitative are said to be **qualitative**.

Interest rate, employment rate, numbers of employees, and number of machines are four examples of quantitative independent variables. In contrast, suppose you own four similar plants manufacturing the same product and that the response of interest is the plant profit per unit of time. Four different plant supervisors, call them A, B, C, and D, manage the four plants, one assigned to each plant. Certainly, plant supervisor is an independent variable that may affect plant profit. Consequently, plant supervisor is a qualitative independent variable that we enter into a prediction equation to model the plant profit y.

Definition

The intensity setting of an independent variable is called a **level**.

For quantitative independent variables the levels correspond to the values that these independent variables may assume and therefore correspond to points on a line. For example, if you think that interest rates may affect your response, and the response y is recorded for three interest rates, 6%, 8%, and 9.2%, then you have observed the independent variable "interest rate," at three levels, 6%, 8%, and 9.2%.

The levels for qualitative independent variables are not quantifiable and therefore do not correspond to points on a line. They can only be defined by describing them. For the manufacturing profit example given above, the independent variable "plant supervisor" is observed at four levels, each level corresponding to one of the supervisors A, B, C, or D.

A good way to portray a model for a response y that is a function of a single quantitative predictor variable is to graph $E(y)$ (or $\hat{y}$) as a function of x, as we did in chapter 11. The resulting curve (or line) is called a **response curve**.

Just as the equation expressing $E(y)$ (or $\hat{y}$) as a function of a single quantitative predictor variable traces a curve on a sheet of paper, the corresponding equation involving two (or more) quantitative predictor variables graphs a **response surface** in a three- (or higher) dimensional space.

The two most commonly used models employing quantitative predictor variables are called first-order and second-order linear models. A first-order model, given by the following equation, graphs as a response plane.

Figure 12.7 *The response surface for a first-order linear model*

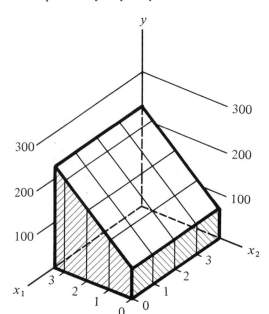

A First-Order Linear Model

$$y = \beta_0 + \beta_1 x_1 + \beta_2 x_2 + \cdots + \beta_k x_k + \epsilon$$

where x_1, x_2, ..., x_k are quantitative predictor variables and ϵ is a random error.

A fitted first-order response surface depicting the relationship between the price y of a stock and two quantitative predictor variables

$$x_1 = \text{annual dividends of the stock}$$

$$x_2 = \text{earnings per share of the stock}$$

is shown in figure 12.7.

Second-order linear models in k quantitative predictor variables x_1, x_2, ..., x_k include all the terms contained in a first-order model plus all two-way cross product terms $x_1 x_2$, $x_1 x_3$, ..., $x_2 x_3$, ... and all terms involving the squares, x_1^2, x_2^2, ..., x_k^2. Thus a second-order model in two predictor variables is given by the following equation:

A Second-Order Linear Model in Two Predictor

Variables

$$y = \beta_0 + \beta_1 x_1 + \beta_2 x_2 + \beta_3 x_1 x_2 + \beta_4 x_1^2 + \beta_5 x_2^2 + \epsilon$$

where x_1 and x_2 are quantitative predictor variables and ϵ is a random error.

The response surface for a second-order model can be curved (induced primarily by the terms involving x_1^2, x_2^2, ..., x_k^2) and it also can be warped (or twisted). Warping of the surface is caused by the terms (called interaction terms) containing the cross products $x_1 x_2$, $x_1 x_3$, ... A second-order response surface, again reflecting the relationship between stock price y and the two predictor variables x_1, the annual dividends, and x_2, the earnings per share, is shown in figure 12.8.

Qualitative independent variables are entered into a model using dummy variables, the number of dummy variables always being one less than the number of levels (categories) associated with the independent variable. For example, if the sales y of a retailing firm depends on the variable "location," and if there are three locations, A, B, and C, the first few terms of the model are

$$y = \beta_0 + \beta_1 x_1 + \beta_2 x_2 + \left\{ \begin{array}{l} \text{terms associated with} \\ \text{other predictor variables} \end{array} \right\} + \epsilon$$

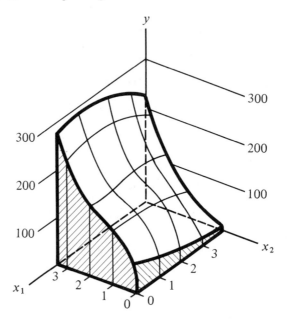

Figure 12.8 *The response surface for a second-order linear model*

where

$$x_1 = 1 \text{ if the response is at location B} \qquad x_1 = 0 \text{ if not}$$
$$x_2 = 1 \text{ if the response is at location C} \qquad x_2 = 0 \text{ if not}$$

LOCATION	x_1	x_2
A	0	0
B	1	0
C	0	1

Then the codings for the three locations are as shown in the table. Thus when a response measurement is taken at location A, we let $x_1 = 0$ and $x_2 = 0$. A dummy variable was used in example 12.2 to account for marital status in the prediction of television-viewing time, and dummy variables also appear in other worked examples in this chapter.

The formulation of the probabilistic model is perhaps the most important part of a regression analysis. Why? Because even if you have all the information contributing to the predictor variables in the model, you may obtain a very poor fit to a set of data if you have not properly formulated the model.

For example, suppose you wish to fit a linear model to the data points shown in figure 12.9(a) and you think one predictor variable x contributes most of the information for the prediction of y. If you fit the first-order model

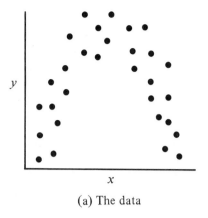

(a) The data

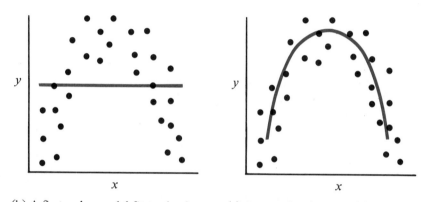

(b) A first-order model fit to the data   (c) A second-order model fit to the data

Figure 12.9 *A comparison of first-order and second-order models fit to the same set of data*

$$y = \beta_0 + \beta_1 x + \epsilon$$

to the data, you can see in figure 12.9(b) that you obtain a very poor fit. In contrast, a second-order model

$$y = \beta_0 + \beta_1 x + \beta_2 x^2 + \epsilon$$

provides a good fit to the data [figure 12.9(c)].

Similarly, if you fit a first-order linear model to a set of data associated with a curved or warped response surface, you will obtain a poor fit. The only way you can improve the fit is to use a second- (or higher) order linear model that will adjust to the curvatures and warpages of the actual response surface.

Learning to formulate the appropriate linear model for a given application takes experience. You will gain some insight into model building by studying the worked examples in this chapter. More information on this important topic can be found in the texts listed in the references.

Exercises

12.15.   To better understand the models you might construct for the mean response $E(y)$, graph the following second-order polynomials:

a. $E(y) = 2x^2$ b. $E(y) = -2x^2$ c. $E(y) = 1 - 2x + x^2$
d. $E(y) = 1 + 2x + x^2$ e. $E(y) = 5 + 2x + x^2$

12.16.   Graph these third-order polynomials:

a. $E(y) = 1 - 2x + x^2 - 3x^3$ b. $E(y) = 1 - 2x - x^2 + 3x^3$

12.17.   Suppose that the mean response for given values of x_1 and x_2 is given by the equation

$$E(y) = 3 - x_1 + 2x_2$$

Graph the contour curves (lines) of $E(y)$ as a function of x_1 for values of x_2 equal to 0, 1, and 2.

12.18.   Suppose that the mean response for given values of x_1 and x_2 is given by the equation

$$E(y) = 4 - 2x_1 + x_2 + 4x_1^2 + 2x_2^2$$

Graph the contour curves of $E(y)$ as a function of x_1 (in the interval $0 \le x_1 \le 5$) for values of x_2 equal to 0, 1, and 2. Note that except for vertical shifts, the contour curves are portions of identical parabolas.

12.19.   Continuing exercise 12.18, add a cross product term to the model. Note that the shapes of the contour curves depend on the value assigned to x_2. For example, let

$$E(y) = 4 - 2x_1 + x_2 - 3x_1 x_2 + 4x_1^2 + 2x_2^2$$

Graph $E(y)$ for $x_2 = 0$, 1, and 2.

12.20.   The relationship between interest rates and the residential housing industry has been observed for some time. High interest rates cause the monthly payment on a home mortgage to be noticeably increased. For example, a 1% increase in the mortgage rate causes a $20.83 increase in the monthly payment on a 30-year $25,000 mortgage. Interest rates, in turn, are affected by economic factors such as the total money supply and the yield on U.S. government securities. Typically, interest rates rise when the money supply is tightened and when treasury bill yields rise.

Suppose you have been assigned the task of investigating the relationship between new residential housing starts y and the interest rates x_1 on conventional 30-year first mortgages, the total money supply x_2, and the yield x_3 on U.S. government securities for each of the past 36 months. Write a second-order model to represent this relationship. Would you expect the interaction terms to be important factors in this analysis? Explain.

12.11 Model Building: Testing Portions of a Model

This section is concerned with a test of an hypothesis that one or more β parameters in the model equal 0. For example, one of the tests described in the computer printout for the television viewer data of section 12.4 used the F statistic to test the hypothesis that $\beta_1 = \beta_2 = \beta_3 = 0$. The F statistic was also used to test an hypothesis that an individual β parameter equals 0. In this section we will explain in greater detail the reasoning behind the test and describe several situations in which the test can be applied.

Suppose you have a model

$$y = \beta_0 + \beta_1 x_1 + \beta_2 x_2 + \cdots + \beta_k x_k + \epsilon$$

or, equivalently,

$$E(y) = \beta_0 + \beta_1 x_1 + \beta_2 x_2 + \cdots + \beta_k x_k$$

and you want to know whether certain variables contribute information for the prediction of y. Or putting it another way, you are asking whether those terms should be included in the model. If a set of x's in the model contributes no information for the prediction of y, then the β parameters associated with those x's should equal zero. Consequently, testing to determine whether a set of terms should be included in the model is a test of an hypothesis that a set of β parameters equals zero.

Suppose we have two models for $E(y)$, one which we will call the "complete model" (call this model 2) and another which we will call the "reduced model" (call this model 1). The reduced model includes only a portion of the terms in the complete model. Thus the complete model contains all the terms in the reduced model plus some additional terms. The purpose of the test is to test the hypothesis that the β parameters associated with the additional terms equal 0. Or putting it another way, we test to see whether the additional terms contribute information for the prediction of y.

We represent the reduced and complete models as follows:

model 1 (reduced model): $\quad E(y) = \beta_0 + \beta_1 x_1 + \beta_2 x_2 + \cdots + \beta_g x_g$

model 2 (complete model): $\quad E(y) = \beta_0 + \beta_1 x_1 + \beta_2 x_2 + \cdots + \beta_g x_g$
$$+ \beta_{g+1} x_{g+1} + \cdots + \beta_k x_k$$

Note that the complete model contains, in addition to all the terms in the reduced model, the terms $\beta_{g+1} x_{g+1}, \beta_{g+2} x_{g+2}, \ldots, \beta_k x_k$.

The test we describe here is intuitive. We use the method of least squares to fit the reduced model and calculate the sum of squares for error SSE_1 (the sum of squares of the deviations between the y-values and the fitted model). Then we fit the complete model and find the sum of squares for error SSE_2. Then we compare the two sums of squares for error, SSE_1 and SSE_2. If the variables $x_{g+1}, x_{g+2}, \ldots, x_k$ contribute information for the prediction of y, then SSE_2 should be significantly smaller than SSE_1. That is, retaining these variables in the model should reduce the sum of squares of the errors of prediction. Consequently, the larger the difference $(SSE_1 - SSE_2)$, the greater the weight of evidence to indicate that the terms should be included in the model. Or, the greater will be the weight of evidence indicating that at least one of the parameters $\beta_{g+1}, \beta_{g+2}, \ldots, \beta_k$ differs from 0.

It can be shown that whenever we add terms to a model, we reduce the sum of squares for error. The question is whether the reduction in the sums of squares for error is just due to chance or whether it is due to information contributed by $x_{g+1}, x_{g+2}, \ldots, x_k$.

To test the hypothesis that these x variables contribute no information for the prediction of y (i.e., $\beta_{g+1} = \beta_{g+2} = \cdots = \beta_k = 0$), we use the test statistic

$$F = \frac{(SSE_1 - SSE_2)/(k - g)}{SSE_2/(n - k - 1)} = \frac{(\text{drop in SSE})/(k - g)}{s^2_{\text{complete model}}}$$

When the assumptions of the earlier sections of this chapter are satisfied—that is, the y-values are independent and normally distributed, with mean $E(y)$ and variance σ^2—the F statistic has an F distribution, with $v_1 = (k - g)$ and $v_2 = (n - k - 1)$ degrees of freedom. Note that $v_1 = (k - g)$ is equal to the difference in the number of parameters in the complete and reduced models. Also, $v_2 = n - (k + 1)$ is equal to the number n of data points less the number of parameters in the complete model.

As we stated earlier, the larger the drop in SSE (the numerator of the F statistic), the greater the weight of evidence favoring rejection of the null hypothesis and acceptance of the alternative hypothesis that at least one or more of the parameters $\beta_{g+1}, \beta_{g+2}, \ldots, \beta_k$ differ from zero. Consequently, we will reject the null hypothesis

$$H_0: \beta_{g+1} = \beta_{g+2} = \cdots = \beta_k = 0$$

when F is too large. That is, we use a one-tailed test and reject H_0 when F exceeds some critical value F_α, as shown in figure 12.10.

Example 12.10 Refer to example 12.9. The second-order model

$$y = \beta_0 + \beta_1 x_1 + \beta_2 x_2 + \beta_3 x_1^2 + \beta_4 x_1 x_2 + \epsilon$$

Figure 12.10 *Rejection region for the F test for* $H_0: \beta_{g+1} = \beta_{g+2} = \cdots = \beta_k = 0$

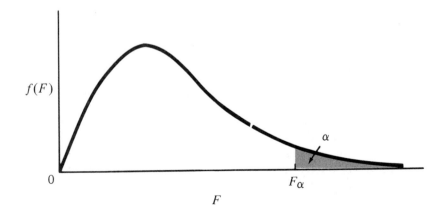

was fitted to the data relating the profit y per sales dollar to the size x_1 of the construction contract and the years x_2 of experience of the construction superintendent for $n = 18$ different projects undertaken by a construction firm. Test the hypothesis

$$H_0 : \beta_3 = \beta_4 = 0$$

that is, that the second-order model does not provide an improvement over the first-order model for the prediction of profit y per sales dollar.

Solution A standard regression analysis program was used to fit the first-order (reduced) model

$$y = \beta_0 + \beta_1 x_1 + \beta_2 x_2 + \epsilon$$

to the data set of example 12.9. The results were as follows:

```
MULTIPLE R           .5731
R SQUARE             .3284
STD. ERROR OF EST.   1.9993
```

ANALYSIS OF VARIANCE

	DF	SUM OF SQUARES	MEAN SQUARE	F RATIO
REGRESSION	2	29.321	14.660	3.668
RESIDUAL	15	59.957	3.997	

INDIVIDUAL ANALYSIS OF VARIABLES

VARIABLE	COEFFICIENT	STD. ERROR	F VALUE
(CONSTANT	8.0136)		
CONT. SIZE	−1.3548	.5530	6.0012
EXPER.	.4626	.4346	1.1333

The computer printout for the *reduced* model shows that

$$SSE_1 = 59.957$$

From the computer printout for the *full* model, shown in the solution to example 12.9, we find

$$SSE_2 = 12.252$$

Since $n = 18$, $k = 4$, and $g = 2$, our test statistic is

$$F = \frac{(SSE_1 - SSE_2)/(k - g)}{SSE_2/(n - k - 1)} = \frac{(59.957 - 12.252)/(4 - 2)}{12.252/13}$$

$$= \frac{23.853}{.942} = 25.322$$

This calculated value of F exceeds the critical value $F_{.05} = 3.81$ (with $v_1 = 2$ and $v_2 = 13$ degrees of freedom). So we reject the null hypothesis and conclude that the second-order model does in fact provide an improvement over the first-order model for the prediction of profit y per sales dollar as a function of the size x_1 of the construction contract and the experience x_2 of the construction superintendent.

Exercises

12.21. The general sales manager of a firm that sells hotel and restaurant supplies is interested in the predictive relationship of advertising and the size of the sales force on the monthly sales volume. To study this issue the manager recorded the sales volume y, the amount x_1 spent in direct mail advertising, and the number x_2 of sales representatives in each of 12 randomly selected sales territories for the past year. The data are shown in the table.

y ($\times$ \$10,000)	x_1 ($\times$ \$1,000)	x_2	y ($\times$ \$10,000)	x_1 ($\times$ \$1,000)	x_2
25.9	9.75	3	30.3	12.20	3
27.1	9.20	2	31.0	13.86	4
27.9	10.00	3	31.7	13.50	3
29.0	11.04	3	33.0	14.20	4
29.8	10.80	2	34.1	16.30	4
30.2	12.00	3	36.2	17.50	4

The model

$$y = \beta_0 + \beta_1 x_1 + \beta_2 x_2 + \beta_3 x_1 x_2 + \epsilon$$

was then fitted to the data by using the BMD-P1R program (see section 12.12), providing the results that follow.

```
REGRESSION TITLE . . . . . . . . . . . . . . BMD REGRESSION #3
DEPENDENT VARIABLE . . . . . . . . . . . . 4
TOLERANCE . . . . . . . . . . . . . . . . . . .0100
```

ALL DATA CONSIDERED AS A SINGLE GROUP

MULTIPLE R	.9809	STD. ERROR OF EST.	.6734
MULTIPLE R-SQUARE	.9621		

ANALYSIS OF VARIANCE

	SUM OF SQUARES	DF	MEAN SQUARE	F RATIO	P(TAIL)
REGRESSION	92.110	3	30.703	67.717	.0001
RESIDUAL	3.627	8	.453		

VARIABLE		COEFFICIENT	STD. ERROR	STD. REG. COEFF.	T	P(2TAIL)
INTERCEPT		14.9600				
X1	1	1.5321	.5910	1.3582	2.5925	.0345
X2	2	−.4323	1.7964	−.1052	.2407	.8870
X1X2	3	−.0553	.1554	−.3165	.3557	.7146

a. Provide a complete analysis of the printout.
b. Use the methods of chapter 11 to fit the simple first-order model

$$y = \beta_0 + \beta_1 x_1 + \epsilon$$

to the data for y and x_1. Compute SSE for this first-order model.

c. Use the computer analysis and the results from part b to determine if sales are adequately described by the amount spent on advertising. That is, for the model

$$y = \beta_0 + \beta_1 x_1 + \beta_2 x_2 + \beta_3 x_1 x_2 + \epsilon$$

test the hypothesis $H_0 : \beta_2 = \beta_3 = 0$.

12.22. Suppose we wish to test an hypothesis that certain variables in a regression model are insignificant in their collective ability to predict the dependent variable y in the presence of the other variables. Why do misleading conclusions sometimes result if we test this hypothesis solely on the basis of an examination of the F-value (or t-values) associated with each individual parameter?

12.12 A Summary of Tests and Estimation and Prediction Procedures for Multiple Regression

The preceding sections indicate that except for one additional test procedure (discussed in section 12.11), the same tests and estimation and prediction procedures are available for multiple regression as for simple linear regression. For a general linear model of the type

$$y = \beta_0 + \beta_1 x_1 + \beta_2 x_2 + \cdots + \beta_k x_k + \epsilon$$

the tests and estimation and prediction procedures are as follows:

1. A test of an hypothesis that one of the parameters, say β_i, equals 0. This can be a t test of the type used in chapter 11, where t is based on $(n - k - 1)$ degrees of freedom. (Note: n is the number of data points and k is the number of independent variables in the model.) Or you can use an F test as indicated in the computer printout for the television viewer data. The F statistic is based on $v_1 = 1$ and $v_2 = (n - k - 1)$ degrees of freedom (see the cautionary comments about interpretation in section 12.6).

2. A confidence interval for an individual regression parameter, say β_i. This confidence interval will be of the form

$$\hat{\beta}_i \pm t_{\alpha/2} s_{\hat{\beta}_i}$$

where the t statistic is based on $(n - k - 1)$ degrees of freedom and $s_{\hat{\beta}_i}$ is the estimated standard deviation of $\hat{\beta}_i$. This latter statistic is shown in the output of all common regression computer programs.

3. A confidence interval for the mean value of y for given values of x_1, $x_2, \ldots, x_k$. The estimate of $E(y)$, given by $\hat{y}$, is obtained by inserting the given values $x_1, x_2, \ldots, x_k$ into the fitted regression equation. A confidence interval for $E(y)$ is determined by a complicated formula, a topic beyond the scope of this text. Very few regression computer programs provide this important confidence interval as an option.

4. A prediction interval for some value of y to be observed in the future. The predicted value of y for specific values of $x_1, x_2, \ldots, x_k$ is obtained

Table 12.7 *Available options for some common regression programs*

PROGRAM AND SUPPLIER	STEPWISE OR STANDARD OUTPUT	t or F TESTS (see paragraph 1)	CONFIDENCE INTERVALS FOR MEAN OF y (see paragraph 3)	PREDICTION INTERVALS FOR PARTICULAR VALUES OF y (see paragraph 4)
BMD-P1R (UCLA Biomedical Computing Facilities)	standard	t	no	no
BMD-P2R (UCLA Biomedical Computing Facilities)	stepwise	t	no	no
BMD-O2R (UCLA Biomedical Computing Facilities)	stepwise	F	no	no
SSP-MULTR (IBM Corp.)	standard	t	no	no
SPSS-REGRESSION (University of Chicago)	optional	F	no	no
SAS (SAS Institute, Raleigh, N.C.)	optional	t	yes	yes
MINITAB (Pennsylvania State University)	standard	t	yes	yes

by substituting these values into the fitted regression equation. As noted in paragraph 3, the regression equation is computed by most packaged programs. The formula for a prediction interval for y is similar to the formula for the confidence interval (paragraph 3) and is very complex. Unfortunately, very few computer program packages provide this computation as an option.

5. **A test of the hypothesis that a set of one or more β parameters simultaneously equals 0.** This F test, which can be used to test an hypothesis that a single β parameter equals 0 (see paragraph 1), is primarily used in model building. It was presented in section 12.11.

A number of packaged regression analysis computer programs are available for computing the quantities listed above, and all of them are very easy to use. You should find out which regression computer programs are available through your campus computing center and familiarize yourself with the use of those programs. The outputs from the various packages are not identical, however. Therefore, you will need to exercise good judgment in directing the package to make the inferences that you desire. Some print a t-value for the test of an hypothesis that one of the parameters β_i equals 0, while some print an F-value; some print a confidence interval for the mean value of y for given values of x_1, x_2, ..., x_k and some do not. Table 12.7 provides a summary of the options available from some of the more common packaged regression computer programs.

12.13 Worked Examples

To give you additional practice in interpreting the results of a regression analysis, we include in this section several worked examples. The analyses (and hence the computer printouts) have been performed using different computer regression programs in order to acquaint you with a variety of computer outputs that you might encounter in practice.

Example 12.11 The sales manager for a pharmaceutical company is concerned about the complacency which seems to exist among the more experienced sales personnel employed by the firm. She has noticed that, up to a point, sales tend to level off and in some cases even tend to decrease with experience. To investigate this phenomenon the sales manager has recorded the territory sales over the past three months and the work experience for 10 of the firm's sales personnel. These data are shown in the table.

SALES, y ($\times$ $1000)	EXPERIENCE, x (years)	SALES, y ($\times$ $1000)	EXPERIENCE, x (years)
36.7	2.0	41.2	4.5
22.9	1.5	18.5	1.0
30.5	4.5	43.4	3.0
9.2	.8	25.5	2.3
38.4	3.5	28.4	5.5

a. Plot the relationship between territory sales and years of work experience.

b. Fit y to x using a second-order polynomial model. Does the second-order model appear to provide a good fit to these data?

Solution a. A plot depicting the relationship between sales y and experience x is shown in figure 12.11. This plot suggests that a second-order model of the form

Figure 12.11 *A plot of the sales and work experience data of example 12.11, with the fitted second-order model superimposed*

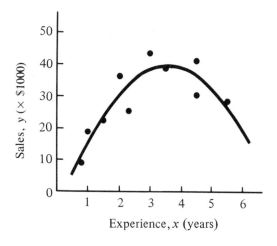

$$y = \beta_0 + \beta_1 x + \beta_2 x^2 + \epsilon$$

should provide an adequate fit to the data.

b. The BMD-P1R regression program was used to fit the second-order model to the sales and work experience data. The results follow.

```
REGRESSION TITLE . . . . . . . . . . . . . .  BMD REGRESSION #2
DEPENDENT VARIABLE . . . . . . . . . . . .  3Y
TOLERANCE . . . . . . . . . . . . . . . . . . .  .0100
```

ALL DATA CONSIDERED AS A SINGLE GROUP

MULTIPLE R	.8961	STD. ERROR OF EST.	5.4526
MULTIPLE R-SQUARE	.8029		

ANALYSIS OF VARIANCE

	SUM OF SQUARES	DF	MEAN SQUARE	F RATIO	P(TAIL)
REGRESSION	847.880	2	423.940	14.259	.00340
RESIDUAL	208.120	7	29.731		

VARIABLE		COEFFICIENT	STD. ERROR	STD. REG. COEFF.	T	P(2TAIL)
INTERCEPT		−6.173				
X1	1	25.332	5.439	3.770	4.657	.002
X2	2	−3.499	.872	−3.249	−4.014	.005

The most important quantities appearing in the output may be interpreted as shown in the paragraphs that follow.

R^2 *(coefficient of determination).* The proportion of variation in sales explained by the second-order model involving the explanatory variable "work experience" is

$$R^2 = .8029$$

Thus approximately 80% of the total variation in sales is explained by this model.

F ratio. The F ratio is 14.259 and the associated P(TAIL) probability is .0034. This is the probability of observing a value of F as large as 14.259 or larger; consequently, the P(TAIL) probability is the significance level of the observed F-value. Then we would reject the null hypothesis

$$H_0: \beta_1 = \beta_2 = 0$$

at the $\alpha = .0034$ level of significance (or any higher level—for example, $\alpha = .05$). This provides strong evidence that at least one, or both, of the terms $\beta_1 x$ and $\beta_2 x^2$ contributes information for the prediction of y.

Variable coefficients. The estimated second-order model relating sales y to work experience x is

$$\hat{y} = -6.173 + 25.332x - 3.499x^2$$

t-value. The listed t-values give the values of the Student's t and the associated P(2TAIL) probabilities (significance levels) for tests of the significance of the individual β parameters in the model. The first values, $t = 4.657$ and P(2TAIL) = .002, indicate that we would reject the hypothesis

$$H_0: \beta_1 = 0$$

at a level of significance of $\alpha = .002$. The second t-value has an analogous interpretation for β_2.

As noted previously, the most important portion of the computer output is the analysis of variance F test. This test tells us that the model used in this analysis contributes information for the prediction of y. At the same time, a value of R^2 as low as .8029 indicates there is much room for improving the model. It is doubtful that the addition to the model of a few higher-order terms in x will greatly increase R^2 (and improve the fit). It appears that it will be necessary to add to the model some of the many other variables that are related to sales.

Example 12.12 A county assessor wishes to develop a model to relate the market value of single-family residences in a community to the size in square feet of each residence and the number of bedrooms contained in the residence. The resulting prediction equation will then be used for assessing the values of single-family residences in the county to establish the amount each homeowner owes in property taxes. The market sales price y, the square feet x_1 of floor space, and the number x_2 of bedrooms were recorded for 20 single-family residences recently sold at fair market value. The data are given in the table.

y ($\times$ \$1000)	x_1 ($\times$ 100)	x_2	y ($\times$ \$1000)	x_1 ($\times$ 100)	x_2
20.9	8.40	2	32.6	14.75	3
21.7	8.65	2	34.7	17.50	4
22.5	10.25	3	35.8	19.20	4
24.0	9.60	2	38.9	18.70	4
26.1	10.50	2	40.0	18.50	3
27.3	11.25	3	42.1	19.00	4
27.5	14.70	3	44.6	17.50	3
29.2	12.80	2	47.9	19.50	4
30.4	13.85	3	49.3	19.00	4
30.5	15.00	4	52.4	19.45	4

An SAS regression program was used to fit the second-order model

$$y = \beta_0 + \beta_1 x_1 + \beta_2 x_2 + \beta_3 x_1^2 + \beta_4 x_2^2 + \beta_5 x_1 x_2 + \epsilon$$

to the data set. The printout obtained from the analysis is shown here.

SAS
MULTIPLE REGRESSION
GENERAL LINEAR MODELS PROCEDURE

DEPENDENT VARIABLE: SALES PRICE

SOURCE	DF	SUM OF SQUARES	MEAN SQUARE	F VALUE
MODEL	4	1519.791	379.948	23.495
ERROR	15	242.569	16.171	
CORRECTED TOTAL	19	1762.360		

R-SQUARE	C.V.	STD. DEV.	S. PRICE MEAN
.8623	11.8555	4.0214	33.9200

PARAMETER	ESTIMATE	T FOR HO: PARAMETER = 0	PR > |T|	STD. ERROR OF ESTIMATE
INTERCEPT	24.6970			
FL. SPACE (X1)	−1.2520	−.5096	.5873	2.4569
X1 SQUARE	.0933	.8771	.4250	.1064
X2 SQUARE	−1.1157	−1.1225	.2465	.9939
(X1)(X2)	.3565	.8317	.4345	.4286

VARIABLE NO. BDRMS. (X2) REDUNDANT AND DELETED FROM PRINTOUT.

Give an analysis of the printout and construct a contour diagram for the second-order model fitted to the same data by the computer.

Solution It is interesting to note that the user can control the SAS program (as well as most other regression programs) so that variables that are clearly redundant are not included in the model. In this program variable x_2, number of bedrooms, is eliminated from the model since its associated t-value did not exceed a predetermined cutoff value, $t = .1$. Although the selection of a control value for the individual t- or F-values is arbitrary, the values $t = .1$ and $F = .01$ are often selected to identify as redundant variables those for which the individual t- or F-values do not exceed the control value.

Even though x_2 is not included as an independent variable in the model, second-order terms involving x_2, namely x_2^2 and the interaction term $x_1 x_2$, are included as nonredundant predictors.

Other interesting observations from the printout are listed in the paragraphs that follow.

R-SQUARE *(coefficient of determination).* Since the coefficient of determination R^2 is .8623, we can say that 86.23% of the variability in sales price for the 20 residences is explained by a second-order model involving x_1 = square footage and x_2 = the number of bedrooms in the residence.

F-value. The F VALUE = 23.495 indicates that we must reject the hypothesis that all the parameters $\beta_1 = \beta_2 = \cdots = \beta_5$ in the second-order model

$$y = \beta_0 + \beta_1 x_1 + \beta_2 x_2 + \beta_3 x_1^2 + \beta_4 x_2^2 + \beta_5 x_1 x_2 + \epsilon$$

equal 0 since the F-value exceeds 4.89, the tabulated F statistic with 4 and 15 degrees of freedom ($\alpha = .01$). Thus at least one of the parameters is significantly different from zero.

Parameter estimates. The fitted second-order model relating sales price to square footage and number of bedrooms for the 20 residences is

$$\hat{y} = 24.6970 - 1.2520 x_1 + .0933 x_1^2 - 1.1157 x_2^2 + .3565 x_1 x_2$$

T FOR HO: PARAMETER = 0. Each t-value tests the hypothesis

$$H_0 : \beta_j = 0$$

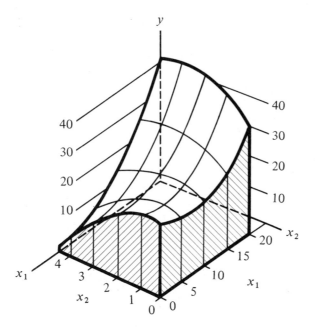

Figure 12.12 *Contour diagram for the second-order model fitted to the data of example 12.12*

for its associated regression parameter β_j. Recall, however, that this is not simply a test of significance of the associated term x_j as a predictor of y. Since the tabulated t with 15 degrees of freedom is 2.602 ($\alpha = .01$), we would not reject H_0 for any of the four parameters of the model, a clear contradiction of the result suggested by an analysis of variance F test. What is implied is that none of the terms x_1, x_1^2, x_2^2, and $x_1 x_2$ is significant as a predictor of y *in the presence of the other three terms* (probably because of a duplication of information content). However, when considered individually any or all of the four terms may be valuable predictors of market sales price y. A contour diagram for the fitted second-order model

$$\hat{y} = 24.6970 - 1.2520x_1 + .0933x_1^2 - 1.1157x_2^2 + .3565x_1 x_2$$

relating sales price to square footage and the number of bedrooms is shown in figure 12.12.

12.14 Summary

A multiple regression analysis is an extension of the simple regression analysis of chapter 11 to the case in which the response variable y is related to a number of predictor variables x_1, x_2, ..., x_k. All the tests and estimation and prediction procedures of chapter 11 are applicable to the general linear model of chapter 12. Even the multiple coefficient of correlation R and the multiple coefficient of determination R^2 have meanings similar to those for the simple coefficient of correlation r and coefficient of determination r^2 encountered in chapter 11.

The major difference between simple and multiple regression analyses is the utility of the latter. Very few response variables in business are adequately modeled by the simple linear probabilistic model

$$y = \beta_0 + \beta_1 x + \epsilon$$

of chapter 11. In contrast, the multiple regression model of chapter 12, when properly constructed, provides a good model for many business response variables. The resulting prediction equation often provides a good estimator of the mean response for given values of the predictor variables or provides a good predictor of values of y that might be observed in the future.

Supplementary Exercises

12.23. Discuss the following terms as they relate to the multivariable predictor model:
a. the least squares equations
b. quantitative independent variable
c. qualitative independent variable
d. interaction terms in a regression model
e. a second-order regression model

12.24. What is meant by model building in regression analysis? Why is model building important?

12.25. The stock market has undergone considerable change over the past ten years. As a result, novel approaches have been developed to create more interest in markets by providing more efficient dealer services to investors. One important issue concerns the rate of dealer participation to variations in bid-ask spreads. Some researchers* examined this issue by regressing

y = dealer participation measured by (dealer purchases + sales)

on

$$x_1 = \text{bid-ask spread} \times 1000$$

$$x_2 = \text{average price of security in 1961}$$

$$x_3 = (\text{high price} - \text{low price})/x_2$$

$$x_4 = \text{activity factor}$$

for $n = 65$ equity issues sold in the over-the-counter market. Their regression analysis provided the results shown in the table.
a. Does dealer participation appear to be responsive to the bid-ask spread in the presence of the other variables in the model? (State and test the appropriate hypothesis.)

*S. Tinic and R. West, "Competition and the Pricing of Dealer Service in the Over-the-Counter Stock Market," *Journal of Financial and Quantitative Analysis* (June 1972).

VARIABLE	COEFFICIENT	STANDARD DEVIATION	t-VALUE
constant	12759.050		
x_1	−.905	.354	−2.557
x_2	33.227	6.893	4.821
x_3	298.117	1181.811	.252
x_4	706.343	47.621	14.833

$R^2 = .812$ *Standard Error* $= 2730.070$ $F = 64.760$

b. Does the first-order model appear to exhibit a good fit to the data examined by the researchers?

c. Provide an analysis of the remaining entries contained in the regression printout.

12.26. A labor union representative is interested in creating a multivariable prediction model to predict the hourly wage of union clerical workers using the employee's age, her or his years of work experience, and the number of years the employee has been a union member. If data are available on 75 male and female clerical workers employed by five different firms in two different states, discuss the type of independent variables you would recommend to use in the study and describe the nature of the levels which will be measured on each.

12.27. Does product competition always result in lower prices to the consumer? Some marketing specialists have suggested that the relationship between price and the number of competing products is of second order—that is, that prices are high in the presence of either few or many competitors and lowest when there is a moderate number of competitors. To study this issue a record was made of the average price for a six-pack of beer and the number of competing brands sold in a certain chain of supermarkets in 12 different cities. The data are shown in the table.

PRICE, y	NUMBER OF COMPETING BRANDS, x	PRICE, y	NUMBER OF COMPETING BRANDS, x
$1.25	4	$1.56	10
1.40	7	1.34	5
1.60	2	1.45	8
1.52	9	1.41	2
1.40	3	1.27	7
1.35	6	1.55	8

Fit the second-order model $y = \beta_0 + \beta_1 x + \beta_2 x^2 + \epsilon$ to these data by solving the least squares equations, as was done in example 12.1, or by using an available regression analysis program. Carefully interpret your results. Do these data suggest that the number x of competing brands can be used to predict the price y for the product involved in this study?

12.28. New product research provides the lifeblood for most manufacturing organizations. But how much should the firm spend to develop and promote new products? The answer depends on the rate of return expected from new ventures and the amount invested in the development and promotion of the new product. The marketing research director for a food products company had recorded the rate of return and the amount invested in the development and promotion of 10 new

RATE OF RETURN, y	INVESTMENT, X ($\times$ \$100,000)	RATE OF RETURN, y	INVESTMENT, X ($\times$ \$100,000)
4.0	1.1	5.6	2.5
9.2	3.5	8.0	3.1
6.0	4.7	6.1	5.9
1.8	1.0	3.2	2.1
7.7	5.2	7.8	4.6

snack food products introduced by his firm over the past 10 years. The data are shown in the table. Fit the second-order model $y = \beta_0 + \beta_1 x + \beta_2 x^2 + \epsilon$ by solving the least squares equations, as was done in example 12.1, or by using an available packaged regression analysis program. Carefully interpret the results of your analysis.

12.29. The production manager of a plant that manufactures a chemical fertilizer recorded the marginal cost of production at various levels of output for 12 randomly

COST, y (per 100 lb)	OUTPUT, X ($\times$ 1000 lb)	COST, y (per 100 lb)	OUTPUT, X ($\times$ 1000 lb)
30.0	1.9	40.5	5.5
29.6	4.5	23.4	3.9
15.1	3.5	15.0	2.4
41.2	7.1	34.9	6.0
22.6	2.6	20.5	1.1
21.0	1.9	34.6	4.5

selected months. The results are shown in the table. A third-order model

$$y = \beta_0 + \beta_1 x + \beta_2 x^2 + \beta_3 x^3 + \epsilon$$

was selected to describe the marginal cost y as a function of output x. A BMD-P1R analysis follows.

```
REGRESSION TITLE . . . . . . . . . . . . . . BMD REGRESSION #3
DEPENDENT VARIABLE . . . . . . . . . . . . 4 Y
TOLERANCE . . . . . . . . . . . . . . . . . . . . .0100
```

ALL DATA CONSIDERED AS A SINGLE GROUP

MULTIPLE R	.7840	STD. ERROR OF EST.	6.2573
MULTIPLE R-SQUARE	.6146		

ANALYSIS OF VARIANCE

	SUM OF SQUARES	DF	MEAN SQUARE	F RATIO	P(TAIL)
REGRESSION	562.002	2	281.001	7.177	.01369
RESIDUAL	352.383	9	39.154		

VARIABLE		COEFFICIENT	STD. ERROR	STD. REG. COEFF.	T	P(2TAIL)
INTERCEPT		21.495				
X1	1	.019	2.466	.004	.008	.994
X2	2			REDUNDANT VARIABLE		

| X3 | 3 | .065 | .044 | .780 | 1.482 | .172 |

Does the third-order model appear to provide a good fit to these data? (In your answer address the issues discussed in the preceding sections.)

12.30. Sales force management requires the proper adjustment of sales territories to equalize sales potential among territories as well as the establishment of sales force goals and compensation plans. To examine this issue the sales manager for a company that sells office machines and supplies has recorded the territory sales y for the past month, the number x_1 of accounts, and the number x_2 of years of work experience for a random sample of $n = 25$ of her firm's sales personnel. The data are shown in the table.

y (× $1000)	x_1 (× 100)	x_2 (years)	y (× $1000)	x_1 (× 100)	x_2 (years)
36.70	15	1.7	23.37	10	1.5
34.74	14	1.7	45.87	19	2.3
22.95	12	1.5	27.29	12	1.6
46.76	18	2.6	32.89	14	1.8
61.26	24	4.7	28.01	13	1.6
21.35	9	1.3	32.64	13	1.9
50.32	22	2.5	34.54	15	1.7
33.67	14	1.9	17.41	7	1.3
65.19	25	4.3	20.36	9	1.4
48.76	21	2.4	15.78	6	1.2
24.68	11	1.7	41.68	16	2.0
25.33	11	1.6	28.00	11	1.6
24.08	12	1.4			

It was assumed that a second-order model for $k = 2$ independent variables appropriately describes the relationship between sales potential and the independent variables "number of accounts" and "sales potential." An SAS regression program was used to fit the second-order model to the data set. The results follow.

SAS
MULTIPLE REGRESSION
GENERAL LINEAR MODELS PROCEDURE

DEPENDENT VARIABLE: SALES

SOURCE	DF	SUM OF SQUARES	MEAN SQUARE	F VALUE
MODEL	4	4065.707	1016.427	280.438
ERROR	20	72.488	3.624	
CORRECTED TOTAL	24	4138.195		

R-SQUARE	C.V.	STD. DEV.	SALES MEAN
.9825	5.6417	1.9038	33.7451

PARAMETER	ESTIMATE	T FOR H0: PARAMETER = 0	PR > |T|	STD. ERROR OF ESTIMATE
INTERCEPT	−12.4366			
NO. ACCTS. (X1)	1.3251	2.1301	.0450	.6221

EMPLOYMT. (X2)	17.9296	2.9937	.0055	5.9891
X1 SQUARE	.0075	.3782	.8215	.0199
X2 SQUARE	−2.0789	2.4489	.0240	.2489

VARIABLE (X1)(X2) REDUNDANT AND DELETED FROM PRINTOUT.

Carefully interpret the results of this analysis.

12.31.   An investment officer is interested in explaining how investors value securities and measure the risk and return inherent in different investment opportunities before making an investment decision. In his study the investment officer recorded the closing price y of 20 different utility company securities on 31 December 1976, the retained earnings per share x_1 as a measure of return, and the percentage x_2 of debt to assets as a measure of financial risk. A second-order model for $k = 2$ independent variables was fitted to the data set by using a common regression computer program. The program provided the output shown.

MULTIPLE R	.8751
R SQUARE	.7658
STD. ERROR OF EST.	6.2569

ANALYSIS OF VARIANCE

	DF	SUM OF SQUARES	MEAN SQUARE	F RATIO
REGRESSION	3	2048.054	682.685	17.438
RESIDUAL	16	626.388	39.149	

INDIVIDUAL ANALYSIS OF VARIABLES

VARIABLE	COEFFICIENT	STD. ERROR	F VALUE
(CONSTANT	12.8965)		
EPS (X1)	9.1856	15.2926	.3608
D/A (X2)	REDUNDANT VARIABLE		
X1 SQ.	10.1441	8.2444	1.4139
X2 SQ.	.0006	.0024	.0498
(X1)(X2)	REDUNDANT VARIABLE		

Interpret the output supplied by the program. Construct a contour diagram of the second-order model that appears in the output.

12.32.   The sales manager for a firm that sells packaged soybeans through a nationwide supermarket chain is interested in examining the relationship of the retail price charged for the product and advertising on retail sales. To study this issue the sales manager recorded the sales y (in thousands of units), the average unit price x_1 per package, and the proportion x_2 of total expenses devoted to advertising for the past year in each of $n = 25$ sales regions. An SPSS regression program was used to fit the second-order model

$$y = \beta_0 + \beta_1 x_1 + \beta_2 x_2 + \beta_3 x_1^2 + \beta_4 x_2^2 + \beta_5 x_1 x_2 + \epsilon$$

to the data. The results follow.*

*In the SPSS regression program, the column headed by B lists the least squares regression coefficients, while the column headed by BETA lists the standardized regression coefficients. For example, the standardized coefficient associated with the jth independent variable is

$$\hat{\beta}_j \left(\frac{s_{x_j}}{s_y} \right)$$

where s_{x_j} and s_y are, respectively, the standard deviations of x_j and y. SPSS is the trademark of SPSS, Inc., for its proprietary software product. Output reproduced with the permission of SPSS, Inc.

SPSS REGRESSION

VARIABLE	MEAN	STANDARD DEVIATION
1 (Y)	30.6519	7.9602
2 (X1)	.3604	.0357
3 (X2)	6.3600	1.7049
4 (X1)SQ.	.1311	.0251
5 (X2)SQ.	43.2400	22.5153
6 (X1)(X2)	2.2840	.6285

		ANAL. OF	DF	SS	MS	F
MULTIPLE R	.8396	VARIANCE				
R SQUARE	.7049	REGRESSION	5	1072.065	214.413	9.080
STD. ERROR	4.8598	RESIDUAL	19	448.678	23.615	

VARIABLES IN THE EQUATION

VARIABLE	B	BETA	STD. ERROR B	F
(CONSTANT)	134.0779			
X1	−628.4045	−2.8183	667.1304	.8873
X2	2.4161	.5175	7.5804	.1016
X1SQ.	841.6853	2.6540	907.5220	.8602
X2SQ.	.2286	.6466	.3071	.5541
X3SQ.	−5.4993	−.4342	21.9262	.0629

Give a complete analysis of the printout and construct a contour diagram for the second-order model fitted to the sample data.

12.33.   An SPSS regression program was used to fit the first-order model

$$y = \beta_0 + \beta_1 x_1 + \beta_2 x_2 + \epsilon$$

to the data of exercise 12.32. Test to see if the second-order model fitted in exercise 12.32 provides an improvement over the first-order model. That is, for the second-order model of exercise 12.32, test the hypothesis

$$H_0 : \beta_3 = \beta_4 = \beta_5 = 0$$

SPSS REGRESSION

		ANAL. OF	DF	SS	MS	F
MULTIPLE R	.8281	VARIANCE				
R SQUARE	.6857	REGRESSION	2	1042.917	521.458	24.009
STD. ERROR	4.6604	RESIDUAL	22	477.826	21.719	

VARIABLES IN EQUATION

VARIABLE	B	BETA	STD. ERROR B	F
(CONSTANT)	35.6170			
X1	−72.8205	−.3271	26.8613	7.3494
X2	3.3458	.7166	.5635	35.2592

12.34.   Refer to exercise 12.32 and exercise 12.33.

a. Use each model to separately estimate the expected annual sales of the firm's product in a sales region where the average unit price is $.40 and 7% of total expenses are devoted to advertising. How do you account for differences between your two estimates?

b. If your computer package will calculate confidence intervals for $E(y)$, find 95% confidence intervals for the estimates of mean response in part a.

Experiences with Real Data

From your library, from personal interviews, or from some other source, obtain data on a dependent variable and at least three variables that are logically related to the dependent variable. Your data set can be sequenced over time—selecting, perhaps, corporate data at many consecutive points over time—or it can be cross-sectional—the selection of data sets from many different people, corporations, or market areas at one point in time. Some possible applications you may wish to pursue are the following:

1. Predict new housing starts in the United States by regressing housing starts for each of the last several years on the prime interest rate, the inflation rate, per capita income, the unemployment rate, and other related indicators. Data are available in the government documents section of your library.

2. Develop a model to assess values on single-family residences of the city in which you reside. Note the "asking price" of several houses currently for sale in your city and use these values as the dependent variable. For each listed house you have recorded, find the square footage, the total number of rooms, the number of bedrooms, the age of the house, and whether or not the house has a good view. These data could be obtained by personal inspection or by asking the owner or realtor listing each house. Then regress asking price on the five or six independent variables recorded for each house.

3. Through personal interviews with students, develop a model to relate grade point average (GPA) to a number of predictive factors, such as age, high school GPA, hours spent studying per week, whether or not the student is a business major, and whether or not the student intends to go on to graduate school. Then regress current college grade point average on the independent variables you have chosen to develop a predictive model.

4. From company reports available in your library, extend the exercise discussed in the Experiences with Real Data of chapter 11 by regressing net sales of a company on advertising expenditures, net earnings per share, number of common stock shares outstanding, capital additions in the previous year, and per capita income in the primary market area. (You may choose to delete some of the independent variables mentioned here and add others that your logic suggests should be included.)

If possible, select a sample size of at least $n = 25$. But in any case, make certain that n is large enough to leave a sufficient (say 10 or more) degrees of freedom for SSE. Also make certain that your data set is "complete." That is, each case must contain a measurement on y and every predictor variable—there must not be any missing data.

When performing your analysis, you will almost surely use an electronic computer. Packaged computer programs such as BMD-O2R, BMD-P1R, BMD-P2R (developed by the UCLA Health Sciences Computing Facility), MULTR (a multiple regression program written for use on their 360 series computers by IBM), SPSS-H (part of the Statistical Package for the Social Sciences series developed by the University of Chicago), or the GLM procedure of the SAS (Statistical Analysis System) computer package are widely available, easy-to-use-and-interpret regression programs. Select one of these programs or some other multiple regression program available in your computer center and perform a regression analysis on the data you have gathered. After obtaining your printout from the computer, interpret the results of your analysis. In the process of your interpretation, give the percentage of variation in y explained by the predictor variables used in the model, identify the significant and the redundant predictor variables in the model, give the equation for predicting y from the predictor variables, and provide any other interpretations from the output that you believe are enlightening.

References

DRAPER, N., and H. SMITH. *Applied Regression Analysis.* New York: Wiley, 1966.

MENDENHALL, W. *An Introduction to Linear Models and the Design and Analysis of Experiments.* Belmont, Calif.: Wadsworth, 1968. Chapters 6 and 7.

NETER, J., and W. WASSERMAN. *Applied Linear Statistical Models.* Homewood, Ill.: Richard D. Irwin, 1974. Chapters 2–12.

WONNACOTT, R. J., and T. H. WONNACOTT. *Econometrics.* New York: Wiley, 1970. Chapters 3–20.

chapter objectives

GENERAL OBJECTIVES Chapter 12 presented a technique, multiple regression, that can be used to analyze response data, where the response is a function of a set of independent variables. The technique was applicable to data generated by a designed experiment or to the case in which the data were collected on uncontrolled independent variables. Chapter 13 presents another way of analyzing these experiments for the case in which the experiment is designed. The technique, which is computationally much simpler than multiple regression, is known as an analysis of variance.

SPECIFIC OBJECTIVES

1. To explain the logic of an analysis of variance.
>Section 13.2

2. To present the analysis of variance for a comparison of two treatment means and to relate the resulting F test to the t test of chapter 9.
>Section 13.2

3. To describe a completely randomized experimental design and to give the analysis of variance for more than two treatment means.
>Sections 13.3, 13.4, 13.5

4. To describe a randomized block design, to explain how it increases the amount of information in an experiment, and to give the analysis of variance for a comparison of two or more treatments.
>Sections 13.6 through 13.9

5. To give applications of the analysis of variance to the solution of business problems.
>Sections 13.3 through 13.8

6. To clearly specify the assumptions upon which an analysis of variance is based.
>Section 13.10

chapter thirteen

The Analysis of Variance

13.1 Introduction

Many experiments are conducted to determine the effect of one or more variables on a response. A linear relationship between a response y and an independent variable x was studied in chapter 12, and, in a more general sense, the comparison of means in chapter 9 was a similar problem. For example, the comparison of the difference in the mean monthly sales volume for two different amounts spent on advertising is a study of the independent variable "advertising expenditures," which is a type of "stimulus," on a response y, the mean rate of sales.

The preceding comments indicate that independent experimental variables may be one of two types, quantitative or qualitative. The former, as the name implies, are those familiar variables that can be measured. Qualitative variables cannot be measured. Thus monthly sales volume, annual income, number of employees, and daily demand represent four quantitative variables. In contrast, the "manufacturer" of automobile tires is a variable that is likely to be related to the wearing characteristics of tires, but there is no way to order or arrange five "manufacturers" in some continuous way. Thus "manufacturers" is a qualitative independent variable. Similarly, we may wish to determine whether the mean yield of a chemical in a manufacturing operation varies depending on the foreman managing the process and also whether it varies from one working shift to another. Four foremen, Jones, Smith, Adams, and Green, would represent the levels, or settings, of a single qualitative variable "foremen"; the three working shifts would identify the levels of a second qualitative variable "shifts."* The complete experiment would involve, therefore, a study of the effect of two independent qualitative variables, "foremen" and "shifts," on the yield of a chemical.

In this chapter we consider the comparison of more than two means, which, in the context of the previous discussion, implies a study of the effect of a single independent variable (quantitative or qualitative) on a response. We will then extend the procedure to include the analysis of designed experiments involving two independent variables and discuss a generalization of the procedure for more than two independent variables.

13.2 The Analysis of Variance

The methodology for the analysis of experiments involving several independent variables can best be explained in terms of the linear probabilistic model of chapter 12. This general method of analysis (chapter 12) can be used to analyze data from designed experiments or from data-collecting situations where the independent variables cannot be, or have not been, controlled. However, when the experiment has been designed, the quantities needed to test hypotheses concerning sets of parameters (described in section 12.11) can be expressed as relatively simple formulas involving the response measurements. The use of these formulas in analyzing data from multivariable designed experiments is called an **analysis of variance.** The rationale underlying an analysis of variance can be explained in terms of the linear probabilistic model of chapter 12 or it can be explained in an intuitive manner. We will use the latter approach in this chapter. For the general approach see the text by Mendenhall (1968), chapter 8, which is listed in the references.

As the name implies, **the analysis of variance procedure attempts to analyze the variation of a response and to assign portions of this variation to each of a set of independent variables.** The reasoning behind the procedure is that response variables vary only because of a variation in a set of unknown independent variables. Since the experimenter will rarely, if ever, include all the variables affecting the response in the experiment, random variation in the response is observed even though all independent variables considered are held constant. **The objective of the analysis of variance is to locate important independent variables in a study and to determine how they interact and affect the response.**

The rationale underlying the analysis of variance can be indicated best with a symbolic discussion. The actual analysis of variance—that is, "how to do it"—can be illustrated with an example.

Recall that the variability of a set of n measurements is proportional

---

*The word "level" is usually used to denote the intensity for a quantitative independent variable. We use it here to refer to the settings for either quantitative or qualitative independent variables.

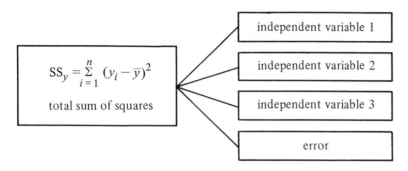

Figure 13.1 *Partitioning of the total sum of squares of deviations*

to the sum of squares of deviations

$$SS_y = \sum_{i=1}^{n} (y_i - \bar{y})^2$$

and that this quantity is used to calculate the sample variance. The analysis of variance partitions the sum of squares of deviations, called the **total sum of squares of deviations**, into parts, each of which is attributed to one of the independent variables in the experiment, plus a remainder that is associated with random error. This partitioning is shown diagrammatically in figure 13.1 for three independent variables.

   If a multivariable linear regression model were written for the response y, as suggested in chapter 12, the portion of the total sum of squares of deviations assigned to error would be the sum of squares of deviations of the y-values about their respective predicted values obtained from the prediction equation $\hat{y}$. You will recall that this quantity, represented by the sum of squares of deviations of the y-values about a straight line, was denoted as SSE in chapter 12.

   For the cases we consider, when the independent variables are unrelated to the response, it can be shown that each of the pieces of the total sum of squares of deviations, divided by an appropriate constant, provides an independent and unbiased estimator of σ^2, the variance of the experimental error. When a variable is highly related to the response, its portion (called the "sum of squares" for the variable) will be inflated. This condition can be detected by comparing the estimate of σ^2 for a particular independent variable with that obtained from SSE using an F test (see section 9.7). If the estimate for the independent variable is significantly larger, the F test will reject an hypothesis of "no effect for the independent variable" and produce evidence to indicate a relation to the response.

   The logic behind an analysis of variance can be illustrated by considering a familiar example, the comparison of two population means for an unpaired experiment, which was analyzed in chapter 9 using a Student's t statistic. We begin by giving a graphic and intuitive explanation of the procedure.

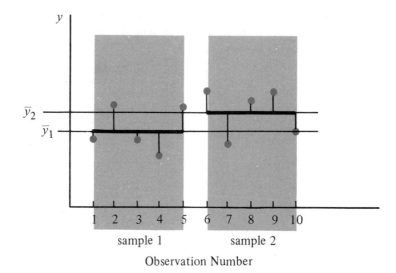

Figure 13.2 *Graphic portrayal of the deviations of the y-values about their means*

Suppose that we have selected random samples of five observations each from two populations, 1 and 2, and that the y-values are plotted as shown in figure 13.2. Note that the $n_1 = 5$ observations from population 1 lie to the left; the $n_2 = 5$ observations from population 2 lie to the right. The sample means $\bar{y}_1$ and $\bar{y}_2$ are shown as horizontal lines in the figure and the deviations of the y-values about their respective means are the vertical line segments. Now examine figure 13.2. Do you think the data provide sufficient evidence to indicate a difference between the two population means? Before we explain, let us look at another figure.

The same two sets of five points are plotted in figure 13.3 except that the distance between the two sets (as measured by the distance between $\bar{y}_1$ and $\bar{y}_2$) is greater in figure 13.3(b) than in figure 13.3(a) and even greater in figure 13.3(c). Therefore, the distance between $\bar{y}_1$ and $\bar{y}_2$ increases as you move from figure 13.3(a) to figure 13.3(b) and then to figure 13.3(c), but the variation within each set is held constant.

Now view the three plots of figure 13.3, (a), (b), and (c), and decide which situation, (a), (b), or (c), provides the greatest evidence to indicate a difference between μ_1 and μ_2. We think you will choose figure 13.3(c) because that plot shows the greatest difference between sample means in comparison with the variation of the points about their respective sample means. This latter variation was held constant for all three plots.

Note that the population means actually may differ for figure 13.3(a) but this fact would not be apparent, because the variation of the points about their respective sample means is too large in comparison with the difference between $\bar{y}_1$ and $\bar{y}_2$. Figure 13.4 shows the same difference between sample means as for figure 13.3(a), but the variation within samples has

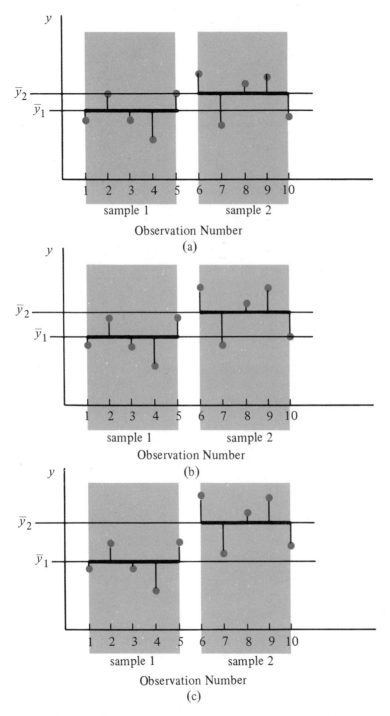

Figure 13.3 *Three fictitious sets of measurements, $n_1 = n_2 = 5$ (the relative positions of points within each set is held constant)*

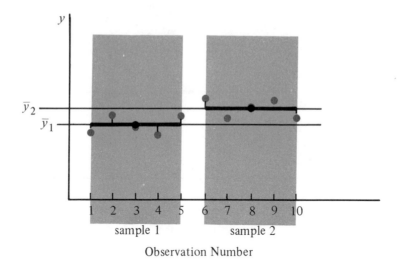

Figure 13.4 *Data showing the same difference between means as shown in figure 13.3(a) but with less within-sample variation*

been reduced. Now it appears that a difference does exist between μ_1 and μ_2.

Now let us leave our intuitive discussion and consider the two sample comparison of means for sample sizes n_1 and n_2. Particularly, we are interested in determining how the total sum of squares of deviations can be partitioned into portions corresponding to the difference between the means and another to the variation within the two samples.

The total sum of squares of deviations of all $(n_1 + n_2)$ y-values about the general mean is

$$\text{Total SS} = \sum_{i=1}^{2} \sum_{j=1}^{n_i} (y_{ij} - \bar{y})^2$$

where $\bar{y}$ is the average of all $(n_1 + n_2)$ observations contained in the two samples. Then with a bit of algebra we can show that

$$\text{Total SS} = \sum_{i=1}^{2} \sum_{j=1}^{n_i} (y_{ij} - \bar{y})^2 = \underbrace{\sum_{i=1}^{2} n_i(\bar{y}_i - \bar{y})^2}_{\text{SST}} + \underbrace{\sum_{i=1}^{2} \sum_{j=1}^{n_i} (y_{ij} - \bar{y}_i)^2}_{\text{SSE}}$$

where $\bar{y}_i$ is the average of the observations in the ith sample, $i = 1, 2$.

The first quantity to the right of the equal sign, called the **sum of squares for treatments** and denoted by the symbol SST, can be shown (with a bit of algebra) to equal

$$\text{SST} = \frac{n_1 n_2}{n_1 + n_2} (\bar{y}_1 - \bar{y}_2)^2$$

Thus SST, which increases as the difference between $\bar{y}_1$ and $\bar{y}_2$ increases, measures the variation between the sample means. Consequently, SST would be larger for the data of figure 13.3(c) than for figure 13.3(a).

The second quantity to the right of the equal sign is the familiar pooled sum of squares of deviations computed in the t test of section 9.4. It is the sum of the sum of squares of deviations of the y-values about their respective sample means. This pooled sum of squares measures within-sample variation. Within-sample variation is usually attributed to experimental error and is consequently called sum of squares for error (denoted by the symbol SSE).

The quantities SST and SSE measure the two kinds of variation that we viewed in the graphic representation of figure 13.3, the variation between means and the variation within samples. The greater the variation between means (the larger SST) in comparison with the variation within samples (SSE), the greater the weight of evidence to indicate a difference between μ_1 and μ_2. How large is large? When will SST be large enough (relative to SSE) to indicate a real difference between μ_1 and μ_2? We will answer these questions in the discussion that follows.

As indicated in chapter 9,

$$s^2 = \text{MSE} = \frac{\text{SSE}}{n_1 + n_2 - 2}$$

provides an unbiased estimator of σ^2. In the language of analysis of variance, s^2 is usually denoted as MSE, meaning mean square for error. Also, when the null hypothesis is true (that is, $\mu_1 = \mu_2$), SST divided by an appropriate number of degrees of freedom yields a second unbiased estimator of σ^2, which we denote as MST (mean square for treatments). For this example, the number of degrees of freedom for MST is equal to 1.

When the null hypothesis is true (that is, $\mu_1 = \mu_2$), MSE (the mean square for error) and MST (the mean square for treatments) estimate the same quantity and should be "roughly" of the same magnitude. When the null hypothesis is false and $\mu_1 \neq \mu_2$, MST will probably be larger than MSE.

The preceding discussion, along with a review of the variance ratio given in section 9.7, suggests the use of

$$\frac{\text{MST}}{\text{MSE}}$$

as a test statistic to test the hypothesis $\mu_1 = \mu_2$ against the alternative $\mu_1 \neq \mu_2$. Indeed, when both populations are normally distributed, it can be shown that MST and MSE are independent in a probabilistic sense and that they can be used in a test statistic, as shown in the box.

Test Statistic for the Null Hypothesis H_0: $\mu_1 = \mu_2$

$$F = \frac{MST}{MSE}$$

The test statistic F follows the F probability distribution of section 9.7. Disagreement with the null hypothesis is indicated by a large value of F, and hence the rejection region for a given α is

$$F \geq F_\alpha$$

Thus the analysis of variance test results in a one-tailed F test. The degrees of freedom for F will be those associated with MST and MSE, which we denote as v_1 and v_2, respectively. Although we have not indicated how we determine v_1 and v_2, in general, $v_1 = 1$ and $v_2 = (n_1 + n_2 - 2)$ for the two-sample experiment described.

Example 13.1 The coded values for the hours of life of two brands of light bulbs are given in table 13.1 for samples of six bulbs drawn randomly from each of the two brands. The true values, in hundreds of hours,

Table 13.1 *Hours of life for two brands of light bulbs, example 13.1*

BULB	BRAND	
	A	B
1	6.1	9.1
2	7.1	8.2
3	7.8	8.6
4	6.9	6.9
5	7.6	7.5
6	8.2	7.9
Totals	43.7	48.2

were coded by multiplying each by $1/100$ to give the coded values shown in table 13.1. Do the data present sufficient evidence to indicate a difference in mean lifetime for the two brands of light bulbs?

Solution Although the Student's t could be used as the test statistic for this example, we will use our analysis of variance F test, since it is more general and can be used to compare more than two means.

The two desired sums of squares of deviations* are

---

*Note: This formula for SST applies to the special case where $n_1 = n_2$.

$$SST = n_1 \sum_{i=1}^{2} (\bar{y}_i - \bar{y})^2 = \left(\frac{n_1 n_2}{n_1 + n_2}\right)(\bar{y}_1 - \bar{y}_2)^2$$

$$= \frac{(6)(6)}{(6+6)}\left(\frac{43.7}{6} - \frac{48.2}{6}\right)^2 = 1.6875$$

$$SSE = \sum_{i=1}^{2} \sum_{j=1}^{6} (y_{ij} - \bar{y}_i)^2 = \sum_{j=1}^{6} (y_{1j} - \bar{y}_1)^2 + \sum_{j=1}^{6} (y_{2j} - \bar{y}_2)^2 = 5.8617$$

(You can verify that SSE is the pooled sum of squares of the deviations for the two samples discussed in section 9.4. Also, note that Total SS = SST + SST.) The mean squares for treatment and error are

$$MST = \frac{SST}{1} = 1.6875$$

$$MSE = \frac{SSE}{2n_1 - 2} = \frac{5.8617}{10} = .5862$$

To test the null hypothesis $\mu_1 = \mu_2$, we compute the test statistic

$$F = \frac{MST}{MSE} = \frac{1.6875}{.5862} = 2.88$$

The critical value of the F statistic for $\alpha = .05$ is 4.96. Although the mean square for treatments is almost three times as large as the mean square for error, it is not large enough to reject the null hypothesis. Consequently, there is not sufficient evidence to indicate a difference between μ_1 and μ_2.

As noted, the purpose of the preceding example was to illustrate the computations involved in a simple analysis of variance. The F test for comparing two means is equivalent to a Student's t test, because an F statistic with one degree of freedom in the numerator is equal to t^2. Had the t test been used for this example, we would have found $t = -1.6967$, which we see satisfies the relationship $t^2 = (-1.6967)^2 = 2.88 = F$. This relationship also holds for the critical values. You can verify that the square of $t_{.025} = 2.228$ (used for the two-tailed test with $\alpha = .05$ and $v = 10$ degrees of freedom) is equal to $F_{.05} = 4.96$.

Of what value is the Total SS? The answer is that it provides an easy way to compute SSE. Since the Total SS partitions into SST and SSE, that is,

$$\text{Total SS} = \text{SST} + \text{SSE}$$

then we have

$$\text{SSE} = \text{Total SS} - \text{SST}$$

Both the Total SS and SST are easy to compute. Hence we can easily find SSE by substituting into the expression above. For example 13.1, we have

$$\text{Total SS} = \sum_{i=1}^{2} \sum_{j=1}^{6} (y_{ij} - \bar{y})^2 = \sum_{i=1}^{2} \sum_{j=1}^{6} y_{ij}^2 - \frac{\left(\sum_{i=1}^{2} \sum_{j=1}^{6} y_{ij} \right)^2}{12}$$

$$= (\text{sum of squares of all } y\text{-values}) - \frac{(\text{total of all } y\text{-values})^2}{12}$$

$$= 711.35 - \frac{(91.9)^2}{12} = 7.5492$$

Then

$$\text{SSE} = \text{Total SS} - \text{SST} = 7.5492 - 1.6875 = 5.8617$$

This is exactly the same value obtained by the tedious computation and pooling of the sums of squares of deviations from the individual samples.

Exercises

13.1.   Analyze the data of exercise 9.47 by the procedure outlined in section 13.2. Determine whether there is evidence of a difference in mean recognition time for the two advertising layouts. Test at the $\alpha = .05$ level of significance. If you have not worked exercise 9.47, analyze the same data using the Student's t test for independent samples, as described in section 9.4. Note that both methods lead to the same conclusion and that the computed values of F and t for the two methods are related. That is, $F = t^2$ (this will hold true only for the comparison of *two* population means).

13.2.   To the investor, risk and the rate of return are the elements of concern when selecting an investment. Suppose an investor is interested in developing a portfolio of either bank issues or industrial bonds. She selects a sample of $n_1 = 7$ recent bank issues and $n_2 = 5$ recent industrial bond issues and records the yield to maturity

Bank	9.14	8.85	9.52	10.16	8.90	9.65	9.85
Industrial	9.69	8.94	8.85	9.45	9.15		

on each. The results are shown in the table. Do these data present sufficient evidence to indicate a difference in average yield to maturity between bank and industrial bond issues? Use $\alpha = .05$.

a. Use the Student's t test from section 9.4.
b. Use the procedure outlined in section 13.2.

13.3 A Comparison of More Than Two Means

An analysis of variance to detect a difference in a set of more than two population means is a simple generalization of the analysis of variance of section 13.2. The random selection of independent samples from p populations is known as a **completely randomized experimental design.**

Assume that independent random samples have been drawn from p normal populations with means μ_1, μ_2, ..., μ_p, respectively, and with variance σ^2. Thus all populations are assumed to possess equal variances. And, to be completely general, we allow the sample sizes to be unequal and we let n_i, $i = 1, 2, ..., p$, be the number in the sample drawn from the ith population. The total number of observations in the experiment is $n = n_1 + n_2 + \cdots + n_p$.

Let y_{ij} denote the measured response on the jth experimental unit in the ith sample and let T_i and $\bar{T}_i$ represent the total and the mean, respectively, for the observations in the ith sample. (The modification in the symbols for sample totals and averages will simplify the computing formulas for the sums of squares.) Then, as in the analysis of variance involving two means,

$$\text{Total SS} = \text{SST} + \text{SSE}$$

where

$$\text{Total SS} = \sum_{i=1}^{p} \sum_{j=1}^{n_i} (y_{ij} - \bar{y})^2 = \sum_{i=1}^{p} \sum_{j=1}^{n_i} y_{ij}^2 - \text{CM}$$

$$= (\text{sum of squares of all } y\text{-values}) - \text{CM}$$

$$\text{CM} = \frac{(\text{total of all observations})^2}{n} = \frac{\left(\sum_{i=1}^{p} \sum_{j=1}^{n_i} y_{ij}\right)^2}{n} = n\bar{y}^2$$

(the term CM denotes "correction for the mean"),

$$\text{SST} = \sum_{i=1}^{p} n_i(\bar{T}_i - \bar{y})^2 = \sum_{i=1}^{p} \frac{T_i^2}{n_i} - \text{CM}$$

$$= \left\{ \begin{array}{l} \text{sum of squares of treatment totals, with each square divided} \\ \text{by the number of observations in that particular total} \end{array} \right\} - \text{CM}$$

$$\text{SSE} = \text{Total SS} - \text{SST}$$

Although the easy way to compute SSE is by subtraction as shown above, it is interesting to note that SSE is the pooled sum of squares for all p samples and is equal to

$$\text{SSE} = \sum_{i=1}^{p} \sum_{j=1}^{n_i} (y_{ij} - \bar{T}_i)^2$$

The unbiased estimator of σ^2, based on $(n_1 + n_2 + \cdots + n_p - p)$ degrees of freedom, is

$$s^2 = \text{MSE} = \frac{\text{SSE}}{n_1 + n_2 + \cdots + n_p - p}$$

The mean square for treatments possesses $(p - 1)$ degrees of freedom, that is, one less than the number of means, and is

$$MST = \frac{SST}{p - 1}$$

To test the null hypothesis

$$H_0: \mu_1 = \mu_2 = \cdots = \mu_p$$

against the alternative that at least one of the equalities does not hold, MST is compared with MSE using the F statistic, based upon $v_1 = p - 1$ and

$$v_2 = \sum_{i=1}^{p} n_i - p = n - p$$

degrees of freedom. The null hypothesis will be rejected if

$$F = \frac{MST}{MSE} > F_\alpha$$

where F_α is the critical value of F, based on $(p - 1)$ and $(n - p)$ degrees of freedom, for a probability α of a type I error.

Intuitively, the greater the difference between the observed treatment means $\bar{T}_1, \bar{T}_2, \ldots, \bar{T}_p$, the greater is the evidence to indicate a difference between their corresponding population means. It can be seen from the formula for SST that SST = 0 when all the observed treatment means are identical because then $\bar{T}_1 = \bar{T}_2 = \cdots = \bar{T}_p = \bar{y}$, and the deviations appearing in SST, $(\bar{T}_i - \bar{y})$, $i = 1, 2, \ldots, p$, equal zero. As the treatment means get farther apart, the deviations $(\bar{T}_i - \bar{y})$ increase in absolute value and SST increases in magnitude. Consequently, the larger the value of SST, the greater is the weight of evidence favoring a rejection of the null hypothesis. This same line of reasoning applies to the F tests employed in the analysis of variance for all designed experiments.

The test is summarized as follows:

F Test for Comparing p Population Means

$H_0: \mu_1 = \mu_2 = \cdots \mu_p$.
H_a: One or more pairs of population means differ.

Test statistic:
$$F = \frac{MST}{MSE}$$

where F is based on $v_1 = (p - 1)$ and $v_2 = (n - p)$ degrees of freedom.

Rejection region: Reject H_0 if $F > F_\alpha$, where F_α lies in the upper tail of the F distribution (with $v_1 = p - 1$ and $v_2 = n - p$) and satisfies the expression

$$P(F > F_\alpha) = \alpha$$

The assumptions underlying the analysis of variance F tests deserve particular attention. The samples are assumed to have been randomly selected from the p populations in an independent manner. The populations are assumed to be normally distributed, with equal variances σ^2 and means μ_1, μ_2, ..., μ_p. Moderate departures from these assumptions will not seriously affect the properties of the test. This is particularly true of the normality assumption.

Example 13.2 Four groups of salespeople for a magazine sales agency were subjected to different sales training programs. Because there were some dropouts during the training programs, the number of trainees varied from group to group. At the end of the training programs, each salesperson was randomly assigned a sales area from a group of sales areas that were judged to have equivalent sales potentials. The number of sales made by each person in each of the four groups of salespeople during the first week after completing the training program is listed in table 13.2. Do the data present sufficient evidence to indicate a difference in the mean achievement for the four training programs?

Table 13.2 *Number of sales made by each person in each training group*

	TRAINING GROUP			
	1	2	3	4
	65	75	59	94
	87	69	78	89
	73	83	67	80
	79	81	62	88
	81	72	83	
	69	79	76	
		90		
T_i	454	549	425	351
$\bar{T}_i$	75.67	78.43	70.83	87.75

Solution We must first compute the quantities listed on page 475.

$$CM = \frac{\left(\sum_{i=1}^{4}\sum_{j=1}^{n_i} y_{ij}\right)^2}{n} = \frac{(\text{total of all observations})^2}{n}$$

$$= \frac{(1{,}779)^2}{23} = 137{,}601.8$$

$$\text{Total SS} = \sum_{i=1}^{4}\sum_{j=1}^{n_i} y_{ij}^2 - CM = (\text{sum of squares of all } y\text{-values}) - CM$$

$$= (65)^2 + (87)^2 + (73)^2 + \cdots + (88)^2 - CM$$

$$= 139{,}511 - 137{,}601.8 = 1{,}909.2$$

$$SST = \sum_{i=1}^{4} \frac{T_i^2}{n_i} - CM$$

$$= \left\{ \begin{array}{l} \text{sum of squares of treatment totals, with each square} \\ \text{divided by the number of observations in that} \\ \text{particular total} \end{array} \right\} - \text{CM}$$

$$= \frac{(454)^2}{6} + \frac{(549)^2}{7} + \frac{(425)^2}{6} + \frac{(351)^2}{4} - \text{CM}$$

$$= 138,314.4 - 137,601.8 = 712.6$$

$\text{SSE} = \text{Total SS} - \text{SST} = 1,196.6$

The mean squares for treatment and error are

$$\text{MST} = \frac{\text{SST}}{p-1} = \frac{712.6}{3} = 237.5$$

$$\text{MSE} = \frac{\text{SSE}}{n_1 + n_2 + \cdots + n_p - p} = \frac{\text{SSE}}{n-p} = \frac{1,196.6}{19} = 63.0$$

The test statistic for testing the hypothesis $\mu_1 = \mu_2 = \mu_3 = \mu_4$ is

$$F = \frac{\text{MST}}{\text{MSE}} = \frac{237.5}{63.0} = 3.77$$

where

$$\nu_1 = (p-1) = 3 \qquad \text{and} \qquad \nu_2 = \sum_{i=1}^{p} n_i - 4 = 19$$

The critical value of F for $\alpha = .05$ is $F_{.05} = 3.13$. Since the computed value of F, 3.77, exceeds $F_{.05} = 3.13$, we reject the null hypothesis and conclude that the evidence is sufficient to indicate a difference in mean achievement for the four training programs.

You may feel that the above conclusion could have been made on the basis of visual observation of the treatment means. However, it is not difficult to construct a set of data that will lead the "visual" decision maker to erroneous results.

13.4 An Analysis of Variance Table for a Completely Randomized Design

The calculations of the analysis of variance are usually displayed in an analysis of variance (ANOVA or AOV) table. The table for the design of section 13.3 involving p treatment means is shown in table 13.3. Column 1 shows the source of each sum of squares of deviations; column 2 gives the respective degrees of freedom; columns 3 and 4 give the corresponding

Table 13.3 *ANOVA table for a comparison of means, completely randomized design*

SOURCE	d.f.	SS	MS	F
treatments	$p - 1$	SST	$MST = SST/(p - 1)$	MST/MSE
error	$n - p$	SSE	$MSE = SSE/(n - p)$	
Totals	$n - 1$	Total SS		

sums of squares and mean squares, respectively. A calculated value of F, comparing MST and MSE, is usually shown in column 5. Note that the degrees of freedom and sums of squares add to their respective totals.

The ANOVA table for example 13.2, shown in table 13.4, gives a compact presentation of the appropriate computed quantities for the analysis of variance.

Table 13.4 *ANOVA table for example 13.2*

SOURCE	d.f.	SS	MS	F
treatments	3	712.6	237.5	3.77
error	19	1,196.6	63.0	
Totals	22	1,909.2		

13.5 Estimation for the Completely Randomized Design

Confidence intervals for a single treatment mean and the difference between a pair of treatment means (section 13.3) are identical to those given in chapter 9. The confidence interval for the mean of treatment i or the difference between treatments i and j are given in the box.

Completely Randomized Design: $(1 - \alpha)100\%$

Confidence Intervals

A single treatment mean:

$$\bar{T}_i \pm t_{\alpha/2} \frac{s}{\sqrt{n_i}}$$

The difference between two treatment means:

$$(\bar{T}_i - \bar{T}_j) \pm t_{\alpha/2}\, s \sqrt{\frac{1}{n_i} + \frac{1}{n_j}}$$

---

where

$$s = \sqrt{s^2} = \sqrt{MSE} = \sqrt{\frac{SSE}{n - p}}$$

$n = n_1 + n_2 + \cdots + n_p$ and $t_{\alpha/2}$ is based on $(n - p)$ degrees of freedom.

Note that the confidence intervals stated above are appropriate for single treatment means or a comparison of a pair of means selected prior to observing the data. The stated confidence coefficients are based on random sampling. If you were to look at the data and then compare the largest and smallest sample means, the assumption of randomness would be disturbed. Certainly the difference between the largest and smallest sample means is expected to be larger than for a pair selected at random.

Example 13.3 Find a 95% confidence interval for the mean number of sales for those trained in training program 1 of example 13.2.

Solution The 95% confidence interval for the mean number of sales is

$$\bar{T}_1 \pm t_{.025}\frac{s}{\sqrt{6}} = 75.67 \pm \frac{(2.093)(7.94)}{\sqrt{6}} = 75.67 \pm 6.78$$

where $t_{.025}$ is based on $(n - p) = 19$ degrees of freedom, and $s = \sqrt{MSE} = \sqrt{63} = 7.94$. Then we estimate the mean number of sales to be contained in the interval from 68.89 to 82.45.

Example 13.4 Find a 95% confidence interval for the difference in mean sales for training programs 1 and 4 of example 13.2.

Solution The 95% confidence interval for $(\mu_1 - \mu_4)$ is

$$(\bar{T}_1 - \bar{T}_4) \pm (2.093)(7.94)\sqrt{\frac{1}{6} + \frac{1}{4}} = -12.08 \pm 10.73$$

Therefore, we estimate that the interval from -22.81 to -1.35 encloses the difference in mean sales for training programs 1 and 4. Because all points in the interval are negative, we infer that μ_4 is larger than μ_1. Also note that the variability in the number of sales for a given salesperson is rather large. Consequently, the sample sizes should be increased if we want to reduce the width of the confidence interval.

Exercises

13.3.   What types of television advertising best hold the attention of children? Goldberg and Gorn * confirmed an earlier study by Ward, who found that advertisements

*M. E. Goldberg and G. J. Gorn, "Children's Reactions to Television Advertising: An Experimental Approach," *Consumer Research*, September 1974.

for foods and gum appear to best capture the attention of young viewers. In an attempt to replicate their experiment, suppose a researcher observes 15 children; 5 children are observed during the showing of an advertisement featuring toys and games, 5 during the showing of an advertisement featuring food and gum, and 5 during the showing of an advertisement directed toward children's clothing. All advertisements are exactly 60 seconds in length. Recorded in the accompanying table are the times of attention to the advertisements for the 15 children.

ADVERTISEMENT	TIME OF ATTENTION
toys, games	45 40 30 25 45
food, gum	50 25 55 45 50
clothing	30 45 40 50 35

   a. Perform an analysis of variance for this experiment.

   b. Do these data provide sufficient evidence to indicate a difference in mean time of attention by the children for the three classes of advertisements?

   c. Do these data support the claims of Goldberg, Gorn, and Ward?

   13.4.   Refer to exercise 13.3. Let μ_A and μ_B denote, respectively, the mean attention time of the children to advertisements directed toward toys and games (A) and food and gum (B).

   a. Find a 95% confidence interval for μ_A.

   b. Find a 95% confidence interval for μ_B.

   c. Find a 95% confidence interval for $(\mu_A - \mu_B)$.

   13.5.   An experiment was conducted to compare the price of a loaf of bread (a particular brand) in four city locations. Eight stores were randomly sampled in locations 1, 2, and 3, but due to an omission only seven were selected from location 4. A completely randomized design was employed. Conduct an analysis of variance for these data.

LOCATION	BREAD PRICE (CENTS)
1	59 63 65 61 64 58 60 61
2	58 61 64 63 57 60 63 60
3	54 59 55 58 59 56 60 55
4	69 70 68 70 66 71 69

   a. Do these data provide sufficient evidence to indicate a difference in the mean price of the bread in stores located in the four areas of the city?

   b. Suppose that prior to seeing the data we wished to compare the mean prices between locations 1 and 4. Estimate the difference in mean prices, using a 95% confidence interval.

   13.6.   The Environmental Protection Agency and certain state agencies have established rigid regulations on the output of polluting effluents by manufacturing plants. A wood products firm has four branch plants located in a certain western state. One plant, plant A, has recently passed an EPA investigation of its water effluents, but management is uncertain about how the other three plants will fare in similar investigations. To study this issue five samples of liquid waste are selected from each plant and the polluting effluents measured in each sample. The results of the experiment are shown in the table.

PLANT	POLLUTING EFFLUENTS (LB/GAL OF WASTE)
A	1.65 1.72 1.50 1.37 1.60
B	1.70 1.85 1.46 2.05 1.80
C	1.40 1.75 1.38 1.65 1.55
D	2.10 1.95 1.65 1.88 2.00

a. Do these data present sufficient evidence to indicate a difference in the mean amount of polluting effluents per gallon of waste discharge for the four plants?

b. Use a 90% confidence interval to estimate the difference in the mean amount of polluting effluents per gallon of waste discharge between plants A and B. Between plants A and C. Between plants A and D.

13.7.   Precision assembly operations, such as those required in the assembly of electronic circuitry in the manufacture of television components and computers, require a high degree of specialized training. The objective of all such training programs is to prepare assemblers to perform their tasks properly in as short a time as possible. A manufacturer of computing machinery has proposed three different training programs for its employees involved in circuitry assembly operations. Fifteen assembly employees were divided equally among the three training programs. After training, the average time for the proper assembly of a circuit board was recorded for each employee, but three employees resigned from the firm during the course of the training program. The data are shown in the table.

TRAINING PROGRAM	AVERAGE ASSEMBLY TIME (MIN)
A	59 64 57 62
B	52 58 54
C	58 65 71 63 64

a. Do these data present sufficient evidence to indicate a difference in mean assembly time for employees trained under the three training programs?

b. Compare the mean assembly time for employees trained under program B with the mean assembly time for those trained under program C, using a 95% confidence interval.

13.6 A Randomized Block Design

The **randomized block design** is a generalization of the paired-difference design of section 9.5. The purpose of this design is to make comparisons between a set of treatments within blocks of relatively homogeneous experimental material. In section 9.5 two treatments (types of automobile tires) were compared within the relative homogeneity of a single automobile, which eliminated auto-to-auto variability.

The difference between a randomized block design and the completely randomized design can be demonstrated by considering an experiment to compare the effect of product display (treatments) on sales in a marketing analysis. Product display may be defined as packaging variations or arrangements within the marketplace.

Suppose that four package designs are selected as the treatments and we wish to study their effect on sales by using 12 supermarkets. Each supermarket would display only one of the four package designs. Then the 12 supermarkets could be randomly assigned as distributors, 3 to each of the four package designs, as shown in accompanying table. Random assignment of the supermarkets to treatments (or vice versa) randomly distributes errors due to variability of the supermarkets to the four treatments and yields four samples that are, for all practical purposes, random and independent. This would be a completely randomized experimental design and would require the analysis of section 13.3.

Treatments	1	2	3	4
Supermarkets	4, 9, 1	11, 3, 5	2, 6, 12	8, 7, 10

The confidence interval for a comparison of means using the completely randomized design depends on s, the estimated standard deviation of the experimental error. If you were able to reduce σ, the value of s would probably decrease and the confidence interval would narrow, indicating an increase in information in the experiment. This would imply that a reduction is required in the magnitude of the random experimental errors that occur when making repeated measurements for a given treatment.

The random experimental error is composed of a number of components. Some of these are due to the differences between the supermarkets, to the failure of consecutive sales records for the product within a supermarket to be identical (due to variation in the supermarket's advertising and changing patterns of its customers' buying behavior), to the failure of the marketing research director to administer the program consistently from supermarket to supermarket, and, finally, to errors in recording the actual sales records in each store. Reduction of any of these causes of error will increase the information in the experiment.

The store-to-store variation in the experiment above can be eliminated by using the supermarkets as blocks. Thus each supermarket would receive each of the four package designs (treatments) assigned in a random sequence. The resulting randomized block design would appear symbolically as shown in figure 13.5. Each display in a single store is an experimental unit.

We see four units (squares) assigned to each store in figure 13.5, where it is assumed that the displays are run in a time sequence in the order shown from top to bottom. The circled number appearing in a given square is the package design (treatment) to be used for the display. Note that each treatment occurs exactly once in each block. In Store 1 the package design displays are run in the order 2, 1, 4, 3. With the randomized block design, only 12 supermarkets are required to obtain 12 sales measurements per treatment, whereas 48 supermarkets would be required to obtain 12 measurements per

Supermarkets (blocks)

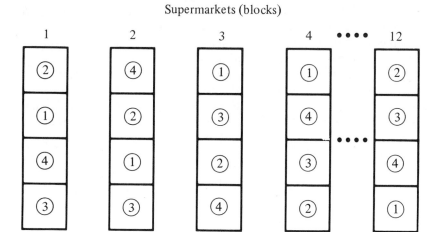

Figure 13.5 *Randomized block design*

treatment using the completely randomized design.

The word "randomized" in the name of the design implies that the treatments are randomly assigned within the block. For our experiment the position within the block pertains to the position in the sequence when assigning a particular package design to a given supermarket over time. For instance, the experiment may be conducted over a period of four weeks, with each store being randomly assigned a different package design to sell each week. The purpose of the randomization (that is, the position in the block) is to eliminate bias due to time. That is, in case the demand for the product in any one week is higher or lower than in another week, it gives every package an equal chance of being assigned to that week.

Blocks may represent time, location, or experimental material. If three treatments are to be compared and there is a suspected trend in the mean response over time, a substantial part of the time variation may be removed by blocking. All three treatments would be randomly applied to experimental units in one small block of time. This procedure would be repeated in succeeding blocks of time until the required amount of data is collected.

As we have seen, a comparison of the sale of competitive or differently designed products in supermarkets should be made within supermarkets, thus using the supermarkets as blocks and removing store-to-store variability. An experiment designed to test and compare subject response to a set of stimuli uses the subjects as blocks to minimize the chance of bias due to fatigue or learning. That is, each person is subjected to the complete set of stimuli, with the stimuli spaced in time (to reduce the possibility of a residual effect from the previous stimulus) and assigned in random order. Experiments in medicine often utilize children within a single family as blocks, applying all the treatments, one each, to children within a family. Because of heredity, children within a family are more homogeneous than those between families. This type of blocking removes the family-to-family variation, just as the stimulus-response experiment is designed to remove the subject-to-subject variation and the market preference experiment is designed to remove the store-to-store variation.

13.7 The Analysis of Variance for a Randomized Block Design

The randomized block design implies the presence of two qualitative independent variables, "blocks" and "treatments." Consequently, the total sum of squares of deviations of the response measurements about their mean can be partitioned into three parts: the sums of squares for blocks, treatments, and error.

Denote the total and average of all observations in block i as B_i and $\bar{B}_i$, respectively. Similarly, let T_j and $\bar{T}_j$ denote the total and mean of all observations receiving treatment j. Then for a randomized block design involving b blocks and p treatments, we have

Total SS = SSB + SST + SSE

$$= \sum_{i=1}^{b} \sum_{j=1}^{p} (y_{ij} - \bar{y})^2 = \sum_{i=1}^{b} \sum_{j=1}^{p} y_{ij}^2 - CM$$

$$= (\text{sum of squares of all } y\text{-values}) - CM$$

$$SSB = p \sum_{i=1}^{b} (\bar{B}_i - \bar{y})^2 = \frac{\displaystyle\sum_{i=1}^{b} B_i^2}{p} - CM$$

$$= \frac{\text{sum of squares of all block totals}}{\text{number of observations in a single total}} - CM$$

$$SST = b \sum_{j=1}^{p} (\bar{T}_j - \bar{y})^2 = \frac{\displaystyle\sum_{j=1}^{p} T_j^2}{b} - CM$$

$$= \frac{\text{sum of squares of all treatment totals}}{\text{number of observations in a single total}} - CM$$

where

$$\bar{y} = (\text{average of all } n = bp \text{ observations}) = \frac{\displaystyle\sum_{i=1}^{b} \sum_{j=1}^{p} y_{ij}}{n}$$

$$CM = \frac{(\text{total of all observations})^2}{n} = \frac{\left(\displaystyle\sum_{i=1}^{b} \sum_{j=1}^{p} y_{ij}\right)^2}{n}$$

Table 13.5 *ANOVA table for a randomized block design*

SOURCE	d.f.	SS	MS
blocks	$b - 1$	SSB	$MSB = SSB/(b - 1)$
treatments	$p - 1$	SST	$MST = SST/(p - 1)$
error	$n - b - p + 1$	SSE	$MSE = \dfrac{SSE}{n - b - p + 1}$
Totals	$n - 1$	Total SS	

The analysis of variance for the randomized block design is presented in table 13.5. The degrees of freedom associated with each sum of squares are shown in the second column. Mean squares are calculated by dividing the sums of squares by their respective degrees of freedom. Note that the degrees of freedom for blocks, treatments, and error always sum to $(n - 1)$.

To test the null hypothesis "there is no difference in treatment means," we use the following test:

F Test for Comparing p Treatments, Using a Randomized Block Design

H_0: The population treatment means are equal.

H_a: One or more pairs of population means differ.

Test statistic:
$$F = \frac{MST}{MSE}$$

where F is based on $v_1 = (p - 1)$ and $v_2 = (n - b - p + 1)$ degrees of freedom.

Rejection region: Reject H_0 if $F > F_\alpha$.

Blocking not only may reduce the experimental error, it also may provide an opportunity to see whether evidence exists to indicate a difference in the mean response for blocks. Under the null hypothesis that there is no difference in mean response for blocks, MSB provides an unbiased estimator of σ^2, based on $(b - 1)$ degrees of freedom. Where a real difference exists in the block means, MSB will probably be inflated in comparison with MSE, and then

$$F = \frac{MSB}{MSE}$$

provides a test statistic. As in the test for treatments, the rejection region is

$$F > F_\alpha$$

Supermarkets

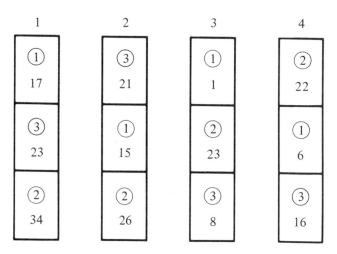

Figure 13.6 *Number of units sold for each package design within each supermarket, example 13.5*

based on $v_1 = (b - 1)$ and $v_2 = (n - b - p + 1)$ degrees of freedom.

Example 13.5 A consumer preference study involving three different package designs (treatments) was laid out in a randomized block design among four supermarkets (blocks). The data shown in figure 13.6 represent the number of units sold for each package design within each supermarket during each of three given weeks. Do the data present sufficient evidence to indicate a difference in the mean sales for each package design (treatment)? Do they present sufficient evidence to indicate a difference in mean sales for the supermarkets?

Solution The treatment and block totals are as follows:

$$T_1 = 39 \qquad T_2 = 105 \qquad T_3 = 68$$
$$B_1 = 74 \qquad B_2 = 62 \qquad B_3 = 32 \qquad B_4 = 44$$

The sums of squares for the analysis of variance are shown individually below and jointly in the analysis of variance table (table 13.6).

$$\text{CM} = \frac{(\text{total})^2}{n} = \frac{(212)^2}{12} = 3745.33$$

$$\text{Total SS} = \sum_{i=1}^{4} \sum_{j=1}^{3} (y_{ij} - \bar{y})^2 = \sum_{i=1}^{4} \sum_{j=1}^{3} y_{ij}^2 - \text{CM}$$

$$= (\text{sum of squares of all } y\text{-values}) - \text{CM}$$

$$= (17)^2 + (23)^2 + \cdots + (16)^2 - \text{CM}$$

$$= 4686 - 3745.33 = 940.67$$

$$\text{SSB} = \frac{\sum\limits_{i=1}^{4} B_i^2}{3} - \text{CM}$$

$$= \frac{\text{sum of squares of all block totals}}{\text{number of observations in a single total}} - \text{CM}$$

$$= \frac{(74)^2 + (62)^2 + (32)^2 + (44)^2}{3} - \text{CM}$$

$$= 4093.33 - 3745.33 = 348.00$$

$$\text{SST} = \frac{\sum\limits_{j=1}^{3} T_j^2}{4} - \text{CM}$$

$$= \frac{\text{sum of squares of all treatment totals}}{\text{number of observations in a single total}} - \text{CM}$$

$$= \frac{(39)^2 + (105)^2 + (68)^2}{4} - \text{CM} = 4292.5 - 3745.33 = 547.17$$

$$\text{SSE} = \text{Total SS} - \text{SSB} - \text{SST} = 940.67 - 348.00 - 547.17 = 45.50$$

Table 13.6 *ANOVA table for example 13.5*

SOURCE	d.f.	SS	MS	F
blocks	3	348.00	116.00	15.30
treatments	2	547.17	273.58	36.09
error	6	45.50	7.58	
Totals	11	940.67		

We use the ratio of mean square for treatments to mean square for error to test an hypothesis of no difference in the expected response for treatments. Thus

$$F = \frac{\text{MST}}{\text{MSE}} = \frac{273.58}{7.58} = 36.09$$

The critical value of the F statistic ($\alpha = .05$) for $v_1 = 2$ and $v_2 = 6$ degrees of freedom is $F_{.05} = 5.14$. Since the computed value of F exceeds the critical value, there is sufficient evidence to reject the null hypothesis and conclude that a real difference does exist in the expected sales for the three package designs.

A similar test can be conducted for the null hypothesis that no difference exists in the mean sales for supermarkets. Rejection of this hypothesis would imply that store-to-store variability does exist and that blocking is desirable. The computed value of F based on $v_1 = 3$ and $v_2 = 6$ degrees of freedom is

$$F = \frac{\text{MSB}}{\text{MSE}} = \frac{116.00}{7.58} = 15.30$$

Since this value of F exceeds the corresponding tabulated critical value, $F_{.05} = 4.76$, we reject the null hypothesis and conclude that a real difference exists in the expected sales in the four supermarkets; that is, the data present sufficient evidence to support our decision to block with respect to supermarkets.

13.8 Estimation for the Randomized Block Design

The confidence interval for the difference between a pair of means is similar to that for the completely randomized design given in section 13.5. It is as given in the box.

Randomized Block Design: A $(1 - \alpha)100\%$ Confidence Interval for the Difference Between Two Treatment Means

$$(\bar{T}_i - \bar{T}_j) \pm t_{\alpha/2} s \sqrt{\frac{2}{b}}$$

where $n_i = n_j = b$, the number of observations contained in a treatment mean, p is the number of treatments, and $t_{\alpha/2}$ is based on $n - b - p + 1$ degrees of freedom.

The difference between the confidence intervals for the completely randomized and the randomized block designs is that s, appearing in the expressions above, will probably be smaller for the randomized block design.

Similarly, we can construct a $(1 - \alpha)100\%$ confidence interval for the difference between a pair of block means. Each block contains p observations corresponding to the p treatments. Therefore, the confidence interval is

$$(\bar{B}_i - \bar{B}_j) \pm t_{\alpha/2} s \sqrt{\frac{2}{p}}$$

Example 13.6 From the problem described in example 13.5, construct a 90% confidence interval for the difference between the expected sales from package designs 1 and 2, that is, for the difference between treatments 1 and 2.

Solution From example 13.5 we have

$$\bar{T}_1 = \frac{T_1}{b} = \frac{39}{4} = 9.75 \qquad \bar{T}_2 = \frac{T_2}{b} = \frac{105}{4} = 26.25$$

$$s^2 = \frac{SSE}{n - b - p + 1} = \frac{45.50}{6} = 7.58333 \qquad \text{and} \qquad s = 2.75$$

The confidence interval for the difference in mean response for a pair of treatments is

$$(\bar{T}_i - \bar{T}_j) \pm t_{\alpha/2}\, s \sqrt{\frac{2}{b}}$$

where, for our example, $t_{.050}$ is based on $(n - b - p + 1) = (12 - 4 - 3 + 1) = 6$ degrees of freedom. Then substituting the appropriate quantities into this formula, we have

$$(26.25 - 9.75) \pm (1.943)(2.75) \sqrt{\frac{2}{4}} = 16.50 \pm 3.78$$

Thus we estimate that the difference between mean sales for the two package designs lies in the interval from 12.72 to 20.28. A much smaller confidence interval can be acquired by increasing b, the number of supermarkets included in the experiment.

Exercises

13.8. Benson reported the average annual overhead per loan, excluding advertising (in 1970 dollars), for the three leading nationwide consumer finance companies.* His findings were as given in the table.

| | | | | | YEAR | | | | | |
COMPANY	1962	1963	1964	1965	1966	1967	1968	1969	1970	1971
A	9.68	9.80	12.11	12.13	12.46	14.31	15.37	16.64	19.50	18.92
B	14.26	14.10	13.95	14.57	14.04	14.76	15.20	15.62	15.01	15.67
C	9.73	10.12	11.30	12.05	12.59	13.00	12.83	13.94	13.53	16.05

a. Do these data provide sufficient evidence to indicate a difference in mean overhead cost per loan for the three finance companies?

b. Use a 90% confidence interval to estimate the difference in mean overhead cost per loan between consumer finance companies A and C.

*G. J. Benson, "The Cost to Consumer Finance Companies of Extending Consumer Credit," *Report of the National Committee on Consumer Finance,* Technical Statistics, vol. II (1975).

13.9.   A study was undertaken to determine the relative typing speeds that could be obtained when using four different brands of typewriters. Each typewriter was assigned to each of eight secretaries, the order of assignment conducted in a random manner. The typing speed, in words per minute, for 10 minutes of typing was recorded for each secretary-typewriter combination. The data obtained are given in the table.

TYPEWRITER BRANDS	SECRETARY							
	1	2	3	4	5	6	7	8
A	79	80	77	75	82	77	78	76
B	74	79	73	70	76	78	72	74
C	82	86	80	79	81	80	80	84
D	79	81	77	78	82	77	77	78

a. Identify the design used for this experiment and justify your diagnosis.

b. Perform an analysis of variance on the data.

c. Do the data provide sufficient evidence to indicate that the mean typing speed for the secretaries varies from brand to brand of typewriters? Test by using $\alpha = .05$.

d. Why was the order of assignment of typewriters to each secretary conducted in a random manner? In general, what is the advantage of randomly assigning the treatments to the blocks?

13.10.   Refer to exercise 13.9. Let μ_C and μ_D, respectively, denote the mean typing speeds for using typewriter C and typewriter D. Find a 99% confidence interval for $(\mu_C - \mu_D)$. Interpret this interval.

13.11.   Reinmuth and Barnes studied the bidding behavior of three drilling contractors who opposed one another on a number of invitations to bid.* Their studies suggest that bidding contractors must not be too cautious, since conservative bidders, whose bid prices range on the high side, win few contract awards and thus take a chance of draining their revenues. The bid data provided by Reinmuth and Barnes are the bids recorded on five randomly selected invitations to bid. Each recorded bid represents the bid price for the hourly operating costs when using a 4000-foot-capacity drilling rig with a four-man crew. The data are shown in the table.

BIDDING CONTRACTOR	BIDDING TRIAL				
	1	2	3	4	5
A	$45.00	$46.00	$43.75	$44.50	$46.50
B	42.50	45.50	43.50	40.90	47.55
C	39.75	40.00	40.20	43.75	47.50

a. Do these data provide sufficient evidence to indicate a difference in the mean bid prices submitted by the three contractors?

*J. E. Reinmuth and J. D. Barnes, "A Strategic Competitive Bidding Approach to Pricing Decisions for Petroleum Industry Drilling Contractors," *Journal of Marketing Research*, August 1975.

b. Is there evidence to suggest a difference in mean bid price over the bid trials?

c. Suppose that prior to looking at the data, we had decided to compare the mean bids submitted by contractors A and C. Find a 90% confidence interval for this difference.

13.9   Some Cautionary Comments on Blocking

There are two major steps in designing an experiment and you need to be careful to distinguish between the two. The first step is deciding what treatments you wish to include in an experiment and how many observations to select per treatment. The treatments often may be the levels of a single qualitative or quantitative variable. For example, you might wish to compare the mean sales per store for three different brands of coffee. Then the three brands of coffee represent three levels of the qualitative variable "brands" and are the three treatments employed in the experiment. Or suppose you wish to see if the mean sales per brand varies with the type of product display in a store. If you had two types of product displays, say A_1 and A_2, and three brands, B_1, B_2, and B_3, then the six treatments would be the combinations of product displays and brands: A_1B_1, A_1B_2, A_1B_3, A_2B_1, A_2B_2, A_2B_3. The best strategies to employ in selecting treatments for an experiment and deciding on the number of observations per treatment are contained in a course on the design of experiments. You will find more information on this topic in the texts listed in the references.

The second step in designing an experiment is deciding how to apply the treatments to the experimental units. If the experimental units are one day of operation in a particular store, you could randomly assign the treatments to a group of stores (a completely randomized design that would yield independent random samples for each of the six treatments), or you could treat each store as a block and run all six of the brand-display combinations in each store. This would tend to block out store-to-store variation and, it is hoped, would reduce σ^2.

Sometimes people confuse these two steps in the design of an experiment and mistakenly use blocking patterns (intended for step 2) as a guide to selecting treatments and selecting the number of observations per treatment (step 1). For example, you should not use the configuration of a randomized block design (for example, figure 13.5) to investigate the effect of two variables on a response (letting the levels of one variable correspond to "treatments" and the levels of the second variable to "blocks"). Actually there is nothing wrong with *selecting* the treatments in this manner. The difficulty with this approach, however, is that the randomized block configuration provides only one observation per combination of variable levels. You can see that there is no way that you could compare the mean response for these treatments (various variable-level combinations) using only one observation per treatment. There would be no degrees of freedom available for estimating σ^2.

The correct way to design a multivariable investigation is to choose the various variable-level combinations to be included in the experiment (the treatments), decide approximately the number of observations needed per treatment, and then decide how to apply (a completely randomized design, randomized block design, etc.) the treatments to the experimental units.

A second point to note is that blocking is not always beneficial. Blocking produces a gain in information if the between-block variation is larger than the within-block variation. Then blocking removes this larger source of variation from SSE, and s^2 assumes a smaller value (as does the population variance σ^2) because of the design. At the same time, you lose information, because blocking reduces the number of degrees of freedom associated with SSE (and s^2). (To see this compare the ANOVA tables for the completely randomized design, table 13.3, and the randomized block design, table 13.5.)

Consequently, if blocking is to be beneficial, the gain in information due to the elimination of block variation must outweigh the loss due to a reduction in the number of degrees of freedom associated with SSE. Unless blocking leaves you with only a small number of degrees of freedom for SSE, the loss in degrees of freedom causes only a small reduction in information in the experiment. Consequently, if you have a reason for suspecting that there is block-to-block variation, it will usually pay to block.

13.10 The Analysis of Variance Assumptions

The assumptions about the probability distribution of the random response y are the same for the analysis of variance as for the regression models of chapters 11 and 12. That is, we assume that the probability distribution of y for any given treatment-block condition is normal, with common variance σ^2, and that the random errors associated with any pair of observations are independent.

In a practical situation you can never be certain that the assumptions are satisfied, but you will often have a fairly good idea whether the assumptions are reasonable for your data. To illustrate, it has been shown that the inferential methods of chapters 11, 12, and 13 are not seriously affected by moderate departures from the assumptions of normality, but you would want the probability distribution of y to be at least mound-shaped. So if y is a discrete random variable that can only assume three values, say $y = 10, 11, 12$, then it is *unreasonable* to assume that the probability distribution of y is approximately normal.

The assumption of a constant variance for y for the various experimental conditions should be approximately satisfied, although violation of this assumption is not too serious if the sample sizes for the various experimental conditions are equal. However, suppose the response is binomial, say the proportion p of people who favor a particular type of investment. We know that the variance of a proportion (section 6.4) is

$$\sigma_{\hat{p}}^2 = \frac{pq}{n} \qquad \text{where} \qquad q = 1 - p$$

and therefore that the variance is dependent on the expected (or mean) value of $\hat{p}$, namely, p. Then the variance of $\hat{p}$ will change from one experimental setting to another, and the assumptions of the analysis of variance have been violated.

A similar situation occurs when the response measurements are Poisson data (say the number of industrial accidents per month in a manufacturing plant). If the response possesses a Poisson probability distribution, the variance of the response equals the mean (section 6.5). Consequently, Poisson response data also violate the analysis of variance assumptions.

Many kinds of data are not measurable and hence are unsuitable for an analysis of variance. For example, many responses cannot be measured but can be ranked. Product preference studies yield data of this type. You know you like Product A better than B and B better than C but you have difficulty assigning an exact value to the strength of your preferences.

So what do you do when the assumptions of an analysis of variance are not satisfied? For example, suppose the variances of the responses for various experimental conditions are not equivalent. This situation can be remedied sometimes by transforming the response measurements. That is, instead of using the original response measurements, we might use their square roots, logarithms, or some other function of the response y. Transformations that tend to stabilize the variance of the response have been found to make the probability distributions of the transformed responses more nearly normal. See the texts listed in the references for discussions of these topics.

When nothing can be done to satisfy (even approximately) the assumptions of the analysis of variance, or if the data are rankings, you should use nonparametric testing and estimation procedures. These procedures, which rely on the comparative magnitudes of measurements (often ranks), are almost as powerful in detecting treatment differences as the tests presented in this chapter. Nonparametric tests, what they are and how to use them, are presented in chapter 18.

13.11 Summary

The completely randomized and the randomized block designs are illustrations of two experiments involving one and two qualitative independent variables, respectively. The analysis of variance partitions the total sum of squares of deviations of the response measurements about their mean into portions associated with each independent variable and the experimental error. The former may be compared with the sum of squares for error, using mean squares and the F statistic, to determine whether the mean squares for the independent variables are unusually large and thereby indicate an effect on the response.

This chapter has presented a brief introduction to the analysis of variance and its associated subject, the design of experiments. Experiments can be designed to investigate the effect of many quantitative and qualitative variables on a response. These may be variables of primary interest to us as well as nuisance variables, such as blocks, which we attempt to separate from the experimental error. The experiments are subject to an analysis of variance when properly designed. A more extensive coverage of the basic concepts of experimental design and the analysis of experiments can be found in the texts listed in the references.

Supplementary Exercises

13.12. State the assumptions underlying the following experimental design models:
a. a completely randomized design   b. a randomized block design

13.13. When choosing an experimental design:
a. Discuss the advantages of blocking.
b. What happens to these advantages as you increase the size of the blocks (the number of experimental units per block)?

13.14. A study was undertaken to compare the productivity of the operators of four identical assembly machines. Production records were examined for three randomly selected days, where the days of record were not necessarily the same for any two assembly machine operators. The data are given in the table. Assuming

OPERATOR 1	OPERATOR 2	OPERATOR 3	OPERATOR 4
230	220	215	225
220	210	215	215
225	220	220	225

that the requirements for a completely randomized design are met, analyze the data. State whether there is statistical support at the $\alpha = .05$ level of significance for the conclusion that the four assembly machine operators differ in average daily productivity.

13.15. Refer to exercise 13.14. Let μ_1 and μ_2 denote the mean production rates of operators 1 and 2, respectively.
a. Find a 90% confidence interval for μ_1. Interpret this interval.
b. Find a 95% confidence interval for $(\mu_1 - \mu_2)$. Interpret this interval.

13.16. A trucking company wished to compare three makes of trucks before ordering an entire fleet of one of the makes. Purchase prices for each of the makes were about the same and thus were ignored in the comparison. Five trucks of each make were run 5000 miles each and the average variable cost of operation per mile was noted for each truck. However, because of tire failure, accidents, and driver illness, two make B and two make C trucks did not complete the 5000-mile test. For those that did finish, the results are as given in the table.

MAKE A	MAKE B	MAKE C
17.3	15.4	17.9
18.3	17.4	19.5
17.6	17.1	18.7
16.8		
18.0		

    a. Perform an analysis of variance for this experiment.

    b. Do these data provide sufficient evidence to indicate a difference in the average variable cost per mile of operation for the three makes of trucks? Use α = .05.

    c. Is there an advantage in having the same number of measurements within each treatment in a completely randomized design? Explain.

    13.17.   Refer to exercise 13.16. Let μ_A and μ_B denote, respectively, the mean variable cost per mile of operating a truck of make A and make B.

    a. Find a 95% confidence interval for μ_A.

    b. Find a 95% confidence interval for μ_B.

    c. Find a 95% confidence interval for $(\mu_A - \mu_B)$.

    d. Is it correct to assume that the confidence interval computed in part c can be obtained as the difference between the confidence intervals found in parts a and b? Explain.

    13.18.   One researcher describes an application of an experimental design to a pricing experiment conducted at the retail store level.* The study involved a product line of the Quaker Oats Company, which manufactures three products, A, B, and C. Product C was considered adequately profitable, but price increases of 4 cents per package were sought for products A and B. To explore the effects of increasing the price on one or both of the products, Quaker chose four retail stores where the sales for A and B were observed to be relatively constant and approximately equal for each product for a period of time. In one store the prices on A and B were not changed; in another the price of A was increased by 4 cents; in the third the price of B was increased by 4 cents; and in the fourth the prices on both A and B were increased by 4 cents. Sales were recorded for each store for five randomly selected weeks after initiating the pricing experiment. The results are shown in the table. Do these data provide sufficient evidence to indicate a difference in the average combined sales of products A and B for any of the four price combinations?

PRICE COMBINATION	SALES (number of units sold)
A and B unchanged	54 46 68 62 57
A increased, B not	53 50 55 59 54
B increased, A not	57 51 58 59 56
both increased	51 45 52 50 45

    *Adopted from W. D. Barclay, "Factorial Design in a Pricing Experiment," *Journal of Marketing Research*, November 1969.

13.19.   Refer to exercise 13.18. Using a 95% confidence interval, estimate the average gain (loss) in revenue for products A and B in each store per week for the following conditions:

a. The price of A and B are both increased by 4 cents.

b. The price of A is left unchanged while the price of product B is increased by 4 cents per package.

13.20.   A cab company is conducting a study of three brands of tires before determining which brand to order for all their cabs. The study involved selecting four different tires from each brand and randomly assigning them to the left front wheel of 12 different cabs. The wear is recorded after 10,000 miles of use. The wear noted at the end of the test period is given in terms of the millimeters of tread wear.

BRAND A	BRAND B	BRAND C
462	250	319
421	336	425
470	322	460
411	268	380

Assume that the requirements for a completely randomized design are met and analyze the data. State whether there is statistical support at the $\alpha = .05$ level for the conclusion that the three brands of tires differ in resistance to wear.

13.21.   A study has been initiated to investigate the cleaning ability of three laundry detergents. Four different brands of automatic washing machines are to be used in the experiment, with each of three laundry detergents tested in each of the four washers. Thus 12 combinations exist within the experiment. Twelve stacks of laundry, containing an equal amount of soil, are to be laundered. At the completion of each wash load, the laundry is to be tested by a meter for "whiteness" and the results are to be recorded.

a. Is this a randomized block design? Explain.

b. Suppose that *two* stacks of soiled laundry are to be subjected to each of the three detergents in each of the four washing machines. What type of experimental design is this?

13.22.   Suppose that a marketing executive undertook a study to examine the comparative effect of three different promotional techniques (treatments) in four different sales areas (blocks) and obtained the results shown in the table. State precisely the conclusions that you would derive from the analysis of variance table. Perform all tests at the 5% level of significance.

SOURCE	d.f.	SS
blocks	3	.03
treatments	2	7.48
error	6	3.90
Totals	11	11.41

13.23.   A study was conducted to compare automobile gasoline mileage for three brands of gasoline, A, B, and C. Four automobiles, all of the same make and model, were employed in the experiment and each gasoline brand was tested in each automobile. Using each brand within the same automobile has the effect of eliminating (blocking out) automobile-to-automobile variability. The data, in miles per gallon, are given in the table.

GASOLINE BRAND	AUTOMOBILE			
	1	2	3	4
A	15.7	17.0	17.3	16.1
B	17.2	18.1	17.9	17.7
C	16.1	17.5	16.8	17.8

a. Do the data provide sufficient evidence to indicate a difference in mean mileage per gallon for the three gasolines?

b. Is there evidence of a difference in mean mileage for the four automobiles?

c. Suppose that prior to looking at the data, we had decided to compare the mean mileage per gallon for gasoline brands A and B. Find a 90% confidence interval for this difference.

13.24. Refer to exercise 13.23. Suppose that gasoline mileage is unrelated to the automobile in which it is used. Carry out an analysis of the data appropriate for a completely randomized design with three treatments.

a. Should a customer conclude that there is a difference in mean gasoline mileage for the three brands of gasoline? Test at the $\alpha = .05$ level.

b. Comparing your answer for part a in exercise 13.23 with your answer for part a above, can you suggest a reason why blocking may be unwise in certain cases?

13.25. A zoning commission has been formed to estimate the average appraisal value of houses in a residential suburb of a city. The commission is considering using one of three different appraisal models in their appraisal efforts. To test for consistency among the three appraisal models, each model is separately used to generate an appraisal value for each of five different residential dwellings. The results are given in the table. Without any specific directives, perform an analysis of the data.

APPRAISAL MODEL	DWELLING				
	1	2	3	4	5
A	$21,000	$37,000	$28,000	$37,000	$30,000
B	22,500	40,000	27,500	39,000	33,000
C	19,000	35,000	27,500	36,000	31,000

What are your conclusions? What recommendations would you make to the zoning commission about the relative merits of the three appraisal models? (Hint: Code the data.)

13.26. A portion of a questionnaire was constructed to enable judges to evaluate four proposed site locations. The judges, selected from the executive group and upper and middle management of the company, were asked to respond concerning their perceptions of accessibility of each site to primary markets, transportation facilities, state corporation regulations, and living desirability of the area relative to each proposed site. Their responses were then collated and coded on a 0-to-20 scale. The data obtained are shown in the table.

	JUDGES							
SITES	1	2	3	4	5	6	7	8
1	9	10	7	5	12	7	8	6
2	4	9	3	0	6	8	2	4
3	12	16	10	9	11	10	10	14
4	9	11	7	8	12	7	7	8

a. The primary objective of this experiment was to compare site locations. Give the type of design employed for this experiment and justify your diagnosis.

b. Perform an analysis of variance on the data.

c. Do the data provide sufficient evidence to indicate that the mean coded scores vary from site to site? Test by using $\alpha = .05$.

d. Suppose that the data did provide sufficient evidence to indicate differences among the mean coded questionnaire scores associated with the four sites. Would this imply that the questionnaire was able to detect a difference in preferences for the building sites?

13.27. Paper machines distribute a thin mixture of wood fibers and water to a wide wire mesh belt that is traveling at a very high speed. Thus it is conceivable that the distribution of fibers, thickness, porosity, and so on will vary along the belt and produce variations in the strength of the final paper product. A paper company designed an experiment to compare the strength of four coatings intended to improve the appearance of packaging paper. Because the uncoated paper strength could vary down a roll, the experiment was conducted as a randomized block experiment. The strength measurements for the four coatings A, B, C, and D were as shown in the table.

POSITION DOWN THE ROLL	COATING			
	A	B	C	D
1	10.4	12.4	13.1	11.8
2	10.9	12.4	13.4	11.8
3	10.5	12.3	12.9	11.4
4	10.7	12.0	13.3	11.4

a. Do these data present sufficient evidence to indicate a difference in mean strength for paper treated with the four paper coatings?

b. Do these data present sufficient evidence to indicate a difference in mean strength for locations down the roll?

c. Suppose that prior to seeing our data, we wished to compare the mean strength between paper coated with coatings A and C. Estimate the difference, using a 95% confidence interval.

13.28. A completely randomized design was employed to compare the effect of five different advertising layouts on product recognition time. Twenty-seven people were employed in the experiment. Regardless of the results of the analysis of variance, it is desired to compare layouts A and D. The results of the experiment were as shown in the table (time, in seconds).

	LAYOUT				
	A	B	C	D	E
	.8	.7	1.2	1.0	.6
	.6	.8	1.0	.9	.4
	.6	.5	.9	.9	.4
	.5	.5	1.2	1.1	.7
		.6	1.3	.7	.3
		.9	.8		
		.7			
Totals	2.5	4.7	6.4	4.6	2.4
Mean	.625	.671	1.067	.920	.48

a. Conduct an analysis of variance and test for a difference in mean recognition time resulting from the five advertising layouts.

b. Compare layouts A and D to see if there is a difference in mean recognition time.

13.29. The experiment in exercise 13.28 might have been more effectively conducted by using a randomized block design with subjects as blocks, since we would expect mean recognition time to vary from one person to another. Four people were used in a new experiment, and each person was shown each of the five advertising layouts in a random order. The results were as shown in the table (time, in seconds).

	LAYOUT				
SUBJECT	A	B	C	D	E
1	.7	.8	1.0	1.0	.5
2	.6	.6	1.1	1.0	.6
3	.9	1.0	1.2	1.1	.6
4	.6	.8	.9	1.0	.4

Conduct an analysis of variance and test for differences in treatments (advertising layouts). Test at the $\alpha = .05$ level.

Experiences with Real Data

Listed below are two exercises which require, respectively, the use of a randomized block experimental design and a completely randomized experimental design. For each exercise perform a complete analysis of variance of the experimental data you have collected. Although you can easily perform the analyses with the assistance of an electronic desk calculator, you may wish to use an available computer library program and an electronic computer to perform the calculations. Check with your campus computing center about the availability of library programs involving the randomized block and completely randomized experimental designs.

1. How do the performances of securities from different investment classes

compare with regard to the stability of average yield over time? Certainly this is one of the first questions that would logically be posed by an investor who is considering a number of different investment portfolios. To study this issue randomly select a list of five companies in the three classes industrial, transportation, and utilities. For instance, you might select the companies shown in the table. From your campus

INDUSTRIAL	TRANSPORTATION	UTILITIES
AMF	Chessie System	Allegheny Power System
Bethlehem Steel	Greyhound	Cleveland Electric
Eaton	Illinois Central	Consolidated Edison
Talley Industries	Norfolk & Western	Long Island Lighting
Texaco	Transway International	Pacific Power and Light

library, record the average annual yield (dividend as a percentage of price) for each company you have chosen for each of the past four years. Within each investment class, compute the average annual yield for each year by averaging the annual yields for the five companies included in that class. This information can be obtained from published sources such as those provided by Dun and Bradstreet and Standard and Poor. This will give you a 3 by 4 table, with investment classes as treatments and years as blocks. Perform an analysis of variance on these data.

    a. Do your data provide sufficient evidence to assume that the average yield differs according to the class in which the issuing company belongs?

    b. Does the average yield differ from year to year?

    c. Discuss the advantage of blocking on years in this problem. That is, discuss the benefits of performing this analysis as a randomized block experiment instead of ignoring the yearly differences and performing the analysis as a completely randomized experiment.

    2. Compare the cost of a bag of groceries in at least three different grocery chains located in your region. Make a list of common market basket items, such as a pound of ground beef, a pound of pork chops, a gallon of milk, one dozen large AA eggs, a loaf of whole wheat bread, and so on. Record the price for each of the items in n_1 stores from grocery chain A, for each item in n_2 stores from chain B, and so forth. You may wish to consider small neighborhood stores and independent supermarkets as separate "chains" in your study. Then find the total cost for the groceries on your list for each store. Do your recorded prices indicate that the various grocery chains considered in your study differ in their pricing of commonly purchased grocery items?

References

GUENTHER, W. C. *Analysis of Variance.* Englewood Cliffs, N.J.: Prentice-Hall, 1964.

HICKS, C. R. *Fundamental Concepts in the Design of Experiments.* New York: Holt, Rinehart and Winston, 1964.

LI, J. C. R. *Introduction to Statistical Inference.* Ann Arbor, Mich.: J. W. Edwards, Publisher, 1961.

MENDENHALL, W. *An Introduction to Linear Models and the Design and Analysis of Experiments.* Belmont, Calif.: Wadsworth, 1968. Chapters 10-13.

NETER, J., and W. WASSERMANN. *Applied Linear Statistical Models.* Homewood, Ill.: Richard D. Irwin, 1974. Chapters 13-24.

chapter objectives

GENERAL OBJECTIVES In this chapter we examine the effect of time as a variable in business decision-making problems. First, we focus on the components of a time series and how certain components suggest the presence or absence of certain implied characteristics of the process that is generating the time series. Smoothing methods are offered to assist in the discovery of the hidden components. Second, we discuss the use of index numbers for making meaningful comparisons over time. We will show how to derive and interpret price indexes and will devote particular attention to the interpretation of some of the more common published indexes.

SPECIFIC OBJECTIVES

1. To define a time series and to interpret and illustrate the components of a time series.

Sections 14.1, 14.2

2. To illustrate the use of smoothing methods to aid in identifying hidden components and in removing the effect of a seasonal component in a time series.

Sections 14.3, 14.4

3. To explain the use of index numbers for making meaningful monetary comparisons over time.

Section 14.5

4. To discuss the usefulness and limitations of the more common published indexes, such as the consumer price index, the wholesale price index, and the Dow-Jones industrial average.

Section 14.5

chapter fourteen

Elements of Time Series Analysis

14.1 Introduction

Businesspeople are constantly faced with variables whose values are random over time. Time, either as an independent experimental variable or as an added dimension to other variables of interest, may help or hinder them in their decision-making processes. We have seen in previous chapters that many unmeasured and uncontrolled variables may cause a response to vary over time and thereby inflate the experimental error. The undesirable effect of time can be reduced by using blocking designs (such as the randomized block design of chapter 13) and by making experimental treatment comparisons within relatively homogeneous blocks of time.

In other decision-making situations, time may be one of the most important variables. This usually occurs in problems of the type described in chapter 12, where we wish to estimate the expected value of y or to predict a new value of y at a future point in time after having observed the pattern of the values of y during past and present time periods. For example, an investor is interested in predicting future security prices, a store manager is interested in the effect of time on demand for products, and the marketing manager is interested in the pattern of sales over time. In a sense, everyone who plans for the future by attempting to budget time and resources is concerned with processes that are random over time. If we believe that interest rates will drop in the near future, we may be wise to rent now and buy a home later. The skier plans vacations during the winter because the seasonal weather pattern calls for the greatest snowfall during the winter.

Any sequence of measurements taken on a response that is variable over time is called a time series. The time series is usually represented by the

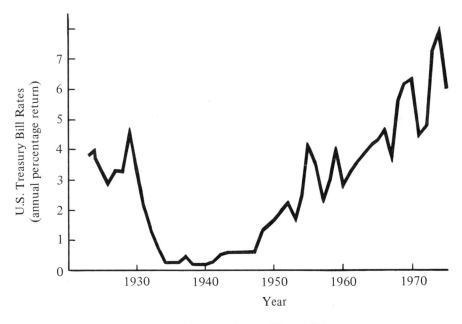

Figure 14.1 *U.S. Treasury bill rates from 1925 to 1976*

mathematical equation listing the values of the response as a function of time, or, equivalently, as a curve on a graph whose vertical coordinate gives the value of the random response plotted against time on the horizontal axis. In figure 14.1 we show a time series that plots U.S. Treasury bill rates as a function of time between the years 1925 and 1976. The same information can be shown in tabular form, but the pattern of change over time would be much more obscure. It is the **pattern** generated by the time series and not necessarily the individual values that offers the planning device. An individual who is planning a long-term investment portfolio will want to buy treasury bills only at times when the rates are projected to be high. Projecting the treasury bill rates into the future implies extending the trends, cycles, and other elements of the pattern from the time series of figure 14.1.

Applications of time series analytical techniques are not limited to business problems concerned solely with economic data. Demand analyses are very common in marketing research problems. Quality control studies are also common business applications of time series analyses. By employing computer feedback systems, modern industrial process control theory utilizes multiple time series to keep the industrial manufacturing process "in control" and to correct and adjust the process when it is found to be "out of control." Furthermore, such a study enables the manufacturer to move to positions of higher product quality and yield and to greater manufacturing profits.

The analysis of time series—that is, the utilization of sample data for purposes of inference (estimation, decision making, and prediction)—is a complicated and difficult subject. Response measurements appearing in a time series are usually correlated, with the correlation increasing as the time interval between a pair of measurements decreases. Consequently, time series data often defy the basic assumptions of independence required for the methods described in previous chapters. Indeed, the methodology of time series analysis is in an embryonic state in comparison with that for the static (time-independent) case with which we are so familiar. For this reason published time series analyses often appear to be based on highly subjective methods, and predictions (forecasts) are most often unaccompanied by a measure of goodness. Perhaps a second reason for the rather primitive techniques often employed in the analysis of time series is the mathematical complexity of the theory underlying the more sophisticated and newer methodologies. The mathematical background required for understanding some of the more powerful methods of time series analysis places the subject beyond the grasp of many nonmathematically trained forecasters.

The preceding comments are intended to introduce the subject of time series analysis and also to explain the absence of a single method of analysis. We will begin by first exploring the components of a time series, which together determine its pattern. In chapter 15 we will examine some of the available techniques for extending the pattern of the time series and forecasting future values. Because of the importance of forecasting in modern business decision making, our emphasis in the study of time series methods will be placed on the analytical methods of chapter 15 rather than on the descriptive methods of this chapter.

14.2 Components of Time Series

Statisticians often think of a time series as the addition of four meaningful *component series:*

Components of Time Series

1. long-term trend
2. cyclical effect
3. seasonal effect
4. random variation

Long-term trends are often present in time series because of the steady increase in population, gross national product, the effect of competition, or other factors, that fail to produce sudden changes in response but produce a steady and gradual change over time. A time series with an upward long-term trend is similar to the increase in the operating revenues of U.S. oil pipeline companies from 1955 to 1976, as shown in figure 14.2.

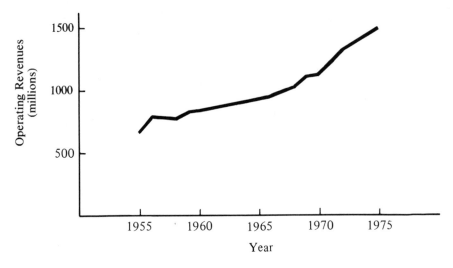

Figure 14.2 *Operating revenues of U.S. oil pipeline companies from 1955
to 1976*

Cyclical effects in a time series are apparent when the response rises
and falls in a gentle, wavelike manner about a long-term trend curve, like
the unemployment rate in the United States from 1900 to 1976, illustrated
in figure 14.3. The unemployment rate does not seem to follow an increasing
or decreasing trend over time but appears to fluctuate with, say, general
conditions of the economy, demands from the military service, and the onset

Figure 14.3 *Unemployment rate in the United States from 1900 to 1976*

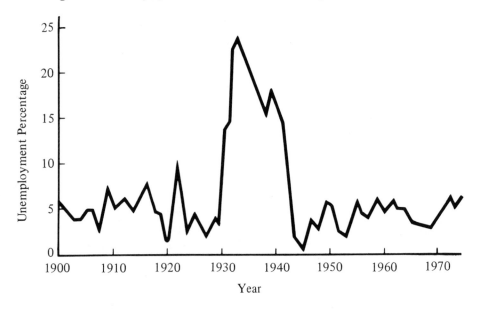

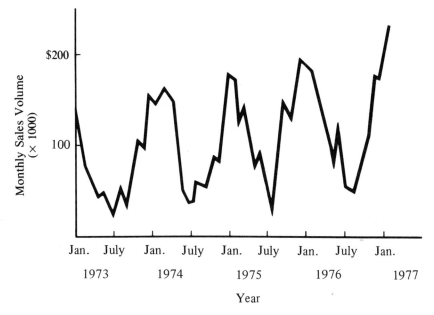

Figure 14.4 *Monthly retail sales volume of an urban department store from 1973 to 1977*

of automation. Generally, cyclic effects in a time series can be caused by changes in the demand for a product, business cycles, stockpiling, and, particularly, the inability of supply to meet exactly the requirements of customer demand when the time series is a plot of economic values over time. For a noneconomic time series, cyclic effects are usually the effect of governmental, economic, or political policies.

Seasonal effects in time series are those rises and falls that always occur at a particular time of the year. For example, auto sales and earnings tend to decrease during August and September because of the changeover to new models, while the sales of television sets rise in the month of December. The essential difference between seasonal and cyclic effects is that seasonal effects are predictable, occurring at a given interval of time from the last occurrence, while cyclic effects are completely unpredictable. A time series component suggesting seasonal effects is illustrated by the sales of a department store, shown in figure 14.4.

The fourth component of a time series is random variation. This component represents the random upward and downward movement of the series after adjustment for the long-term trend, the cyclic effect, and the seasonal effect. Random variation, which might appear as the variation from the increasing linear trend shown in figure 14.5, is the unexplained shifting and bobbing of the series over the short-term period. Political events, weather, and an amalgamation of many human actions tend to cause random and unexpected changes in a time series.

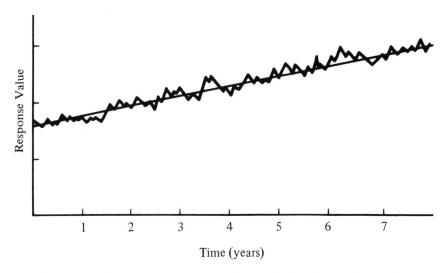

Figure 14.5 *Time series that shows random variation from an increasing linear trend*

All time series contain random variation. In addition, a time series may contain none, one, two, or all of the three components long-term trend, cyclic effect, and seasonal effect. The objective of a time series analysis is to identify the components which do exist in order to identify their causes and to forecast future values of the time series.

With most time series processes it is not easy to distinguish between

Figure 14.6 *Monthly retail sales volume of an urban department store from 1973 to 1977*

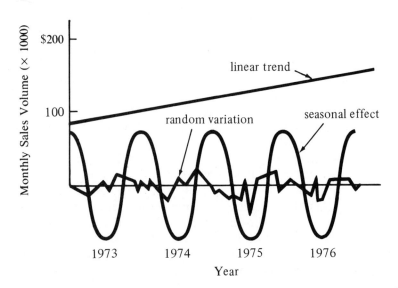

the components. Often seasonal and cyclic effects, or the three components long-term trend and cyclic and seasonal effects, have become so integrated that they are inseparable. On the other hand, if the components appear to be distinguishable, it is not difficult to separate them. For instance, the monthly sales of the urban department store shown in figure 14.4 illustrate a seasonal effect and a long-term trend with superimposed random variation. The long-term trend and seasonal effect, when identified, can be subtracted from the response values. The remainder is attributable to random variation. An illustration of how this might be accomplished is shown in figure 14.6.

The communications engineer refers to the long-term trend and the cylic and seasonal effects as the signal of the time series. We will use this terminology for purposes of discussion. Since the signal is the part of the time series that is deterministic, it is important for purposes of prediction to be able to separate the signal from random variation, called noise by the communications engineer. Determination of future values would simply require the statistician to add the extended patterns of the known components of the signal to the projected random variation.

Since the random variation is at best probabilistic, accurate estimation of future values can be expected only when the magnitude of the random variation, as measured by its variance, is small. Otherwise, the oscillations (fluctuations) of the random variation over time may overwhelm the effect of the signal components or even cancel them out entirely. Such a process is illustrated in figure 14.7, where the seasonal effect and long-term trend from figure 14.6 are combined with a random variation with a large variance. The resultant time series illustrates only the long-term trend, the seasonal

Figure 14.7 *Time series whose seasonal effect has been hidden by the presence of random variation with excessively large variance*

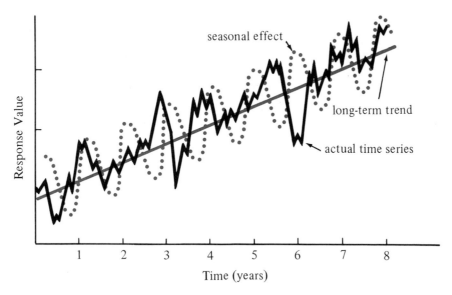

effect having been "hidden" by the random variation.

Even if the proper signal is discovered and projected, the predicted values may be inaccurate if the magnitude of the random variation is great. In that case the best we can do is to give a probability interval for the predicted value, where the probability interval is based on the supposed probability distribution of the random variation.

Exercises

14.1.   List and define the components of a time series.

14.2.   Which of the time series components would you expect to be present in each of the following series?

a. The interest rate on 30-year first mortgages issued by the Seattle First National Bank from January 1967 to January 1977.

b. The quarterly earnings of the Exxon Corporation for the years 1965 through 1976.

c. The monthly sales of skiing equipment by the sporting goods division of a large retail department store chain over the years 1965 through 1976.

d. The consumer price index as computed by the Bureau of Labor Statistics for the months from January 1962 through December 1977.

e. The number of tourists visiting Hawaii during each month from January 1966 to January 1976.

14.3.   The time series in figure 14.8 shows sugar prices at world market (in dollars per 100 pounds) for the years 1966 through 1973. Which time series components appear to exist within this time series? Break down the sugar price time series into

Figure 14.8 *Sugar price at world market, exercise 14.3*

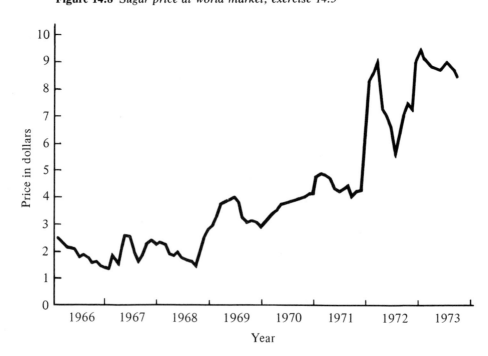

its components as was done in figure 14.6 for the sales of an urban department store.

14.4. The data given in the table represent the number of tourists (in thousands) visiting the state of Hawaii for each month of two consecutive years.

| | NUMBER OF TOURISTS | |
MONTH	Year 1	Year 2
January	15.592	18.024
February	15.169	18.806
March	15.469	21.707
April	12.085	13.463
May	13.037	14.930
June	11.584	12.287
July	12.798	16.248
August	11.782	11.466
September	12.258	12.672
October	14.808	13.630
November	12.472	14.422
December	13.069	14.441

Source: M. D. Geurts and J. B. Ibrahim, "Comparing the Box-Jenkins Approach with the Exponentially Smoothed Forecasting Model: Application to Hawaii Tourists," *Journal of Marketing Research,* May 1975.

a. Plot the Hawaii tourist traffic against time and construct the time series.
b. Which time series components appear to exist in the sales pattern?

14.5. Suppose a time series exhibits a sudden shift from its normal pattern. How can the decision maker determine whether the shift is an indication of a lasting change in the time series (a new or revised signal) or is simply random variation? Which component was the cause of the drastic increase in sugar prices noticed in figure 14.8?

14.3 Smoothing Methods

Traditional methods of time series analysis have rested heavily on **smoothing techniques** that attempt to cancel out the effect of random variation and presumably reveal the components that are being sought. Smoothing can be accomplished by utilizing a **moving average** of the response measurements over a fixed number of time periods. Thus we might average the monthly sales of a company over a four-month period and plot the average at the midpoint of the four-month time interval. The next point in the series would be obtained by adding to the four-month total the next month's sales in the series, dropping the sales from the earliest month of the previous four, and computing a new four-month average. Thus the time series of moving averages shows a point for each month that is the calculated average response for a specified time interval below and above the given month. The net effect is to transform the original sales time series to a moving-average series that is smoother (less subject to rapid oscillations) and more likely to reveal the underlying trend or cycles in the pattern of sales over time.

At time point t the moving average $\bar{y}_t$ of the response measurements over M time periods is found by computing

$$\bar{y}_t = \frac{y_{t-(M-1)/2} + y_{t+1-(M-1)/2} + y_{t+2-(M-1)/2} + \cdots + y_{t+(M-1)/2}}{M}$$

where M is an odd number and y_t is the process response at time t, y_{t-1} is the process response at time $t - 1$, and so forth. For instance, if the moving average is to be computed over three period intervals ($M = 3$), the first few moving averages are computed as follows:

$$\bar{y}_2 = \frac{y_1 + y_2 + y_3}{3}$$

$$\bar{y}_3 = \frac{y_2 + y_3 + y_4}{3}$$

$$\bar{y}_4 = \frac{y_3 + y_4 + y_5}{3}$$

and so forth. The moving-average formula can be simplified by rewriting it in the form shown here (called a recursive form):

$$\bar{y}_t = \bar{y}_{t-1} + \frac{\text{next observation} - \text{most remote observation}}{M}$$

since all we are doing at each step is to recompute the average by adding in the next observation and dropping the observation that occurred M periods in the past.

If M is an even number, the moving averages will occur *between* the time points instead of *at* the time points. Thus it is often constructive to compute the moving average over an odd number of time periods so that we have values of comparison that are actual values.

The primary disadvantage of using a moving average for smoothing is that unless $M = 1$, we do not have a smoothed value corresponding to each response value. For instance, if we compute the moving averages over each $M = 5$ consecutive responses of the time series, there would be no smoothed value corresponding to the first two or to the last two values. With a large number of response measurements, this is no problem, but it may be a serious consideration when the number of response measurements is small.

Another smoothing scheme, called **exponential smoothing**, is more efficient than the moving average in the sense that it computes a smoothed value corresponding to each response measurement. The exponentially smoothed response value at time period t is denoted by S_t. The smoothing scheme begins by assigning $S_1 = y_1$ at the first period. For the second time period,

$$S_2 = \alpha y_2 + (1 - \alpha)S_1$$

and for any succeeding time period t, the smoothed value S_t is found by computing

$$S_t = \alpha y_t + (1 - \alpha)S_{t-1} \qquad 0 \le \alpha \le 1$$

This equation is called the **basic equation of exponential smoothing** and the constant α is called the **smoothing constant.**

The symbol α is used exclusively throughout chapters 14 and 15 to represent the smoothing constant, in spite of the fact that α was defined as the probability of making a type I error, or the significance level of a statistical test, in chapter 8. Since smoothing is discussed only in chapters 14 and 15, this double usage of the symbol α should not present a problem. We have chosen not to use a symbol different from α in either case since both uses of α are traditional in statistical literature.

The moving-average smoothing scheme forms averages over M time periods, but S_t computes an average from all past values y_t, y_{t-1}, ..., y_1, where y_t is the value at the time period t, y_{t-1} is the value at time period $t - 1$, and y_1 is the value from the first time period in which data are available. This can be seen if we expand the basic equation by first substituting

$$S_{t-1} = \alpha y_{t-1} + (1 - \alpha)S_{t-2}$$

into the equation for S_t to obtain

$$S_t = \alpha y_t + (1 - \alpha)\alpha y_{t-1} + (1 - \alpha)^2 S_{t-2}$$

By substituting for S_{t-2}, then for S_{t-3}, and so forth, until we substitute y_1 for S_1, it can be shown that the expanded equation can be written as

$$S_t = \alpha \sum_{i=0}^{t-2} (1 - \alpha)^i y_{t-i} + (1 - \alpha)^{t-1} y_1$$

Even though remote responses are not dropped in an exponential smoothing scheme as they are in a moving average, their contribution to the smoothed value S_t becomes less at each successive time point. The speed at which remote responses are dampened (smoothed) out is determined by the selection of the smoothing constant α. For values of α near 1, remote responses are dampened out quickly; for α near 0, they are dampened out slowly.

The theory of selecting the "best" smoothing constant α is omitted here because it requires more than an elementary knowledge of mathematics. Although it is beyond the scope of this text, we will make some general remarks about its selection. When the underlying response is quite "volatile" (the magnitude of the random variation is large), we would like to average out the effects of the random variation quickly. Thus we would select a small smoothing constant so that the smoothed value S_t will reflect S_{t-1}, the averaged values from the first $(t - 1)$ time periods, to a greater extent than it reflects the "noisy" measurement y_t. Similarly, for a moderately stable process, a large smoothing constant would be selected.

The following example illustrates the use of three different smoothing models on a rather volatile time series.

Example 14.1 The week's end closing prices for the securities of the Color-Vision Company, a manufacturer of color television sets, have been recorded over a period of 30 consecutive weeks. Find the 5-week moving-average time series and the exponentially smoothed time series, using smoothing constants $\alpha = .1$ and $\alpha = .5$.

Solution The original process values y_t (week's end closing prices for each of the 30 weeks), the 5-week moving averages $\bar{y}_t$, and the exponen-

Table 14.1 *Original and smoothed week's end closing prices for the securities of the Color-Vision Company over 30 consecutive weeks**

t	y_t	$\bar{y}_t$	$S_t(\alpha = .1)$	$S_t(\alpha = .5)$
1	71		71.0	71.0
2	70		70.9	70.5
3	69	68.4	70.7	69.8
4	68	67.2	70.4	68.9
5	64	67.6	69.8	66.5
6	65	69.4	69.3	65.8
7	72	70.8	69.6	68.9
8	78	73.0	70.4	73.5
9	75	75.0	70.9	74.3
10	75	74.6	71.3	74.7
11	75	74.0	71.7	74.9
12	70	74.0	71.5	72.5
13	75	73.8	71.9	73.8
14	75	74.4	72.2	74.4
15	74	77.6	72.4	74.2
16	78	79.0	73.0	76.1
17	86	79.0	74.3	81.1
18	82	78.8	75.1	81.6
19	75	77.6	75.1	78.3
20	73	75.2	74.9	75.7
21	72	73.0	74.6	73.4
22	73	73.4	74.4	73.5
23	72	75.4	74.2	72.8
24	77	77.2	74.5	74.9
25	83	78.8	75.4	79.0
26	81	81.4	76.0	80.0
27	81	83.0	76.5	80.5
28	85	83.2	77.4	82.8
29	85		78.2	83.9
30	84		78.8	84.0

*All smoothed values have been rounded to the nearest tenth of a unit.

tially smoothed time series S_t for $\alpha = .1$ and $\alpha = .5$ are listed in table 14.1. The moving averages were found by computing

$$\bar{y}_t = \frac{y_{t-2} + y_{t-1} + y_t + y_{t+1} + y_{t+2}}{5}$$

for each of the time periods $t = 3, 4, ..., 28$. For instance, the seventh moving-average value $\bar{y}_7$ was found by computing

$$\bar{y}_7 = \frac{y_5 + y_6 + y_7 + y_8 + y_9}{5} = \frac{64 + 65 + 72 + 78 + 75}{5} = 70.8$$

The original time series and the moving-average time series are shown together in figure 14.9.

The exponentially smoothed time series employing a smoothing constant $\alpha = .1$ was computed by first setting

$$S_1 = 71.0$$

and then computing

$$S_2 = (.1)(70) + (1 - .1)(71.0) = 70.9$$

$$S_3 = (.1)(69) + (1 - .1)(70.9) = 70.7$$

and so forth.

Figure 14.9 *Week's end closing security prices for the Color-Vision Company over 30 weeks with the 5-week moving-average process superimposed*

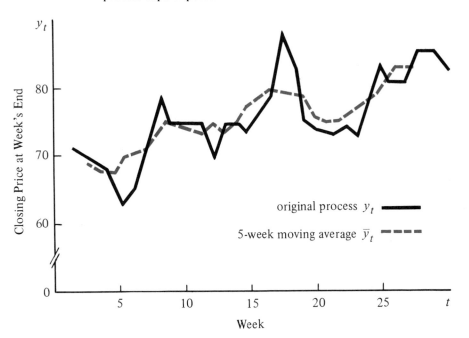

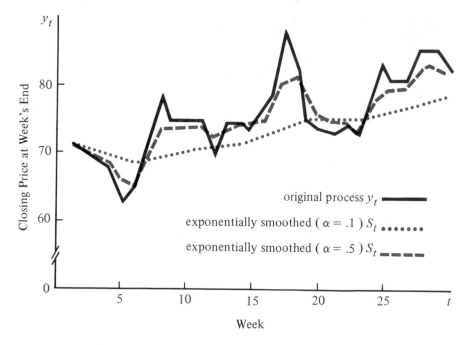

Figure 14.10 *Week's end closing security prices for the Color-Vision Company over 30 weeks, with the two exponentially smoothed processes superimposed*

Similarly, each of the values for the exponentially smoothed time series with $\alpha = .5$ are found by first setting

$$S_1 = 71.0$$

and then computing

$$S_t = (.5)y_t + (1 - .5)S_{t-1}$$

for each of the time periods $t = 2, 3, 4, ..., 30$. The original time series is plotted together with both of the exponentially smoothed time series in figure 14.10.

Observe that in both figures 14.9 and 14.10, the smoothed time series appear more stable than the original series. However, the 5-week moving-average series and the exponentially smoothed time series with $\alpha = .5$ appear to be much less stable than the exponentially smoothed series with $\alpha = .1$. The latter series, although undershooting the original series much of the time, appears to suggest the presence of a linear trend with a cyclic effect. Hence the true components of the original series (if a linear trend and a cyclic effect are the true components) most readily become apparent when the original series is smoothed by an exponential smoothing scheme employing a small smoothing constant. This is not a generalization, however; the small smoothing constant happens to yield the best results for the data used here.

Perhaps the major advantage of smoothing techniques is typified by the old saying that a picture is worth a thousand words. Moving averages and exponentially smoothed time series sometimes make trends, cycles, and seasonal effects more visible to the eye and consequently lead to a simple and useful description of the time series process for the businessperson or economist. When presented simply as a description of the time series (as is often the case), smoothing techniques often ignore the basic objective of statistics—inference—and leave this difficult task to the reader. Thus we still seek techniques for estimation and prediction that are accompanied by measures of goodness. We will investigate such methods in chapter 15.

Exercises

14.6.   Refer to exercise 14.4. Smooth the monthly Hawaii tourist traffic time series by computing a three-month moving average. Plot the smoothed series and the original series on the same sheet of graph paper. Now repeat this process for a five-month moving average. What conclusions do you draw concerning the inherent components within the time series? Which moving-average model was most useful in detecting these inherent components?

14.7.   What is the main disadvantage of the moving-average smoothing technique? How can the effect of this disadvantage be minimized?

14.8.   The accompanying table shows the number of miles flown by air carriers on international routes for each month over five consecutive years. To simplify recording, the data have been coded in millions of passenger miles.

	YEAR				
MONTH	1	2	3	4	5
January	284.0	315.1	340.1	359.9	416.9
February	276.9	300.9	317.9	342.0	391.1
March	317.0	356.0	362.1	405.8	419.0
April	312.9	347.9	347.9	395.8	460.8
May	317.9	354.9	362.8	419.8	472.0
June	373.9	421.9	434.8	472.0	534.8
July	412.8	466.8	489.8	547.8	622.0
August	405.0	466.8	505.2	558.9	606.0
September	354.9	403.8	403.8	463.1	507.7
October	306.1	346.8	358.8	407.0	460.8
November	270.9	304.9	310.1	362.1	389.9
December	306.1	335.9	336.9	405.0	431.8

a. Smooth the monthly international airline traffic data by computing a five-month moving-average series.

b. Smooth the monthly international airline traffic data by computing an exponentially smoothed series, employing the smoothing constant $\alpha = .1$.

c. Smooth the monthly international airline traffic data by computing an exponentially smoothed series, employing the smoothing constant $\alpha = .4$.

d. On a sheet of graph paper, superimpose the three smoothed series on the original series. What conclusions do you draw?

14.4 Adjustment of Seasonal Data

Suppose we are interested in examining short-term trends or the effect of an assumed business cycle on the time series representing business activity. This is a difficult, if not impossible, task if the time series exhibits a pronounced seasonal component, since the seasonal fluctuations tend to overwhelm the other components. If the seasonal component can be removed from the time series, identification, examination, and interpretation of trends and cycles is greatly simplified.

The most common method for removing the seasonal component from a time series containing a seasonal component is by employing a moving-average model, with M equal to the number of time points in one complete seasonal period. In most seasonal time series, the seasonal period is either 4 or 12 time periods. The former are time series consisting of **quarterly** data, like quarterly jewelry or toy sales, typically relatively constant during the first three quarters of each year but much higher during the fourth quarter owing to the effect of the Christmas market. When the time points are months and the time series is seasonal, the seasonal period is almost always 12 months. Sales of soft drinks, beer, and golf equipment follow the seasons—sales are high during the warm months, low during the cool months. The demand and sales for air conditioners, clothing, skiing equipment, gardening equipment, airline travel, and many other items also follow a 12-month seasonal pattern.

The moving-average smoothing technique is used to remove the seasonal component from a time series by performing the following computations:

1. Let M equal the number of time periods in one complete seasonal period.
2. Compute the M-period moving average for the time series.
3. If M is an even number, center the moving-average values by averaging adjacent moving averages.

The obvious disadvantage of the moving-average deseasonalization method is that $M/2$ observations are lost at the beginning and at the end of the time series. Typically this is not a serious problem if the time series contains a sufficient number of measurements. A rule of thumb is that we should have **at least three complete seasons** represented within the time series before the moving-average deseasonalization methods are applied.

As an illustration of the moving-average deseasonalization technique, consider the time series of monthly sales of a regional brewing company for the years 1974–1976, as shown in the first column of table 14.2. Looking ahead to figure 14.11, we can see that this time series follows a 12-month seasonal pattern. Thus we let $M = 12$.

The first moving-average value falls between June and July of 1974 since $M = 12$ is an even number. We find this first moving-average value to be

Table 14.2 *Actual sales and 12-month moving-average time series of sales of a brewing company, 1974–1976 (thousands of barrels)*

TIME	ACTUAL SALES	12-MONTH MOVING AVERAGE	CENTERED MOVING AVERAGE
1974			
January	19.6		
February	18.6		
March	23.2		
April	24.5		
May	27.7		
June	30.0		
July	28.7	24.667	24.821
August	33.8	24.975	25.038
September	25.1	25.100	25.304
October	22.1	25.508	25.596
November	21.8	25.683	25.721
December	20.9	25.758	25.896
		26.033	
1975			
January	23.3	26.500	26.267
February	20.1	26.100	26.300
March	28.1	26.208	26.152
April	26.6	26.458	26.333
May	28.6	26.500	26.479
June	33.3	26.450	26.475
July	34.3	26.558	26.504
August	29.0	26.783	26.671
September	26.4	26.808	26.796
October	25.1	26.858	26.833
November	22.3	26.858	26.858
December	20.3	26.525	26.692
1976			
January	24.6	26.858	26.692
February	22.8	27.108	26.983
March	28.4	26.983	27.046
April	27.2	27.200	27.092
May	28.6	27.192	27.196
June	29.3	27.292	27.242
July	38.3		
August	32.0		
September	24.9		
October	27.7		
November	22.2		
December	21.5		

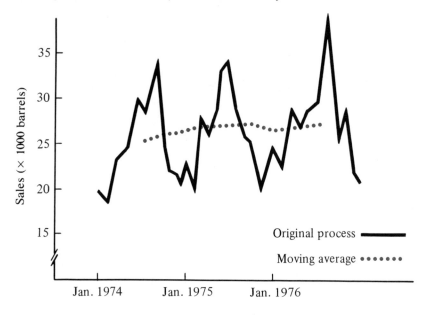

Figure 14.11 *Sales volume for the brewing company for the years 1974, 1975, and 1976, with a superimposed 12-month moving average*

$$\bar{y}_{\text{June-July}} = \frac{\begin{aligned}19.6 + 18.6 + 23.2 + 24.5 + 27.7 + 30.0 + 28.7 + 33.8 \\ + 25.1 + 22.1 + 21.8 + 20.9\end{aligned}}{12}$$

$$= \frac{296}{12} = 24.667$$

Similarly, we find

$$\bar{y}_{\text{July-Aug}} = \frac{296}{12} + \frac{23.3 - 19.6}{12} = \frac{299.7}{12} = 24.975$$

Notice that the July–August value is computed by using the shortcut procedure introduced in section 14.3 (p. 513). The shortcut procedure cannot be used to compute the first moving average but can be used to compute every moving average after the first.

We wish to have a moving average uniquely associated with each observation in the time series (except the first six and last six observations, which are lost by employing the moving average). Thus we **center** the moving-average values by computing

$$\bar{y}_{\text{July}} = \frac{24.667 + 24.975}{2} = 24.821$$

Taking the process one step further, we find

$$\bar{y}_{\text{Aug-Sept}} = \frac{299.7}{12} + \frac{20.1 - 18.6}{12} = \frac{301.2}{12} = 25.1$$

$$\bar{y}_{\text{Aug}} = \frac{\bar{y}_{\text{July-Aug}} + \bar{y}_{\text{Aug-Sept}}}{2} = \frac{24.975 + 25.1}{2} = 25.038$$

The remaining 22 moving averages have been computed and are listed in table 14.2.

In figure 14.11 we have plotted the original time series of sales of the brewing company for the years 1974 through 1976 and have superimposed the 12-month moving average. Having smoothed out the seasonal fluctuations, the moving average exhibits only a linear growth trend. No short-term trends or business cycles appear for the period 1974 through 1976. In the final analysis, the brewing company's sales manager can assume that sales of the firm's products are increasing relatively over time but fluctuate seasonally over the year, exhibiting greater than average sales during the warm months, and lesser than average sales during the cooler months. The only other factors affecting the firm's time series are of minimal importance and very short-lived, allowing for their categorization as random variation.

Exercises

14.9.   Refer to exercise 14.8. Remove the seasonal component from these data by applying a 12-month moving average to the data. Plot the original series and the deseasonalized series together on the same piece of graph paper. After removing the seasonal component from the series, what inherent components do these data appear to possess?

14.10.   Refer to the two exponentially smoothed series computed in exercise 14.8. Did either of these smoothed series appear to eliminate the seasonal component from the airline traffic time series?

14.11.   Texas Chemical Products manufactures an agricultural chemical that is applied to farmlands after crops have been harvested. Since the chemical tends to deteriorate in storage, Texas Chemical cannot stockpile quantities in advance of the winter season demand for the product. Sales of the product over four consecutive

| | YEAR | | | |
MONTH	1	2	3	4
January	123.327	133.708	143.747	145.151
February	129.585	146.156	159.360	146.323
March	157.480	174.000	168.129	164.262
April	155.027	162.574	152.642	157.848
May	161.040	176.280	178.682	181.914
June	169.076	154.033	164.432	169.352
July	142.196	165.715	160.469	165.623
August	156.731	167.835	169.940	174.069
September	169.057	165.715	160.469	165.623
October	185.070	223.205	208.081	215.474
November	208.645	238.217	220.516	212.594
December	238.468	251.588	243.519	258.063

years are shown in the accompanying table (the recorded sales values are listed in thousands of pounds).

Remove the seasonal component from these data by applying a 12-month moving average to the data. Plot the original series and the deseasonalized series together on the same piece of graph paper. After removing the seasonal component from the series, what inherent components does the time series appear to possess?

14.5 Index Numbers

Because of the variability in the buying power of the dollar over time, it is necessary to deflate some values and inflate others in order to make meaningful comparisons. For example, to compare the relative cost of a four-year college education today with its cost in 1940, we must first determine the buying power of the dollar today as compared with the buying power of a 1940 dollar. *Index numbers* are computed for such purposes and are used every day by businesspeople and economists to make meaningful comparisons over time. Their use is not limited strictly to monetary comparisons, but for business problems application of index numbers to other than monetary processes is uncommon.

Definition

An **index number** is a ratio or an average of ratios expressed as a percentage. Two or more time periods are involved, one of which is the base time period. The value at the base time period serves as the standard point of comparison, while the values at the other time periods are used to show the percentage change in value from the standard value of the base period.

The concept of an index number is best illustrated by an example.

Example 14.2 Suppose that we wish to compare the average hourly wages for a journeyman electrician in 1950, 1955, 1960, 1965, 1970, and 1975, using 1950 as the base year. The average hourly wages and their computed index numbers for each of the six years are listed in table 14.3.

The wage index for each year is computed by evaluating the ratio

$$I_k = \frac{(\text{average hourly wages in year } k)(100)}{\text{average hourly wages in 1950}}$$

Each wage index is a percentage that indicates the percentage of 1950 wages that were earned in the year of interest. For example, the wage index for 1975 indicates that the hourly wage in 1975 was 442.5% of the 1950 hourly wage.

Table 14.3 *Average hourly wages for journeyman electricians for six years*

YEAR	AVERAGE HOURLY WAGES	WAGE INDEX (1950 base)
1950	$2.00	100
1955	2.85	142.5
1960	3.90	195
1965	5.25	262.5
1970	6.00	300
1975	8.85	442.5

Definition

An **index time series** is a list of index numbers for two or more periods of time, where each index number employs the same base year.

The preceding example is an index time series for the years 1950, 1955, 1960, 1965, 1970, and 1975. An index time series is simply a transformation of the original time series to one giving each year's value as a percentage of the value for the base year.

A commonly used index for comparing two sets of prices from a wide variety of items is called a *simple aggregate index.*

Definition

A **simple aggregate index** is the ratio of an aggregate (sum) of commodity prices for a given year k to an aggregate of the prices of the same commodities in some base year, expressed as a percentage.

This index is computed by evaluating the formula given in the box.

Simple Aggregate Index

$$I_k = \frac{\displaystyle\sum_{i=1}^{n} p_{ki}}{\displaystyle\sum_{i=1}^{n} p_{oi}} (100)$$

where the p_{ki}'s are the prices in year k of i items, i ranging from 1 to n, and the p_{oi}'s are the base year prices of the same i items.

Table 14.4 *Average consumer prices (cents/pound) for 1955 and 1975*

ITEM	1955	1975
sugar	10	40
wheat flour	11	20
butter	71	99
sirloin steak	91	186
ground beef	39	88
frying chicken	51	62

Example 14.3 The average consumer prices, in cents per pound, for certain staple food items in 1955 and 1975 are given in table 14.4. Find the aggregate price index for 1975.

Solution The aggregate price index for 1975 is

$$I_{1975} = \frac{40 + 20 + 99 + 186 + 88 + 62}{10 + 11 + 71 + 91 + 39 + 51}(100) = \frac{495}{273}(100) = 181$$

which implies that the prices of these six items in 1975 are 81% higher than they were in 1955.

The greatest weakness of the simple aggregate index is that changes in the measuring units may drastically affect the value of the index. In example 14.3, suppose that we had considered the prices of 10-pound bags of sugar and wheat flour instead of the prices of these items per pound. If the 10-pound prices were $1.00 and $4.00 for the sugar and $1.10 and $2.00 for the flour, then our price index would be

$$I_{1975} = \frac{1035}{462}(100) = 224$$

an increase of 43 percentage points over our former index value. By changing the scale of measurement of some of the units being measured within the simple aggregate index, the statistician could derive almost any index value at his or her discretion. This lack of objectivity of the simple index tends to lessen its usefulness for making meaningful comparisons.

An index that gives a more uniform measure of comparison is called a *weighted aggregate index.*

Definition

A **weighted aggregate index** is the ratio of an aggregate of weighted commodity prices for a given year k to an aggregate of the weighted prices of the same commodities in some base year, expressed as a percentage.

In a weighted aggregate index, the prices do not necessarily contribute equally to the value of the index. Each price is weighted (multiplied) by the quantity of the item produced or the number of units purchased or consumed. Thus each item is included according to its importance in the aggregate of prices of the items being described by the index. The index is found by computing the formula given in the box.

Weighted Aggregate Index

$$I_k = \frac{\sum\limits_{i=1}^{n} p_{ki} q_{ki}}{\sum\limits_{i=1}^{n} p_{0i} q_{0i}} (100)$$

where the q_{ki}'s and the q_{0i}'s are the quantities associated with the n prices for the reference year and the base year, respectively.

The U.S. Department of Labor uses a special form of a weighted aggregate index for several of its published indexes. This index is called the **Laspeyres index** and is found by computing

$$L = \frac{\sum\limits_{i=1}^{n} p_{ki} q_{0i}}{\sum\limits_{i=1}^{n} p_{0i} q_{0i}} (100)$$

The rationale for using base year quantities as the weights for the reference year prices is that the base year quantities do not change from year to year. Thus we can make more meaningful comparisons of the change in prices and buying power over time, since we are considering only the change in price per given number of units and not changing the number of units. By using the base year quantities, though, we tend to give too much weight to the commodities whose prices have increased, since an increase in price will often be accompanied by a decrease in the quantity consumed or purchased. However, if essential or staple items are being considered, the effect of using only the base year weights on the value of the index should be small. Since it is often difficult and expensive to obtain the quantities for each time period, it may be a worthwhile trade-off to give up some accuracy of the computed index value by employing the Laspeyres index.

Example 14.4 The average consumer prices for certain staple food items from example 14.3 are listed in table 14.5. Also, we are given the average amount of each item recommended as necessary to sustain a fam-

Table 14.5 *Average consumer prices (cents/pound) and amount recommended for a family of four*

ITEM	1955 PRICE, p_0	1975 PRICE, p_k	1955 QUANTITY (LB), q_0
sugar	10	40	25
wheat flour	11	20	60
butter	71	99	50
sirloin steak	91	186	25
ground beef	39	88	120
frying chicken	51	62	40

ily of four in 1955. Find the Laspeyres index to measure the amount of change of the 1975 prices on these items from 1955 to 1975.

Solution The Laspeyres index is

$$L = \frac{40(25) + 20(60) + 99(50) + 186(25) + 88(120) + 62(40)}{10(25) + 11(60) + 71(50) + 91(25) + 39(120) + 51(40)} (100)$$

$$= \frac{24,840}{13,455} (100) = 185$$

Thus the cost for these staple food items in 1975 is 85% higher than the cost for these items in 1955, when considering the total annual food expenditures.

Other less frequently used weighted aggregative price indexes are the **Paasche index** and **Fisher's ideal index**. The Paasche index uses the reference year quantities rather than the base year quantities as weights for the weighted index. Otherwise, the computational procedure is the same for the Laspeyres and the Paasche indexes. Previously we mentioned that the Laspeyres index tends to "overweigh" commodities whose prices have increased. Analogously, the Paasche index tends to "underweigh" commodities whose prices have increased. Hence we might suspect that the price index should be somewhere between these two indexes. This is the logic behind the use of Fisher's ideal index.

The Fisher index is computed from the Laspeyres and the Paasche indexes as follows:

$$\text{Fisher's index} = \sqrt{(\text{Laspeyres index})(\text{Paasche index})}$$

$$= 100 \sqrt{\frac{\sum p_{ki} q_{0i}}{\sum p_{0i} q_{0i}} \frac{\sum p_{ki} q_{ki}}{\sum p_{0i} q_{ki}}}$$

Although the Fisher index might seem to measure the price index more accurately than the Laspeyres or the Paasche indexes, it is seldom used in practice. The Fisher index, since it is a function of the Paasche index, requires a new set of quantities at each time period. These are often difficult and expensive to obtain. Also the Fisher index does not give a uniform index for purposes of comparison in an index time series. That is, like the Paasche index, the Fisher index does not hold the quantity measure constant, as does the Laspeyres index. Thus the Laspeyres index, or some form of the Laspeyres index, is used at the practical exclusion of other types of indexes when the business statistician chooses to use a weighted aggregate index.

Two important price indexes computed regularly by the Bureau of Labor Statistics are the **consumer price index** and the **wholesale price index**. Both may be somewhat misleading and deserve some explanation. The consumer price index (CPI) is computed and published monthly. It is computed from data gathered from a random sample of clerical workers' and wage earners' family expenditures from across the country. About 300 different items from food products to clothing to rent to luxuries to fees paid to doctors and dentists are sampled for each family selected. A weighted aggregate index is computed using the information gathered in the year 1967 as the base information. The computed index is then published as the consumer price index. The CPI is used as a factor to deflate gross national product and other measures of economic wealth to "cancel out" the effect of inflation or deflation. Also, other price indexes, such as the retail price index, are computed as a function of the CPI.

Now let us interpret the CPI as a price deflator or as a cost-of-living indicator. The CPI is not representative of all American families but only those from which the sample data have been collected—clerical workers and wage earners. The CPI says little or nothing about the cost of living for a professional person or a farmer. In addition, the CPI is restricted in attention to clerical workers and wage earners of moderate income, since those included in the sample have moderate incomes. Thus the CPI "market basket index" does not reflect purchases of families at the lowest and highest extremes of the income scale, even though they may be families of clerical workers and wage earners. A further limiting factor of the CPI is that it ignores taxation and product quality changes over time. It is almost impossible to select an index that would be perfectly representative in every way, so, recognizing its limitations, we use the CPI as a measure, if not a valid measure, of the changes in prices of goods and services for American families.

The wholesale price index (WPI) is computed from information obtained by sampling the producers' selling prices of about 2000 different items in the primary markets. Agricultural commodities, raw materials, fabricated products, and many manufactured items are included. The WPI is a weighted aggregate index computed and published monthly by the Bureau of Labor Statistics. Its base period is 1967–1969.

Many industrial contracts, especially those with the Department of Defense, allow for an adjustment of the contract price and payments according to the change in the WPI. This allows the contractor and the buyer to negotiate and plan in terms of some constant dollar amounts. The primary difficulty is that the WPI is not really an indicator of wholesale prices at all but represents the *change* in producers' selling prices. In many instances this makes no difference, but the name of the index, the wholesale price index, may in itself be misleading and cause the businessperson to form unwarranted conclusions.

Security market indexes, such as the Dow-Jones industrial average, are computed differently from the wage and price indexes previously described. They might better be described as averages of a group of individual time series rather than as indicators of change of value or price.

The Dow-Jones industrial average purports to compute the average of the daily closing prices of the securities of 30 predetermined industrial firms. The computations become involved, though, when financial occurrences such as mergers and stock splits take place. For the sake of discussion, suppose that the Dow-Jones average consists of the securities of only two firms, one whose stock is valued at $20 per share, the other valued at $30. The Dow-Jones average is

$$\frac{1}{2} (\$20 + \$30) = \$25$$

During the next day, suppose that a stock split occurs for the firm whose securities were valued at $20 per share. Now each security from the first firm is worth only $10, and the Dow-Jones average, ignoring the stock split, is

$$\frac{1}{2} (\$10 + \$30) = \$20$$

To consider the effect of the split, the Dow-Jones average finds the divisor d such that

$$\frac{1}{d} (\$10 + \$30) = \$25$$

where $25 was the average before the split, Here $d = 1.6$. If during the day of the stock split, the first security increased in price to $12 per share while the other remained at $30, the new Dow-Jones average is

$$\frac{1}{1.6} (\$12 + \$30) = \$26.25$$

Each time stock dividends are declared, a merger occurs, or a stock split takes place, the Dow-Jones industrial average becomes less meaningful because of the continual adjustment of the divisor term d. Although the divisor started out as 30 (since the Dow-Jones average is computed from the security prices of 30 firms), it is now about 3.9 and will continue to decrease as more stock splits, mergers, and other financial happenings occur. Actually the Dow-Jones average is not an average at all. Its usefulness as a measure of market value has diminished and will continue to diminish with time.

Exercises

14.12.   For each of the following decision-making situations, explain how an individual might construct and use an index number as an aid in arriving at a conclusion.

a. A public utility company wishes to increase its rates by the amount of inflation in the economy of its area of service since rates were last set. The current rates were established in 1973.

b. An individual would like to know the real buying power of a dollar she currently earns in comparison with the buying power of a dollar she earned 10 years ago.

c. A homeowner wishes to establish a selling price for his home, which cost him $25,000 in 1960.

d. Suppose the homeowner mentioned in part c sells his home for $55,000. He would like to estimate his capital gain on the sale of his house in terms of 1960 dollars.

e. The owner of a small hardware store would like to know how the growth in sales of her store over the past five years compares with the growth in sales of a large competing hardware chain.

14.13.   The cost indexes for the five primary categories of consumer expenditures are shown in the table for the years 1970 and 1975 (1967 = 100).

YEAR	HOUSING	FOOD	APPAREL AND UPKEEP	TRANSPORTATION	HEALTH AND RECREATION
1970	140.1	132.8	135.9	135.5	147.4
1975	172.2	180.7	145.2	157.6	157.5

Source: U.S. Department of Labor, Bureau of Labor Statistics, *CPI Detailed Report*, December 1970 and December 1975.

a. Compute the percentage gain in the prices of items within each of the five main categories of consumer expenditures from 1970 to 1975.

b. Suppose the average family of four spent the following amounts in each of the five primary categories of consumer expenditures in 1970: housing, $3300; food, $2380; apparel and upkeep, $940; transporation, $1150; and health and recreation, $1470. Estimate their expenditures in each of these five categories for the year 1975.

14.14.   Listed in the table are the prices in five different years of five common family market basket items.

| ITEM | YEAR | | | | |
	1955	1960	1965	1970	1975
milk (qt)	22.4	25.1	26.0	33.8	39.6
wheat flour (lb)	10.7	11.2	12.3	14.1	20.3
hamburger (lb)	39.0	51.3	54.2	78.0	88.1
coffee (lb)	93.0	74.6	70.7	76.5	122.2
potatoes (lb)	4.7	6.7	11.1	27.3	20.5

Source: U.S. Department of Labor, Bureau of Labor Statistics, CPI Detailed Report, December 1970 and December 1975.

a. Using 1955 as the base year, compute the simple price index for each food product for each of the five years.

b. Using 1955 as the base year, compute the simple aggregate index of food prices for each of the five years.

14.15.   Refer to exercise 14.14. In 1955 the average family of four bought 350 quarts of milk, 60 pounds of wheat flour, 120 pounds of hamburger, 40 pounds of coffee, and 140 pounds of potatoes. Compute the Laspeyres index for the prices of this market basket for each of the five years under study, using 1955 as the base year.

14.6 Summary

A time series is a sequence of measurements taken on a response that varies over time. It is usually represented in graphical form, with its response values listed on the vertical axis plotted against time on the horizontal axis. We may think of a time series as consisting of components such as a long-term trend, a cyclical effect, or a seasonal effect. Any or none of these components may be present in a time series. In addition, all time series possess random variation, which tends to obscure the nonrandom components of the series. Smoothing methods may sometimes reveal the underlying components in a time series by canceling out the effects of random variation.

The values in a time series are often misleading, especially those which occur in monetary units. Index numbers are used to deflate some monetary values and inflate others so that meaningful monetary comparisons can be made over time. Index numbers measure the change in price or value of a single or an aggregate of commodities from some base time period to a reference time period. Weighted indexes, which multiply the commodity prices by a predetermined factor, are the most commonly used type of index number. Most price indexes computed regularly by the Bureau of Labor Statistics are weighted indexes.

Care must be taken in interpreting an index number or a series. Many commonly used indexes, such as the consumer price index, the wholesale price index, and the Dow-Jones industrial average, do not really indicate what their titles might imply. We must examine the exact intent of an index, the identity of the commodities that are included within the index, and the weights associated with each commodity included within the index before meaningful conclusions can be made.

Supplementary Exercises

14.16.   Which of the time series components would you expect to be present in each of the following series?

a. The week's end closing values of the Dow-Jones industrial average from 1 January 1965 to 1 January 1978.

b. The gross monthly sales volume of Sears, Roebuck and Company from January 1960 to January 1978.

c. The gross monthly sales volume of the Miller Brewing Company from June 1960 to June 1977.

d. The annual sales of life insurance in the United States from 1950 to 1978.

e. The monthly farm employment in the United States from January 1955 to January 1978.

14.17. A recent opinion by the Accounting Principles Board allows companies with considerable investment in securities to overcome the problem of earnings and income fluctuation by smoothing income and spreading market fluctuation over a wider period of time.* Using smoothing methods, realized and unrealized gains and losses are combined so that each period's income includes a pro rata portion of the gains or losses arising in prior periods. The data in the table represent the gross monthly sales volume (in thousands of dollars) of a pharmaceutical company from January 1976 to January 1978.

1976		1977	
Month	Sales	Month	Sales
January	18.0	January	23.3
February	18.5	February	22.6
March	19.2	March	23.1
April	19.0	April	20.9
May	17.8	May	20.2
June	19.5	June	22.5
July	20.0	July	24.1
August	20.7	August	25.0
September	19.1	September	25.2
October	19.6	October	23.8
November	20.8	November	25.7
December	21.0	December	26.3

a. Plot the sales values against time and construct the time series.

b. Which time series components appear to exist in the sales pattern?

c. Smooth the monthly sales values by computing a three-month moving-average series. Plot the smoothed series and the original series together on the same sheet of graph paper. What conclusions do you draw?

14.18. Refer to exercise 14.17.

a. Smooth the monthly sales volumes by computing an exponentially smoothed series, employing the smoothing constant $\alpha = .1$.

b. Smooth the monthly sales volumes by computing an exponentially smoothed series, employing the smoothing constant $\alpha = .25$.

c. On a sheet of graph paper, superimpose the two smoothed series on the original series. What conclusions do you draw?

14.19. Refer to exercise 14.17. Break down the original series into its components, as was done in figure 14.6 for the sales of the department store (p. 508). Is it always possible to separate the components of a time series? Explain.

14.20. The accompanying table lists the monthly credit sales (in thousands of dollars) for a department store over five consecutive years. Remove the seasonal component from these data by applying a 12-month moving average to the data. Plot the original series and the deseasonalized series together on the same piece of graph paper. After removing the seasonal component, what inherent components do these data appear to possess?

*See T. E. Lynch, "Accounting for Investments in Equity Securities by the Equity and Market Value Methods," *Financial Analysts Journal*, February 1975.

MONTH	YEAR				
	1	2	3	4	5
January	12.3327	13.3708	14.3747	14.5151	15.1144
February	12.9585	14.6156	15.9360	14.6323	15.3555
March	15.7480	17.4000	16.8129	16.4262	18.6998
April	15.5027	16.2574	15.2642	15.7848	17.1827
May	16.1040	17.6280	17.8682	18.1914	18.2053
June	16.9076	16.4033	16.4432	16.9352	18.6406
July	14.2196	14.8934	15.1211	16.3979	17.2713
August	15.6731	16.7835	16.9940	17.4069	17.1966
September	16.9057	16.5715	16.0469	16.5623	17.4080
October	18.5070	22.3205	20.8081	21.5474	21.2315
November	20.8645	23.8217	22.0516	21.2594	21.5733
December	23.8468	25.1588	24.3519	25.8063	27.5733

Source: J. E. Reinmuth and D. R. Wittink, "Recursive Models for Forecasting Seasonal Processes," *Journal of Financial and Quantitative Analysis*, September 1974.

14.21.   Refer to the credit sales data listed in exercise 14.20.

a. Smooth the monthly sales volumes over the five-year period by computing an exponentially smoothed series, employing the smoothing constant $\alpha = .1$.

b. Smooth the monthly sales volumes over the five-year period by computing an exponentially smoothed series, employing the smoothing constant $\alpha = .4$.

c. On a sheet of graph paper, superimpose the two smoothed series on the original series. What conclusions do you draw?

14.22.   Refer to exercises 14.20 and 14.21. Does the 12-month moving-average model used in exercise 14.20 appear to more effectively eliminate the seasonal pattern from the credit sales data than either of the exponential smoothing models from exercise 14.21? Would a moving-average model with an M other than 12 have eliminated the seasonal component as well? Explain.

14.23.   Criticize the following uses of index numbers:

a. The portfolio manager for a pension fund must decide whether to invest a large sum in a corporate bond account or in a mutual fund. If the Dow-Jones industrial average drops by more than five points in the next week, he will invest in the corporate bond account. Otherwise, he will elect the mutual fund investment.

b. A business economist observes that the consumer price index has risen from 124 in 1970 to 172 in 1975. She concludes that living costs have increased by 48% during that time.

c. A government contractor notices that the wholesale price index has increased by 13% over the past two years. Thus she asks the government for a 13% increase in the contractual payoff.

d. A lumber wholesaler is interested in the real increase in sales he has incurred over the past five years. To measure the increase he computes a Laspeyres index, considering all types of lumber he handles, using the current and five-year-old board foot selling prices and the quantities of each type of lumber he sold five years ago.

14.24.   The consumer price index, computed monthly by the U.S. Department of Labor as a measure of the cost of consumer goods and services, was 162.8 for October 1975, using 1967 as the base year. Over this time, suppose that Bob Owen's salary rose from $12,500 in 1967 to $17,850 by October 1975. Has Owen's "real" income actually risen? Explain.

14.25.   The costs, in cents per pound, of six common meat products for five consecutive years are listed in the table.

YEAR	GROUND BEEF	ROUND STEAK	RIB ROAST	HAM	FRYING CHICKEN	PORK CHOPS
1971	68.4	135.8	115.8	66.1	40.6	108.5
1972	74.5	147.6	126.7	73.1	41.1	124.9
1973	94.1	174.8	148.7	96.8	59.1	156.2
1974	95.4	180.4	155.0	98.4	55.4	156.9
1975	88.1	194.9	174.2	117.6	62.4	185.3

Source: U.S. Department of Labor, Bureau of Labor Statistics, *CPI Detailed Report*, selected issues 1971–1975.

a. Using 1971 as the base year, compute the simple price index for each type of meat for each of the five years.

b. Using 1971 as the base year, compute the simple aggregate index of meat prices for each of the five years.

14.26. Refer to exercise 14.25. In 1971 the average family of four bought 115 pounds of ground beef, 75 pounds of round steak, 50 pounds of rib roast, 50 pounds of ham, 100 pounds of frying chicken, and 80 pounds of pork chops. Compute the Laspeyres index for the meat prices over the past five years.

14.27. Refer to exercise 14.25. In 1975 the average family of four consumed 135 pounds of ground beef, 70 pounds of round steak, 50 pounds of rib roast, 35 pounds of ham, 110 pounds of frying chicken, and 75 pounds of pork chops. Compute the Paasche index for the meat prices in 1975, using 1971 as the base year.

14.28. Refer to exercises 14.26 and 14.27. Compute Fisher's ideal index for the meat prices in 1975, using 1971 as the base year. Discuss the significance of this index.

14.29. A power company was interested in examining the relative costs of heating fuels over a four-year period. In their study the power company contacted a sample of householders who heat by electric heat, use an oil furnace, or heat by natural gas and obtained the average quantity each householder used per month. (Quantities were obtained from utility company records.) The results are given in the table.

FUEL	UNIT COST				AVERAGE MONTHLY USAGE			
	1973	1974	1975	1976	1973	1974	1975	1976
electricity	1.70	1.85	2.05	2.05	67	75	68	70
oil	.32	.39	.41	.42	230	241	225	256
gas	8.20	9.05	9.70	9.90	7.2	6.9	6.8	7.0

a. Find the Laspeyres index for the 1974, 1975, and 1976 average monthly heating costs, using 1973 as the base year.

b. Find the Paasche index for the 1974, 1975, and 1976 average monthly heating costs, using 1973 as the base year.

14.30. Refer to exercise 14.29. Compute Fisher's ideal index for the average monthly heating costs during 1974, 1975, and 1976, using 1973 as the base year. What conclusions do you draw about the relative heating costs over the four-year period from 1973 through 1976, based on the Laspeyres, Paasche, and Fisher indexes? If the results were contradictory, which index would you consider the most reliable? Explain.

14.31.   Suppose that the security price movements of the common stock of four companies, American Biscuit Corporation, National Motors, Southwestern Oil, and Texas Chemical Products, are considered to be representative of the behavior of all industrial securities traded on the New York stock exchange. The closing prices of the securities of the four companies for four consecutive market days in October are given in the table. Using the procedure discussed in section 14.5, compute a

	AMERICAN BISCUIT	NATIONAL MOTORS	SOUTHWESTERN OIL	TEXAS CHEMICAL
October 10	27	30	62	14
October 11	26	15*	60	13
October 12	28	17	30*	15
October 13	29	18	32	14

*National Motors declared a 2-for-1 stock split after the close of the market on October 10, while Southwestern Oil declared a 2-for-1 stock split on October 11.

Dow-Jones-type market average for each of the four listed market days, using the security prices of these four representative companies.

Experiences with Real Data

Consumers have been distressed by the precipitous rise in retail food prices and the cost of heating fuels over the past few years. Food prices are now rising twice as fast as the increase in all prices, while for the twenty-year period 1951-1971, food prices advanced a total of 43%, considerably lower than the 62% for nonfood items. Meanwhile, energy costs have almost tripled since 1971.

Perhaps, then, the energy crisis and the resultant spiraling costs of fuels is the primary reason for the rise in food prices. Petroleum derivatives are needed in growing, processing, transporting, storing, and preparing food items. Thus it seems only natural to expect food prices to continue to soar as long as energy costs continue to rise.

To note the parallel rise in food and energy costs over the past few years, construct a time series of the tabulated price index over the past 36 months (three years of monthly data) for each of the following items: diesel fuel, heating oil, natural gas, regular gasoline, and food (as an overall category of consumer expenditures). These indexes are tabulated monthly and published in the *CPI Detailed Report*, U.S. Department of Labor, Bureau of Labor Statistics.

Select one of the energy resource time series and the food index time series and smooth each series, using either a moving-average or exponential smoothing procedure. Use your own judgment in the selection of the constant M for the moving-average process or the selection of a smoothing constant α for the exponential smoothing procedure. Do the smoothed series exhibit a noticeable seasonal effect? Are any other time series components evident after smoothing? What does the presence of the components you have detected suggest about the flow of food and energy costs from one month to another within a particular year? Is there a noticeable relationship between energy costs and food costs over the period of your analysis.

References

DOODY, F. S. *Introduction to the Use of Economic Indicators.* New York: Random House, 1965.

FREUND, J. E., and F. J. WILLIAMS; revised by B. Perles and C. Sullivan. *Modern Business Statistics*. Englewood Cliffs, N.J.: Prentice-Hall, 1969. Chapters 15 and 16.

SPURR, W. A., and C. P. BONINI. *Statistical Analysis for Business Decisions*. Rev. ed. Homewood, Ill.: Richard D. Irwin, 1973. Chapter 18.

U.S. DEPARTMENT OF COMMERCE. *Business Statistics*, biennial supplement to the *Survey of Current Business*. Washington, D.C.: U.S. Government Printing Office, 1967 et seq.

chapter objectives

GENERAL OBJECTIVES In chapter 14 we introduced smoothing as a procedure to aid in identifying the inherent components of a time series. In this chapter we show that smoothing methods can also be used to extend the pattern of a time series to forecast future business activity. Our purpose in this chapter is to introduce a number of different econometric and time series forecasting models and to illustrate their use in practice. We do not devote attention to the derivation of the various models or to the appropriate tailoring of a model to a particular time series.

SPECIFIC OBJECTIVES

1. To discuss the need for forecasting and to discuss the incorporation of judgment into the forecast.
Section 15.1

2. To discuss various measures of forecast accuracy to aid in the selection of a best forecasting model.
Section 15.2

3. To introduce and illustrate some common econometric forecasting models.
Sections 15.2, 15.3, 15.4

4. To introduce and illustrate two useful time series forecasting models.
Sections 15.5, 15.6

5. To briefly discuss the growth model, the Box-Jenkins procedures, and qualitative forecasting models.
Section 15.7

chapter fifteen

Forecasting Models

15.1 Introduction

We all exist in an environment governed by time. It is thus a common goal for business organizations, public organizations, and individuals to allocate available time among competing resources in some optimal manner. This is accomplished by making forecasts of future activities and taking the proper actions as suggested by these forecasts.

In business and public administration, the organization is concerned with both short-term and long-term forecasts. The short-term forecast is usually planned for looking no more than one year into the future and involves forecasting sales, price changes, and customer demand, which, in turn, reflect the need for seasonal employment, short-term capital expenditures, and inventory management procedures. The long-term forecast usually looks from two to ten years into the future and is used as a planning model for product line and capital investment decisions, as indicated by changing demand patterns.

Naturally the further a forecast is projected into the future, the more speculative it becomes. But since the future is *always* uncertain, we cannot expect complete accuracy for any given forecast. The time series underlying the process to be forecast is bound to be influenced by many causal factors—some forcing the time series up while conflicting factors act to force the series down. Nevertheless, it is essential for businesspeople to make forecasts of future business activity in order to efficiently budget their time and resources. They cannot hope to account for every possible factor that may cause the response of interest to rise or fall over time. All that can be expected is that the benefits gained by forecasting offset the opportunity cost for not forecasting. It is important to note that such benefits are not limited to real monetary savings but may imply a sharpening of the business-person's thinking to consider the interplay of the events that affect the movement of the time series.

We recall from section 14.3 that smoothing methods sometimes reveal trends and seasonal and cyclic effects in the time series. Forecasting by extending these patterns is a very speculative procedure. We must first assume that the past is a mirror of the future—that past trends and cycles will continue into the future. This is seldom the case. In the end, mathematical forecasting procedures and judgment must work hand in hand. Thus we must not only smooth the data and try to extend the signal components into the future but also predict the impact of unknown factors, such as political events, research, changing buyer behavior, and new product development. These subjective evaluations must, in turn, be used to condition the forecast obtained from the mathematical forecasting model. As long as uncertainty is involved with future business and economic activity, forecasting must be recognized as an art that becomes more perfect as the forecaster gains experience and the ability to adapt procedures to meet the changing environment of the firm.

Forecasting models are generally classified as **econometric models, time series models, or qualitative forecasting models.** The first two are projection techniques, which initially involve the fitting of a theoretical model to a sample data set. Qualitative models are designed for forecasting the sales of a newly introduced product and for other cases where a relevant historical sample data base is not available. We will devote most of our attention to econometric and time series models and give only a few brief passing remarks to qualitative forecasting models.

Our purpose throughout this chapter will be to illustrate the use of various forecasting models. As a result there may be instances where an example or an exercise shows that a particular model provides a poor fit to a sample data set. We believe that if you are aware of the characteristics of the various models and know how to use them in practice, then you will be able to properly tailor an appropriate model to an underlying time series in order to obtain an adequate level of forecast accuracy.

15.2 Econometric Forecasting Models

An econometric model is a system of one or more equations that describe the relationship among a number of economic and time series variables. Econometric models are probabilistic models and capitalize on the probabilistic relationship that exists between a dependent variable representing the time series and any of a number of independent variables.

The linear regression model of chapters 11 and 12 is an econometric

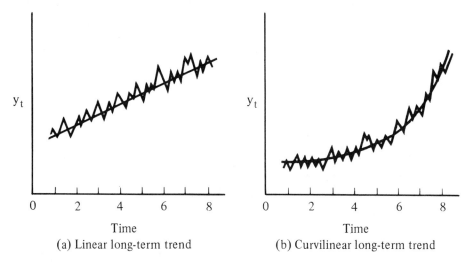

Figure 15.1 *Time series with linear and curvilinear long-term trends*

model that sometimes provides a suitable probabilistic model for establishing the long-term trend for a time series. For example, a long-term upward or downward trend similar to that shown in figure 15.1(a) might be isolated by using the procedures of chapter 11 to fit the straight line

$$y = \beta_0 + \beta_1 x + \epsilon$$

where the independent variable x represents time. A curvilinear long-term trend, as shown in figure 15.1(b), could be modeled by using a second-order function, such as

$$y = \beta_0 + \beta_1 x + \beta_2 x^2 + \epsilon$$

The corresponding prediction equation, $\hat{y} = \hat{\beta}_0 + \hat{\beta}_1 x + \hat{\beta}_2 x^2$, could be determined by using the method of least squares, as described in chapter 12.

   The assumption of independence of the random error ϵ associated with successive measurements will not usually be satisfied. Consequently, the probabilistic statements associated with the estimation of $E(y)$ or the prediction of y will be incorrect. We would suspect that they would be conservative and that knowledge of the actual pattern of correlation would permit more accurate estimation and prediction. If the response is an average over a period of time, the correlation of adjacent response measurements will be reduced and will quite possibly satisfy adequately the assumption of independence implied in the least squares inferential procedures.

   Other linear models (linear in the unknown terms, or "weights," β_0, β_1, β_2, etc.) can be constructed and fitted to data generated from economic time series by using the method of least squares. For example, the yearly production y of steel is a function of its price, the price of competitive structural materials, the production of competitive products during the preceding year, the amount of steel purchased during the immediately preceding years (to measure current inventory), and of many other variables. A linear model relating these independent variables to steel production might be

$$y = \beta_0 + \beta_1 x_1 + \beta_2 x_2 + \beta_3 x_3 + \cdots + \beta_k x_k + \epsilon$$

where

x_1 = time

x_2 = price of steel

x_3 = x_2^2 (allowing curvature in the response curve as a function of price)

x_4 = production of aluminum during previous year

x_5 = price of aluminum

x_6 = steel production during previous year

$\vdots$

x_k = $x_2 x_5$ (an interaction effect between steel and aluminum prices)

We could even include variables of the type, say, $x_7 = \sin(2\pi x_1/3)$, which would reflect a cyclic effect due to time, with a period of 3 years. In other words, the model builder has unlimited room for ingenuity in constructing the linear model. The test of the model is how well it agrees with the sample data and then, more importantly, how well it continues to forecast the future.

The primary feature that distinguishes econometric forecasting models from time series models is their use of economic and demographic variables which are thought to be causally related to y. Econometric models attempt to describe the relationship among such variables by use of one or more regression equations, while time series models ignore these causal variables and rely on a projection of the time series components inherent in y.

In building an econometric forecasting model, we usually begin with a large number of variables that might be closely related to the response. We then combine these variables to form models that are fitted to the sample data by using the method of least squares (see chapters 11 and 12). However, a model that fits past data very well may be insensitive to the uncertainties associated with future events and may lead to inaccurate forecasts. Since forecasting is concerned with *future* events, we should select a forecasting model that demonstrates the best ability to forecast the future, not fit the past.

Speaking in a very general sense, we try different combinations of variables and continue this procedure until we obtain a model that predicts future values with a relatively small error of prediction, $(y_t - \hat{y}_t)$, where y_t and $\hat{y}_t$ are the actual and forecast values, respectively, for time period t. The model is tested by computing the sum of squares for forecast errors (SSE), as shown in the box.

SSE is computed for observed and predicted response measurements collected at **future points in time.** The model yielding the smallest SSE is the one providing the greatest measure of forecast accuracy.

Computation of SSE

$$\text{SSE} = \sum_{t=1}^{n} (y_t - \hat{y}_t)^2$$

Exercises

15.1. Since 1972, enrollment in schools of business has increased at a much more rapid rate than the overall growth in college enrollment in the United States. To properly allocate resources to accommodate the students and to adjust for other effects of enrollment, college and university administrators must be able to predict the anticipated number of students who will enroll in business courses as well as the number who will enroll in other areas. Design an econometric forecasting model to predict the enrollment in all undergraduate business courses at your college or university during each of the next two terms (or semesters). In particular, suggest the predictor variables you believe are related to business enrollment and indicate where you could obtain measurements for these variables and for business enrollment for each of the past several terms.

15.2. To create better awareness of the potential of fire damage and to better protect their customers, a large fire insurance company has offered home smoke detection sensors to policyholders for a price of $50. The number of smoke sensors sold to policyholders over the first eight months of the offer is shown in the table.

Month, t	1	2	3	4	5	6	7	8
Number Sold, y_t	410	290	420	310	410	580	600	990

The sales manager for the insurance company claims that the simple linear econometric model $\hat{y}_t = 200 + 66.67t$, where t is the month as numbered since the beginning of the sales of the smoke sensors, describes the sales of smoke sensors by the insurance company over time.

a. On a piece of graph paper, plot the time series and superimpose the linear model $\hat{y}_t = 200 + 66.67t$. Does this model appear to provide an adequate fit to the time series?

b. Use the procedures of chapter 12 to fit the second-order model $\hat{y}_t = \hat{\beta}_0 + \hat{\beta}_1 t + \hat{\beta}_2 t^2$ to the time series for the smoke sensor sales. Compare the fit of the second-order model with that of the simple linear model $\hat{y}_t = 200 + 66.67t$ to the time series.

c. Use both the simple linear model $\hat{y}_t = 200 + 66.67t$ and the second-order model you have derived to forecast the sales for the smoke sensors during months $t = 9$ and $t = 10$.

15.3. One way of determining whether a model provides an adequate fit to an underlying data set is to examine the residuals $(y_t - \hat{y}_t)$. If a plot of the residuals against time appears to be random (i.e., scattered, or not patterned), the model is said to provide a good fit to the underlying time series. Refer to exercise 15.2 and plot the residuals against time for both the linear model $\hat{y}_t = 200 + 66.67t$ and the second-order model derived as part of the exercise. What do these plots suggest?

15.3 A Least Squares Sinusoidal Model

In figure 14.9 the week's end security prices for the Color-Vision Company were plotted against time. (These data are actually the week's end closing prices of a well-known U.S. computing machine manufacturer recorded during a period in the mid 1960s.) You can notice the "peaks" in the time series at the 8th, 17th, 28th, and 29th time points and "valleys" at the 5th, 12th, and 23d time points. Very roughly, you could infer that these peaks and valleys suggest that the time series follows a cyclic pattern, with 10 time points as the period of one complete cycle. Assuming that the security prices for the Color-Vision Company follow this pattern, we will construct an appropriate least squares model to forecast prices over time.

The model we will use is*

$$y_t = \beta_0 + \beta_1 t + \beta_2 \cos\left(\frac{2\pi t}{10}\right) + \beta_3 \sin\left(\frac{2\pi t}{10}\right) + \beta_4 t \cos\left(\frac{2\pi t}{10}\right)$$
$$+ \beta_5 t \sin\left(\frac{2\pi t}{10}\right) + \epsilon$$

Notice that this model is a special form of the multiple linear regression model (see chapter 12)

$$y = \beta_0 + \beta_1 x_1 + \beta_2 x_2 + \beta_3 x_3 + \beta_4 x_4 + \beta_5 x_5 + \epsilon$$

where

$$x_1 = t, \text{ the index of time} \quad x_4 = t\cos\left(\frac{2\pi t}{10}\right)$$

$$x_2 = \cos\left(\frac{2\pi t}{10}\right) \quad x_5 = t\sin\left(\frac{2\pi t}{10}\right)$$

$$x_3 = \sin\left(\frac{2\pi t}{10}\right)$$

The linear term $\beta_1 t$ should reflect the linear trend of the series with time, while the terms $\cos(2\pi t/10)$ and $\sin(2\pi t/10)$ give a cyclic effect with a period of 10 weeks to the forecasted prices. The inclusion of the terms $t\cos(2\pi t/10)$ and $t\sin(2\pi t/10)$ allows the amplitudes (heights) of

---

*The notation $(2\pi t/10)$ refers to the number of radians at which the associated trigonometric term must be evaluated. Since one complete cycle equals 2π radians, $(2\pi t/10)$ at $t = 3$ would imply that the cosine must be evaluated at a point 3/10 of the way through its complete cycle. Thus $\cos(2\pi 3/10) = -.30$.

the cyclic function to change over time, a necessary requirement since the peaks in the time series of figure 14.9 have varying heights. A multiple linear regression model with periodic (trigonometric) terms for the independent variables, such as the one suggested above, is called a sinusoidal model.

A sinusoidal model is merely one of many types of econometric models that are adaptable to the method of least squares. It is especially useful here because of the cyclic effects that appear to exist in the Color-Vision prices over time, but it is primarily used as an illustration of how nonlinear functions of the independent variables may be combined linearly to produce a least squares econometric forecasting equation.

The forecasting model computed by the method of least squares as the best-fit equation to the Color-Vision data is

Table 15.1 *Actual and forecasted week's end security prices for the Color-Vision Company; forecasts determined by a least squares sinusoidal forecasting model*

WEEK, t	COS	SIN	t COS	t SIN	ACTUAL VALUES, y_t	FORECAST, $\hat{y}_t$	FORECAST ERROR, $e_t = y_t - \hat{y}_t$
1	.80	.58	.80	.58	71	70.65	.35
2	.30	.95	.60	1.90	70	67.81	2.19
3	−.30	.95	−.90	2.85	69	66.01	2.99
4	−.80	.58	−3.20	2.32	68	66.00	2.00
5	−1.00	.00	−5.00	.00	64	67.83	−3.83
6	−.80	−.58	−4.80	−3.48	65	70.92	−5.92
7	−.30	−.95	−2.10	−6.65	72	74.07	−2.07
8	.30	−.95	2.40	−7.60	78	76.09	1.91
9	.80	−.58	7.20	−5.22	75	76.44	−1.44
10	1.00	.00	10.00	.00	75	75.28	−.28
11	.80	.58	8.80	6.38	75	73.29	1.71
12	.30	.95	3.60	11.40	70	71.54	−1.54
13	−.30	.95	−3.90	12.35	75	71.03	3.97
14	−.80	.58	−11.20	8.12	75	72.06	2.94
15	−1.00	.00	−15.00	.00	74	74.27	−.27
16	−.80	−.58	−12.80	−9.28	78	76.89	1.11
17	−.30	−.95	−5.10	−16.15	86	78.95	7.05
18	.30	−.95	5.40	−17.10	82	79.69	2.31
19	.80	−.58	15.20	−11.02	75	79.00	−4.00
20	1.00	.00	20.00	.00	73	77.46	−4.46
21	.80	.58	16.80	12.18	72	75.93	−3.93
22	.30	.95	6.60	20.90	73	75.28	−2.28
23	−.30	.95	−6.90	21.85	72	76.04	−4.04
24	−.80	.58	−19.20	13.92	77	78.11	−1.11
25	−1.00	.00	−25.00	.00	83	80.71	2.29
26	−.80	−.58	−20.80	−15.08	81	82.86	−1.86
27	−.30	−.95	−8.10	−25.65	81	83.83	−2.83
28	.30	−.95	8.40	−26.60	85	83.29	1.71
29	.80	−.58	23.20	−16.82	85	82.04	2.96
30	1.00	.00	30.00	.00	84	79.63	4.37

$$\hat{y}_t = 68.85 + .43t + 4.25 \cos\left(\frac{2\pi t}{10}\right) - 3.23 \sin\left(\frac{2\pi t}{10}\right)$$
$$- .21t \cos\left(\frac{2\pi t}{10}\right) + .01t \sin\left(\frac{2\pi t}{10}\right)$$

where $\hat{y}_t$ is the forecast price at the end of week t, for $t = 1, 2, 3, \ldots,$ 30. The original data, the forecast prices, and the forecast errors are shown in table 15.1.

How well does this model forecast the week's end security prices of the Color-Vision Company? The observed and forecasted prices listed in table 15.1 are plotted against time in figure 15.2. It appears that the least squares forecasting equation we have selected is not too responsive to the sudden changes of value in the original series. It must be remembered that the observed and forecasted security prices should agree to some extent because the observed prices were used in determining the best-fitting model according to the least squares theory. In spite of this we would regard the agreement to be good, considering the simplicity of the econometric model used.

The sinusoidal model has been introduced simply to illustrate the usefulness of trigonometric variables as predictors in an econometric forecasting model. In a real application these artificial variables would always be supplemented by measurable predictors, such as last week's sales or a market

Figure 15.2 *Least squares forecasts for the week's end security prices of the Color-Vision Company*

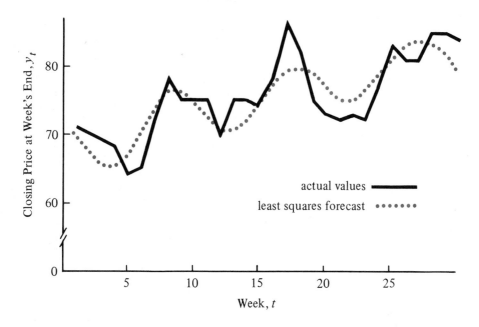

index, resulting in a more accurate forecasting equation than the one we have obtained.

We should note that the addition of extra terms in a least squares model can only give us a greater fit to the underlying time series. We must be careful, however, to avoid the problems resulting from "overfitting" in a least squares model. It was mentioned in chapter 12 that these problems are usually negligible if the sample size $n \le 4k$, where k is the number of predictor variables used in the model. If a term introduced as a predictor variable is redundant, it will simply be "weighted out" by having a small regression coefficient in the least squares forecasting equation. [Note that the coefficient of the $t \sin (2\pi t / 10)$ term in our preceding model was only .01. Thus $t \sin (2\pi t / 10)$ does not contribute much to the value of $\hat{y}_t$.] The ultimate criterion of goodness of the forecasting model, of course, is how well the model forecasts the future.

Forecasting always requires extrapolation beyond the interval of time in which the sample data were collected, a procedure that can lead to errors of prediction considerably larger than those expected according to the least squares theory. This is occasionally caused by a change in the basic form of the linear model relating y to the original independent variables but is most often due to the omission of one or more independent variables that may well have remained stable over the interval of time in which the sample data were collected. Taxation policies, research expenditures, political developments, and buyer behavior are examples of variables that may remain stable over the sample collection period but may cause abrupt and relatively large forecast errors when their values exhibit sudden changes during the forecast period.

We almost always fail to include in the model one or more "dormant" variables, with the result that forecasts far into the future may be subject to large errors. In spite of this, the least squares forecasting equation, with the relatively crude least squares bound on the error of prediction, is a very useful tool. This is particularly true for short-term forecasting, where relative stability persists in the system for at least a short period of time. The important point to keep in mind is that large errors can occur when extrapolating.

Exercises

15.4.   Refer to chapter 14, exercise 14.17. A statistician claims that the sinusoidal forecasting model

$$\hat{y}_t = 17.0 + .365t + 1.1 \sin \left(\frac{2\pi t}{6} \right)$$

which indicates the presence of a six-month cycle in the sales pattern, is an efficient model for forecasting the sales of the Pharmaceutical Company. Let January 1976 be designated as time period $t = 1$, February 1976 as $t = 2$, and so forth. Using the forecasting model given above, plot the forecast and actual sales of the Pharmaceutical Company. Does the suggested sinusoidal model appear to be a good forecasting equation for these sales data?

15.5. Employing the methods of chapter 11, construct a simple linear forecasting model to fit the monthly sales volume data of the pharmaceutical company (the data of exercise 14.17).

a. Compute SSE as a measure of the fit of your derived linear model to the actual sales data.

b. Compute SSE for the sinusoidal model of exercise 15.4. Which model appears to be better for forecasting the sales of the pharmaceutical company?

c. Superimpose the linear model derived in this exercise with the sinusoidal model of exercise 15.4 on a plot of the actual sales data. Which model seems more responsive to periodic cycles inherent in the actual sales data?

15.6. What purpose is served by the use of terms such as $t \cos$ and $t \sin$ as well as $\cos$ and $\sin$ as independent variables in a sinusoidal econometric forecasting model? Give an example of a time series in which the use of the terms $t \cos$ and $t \sin$ as predictors would be appropriate.

15.7. The director of the aid to dependent children (ADC) program of a state welfare agency, realizing the uneven demand for state support due to seasonal employment opportunities, has devised a sinusoidal model for forecasting the number of children supported under the ADC program. Her model for forecasting the number requiring support each quarter is

$$\hat{y}_t = 1.75 + .10t + 1.20 \cos\left(\frac{2\pi t}{4}\right) + .10t \cos\left(\frac{2\pi t}{4}\right)$$

where $\hat{y}_t$ is measured in tens of thousands and t is the quarter number for each of the next few three-month segments of time. Use this model to forecast the quarterly demand to the ADC program over the next two years (quarters $t = 1, 2, ..., 8$).

15.4 The Autoregressive Forecasting Model

Econometric models accounting for correlations between adjacent observations in a time series are called **autoregressive models**. Thus if y_t and y_{t-1} are the responses at times t and $(t-1)$, respectively, we might write

$$y_t = \beta_0 + \beta_1 x + \epsilon \qquad \text{where} \qquad x = y_{t-1}$$

Thus we are using knowledge of the response at time period $(t-1)$ to predict the response at time t. We refer to the model above as the **first-order autoregressive model**. The general p**th-order** autoregressive equation is

$$y_t = \beta_0 + \beta_1 y_{t-1} + \beta_2 y_{t-2} + \beta_3 y_{t-3} + \cdots + \beta_p y_{t-p} + \epsilon$$

An autoregressive model of order greater than one is necessary for including information in the model due to the correlation between observations separated by more than one unit of time. For instance, the sales volume of the urban department store, illustrated in figure 14.4, seems to indicate an obvious annual (12-month) cycle. Thus a "good" autoregressive model for forecasting the future sales of the urban department store might be

$$y_t = \beta_0 + \beta_1 y_{t-1} + \beta_2 y_{t-6} + \beta_3 y_{t-12} + \epsilon$$

where $\beta_1 y_{t-1}$ is the effect on the next month's sales caused by last month's sales volume, $\beta_2 y_{t-6}$ is the 6-month effect, and $\beta_3 y_{t-12}$ is the 12-month effect. The 6-month effect is chosen because of an assumed significant strong negative relationship of sales for months separated by 6 months—high sales in December, low sales in June, moderately high in November and January, moderately low in May and July. The 12-month effect is more apparent. Sales are always high in December, always low in June, always moderately high in November and January, and always moderately low in May and July. Thus there is an apparent high positive correlation between the sales for the same month in different years.

The autoregressive model need not be restricted to a linear relationship. For example, we could use

$$y_t = \beta_0 + \beta_1 x + \beta_2 x^2 + \epsilon \qquad \text{where} \qquad x = y_{t-1}$$

if we have reason to believe that each response is a quadratic function of the most recent response.

It is not easy to determine the equation that best forecasts the future values of a random process from its past values or other related variables. The autoregressive model suggests that if a high correlation exists between the values of a random process at constant intervals of time, then the autoregressive model may be appropriate for forecasting.

The only way we can determine whether the autoregressive model, or any other forecasting model, is a good forecasting model is to use each model to develop forecasts. The model that should be best for our purposes is the one that yields the most accurate forecasts, as measured by the smallness of the SSE.

Autoregressive models can be fitted to time series data by using the method of least squares. Essentially this amounts to letting y_{t-1}, y_{t-2}, y_{t-3}, ..., y_{t-p}, called the lagged process responses, assume the roles of the independent variables in the linear regression model. Then we solve for estimates of the regression parameters β_0, β_1, ..., β_{p+1} by the least squares methods outlined for the general linear regression model in chapter 12. The theory of inference associated with autoregressive models is beyond the scope of this text but can be found in the text by Box and Jenkins (1969) listed in the references.

We have fitted the autoregressive model

$$y_t = \beta_0 + \beta_1 y_{t-1} + \beta_2 y_{t-2} + \epsilon$$

to the week's end security prices of the Color-Vision Company. A second-order autoregressive model was selected only to motivate the use of an autoregressive model, not to suggest that the second-order model is the "best" autoregressive model for forecasting the Color-Vision prices. The second-order autoregressive equation for forecasting the security prices one week ahead of the available data is

$$\hat{y}_t = 18.74 + 1.03y_{t-1} - .28y_{t-2}$$

where y_t is the most recent week's end security price and y_{t-1} is the next most recent week's end price.

To see how well the model fits the data, we compute the estimated price $\hat{y}_t$ for each week when actual observed closing prices exist for the two preceding weeks. For instance, the predicted price at the end of the tenth week is

$$\hat{y}_{10} = 18.74 + 1.03y_9 - .28y_8 = 18.74 + 1.03(75) - .28(78) = 74.15$$

where the actual price at the end of the ninth week was 75 and for the eighth week was 78. Table 15.2 lists the actual and autoregressive fitted values for the Color-Vision security prices, which are plotted in figure 15.3.

Table 15.2 *Actual and fitted week's end security prices for fitted values determined by a second-order autoregressive model*

WEEK, t	y_{t-1}	y_{t-2}	TRUE VALUE, y_t	ESTIMATES, $\hat{y}_t$	ERROR, $e_t = y_t - \hat{y}_t$
1	70†	69†	71	71.82	−.82
2	71	70†	70	72.57	−2.57
3	70	71	69	70.26	−2.26
4	69	70	68	70.51	−2.51
5	68	69	64	69.75	−5.75
6	64	68	65	65.90	−.90
7	65	64	72	68.04	3.96
8	72	65	78	74.99	3.01
9	78	72	75	79.24	−4.24
10	75	78	75	74.48	.52
11	75	75	75	75.31	−.31
12	75	75	70	75.31	−5.31
13	70	75	75	70.15	4.85
14	75	70	75	76.70	−1.70
15	75	75	74	75.31	−1.31
16	74	75	78	74.28	3.72
17	78	74	86	78.68	7.32
18	86	78	82	85.83	−3.83
19	82	86	75	79.48	−4.48
20	75	82	73	73.37	−.37
21	73	75	72	73.25	−1.25
22	72	73	73	72.77	.23
23	73	72	72	74.08	−2.08
24	72	73	77	72.77	4.23
25	77	72	83	78.21	4.79
26	83	77	81	83.01	−2.01
27	81	83	81	79.28	1.72
28	81	81	85	79.84	5.16
29	85	81	85	83.96	1.04
30	85	85	84	82.85	1.15

†The values of y_t for the two weeks preceding week 1 were 69 and 70, respectively.

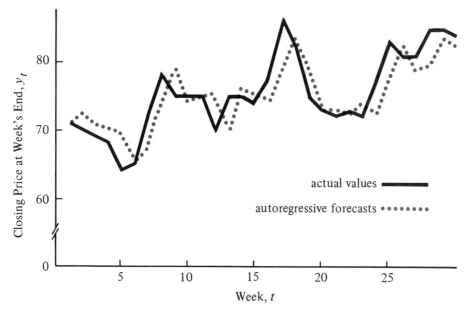

Figure 15.3 *Autoregressive forecasts for the week's end security prices of the Color-Vision Company*

Figure 15.3 exhibits a very noticeable lag-one effect. That is, the fitted prices appear to be practically identical with the actual price from the previous time period. This relationship is due to the fact that the autoregressive model we have selected estimates future security prices as a function of only the two latest security prices. But since the correlation between prices separated by two time periods is small, the forecasted price is essentially a function of only the most recent observation. Thus a logical conclusion might be that the autoregressive model, at least the second-order autoregressive model we have chosen, is not a good model to use to forecast future Color-Vision security prices since the model does not appear to adequately fit the available past prices. However, in all 30 cases, the forecast error is less than 10% of the true value, a standard of accuracy commonly required by management.

If the actual process we are attempting to forecast is quite stable, a first-order or second-order autoregressive model should give fairly accurate forecasts. When the actual process is rather volatile, autoregressive forecasts are typically not very accurate. But we could hardly expect any forecasting model to predict accurately the future behavior of a volatile process. Some additional forecasting accuracy can be expected for each additional autoregressive term added to the autoregressive forecasting equation. However, inclusion of y_{t-2} into the autoregressive model

$$y_t = \beta_0 + \beta_1 y_{t-1} + \beta_2 y_{t-2} + \epsilon$$

may not add enough in predictive accuracy to offset the cost and effort necessary to incorporate the extra term, and its values over time, into the model. Only the terms that add significantly to predictive accuracy should be incorporated into the model, where significance is measured by the relative sizes of the correlations between response values separated by constant intervals of time.

Time lags at which significant correlations exist can seldom be determined by only a visual inspection of the plot of the time series. In general, we determine whether an autoregressive model is appropriate by computing a **correlogram**, a graphical display of the correlations existing between observations spaced by a constant interval of time. These time-lag correlations, called **autocorrelations**, are computed in a manner similar to the method of section 11.8. If y_1, y_2, ..., y_n are the values for n consecutive time periods in a time series, then the autocorrelation between the values separated by k time periods is

$$r_k = \frac{\sum\limits_{t=1}^{n} (y_t - \hat{y}_t)(y_{t+k} - \hat{y}_{t+k})}{\sum\limits_{t=1}^{n} (y_t - \hat{y}_t)^2} \quad \text{and} \quad -1 \le r_k \le 1$$

where $\hat{y}_t$ is the value at time t of the linear trend equation fitted to the time series. That is, by the linear regression methods of chapter 11, the linear equation $\hat{y}_t = \hat{\alpha}_0 + \hat{\alpha}_1 t$ is fitted to the response values over time. The terms $\hat{y}_t$ are used in the computation of r_k to "cancel out" the effects that may exist in a linear trend over time.

If the inherent trend is not linear, then $\hat{y}_t$ is determined from some nonlinear model that appropriately fits the underlying time series. However, studies show that linear detrending is usually satisfactory for removing the trend inherent in a time series.

A correlogram for a time series may appear as shown in figure 15.4.

Figure 15.4 *Correlogram for a time series*

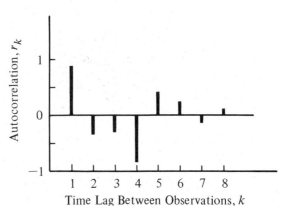

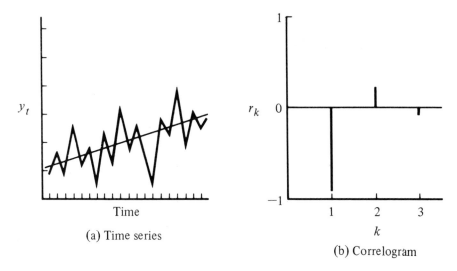

Figure 15.5 *Plot and correlogram of a time series that is very unstable over time*

The autocorrelations for response values separated by one and four time periods appear to be greater than the other correlations. Thus an appropriate autoregressive forecasting function for this process might be

$$y_{t+1} = \beta_0 + \beta_1 y_t + \beta_2 y_{t-3} + \epsilon$$

We should note before concluding our discussion of autoregressive models that a negative autocorrelation is just as meaningful as a positive one. We

Figure 15.6 *Plot and correlogram of a time series that is rather stable over time*

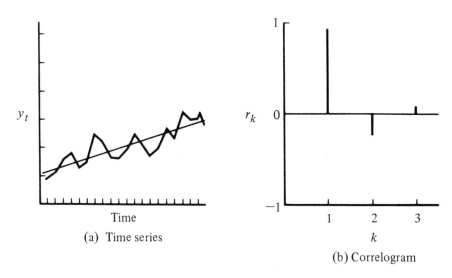

are interested in the absolute magnitude $|r_k|$ and not in its sign. A negative r_k implies that when a response y_t is greater than the average trend, the response y_{t+k} at k time periods ahead of y_t (or y_{t-k} at k time periods behind y_t) tends to be less than the average trend. If significant, such a relationship may be just as useful for purposes of prediction as when both y_t and y_{t+k} are simultaneously either greater or less than the average trend. As an illustration, compare figures 15.5 and 15.6. In figure 15.5 we note a process that is highly unstable, giving a value for the time series on a different side of the average trend line each successive time period. Its autocorrelation value r_1 is negative. In figure 15.6 the time series is more stable, does not alternate around the average trend as often, and has a positive autocorrelation value r_1. In either case, y_{t-1} would probably be a good predictor of y_t in a first-order autoregressive model.

Exercises

15.8.   Refer to chapter 14, exercise 14.17. Fit a first-order autoregressive model

$$y_t = \beta_0 + \beta_1 y_{t-1} + \epsilon$$

to the sales data from the pharmaceutical company by letting the one-month lagged sales value y_{t-1} assume the role of the independent variable x and by employing the linear regression methods of chapter 11. Compute the estimated autoregressive values and plot them against the true monthly sales values.

15.9.   Assume the model you have derived in exercise 15.8 is a forecast model derived independent of the 1977 sales values. Use the autoregressive model to forecast the monthly 1977 sales values and measure the forecast accuracy by computing SSE. Compare the forecast accuracy of the autoregressive model for forecasting the monthly 1977 sales values against those obtained by using the sinusoidal model in exercise 15.4.

15.10.   The sales manager for a firm that manufactures energy-saving thermostats and rheostats claims that the following second-order autoregressive model can be used to provide adequate forecasts for his firm's monthly gross sales volume (in thousands of dollars):

$$\hat{y}_t = 17.8 + .9y_{t-1} - .2y_{t-2}$$

Suppose that the gross monthly sales volume (in thousands of dollars) over the next 12 months are as given in the table. Beginning with March, use the suggested

Month	Jan.	Feb.	Mar.	Apr.	May	June	July	Aug.	Sept.	Oct.	Nov.	Dec.
Sales	51.2	50.7	52.6	54.9	55.5	56.2	57.1	56.9	58.3	58.6	59.4	60.0

autoregressive forecasting equation to estimate the monthly sales volume based on past sales data. Compare the forecasted sales volumes with the actual sales volumes by computing SSE as a measure of forecast accuracy.

15.11.   How can autoregressive terms be used in an econometric forecasting model to assist in forecasting future values of a time series with a distinct seasonal component?

15.5 An Exponential-smoothing Forecasting Model

In section 14.3 we examined the use of smoothing techniques for "averaging out" the effects of random variation. It was mentioned that the smoothed series is a series of averages, each computed from available past data. In the case of the moving-average series, only the data from the past M periods are used for each value; with the exponentially smoothed series, all available past data are used for the computation of each smoothed value.

Smoothing methods can also be used for forecasting. We will present two time series forecasting models based on smoothing techniques. Other time series models are left to you to explore.

The first model we introduce is the **multiple exponential-smoothing forecasting model** developed by R. G. Brown. Suppose that we have the observations y_1, y_2, ..., y_t from a time series. Our objective is to compute a forecast of the process value y_{t+T}, which is T time points ahead of our available data. (This is sometimes referred to as a forecast lead period of T.)

Brown's method gives a convenient way of expressing the forecast y_{t+T} in terms of the exponentially smoothed statistics discussed in section 14.3. If the time series appears **constant over time**, such as perhaps the average annual rainfall in Portland, Oregon, over the past 25 years, then we use the simple exponentially smoothed value S_t to forecast y_{t+T}.

First-Order Exponential-smoothing Forecasting Model

$$\hat{y}_{t+T} = S_t = \alpha y_t + (1 - \alpha)S_{t-1}$$

If the process is **linear over time**, as was the illustration in figure 14.2, the forecast is as given in the next box.

Second-Order Exponential-smoothing Forecasting Model

$$\hat{y}_{t+T} = \left(2 + \frac{\alpha T}{1 - \alpha}\right) S_t - \left(1 + \frac{\alpha T}{1 - \alpha}\right) S_t(2)$$

where $S_t(2) = \alpha S_t + (1 - \alpha)S_{t-1}(2)$.

The statistic $S_t(2)$ is called the **double-smoothed statistic** and is a smoothing of the smoothed values. That is, the series of $S_t(2)$-values is a smoothing of the series of S_t-values, where the observations, the y_t-values, in the S_t series are the counterparts of the S_t-values in the $S_t(2)$ series. The statistic $S_t(2)$ gives an indication of the trend of the averages S_t over time. Hence its inclusion in the model will account for a linear trend of y_t with time.

If the time series is neither constant nor linear over time, it is best to use a triple exponential-smoothing forecasting model. Higher-order exponential-smoothing forecasting models exist, but the computational difficulties in computing the forecasting equation for models of an order higher than the triple exponential-smoothing models are considerable. (See R. G. Brown in the references.) Unless a time series is extremely volatile, triple exponential-smoothing forecasting models work quite well.

The suggested forecasting equation when the time series is **neither constant nor linear with time** is given in the box.

Third-Order Exponential-smoothing Forecasting Model

$$\hat{y}_{t+T} = [6(1-\alpha)^2 + (6-5\alpha)\alpha T + \alpha^2 T^2]\frac{S_t}{2(1-\alpha)^2}$$

$$- [6(1-\alpha)^2 + 2(5-4\alpha)\alpha T + 2\alpha^2 T^2]\frac{S_t(2)}{2(1-\alpha)^2}$$

$$+ [2(1-\alpha)^2 + (4-3\alpha)\alpha T + \alpha^2 T^2]\frac{S_t(3)}{2(1-\alpha)^2}$$

The **triple-smoothed statistic** $S_t(3)$ is, in a sense, describing the average rate of change of the average rates of change. It is found by computing

$$S_t(3) = \alpha S_t(2) + (1-\alpha)S_{t-1}(3)$$

The forecasting equations listed earlier for the constant model, the linear model, and the second-order model were presented without an explanation of their derivation. The derivations, in each case, are very involved and are eliminated from our discussion. Those interested in the derivation of these forecasting equations are referred to chapter 9 in the text by Brown (1963).

The purpose of using the smoothed statistics S_t, $S_t(2)$, and $S_t(3)$ in Brown's method is to develop estimates for the coefficients of a model that adequately describes the relationship of the value y_t with time. This is performed recursively by continually updating the coefficients in the forecasting model as more data become available. That is, a different forecasting equation is acquired at every point in time, each based on all the available past and present response values. The recursions are initiated by assuming some value for $S_t(2)$ and $S_t(3)$ at the first time period $t = 1$. A wise choice for these initial values is the first observation y_1. The values selected in the beginning are not critical, though, since their contribution to the forecasting equation will decrease as more data are incorporated into the model.

The selection of the smoothing constant α is arbitrary, as explained in the discussion of smoothing methods in section 14.3. The selection rule suggested for data smoothing is also recommended for the use of smoothing methods for forecasting. That is, if the process is volatile, a small value of α is selected; if the process is stable, a large α will probably give the most accurate forecasts. Some authors have suggested that the smoothing constant α change with time to reflect any new trends in the process. The benefits of such a plan, though, would probably be outweighed by the additional computational difficulties.

You might ask why Brown's method should be used when a least squares model can be fitted to the data to represent the same polynomial relationship between y_t and time that is being assumed for Brown's method. The advantage of Brown's method is that it is a recursive method which develops a new forecasting model each time an additional observation is observed. We could also compute a new least squares model with each new observation but the computational difficulties would be enormous. Least squares models are usually of the type described in sections 15.2 and 15.3, where one forecasting model is constructed from which forecasts are made, without updating the model each time new observational information is obtained. Thus the multiple smoothing forecasting method of Brown is more efficient since it incorporates more information into the forecasting model.

Care should be taken when we are attempting to forecast further than one time point ahead $(T > 1)$. Unforeseeable events may cause the time series to react differently than it has reacted before. Thus, regardless of how much past data are incorporated into the model, we may not be able to forecast future values accurately. If you are interested in the details of Brown's method of forecasting, we again refer you to Brown (1963).

Use Brown's method of multiple smoothing to forecast the week's end security prices of the Color-Vision Company. Assume that the relationship of the security prices to time is neither constant nor linear. Let $T = 1$.

Since we are assuming that our process is neither constant nor linear with time, we will employ the triple smoothing method to forecast the prices one time point ahead of available data. Since the process is rather volatile, as illustrated in figure 14.9, we select $\alpha = .2$. Since $T = 1$, our forecasts are computed from the equation

$$\hat{y}_{t+1} = [6(1 - .2)^2 + \{6 - 5(.2)\}(.2) + (.2)^2]\frac{S_t}{2(1 - .2)^2}$$
$$- [6(1 - .2)^2 + 2\{5 - 4(.2)\}(.2) + 2(.2)^2]\frac{S_t(2)}{2(1 - .2)^2}$$

$$+ [2(1 - .2)^2 + \{4 - 3(.2)\} (.2) + (.2)^2] \frac{S_t(3)}{2(1 - .2)^2}$$

$$= 3.8125 S_t - 4.375 S_t(2) + 1.5625 S_t(3)$$

At $t = 1$ we have $S_1 = 71$, $S_1(2) = 71$, and $S_1(3) = 71$, since $y_1 = 71$. At each succeeding time period $t = 2, 3, \ldots, 30$, the smoothed statistics are found by computing

$$S_t = (.2)y_t + (1 - .2)S_{t-1}$$

$$S_t(2) = (.2)S_t + (1 - .2)S_{t-1}(2)$$

$$S_t(3) = (.2)S_t(2) + (1 - .2)S_{t-1}(3)$$

The smoothed statistics are then reentered into the forecasting equation to obtain the forecast for the next time period.

Table 15.3 *Actual and forecasted week's end security prices for the Color-Vision Company; Brown's triple-smoothing forecasting method was used to compute the forecasts*

TIME, t	ACTUAL PRICE, y_t	FORECAST, $\hat{y}_t$	FORECAST ERROR, $e_t = y_t - \hat{y}_t$
1	71		
2	70	71.00	−1.00
3	69	70.40	−1.40
4	68	69.44	−1.44
5	64	68.28	−4.28
6	65	65.22	−.22
7	72	64.06	7.94
8	78	67.70	10.30
9	75	73.64	1.36
10	75	75.45	−.45
11	75	76.42	−1.42
12	70	76.84	−6.84
13	75	73.93	1.07
14	75	75.01	−.01
15	74	75.60	−1.60
16	78	75.26	2.74
17	86	77.36	8.64
18	82	88.35	−1.35
19	75	84.42	−9.42
20	73	80.59	−7.59
21	72	76.82	−4.82
22	73	73.83	−.83
23	72	72.61	−.61
24	77	71.35	5.65
25	83	73.68	9.32
26	81	78.81	2.19
27	81	80.73	.27
28	85	81.79	3.21
29	85	84.70	.30
30	84	86.29	−2.29

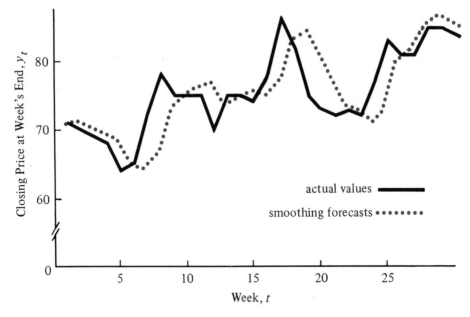

Figure 15.7 *Exponential-smoothing forecasts for the week's end security prices of the Color-Vision Company*

The forecasts obtained by Brown's method and the true values are listed in table 15.3 and illustrated in figure 15.7. As with the autoregressive model, there is a noticeable one-period lag effect to the multiple exponential smoothing forecasts. The forecasts are, however, responsive to changes in the direction of the security price time series. In fact, in only 5 of the 30 time periods, the forecast error percentage exceeds 10% of the actual price y_t.

The multiple exponential-smoothing model is often used as a tracking model to detect turning points in a time series. Identification of turning points is very useful when studying such time series as price movements, consumer buying habits, economic indicators like the floating value of currencies, or security price movements over time. The multiple exponential-smoothing model is generally interpreted as having noted a time series that has bottomed out when the true values *cut over* the smoothing forecasts, as illustrated in figure 15.8(a). A process that has peaked out is noted when the true values *cut under* the smoothing model forecasts. [See figure 15.8(b).]

In figure 15.7 the multiple exponential-smoothing model indicates that the time series of Color-Vision prices has reached a low point (bottoming out) at periods 6, 13, 16, and 24. When prices have reached a high point and should be on the way down, peaks are noted at periods 10, 14, 18, and 29. If an investor with a typical buy-low-and-sell-high objective had purchased Color-Vision securities each time a low point was indicated by the model and had sold the securities each time the prices peaked out, he

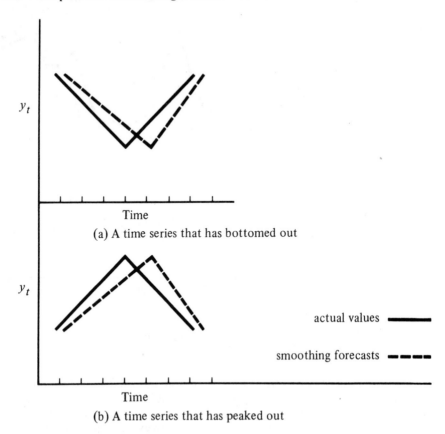

(a) A time series that has bottomed out

actual values ▬▬▬▬▬

smoothing forecasts ▬ ▬ ▬ ▬

(b) A time series that has peaked out

Figure 15.8 *Detection of turning points in a time series by the multiple exponential-smoothing model*

would have made considerable profit over the period indicated. In fact, his average gain per share would have been $25 over the 24-week period $t = 5$ to $t = 30$.

Used simply as a forecasting model, the multiple exponential-smoothing model with smoothing constant $\alpha = .2$ appears less reliable than the sinusoidal model for forecasting the security prices of the Color-Vision Company. This conclusion applies only to the security prices of the Color-Vision Company and should not be taken as a general rule to apply when selecting a forecasting model for a time series. A forecasting model that is appropriate for one time series may be totally inappropriate for generating forecasts of another time series.

Exercises

15.12. Refer to exercise 15.10. Assume that the true pattern of the monthly sales of the manufacturing firm is linear over time and use Brown's multiple exponential-smoothing forecasting model to forecast the monthly sales one month ahead of available data.

a. Compute the monthly forecasts using a smoothing constant of $\alpha = .1$.

b. Compute the monthly forecasts using a smoothing constant of $\alpha = .3$.

c. Compare the accuracy of the two forecast equations in parts a and b by computing SSE as a measure of forecast accuracy for each.

15.13.   Explain how the multiple exponential-smoothing forecasting model can be adjusted to accommodate sudden shifts in a time series. For example, consider the case of a candy manufacturer who uses the multiple exponential-smoothing model to forecast future monthly sugar prices. As we can see from the time series shown in exercise 14.3, sugar prices fluctuated gently until 1972, when they suddenly skyrocketed. How can the candy manufacturer adjust the multiple exponential-smoothing model to accommodate the sudden shifts in sugar prices?

15.14.   You may have noticed from the text that the various multiple exponential-smoothing forecasting models are simply linear equations equating the forecast value $\hat{y}_{t+T}$ to the smoothed statistics S_t, $S_t(2)$, and $S_t(3)$. The coefficients of each equation are determined by the selection of the smoothing constant α and the forecast lead time T. For both the double-smoothed and triple-smoothed forecasting models, compute the coefficients relating $\hat{y}_{t+T}$ to the smoothed statistics for the following:
a. $\alpha = .1$, $T = 1$ b. $\alpha = .1$, $T = 2$
c. $\alpha = .25$, $T = 1$ d. $\alpha = .25$, $T = 3$

15.15.   As was mentioned in the text, the multiple exponential-smoothing forecasting model is sometimes used to detect turning points in a time series. In production management this means that when a peak in demand is observed, we should curtail production and reduce inventories; when a bottom is observed, inventories

MONTH	YEAR 1	YEAR 2	YEAR 3
January	57	63	58
February	59	62	64
March	62	66	66
April	60	69	63
May	63	64	65
June	66	62	68
July	63	70	66
August	67	71	72
September	70	59	69
October	68	59	62
November	64	61	65
December	66	61	65

should be replenished. The table shows the monthly demand for an air purification unit produced by a small manufacturing firm over a three-year period. Use the triple exponential-smoothing forecasting model, with $\alpha = .2$ and $T = 1$, to find the months during which production should have either been increased or curtailed.

15.6 The Exponentially Weighted Moving-Average Forecasting Model

One time series forecasting method has proven to be especially effective for generating forecasts of a process with a pronounced **seasonal component**. This method, called the **exponentially weighted moving-average (EWMA)** fore-

casting model, operates by separately estimating, at each point in time, the smoothed average, the average trend gain, and the seasonal factor, and then combining these three components to compute a forecast.

As we noted in section 14.2, the seasonal effect is independent of a long-term trend and cyclic effects. Furthermore, since the seasonal effect is recurrent and periodic, it is predictable. Seasonality is a very common effect in business forecasting problems, especially as a component of the sales pattern of a firm. In most retailing enterprises sales tend to be high around December; breweries and soft drink companies tend to have a sales pattern that follows the temperature—high sales when the weather is warm and low sales during cool weather; ski manufacturers have a sales pattern that peaks every winter. We could create an endless list of examples of time series in business that exhibit a seasonal pattern. Thus the exponentially weighted moving-average model is an extremely valuable aid to the business forecaster.

At each point in time, the separate components of the time series that are estimated by the EWMA model are as follows:

1. The **smoothed average** at time t:

$$S_t = (\alpha)\frac{y_t}{F_{t-L}} + (1 - \alpha)(S_{t-1} + R_{t-1})$$

2. The updated **trend gain** at time t:

$$R_t = (\beta)(S_t - S_{t-1}) + (1 - \beta)R_{t-1}$$

3. The updated **seasonal factor** at time t:

$$F_t = (\gamma)\frac{y_t}{S_t} + (1 - \gamma)F_{t-L}$$

In these equations, α, β, and γ are **smoothing constants**, arbitrarily selected by the statistician and satisfying the properties

$$0 \leq \alpha \leq 1 \qquad 0 \leq \beta \leq 1 \qquad 0 \leq \gamma \leq 1$$

The index L in the seasonal factor F_{t-L} is the seasonal period—the number of time points required for a repeat of the seasonal effect present in the time series—y_t is the true value at time t, and the statistics S_{t-1} and R_{t-1} are the smoothed average and trend gain, respectively, which are estimated at time period $(t - 1)$. These component equations are combined to give the forecast of the response value T time periods ahead of the most recent data (time period t).

EWMA Forecasting Equation

$$\hat{y}_{t+T} = [S_t + (T)R_t]F_{t-L+T}$$

Table 15.4 *Using the EWMA forecasting equation*

COMPONENT (FACTOR)	ESTIMATE BASED ON MOST RECENT RESPONSE VALUE	ESTIMATE BASED ON REMOTE VALUE
smoothed average	y_t / F_{t-L}	$S_{t-1} + R_{t-1}$
trend gain	$S_t - S_{t-1}$	R_{t-1}
seasonal factor	y_t / S_t	F_{t-L}

In each of the component equations, the model computes an exponentially weighted moving average—from which the name of the model arises. Each component equation estimates a factor that will be used in the forecasting equation by weighting an estimate of that factor based on the most recent response value with the previous value of the factor. We can see how this is performed by examining table 15.4. Educational psychologists refer to such updating schemes as those illustrated by the three component equations of our model as **learning curves,** since we are learning more about each component by incorporating new information into the estimation equations while still utilizing the available past information.

Initial values must be determined for the statistics S and R so that we can compute S_t and R_t at time $t = 1$. Furthermore, initial values are needed for the seasonal factor F_t **at each time point** (each month, each week, etc.) **of the entire seasonal period.** The initial values are usually improved by using two or more complete seasonal periods of the data as a "warm-up" period before any forecasts are computed. For instance, if we are interested in analyzing and then forecasting the monthly sales of a retail department store, initial values for S and R are selected and 12 seasonal factors (one for each month) are selected. These values are used to initiate the recursion for each equation. After two or more complete seasonal periods, the erroneous effects of the initial estimates for S, R, and the 12 values of F should be sufficiently dampened out. The rate at which the initial values are dampened out depends on the selection of smoothing constants α, β, and γ. **As a general rule, α and β are small, usually about .1, while the seasonal factor smoothing constant γ is best set at about .4.** The statistician should not arbitrarily select those values, though, but should try many combinations of values for α, β, and γ until a combination is found that generates sufficiently accurate forecasts. Note, however, that the number of combinations of possible values for α, β, and γ is exceedingly large.

Although the smoothing constants are usually not changed as the EWMA model is continuously applied over time, they can be adjusted to accommodate sudden shifts in the time series. For instance, if a temporary shift in the level of a time series is expected due to a sales promotion campaign, a labor strike, the weather, or some other atypical situation, the constants α and β can be temporarily increased to some very large value, say .8, and then returned to their original values when the atypical situation has subsided. If some effect has dampened the seasonal effect or increased its amplitude, the smoothing constant γ should likewise be temporarily increased.

In short, the forecaster should use ingenuity and experience to continuously adapt the model to its environment in order to guarantee that the resulting forecasts are as accurate as possible.

Example 15.2 The monthly sales volumes for a regional brewing company are listed in table 15.5 for the period of time from 1 January 1972 to 1 January 1977. Use the sales data from 1972 and 1973 to generate estimates for the smoothed average, the trend factor, and the monthly seasonal factors. Beginning on 1 January 1974, use the exponentially weighted moving-average method to forecast the sales of the brewery one month ahead of available data.

Table 15.5 *Monthly sales (thousands of barrels) of a regional brewing company from 1 January 1972 to 1 January 1977*

| | SALES VOLUME | | | | |
MONTH	1972	1973	1974	1975	1976
January	18.7	18.3	19.6	23.3	24.6
February	15.6	17.6	18.6	20.1	22.8
March	18.3	24.1	23.2	28.1	28.4
April	19.6	21.8	24.5	26.6	27.2
May	21.4	23.3	27.7	28.6	28.6
June	28.9	28.7	30.0	33.3	29.3
July	24.5	30.0	28.7	34.3	38.3
August	24.5	29.1	33.8	29.0	32.0
September	21.9	23.5	25.1	26.4	24.9
October	20.1	21.6	22.1	25.1	27.7
November	17.9	21.6	21.8	22.3	22.2
December	17.9	19.8	20.9	20.3	21.5

Solution We will use the previously suggested smoothing constants, $\alpha = .1$, $\beta = .1$, and $\gamma = .4$, in the three recursive component equations. Initial estimates for the factors S and R and the 12 seasonal factors may be obtained as follows:

a. Let S_0, the initial value for smoothed average, be represented by the first actual value y_1. Thus $S_0 = 18.7$.

b. On a sheet of graph paper, plot the available data for the time series. (In our problem we are assuming that the monthly data from 1972 and 1973 are available as sample data.) With the aid of a ruler, draw a line through a plot of the available time series data which, with eyeball accuracy, appears to depict the linear trend of the data. Such a line is illustrated for the brewing company sales data in figure 15.9.

c. Let R_0, the initial value for the trend gain, be represented by the slope of the arbitrarily drawn trend line; that is, the average gain in trend (sales) per unit of time. The value of the trend line at $t = 1$ is 21; the value at $t = 24$ is 24. Thus

$$R_0 = \frac{24 - 21}{24 - 1} = \frac{3}{23} = .13$$

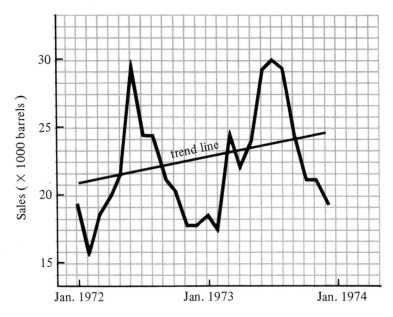

Figure 15.9 *Sales volume of a regional brewing company for the years 1972 and 1973, illustrating an arbitrarily drawn trend line*

d. The initial monthly seasonal factors are found by evaluating the ratio

$$\frac{\text{actual sales volume for the month}}{\text{value of the trend for the month}}$$

for each month of the first entire seasonal pattern. For the Bush Brewing Company, the first three seasonal indexes are

$$F_{\text{January}} = \frac{18.7}{21.0} = .89$$

$$F_{\text{February}} = \frac{15.6}{21.13} = .74$$

$$F_{\text{March}} = \frac{18.3}{21.26} = .86$$

The other monthly indexes are calculated in a similar fashion.

Using the initial values for the smoothed average, the trend factor, and the 12 seasonal factors, the recursive component equations are initiated. At the first time point $t = 1$, we compute these statistics:

$$S_1 = \alpha \left(\frac{y_1}{F_{\text{Jan}}} \right) + (1 - \alpha)(S_0 + R_0)$$

$$= .1 \left(\frac{18.7}{.89} \right) + .9(18.7 + .13) = 19.05$$

$$R_1 = \beta(S_1 - S_0) + (1 - \beta)R_0$$

$$= .1(19.05 - 18.7) + .9(.13) = .15$$

Table 15.6 *Actual and forecasted monthly sales of the brewing company; forecasting method is an exponentially weighted moving average (listed values in thousands of barrels)*

TIME, t	ACTUAL SALES, y_t	FORECAST, $\hat{y}_t$	FORECAST ERROR, $e_t = y_t - \hat{y}_t$
1974			
January	19.6	22.7	−3.1
February	18.6	19.7	−1.1
March	23.2	24.2	−1.0
April	24.5	24.0	.5
May	27.7	26.0	1.7
June	30.0	34.2	−4.2
July	28.7	30.9	−2.2
August	33.8	30.1	3.7
September	25.1	26.2	−1.1
October	22.1	23.9	−1.8
November	21.8	22.0	−.2
December	20.9	21.2	−.3
1975			
January	23.3	22.6	.7
February	20.1	20.5	−.4
March	28.1	25.4	2.7
April	26.6	26.2	.4
May	28.6	28.7	−.1
June	33.3	35.0	−1.7
July	34.4	32.5	1.9
August	29.0	34.4	−5.4
September	26.4	27.4	−1.0
October	25.1	24.6	.5
November	22.3	23.4	−1.1
December	20.3	22.4	−2.1
1976			
January	24.6	24.1	.5
February	22.8	21.4	1.4
March	28.4	27.9	.5
April	27.2	27.5	−.3
May	28.6	29.8	−1.2
June	29.3	35.5	−6.2
July	38.3	33.7	4.6
August	32.0	33.1	−1.1
September	24.9	27.9	−3.0
October	27.7	25.3	2.4
November	22.2	23.7	−1.5
December	21.5	22.2	−.7

$$F_1 = \gamma\left(\frac{y_1}{S_1}\right) + (1 - \gamma)F_{Jan} = .4\left(\frac{18.7}{19.05}\right) + .6(.89) = .93$$

At each succeeding month of the smoothing period (the first 24 months of data), we compute these statistics:

$$S_t = .1\left(\frac{y_t}{F_{t-12}}\right) + .9(S_{t-1} + R_{t-1})$$

$$R_t = .1(S_t - S_{t-1}) + .9(R_{t-1})$$

$$F_t = .4\left(\frac{y_t}{S_t}\right) + .6(F_{t-12})$$

Forecasting is begun at the 24th time period by computing the forecast $\hat{y}_{25} = (S_{24} + R_{24})F_{13}$ of the sales during the 25th time period. Thereafter, forecasts are made one month ahead of available data by the forecasting equation

$$\hat{y}_{t+1} = (S_t + R_t)F_{t-11}$$

for the time periods $t = 25, 26, ..., 60$. The 1-month-ahead forecasts and the actual monthly sales values for the brewing company for the years 1974, 1975, and 1976 are listed in table 15.6 and are plotted in figure 15.10.

Figure 15.10 *Weighted moving-average forecasts for the monthly sales of the brewing company*

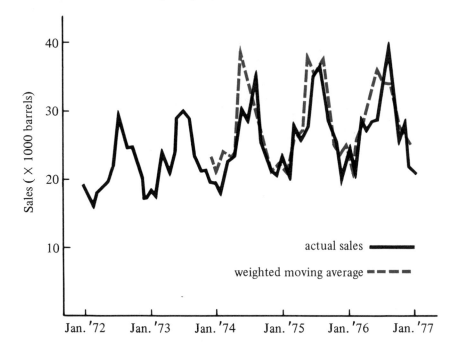

From figure 15.10 and the forecasting equation, we can note the dependence of the EWMA forecasting model on the seasonal factors. Notice, for example, the erratic up-and-down behavior of the true sales pattern during the summer of 1974. This same behavior is exhibited by the sales forecasts for the summer months of 1975, since the seasonal indexes used in the forecast equations for the summer months of 1975 were those computed for the summer months of 1974.

The brewing company example emphasizes an earlier remark. The EWMA forecasting model is an appropriate forecasting model only if the seasonal pattern is pronounced, regular, and predictable. Otherwise, because of the inclusion of the seasonal factor as a multiplication term in the forecasting equation, an erratic behavior in the seasonal pattern will cause a similarly erratic and possibly erroneous response in the pattern of the forecasts during the following season.

Exercises

15.16.   Discuss the selection of initial values for the three recursive equations involving S, R, and the L-values of F. Why is a warm-up period needed for the values of S, R, and F before any forecasts are generated?

15.17.   Refer to exercise 14.4 and the Hawaii tourist traffic data. Assume that tourist traffic data from years prior to those listed have provided the following initiating values for S_0, R_0, and the 12 values of F:

$$S_0 = 13.20 \qquad R_0 = .17$$

$$F_{Jan} = 1.23 \qquad F_{Apr} = .89 \qquad F_{July} = .98 \qquad F_{Oct} = .93$$

$$F_{Feb} = 1.24 \qquad F_{May} = .91 \qquad F_{Aug} = .83 \qquad F_{Nov} = .87$$

$$F_{Mar} = 1.33 \qquad F_{June} = .78 \qquad F_{Sept} = .87 \qquad F_{Dec} = .89$$

Label January of year 1 as month $t = 1$ and then use the EWMA forecasting model to forecast the tourist traffic one month ahead of available data for months $t = 1$ through $t = 24$. Use $\alpha = .1$, $\beta = .1$, and $\gamma = .4$.

15.18.   The data in the table represent the quarterly earnings per share for the shareholders of a dry goods manufacturer for the years 1970 through 1976. Use

	1970	1971	1972	1973	1974	1975	1976
Q1	.65	.63	.71	.88	1.01	1.02	.90
Q2	.43	.39	.51	.54	.53	.48	.43
Q3	.89	.85	1.03	1.18	1.12	.98	.89
Q4	.56	.61	.66	.75	.64	.77	.70

the years 1970 through 1973 as assumed past data to develop initiating values for the model. Then use the EWMA forecasting model to forecast the quarterly earnings one quarter ahead of available data for the years 1974, 1975, and 1976. Use $\alpha = .1$, $\beta = .1$, and $\gamma = .4$.

15.19.   Refer to exercise 15.18. Use the EWMA model to forecast the quarterly earnings two quarters ahead of available data for the years 1974, 1975, and 1976.

Use α = .1, β = .1, and γ = .4. Compare the forecast accuracy of the one-period lead and the two-period lead forecasting models by computing SSE for each model. What does this suggest about the relationship between forecast accuracy and the length of the forecast lead period?

15.7 Other Forecasting Models and Procedures

The forecasting models used by the practicing forecaster are numerous and varied. We have introduced only some of the most commonly used models in this chapter. Other forecasting models and procedures should at least be mentioned to give you a complete picture of the available set of methods. In this section we present brief looks at some other forecasting methods.

The Growth Model

Growth processes often exhibit little or no cyclic or seasonal effect and the trend appears to be exponential. The revenue per barrel from equity oil sold by Saudi Arabia for the years 1964 through 1975 (figure 15.11) exhibits this type of growth trend.

An exponential growth model suitable for data of the type exhibited in figure 15.11 is

$$g = ae^{bx}\epsilon'$$

where

$$x = \text{time} \qquad \epsilon' = \text{random error}$$

Figure 15.11 *Revenue per barrel from equity oil sold by Saudi Arabia, 1964–1975*

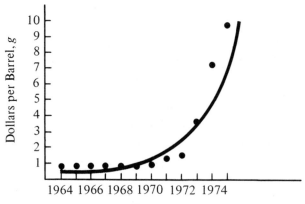

Note that the larger the general level of earnings, as indicated by ae^{bx}, the larger will be the random error, since ϵ' is multiplied by this quantity.

The parameters a and b can be estimated by the method of least squares by taking the natural logarithm of the growth model. That is,

$$\ln g = \ln (ae^{bx}\epsilon') = \ln a + bx + \ln \epsilon'$$

If we let

$$y = \ln g \qquad \beta_1 = b$$
$$\beta_0 = \ln a \qquad \epsilon = \ln \epsilon'$$

we have the straight-line linear model of chapter 11,

$$y = \beta_0 + \beta_1 x + \epsilon$$

This logarithmic transformation of the response measurements has the additional advantage of causing a relatively larger reduction in the random errors for large x and hence has a tendency to stabilize the variance of the random errors as required by the least squares inferential procedures.

The estimator of the parameter that measures the rate of exponential growth is β_1, since $\beta_1 = b$. Confidence intervals for b can be obtained by using the methods of section 11.5.

Note that the expected value of y is equal to the expected value of $\ln g$. Confidence intervals can be obtained for $E(y)$ and prediction intervals for y by using the methods of sections 11.6 and 11.7, respectively, when the least squares assumptions are satisfied. This procedure is followed by taking the antilogarithms of the limits in the confidence and prediction intervals of sections 11.6 and 11.7. Unfortunately, we really wish to estimate $E(g)$, but $e^{E(y)}$ is not equal to $E(g)$. This difference will not be too great, and consequently confidence intervals obtained for $E(y)$ can be used to obtain approximate confidence intervals for $E(g)$. Prediction intervals for g can be approximated by transforming the prediction interval for y (section 11.7), say

$$k_1 < y < k_2 \qquad \text{to} \qquad e^{k_1} < g < e^{k_2}$$

The Box-Jenkins Forecasting Procedure

The Box-Jenkins forecasting procedure has recently been shown to be a highly efficient technique for forecasting in situations where the inherent pattern in an underlying series is very complex and difficult to discern. This technique relies on a careful examination of the autocorrelations derived from a time series. Then a forecasting model is created that combines an autoregressive component of the form

$$y_t = \phi_1 y_{t-1} + \phi_2 y_{t-2} + \phi_3 y_{t-3} + \cdots + \phi_p y_{t-p} + e_t$$

with a moving-average component involving current and past error terms of the form

$$y_t = e_t - \theta_1 e_{t-1} - \theta_2 e_{t-2} - \cdots - \theta_q e_{t-q}$$

To remove any trends that may exist in the underlying time series, differencing is applied to the time series. That is, we compute the difference between observations recorded at adjacent periods of time. Then an analysis is performed on the first differences of successive observations instead of the observations themselves. In shortcut notation, the general class of Box-Jenkins models is represented as (p, d, q), where p represents the order of involvement of autoregressive terms in the model, q represents the order of moving-average terms, and d is the number of regular differences of the original series necessary to remove any inherent trends. Box and Jenkins (1969) [see the references] note that only rarely do any of these variables assume a value greater than two.

There are four stages of analysis involved in the application of the Box-Jenkins forecasting procedure.

1. *Model identification.* Since the Box-Jenkins procedure encompasses a broad range of different models, we must identify the particular (p, d, q) combinations that seem to provide an adequate fit to the underlying time series. This is done by matching sample autocorrelations computed from the data against theoretical autocorrelation functions for various (p, d, q) autoregressive moving-average models.

2. *Estimation of the model parameters.* For the particular (p, d, q) combination identified in stage 1, the method of least squares is used to fit the tentatively entertained model to the underlying time series. Coefficients are thus obtained for the autoregressive or moving-average components of the model.

3. *Diagnostic checking.* A diagnostic check is made of the adequacy of fit of the estimated model by analyzing the residuals it generates. If these residuals do not appear to exhibit any distinct pattern, but appear random over time, the fitted model is assumed to provide an adequate fit to the underlying time series. If at this stage the fit is found to be less than adequate, we return to the first stage and entertain a new class of models.

4. *Forecasting.* The acceptable forecasting model is used to generate forecasts of future values.

Unlike the EWMA model and regression procedures, which assume that a certain fixed pattern exists in the time series, the Box-Jenkins procedures are adaptive and able to accommodate any inherent pattern. This fact should be apparent from the four-stage analysis presented above.

Although the Box-Jenkins procedures have been found to be very efficient for handling time series with complex inherent patterns, recent studies have indicated some shortcomings in this new methodology. Computer programs designed to facilitate the Box-Jenkins procedure are often very expensive to apply. In addition, Newbold and Granger (1974) [see the references] have noted that the Box-Jenkins procedures are seldom capable of outperforming the EWMA model when applied to a time series with a distinct seasonal component. An additional difficulty arises in the model identification stage: Matching of the sample autocorrelations with theoretical autocorrelation functions is done subjectively and thus relies heavily on the experience of the forecaster. Further developments in the theory of the Box-Jenkins procedure may perhaps overcome this latter shortcoming.

Qualitative Forecasting Models*

When a relevant data base is not available, such as in the case of a new product being considered for market introduction, the methods and models we have introduced up to this point cannot be used. The forecaster must then rely upon other evidence, usually in the form of subjective or judgmental information, as the basis for a market forecast. Forecasting models that do not rely on a historical data base are called qualitative forecasting models. We briefly introduce four of the more common qualitative models here and discuss their use in the context of sales forecasting for a new product introduction.

The panel consensus method assumes that the organization or firm contains experts who have special knowledge or experience which enables them to effectively evaluate the uncertain effects of the future. It further assumes that these experts will recognize one another's special areas of expertise and, by supplementing each other's knowledge, arrive at a consensus as to the appropriate sales forecast for the firm's product.

One obvious difficulty with the panel consensus method is that certain social factors may make a consensus agreement impossible. Perhaps certain members of the panel are simply not willing to compromise. Also, a hierarchical bias may exist within the group, causing lower-level experts to be very reluctant in their criticism of superiors even when they believe their opinions to be of more value than that adhered to by their superiors.

The delphi method attempts to eliminate the bandwagon effect of majority opinion by employing a sequence of questionnaires that are designed to filter out the factors that can best be of assistance in focusing attention on the resultant forecast. The method employs the services of a panel of experts, either in-house experts or experts hired as consultants from the outside. The responses of the experts to a first questionnaire are used to generate a second questionnaire. These same experts or new experts may be used to respond to the second questionnaire. Their responses are then used to generate questions for a third questionnaire, and so forth, until the experts have sufficient information to sharpen their focus on the expected level of sales.

Expert information is passed on from one expert or set of experts to another by using the delphi method. Successive experts may modify previous expert opinion, but as their evaluation is made separately from the expression of opinion of earlier experts, snowballing and hierarchical influencing are not likely to occur.

The historical analogy is a qualitative model for forecasting the sales

*See the article by Reinmuth (1974), listed in the references.

of a new product which assumes that we can use the sales history of some product previously introduced to gauge the success of our current product. A natural assumption underlying this approach is that the earlier product must have had an economic and market environment during its introductory stage that is similar to the environment for the current product. If this assumption is not valid, we have no justification for the analogy.

A sample survey of opinions from those best able to assess the potential sales of a product—the potential buyers—seems to be the most logical approach for obtaining an accurate new product sales forecast. Typical **market research methods** identify the population of prospective buyers of the product, select a representative sample of size n from this population, then find the proportion p of this sample who indicate that they would buy the product if presented the opportunity. The sales forecast is then Np, where N is the number of prospective buyers in the population.

Realizing the difficulty with a required yes-no response to a consumer survey on purchase intentions, Juster (1966) [see the references] has devised a response form that allows the market research survey to gather probabilistic information from the respondents. The Juster approach requires that the respondents select a descriptive word (certain, good possibility, very slight possibility, etc.) which describes his or her intention regarding the purchasing of the new product. Juster associates each descriptive word with a "purchase intent probability" determined from numerous follow-up surveys. The average of the individual purchase intent probabilities is then used as the estimate of p, the proportion of those in the population of buyers who will buy the product.

Estimation of the population proportion p in a market research survey is an application of survey sampling. An appropriate survey design should be selected according to the characteristics of the population being sampled (see chapter 16).

Exercises

15.20.   The data in the table represent the earnings to the shareholders of Petroleum Explorations, Inc., for the years 1970 through 1976. Fit an exponential

Earnings, g	$.25	$.40	$.62	$.86	$1.20	$1.78	$2.35
Year, x	1970	1971	1972	1973	1974	1975	1976

model to the earnings data by using the logarithmic transformation and the method of least squares. Estimate the earnings to the shareholders of Petroleum Explorations, Inc., for the year 1977 by using the model you have derived.

15.21.   Assume that the logarithms of the earnings in exercise 15.20 are independent and that the least squares assumptions hold. Further assume that the growth process will remain fairly stable during 1977. Find a 95% prediction interval for the earnings in 1977.

15.22.   Based on the information provided in section 15.7, under what conditions should you use a Box-Jenkins forecasting procedure?

15.23.   If you were assigned the task of generating forecasts of future values of a process with a distinct seasonal component, such as the sales of skiing equipment, soft drinks, beer, or recreation equipment, would you use a Box-Jenkins forecasting procedure or the EWMA forecasting model? Explain.

15.24.   For years airplane manufacturing companies have sought to develop an economical and efficient airplane for transporting large numbers of people between short city pairs, such as San Francisco–Los Angeles, Dallas-Houston, Washington-New York, and London-Paris. Large jet-powered air buses, such as those developed by the Lockheed Corporation, seem to be the best answer so far, but recent developments suggest that helicopters may prove to be even more economical and efficient for short city pairs, such as San Francisco–Los Angeles, Dallas–Houston, Washington–New military and commercial helicopters to forecast the two-year sales of a new 300-passenger, 500-mile-range helicopter it has just developed for the commercial market. Explain exactly how you might conduct this forecasting exercise by using each of the following forecasting methodologies:

a. the panel consensus method   b. the delphi method
c. a historical analogy   d. a market research survey

15.8 Summary

Forecasting is a necessary task for the businessperson who wishes to budget time and resources. Forecasts provide a plan, however tentative, for the businessperson to follow in order to achieve objectives and still remain competitive.

Forecasting methods are many and varied. Traditionally forecasting has been quite subjective, without relying on rigorous mathematical forecasting models. The use of mathematical forecasting models has been suggested in this chapter for these reasons:

1. Mathematical models can be designed to track the specific components (long-term trend, cyclic and seasonal effects, etc.) in the time series under study.
2. Mathematical models can be adapted to conform to the intuitive knowledge of the businessperson about the time series.

The last point is especially important; it implies that the ingenuity of the business statistician is an essential factor in the selection of an appropriate forecasting model.

The ultimate criterion for the value of a forecasting model is how well it forecasts the future. Inherent in this criterion is the assumption that past behavior of the time series mirrors its future behavior. Dormant variables and other factors may cause a time series to react differently in the future than it has reacted before. Thus a model that fits the sample data well might not be a good model for forecasting the future behavior of the time series.

Statistics provides only a starting point in the analysis of a time series and the adoption of a forecasting procedure. Statistical and mathematical models are limited and cannot provide a complete solution to forecasting problems as long as uncertainty exists in the problem. Personal judgment and subjective evaluations concerning the forecast environment must be used to condition the forecast obtained from a mathematical forecasting model.

Supplementary Exercises

The exercises in this supplementary set are divided into two parts. The first part contains exercises that, for the most part, require descriptive answers and do not involve extensive amounts of computation. The second part contains exercises that require extensive computational effort for their solution. The student with access to an electronic computer and the ability to write computer programs may find it helpful to write programs to facilitate the solutions of these exercises. This latter set may be eliminated if the student desires only an overview of forecasting methods.

Part I   Descriptive Exercises

15.25.   Discuss the importance of the following factors as they relate to our discussion of forecasting concepts and methods:

a. sinusoidal model
b. seasonal factor
c. period of a cycle or seasonal factor
d. curvilinear long-term trend
e. correlogram
f. smoothing constant
g. Box-Jenkins procedures
h. qualitative forecasting methods

15.26.   Spurred by the recession of the early 1970s and the legalization of the sale of gold to the public by the U.S. government, gold prices rose dramatically during the period from 1973 through 1974. Speculators and investment counselors, noticing this growth trend, forecast a continued rise in gold prices for the years

Figure 15.12 *U.S. dollars per troy ounce of gold on the London market, 1973–1976*

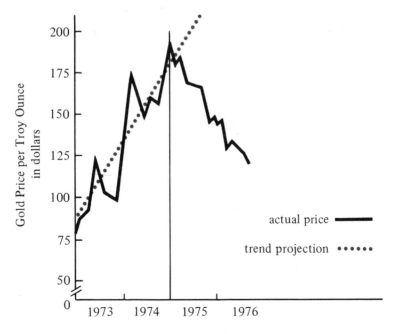

following 1974. However, the trend did not continue, as shown in figure 15.12. Why were the trend projection forecasting procedures employed by the speculators and investment counselors especially risky? In general, discuss the possible dangers of forecasting by extrapolating beyond the interval of time in which the sample data were collected.

15.27.   For each of the following forecasting problems, suggest the forecasting method you would recommend, defend the selection of your chosen method, and suggest the relevant data you would use in the forecast.

a. The unemployment rate in the United States for each of the next 12 months.

b. The total number of housing starts in the United States in the next calendar year.

c. The demand for electrical power for each of the next 20 years by the resident and commercial users in Topeka, Kansas.

d. The sales of skis and skiing equipment by a large sporting goods chain during each of the next 24 months.

e. The sales over the next 6 months of a new diet drink being offered for the first time by the Coca-Cola Company.

f. The sales of term life insurance in the state of Georgia during the next calendar year.

15.28.   Explain the difference in the conditions underlying a forecasting problem that would suggest the use of a forecasting method from each of the following classes of forecasting methods: econometric methods, time series methods, and qualitative forecasting methods.

15.29.   In section 15.6 it was mentioned that the smoothing constants in the EWMA model should be appropriately adjusted to accommodate sudden shifts in a time series due to atypical situations such as labor strikes, price cut promotions, and the weather. Adjustments to accommodate atypical situations should not, however, be limited to the EWMA model but should be applied to all forecasting models. Explain how relevant subjective information can, in a general sense, be incorporated into a total forecasting model to adapt the model to its total environment.*

15.30.   To conserve the use of natural gas, a public utility has offered energy-saving thermostats to its commercial subscribers over the past few years. The data in the table represent the number of thermostats sold each month of two

MONTH	YEAR 1	YEAR 2
January	19	23
February	19	20
March	17	21
April	15	17
May	15	19
June	16	20
July	15	19
August	17	20
September	17	22
October	19	25
November	23	24
December	22	26

*See J. E. Reinmuth and M. D. Geurts, "A Bayesian Approach to Forecasting Effects of Atypical Situations," *Journal of Marketing Research*, August 1972.

consecutive years by the utility to its commercial subscribers. A staff assistant for the public utility claims that the sinusoidal forecasting model

$$\hat{y}_t = 16.10 + .33t + 3.20 \cos\left(\frac{2\pi t}{12}\right)$$

which indicates the presence of a 12-month (annual) cycle in the sales pattern, is an efficient model for forecasting the monthly sales of the energy-saving thermostats. Let January of year 1 be designated as time period $t = 1$, February of year 2 as $t = 2$, and so forth. Using the sinusoidal model, forecast the sales for the two-year period and compare your forecasts with the actual sales values. Does the suggested sinusoidal model appear to be a good forecasting equation for these data?

15.31.   Employing the methods of chapter 11, construct a linear forecasting model to fit the monthly sales data of exercise 15.30. Compare the fitted values obtained from this model with the forecast values obtained from the sinusoidal model of exercise 15.30 by computing SSE for the values generated by each model.

15.32.   When developing a forecasting model, why is it advisable to test the model using "hold-back" data? That is, why should we see how well the model forecasts during a period outside the data set used to derive the model, and why is it insufficient to rely on a model that simply fits the sample data set well?

15.33.   In some cases a historical data base may exist for a product, but due to a change in the market environment of the product, the data base becomes worthless. For instance, the demand for large station wagons was unalterably changed after the energy crisis of 1973. Suppose you were acting in an advisory capacity to the Ford Motor Company (for example) in an attempt to forecast future demand for full-sized station wagons. How would you recommend Ford approach this problem? Would you suggest they use past sales data and modify it to accommodate new trends in buying behavior or would you recommend they discard past data and use a qualitative forecasting procedure? If the latter, which method(s) would you recommend they explore?

15.34.   For the exponentially weighted moving-average forecasting model (EWMA), discuss the selection of initial values for the three recursive equations involving S, R, and the L-values of F. Why is a warm-up period needed for the values of S, R, and F before any forecasts are generated?

Part II   Computational Exercises

15.35.   Refer to exercise 15.30. Fit a first-order autoregressive model

$$y_t = \beta_0 + \beta_1 y_{t-1} + \epsilon$$

to the sales data by letting the one-month lagged sales value y_{t-1} assume the role of the independent variable x and by employing the linear regression methods of chapter 11. Compute the estimated autoregressive values and compute SSE as a measure of the fit of the first-order model to the data. How does the fit of the autoregressive model compare with, respectively, the fit of the sinusoidal model and the linear model of exercises 15.30 and 15.31?

15.36.   Refer to exercise 15.30. Assume that the true pattern of monthly sales is approximately linear over time and use Brown's multiple exponential-smoothing forecasting method to forecast the monthly sales one month ahead of available data. Plot the multiple exponential-smoothing forecasts against the true monthly sales volumes. Let the smoothing constant be $\alpha = .1$.

15.37.   The data in the table represent the monthly revenue, in thousands of dollars, of a resort restaurant. Assume that the pattern of revenue over time

MONTH	1975	1976	1977
January	12.0	13.3	15.3
February	13.2	14.8	16.4
March	14.5	17.0	18.9
April	15.6	18.1	18.4
May	20.0	21.8	21.2
June	20.5	21.1	23.8
July	21.9	24.0	25.4
August	19.2	21.2	24.5
September	16.1	20.7	23.4
October	16.7	17.4	20.4
November	14.8	16.8	18.0
December	14.2	14.7	16.6

is neither constant nor linear. Use Brown's multiple exponential-smoothing forecasting method to forecast the monthly revenue one month ahead of available data. Plot the forecast revenue against the true monthly revenue. Use the smoothing constant of $\alpha = .1$.

15.38.   Refer to exercise 15.37. Use the exponentially weighted moving-average forecasting method to forecast the monthly revenue one month ahead of available data for the year 1977. Use the sales data from 1975 and 1976 to recursively generate values for the smoothed statistics S, R, and the 12 values of F. Plot the forecast revenues for 1977 against the true revenues. Use the smoothing constants $\alpha = .1$, $\beta = .1$, and $\gamma = .4$.

15.39.   Repeat exercise 15.38 by using the exponentially weighted moving-average forecasting method to forecast the monthly revenue for 1977 three months ahead of available data. Compare the accuracy of the one-month lead forecasts generated in exercise 15.38 with the accuracy of the three-month lead forecasts of this exercise by computing SSE for each model. What does this suggest about the relationship between forecast accuracy and the length of the forecast lead time?

15.40.   Refer to chapter 14, exercise 14.20. Use the exponentially weighted moving-average forecasting model to forecast the monthly sales for the department store one month ahead of available data for year 3, year 4, and year 5. Use the sales data from year 1 and year 2 to recursively generate values for the smoothed statistics S, R, and the 12 values of F. Plot the forecast sales for year 3, year 4, and year 5 against the true sales volumes. Use the smoothing constants $\alpha = .1$, $\beta = .1$, and $\gamma = .4$. (You may wish to compare your results with those derived by Reinmuth and Wittink; see the reference given at exercise 14.20.)

15.41.   The data in the table represent the quarterly earnings per share for the shareholders of the Coca-Cola Company for the years 1960 through 1969. An

	1960	1961	1962	1963	1964	1965	1966	1967	1968	1969
Q1	.13	.15	.16	.18	.22	.26	.31	.35	.38	.42
Q2	.21	.21	.24	.26	.31	.36	.42	.47	.51	.57
Q3	.26	.27	.30	.33	.39	.45	.52	.59	.66	.74
Q4	.12	.14	.16	.18	.22	.26	.31	.35	.38	*

*Earnings from Q4 of 1969 were not available.

executive of Coca-Cola suggested that the sinusoidal regression model

$$y_t = \beta_0 + \beta_1 t + \beta_2 \sin\left(\frac{2\pi t}{4}\right) + \beta_3 \cos\left(\frac{2\pi t}{4}\right) + \epsilon$$

where y_t is the amount of earnings from quarter t and t is the index of time ranging from $t = 1$ for Q1 of 1960 to $t = 39$ for Q3 of 1969, provides a good model for forecasting Coke's quarterly earnings.

a. Fit the sinusoidal model to the earnings data. What are the estimates for the parameters in the forecast equation?

b. How well does the sinusoidal model fit the earnings data? That is, how much of the variation in earnings can be explained by the model? (See section 11.9.)

15.42.   In examining a plot of the time series of quarterly earnings of Coca-Cola (see exercise 15.41), you can notice that the amplitudes of the seasonal patterns appear to be increasing over time. Thus a sinusoidal model such as

$$y_t = \beta_0 + \beta_1 t + \beta_2 \sin\left(\frac{2\pi t}{4}\right) + \beta_3 \cos\left(\frac{2\pi t}{4}\right) + \beta_4 t \sin\left(\frac{2\pi t}{4}\right)$$
$$+ \beta_5 t \cos\left(\frac{2\pi t}{4}\right) + \epsilon$$

may provide a better fit to the Coca-Cola quarterly earnings.

a. Fit the sinusoidal model to the data of exercise 15.41. What are the estimates for the parameters in the forecasting equation?

b. What greater proportion of variation is explained by the model of exercise 15.42 over the model of exercise 15.41?

15.43.   Refer to exercise 15.41. Use the exponentially weighted moving-average forecasting model to forecast quarterly earnings for Coca-Cola one quarter ahead of available data for the years 1966 through 1969. Use the earnings data from 1960 through 1965 to recursively generate values for the smoothed statistics S, R, and the four values of F. Use the smoothing constants $\alpha = .1$, $\beta = .1$, and $\gamma = .4$.

15.44.   Refer to exercises 15.42 and 15.43. Compare the accuracy of the forecasts generated by the EWMA model of exercise 15.43 with the accuracy of the fitted values suggested by the sinusoidal model of exercise 15.42 over the years 1966 through 1969 by computing SSE for each model. Does this provide a valid comparison of the two models to forecast the earnings data? Explain.

15.45.   Deposits are the lifeblood of commercial banks. They are the chief source of bank funds and provide the base for loans and other financial transactions

MONTH	1965	1966	1967	1968	1969
January	13.48	14.50	15.29	15.94	16.04
February	13.08	13.80	14.57	15.11	15.19
March	13.37	13.94	14.63	15.11	15.15
April	13.88	14.55	15.15	15.46	15.88
May	13.16	14.05	14.55	14.93	15.13
June	13.63	14.53	15.07	15.39	15.53
July	13.80	14.58	15.45	15.69	16.00
August	13.76	14.33	15.04	15.49	15.44
September	14.18	14.97	15.70	15.69	15.60
October	14.27	14.69	15.06	15.66	16.06
November	14.65	15.11	15.57	15.95	16.15
December	15.09	15.49	16.12	16.29	16.33

Source: U.S. Board of Governors of Federal Reserve System, *Deposits, Reserves, and Borrowings of Member Banks.*

of the bank. Thus if a bank can effectively forecast deposits, it can accurately manage its investment activities to maintain a high degree of both liquidity and profitability. A San Francisco banking executive obtained the data shown in the table, representing the level of demand deposits for the San Francisco city banks over the period 1965 through 1969. All figures are in billions of dollars.

a. Use the monthly data from 1965, 1966, and 1967 to develop initial estimates for the smoothing contants S, R, and the 12 values of F in the EWMA forecasting model as outlined in example 15.2, section 15.6. Use $\alpha = .1$, $\beta = .1$, and $\gamma = .4$.

b. Using the monthly data from 1965, 1966, and 1967 as assumed past data, develop the EWMA forecasting model over this period and generate monthly forecasts for the 24 months of 1968 and 1969. Let the forecast lead period be $T = 1$. (That is, develop forecasts one period in advance of available data.)

 15.46.   The accompanying table gives the annual gross revenue for a manufacturer of a water purification unit for the years 1969 through 1976.

Year	1969	1970	1971	1972	1973	1974	1975	1976
Revenue ($\times$ 1000)	21.1	19.4	23.0	26.9	29.1	40.5	64.3	101.2

a. Using a log transformation on the annual revenues, fit a growth curve to the data. Plot the fitted values against the actual gross revenues.

b. Estimate the gross revenue for the company for 1977.

15.47.   In an attempt to discover the best model to forecast future values of a time series, it is typical to try many models and then select the one returning the greatest measure of forecast accuracy. However, Bates and Granger note that discarded forecasts nearly always contain some useful information independent of that provided by the chosen model.* They suggest a forecast that combines the forecasts generated by different forecasting models. One way of doing this is to create a regression model, where

$$y_t = \text{the actual value at time } t$$
$$x_1 = \text{the forecast of } y_t \text{ provided by model 1}$$
$$x_2 = \text{the forecast of } y_t \text{ provided by model 2}$$

Then the combined model is

$$\hat{y}_t = \beta_0 + \beta_1 x_1 + \beta_2 x_2$$

providing a forecast $\hat{y}_t$ for the outcome at time period t that is a combination of two other forecasts, x_1 and x_2. Now refer to exercises 15.37 and 15.38, which provide one-month lead forecasts of the revenues of a restaurant. Use the 1977 one-month lead forecasts to develop a combined forecast model, where x_1 is the forecast derived by using Brown's method (exercise 15.37) and x_2 is the EWMA forecast (exercise 15.38). Regress the actual 1977 revenues on x_1 and x_2 and then use the combined model to forecast the 1977 revenues. Compare the accuracy of the forecasts generated by the combined model with the accuracy of the forecasts generated by the models of exercises 15.37 and 15.38 for the 12 months of 1977 by computing SSE for each model.

*J. M. Bates and C. W. J. Granger, "The Combination of Forecasts," *Operational Research Quarterly* (1969).

Experiences with Real Data

1. In exercise 15.1 we noted that enrollment in business courses is increasing at a more rapid rate than the overall college and university enrollment. However, due to course sequencing and the personal preference of students not to enroll during certain terms, enrollment and its growth fluctuates from term to term. It would thus seem appropriate that an exponentially weighted moving-average forecasting model, which allows for periodic surges and declines within each year, be used as a planning model by colleges and universities to forecast student enrollment on a term-by-term basis.

Use the EWMA model to forecast the enrollment in a certain segment of the business program offered at your college or university. For instance, you may choose to forecast enrollment for accounting courses, finance courses, or marketing courses. Perhaps your class can be divided into groups, with each group selecting a different segment of the business program for which to forecast enrollment. From the registrar obtain the enrollment in the segment you have chosen for each term (semester) of the past five school years. Use the first three years of enrollment data to recursively generate values for the smoothed statistics S, R, and F, and then forecast the enrollment one term ahead of available data for each term of years four and five. Remember that a separate seasonal factor (F) will be required for each term or semester in your school's academic year. (Initially use the smoothing constants $\alpha = .1$, $\beta = .1$, and $\gamma = .4$, but you may wish to adjust these values until you obtain a forecast model that is responsive to your enrollment data.) How accurate are your forecasts?

2. Exercises 14.8 and 14.11 both provide a substantial amount of real data derived, respectively, from government documents and company reports. Select one of these two data sets, divide the set into two parts, and fit what you consider to be an appropriate forecasting model to the assumed past data. You may wish to fit many different models to the assumed past data, comparing the fit of each by the size of its computed SSE. In addition, you may incorporate two or more forecasts together in a combined model, as discussed in exercise 15.47. Use the model you consider to be most appropriate for the data set you have selected to forecast over the data set you have held back as assumed future data. Use a forecast lead period of $T = 1$. How accurate are your forecasts?

References

Box, G. E. P., and G. M. Jenkins. *Time Series Analysis, Forecasting and Control.* San Francisco: Holden-Day, 1969.

Brown, R. G. *Smoothing, Forecasting, and Prediction of Discrete Time Series.* Englewood Cliffs, N.J.: Prentice-Hall, 1963.

Chambers, J.; S. Mullick; and D. Smith. "How to Choose the Right Forecasting Model." *Harvard Business Review* 49 (July–August 1971): 45–74.

Chisholm, R. K., and G. R. Whitaker, Jr. *Forecasting Methods.* Homewood, Ill.: Richard D. Irwin, 1971.

Juster, F. T. "Consumer Buying Intentions and Purchase Probability: An Experiment in Survey Design." *Journal of the American Statistical Association* 61 (September 1966): 658–696.

Newbold, P., and C. W. J. Granger. "Experience with Forecasting Invariant Time Series and the Combination of Forecasts." *Journal of the Royal Statistical Society* (A) 137, pt. 2 (1974): 131–164.

Reinmuth, J. E. "Forecasting the Impact of a New Product Introduction." *Journal of the Academy of Marketing Science* 2 (Spring 1974): 391–400.

REINMUTH, J. E., and D. WITTINK. "Recursive Models for Forecasting Seasonal Processes." *Journal of Financial and Quantitative Analysis* 9 (September 1974): 659–683.

WHEELWRIGHT, S. C., and S. MAKRIDAKIS. *Forecasting Methods for Management.* New York: Wiley, 1973.

chapter objectives

GENERAL OBJECTIVES In the preceding chapters we presented various sample statistics without explaining how a sample is selected. In this chapter we offer a variety of methods for selecting a sample and provide guidelines for the selection of an appropriate sampling design. We will stress the importance of using a sampling design that provides for the selection of samples in a random manner, since inferential statements about the population are not possible when using nonrandom sampling.

SPECIFIC OBJECTIVES

1. To discuss the advantages of sampling over a complete census and to define the terms used in survey sampling.
Section 16.1

2. To discuss and illustrate the errors that can result from survey sampling.
Section 16.2

3. To explain how to select a random sample by using the table of random numbers.
Section 16.3

4. To explain how to use the results of a simple random sample to make inferences about a population.
Section 16.4

5. To discuss the importance of stratified random sampling and to explain how to use the results of a stratified random sample to make inferences about a population.
Section 16.5

6. To discuss the importance of cluster sampling and to explain how to use the results of a cluster sample to make inferences about a population.
Section 16.6

7. To show how to find an appropriate sample size when using simple random sampling, stratified random sampling, or cluster sampling.
Section 16.7

8. To discuss the use and interpretation of the systematic sampling design, ratio estimation, the two-stage cluster-sampling design, and randomized response sampling.
Section 16.8

chapter sixteen

Survey Sampling

16.1 Introduction

As we have repeated a number of times throughout the text, the main objective of statistics is to make inferences about a large body of data, the population, based on information contained in a sample. In previous chapters we have discussed a number of statistical procedures that can be used to analyze our available sample data set and to make inferences about certain characteristics of the population (e.g., the population mean μ, the population proportion p). In this chapter we offer a variety of methods for selecting the sample, called **sampling designs,** which can be used to generate our sample data set.

The main objective of any sampling design is to provide guidelines for selecting a sample that is *representative* of its underlying population, thus providing a specified amount of information about the population at a minimum cost. If the underlying population is uniform in the characteristics to be measured, almost any sample provides acceptable results. For example, the Environmental Protection Agency (EPA) bases its diagnosis of the purity of a city's water supply on the analysis of a few pints of water. This is possible because the EPA assumes that one drop of water provides more or less the same proportionate amount of impurities as another drop. Consider, on the other hand, a nationwide survey to determine attitudes of adults toward renting versus buying a home. Just due to pure chance, it is possible that only residents of New York City are selected as members of the sample. However, such a sample may not accurately reflect the attitudes of all adults, since residents of New York City are much less likely to own their own home than, say, residents of Los Angeles. A more representative sample could be obtained by randomly selecting individuals from different regions of the country, perhaps by first dividing the population into states, and then by randomly selecting a specified number of individuals from each state. The information provided by the random selection of adults chosen from each state would be combined to enable us to make inferences about the rent-versus-buy attitudes of the entire population of adults of the country.

It might appear that the only way to *guarantee* that our experimental data set truly represents the population is by conducting a **census,** a recording of every element contained in the population. Why, then, do we typically conduct a sample investigation instead of a complete census? Samples are used because in the majority of cases their advantages outweigh the advantages of a complete census.

Some advantages of sampling, like economy and practicability, are obvious; others are more subtle. For instance, sampling provides more timely results since it gives us the opportunity to gain rapid information retrieval on a varying process and, in a sense, to determine its state at some fixed time. Another situation in which sampling is more efficient than a complete census is in an experiment that results in the destruction of the elements of the sample. For example, in quality control studies, such as the testing of flashbulbs, testing destroys the product. Thus sampling must be used because a census would leave nothing to market. Sometimes sampling may actually provide more accurate results than a census, since the heavy work load of a census may cause researcher fatigue, which, in turn, may be responsible for increasingly careless sampling habits by the researchers. Also, the population itself may be dynamic and never in one state long enough to allow for a complete measurement of its characteristics.

When using sampling we should take every precaution to ensure that our sampling is conducted in a "random manner" and that our sample is a *random sample.* As you will recall, in section 7.3 we found that **a random sample is selected in such a way that every combination of *n* measurements from the population has an equal probability of being selected.** The primary advantage of using random sampling designs is that when the samples are random, we know the probabilities of including various observations in the sample. Hence we can make *probabilistic* statements about the underlying population. If the samples are selected in a deterministic (nonrandom) manner, the probabilities of observing various sample measurements are unknown and we can only make *descriptive* statements about the sample. As you will subsequently see, most survey designs employ some restrictions in selecting the sample, but ultimately random sampling is employed to obtain the sample observations.

In this chapter we will consider the problem of sampling from a **finite population** of measurements and only occasionally will we refer to sampling from an infinite population. Our study will focus on the selection of an appropriate sample design, which, as you will see, involves a balance between the quantity of information obtained and the cost of sampling. The ultimate objective of the sample survey—an inference concerning the underlying population—is based on estimates of the population mean, total, or proportion. For each of these estimates, we will give a bound on the error of estimation. Naturally the estimation formulas differ according to the particular sampling design used in the survey.

Before proceeding with our discussion, we introduce some terms commonly used in survey sampling. These terms are introduced in the order in which they are involved in a sample survey.

Definition

The **sampling design** or **survey design** specifies the method of collecting the sample.

The design does not specify a method of collecting or measuring the actual data. It specifies only a method for collecting the objects that contain the required information. These objects are called *elements.*

Definition

An **element** is an object on which a measurement is taken.

The elements may occur individually or in groups in the population. A group of elements, like a household of community residents or a carton of light bulbs, is called a *sampling unit.*

Definition

Sampling units are nonoverlapping collections of elements from the population. In some cases a sampling unit is an individual element.

To select a random sample of sampling units, we need a list of all sampling units contained in the population. Such a list is called a *frame.*

Definition

A **frame** is a list of sampling units.

When conducting a sample survey, our first task is to identify the sampling units and to construct a frame that provides a list of the sampling units.

According to our sampling design, a predetermined number of sampling units are then randomly selected from the frame. Our sample then consists of the elements contained in the chosen sampling units. We can then use the information obtained from the sample to make inferences about certain characteristics of the population.

Exercises

16.1. What is meant by a random sample? Why is it necessary to use random, as opposed to nonrandom, sampling?

16.2. Why is sampling usually preferable to conducting a census of the population?

16.3. A commonly used sampling technique in public opinion polling is called *quota sampling*. Using this technique the interviewer selects, according to his own discretion, a predetermined number (a quota) of individuals from each of several segments of the population. For instance, he may be requested to interview 10 mechanics, 36 housewives, or 7 lawyers.

a. Under what conditions does quota sampling provide a random sample?

b. Under what conditions does it provide a nonrandom sample?

16.4. In each of the following sample surveys, discuss the selection of an appropriate sampling unit and frame.

a. An economist wishes to conduct a survey for estimating the average weekly grocery expenditures per family in a city.

b. An administrative assistant to the governor of a midwestern state wishes to estimate the proportion of the voting public in the state who favor legislation prohibiting strip-mining.

c. A supermarket chain that operates retail grocery stores throughout the United States wishes to survey the opinions of their employees about a company-sponsored health insurance plan.

d. A marketing representative for a large merchandising chain wishes to survey buyers to determine their attitudes toward a new product line.

16.2 Bias and Error in Sampling

There are two major types of statistical inference: estimation and decision making. And, as we have seen, each type has an associated error. Decision-making errors, usually referred to as type I and type II errors, were defined in chapters 6 and 8. The error of estimation, discussed in chapters 8 and 9, is defined here.

Definition

Let $\hat{\theta}$ be a sample estimator of the population parameter θ. The **error of estimation** is the absolute difference $|\hat{\theta} - \theta|$.

When choosing a sample size, the experimenter should specify a bound
B on the error of estimation that he or she is willing to tolerate (see sections
8.12 and 11.7). That is, the experimenter should specify the value B and
then choose the sample size so that $\hat{\theta}$ and θ differ by more than B only
a very small fraction of the time. For example, an auditor may wish the
chances to be very small that the average account balance obtained from
a sample of a firm's accounts receivable differ from the true account balance
by more than some specified amount, say $2.00. As you might expect, B
and n are inversely related; the smaller the tolerable error of estimation,
the greater is the quantity of information required to satisfy the specified
bound.

There are three sources of errors in sample surveys. The most common
source is **random variation.** For example, suppose a retailing organization
is interested in estimating the average income per household in a particular
community. In the selection of a random sample of households, by chance
all those selected may be in the higher income brackets. Common sense
and intuition would suggest the presence of an error of estimation if, for
instance, an average household income of $49,000 is obtained for the communi-
ty. However, when the estimation error is moderate, it may go undetected,
leading to erroneous inferences and, perhaps, to faulty decision making. If
a slightly inflated, but believable, average household income of $25,000 is
obtained, the retailing organization may decide not to market lower-priced
economy product lines that are thought to appeal more to those in moderate-to-
lower-income communities. On the basis of the inflated average household
income, the firm may be erroneously assuming that the community contains
very few moderate-to-lower-income families.

Another source of error in sample surveys is a **misspecification** of the
population. Such errors are fairly common in public opinion polling for election
surveys. The true population of interest in an election survey consists of
those who will vote on election day. However, typical election surveys deal
with the opinions of registered voters, many of whom will not vote on election
day.

Errors due to a misspecification of the population may also arise from
sources such as an incorrect list of the population elements, incorrect
information recorded on an inventory ledger, erroneous selection of sample
elements (such as substituting a neighbor when a respondent is not found
at home), question sensitivity, mistakes in the collection of information from
the sample due to intentional or unintentional interviewer bias, or errors
in processing the sample information. For the most part, such causes are
controllable; in other cases, such as in the measurement of the dimensions
of timber or lumber which swells with an accumulation of moisture, the
causes are uncontrollable.

The classic example of a misspecification error in an election survey
is the 1936 *Literary Digest* poll. The *Literary Digest* obtained its sample from
telephone directories and magazine subscription rolls and, on the basis of
the sample responses, predicted that Republican Alfred M. Landon would

soundly defeat incumbent Democratic President Franklin D. Roosevelt. As we know, the results were exactly the opposite—Roosevelt defeated Landon by one of the greatest pluralities ever, causing the *Literary Digest* to lose much of its credibility. What went wrong? During the economically depressed times of the thirties, only the affluent, who were typically Republicans, could afford telephones and magazine subscriptions. The sample did not provide a representative cross section of the voting public.

Errors due to misspecification of the population are also common in consumer research surveys, where the sample typically consists entirely of housewives, while excluding men, working women, and students because of their relative inaccessibility. In addition, we noticed in chapter 14 that the U.S. Bureau of Labor Statistics is guilty of a certain amount of misspecification in their published consumer price index. Such surveys cannot expect to reflect the unique buying habits of segments of the population not represented in the sample.

An additional source of error in sample surveys is due to **nonresponse** by some members of the sample. It is common for the researcher to assume that respondents and nonrespondents provide similar cross sections of the population when, in fact, this is seldom the case. In consumer surveys the nonrespondents are typically working people and the respondents are usually housewives; in public opinion surveys nonrespondents (those registering "no opinion") are usually the very contented members of the sample who generally prefer things as they are.

The researcher can only minimize the chances of errors due to random variation by selecting a proper sampling design. The researcher can have a much more direct effect on errors due to nonresponse. Continued efforts can be made to reach nonrespondents, or, in some cases, nonrespondents can be replaced by randomly selected alternates.

To minimize the chances of incurring errors due to misspecification, a very careful statement of the survey objective can be made in advance of the study, thus providing a clear image of the elements that comprise the population. Most important of all, the researcher should take great care in phrasing inferences in terms of the actual population from which the sample information was derived and not in terms of some other, perhaps conceptual, population of greater appeal.

Experience is the best guide to use for controlling sources of error in survey sampling. Individuals or agencies who have designed or conducted numerous surveys of a particular type (e.g., public opinion, market research, account audits, inventory audits) develop a reputation for anticipating certain possible pitfalls in the survey. They are then able to design the sample and the survey methods so as to avoid most common controllable sources of bias and error while minimizing the impact of uncontrollable sources of error.

Exercises

16.5.   When conducting an audit of a firm's accounts receivable, it is typical for the auditor to select a random sample of accounts and then to verify the account balance with each account holder. Suppose, for example, that an accounting firm has undertaken an audit of the accounts receivable held by a large municipal hospital. Suggest some possible controllable and uncontrollable sources of error in the audit of the hospital's accounts.

16.6.   Refer to exercise 16.5. Suggest some auditing guidelines that may help to control the bias and error that may occur in the auditing process. How might the auditor minimize the effect of random variation?

16.7.   Most prime-time television shows in the United States are at the mercy of weekly ratings issued by the A. C. Nielsen organization. Each week the Nielsen organization obtains nationwide samples randomly selected from residential directories (not telephone directories). In a mail survey the occupant of each residence selected is asked to keep a weekly tabulation of the channel selection during the evening prime-time hours and to return the results immediately at the week's end. Suggest some possible sources of error in the Nielsen survey process.

16.8.   Suppose you represent an advertising agency that buys considerable network television advertising time during the evening prime-time hours. Would you be willing to rely solely on the Nielsen ratings to select a network program and time slot for your advertisements? Explain. What additional information would you request from the Nielsen organization other than their widely published ratings?

16.3 How to Select a Random Sample

In this section we explain how to conduct the most basic sample survey design, a *simple random sample.*

Definition

Suppose a sample of n measurements is selected from a *finite* population of N measurements. If the sampling is conducted in such a way that every possible sample of size n has an equal probability of being selected, the sampling is said to be **random** and the result is said to be a **simple random sample.**

As we noted in section 7.3, perfect random sampling is difficult to achieve in practice. If the population is not too large, each of the N measurements can be written on a slip of paper or on a poker chip and then placed in a bowl. A random sample of n measurements can then be drawn from the bowl.

The best way to ensure that we are employing random sampling is to use a table of random numbers. One such table is table 14 in the appendix. This table has been constructed so that the integers from 0 through 9 occur randomly and with equal frequency.

We illustrate the use of the random number table with an example.

Example 16.1 The effective management of cash flows by business organizations is necessary for the proper budgeting and control of their present and future resources. When cash flows are high, the firm is in a position to purchase inventories and capital goods on short notice, thereby taking advantage of price cuts offered by suppliers. When the firm is short of cash, it cannot buy ahead and therefore usually ends up paying more for goods and supplies.

One of the best measures of the cash position of a retail merchandise organization is provided by the short-term accounts receivable held by the firm. In an analysis of the cash position of a department store, an accounting firm decides to select a simple random sample of $n = 15$ monthly retail accounts receivable from among the $N = 1000$ current monthly retail accounts of the department store in order to estimate the total amount due on all outstanding accounts receivable. We know that a simple random sample will be obtained if every possible sample of $n = 15$ accounts has the same chance of being selected. We will determine which accounts are to be included in the sample of size $n = 15$ by using the table of random numbers, table 14 in the appendix.

Solution We can think of the $N = 1000$ accounts receivable as being numbered 001, 002, ..., 999, 000. That is, we have 1000 three-digit numbers, where 001 represents the first account, 999 the 999th account, and 000 the 1000th.

Refer to table 14 of the appendix and arbitrarily select a starting point. Suppose our starting point is the first number in the fifth column. If we drop the last two digits of each five-digit number, we see that the first three-digit number formed is 816, the second is 309, the third is 763, and so on. If a random number occurs twice, the second occurrence is omitted and another number is selected as its replacement. Taking a random sample consisting of the first 15 nonrepeated three-digit numbers from column 5, we obtain the following numbers:

816	277	709
309	988	496
763	188	889
078	174	482
061	530	772

If the accounts receivable are numbered, we merely choose the accounts with the corresponding numbers and form our simple random sample of $n = 15$ from $N = 1000$. If the accounts receivable are not numbered, we can refer to a list of the accounts and select the 61st, the 78th, the 174th, and so on until $n = 15$ accounts have been selected.

In example 16.1 the population size $N = 1000$ made it possible to uniquely associate each element in the population with a different three-digit number. What would we do if, say, $N = 964$? Clearly we can associate the three-digit numbers 001, 002, ..., 963, 964 with the elements of the population. All the remaining three-digit numbers, 965, 966, ..., 999, 000, are simply ignored when selecting our sample of n three-digit numbers from the table of random numbers.

Sometimes an experimenter uses his or her judgment to select a representative sample or applies some intuitive means to "randomly" select the sample. Both procedures are subject to experimenter bias and should be avoided when seeking a simple random sample.

Exercises

16.9.   The number and size of delinquent accounts are of vital concern to companies offering consumer credit. Excessive delinquency can seriously impair the cash position of the company and possibly entail costly account collection activities. Table 16.1 lists the account receivable balances of the $N = 100$ regular customers of a wholesale hardware company, with delinquent accounts denoted by an asterisk.

Table 16.1 *The population of accounts receivable for a wholesale hardware company*[†]

ACCT. NO.	ACCT. BAL.	ACCT. NO.	ACCT. BAL.	ACCT. NO.	ACCT. BAL.	ACCT. NO.	ACCT. BAL.
1	$136	26	$152	51	$146	76*	$233
2	216	27*	858	52	173	77	85
3	520	28	130	53	235	78	162
4	312	29	416	54*	220	79	165
5	180	30*	144	55*	600	80	483
6	250	31	320	56	405	81	342
7*	235	32	210	57	135	82	290
8	345	33	275	58	160	83*	95
9	260	34	540	59	325	84*	338
10	185	35	350	60	290	85	710
11	285	36	560	61	302	86	155
12	310	37*	365	62	250	87	115
13*	430	38	125	63*	280	88	200
14	605	39	312	64	235	89	450
15*	310	40	165	65	560	90*	245
16	60	41*	360	66	320	91	260
17	155	42*	450	67	305	92	530
18*	190	43	235	68	160	93	200
19	425	44	190	69	255	94	216
20	75	45	345	70*	204	95*	195
21	315	46	240	71	120	96	395
22	240	47	389	72	403	97	126
23*	209	48	80	73	265	98	250
24	178	49	205	74*	322	99	405
25	313	50*	215	75	602	100	196

[†] All account balances have been rounded to the nearest whole dollar.
*Delinquent accounts.

Suppose you represent an auditing firm that has been hired by the company to examine its accounts receivable. Sufficient time and money are available to investigate only $n = 15$ of the accounts. Using the table of random numbers, table 14 in the appendix, select a simple random sample of $n = 15$ accounts from all $N = 100$ accounts receivable of the wholesale hardware company.

16.10.   Suppose you have been hired as a marketing representative for a clothing manufacturer that has developed a new line of casual clothing designed to appeal to college students. You are requested to select a representative sample of the students at your college (or university) and to survey their attitudes toward the new line of clothing.

a. Using your college's student directory as a frame, use the table of random numbers to select a simple random sample of $n = 50$ students.

b. How well does your sample of $n = 50$ students reflect the underlying population of all students at your college (or university)? (Hint: Using figures provided by the registrar's office, compare your sample distribution with the actual distribution of students according to the male-female ratio, according to academic class, and according to major.)

16.4 Estimation Based on a Simple Random Sample

The selection of a simple random sample, the most elementary survey design, was presented in section 16.3. After gathering the sample observations, our next objective is to estimate certain population parameters of interest. Most often we are interested in estimating the population mean μ or the population total τ (Greek letter tau). For example, the accounting firm in example 16.1 might be interested in the mean dollar value for the accounts receivable as well as the total dollar amount of these accounts.

The computational formulas for estimating the population mean μ and the population total τ, based on simple random sampling, are shown in the boxes. Recall, however, that a point estimate such as $\hat{\mu}$ or $\hat{\tau}$ tells us nothing about the *goodness* of our estimation. Therefore, variance formulas are given so that we can place bounds on the error of estimation of μ and τ.

When using simple random sampling to estimate the population mean μ, the estimator is as follows:

Estimation of the Population Mean for a Simple Random Sample

Estimator:

$$\hat{\mu} = \bar{y} = \sum_{i=1}^{n} \frac{y_i}{n}$$

Variance of the estimator:

$$\hat{\sigma}_{\bar{y}}^2 = \left(\frac{s^2}{n}\right)\left(\frac{N-n}{N}\right) \quad \text{where} \quad s^2 = \sum_{i=1}^{n} \frac{(y_i - \bar{y})^2}{n-1}$$

Bounds on the error of estimation:

$$\bar{y} \pm 2\hat{\sigma}_{\bar{y}}$$

Recall from section 8.4 that the quantity $2\sigma_{\bar{y}}$ is an approximate bound on the error of estimation. This bound is taken to imply that at least 75%, and most likely 95%, of the estimates will deviate from the mean by less than $2\sigma_{\bar{y}}$. Throughout this chapter we will use the two-standard-deviation approximate bound to describe an interval estimator for μ and τ when using each of several different sampling designs.

When the survey objective is to use simple random sampling to estimate the population total τ, we have the following formulas:

Estimation of the Population Total for a Simple Random Sample

Estimator:
$$\hat{\tau} = N\bar{y}$$

Variance of the estimator:
$$\hat{\sigma}_{\hat{\tau}}^2 = N^2 \, \hat{\sigma}_{\bar{y}}^2$$

Bounds on the error of estimation:
$$N\bar{y} \pm 2\hat{\sigma}_{\hat{\tau}}$$

Note that the variance of the population mean estimator, $\hat{\sigma}_{\bar{y}}^2$, is the same as that given in chapter 8 except that it is multiplied by a finite population correction (fpc) factor, $(N - n)/N$, to adjust for sampling from a finite population. When n is small relative to the population size N, the fpc, $(N - n)/N$, is close to 1. Practically speaking, the fpc can be ignored if $n \leq N/20$. In that case $\hat{\sigma}_{\bar{y}}^2$ reduces to the more familar quantity s^2/n.

Example 16.2 Refer to the accounts receivable audit of example 16.1. Suppose the simple random sample of $n = 15$ accounts provided the 15 account balances listed in table 16.2.

Table 16.2 *Account balances for example 16.2*

$14.50	$23.40	$42.00
30.20	15.50	13.30
17.80	27.50	23.70
10.00	6.90	18.40
8.50	19.50	12.10

a. Estimate the average μ due for all $N = 1000$ accounts receivable of the department store, and place a bound on the error of estimation.

b. Estimate the total τ due to all outstanding accounts receivable, and place a bound on the error of estimation.

Solution It will help us in our computations to list the sample data as shown in the accompanying table.

y_i	y_i^2
$14.50	210.25
30.20	912.04
17.80	316.84
10.00	100.00
8.50	72.25
23.40	547.56
15.50	240.25
27.50	756.25
6.90	47.61
19.50	380.25
42.00	1764.00
13.30	176.89
23.70	561.69
18.40	338.56
12.10	146.41

$$\sum_{i=1}^{15} y_i = 283.30 \qquad\qquad \sum_{i=1}^{15} y_i^2 = 6570.85$$

a. Our estimate of the mean account balance μ is

$$\bar{y} = \frac{\sum_{i=1}^{15} y_i}{15} = \frac{283.30}{15} = \$18.89$$

To find a bound on the error of estimation of μ, we must first compute

$$s^2 = \frac{\sum_{i=1}^{15} (y_i - \bar{y})^2}{14} = \frac{\sum_{i=1}^{15} y_i^2 - \left(\sum_{i=1}^{15} y_i\right)^2 \Big/ 15}{14}$$

$$= \frac{1}{14}\left[6570.85 - \frac{(283.30)^2}{15}\right] = \frac{1}{14}\,[6570.85 - 5350.59] = 87.16$$

The estimated variance of $\bar{y}$ is therefore

$$\hat{\sigma}_{\bar{y}}^2 = \left(\frac{s^2}{n}\right)\left(\frac{N-n}{N}\right) = \left(\frac{87.16}{15}\right)\left(\frac{1000-15}{1000}\right) = 5.72$$

An estimate of the mean account balance μ, with a bound on the error of estimation, is

$$\bar{y} \pm 2\hat{\sigma}_{\bar{y}} = \$18.89 \pm 2\sqrt{5.72} = \$18.89 \pm \$4.78$$

b. An estimate of the total amount due on outstanding accounts receivable is provided by

$$\hat{\tau} = N(\bar{y}) = 1{,}000(\$18.89) = \$18{,}890$$

Since the estimated variance of $\hat{\tau}$ is $\hat{\sigma}_{\hat{\tau}}^2 = N^2\,\hat{\sigma}_{\bar{y}}^2$, an estimate of the total due on all $N = 1,000$ accounts, with a bound on the error of estimation, is

$$\hat{\tau} \pm 2\hat{\sigma}_{\hat{\tau}} = N\bar{y} \pm 2N\hat{\sigma}_{\bar{y}} = \$18,890 \pm 2(1,000)\sqrt{5.72}$$
$$= \$18,890 \pm \$4,783$$

In an experimental investigation it may be of interest to estimate the **proportion** of the population that possesses a specified characteristic. An auditor may be interested in the proportion of delinquent accounts; a market researcher may be interested in the firm's proportionate share of total market sales; or a corporate executive may be interested in the proportion of shareholders favoring a particular company policy decision.

Recall that we first studied the estimation of the population proportion p in section 8.10. There it was assumed that our estimation was based on a simple random sample selected from an *infinite* population. When the population size N is **finite**, the population proportion p is estimated as follows:

Estimation of the Population Proportion for a Simple Random Sample

Estimator:

$$\hat{p} = \frac{y}{n}$$

Variance of the estimator:

$$\hat{\sigma}_{\hat{p}}^2 = \left(\frac{\hat{p}\hat{q}}{n-1}\right)\left(\frac{N-n}{N}\right) \qquad \text{where} \qquad \hat{q} = 1 - \hat{p}$$

Bounds on the error of estimation:

$$\hat{p} \pm 2\hat{\sigma}_{\hat{p}}$$

In this case y is the total number of the n sample elements that possess the specified characteristic.

Example 16.3 Manufacturing organizations often resort to short-term price discounts in order to encourage their customers to increase their order size and buy ahead, thus enhancing the manufacturer's cash position. Consistent with this basic intent, a manufacturer and wholesale distributor of frozen food products is considering discounting by 20% the price of frozen dinners to buyers who double their monthly order. As frozen foods require costly storage, it is not certain that the buyers will take advantage of the discount offer. A random sample of $n = 50$ of the firm's $N = 430$ buyers was contacted, with 15 of the 50 indicating they would accept the price discount offer and double their monthly order. Estimate the proportion p of all $N = 430$ of the firm's buyers who will take advantage of the offer, and place a bound on the error of estimation.

Solution An estimate of the proportion p of all the firm's buyers who will take advantage of the firm's price discount is

$$\hat{p} = \frac{y}{n} = \frac{15}{50} = .30$$

To place bounds on the error of estimation, we must first compute the variance $\hat{\sigma}^2_{\hat{p}}$. We find

$$\hat{\sigma}^2_{\hat{p}} = \left(\frac{\hat{p}\hat{q}}{n-1}\right)\left(\frac{N-n}{N}\right) = \left[\frac{(.30)(.70)}{49}\right]\left(\frac{430-50}{430}\right)$$

$$= \left(\frac{.21}{49}\right)(.88) = .003771$$

An estimate of p, with a bound on the error of estimation, is

$$\hat{p} \pm 2\hat{\sigma}_{\hat{p}} = .30 \pm 2\sqrt{.003771} = .30 \pm .12.$$

That is, we estimate that the proportion of all the firm's buyers who will take advantage of the price discount is .30, with a bound on the error of estimation of .12.

Exercises

16.11. In the face of the energy crisis and rising numbers of highway deaths, Congress in 1974 created legislation imposing a maximum 55-mph speed limit on all public highways. Since then, much debate has occurred on the public's acceptance of this law. To study this issue the California Highway Patrol decided to randomly select $n = 25$ vehicles traveling a certain section of interstate highway and to measure their speeds. The average speed of the 25 vehicles was found to be 57.5 mph with a standard deviation of 9.3 mph. Estimate the average speed μ of all vehicles traveling on the highway, and place a bound on the error of estimation. (Assume that N is sufficiently large so that we can ignore the fpc factor.)

16.12. The Federal Trade Commission (FTC) has proposed legislation to permit pharmacies to advertise prices for retail drugs. Such legislation would allow price competition for prescription drugs among retail pharmacies, thus enabling the consumer to obtain the best available bargain for prescriptions. An FTC staff member undertook a study to examine the disparity in prices charged by retail pharmacies in Cincinnati for the drug raudixin, a high blood pressure remedy. A table of random numbers was used to select $n = 20$ pharmacies from among the $N = 152$ retail pharmacies of Cincinnati. The price charged for 100 tablets of raudixin by each of the selected pharmacies is shown below.

$3.75	$4.10	$10.40	$7.50	$2.95	$5.75	$7.50	$8.90	$ 4.75	$11.75
5.85	7.65	8.10	6.50	7.50	5.50	8.00	4.50	10.25	4.95

Estimate the average price μ charged for 100 tablets of raudixin by all the 152 retail pharmacies of Cincinnati, and place a bound on the error of estimation.

16.13. Labor unions are organized to represent the interests of member employees in their jobs. Wages and salaries comprise only part of these interests.

Fringe benefits, such as retirement and insurance benefits, vacation plans, and working conditions, are often of equal concern. Suppose the Textile Workers Union is interested in determining the proportion of the $N = 352$ employees it represents in a South Carolina textile plant who are satisfied with the company's current retirement and insurance benefits. Using a table of random numbers, a union official selects $n = 40$ of the employees and finds that 23 of them are satisfied with the retirement and insurance benefits currently offered by the company. Estimate the proportion p of all the textile company's employees represented by the union who favor the current benefits offered by the firm. Place a bound on the error of estimation.

16.14. Refer to exercise 16.9. Use your simple random sample of $n = 15$ accounts drawn from all $N = 100$ accounts listed in table 16.1.

a. Estimate the average balance μ for all 100 accounts. Place a bound on the error of estimation.

b. Estimate the total receivable on all accounts, and place a bound on the error of estimation.

c. Estimate the proportion of delinquent accounts, and place a bound on the error of estimation. How does your estimate of p compare with the true p for the population of all $N = 100$ accounts receivable?

d. Estimate the total number of delinquent accounts, and place a bound on the error of estimation. [Hint: If p is the population proportion of delinquent accounts, then Np is the *total* number of delinquent accounts. An estimator of Np is $N\hat{p}$, which has an estimated variance of $N^2 \hat{\sigma}_{\hat{p}}^2$.

16.5 Stratified Random Sampling

A second type of sampling design, which often provides a specified amount of information at less cost than simple random sampling, is called *stratified random sampling*. This design is recommended when the population consists of a set of heterogeneous (dissimilar) groups.

Definition

A **stratified random sample** is a random sample obtained by separating the population elements into nonoverlapping groups, called **strata**, and then selecting a simple random sample within each stratum.

The stratified random sampling design has three major advantages over simple random sampling. First, the *cost* of collecting and analyzing the data is often reduced by stratifying the population into groups, where the elements possess similar characteristics within the groups but are dissimilar between groups. For example, in a survey of industrial buyers, it would cost more to obtain information from those in foreign countries than from domestic buyers. We should, therefore, take small samples from the strata with high sampling costs in order to satisfy our objective of minimizing the total cost of sampling.

The second advantage concerns the *variance* of the estimator of the population mean. This variance is often reduced by using stratified random sampling, because the variation within strata is usually smaller than the overall population variance. For example, the power usage for industrial users is likely to be much more variable than for residential users. Therefore, the power company official who wishes to estimate the average usage for all company subscribers should select larger samples from the less homogeneous industrial sector in order to obtain good estimates of the population parameters.

The third advantage of stratified random sampling is that separate estimates are provided for the parameters of each stratum, without selecting another sample and incurring additional cost. For instance, it may be more useful to know the mean monthly power usage for the residential users and the industrial users of a city than to know only the mean monthly usage for all the city's subscribers. Stratified random sampling allows us to analyze stratum differences, so that those groups within the population that deserve special attention can be more easily noted.

In this chapter we will use a **proportional allocation procedure,** which partitions the sample size among the strata proportional to the size of the strata. The major advantage of using proportional allocation is that it provides a "self-weighting" sample, since the sampling fraction is the same in each stratum. Sampling cost savings are thus obtained when many estimates have to be made. Optimal allocation, which partitions the sample size according to cost, variability, and stratum size, produces estimates with smaller variances than those obtained by using proportional allocation when the strata sampling costs and variances differ widely. However, Cochran (see the references) notes that the superiority of optimal allocation is exaggerated because of errors obtained when estimating strata sampling costs and variances. Furthermore, he suggests that the simplicity and self-weighting feature of proportional allocation should offset up to a 20% increase in the variance of the derived estimators.

The first step in selecting a stratified random sample is to clearly specify the strata, associating each element of the population with one and only one stratum. In some cases this task may not be so easy. In an opinion poll example, would we classify those living in a city of 1000 as urban or rural constituents? In the power usage example, would the residence of an accountant whose office is in his home be placed in the residential or commercial stratum? Resolution of such conflicts does not affect our results as long as it is always done consistently. For example, we could say that cities of under 2500 population will always be considered rural, those larger, urban; combination business-residence units will be classified according to which use occupies the greatest floor space of the dwelling.

Once the strata have been specified, we can use the method of section 16.3 to select a simple random sample within each stratum. The overall sample size n depends on our available budget for sampling and on how precise and accurate we wish the estimate to be (these latter requirements are discussed in section 16.7). Using proportional allocation, the sample size is allocated among the L strata so that $n = n_1 + n_2 + \cdots + n_L$, where each n_i is as given by the formula in the box.

Allocation of the Sample Among the Strata

$$n_i = n\left(\frac{N_i}{N}\right) \qquad i = 1, 2, \ldots, L$$

where N_i is the number of elements in stratum i and

$$N = \sum_{i=1}^{L} N_i$$

is the size of the population.

From information obtained from the sample elements, we can compute the estimated mean $\bar{y}_i$ and the variance s_i^2 of the observations within each stratum by using the formulas given in the next box.

Estimation of the Mean and Variance of Each Stratum

$$\bar{y}_i = \frac{\sum_{j=1}^{n_i} y_{ij}}{n_i}$$

$$s_i^2 = \frac{\sum_{j=1}^{n_i} (y_{ij} - \bar{y}_i)^2}{n_i - 1} \qquad i = 1, 2, \ldots, L$$

where y_{ij} is the jth observation in stratum i.

The variance s_i^2 is an estimate of the corresponding true stratum variance σ_i^2.

The estimator $\bar{y}_{st}$ of the population mean μ, based on stratified random sampling, is given in the box.

Estimation of the Population Mean for a Stratified Random Sample

Estimator:

$$\bar{y}_{st} = \frac{1}{N} \sum_{i=1}^{L} N_i \bar{y}_i$$

Variance of the estimator:

$$\hat{\sigma}^2_{\bar{y}_{st}} = \frac{1}{N^2} \sum_{i=1}^{L} N_i^2 \left(\frac{N_i - n_i}{N_i} \right) \left(\frac{s_i^2}{n_i} \right)$$

Bounds on the error of estimation:

$$\bar{y}_{st} \pm 2 \hat{\sigma}_{\bar{y}_{st}}$$

Example 16.4 The period from 1973 through 1975 witnessed a rapid decline in the number of new housing starts in the United States. This decline was primarily due to a shortage of available home loan funds from banks and other savings institutions. To increase the supply of funds available for home mortgages, a large manufacturing organization instituted a policy to encourage their employees to regularly invest part of their incomes with local savings institutions. The firm later decided to conduct a survey of its employees' savings habits to judge the effectiveness of the savings campaign. It is desired to estimate the average amount invested in savings by the employees from their past month's incomes. Suggest a survey design for this problem.

Solution The employees of the firm can be categorized as clerical workers and laborers, foremen and middle managers, and higher-level executives. A stratified random sample with $L = 3$ strata appears to be the most appropriate sample survey design. Within each stratum the spending and investment habits of the employees should be reasonably homogeneous. Simple random sampling should be used to select a sample of employees from each stratum to question them about their savings investment from their last month's incomes.

Suppose the manufacturing organization employs a total of 5000 people, of which 3500 are clerical workers or laborers, 1000 are foremen or middle managers, and 500 are executives. The research department of the firm has enough time and money to interview only $n = 50$ employees. Using a proportional allocation, we would partition the sample size $n = 50$ as follows:

$$n_1 = n \left(\frac{N_1}{N} \right) = 50 \left(\frac{3500}{5000} \right) = 35$$

$$n_2 = 50 \left(\frac{1000}{5000} \right) = 10 \qquad \text{and} \qquad n_3 = 5$$

An alphabetical list of employees in each category, available through the payroll office, provides a base from which we can select our sample. Arbitrarily beginning in column 8 of table 14, we select the first 35 nonrepeated four-digit random numbers between 0000 and 3501 to identify the 35 clerical workers and laborers to be included in our sample. The results are shown in table 16.3. Therefore, the first member of the sample should be the clerical worker or laborer who is 84th in alphabetical order, the next, the 118th, and so on. Similarly, three-digit random numbers

Table 16.3 *Random numbers to identify the employees to sample from stratum 1, clerical workers and laborers*

1419	0473	0151	1798	0691
2483	2638	0118	3159	2096
1873	2872	2349	2084	3068
0585	1539	3409	0827	0084
1761	2533	3208	2635	1411
2985	0815	1505	2217	2191
1360	3001	1256	0664	1842

from 000 to 999 should be used to select a sample of $n_2 = 10$ foremen and middle managers, and three-digit numbers from 001 to 050 used to select $n_3 = 5$ executives.

Having selected the sample elements (employees), we proceed with the interview. From the responses of the employees, we compute the mean $\bar{y}_i$ and the variance s_i^2 of the observations within each stratum. The computed strata means and variances are shown in table 16.4.

We estimate the average investment $\bar{y}_{st}$ in savings from last month's income, using the data from table 16.4, as

$$\bar{y}_{st} = \frac{1}{N} \sum_{i=1}^{L} N_i \bar{y}_i = \frac{1}{5,000} [(3,500)(10.16) + (1,000)(25.50)$$

$$+ (500)(21.80)]$$

$$= \frac{1}{5,000} (71,960) = \$14.39$$

Therefore, the estimated average amount of last month's income invested in savings by all the employees of the firm is $14.39.

The estimated variance of $\bar{y}_{st}$ is

$$\hat{\sigma}_{\bar{y}_{st}}^2 = \frac{1}{N^2} \sum_{i=1}^{3} N_i^2 \left(\frac{N_i - n_i}{N_i}\right)\left(\frac{s_i^2}{n_i}\right)$$

$$= \frac{1}{(5000)^2} \left[\frac{(3500)^2(.99)(16.81)}{35} + \frac{(1000)^2(.99)(22.09)}{10} \right.$$

$$\left. + \frac{(500)^2(.99)(125.44)}{5}\right]$$

$$= .5688$$

Table 16.4 *Calculations for example 16.4*

STRATUM 1	STRATUM 2	STRATUM 3
$n_1 = 35$	$n_2 = 10$	$n_3 = 5$
$\bar{y}_1 = \$10.16$	$\bar{y}_2 = \$25.50$	$\bar{y}_3 = \$21.80$
$s_1^2 = 16.81$	$s_2^2 = 22.09$	$s_3^2 = 125.44$
$N_1 = 3500$	$N_2 = 1000$	$N_3 = 500$

The estimate of the average savings, with a bound on the error of estimation, is provided by

$$\bar{y}_{st} \pm 2\hat{\sigma}_{\bar{y}_{st}} = \$14.39 \pm 2\sqrt{.5688} = \$14.39 \pm 2(.75) = \$14.39 \pm \$1.50$$

If the survey objective is to use stratified random sampling to estimate the population total τ, then the estimator is as given in the box.

Estimation of the Population Total for a Stratified

Random Sample

Estimator:

$$\hat{\tau} = N\bar{y}_{st}$$

Variance of the estimator:

$$\hat{\sigma}_{\hat{\tau}}^2 = N^2 \hat{\sigma}_{\bar{y}_{st}}^2$$

Bounds on the error of estimation:

$$\hat{\tau} \pm 2\hat{\sigma}_{\hat{\tau}}$$

Example 16.5 Refer to example 16.4. Estimate the total of last month's income invested in savings by the employees of the manufacturing organization. Place a bound on the error of estimation.

Solution From our previous computations $\bar{y}_{st} = \$14.39$. Therefore, an estimate of the total savings is

$$\hat{\tau} = N\bar{y}_{st} = (5,000)(\$14.39) = \$71,950$$

To find bounds on the error of estimation of τ, we must first compute the estimated variance $\hat{\sigma}_{\hat{\tau}}^2$.

$$\hat{\sigma}_{\hat{\tau}}^2 = N^2 \hat{\sigma}_{\bar{y}_{st}}^2 = (5,000)^2(.5688) = 14,220,000$$

The estimate of the total savings, with a bound on the error of estimation, is given by

$$\hat{\tau} \pm 2\hat{\sigma}_{\hat{\tau}} = \$71,950 \pm 2\sqrt{14,220,000} = \$71,950 \pm 2(3,771)$$
$$= \$71,950 \pm \$7,542$$

Therefore, we are approximately 95% certain that the total investment in savings by the employees is contained in the interval from $64,410 to $79,490.

So far we have been concerned with the use of stratified random sampling to estimate the population mean and total. In contrast, suppose the manufacturing organization wishes to estimate the proportion of employees who invested part of their last month's income in a savings account. Using the same strata as defined earlier, the researcher can select a simple random sample from each stratum and find the proportion $\hat{p}_i$ of employees in stratum i who invested part of last month's income in savings. The sample proportions obtained from the strata can be combined to provide an estimate of the population proportion.

Estimation of the Population Proportion for a Stratified Random Sample

Estimator:

$$\hat{p}_{st} = \frac{1}{N} \sum_{i=1}^{L} N_i \hat{p}_i$$

Variance of the estimator:

$$\hat{\sigma}^2_{\hat{p}_{st}} = \frac{1}{N^2} \sum_{i=1}^{L} N_i^2 \left(\frac{N_i - n_i}{N_i}\right)\left(\frac{\hat{p}_i \hat{q}_i}{n_i - 1}\right) \qquad \hat{q}_i = 1 - \hat{p}_i$$

Bounds on the error of estimation:

$$\hat{p}_{st} \pm 2\hat{\sigma}_{\hat{p}_{st}}$$

Example 16.6 Of the $n = 50$ employees interviewed in the savings investment study, the number of those who indicated that they had actually participated is given in the table. Estimate the proportion of all employees

STRATUM	SAMPLE SIZE	NUMBER PARTICIPATING	$\hat{p}_i$
1	$n_1 = 35$	21	.60
2	$n_2 = 10$	7	.70
3	$n_3 = 5$	4	.80

participating in the savings program, and place a bound on the error of estimation.

Solution The desired estimate is given by $\hat{p}_{st}$, where

$$\hat{p}_{st} = \frac{1}{5000} [(3500)(.60) + (1000)(.70) + (500)(.80)] = .64$$

Bounds on the error of estimation can be found by first computing the variance:

$$\hat{\sigma}^2_{\hat{p}_{st}} = \frac{1}{(5000)^2} \left[3500^2 \left(\frac{3500 - 35}{3500}\right)\left(\frac{(.6)(.4)}{34}\right) \right.$$
$$+ 1000^2 \left(\frac{1000 - 10}{1000}\right)\left(\frac{(.7)(.3)}{9}\right) + 500^2 \left.\left(\frac{500 - 5}{500}\right)\left(\frac{(.8)(.2)}{4}\right)\right]$$

$$= .004744$$

The estimate of the proportion of employees participating in the manufacturing organization's savings program, with a bound on the error of estimation, is given by

$$\hat{p}_{st} \pm 2\hat{\sigma}_{\hat{p}_{st}} = .64 \pm 2\sqrt{.004744} = .64 \pm 2(.069) = .64 \pm .14$$

Exercises

16.15.   Manufacturing organizations spend millions of dollars each year in the development, promotion, and marketing of new products. Nevertheless, the rate of success of new products is miniscule. Evidence indicates that fewer than one out of ten new products eventually satisfies the firm's criteria for success. One useful market research activity to measure new product acceptance is to test-market the product in a representative sales region. As a case in point, consider a manufacturer of farm implements who wishes to introduce a new orchard sprayer in three western states, Washington, Oregon, and California. To test the market acceptability of the new sprayer, 30 retail outlets are selected in the three states and the number of sprayers sold over a 12-month period is observed. The 30 retail outlets used in the test-marketing are selected by using stratified random sampling and proportionately allocating the sample among the three states. The results are given in the table.

	WASHINGTON	OREGON	CALIFORNIA
n_i	9	6	15
$\bar{y}_i$	26	23	39
s_i^2	31.2	19.3	38.5

a. Estimate the average number μ of sales for all 250 retail outlets in the three western states, and place a bound on the error of estimation.

b. Estimate the total sales in the three western states if the new sprayer were made available to all 250 retail outlets. Place a bound on the error of estimation.

16.16.   Job dissatisfaction among employees may result in costs to a firm due to poor quality of workmanship or needless employment costs caused by inexcusable employee absence from the job. In an examination of this latter cost, a corporate personnel manager wished to determine the number of workdays lost due to sick leave and leave without pay by her firm's 2700 employees. For administrative convenience the personnel manager decided to use stratified random sampling and proportionately allocate a sample of $n = 27$ among the laborers, technicians, and administrators employed within the company. The data (workdays lost) obtained from sampling 15 laborers, 10 technicians, and 2 administrators are given in the table.

LABORERS			TECHNICIANS		ADMINISTRATORS
8	24	0	4	5	1
0	16	32	0	24	8
6	0	16	8	12	
7	4	4	3	2	
9	5	8	1	8	

a. Estimate the average number μ of days lost due to sick leave and leave without pay by all 2700 employees of the firm, and place a bound on the error of estimation.

b. Estimate the total number of days lost due to sick leave and leave without pay by the employees of the firm. Place a bound on the error of estimation.

16.17.   Chain stores such as Sears, Roebuck and Co., J. C. Penney, and most banking chains process all credit accounts through a central or regional office, not separately through each branch store. This provides more efficient central control of branch management activities. The credit manager of a chain of four wholesale bakeries is concerned about the volume of delinquent accounts currently outstanding. To reduce the cost of sampling, stratified random sampling is used, with each bakery serving as a separate stratum. From records available in his office, the credit manager decides to use proportional allocation to select a stratified random sample of $n = 50$ accounts from all $N = 200$ accounts receivable. Doing so, he notes the results given in the table.

	BAKERY			
	1	2	3	4
Number of Accounts Receivable	$N_1 = 56$	$N_2 = 68$	$N_3 = 40$	$N_4 = 36$
Sample Size	$n_1 = 14$	$n_2 = 17$	$n_3 = 10$	$n_4 = 9$
Number of Delinquent Accounts	$y_1 = 5$	$y_2 = 7$	$y_3 = 5$	$y_4 = 1$

a. Estimate the proportion p of delinquent accounts for the chain of bakeries, and place a bound on the error of estimation.

b. There is reason to believe that the manager of Bakery 3 is too lenient in granting credit to his customers. Estimate the proportion p_3 of delinquent accounts for Bakery 3, and place a bound on the error of estimation.

16.6 Cluster Sampling

Frequently it is easier to sample *clusters* of elements rather than the individual elements themselves.

Definition

A **cluster sample** is obtained by randomly selecting a set of m collections of sample elements, called clusters, from the population and then conducting a complete census within each cluster.

Cluster sampling will usually provide a specified amount of information at a minimum cost when either:

1. a frame listing the elements of the population does not exist or would be very costly to obtain, or
2. the population is large and spread out over a wide area.

As an illustration, consider an economist who wishes to estimate the average weekly expenditures on food per household in a city. To use either simple random sampling or stratified random sampling, the economist must have a list from which the sample elements (households) can be selected. However, a list of all the households in the city may be very costly or even impossible to obtain. Even if such a list is available, the survey costs may still be substantial because, with simple random sampling or stratified random sampling, the households chosen in the sample would probably be scattered over a wide area. As a result, the cost of conducting the survey among the scattered households would be great due to the interviewers' travel times and other related expenses.

Rather than select a sample of households scattered throughout the entire city, the economist could use cluster sampling and divide the city into clusters—political wards, perhaps—and then select a random sampling of clusters. This should be easy to accomplish since a list of wards is readily available. Every household within each chosen ward would then be surveyed. The total survey costs are reduced since the economist has eliminated the need to develop a costly list of all households and since households within each ward are close together geographically, thereby reducing the interviewers' expenses.

While cost savings often result by using cluster sampling, this benefit is not always gained without a concession in return. The cluster design sometimes increases the sampling error, since elements within the same cluster tend to have common characteristics. For example, in surveys of human populations, clusters are often neighborhoods, but neighborhoods typically consist of families of the same age, income, ethnic background, and occupational class. Therefore, when randomly selecting clusters for the survey, there is a chance that certain socioeconomic classes will not be represented at all if it should happen that their neighborhoods are not included. On the other hand, certain other classes of households may be overrepresented.

We can reduce sampling error by selecting many small clusters instead of a few large clusters. The smaller the cluster sizes (e.g., city blocks instead of school districts), the lesser is the chance that we are excluding certain classes of elements from our sample. Therefore, more information about the population can be gained by selecting a larger number of smaller-sized clusters.

Once the clusters have been specified, a list containing all clusters must be prepared. Simple random sampling is then used to select a random sample of m clusters from among the M clusters in the population.

When using cluster sampling, the population mean μ is estimated as shown in the box. In the formulas, n_i is the number of elements in the ith cluster and t_i is the total of the measurements in cluster i.

Estimation of the Population Mean for a Cluster Sample

Estimator:

$$\hat{\mu} = \bar{y}_{cl} = \frac{\sum\limits_{i=1}^{m} t_i}{\sum\limits_{i=1}^{m} n_i}$$

Variance of the estimator:

$$\hat{\sigma}^2_{\bar{y}_{cl}} = \left(\frac{M - m}{Mm\bar{n}^2}\right)\left(\frac{\sum\limits_{i=1}^{m}(t_i - \bar{y}_{cl} n_i)^2}{m - 1}\right)$$

Bounds on the error of estimation:

$$\bar{y}_{cl} \pm 2\,\hat{\sigma}_{\bar{y}_{cl}}$$

where

$$\bar{n} = \frac{1}{m}\sum\limits_{i=1}^{m} n_i \qquad \bar{t} = \frac{1}{m}\sum\limits_{i=1}^{m} t_i$$

M is the number of clusters in the population, and m is the number of sampled clusters.

The estimate of the population total τ is given in the next box.

Estimation of the Population Total for a Cluster Sample

Estimator:

$$\hat{\tau} = \frac{M}{m}\sum\limits_{i=1}^{m} t_i$$

Variance of the estimator:

$$\hat{\sigma}^2_{\hat{\tau}} = M^2\left(\frac{M - m}{Mm}\right)\left(\frac{\sum\limits_{i=1}^{m}(t_i - \bar{t})^2}{m - 1}\right)$$

Bounds on the error of estimation:

$$\hat{\tau} \pm 2\hat{\sigma}_{\hat{\tau}}$$

As we stated before, the number of elements in the *i*th cluster is denoted by n_i, while t_i is the total of the measurements within cluster *i*. Therefore

$$t_i = \sum_{j=1}^{n_i} y_{ij}$$

where y_{ij} is the *j*th observation within cluster *i*. The terms $\bar{n}$ and $\bar{t}$ denote, respectively, the mean cluster size and mean cluster total.

Example 16.7 The objective of product advertising is to increase sales or to create awareness and interest in the company's products. Therefore, it is essential that advertising be placed with the media most likely to reach the buying public. An advertising agent for a firm that sells household products wishes to estimate the monthly expenditures on magazines and newspapers by the households of a certain midwestern city to determine whether such expenditures are sufficient to warrant the use of these media sources for advertising. Since no list of households is available, and to control direct interviewer costs, cluster sampling is used, with voting precincts forming the clusters. A simple random sample of 10 precincts is selected from the 50 precincts within the city. Interviewers then survey every household within the 10 precincts and record the total household expenditures on magazines and newspapers during the past month. The data are shown in table 16.5.

Table 16.5 *Monthly expenditures on magazines and newspapers by the households of example 16.7*

SAMPLED PRECINCT, i	NUMBER OF HOUSE- HOLDS, n_i	TOTAL EXPENDI- TURES, t_i	SAMPLED PRECINCT, i	NUMBER OF HOUSE- HOLDS, n_i	TOTAL EXPENDI- TURES, t_i
1	62	$380	6	69	$403
2	55	517	7	58	555
3	49	480	8	74	486
4	71	613	9	57	450
5	70	540	10	65	395
Sums				$\sum_{i=1}^{10} n_i = 630$	$\sum_{i=1}^{10} t_i = \$4819$

a. Estimate the average monthly household expenditures on magazines and newspapers in the city, and place a bound on the error of estimation.

b. Estimate the total monthly expenditures on magazines and newspapers by all the households in the city, and place a bound on the error of estimation.

Solution a. The population mean μ is estimated by

$$\bar{y}_{cl} = \frac{\sum_{i=1}^{10} t_i}{\sum_{i=1}^{10} n_i} = \frac{\$4819}{630} = \$7.65$$

To calculate $\hat{\sigma}^2_{\bar{y}_{cl}}$ we will first evaluate the sum of squares term

$$\sum_{i=1}^{m} (t_i - \bar{y}_{cl} n_i)^2$$

It can be shown that

$$\sum_{i=1}^{m} (t_i - \bar{y}_{cl} n_i)^2 = \sum_{i=1}^{m} t_i^2 - 2\bar{y}_{cl} \sum_{i=1}^{m} t_i n_i + \bar{y}^2_{cl} \sum_{i=1}^{m} n_i^2$$

Taken individually, we have

$$\sum_{i=1}^{10} t_i^2 = (380)^2 + (517)^2 + \cdots + (395)^2 = 2,374,613$$

$$\sum_{i=1}^{10} t_i n_i = (380)(62) + (517)(55) + \cdots + (395)(65) = 304,124$$

$$\sum_{i=1}^{10} n_i^2 = (62)^2 + (55)^2 + \cdots + (65)^2 = 40,286$$

Substituting these values into the sum of squares equation, we find

$$\sum_{i=1}^{10} (t_i - \bar{y}_{cl} n_i)^2 = 2,374,613 - 2(7.65)(304,124) + (7.65)^2(40,286)$$

$$= 79,153.235$$

The average cluster size is

$$\bar{n} = \frac{1}{m} \sum_{i=1}^{m} n_i = \frac{1}{10} (630) = 63$$

Since the total number of clusters in the population is $M = 50$, we find

$$\hat{\sigma}^2_{\bar{y}_{cl}} = \left(\frac{M - m}{Mm\bar{n}^2} \right) \left(\frac{\sum_{i=1}^{m} (t_i - \bar{y}_{cl} n_i)^2}{m - 1} \right)$$

$$= \left(\frac{50 - 10}{(50)(10)(63)^2} \right) \left(\frac{79,153.235}{9} \right) = .1773$$

Therefore, an estimate of μ, the average monthly household expenditures on magazines and newspapers, with a bound on the error of estimation, is

$$\bar{y}_{cl} \pm 2\hat{\sigma}_{\bar{y}_{cl}} = \$7.65 \pm 2\sqrt{.1773} = \$7.65 \pm \$.84$$

b. An estimate of the total monthly expenditures on magazines and newspapers is

$$\hat{\tau} = \frac{M}{m} \sum_{i=1}^{m} t_i = \frac{50}{10} (\$4,819) = \$24,095$$

which does *not* depend on knowledge of the population size N.

To place a bound on the error of estimation, we first calculate the expression

$$\sum_{i=1}^{m} (t_i - \bar{t})^2 = \sum_{i=1}^{m} t_i^2 - \frac{1}{m} \left(\sum_{i=1}^{m} t_i \right)^2$$

$$= 2,374,613 - \tfrac{1}{10} (4,819)^2 = 52,336.90$$

The estimated variance is

$$\hat{\sigma}_{\hat{\tau}}^2 = M^2 \left(\frac{M - m}{Mm} \right) \left(\frac{\sum_{i=1}^{m} (t_i - \bar{t})^2}{m - 1} \right)$$

$$= 50^2 \left(\frac{50 - 10}{(50)(10)} \right) \left(\frac{52,336.90}{9} \right) = 1,163,042.222$$

The estimate of total monthly expenditures on magazines and newspapers by all the households in the city, with a bound on the error of estimation, is

$$\hat{\tau} \pm 2\hat{\sigma}_{\hat{\tau}} = \$24,095 \pm 2 \sqrt{1,163,042.222}$$

$$= \$24,095 \pm \$2,157$$

Often an experimenter wishes to use cluster sampling to estimate a population proportion p. For instance, in a preelection survey it may be desired to estimate the proportion of residents of a community who favor a particular ballot measure; or it may be desired to estimate the proportion of automobiles in a city that do not pass current emission standards, or the proportion of members of a nationwide union favoring a negotiated salary adjustment. To estimate p when using cluster sampling, we first find a_i, the number of elements in cluster i that possess the characteristic of interest, for each cluster $i = 1, 2, ..., m$. Then an estimate of the proportion of elements in the population possessing the characteristic is given by the formula in the box.

When the cluster sizes $n_1, n_2, ..., n_m$ are equal, $\hat{\sigma}_{\hat{p}_{cl}}^2$ is a good estimator of the true variance for any number m of sample clusters. However, experience has shown that when the cluster sizes are not equal, $\hat{\sigma}_{\hat{p}_{cl}}^2$ is a good estimator only when m is large, say $m \geq 20$.

Estimation of the Population Proportion for a Cluster Sample

Estimator:

$$\hat{p}_{cl} = \frac{\displaystyle\sum_{i=1}^{m} a_i}{\displaystyle\sum_{i=1}^{m} n_i}$$

Variance of the estimator:

$$\hat{\sigma}^2_{\hat{p}_{cl}} = \left(\frac{M - m)}{Mm\bar{n}^2}\right)\left(\frac{\displaystyle\sum_{1=1}^{m}(a_i - \bar{p}_{cl}n_i)^2}{m - 1}\right)$$

Bounds on the error of estimation:

$$\hat{p}_{cl} \pm 2\hat{\sigma}_{\hat{p}_{cl}}$$

Exercises

16.18.   In example 16.7 cluster sampling was used to reduce the cost of the household surveys. Someone might suggest that survey costs could be further reduced in the survey by conducting telephone interviews instead of personal interviews. Provide your comments in support of or opposition to the suggestion to use telephone interviews.

16.19.   A recent news article noted that the rate of increase of charitable contributions lags far behind the rate of inflation during periods of recession. Concerned about a possible decline in voluntary contributions in her region, a regional director for the American Cancer Society is interested in estimating the average contribution per household and the total contribution from all households in her city. At this stage in the fund-raising campaign, neighborhood volunteers have canvassed a random selection of 12 of the 47 voting precincts in the city and have obtained the data given in the table for the amount of donations.

PRECINCT	NUMBER OF HOUSEHOLDS	TOTAL DONATIONS	PRECINCT	NUMBER OF HOUSEHOLDS	TOTAL DONATIONS
1	36	$117	7	29	$165
2	42	105	8	52	105
3	40	210	9	44	121
4	47	142	10	40	103
5	39	235	11	45	136
6	50	96	12	36	190

a. Estimate the average contribution per household in the city, and place a bound on the error of estimation.

b. Estimate the total contribution from all households in the city, and place a bound on the error of estimation.

16.20. An inspector for a hardware chain wishes to estimate the proportion of defective light bulbs shipped to its warehouse by a manufacturer. The bulbs are shipped in cartons containing 12 boxes, with each box containing 6 bulbs. Design a cluster-sampling experiment for the inspector. Should he use cartons of bulbs or boxes of bulbs as clusters? Explain.

16.21. Refer to exercise 16.20. Suppose the inspector decides to use boxes of bulbs as clusters and randomly selects $m = 20$ boxes from among the 100 cartons received in the shipment. The number of defective bulbs found in each of the 20 boxes of bulbs is as follows:

$$0, 2, 0, 0, 1, 1, 0, 1, 2, 1, 0, 0, 0, 1, 0, 0, 3, 0, 2, 1$$

Estimate the proportion p of defective bulbs in the shipment, and place a bound on the error of estimation.

16.7 Finding the Sample Size

One of the first questions asked by someone undertaking a sample survey is, "How many sample elements should I select?" Since sampling is time-consuming and costly, our objective in selecting a sample is to obtain a specified amount of information about a population parameter at a minimum cost. We can accomplish this objective by first deciding on a bound on the error of estimation (which measures our specified information content) and then applying an appropriate sample size estimation formula.

In section 16.1 we saw that when the population is uniform, a small sample provides the same amount of information as a large sample. Thus the physician can base his diagnosis on one drop of the patient's blood; the quality control engineer can judge a large lot of transistors by testing only a few. Selecting a large sample in such instances is a waste of time and money. On the other hand, if the population consists of many highly diverse elements, a small sample may provide a poor reflection of the population. In a study to estimate the average height of the male students attending a particular college, a small sample, say $n = 3$ students, may by chance consist entirely of members of the varsity basketball team. A random sampling of $n = 100$ students should, however, provide a much broader coverage of, and hence more information about, the heights of the male students.

The objective behind the selection of a sampling design and selection of the sample size are the same—to obtain a specified amount of information at a minimum cost. Sampling design decisions are made according to the "lay of the land," that is, how the elements group themselves together in the population, and according to the cost of recording information contained by those elements. Sample size decisions are made according to the inherent variability in the population of measurements and how accurate the experimenter wishes the estimate to be. These two criteria are, of course, inversely related. To obtain greater accuracy, and hence more information about a population, we must select a larger sample size; the greater the inherent variability in the population, the larger is the sample size required to maintain a fixed degree of accuracy in estimation.

Simple Random Sampling

When using simple random sampling, the sample size required to estimate the population mean μ, with a bound B on the error of estimation, is as given in the box.

Sample Size for Estimating μ for a Simple Random Sample

$$n = \frac{N\sigma^2}{(N-1)D + \sigma^2} \qquad D = \frac{B^2}{4}$$

where σ^2 is the population variance, N is the number of elements in the population, and B is the bound on the error of estimation.

When N is very large, the sample size formula reduces to the more familiar formula given in chapter 8.

Sample Size for Estimating μ for a Simple Random Sample When N Is Very Large

$$n = \frac{4\sigma^2}{B^2}$$

When the objective is to estimate the population total τ, with a bound B on the error of estimation, we must substitute $D = B^2/4N^2$ into the sample size formula in the preceding box.

Some students may notice a dilemma in the guidelines provided for finding n. To find n we must know the population variance, but to estimate σ^2 we must have a set of sample measurements from the population. The variance can be estimated by s^2 obtained from a previous sample or by knowledge of the range of measurements, giving the estimate

$$\hat{\sigma}^2 = \tfrac{1}{16}(\text{range})^2$$

The range approximation procedure is derived from the Empirical Rule (section 3.8) and provides a very rough approximation to σ^2.

Example 16.8 The manager of the credit division of a commercial bank would like to know the average amount of credit purchases placed on

Bankcard each month by the customers who hold Bankcards issued by the bank. Since there are currently 20,000 open Bankcard accounts at the bank, time and expense prohibit a complete review of every account. The manager thus proposes selecting a simple random sample of open Bankcard accounts to estimate the average monthly account balance μ, with a bound on the error of estimation of $B = \$10$. Although no prior information is available to estimate the variance σ^2 of monthly Bankcard account levels, it is known that most account levels lie within the range from $50 to $450. Find the sample size necessary to achieve the stated bound.

Solution To estimate the population variance σ^2, we use the rule that says the range is approximately equal to four standard deviations. Therefore,

$$\hat{\sigma}^2 = \tfrac{1}{16}(\text{range})^2 = \tfrac{1}{16}(450 - 50)^2 = \tfrac{1}{16}(400)^2 = 10{,}000$$

Using the formula for finding n when estimating μ, we obtain

$$D = \frac{B^2}{4} = \frac{(10)^2}{4} = 25$$

$$n = \frac{N\sigma^2}{(N-1)D + \sigma^2} = \frac{(20{,}000)(10{,}000)}{(19{,}999)(25) + 10{,}000} = 392.18$$

Therefore, the credit manager must randomly select approximately 393 accounts to estimate μ, the average monthly Bankcard account balance, accurate to within $10 of the true value.

Notice that since N was very large in example 16.8, we could have used the shortcut formula $n = 4\sigma^2/B^2$. Doing so, we find

$$n = \frac{4(10{,}000)}{10^2} = 400$$

a slightly larger sample size than that specified by the exact formula.

When the purpose is to estimate the population proportion p, the variance $\sigma^2 = pq$ depends on p, the population characteristic to be estimated. We must then use a guessed value for p or, taking a conservative approach by allowing the variance to attain its maximum possible value, we can assume p is near $\tfrac{1}{2}$ and let $\sigma^2 = .25$. The sample size required to estimate p, with a bound B on the error of estimation, is then the same as that given earlier for the estimation of μ.

Stratified Random Sampling

With stratified random sampling we select a separate simple random sample within each of the L strata. Therefore, we cannot determine n until we know the relationship between n and the sample allocation to the strata $n_1, n_2, \ldots, n_L$. Although there are many ways to allocate n among the strata, in section 16.5 we used only proportional allocation.

When using stratified random sampling, we must also consider the fact that the variances of the strata, $\sigma_1^2, \sigma_2^2, ..., \sigma_L^2$, may not be equal. We will need approximations for each of these variances, which we can obtain from previous samples or by estimating the range of measurements within each stratum. In the latter case the range approximation

$$\hat{\sigma}_i^2 = \frac{1}{16} (range)^2$$

provides a rough estimate of the variance of the measurements in stratum *i* based on the *range* of measurements within stratum *i*.

Using stratified random sampling with proportional allocation, the sample size required to estimate the population mean μ, with a bound *B* on the error of estimation, is as given in the box.

Sample Size for Estimating μ for a Stratified Random Sample

$$n = \frac{\sum_{i=1}^{L} N_i \sigma_i^2}{ND + \dfrac{1}{N} \sum_{i=1}^{L} N_i \sigma_i^2} \quad \text{and} \quad D = \frac{B^2}{4}$$

where σ_i^2 and N_i are, respectively, the variance and size of the *i*th stratum.

The sample size required to estimate the population total τ, with a bound *B* on the error of estimation, is obtained by substituting $D = B^2/4N^2$ into the equation in the box.

Example 16.9 The dean of a business school is considering canvassing the members of the school's alumni association for the purpose of generating donations to the school's development fund. Currently there are 3,500 members of the alumni association, 2,100 of whom live in state while the remainder live out of state. The dean has decided to select a stratified sample of alumni (stratified according to current residence) to estimate total donations, and on the basis of the sample evidence, he will decide whether to contact all remaining alumni. Find the number *n* of alumni that should be contacted if the dean wishes to estimate the total alumni contributions with a bound on the error of estimation of $10,000. How should this sample size be allocated between in-state and out-of-state alumni? On the basis of prior fund-raising drives, the standard deviations for donations by in-state and out-of-state alumni were found to be $30 and $20, respectively.

Solution We let the in-state alumni constitute stratum 1 and the out-of-state alumni constitute stratum 2. From prior experience we know that $\sigma_1 = \$30$ and $\sigma_2 = \$20$. Therefore,

$$\sigma_1^2 = (30)^2 = 900 \qquad \text{and} \qquad \sigma_2^2 = (20)^2 = 400$$

Since we are finding the sample size for the estimation of a population total, we compute

$$D = \frac{B^2}{4N^2} = \frac{(10,000)^2}{4(3,500)^2} = 2.04$$

Before finding n, we compute

$$\sum_{i=1}^{2} N_i \sigma_i^2 = (2,100)(900) + (1,400)(400) = 2,450,000$$

The required sample size is then

$$n = \frac{\displaystyle\sum_{i=1}^{2} N_i \sigma_i^2}{ND + \dfrac{1}{N}\displaystyle\sum_{i=1}^{2} N_i \sigma_i^2} = \frac{2,450,000}{(3,500)(2.04) + (1/3,500)(2,450,000)} = 312.5$$

Therefore, the dean should select 313 alumni to estimate the total contributions τ by all alumni, with a bound on the error of estimation of $10,000. Using proportional allocation, the sample should be partitioned according to the formula

$$n_i = n\left(\frac{N_i}{N}\right)$$

Therefore, he should randomly select

$$n_1 = 313\left(\frac{2100}{3500}\right) = 187.8$$

or 188 in-state alumni and

$$n_2 = 313\left(\frac{1400}{3500}\right) = 125.2$$

or 125 out-of-state alumni.

When using stratified random sampling to estimate the population proportion p, we must substitute $\sigma_i^2 = p_i q_i$ for the variance of stratum i and let $D = B^2/4$. In this expression p_i is the population proportion for stratum i, which can be estimated either by a judicious guess or by a conservative approach of letting $p_i q_i = .25$ for stratum i. Using proportional allocation and the conservative approach for assessing each p_i, the formula for the sample size to estimate the population proportion p, with a specified bound B on the error of estimation, simplifies to that shown in the box.

Sample Size for Estimating *p* for a Stratified Random
Sample When *N* Is Very Large

$$n = \frac{N}{NB^2 + 1}$$

Cluster Sampling

With stratified random sampling we first partition the population into strata, then we select a random sample from every stratum. The procedure is reversed with cluster sampling. After the population is divided into clusters, a few clusters are randomly selected from the group. Within each chosen cluster, every sample element is recorded. The sampling units are the individual elements of the population when using stratified random sampling, but with cluster sampling the sampling units are clusters of elements.

Finding the sample size when using cluster sampling thus amounts to choosing the number *m* of clusters of elements we will select. The quantity of information provided by a cluster sample is affected not only by *m* but also by the size of the clusters. Finding the number of clusters to select in a cluster sample is a more complex task than the selection of the sample size *n* when using simple random sampling or stratified random sampling. For that reason detailed instructions on the topic are not given in this elementary text. (If you are interested in the topic, review section 7.4 of the text by Mendenhall, Ott, and Scheaffer listed in the references.) The only advice we offer on selecting a sample size when using cluster analysis is to repeat a statement made in section 16.6. When using cluster sampling, more information about the population can be gained by selecting a larger number of smaller-sized clusters. The rare exception to this rule is the case in which the population consists of many small homogeneous (similar) segments.

Exercises

16.22.   A government agency is interested in comparing the salaries of its clerical workers with the salaries of those employed in private industry. The agency director has proposed selecting a simple random sample from among the agency's 317 clerical workers to compute the average employee salary. The variance of clerical workers' salaries is unknown, but records show that the maximum and minimum salaries are $12,500 and $7,420, respectively. Find the number *n* of clerical workers to select in order to estimate μ with a bound on the error of estimation of $200.

16.23.   Timber cruisers typically use a formula based on basal circumference and topped-off height to estimate the board footage in standing timber. Suppose a timber cruiser wishes to estimate the board footage in a stand of timber containing 150 trees that would top off to 30-foot logs, 100 that would top off to 40-foot logs, and 50 that would top off to 50-foot logs. How many trees should he select in each height stratum if he wishes to estimate the total board footage in the stand, with a bound on the error of estimation of 20,000 board feet? From past experience it is known that the standard deviation of the board feet of lumber in 30-foot, 40-foot, and 50-foot logs is 300, 350, and 400, respectively.

16.24.   In the face of a recessed economy, a textile firm is considering reducing its work schedule to a four-day workweek. As an alternative the company would likely choose to close one of its four main manufacturing divisions, terminating the employees in that division. To gain a feeling of employee sentiment, the firm's personnel manager wishes to choose a sample of employees from among the four divisions and estimate the proportion *p* favoring a reduced workweek, with a bound on the error of estimation of *B* = .10. The firm employs 75 individuals each in divisions 1 and 2, 65 in division 3, and 40 in division 4. It is estimated that about 75% of those in division 4 favor a reduced workweek, while the employees in the other three divisions appear equally divided on the issue. Find the sample size and allocation necessary to achieve the stated bound.

16.25.   Refer to exercise 16.24. Use the conservative approach for the estimation of the variance σ_i^2 for each division. How does this affect the required sample size and allocation for estimating *p* with a bound on the error of estimation of *B* = .10?

16.26.   In an effort to reduce gasoline consumption, the director of a state motor pool is considering the installation of an experimental carburetor on each of the 318 automobiles operated out of the motor pool. Currently the motor pool consists of 75 Plymouths, 105 Chevrolets, and 138 Fords. Before deciding whether to buy a new carburetor for each car, the director wishes to select a sample of cars from the motor pool and to note the gain in gasoline economy in terms of the reduced mileage per gallon. Past evidence indicates that the variance of gasoline mileage for Fords is about 4.0, for Chevrolets is about 4.8, and for Plymouths is about 6.1. Find the sample size and allocation among the three brands of cars for estimating the average gasoline mileage using the new carburetor, with a bound on the error of estimation of *B* = 2 mpg.

16.27.   A labor union represents 857 meat cutters employed by a supermarket chain in 214 stores located in 35 cities in four western states. The union wishes to select a sample of the meat cutters in order to estimate their average hourly wage and the proportion who are satisfied with employee benefits currently offered by the chain. Because of the wide geographic distribution of the meat cutters, a union official has suggested using cluster sampling. Would you suggest that they select stores, cities, or states as clusters? Explain.

16.8 Other Sampling Designs and Procedures

The sampling designs we have described so far—simple random sampling, stratified random sampling, and cluster sampling—are those most commonly used in business and economics. Other less frequently used sampling designs and methods are briefly introduced in this section. For a complete discussion of these and other sampling procedures, you are referred to the text by Cochran listed in the references.

Systematic Sampling

A design that avoids the cumbersome data collection requirements of simple random sampling is **systematic sampling.** A systematic sample is obtained by randomly selecting one element from the first *k* elements in the frame and then selecting every kth element thereafter. Since it is easier and less time-consuming to perform than simple random sampling, systematic sampling can provide more information per sampling dollar. It is especially useful in auditing, when the relevant information is recorded in an orderly form, say in computer storage or on file cards. Selecting credit accounts, equipment maintenance records, or sales force data from computer-stored company records can be accommodated easily, cheaply, and efficiently by using systematic sampling.

There are some instances where systematic sampling should not be used. If hidden periodicities exist in the population, the 1-in-k systematic sample may bias results by introducing sampling error resulting from the periodic influence. Sales records and financial data observed over time often have inherent cyclical behavior—restaurant sales are greater on the weekend than during the week, cash levels are highest around the tenth of the month, personal loans are more frequent during the winter months. In addition, production processes often exhibit periodic behavior when certain individuals or machines are responsible for the output of the process at regular intervals. In such instances the 1-in-k systematic sample may either be completely in phase or totally out of phase with the cyclic component, thus misrepresenting its influence in the sample.

Systematic sampling should also be avoided when the population size is unknown. As evidence of this caution, note that the sampling frequency k must be selected to be less than or equal to N/n, where the sample size n is specified in advance. If N is unknown, we cannot accurately specify k. However, this is not a problem if N is assumed to be infinitely large. (See the text by Mendenhall, Ott, and Scheaffer, page 149, listed in the references.)

The estimation formulas for a systematic sample can be found in chapter 8 of *Elementary Survey Sampling*, by Mendenhall, Ott, and Scheaffer. Not surprisingly, these formulas are the same as those offered in section 16.4 for simple random sampling.

Ratio Estimation

Ratio estimation is an estimation procedure based on the relationship between two variables y and x measured on the same set of sampled elements. Ratio estimation, like linear regression (chapter 11), uses information on a variable x to estimate μ_y or τ_y. For example, an auditor may wish to estimate the actual dollar value of inventory from the inventory recorded on computer accounts; a lumber dealer may wish to estimate the total board feet τ_y in a stand of 50-foot Douglas fir trees based on the total of their basal circumferences τ_x; or a corporate executive may be interested in estimating the total sales τ_y for the firm in the next quarter based on the current quarter's sales τ_x.

A ratio estimate requires the measurement of two variables, y and x, on each element in the sample and a computation of the ratio of their sums,

$$\hat{R} = \frac{\sum_{i=1}^{n} y_i}{\sum_{i=1}^{n} x_i}$$

It is assumed that

$$R = \frac{\tau_y}{\tau_x}$$

Then $\hat{R}$ is an estimator of R (not unbiased) and $\tau_y = R\tau_x$.

The use of ratio estimation is based on the assumption that the relationship between the variables y and x is stable over the entire population. Peterson and his coauthors (see the references) report on the use of assessment/sales ratios to monitor the assessment performances of public assessment offices. When y is the assessed value of a property and x is its market sales price, the authors note a much higher ratio for "blighted" neighborhoods than for "upward transitional" neighborhoods. Therefore, if their study reflects assessment practices in most urban communities, we must assume that a single assessment/sales ratio does not adequately portray the quality of assessments for a particular community. In short, ratio estimation is appropriate when the measure of linear correlation between y and x is strong; at least equal to $\frac{1}{2}$ according to Mendenhall, Ott, and Scheaffer. If it is not, the relationship between y and x will be unstable, resulting in estimates for μ_y and τ_y that are less precise than those obtained by using $\bar{y}$ or $N\bar{y}$.

As we have noted, ratio estimation is an estimation procedure; it is not a sampling design. That is, an appropriate sampling design is chosen for the selection of measurements on the variables y and x. Then ratio estimation is used to estimate the ratio R from the sample information. It can be used with any of the survey designs presented earlier in this chapter.

Two-Stage Cluster Sampling

When using cluster sampling, convenient groups or clusters of elements are randomly selected from the population. After the clusters are chosen, *all* the elements from each selected cluster are sampled.

A **two-stage cluster sample** is obtained by choosing a simple random sample of clusters and then selecting a random sample of elements from each cluster. Therefore, when cluster sizes are very large or the elements within each cluster are quite similar, two-stage cluster sampling provides an economical and efficient alternative to cluster sampling.

In recent years commercial banks have become very competitive. Many have sought to attract new depositors and to encourage current depositors to place more money in checking accounts and less in savings accounts by offering free check-writing privileges. Suppose a large branch banking system wishes to survey the opinions of their current depositors on their reaction to a proposal that would offer them free checking accounts. The bank could consider each branch bank as a separate cluster, select a random sample of branch banks, and then draw a random sample of depositors from each branch. Similarly, a nationwide study to examine the total expenditures by families on medical services could be obtained by first selecting a simple random sampling of states and then choosing a sample of families from each selected state.

This latter situation is an example of **area sampling,** where the initial clusters are geographical areas. Area sampling need not be two-stage but can be a multistage analysis. For example, in the medical expenditures study, the states selected in the first stage of the analysis may be partitioned into counties, and then selected counties partitioned into school districts. Then from the selected school districts, the researcher could canvass all families or select a random sampling of families. Public opinion polling agencies tend to rely on area sampling, usually using stratified sampling within selected clusters.

Cost savings is the primary advantage of two-stage or multistage cluster sampling over more conventional sampling procedures. A frame listing all the elements in the population may be very costly, if not impossible, to obtain. It may, however, be easy and inexpensive to obtain a list of clusters. In addition, cluster sampling reduces the cost of sampling by concentrating the sampling effort to smaller areas where the elements are physically close together.

As with cluster sampling, the use of two-stage or multistage cluster sampling may lead to imprecise estimates for μ, τ, and p, since we are intentionally excluding part of the population from our sample. Thus biases may result when using cluster sampling. It is the responsibility of the problem analyst to decide whether the potential cost savings gained by using a clustering design do offset the effect of possible sampling biases.

Randomized Response Sampling

In sampling of human populations, two nonsampling errors that frequently distort the research findings involve a refusal on the part of some of the respondents to answer all the questions or their act of deliberately providing incorrect information. Often such distortions result when the respondent is afraid of losing prestige or becoming embarrassed by truthful responses to sensitive questions. The bias produced by these nonsampling errors is sometimes large enough to make the sample estimates seriously misleading.

Nevertheless, personal opinions, controversial topics, and intimate behavior are frequently the topics the business researcher is asked to survey. For instance, a retailer may be interested in estimating the incidence of shoplifting; a wholesale distributor of sundries may wish to know the frequency of purchase of birth control products in a certain city; the Internal Revenue Service may be interested in determining the frequency of intentional errors on federal income tax forms; or a police agency may be interested in estimating the frequency of drug usage in its jurisdiction. Even such seemingly harmless topics as age, income, or marital status sometimes trigger sensitive reactions in the respondent.

A special sampling procedure has been created to handle surveys dealing with potentially sensitive or embarrassing material. This procedure is called **randomized response sampling**, and it requires that a question on the sensitive topic be paired with an innocuous question. The respondent then answers only one of the two questions which he or she selects at random. For example, the sensitive question

S: "Have you ever willfully shoplifted from Macy's?"

could be paired with the innocuous question

A: "Have you ever visited the state of Florida?"

The interviewer is given an answer but is unaware of which question has been answered by the respondent. Thus the respondent is saved possible embarrassment or incrimination, since his or her answer is camouflaged by the presence of the innocuous question.

To apply randomized response sampling, a randomization device is created so that the respondent selects the sensitive question with a known probability. Commonly the randomization device consists of a black bag containing a predetermined mixture of black and white beads. The respondent simply selects a bead, notes its color (without letting the interviewer see the color of the bead selected), and answers the appropriate question.

It should be noted that randomized response sampling does not *compete* with simple random sampling, stratified random sampling, or cluster sampling. To the contrary, the randomized response design is used for collecting data *within* those designs, in cases when survey questions may be somewhat sensitive, in order to reduce the chances of bias and error due to nonresponse or incorrect responses. Formulas for the analysis of experimental results obtained in a randomized response survey can be found in the article by Greenberg and coauthors listed in the references.

Exercises

16.28.   For each of the following problem settings, indicate whether a systematic sample, a two-stage cluster sample, or ratio estimation is appropriate.

a. A department store chain with a number of branch stores located in four eastern states wishes to estimate the total accounts receivable for all stores. Accounts are kept separately in each store.

b. A local electric utility co-op is interested in determining the effect of an energy conservation program begun last month. They wish to obtain a measure of the average household power usage during the current month.

c. An economist wishes to conduct a consumer survey to estimate the average monthly food expenditures per household for the households of a certain community. Based on previous studies the economist believes that the amount spent on food by a household is correlated with total household income.

d. An advertising firm has undertaken a promotional campaign for a new product.

It wishes to sample potential customers to determine market acceptability of the product in a certain small community.

e. The city council of a large southwestern city wishes to know the proportion of residents of the city who favor the introduction of a city income tax whose proceeds are earmarked for the reduction of property taxes.

16.29. Telephone companies and other utilities are examples of regulated monopolies. They cannot arbitrarily adjust their rate structure, as do unregulated companies, but must make an appeal for a rate change to the public utilities commissioner of their state. To keep pace with increasing costs resulting from a high inflation rate, Pacific Northwest Bell has proposed an increase in all long-distance calls for their subscribers in the state of Washington. However, the Washington public utilities commissioner claims that increased rates will simply dampen use of long-distance by Washington subscribers, thus causing no revenue increase to Pacific Northwest Bell. Design a survey by which Pacific Northwest Bell can survey its Washington subscribers to estimate the company's total monthly revenue gain under the new rate structure.

16.30. The control of highway traffic and problems of traffic intensity are issues of vital concern to planning and law enforcement agencies, as well as to commercial and private users of public highways. The greater the traffic intensity on any highway, the more frequent are the needed repairs and the need for law enforcement agencies to assist in cases of accidents or traffic congestion. For those using the highway for business purposes, traffic intensity introduces the likelihood of opportunity costs due to delivery delays and other factors. To measure the average traffic intensity per day, the state highway commission has installed a mechanical traffic counter along a busy portion of a state highway. Traffic counters are not always reliable, though, since trucks with multiple axles tend to inflate the count. To accommodate this problem, an actual traffic count is obtained and compared with the count provided by the traffic counter. Design a sampling procedure to estimate the average daily traffic on the highway from the count provided by the traffic counter.

16.31. Concerned with untraceable inventory shortages, the manager of a clothing store wishes to estimate the proportion of shoppers entering her store who can be considered shoplifters. The manager figures that she has sufficient resources to interview up to 60 randomly selected shoppers, but she does not know how to set up her sample survey. Place yourself in the position of a consultant to the manager and suggest an appropriate research design for her problem. In particular, write an appropriate questionnaire, develop a randomization device, and explain its use.

16.9 Summary and Concluding Remarks

In section 16.1 we observed that when samples are random, we can make probabilistic statements about the underlying population. On the other hand, when samples are selected in a nonrandom manner, we can make only descriptive statements about the population. Three common nonrandom sampling procedures are convenience sampling, judgment sampling, and quota sampling. *Convenience sampling* refers to the selection of elements which can be obtained simply and conveniently. Selecting the first *n* customers to enter a store on a particular date would comprise a convenience sample.

Judgment sampling involves the selection of elements that, according to the judgment and intuition of the sampler, accurately reflect the population. *Quota sampling* requires that the sampler arbitrarily select a predetermined number of individuals from different population sectors—usually different occupational classes. Quotas are usually predetermined in an attempt to characterize the population, but the actual selection is not necessarily performed in a random manner.

Nonrandom sampling procedures should *never* be used when the objective of the sampling exercise is inference—that is, drawing conclusions about the population based on information contained in the sample. Nonrandom sampling should be used only if we are willing to restrict our study to descriptive statements about the sample and forgo inferential statements about the population.

The sampling designs discussed in this chapter—simple random sampling, stratified random sampling, cluster and two-stage cluster sampling, and systematic sampling—are all designs that provide random samples. Consequently, when using these designs we know the probabilities of including various observations in the sample, which allows us to make probabilistic statements about the population.

Two sampling procedures were introduced in this chapter for handling special types of sampling problems. The randomized response model is useful for collecting survey information when the survey topic is sensitive or potentially embarrassing to the respondent. Ratio estimation provides a model for incorporating the effects of an auxiliary variable x in the estimation of μ_y or τ_y.

Supplementary Exercises

16.32.   What is the objective of sampling? How does this objective relate to the objective of statistics? (Hint: See section 16.1.)

16.33.   What is meant by random sampling?

16.34.   Do the following examples represent applications of random sampling or nonrandom sampling? Explain.

a. The constituents returning a questionnaire on U.S. foreign policy to their Congressman.

b. The apples in an enclosed 5-lb bag purchased at a local supermarket.

c. A 5-lb bag of apples selected by a housewife from a bin of apples at a local supermarket.

d. Department store queries to every tenth credit account customer about new store hours.

e. Warranty cards providing personal and demographic information received by a manufacturer from those who have recently purchased one of the manufacturer's small kitchen appliances.

16.35.   Discuss the advantages of conducting a sample survey instead of a census in each of the following instances:

a. A candidate for governor of the state of Illinois wishes to know the proportion of Illinois voters favoring his candidacy one week prior to the election.

b. A marketing representative for a breakfast food company is interested in determining the total first-year sales of a new packaged breakfast food the company has developed.

c. A local newspaper has adopted a more liberal news editorial policy. To obtain reader reaction to this change, an agent for the newspaper randomly selects ten local subscribers from a subscription list, contacts them by phone, and asks them for their opinion of the change in editorial policy.

d. To determine the proportion of residents favoring a municipal bond levy, the city manager proposes selecting every 25th individual listed in the city's telephone directory and obtaining his or her opinion by a telephone poll.

e. An oil company executive is interested in determining the average price per gallon of low-lead gasoline charged by its retail stations in the state of Kansas. From a list of stations the executive randomly selects 20 of the 249 retail stations and obtains their retail price by telephone.

16.36.   Refer to exercise 16.35. For each sampling problem suggest modifications to the sampling procedure which may lessen the chances of incurring bias and error in the survey.

16.37.   "Performance risk" relates to whether a product will perform as it should. "Psychological risk" concerns the consumer's perceived self-concept when purchasing the product. Advertising cannot affect the former but can affect the latter. A discount store in a city containing 745 households has adopted a new advertising theme designed to alleviate psychological risk associated with its merchandise. From a residential directory a simple random sample of $n = 50$ households is selected. One month after the new advertising campaign, the households are contacted, with 13 heads of the households indicating that they perceive the discount store's merchandise as inferior to that offered by competing stores. Estimate the true proportion of households that perceive psychological risk in the merchandise offered by the discount store, and place a bound on the error of estimation.

16.38.   The starting point for the development of an understanding of consumer behavior is consumer demography, the descriptive measures characterizing the buying public. From company records the manager of an automobile dealership has obtained a simple random sample of 25 customer files of the 582 customers who have purchased a certain small economy car in the past year. The mean and variance of the ages of the 25 customers are found to be $\bar{y} = 27.5$ and $s^2 = 16.81$. Estimate the average age of purchasers of the economy car from the dealership, and place a bound on the error of estimation.

16.39.   Test-marketing provides an opportunity to investigate the nature and degree of customer acceptance of a product by marketing the product in certain selected areas. The sales manager for a typewriter manufacturer wishes to know whether sufficient demand exists in a large city to justify adding a new electric portable typewriter to her stock. Currently she serves four large chains, consisting of 25, 20, 30, and 25 stores. For administrative convenience she decides to use stratified random sampling, with each chain of stores representing a stratum. The sales manager

	STRATUM		
1	2	3	4
16	10	5	17
12	17	18	11
10	12	13	12
13	6	15	15
9		20	18
		12	

has enough time and money to obtain sales data in only 20 stores. Using proportional allocation she randomly selects 5 stores from the first chain, 4 from the second, 6 from the third, and 5 from the fourth. After a month the sales figures are as shown in the table. Estimate the average monthly sales per store, and place a bound on the error of estimation.

16.40.   Refer to exercise 16.39. Estimate the total monthly sales if the new electric portable typewriter were sold in all 100 stores, and place a bound on the error of estimation.

16.41.   An insurance executive is concerned that the high inflation rate may leave many clients with insufficient fire insurance coverage for their homes. He has proposed an automatic escalator clause, which increases the client's coverage (and annual premiums) according to the annual inflation rate. To gain client reaction to the proposed policy, the insurance executive decides to select a stratified random sample of the clients served by his company in the three counties of his jurisdiction.

	COUNTY		
	A	B	C
Total Clients	231	407	187
Number Surveyed	21	37	17
Number Approving			
Escalator Policy	8	20	9

The results are shown in the table. Estimate the proportion p of clients in the three counties favoring the escalator policy, and place bounds on the error of estimation.

16.42.   A manufacturer of chain saws has received complaints from buyers about excessive repair costs. To study this problem the manufacturer wishes to estimate the average repair cost per saw per month for saws he has sold to logging and timber companies. He cannot obtain the repair cost for each saw, but he can find the total repair costs and the number of saws owned by different companies. Thus he decides to use cluster sampling, with each company as a cluster. From the $M = 87$ logging and timber companies that buy chain saws from the manufacturer, he selects a simple random sample of $m = 12$. The data in the table represent the repair costs during

COMPANY	NUMBER OF SAWS	REPAIR COSTS	COMPANY	NUMBER OF SAWS	REPAIR COSTS
1	4	$ 55	7	9	$103
2	7	83	8	1	15
3	5	47	9	8	110
4	11	210	10	11	164
5	15	235	11	7	80
6	6	88	12	10	146

the past month for each company. Estimate the average repair cost per saw for the past month, and place bounds on the error of estimation.

16.43.   Refer to exercise 16.42. Estimate the total amount spent by the 87 logging and timber companies on repair of the manufacturer's chain saws during the past month, and place bounds on the error of estimation.

16.44. Upon checking his sales records, the manufacturer cited in exercise 16.42 notes that he has sold 703 chain saws to the 87 logging and timber companies. Using this additional information, estimate the total amount spent on repairs by the 87 companies, and place a bound on the error of estimation. [Hint: If $\bar{y}_{cl}$ is the mean obtained by cluster sampling and N is the number of elements in the population, then $\hat{\tau} = N\bar{y}_{cl}$ and $\hat{\sigma}_{\hat{\tau}}^2 = N^2\hat{\sigma}_{\bar{y}_{cl}}^2$.] How do these results compare with those obtained in exercise 16.43?

16.45. According to a New York management consulting firm, employees in the United States and Canada presently steal more than $10 million a day in cash and merchandise from their employers. The president of a small manufacturing firm has commissioned a group to study employee theft in his firm. He wants to estimate the proportion of employees who have ever willfully stolen anything of value from the firm. Suppose you represent the group commissioned by the president. Develop an appropriate survey design to investigate the problem. Include an appropriate questionnaire in your analysis.

16.46. In an effort to budget the coming year's expenses, the manager of a company motor pool wishes to estimate the total annual maintenance costs for the 170 company cars currently operated by company personnel. Currently those operating the company cars pay for maintenance themselves and are reimbursed at the year's end. How many company employees with company cars should the manager contact if she wishes to estimate the total annual maintenance costs with a bound of $500 on the error of estimation? Assume that past evidence suggests that the standard deviation of annual maintenance costs is about $100 per car.

16.47. Time and motion studies are often used in industry to equate work loads, to determine bases for compensation, and to design machinery for assisting certain manual operations. The production manager of a large assembly plant is interested in knowing the average time spent on a certain assembly operation by the firm's $N = 225$ assembly employees. He feels that men and women probably show a difference in assembly time, so he wants to stratify on sex. It has been observed from previous studies that the assembly times for the 100 male assemblers range from 8 to 18 minutes, while for the female assemblers the range is from 5 to 12 minutes. Using proportional allocation find the approximate sample size necessary to estimate the average assembly time for all employees to within 1 minute. How many men and how many women should be included in the sample?

16.48. Under what conditions is it desirable to use systematic sampling as an alternative to simple random sampling? When should systematic sampling not be used?

16.49. Select an example from business or government that illustrates an effective application of systematic sampling. Select another example that illustrates a case where simple random sampling, but not systematic sampling, should be used to generate the survey information.

16.50. The assessment/sales (A/S) ratio is used as a standard measure of assessment performance by public and private assessment offices. In a jurisdiction requiring assessments at full market value, the A/S ratio should be near 1.0. When the A/S ratio exceeds 1.0, property taxation based on the assessments is said to be regressive. In a community with $N = 510$ residential properties, a random sample of $n = 10$ properties with recent sales prices is observed. The assessed valuations and sales prices are shown in the table. Assuming that the 10 sample properties represent a random sample of all residential properties from the community, compute the assessment/sales ratio for the community.

PROPERTY	ASSESSED VALUATION	SALES PRICE	PROPERTY	ASSESSED VALUATION	SALES PRICE
1	$27,000	$28,000	6	$46,000	$43,500
2	25,000	24,500	7	29,000	30,000
3	31,000	33,500	8	37,000	34,500
4	38,000	42,000	9	51,000	49,000
5	18,000	19,500	10	33,000	34,500

16.51.   Refer to exercise 16.50. Use ratio estimation to estimate the total assessed valuation of all residential properties in the community. If all residential properties in the community are taxed at the rate of $2\frac{1}{2}\%$ of assessed valuation, estimate the property tax revenue generated by the community from residential properties.

16.52.   What is meant by area sampling? Give an example from business or government where area sampling may prove useful in generating survey information.

16.53.   To improve typing service an executive of a large publishing company wishes to estimate the total volume of pages typed by secretaries in the company in one day. The company contains eight departments, each generating about the same volume of typed material per day. Each department employs approximately 25 secretaries, and the work output varies considerably from secretary to secretary. Using a two-stage cluster-sampling model, design a survey to answer the executive's question. Carefully identify the clusters and elements in the survey and specify whether you would suggest large or small numbers of each.

Experiences with Real Data

In this chapter we have considered a number of specific sample survey designs and their associated methods of estimation. The most basic design, simple random sampling, makes no attempt to reduce sampling costs but only insures that every possible sample of size n selected from the population is equally likely to be drawn. Stratified random sampling divides the population into homogeneous groups called strata, then selects a simple random sample from each stratum. The effect of the design is to minimize the cost of the survey while providing an estimator with a smaller variance than can be acquired by simple random sampling. Cluster sampling is used to reduce the costs of conducting the sample survey and is most appropriate when a frame is not available or when travel costs are considerable. We also briefly discussed systematic sampling, which often provides a less costly alternative to simple random sampling or stratified random sampling when the population elements are aligned, such as on file cards, or people as they pass through a door.

When choosing a sample survey design for a real application, it is seldom clear which design is best in spite of the guidelines for design selection that we have presented in our discussion. Exactly what is meant by a "homogeneous group" when identifying strata? When do travel costs become "considerable" in order to justify cluster sampling? How pronounced do periodicities have to be in the population to nullify systematic sampling? Such questions are often very difficult, if not impossible, to answer satisfactorily before conducting the survey. It is unreasonable, however, to expect that a survey must be conducted before it is determined whether the design used in the survey is appropriate. Experience is the best guide in establishing a policy for the selection of an appropriate sample survey design.

Conduct an experiment to compare the results when using simple random sampling, stratified random sampling, cluster sampling, and systematic sampling in a particular survey application. To conduct this experiment first divide your class into a number of groups, say 10 groups of students. Then select a topic to investigate in the survey. Perhaps you may wish to ask each respondent. "How much did you spend last week for entertainment?" Each different sample survey design should then be applied to the population of students within your college or university by each of the student groups from your class. The procedures for this experiment might be as follows:

1. Using the student directory, randomly select $n = 25$ students, contact them, and record their responses. This would comprise the simple random sample.
2. Consider each academic class (e.g., freshman, sophomore, junior, senior, graduate) as a separate stratum, stratify the population of students, and partition your sample of 25 among the strata. Using the student directory randomly select an appropriate number within each stratum, contact them, and record their responses.
3. Cluster the population of students according to the pages of your student directory. Then randomly select two or three pages from the directory, contact each student on every chosen page, and record their responses.
4. Conduct a systematic sample by choosing every 20th student passing a certain point on your campus. For instance, a student conducting the survey may station himself at the entrance to the campus student union, select each 20th student entering the building, and record the responses from those chosen in the survey. Use a sample size of $n = 25$.
5. Compute estimates of the mean and the variance, and place a bound on the error of estimation for the separate survey results obtained under each of the four research designs.

As a class, compare the estimates obtained under each survey design by each of the class groups. Does one of the survey designs appear to consistently exhibit smaller variation, and hence tighter bounds on the error of estimation, than the other designs? Which design appears to be the least advisable for the survey experiment you have conducted? On the basis of your experiment, can you offer advice on the selection of an appropriate sampling design when conducting a sample survey?

References

COCHRAN, W. G. *Sampling Techniques.* New York: Wiley, 1963.

GREENBERG, B. G.; R. T. KUEBLER, JR.; J. R. ABERNATHY; and D. G. HORVITZ. "Application of Randomized Response Technique in Obtaining Quantitative Data." *Journal of the American Statistical Association* 66 (1971): 243–250.

MENDENHALL, W.; L. OTT; and R. L. SCHEAFFER. *Elementary Survey Sampling.* Belmont, Calif.: Wadsworth, 1971.

PETERSON, G. E.; A. P. SOLOMON; H. MADJID; and W. C. APGAR. *Property Taxes, Housing, and the Cities.* Washington, D.C.: Heath Publishing, 1973.

chapter objectives

GENERAL OBJECTIVES Many types of business surveys and experiments result in the observation of qualitative rather than quantitative response variables. That is, the responses can be classified but not quantified. Consequently, data from these experiments will be the number of response observations falling in each of the data classes included in the experiment. The objective of this chapter is to present methods for analyzing count (or enumerative) data.

SPECIFIC OBJECTIVES

1. To present the properties of the multinomial experiment. Except for minor exceptions, which are noted, all the techniques presented in this chapter will assume that the sampling satisfies the properties of a multinomial experiment.
>> **Section 17.1**

2. To present the chi-square test, which will be used to test hypotheses concerning the class (or cell) probabilities.
>> **Section 17.2**

3. To present a test of an hypothesis concerning specific values of the cell probabilities.
>> **Section 17.3**

4. To present a method of testing the independence of two methods of classification.
>> **Sections 17.4, 17.5**

5. To suggest other applications of the general statistical test of section 17.2.
>> **Section 17.6**

chapter seventeen

Analysis of Enumerative Data

17.1 A Description of the Experiment

Many experiments, particularly in the social sciences, result in enumerative (or count) data. For instance, the classification of people into five income brackets results in an enumeration or count corresponding to each of the five income classes. Or we might be interested in studying the reaction of a sales prospect to a particular product promotional device. If the reactions of a prospect to the promotional device can be classified in one of three ways, and if a large number of prospects were subjected to the device, the experiment would yield three counts, indicating the number of prospects falling in each of the reaction classes. Similarly, a traffic study might require a count and classification of the type of motor vehicles using a section of highway. An industrial process manufactures items that fall into one of three quality classes: acceptable, seconds, and rejects. A student of the arts might classify paintings in one of k categories according to style and period in order to study trends in style over time. We might wish to classify ideas in a philosophical study, or style in the field of literature. The results of an advertising campaign would yield count data that indicate a classification of consumer reaction. Indeed, many observations in the physical sciences are not amenable to measurement on a continuous scale and result in enumerative or classificatory data.

The illustrations in the preceding paragraph exhibit, to a reasonable degree of approximation, the following characteristics, which define a **multinomial experiment**:

The Characteristics of a Multinomial Experiment

1. The experiment consists of n identical trials.
2. The outcome of each trial falls into one of k classes, or cells.
3. The probability that the outcome of a single trial will fall in a particular cell, say cell i, is p_i ($i = 1, 2, ..., k$) and remains the same from trial to trial. Note that

$$p_1 + p_2 + p_3 + \cdots + p_k = 1$$

4. The trials are independent.
5. The experimenter is interested in $n_1, n_2, n_3, ..., n_k$, where n_i ($i = 1, 2, ..., k$) is equal to the number of trials in which the outcome falls in cell i. Note that

$$n_1 + n_2 + n_3 + \cdots + n_k = n$$

A multinomial experiment is analogous to tossing n balls at k boxes (cells), where each ball will fall in one of the boxes. The boxes are arranged so that the probability that a ball will fall in a box varies from box to box but remains the same for a particular box in repeated tosses. Finally, the balls are tossed in such a way that the trials are independent. At the conclusion of the experiment, we observe n_1 balls in the first box, n_2 in the second, ..., and n_k in the kth. The total number of balls is

$$\sum_{i=1}^{k} n_i = n$$

Note the similarity between the binomial and multinomial experiments and, in particular, note that the binomial experiment represents the special case for the multinomial experiment when $k = 2$. The single parameter p of the binomial experiment is replaced by the k parameters $p_1, p_2, ..., p_k$ of the multinomial experiment. The object of this chapter is to make inferences about the cell probabilities $p_1, p_2, ..., p_k$. The inferences will be expressed in terms of a statistical test of an hypothesis concerning their specific numerical values or their relationship to one another.

If we proceeded as in chapter 6, we would derive the probability of the observed sample ($n_1, n_2, ..., n_k$) for use in calculating the probability of the type I and type II errors associated with a statistical test. Fortunately, we have been relieved of this chore by the British statistician Karl Pearson, who proposed a very useful test statistic for testing hypotheses concerning $p_1, p_2, ..., p_k$ and gave its approximate probability distribution in repeated sampling. This test statistic is discussed in the next section.

The Chi-square Test

Suppose that $n = 100$ balls are tossed at the cells (boxes) and that we know that p_1 is equal to .1. How many balls would be expected to fall in the first cell? Referring to chapter 6 and utilizing knowledge of the binomial experiment, we calculate

$$E(n_1) = np_1 = 100(.1) = 10$$

Similarly, the expected number falling in the remaining cells may by calculated by using the formula

$$E(n_i) = np_i \qquad i = 1, 2, ..., k$$

Now suppose that we hypothesize values for $p_1, p_2, ..., p_k$ and calculate the expected value for each cell. Certainly if our hypothesis is true, the cell counts n_i should not deviate greatly from their expected values np_i ($i = 1, 2, ..., k$). Hence it seems intuitively reasonable to use a test statistic involving the k deviations

$$n_i - np_i \qquad i = 1, 2, ..., k$$

In 1900 Karl Pearson proposed the following test statistic, which is a function of the square of the deviations of the observed counts from their expected values, weighted by the reciprocal of their expected value:

Chi-square Test Statistic

$$X^2 = \sum_{i=1}^{k} \frac{[n_i - E(n_i)]^2}{E(n_i)} = \sum_{i=1}^{k} \frac{(n_i - np_i)^2}{np_i}$$

Although the mathematical proof is beyond the scope of this text, it can be shown that when n is large, X^2 possesses, approximately, a chi-square probability distribution in repeated sampling. Experience has shown that the cell counts n_i should not be too small in order that the chi-square distribution provide an adequate approximation to the distribution of X^2. As a rule of thumb, **we will require that all expected cell counts equal or exceed 5**, although Cochran (see the references) has noted that this value can be as low as 1 for some situations.

You will recall from section 9.6 the use of the chi-square probability distribution for testing an hypothesis concerning a population variance σ^2. Particularly, we stated that the shape of the chi-square distribution would vary depending on the number of degrees of freedom associated with s^2. We discussed the use of table 5 in the appendix, which presents the critical values of χ^2 corresponding to various right-hand tail areas of the distribution. Therefore, we must know which χ^2 distribution to use—that is, the number of degrees of freedom—for approximating the distribution of X^2, and we must know whether to use a one-tailed or a two-tailed test in locating the rejection region for the test. The latter problem may be solved directly. Since large deviations of the observed cell counts from those expected tend to contradict the null hypothesis concerning the cell probabilities $p_1, p_2, ..., p_k$, we reject the null hypothesis when X^2 is large and employ a one-tailed statistical test, using the upper-tail values of χ^2 to locate the rejection region.

The determination of the appropriate number of degrees of freedom to be employed for the test can be rather difficult and therefore will be specified for the practical applications described in the following sections. In addition, we state here the principle involved (which is fundamental to the mathematical proof of the approximation) so that you may understand why the number of degrees of freedom changes with various applications. This principle states that the appropriate number of degrees of freedom equals the number k of cells less one degree of freedom for each independent linear restriction placed on the observed cell counts. For example, one linear restriction is always present because the sum of the cell counts must equal n; that is,

$$n_1 + n_2 + n_3 + \cdots + n_k = n$$

Other restrictions will be introduced for some applications because of the necessity for estimating unknown parameters required in the calculation of the expected cell frequencies or because of the method in which the sample is collected. These will become apparent as we consider various practical examples.

17.3 A Test of an Hypothesis Concerning Specified Cell Probabilities

The simplest hypothesis concerning the cell probabilities is one that specifies numerical values for each cell. For example, consider a customer preference study in which three different package designs are used to display the same product. We wish to test the hypothesis that the buyer has no preference in the choice of package design; that is, we wish to test

$$H_0: p_1 = p_2 = p_3 = \tfrac{1}{3}$$

against

$$H_a: \text{At least one } p_i \text{ is different from } \tfrac{1}{3}$$

where p_i is the probability that a customer will choose package design i ($i = 1, 2,$ or 3).

Suppose that the product to be packaged is a food product and that the three packages are displayed side by side in several supermarkets in a particular city. In one day's time it was noted that $n = 90$ customers purchased the product, of which $n_1 = 23$ purchased package design 1, $n_2 = 36$ purchased package design 2, and $n_3 = 31$ purchased package design 3. Thus $n_1 = 23$, $n_2 = 36$, and $n_3 = 31$ represent the observed cell frequencies for cells 1, 2, and 3. The expected cell frequencies are the same for each cell, namely,

$$E(n_i) = np_i = 90(\tfrac{1}{3}) = 30$$

The observed and expected cell frequencies are presented in table 17.1. Noting the discrepancy between the observed and expected cell frequency, we wonder whether the data present sufficient evidence to warrant rejection of the hypothesis of no preference.

Table 17.1 *Observed and expected cell counts for the customer preference study*

| | PACKAGE DESIGN | | |
	1	2	3
Observed Cell Frequency	$n_1 = 23$	$n_2 = 36$	$n_3 = 31$
Expected Cell Frequency	30	30	30

The chi-square test statistic for our example possesses $(k - 1) = 2$ degrees of freedom since the only linear restriction on the cell frequencies is that

$$n_1 + n_2 + \cdots + n_k = n$$

or, for our example,

$$n_1 + n_2 + n_3 = 90$$

Therefore, if we choose $\alpha = .05$, we will reject the null hypothesis when $X^2 > 5.991$ (see table 5 of the appendix).

Substituting into the formula for X^2, we obtain

$$X^2 = \sum_{i=1}^{k} \frac{[n_i - E(n_i)]^2}{E(n_i)} = \sum_{i=1}^{k} \frac{(n_i - np_i)^2}{np_i}$$

$$= \frac{(23 - 30)^2}{30} + \frac{(36 - 30)^2}{30} + \frac{(31 - 30)^2}{30} = 2.87$$

Since X^2 is less than the tabulated value 5.991 of χ^2, the null hypothesis is not rejected. We conclude that the data do not present sufficient evidence to indicate that the buyers of the area have a preference for a particular package design.

Exercises

17.1.   List the characteristics of a multinomial experiment.

17.2.   During a given day the manager of a large supermarket observed the number of shoppers choosing each of the market's six checkout stands. The observed

Stand Number	1	2	3	4	5	6
Frequency	84	110	146	152	61	47

results are given in the table. Do these data present sufficient evidence to indicate that some checkout stands were preferred over others? Use $\alpha = .05$.

17.3.   Over a two-year period the manager of a motel recorded the number of vacant rooms in his motel each evening. The relative frequencies of the occurrence of 0, 1, 2, ... vacant rooms enabled the manager to calculate the approximate probabilities given in the accompanying table. Since the manager recorded these data,

Number of Vacant Rooms	0	1	2	3	≥ 4
Probability	.10	.25	.35	.20	.10

a new motel has been built near the manager's motel. In the first 100 days since completion of the new motel, the manager recorded the numbers of room vacancies per day. These data are shown in the next table. Do these data present sufficient

Number of Vacant Rooms	0	1	2	3	≥ 4
Number of Days	3	16	35	25	21

evidence to indicate that the pattern of vacancies in the manager's motel has changed since the opening of the new motel? Test by using a 5% level of significance.

17.4.   Wolgast noted in 1958* that when it is time to buy a car, in 4% of the cases the wife is the family member who selects the car to be purchased; 31% of the time this decision is shared equally between the husband and the wife; 56% of the time the husband makes the decision; and 9% of the time the decision is made by someone else. To see if Wolgast's findings still hold true in light of the greater awareness and influence of today's woman, a marketing representative randomly selected 200 families that had recently purchased a new automobile. In 18 of the families the wife selected the car to be purchased; in 75 the decision was shared equally between the husband and wife; in 92 cases the husband made the decision; and in the remainder of cases the decision was made by someone else. Do these data contradict the findings reported by Wolgast in 1958? Use $\alpha = .05$.

*E. H. Wolgast, "Do Husbands or Wives Make Purchasing Decisions?" *Journal of Marketing*, October 1958.

17.4 Contingency Tables

A problem frequently encountered in the analysis of count data concerns the independence of two methods of classifying observed events. For example, suppose we wish to classify defects found on furniture produced in a manufacturing plant, first, according to the type of defect and, second, according to the production shift. Ostensibly, we wish to investigate a contingency—a dependence between the two classifications. Do the proportions of various types of defects vary from shift to shift?

A total of $n = 309$ furniture defects were recorded and the defects were classified according to one of four types: A, B, C, or D. At the same time each piece of furniture was identified according to the production shift in which it was manufactured. These counts are presented in table 17.2, which is known as a **contingency table**. (Note: Numbers in parentheses are the expected cell frequencies.)

Let p_A be the unconditional probability that a defect will be of type A. Similarly, define p_B, p_C, and p_D as the probabilities of observing the three other types of defects. Then these probabilities, which we will call the **column probabilities** of table 17.2, will satisfy the requirement

$$p_A + p_B + p_C + p_D = 1$$

In like manner, let $p_i (i = 1, 2,$ or $3)$ be the **row probability** that a defect will have occurred on shift i, where

$$p_1 + p_2 + p_3 = 1$$

Then if the two classifications are independent of each other, a cell probability will equal the product of its respective row and column probabilities in accordance with the multiplicative law of probability.

For example, the probability that a particular defect will occur in shift 1 and be of type A is $p_1 p_A$. Thus we observe that the numerical values of the cell probabilities are unspecified in the problem under consideration. The null hypothesis specifies only that each cell probability will equal the product of its respective row and column probabilities and therefore will imply indepen-

Table 17.2 *Contingency table classifying defects of furniture according to type and production shift*

	TYPE OF DEFECT				
SHIFT	A	B	C	D	TOTALS
1	15(22.51)	21(20.99)	45(38.94)	13(11.56)	94
2	26(22.99)	31(21.44)	34(39.77)	5(11.81)	96
3	33(28.50)	17(26.57)	49(49.29)	20(14.63)	119
Totals	74	69	128	38	309

dence of the two classifications. The alternative hypothesis is that this equality does not hold for at least one cell.

The analysis of the data obtained from a contingency table differs from the problem discussed in section 17.3, because we must *estimate* the row and column probabilities in order to estimate the expected cell frequencies.

If proper estimates of the cell probabilities are obtained, the estimated expected cell frequencies may be substituted for the $E(n_i)$ in X^2, and X^2 will continue to possess a distribution in repeated sampling that is approximated by the chi-square probability distribution. The proof of this statement, as well as a discussion of the methods for obtaining the estimates, is beyond the scope of this text. Fortunately, the procedures for obtaining the estimates, known as the method of maximum likelihood and the method of minimum chi-square, yield estimates that are intuitively obvious for our relatively simple applications.

It can be shown that the estimator of a column probability equals the column total divided by $n = 309$. If we denote the total of column j as c_j, then we have

$$\hat{p}_A = \frac{c_1}{n} = \frac{74}{309} \qquad\qquad \hat{p}_C = \frac{c_3}{n} = \frac{128}{309}$$

$$\hat{p}_B = \frac{c_2}{n} = \frac{69}{309} \qquad\qquad \hat{p}_D = \frac{c_4}{n} = \frac{38}{309}$$

Similarly, the row probabilities p_1, p_2, and p_3 may be estimated by using the row totals r_1, r_2, and r_3:

$$\hat{p}_1 = \frac{r_1}{n} = \frac{94}{309}$$

$$\hat{p}_2 = \frac{r_2}{n} = \frac{96}{309}$$

$$\hat{p}_3 = \frac{r_3}{n} = \frac{119}{309}$$

Denote the observed frequency of the cell in row i and column j of the contingency table by n_{ij}. Then the estimated expected value of n_{11} is

$$\hat{E}(n_{11}) = n[\hat{p}_1 \cdot \hat{p}_A] = n\left(\frac{r_1}{n}\right)\left(\frac{c_1}{n}\right) = \frac{r_1 c_1}{n}$$

where $\hat{p}_1 \cdot \hat{p}_A$ is the estimated cell probability. Similarly, we may find the estimated expected value for any other cell, say $\hat{E}(n_{23})$;

$$\hat{E}(n_{23}) = n[\hat{p}_2 \cdot \hat{p}_C] = n\left(\frac{r_2}{n}\right)\left(\frac{c_3}{n}\right) = \frac{r_2 c_3}{n}$$

Thus we see that the estimated expected value of the observed cell frequency n_{ij} for a contingency table is equal to the product of its respective row and column totals divided by the total frequency.

Estimated Expected Cell Frequency

$$\hat{E}(n_{ij}) = \frac{r_i c_j}{n}$$

The estimated expected cell frequencies for our example are shown in parentheses in table 17.2.

We may now use the expected and observed cell frequencies shown in table 17.2 to calculate the value of the test statistic:

$$X^2 = \sum_{i=1}^{12} \frac{[n_i - \hat{E}(n_i)]^2}{\hat{E}(n_i)}$$

$$= \frac{(15 - 22.51)^2}{22.51} + \frac{(26 - 22.99)^2}{22.99} + \cdots + \frac{(20 - 14.63)^2}{14.63} = 19.18$$

The only remaining obstacle is the determination of the appropriate number of degrees of freedom associated with the test statistic. We will give this as a rule that we will attempt to justify. The degrees of freedom associated with a contingency table possessing r rows and c columns will always equal $(r - 1)(c - 1)$. Thus for our example we will compare X^2 with the critical value of χ^2 with $(r - 1)(c - 1) = (3 - 1)(4 - 1) = 6$ degrees of freedom.

You will recall that the number of degrees of freedom associated with the X^2 statistic equals the number of cells (in this case, $k = rc$) less one degree of freedom for each independent linear restriction placed on the observed cell frequencies. The total number of cells for the data of table 17.2 is $k = 12$. From this we subtract one degree of freedom because the sum of the observed cell frequencies must equal n; that is,

$$n_{11} + n_{12} + \cdots + n_{34} = 309$$

In addition, we used the cell frequencies to estimate three of the four column probabilities. Note that the estimate of the fourth column probability will be determined once we have estimated p_A, p_B, and p_C, because

$$p_A + p_B + p_C + p_D = 1$$

Thus we lose $(c - 1) = 3$ degrees of freedom for estimating the column probabilities.

Finally, we used the cell frequencies to estimate $(r - 1) = 2$ row probabilities and therefore we lose $(r - 1) = 2$ additional degrees of freedom. The total number of degrees of freedom remaining is

$$\text{d.f.} = 12 - 1 - 3 - 2 = 6$$

In general, we see that the total number of degrees of freedom associated with an $r \times c$ contingency table is

$$\text{d.f.} = rc - 1 - (c - 1) - (r - 1) = (r - 1)(c - 1)$$

Therefore, if we use $\alpha = .05$, we will reject the null hypothesis that the two classifications are independent if $X^2 > 12.592$. Since the value of the test statistic, $X^2 = 19.18$, exceeds the critical value of χ^2, we reject the null hypothesis. The data present sufficient evidence to indicate that the proportion of the various types of defects varies from shift to shift. A study of the production operations for the three shifts would probably reveal the cause.

Example 17.1 The introduction of unit pricing in food stores makes it easier for shoppers to choose cheaper items. However, Isakson and Maurizi* have found that lower-income shoppers do not appear to be taking proper advantage of unit pricing, perhaps because they do not understand the unit-pricing labeling system as well as do middle-income and upper-income shoppers.

In a follow-up study to provide a check on the authors' findings, an economist observed the purchase selection of $n = 1000$ shoppers in a large supermarket. The supermarkets were located in three different areas of a city where the buyers were, respectively, of lower-income, middle-income, and upper-income families. Packages of the same brand but with different unit prices were placed adjacent to one another on the store shelves. The data for the $n = 1000$ shoppers, classified according to purchase decision and income class, are shown in table 17.3. Do these data

Table 17.3 *Data for example 17.1*

	INCOME CLASS			
	Lower	Middle	Upper	TOTALS
Understand Labels	249 (259.6)	494 (490.9)	201 (193.5)	944
Do Not Understand Labels	26 (15.4)	26 (29.1)	4 (11.5)	56
Totals	275	520	205	1000

present sufficient evidence to support the findings of Isakson and Maurizi?

Solution The problem asks whether the data provide sufficient evidence to indicate a dependence between a buyer's income classification and his or her ability to properly interpret unit-pricing labels. We therefore analyze the data as a contingency table.

The estimated expected cell frequencies may be calculated by using the appropriate row and column totals in the formula

*H. R. Isakson and A. R. Maurizi, "The Consumer Economics of Unit Pricing," *Journal of Marketing* (1973).

$$\hat{E}(n_{ij}) = \frac{r_i c_j}{n}$$

Thus we have

$$\hat{E}(n_{11}) = \frac{r_1 c_1}{n} = \frac{(944)(275)}{1000} = 259.6$$

$$\hat{E}(n_{12}) = \frac{r_1 c_2}{n} = \frac{(944)(520)}{1000} = 490.9$$

and so on. These estimated values are shown in parentheses in table 17.3.

The value of the test statistic X^2 can now be computed and compared with the critical value of χ^2 possessing $(r - 1)(c - 1) = (1)(2) = 2$ degrees of freedom. Then for $\alpha = .05$, we will reject the null hypothesis when $X^2 > 5.991$. Substituting into the formula for X^2, we obtain

$$X^2 = \frac{(249 - 259.6)^2}{259.6} + \frac{(494 - 490.9)^2}{490.9} + \cdots + \frac{(4 - 11.5)^2}{11.5} = 13.26$$

Observing that X^2 falls in the rejection region, we reject the null hypothesis of independence of the two classifications. A comparison of the percentage of shoppers in each income class who properly interpret the unit-pricing labels suggests that the lower-income classes are not properly taking advantage of unit pricing. Therefore, the Isakson and Maurizi findings have been supported by the economist's survey.

Exercises

17.5.   Are there some companies whose securities should be systematically excluded in an investment portfolio for social, political, or moral reasons? Should the investment manager select securities for his clients' portfolios to further such noneconomic goals? Some researchers have noted that reaction to these questions varies considerably from one fund management group to another.* To explore this issue fund managers employed by four of the major fund management agencies were asked whether they approve or disapprove of selecting securities for a portfolio to further social, political, or moral goals. The number of fund managers falling in each of the eight possible categories is shown in the table. Do these data present sufficient

	FUND MANAGEMENT AGENCY			
	A	B	C	D
Favor Noneconomic Goals	7	6	11	13
Do Not Favor Non-economic Goals	23	31	30	41
Totals	30	37	41	54

*B. G. Malkiel and R. E. Quandt, "Moral Issues in Investment Policy," *Harvard Business Review*, March–April 1971.

evidence to assume that the fraction of fund managers favoring the use of noneconomic goals in portfolio selection differs in the four fund management agencies? Use $\alpha = .10$.

17.6.   Many insurance companies are beginning to question the policy of offering reduced rates to owners of subcompact cars since the companies claim that their rate of serious and fatal accidents is higher than that of owners of larger-sized cars. To investigate this issue an analysis of accident data was made to determine the distribution of numbers of cases in which at least one individual was either fatally or critically injured for automobiles of three sizes. The data for 346 accidents are

	SIZE OF AUTO		
	Subcompact	Compact	Full Size
Fatal or Critical Injury	67	26	16
No Fatal or Critical Injury	128	63	46

shown in the table. Do these data indicate that the frequency of fatal and critical injuries in auto accidents depends on the size of automobiles?

17.7.   To encourage energy conservation many public utility corporations are now offering reduced rates to their consumers who reduce their energy usage by a certain fixed percentage each month.* A public utility that serves three communities has offered reduced rates to its consumers who reduce their monthly electrical consumption by at least 15%. After the first month of the energy conservation program, 510 subscribers in the three communities were selected at random for analysis. The observed results are given in the table. Do these data present sufficient evidence

	COMMUNITY		
	Appleton	Beechwood	Cherryville
Reduced Consumption by More Than 15%	16	63	29
Did Not Reduce Consumption by More Than 15%	72	202	118

to indicate a difference in the proportion of subscribers participating in the energy conservation program in the three communities? Use $\alpha = .05$.

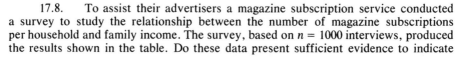

17.8.   To assist their advertisers a magazine subscription service conducted a survey to study the relationship between the number of magazine subscriptions per household and family income. The survey, based on $n = 1000$ interviews, produced the results shown in the table. Do these data present sufficient evidence to indicate

*Environmental Protection Agency, *Environmental News*, July 1975.

NUMBER OF SUBSCRIPTIONS PER HOUSEHOLD	FAMILY INCOME			
	Less Than $8,000	$8,000–$11,999	$12,000–$15,999	At Least $16,000
0	28	54	78	23
1	29	151	301	73
2	10	31	69	57
more than 2	5	19	40	32

that the number of magazine subscriptions per household depends on family income? Test at the $\alpha = .10$ level of significance.

17.5 $r \times c$ Tables with Fixed Row or Column Totals

In the preceding section we described the analysis of an $r \times c$ contingency table by using examples which, for all practical purposes, fit the multinomial experiment described in section 17.1. While the methods of collecting data in many surveys may obviously satisfy the requirements of a multinomial experiment, other methods do not. For example, we might not wish to sample randomly the population described in example 17.1 because we might find that, owing to chance, one income category contains a very small number of people (or even might fail to appear in the sample). Thus we might decide beforehand to interview a specified number of people in each column category, thereby fixing the column totals in advance.

While these restrictions tend to disturb somewhat our visualization of the experiment in the multinomial context, they have no effect on the analysis of the data. As long as we wish to test the hypothesis of independence of the two classifications, and none of the row or column probabilities is specified in advance, we may analyze the data as an $r \times c$ contingency table. It can be shown that the resulting X^2 possesses a probability distribution in repeated sampling that is approximated by a chi-square distribution with $(r - 1)(c - 1)$ degrees of freedom.

Example 17.2 Because of the almost universal presence of TV sets in American households and because of television's unique combination of sight and sound, television offers an ideal media for advertisers whose products require demonstration. This is especially true with advertisements designed for children, as their attention span and reading comprehension is typically not sufficient so that they can be reached effectively by written sources. But how do children react to commercials? Ward conducted a study to determine whether a relationship exists between a child's age and his or her degree of understanding of certain se-

Table 17.4 *Data for example 17.2*

LEVEL OF UNDERSTANDING	5-7	8-10	11-12	TOTALS
	AGE			
I	55 (35.7)	37 (35.7)	15 (35.7)	107
II	35 (48.3)	50 (48.3)	60 (48.3)	145
III	10 (16)	13 (16)	25 (16)	48
Totals	100	100	100	300

lected TV commercials.* Three hundred children were selected in the study, with the children equally divided among three age categories. The results of the study are shown in table 17.4, with level I of understanding implying practically no understanding of the message of the commercial. Do these data present sufficient evidence to indicate that the level of understanding of TV commercials is related to a child's age?

Solution You will observe that the test of an hypothesis concerning a lack of relationship between a child's age and his or her level of understanding of a TV commercial is identical to the test of an hypothesis implying independence of the row and column classifications. Suppose we denote the fraction of children with level I of understanding by p_1 and the fraction with level II of understanding by p_2. If we hypothesize that p_1 and p_2 are the same for all children of ages 5-12, we imply that the fraction of children possessing level III of understanding is $p_3 = 1 - p_1 - p_2$ and that the row probabilities are p_1, p_2, and p_3, respectively. The probability that a child from the sample of $n = 300$ children falls in any one of the three age classes is $\frac{1}{3}$ since this was fixed in advance. Then the cell probabilities for the table are obtained by multiplying the appropriate row and column probabilities under the null hypothesis. Our test is then equivalent to a test of independence of the two classifications.

The estimated expected cell frequencies, calculated using the row and column totals, appear in parentheses in table 17.4. We then find that

$$X^2 = \sum_{i=1}^{9} \frac{[n_i - \hat{E}(n_i)]^2}{\hat{E}(n_i)}$$

$$= \frac{(55 - 35.7)^2}{35.7} + \frac{(37 - 35.7)^2}{35.7} + \cdots + \frac{(25 - 16)^2}{16} = 36.9$$

The critical value of χ^2 for $\alpha = .05$ and $(r - 1)(c - 1) = (2)(2) = 4$ degrees of freedom is 9.488. Since X^2 exceeds this critical value, we reject the null hypothesis and conclude that children's understanding of TV

*S. Ward, "Children's Reactions to Commercials." Reprinted from the *Journal of Advertising Research*. © Copyright 1972, by the Advertising Research Foundation.

commercials is not the same over the age span 5–12. In fact, their under-standing appears to increase with age.

Exercises

17.9.   A common objective of all investors is to select an investment portfolio with minimum possible risk. Typically risk is measured by the fluctuation of security prices—volatile securities are risky, securities with stable prices are not risky. An investor is interested in studying the risk behavior of security prices from different industrial classes. In the course of her study, she randomly selects 50 manufacturing companies, 30 retail chains, and 25 utilities and records their average weekly price change over the past year. The results are shown in the table. Do these data present

AVERAGE PRICE CHANGE	INDUSTRIAL CLASS		
	Manufacturing	Retailing	Utility
less than 2 pts. (low risk)	27	9	5
from 2 to 5 pts. (moderate risk)	16	13	10
more than 5 pts. (high risk)	7	8	10
Totals	50	30	25

sufficient evidence to assume that security price risk varies according to the industrial class of the issuing firm? Test with $\alpha = .05$.

17.10.   By tradition U.S. labor unions have been content to leave the management of the company to the managers and corporate executives. But in Europe worker participation in management decision making is an accepted idea and one that is continually spreading. To study the effect of worker satisfaction with worker participation in managerial decision making, 100 workers were interviewed in each of two separate West German manufacturing plants. One plant had active worker participation in managerial decision making; the other did not. Each selected worker was asked whether he or she generally approved of the managerial decisions made within the firm. The results of the interviews are shown in the table.

	PARTICIPATIVE DECISION MAKING	NO PARTICIPATIVE DECISION MAKING
Generally Approve of the Firms Decisions	73	51
Do Not Approve of the Firm's Decisions	27	49

a. Do the data provide sufficient evidence to indicate that approval or disapproval of management's decisions depends on whether workers participate in decision making? Test by using the X^2 test statistic. Use $\alpha = .05$.

b. Do these data support the hypothesis that workers in a firm with participative decision making more generally approve of the firm's managerial decisions than those employed by firms without participative decision making? Test by using the z test presented in section 8.14. This problem requires a one-tailed test. Why?

17.11.   Usually the small investor attempting to develop a personal investment portfolio considers only stocks or bonds. There is, however, a third possibility that is especially appealing in times preceding an anticipated bull market. Convertible bonds—bonds that can be converted into stocks—offer the stable return of a bond when the securities market is bearish and offer the opportunity to convert to stocks should the issuing firm's security prices begin to rise. To properly advise his clients, an investment banker contacted 50 investment analysts from each of three investment firms that are members of the New York stock exchange. Each was asked whether he or she currently advise investing in stocks, bonds, or convertible bonds. The

	INVESTORS EMPLOYED BY FIRM		
ADVISE INVESTING IN	A	B	C
stocks	13	16	7
bonds	31	24	35
convertibles	6	10	8
Totals	50	50	50

responses are shown in the table. Do these data present sufficient evidence to indicate that the three investment firms differ in their advice for investors? Use $\alpha = .10$.

17.12.   Refer to exercise 17.11. Do these data provide sufficient evidence to indicate that the proportion of investment officers favoring investment in convertible bonds differs significantly for the three investment firms? Use $\alpha = .10$.

17.6   Other Applications

The applications of the chi-square test for analyzing enumerative data described in sections 17.3, 17.4, and 17.5 represent only a few of the interesting classification problems that may be approximated by the multinomial experiment and for which our method of analysis is appropriate. By and large, these applications are complicated to a greater or lesser degree because the numerical values of the cell probabilities are unspecified and hence require the estimation of one or more population parameters. Then, as in sections 17.4 and 17.5, we can estimate the cell probabilities. Although we omit the mechanics of the statistical tests, several additional applications of the chi-square test are worth mentioning.

For example, suppose we wish to test an hypothesis stating that a population possesses a normal probability distribution. The cells of a sample frequency histogram (for example, figure 3.2) correspond to the k cells of the multinomial experiment. The observed cell frequencies are the number of measurements falling in each cell of the histogram. Given the hypothesized normal probability distribution for the population, we could use the areas under the normal curve to calculate the theoretical cell probabilities and hence the expected cell frequencies. The difficulty arises when μ and σ are unspecified for the normal population and must be estimated to obtain the estimated cell probabilities. This difficulty, of course, can be surmounted.

The construction of a two-way contingency table to investigate dependency between two classifications can be extended to three or more classifications. For example, if we wish to test the mutual independence of three classifications, we can employ a three-dimensional "table" or rectangular parallelepiped. The reasoning and methodology associated with the analysis of both the two- and three-way tables are identical, although the analysis of the three-way table is a bit more complex.

A third and interesting application of our methodology is its use to investigate the rate of change of a multinomial (or binomial) population as a function of time. For example, we might study the decision-making ability of a human (or any animal) as he is subjected to an educational program and tested over time. If, for instance, he is tested at prescribed intervals of time and the test is of the yes-no type, yielding a number y of correct answers that follows a binomial probability distribution, we would be interested in the behavior of the probability p of a correct response as a function of time. If the number of correct responses was recorded for c time periods, the data would fall into a $2 \times c$ table similar to that in example 17.1 (section 17.4). We would then be interested in testing the hypothesis that p is equal to a constant—that is, that no learning has occurred—and we would then proceed to more interesting hypotheses to determine whether the data present sufficient evidence to indicate a gradual (say, linear) change over time as opposed to an abrupt change at some time. The procedures we have described can be extended to decisions involving more than two alternatives.

You will observe that our learning example is common to business, to industry, and to many other fields, including the social sciences. For example, we might wish to study the rate of consumer acceptance of a new product for various types of advertising campaigns as a function of the length of time that the campaign has been in effect. Or we might wish to study the trend in the lot fraction defective in a manufacturing process as a function of time. Both of these examples, as well as many others, require a study of the behavior of a binomial (or multinomial) process as a function of time.

The examples we have just described are intended to suggest the relatively broad application of the chi-square analysis of enumerative data, a fact that should be borne in mind by the experimenter concerned with this type of data. The statistical test employing X^2 as a test statistic is often called a "goodness of fit" test. Its application for some of these examples requires care in the determination of the appropriate estimates and the number of degrees of freedom for X^2, which, for some of these problems, may be rather complex.

17.7 Summary

The preceding material concerned a test of an hypothesis regarding the cell probabilities associated with a multinomial experiment. When the number n of observations is large, the test statistic X^2 can be shown to possess, approximately, a chi-square probability distribution in repeated sampling, the number of degrees of freedom being dependent on the particular application. In general, we assume that n is large and that the minimum expected cell frequency is equal to or greater than 5.

Several words of caution concerning the use of the X^2 statistic as a method of analyzing enumerative data are appropriate. The determination of the correct number of degrees of freedom associated with the X^2 statistic is very important in locating the rejection region. If the number is incorrectly specified, erroneous conclusions might result. Also, note that nonrejection of the null hypothesis does not imply that it should be accepted. We would have difficulty in stating a meaningful alternative hypothesis for many practical applications and, therefore, would lack the knowledge of the probability of making a type II error. For example, we hypothesize that the two classifications of a contingency table are independent. A specific alternative would have to specify some measure of dependence, which may or may not possess practical significance to the experimenter. Finally, if parameters are missing and the expected cell frequencies must be estimated, the estimators of missing parameters should be of a particular type in order that the test be valid. In other words, the application of the chi-square test for other than the simple applications outlined in sections 17.3, 17.4, and 17.5 requires experience beyond the scope of this introductory presentation of the subject.

Supplementary Exercises

17.13. A city expressway utilizing four lanes in each direction was studied to see whether drivers preferred to drive on the inside lanes. A total of 1000 automobiles were observed during the heavy early morning traffic and their respective lanes recorded. The results are shown in the table. Do the data present sufficient evidence to indicate

Lane	1	2	3	4
Observed Count	294	276	238	192

that some lanes were preferred over others? Test the hypothesis that $p_1 = p_2 = p_3 = p_4 = 1/4$, using $\alpha = .05$.

17.14. A manufacturer of buttons wishes to determine whether the fraction of defective buttons produced by three machines varies from machine to machine. Samples of 400 buttons are selected from each of the three machines and the number of defectives counted for each sample. The results are shown in the table. Do these data present sufficient evidence to indicate that the fraction of defective buttons varies from machine to machine? Test with $\alpha = .05$.

Machine Number	1	2	3
Number of Defectives	16	24	9

17.15. A manufacturer of floor polish conducted a consumer preference experiment to see whether a new floor polish, A, was superior to those of four of his competitors. A sample of 100 housewives viewed five patches of flooring that had received the five polishes and each indicated the patch that she considered superior in appearance. The lighting, background, and so forth, were approximately the same for all five patches. The results of the survey are given in the table. Do

Polish	A	B	C	D	E
Frequency	27	17	15	22	19

these data present sufficient evidence to indicate a preference for one or more of the polished patches of floor over the others? If you were to reject the hypothesis of no preference for this experiment, would this imply that polish A is superior to the others? Can you suggest a better method for conducting the experiment?

17.16. A printer is interested in examining the relationship between the number of printing errors and the type size used. He selects three different books recently printed by his company, each printed using a different type size. From each book he records the number of pages with printing errors and the number of error-free

	TYPE SIZE		
	A	B	C
Pages with Errors	23	17	41
Pages Without Errors	241	183	210

pages. The results are shown in the table. Do the data indicate a dependence between type size and printing errors?

17.17. A quality control engineer for a factory wishes to examine the operating efficiency of two assembly machine operators. The operators are in charge of the same machine but work during different shifts. During a given week the number of good and defective finished items produced by the assembly machine while each operator was on duty are as shown in the table. Do these data present sufficient

	OPERATOR A	OPERATOR B
Good	551	416
Defective	16	17

evidence to indicate that the operators work with about the same efficiency? (Test the hypothesis that the proportion of defective items produced by each operator is the same.)
  a. Test by using the X^2 statistic. Use $\alpha = .05$.
  b. Use the z test from section 8.14.

17.18.   Following the lead of Oregon and Vermont, the Environmental Protection Agency is supporting the concept of national legislation to control nonreturnable beverage containers.* The enactment of such legislation usually results in reduction of roadside litter, and, in addition, EPA officials estimate that the equivalent of 92 thousand barrels of oil a day could be saved in reduced energy consumption by mandatory deposit legislation. To examine regional differences in opinion toward the EPA proposal, an EPA official randomly chose 100 public officials each from four sectors of the United States. Their responses are recorded in the table. Do

	REGION IN WHICH THE PUBLIC OFFICIAL RESIDES			
	West	South	Midwest	Northeast
Approve of Mandatory Deposit Legislation	54	48	45	39
Disapprove of Mandatory Deposit Legislation	46	52	55	61

these data present sufficient evidence to indicate a difference in the proportion of public officials in the various regions who favor national legislation to control nonreturnable beverage containers? Test by using a 5% level of significance.

17.19.   A survey of the opinions of the stockholders of a corporation about a proposed merger was taken to determine whether the resulting opinion was independent of the number of shares held. Two hundred stockholders were interviewed, with the results shown in the table. Do these data present sufficient evidence to indicate

	OPINION		
SHARES HELD	In Favor	Opposed	Undecided
under 100	37	16	5
100–500	30	22	8
over 500	32	44	6

that the opinions of the stockholders concerning the merger were dependent on the number of shares held by the stockholder? Test with $\alpha = .05$.

17.20.   Suppose the responses for the data in exercise 17.19 were reclassified according to whether the stockholder was male or female, as shown in the table.

	OPINION		
SEX	In Favor	Opposed	Undecided
Female	39	46	9
Male	60	36	10

Do these data present sufficient evidence to indicate that stockholder reaction to

*Environmental Protection Agency, *Environmental News*, October 1975.

the proposed merger depended on whether the stockholder was male or female? Test with $\alpha = .05$.

17.21. A study of the purchase decisions for three stock portfolio managers A, B, and C was conducted to compare the rates of stock purchases that resulted in profits over a time period that was less than or equal to one year. One hundred randomly selected purchases obtained for each of the managers showed the results

	MANAGER		
	A	B	C
Purchases Resulting in a Profit	63	71	55
Purchases Resulting in No Profit	37	29	45
Totals	100	100	100

given in the table. Do these data provide evidence of differences among the rates of successful purchases for the three portfolio managers? Test with $\alpha = .05$.

17.22. A carpet company is interested in comparing the fraction of new home builders favoring carpet over other floor coverings for homes in three different areas of a city. The objective is to decide how to allocate sales effort to the areas. A

	AREA		
	1	2	3
Carpet	69	126	16
Other Materials	78	99	27

survey is conducted, and the resulting data are given in the table. Do the data indicate a difference in the percentage favoring carpet from one region of the city to another?

17.23. A pharmaceutical manufacturer conducted a study to determine the effectiveness of a new drug (serum) for arthritis. The study involved the comparison of two groups, each consisting of 200 arthritic patients. One group was inoculated with the serum, while the other received a placebo (an inoculation that appears to contain serum but actually is nonactive). After a period of time, each person in the study was asked to state whether his or her arthritic condition was improved.

	TREATED	UNTREATED
Improved	117	74
Not Improved	83	126

The observed results are shown in the table. Do these data present sufficient evidence to indicate that the serum was effective in improving the condition of arthritic patients?
a. Test by using the X^2 statistic. Use $\alpha = .05$.
b. Use the z test of section 8.14.

17.24.   A company selling air brushes is about to mount a new advertising campaign. Before doing so it wishes to determine if the campaign should be directed to the male or female market individually, or to both together. A sample cf 1750 users of the air brush revealed the distribution shown in the table of regular and occasional users, by sex. Do these data present sufficient evidence to indicate that

TYPE OF USER	MALE	FEMALE
regular	170	465
occasional	475	640

the use of the air brush on an occasional or regular basis is related to the sex of the user? Use $\alpha = .01$.

17.25.   A study was conducted to determine whether individuals earning over $20,000 per year use the services of an accountant in the preparation of their income tax returns at the same rate in different regions of the United States. Four states were selected as representative of the four regions Northeast, South, Midwest, and West. From each state a random selection of individuals with annual incomes in excess of $20,000 was obtained. Each individual was then asked whether he or she uses the services of an accountant in the preparation of income taxes. The results

	RHODE ISLAND	FLORIDA	IOWA	CALIFORNIA
Use an Accountant	46	121	63	108
Prepares Own Taxes	149	179	178	192

are given in the table. Do the data indicate a difference in the rate of use of the services of an accountant from one region to another? Test with $\alpha = .05$.

17.26.   A survey was conducted by an auto repairman to determine whether various auto ills depended on the make of the auto. His survey, restricted to this year's model, showed the results given in the table. Do these data present sufficient

	TYPE OF REPAIR		
MAKE	Electrical	Fuel Supply	Other
A	12	19	7
B	14	7	9
C	6	21	12
D	33	44	19
E	7	9	6

evidence to indicate a dependency between auto makes and type of repair for these new-model cars? Note that the repairman did not utilize all the information available when he conducted his survey. In conducting a study of this type, what other factors should be recorded?

17.27.   The city manager of a midwestern city conducted a survey to determine whether the incidence of various types of crime varied from one part of the city to another. The city was partitioned into three regions and the crimes classified as homicide, car theft, grand larceny, petty larceny, and other. An analysis of 1599

CITY REGION	HOMICIDE	CAR THEFT	GRAND LARCENY (NEGLECTING AUTO THEFT)	PETTY LARCENY	OTHER
1	12	239	191	122	47
2	17	163	278	201	54
3	7	98	109	44	17

cases showed the results given in the table. Do these data present sufficient evidence to indicate that the occurrence of various types of crime depends on city region?

17.28.   In search of fair treatment and to eliminate discrimination in banking practices against women, female financiers are launching their own savings and lending institutions in many cities across the country.* These savings institutions are largely owned and directed by women and intend to provide women with easier access to credit and banking careers. The president of a women's bank in a large eastern city conducted a survey to determine the attitudes of women of different ages toward the concept of a bank owned and operated by women. The results of her survey of 480 women from five different age groups are shown in the table. Do these data

	AGE GROUP				
	21–30	31–40	41–50	51–60	Over 60
Support Concept of a Women's Bank	52	67	38	34	35
Do Not Support Concept of a Women's Bank	43	43	48	71	49
Totals	95	110	86	105	84

present sufficient evidence to assume that the proportion of women favoring the concept of a bank owned and operated by women differs among women from different age groups? Use $\alpha = .10$.

17.29.   During times of business decline and recession, many suggestions are offered to spur the economy into a turnaround. A survey was conducted among

OPINION	BUSINESS EXECUTIVES	ECONOMISTS	GOVERNMENT OFFICIALS
increase government spending	10	15	39
cut personal income taxes	37	37	33
decrease interest rates	24	34	15
offer tax incentives to business	29	14	13
Totals	100	100	100

*Adapted from "Now a Rush by Women to Start Their Own Banks," *U.S. News and World Report*, 27 October 1975.

100 business executives, 100 economists, and 100 government officials to find the opinion of each regarding the best way of reversing the trend of a business decline. Their responses are shown in the table. Do these data present sufficient evidence to assume that opinion regarding the best way to spur the economy into a turnaround during times of recession differs among business executives, economists, and government officials? Test by using a 5% level of significance.

17.30. Myers and Mount question the usual assumption that ability to buy, as measured by income, and social class identify similar market segments for a wide variety of consumer durables and services.* They note, for instance, that members of lower social classes often tend to purchase large cars and other durables similar to those purchased by members of a higher income and social class. To examine this issue 1000 recent purchasers of new automobiles were categorized according to their social class and to the price class of automobile purchased. The results of

	SOCIAL CLASS				
AUTOMOBILE PRICE CLASS	1	2	3	4	5
low-priced compact	39	169	198	24	11
moderately priced intermediate	25	93	221	42	17
high-priced luxury automobile	15	19	53	51	23
Totals	79	281	472	117	51

the study are shown in the table. Do these data present sufficient evidence to indicate that the price class of automobile purchased depends on social class? Test at a 5% level of significance.

Experiences with Real Data

The age of consumerism has created a new awareness on the part of business toward truth in packaging, honesty in advertising, fair pricing, and consumer protection through expressed or written warranties and product safety testing. The consumer movement also affects service-oriented businesses by demanding quality service at a fair price from professed service experts. The movement clearly calls for responses from business, government, and consumers and provides ample evidence that many improvements need to be made by business firms.

Some have offered the notion that consumerism is a temporary phenomenon—that it is primarily political and not economic in nature. Others see consumerism as the ongoing search of consumers to obtain better values for their money, a search that will continue indefinitely. Most tend to agree, however, that responses to consumerism differ according to the vested interests of the respondent—clearly the housewife views the quality of a laundry detergent from a different perspective than that of the detergent manufacturer.

Conduct separate surveys to determine opinions from different segments of your community on the nature of consumerism and how they visualize consumerism as part of the free enterprise system. Use the procedures of chapter 16 to select

*Adapted from J. H. Myers and J. F. Mount, "More on Social Class vs. Income as Correlates of Buying Behavior," *Journal of Marketing,* October 1973.

a random sample of 60 students from your college or university campus, 60 local professional businesspeople, and 60 local residents who are neither students on your campus or professional businesspeople. Contact half of the people chosen within each group and ask them to respond to the statement,

Consumerism is primarily:

a. political in nature
b. economic in nature
c. social in nature
d. other.

Then record the responses of this group in a table constructed as follows:

OPINION	STUDENTS	BUSINESS	TOWNSPEOPLE
political in nature			
economic in nature			
social in nature			
other			

Contact the remaining people not interviewed in the first survey and ask them to respond to the statement

Consumerism is an attempt to preserve the free enterprise economy by making the market work better.

a. Agree.
b. Disagree.
c. Uncertain.

Display these data in a table constructed as follows:

OPINION	STUDENTS	BUSINESS	TOWNSPEOPLE
agree			
disagree			
uncertain			

After recording your results, compute a chi-square statistic for each data set. Is there sufficient evidence to indicate that opinions about the nature of consumerism vary among students, businesspeople, and townspeople? Is there sufficient evidence to indicate that students, businesspeople, and townspeople differ in their agreement about consumerism's role in a free enterprise economy? You may wish to compare your results with those found by R. M. Gaedeke in "What Business, Government and Consumer Spokesmen Think about Consumerism," from the *Journal of Consumer Affairs* (Summer 1970).

The two studies described above are probably best conducted as a class project rather than as individual projects because of the volume of work required. The calculations can be accomplished on an electronic desk calculator, but you may wish to use an available library computer program to perform your calculations. You should contact your campus computer center to see if a chi-square analysis subroutine is available or to see which software packages are available for use on your campus computer. Useful library programs for performing a chi-square analysis are available in the Biomed (UCLA Health Services Computing Facility), the SAS (Statistical Analysis System), and the SPSS (Statistical Package for the Social Sciences) software packages.

References

ANDERSON, R. L., and T. A. BANCROFT. *Statistical Theory in Research.* New York: McGraw-Hill, 1952.

COCHRAN, W. G. "The X^2 Test of Goodness of Fit." *Annals of Mathematical Statistics* 23 (1952): 315–345.

DIXON, W. J., and F. J. MASSEY, JR. *Introduction to Statistical Analysis.* 3d ed. New York: McGraw-Hill, 1969.

PIERCE, A. *Fundamentals of Nonparametric Statistics.* Belmont, Calif.: Dickenson Publishing, 1970.

chapter objectives

GENERAL OBJECTIVES Chapters 8 and 9 presented statistical techniques for comparing two populations by comparing their respective population parameters. The techniques are applicable to data measured on a continuum and, in chapter 9, to data possessing normal population relative frequency distributions. The purpose of this chapter is to present several statistical test procedures for comparing populations for the many types of business data that do not satisfy these assumptions.

SPECIFIC OBJECTIVES

1. To explain the difference between parametric and nonparametric tests.
Section 18.1

2. To explain why nonparametric tests are useful and, sometimes, essential.
Section 18.1

3. To present a quick and easy method for comparing two populations—the sign test.
Section 18.2

4. To present a nonparametric test for comparing two populations based on independent random samples—the Mann-Whitney U test.
Section 18.3

5. To present a nonparametric test for comparing two populations for a paired-difference experiment—the Wilcoxon rank sum test.
Section 18.4

6. To present a nonparametric test to detect nonrandomness in a sequence—the runs test.
Section 18.5

7. To present a nonparametric test for detecting correlation between two variables—Spearman's r_s.
Section 18.6

chapter eighteen
Nonparametric Statistics

18.1 Introduction

Some experiments yield response measurements that defy quantification. That is, they generate response measurements that can be ordered (ranked), but the location of the response on a scale of measurement is arbitrary. Although experiments of this type occur in almost all fields of study, they are particularly evident in social science research and in studies of consumer preference. For example, suppose that a judge is employed to evaluate and rank the sales abilities of four salespeople, the edibility and taste characteristics of five brands of cornflakes, or the relative appeal of five new automobile designs. Clearly it is impossible to give an exact measure of sales competence, palatability of food, or design appeal. Thus the response measurements here differ markedly from those presented in preceding chapters. The judge's scale of measurement may be a Likert scale,* a semantic differential, or an ordinal scale of his own design.

Nonparametric statistical methods are useful not only for analyzing ranked data but also for the case where only directional differences are available. That is, a judge may indicate his preference between a pair of test items but may not be willing or able to indicate a measure of the magnitude of his preference.

The word "nonparametric" evolves from the type of hypothesis usually tested when dealing with ranked data of the type described above. **Parametric hypotheses** are those concerned with the population parameters. In contrast, many times **nonparametric hypotheses** are concerned with the *form* of the

---

*See the footnote on page 678.

population frequency distribution. Tests of hypotheses concerning the binomial parameter p, tests concerning μ or σ^2, and analysis of variance tests are parametric. On the other hand, an hypothesis stating that a particular population possesses a normal distribution (without specification of parameter values) is nonparametric. Similarly, an hypothesis stating that the distributions for two populations are identical is nonparametric.

The latter hypothesis is pertinent to the three ranking problems previously described. Even though we do not have an exact measure of sales competence, we can imagine that one exists and that in repeated performances a given salesperson would generate a population of such measurements. An hypothesis stating that the four probability distributions for the populations (associated with the four salespeople) are identical would imply no difference in the sales ability of the salespeople. Similarly, we can imagine that a scale of palatability for cornflakes does exist (even if unknown to us) and that a population of responses representing the reactions of a very large set of prospective consumers corresponds to each brand. An hypothesis stating that the distributions of palatability for the five brands are identical implies no difference in consumer preference for these products.

Nonparametric statistical procedures apply to data other than those that are difficult to quantify. They are particularly useful in making inferences in situations where serious doubt exists about the assumptions underlying standard methodology. For example, the t test for comparing a pair of means (presented in section 9.4) is based on the assumption that both populations are normally distributed with equal variances. Now, admittedly, the experimenter never knows whether these assumptions hold in a practical situation, but he will often be reasonably certain that departures from the assumptions are small enough so that the properties of his statistical procedure will be undisturbed. That is, α and β will be approximately what he thinks they are. On the other hand, it is not uncommon for the experimenter to question his assumptions and to wonder whether he is using a valid statistical procedure. This difficulty may be avoided by using a nonparametric statistical test, thereby avoiding reliance on an uncertain set of assumptions.

In this chapter we will discuss only the most common situations in which nonparametric methods are used. For a more complete treatment of the subject, see the texts by Siegel (1956) or Conover (1971) listed in the references.

18.2 The Sign Test for Comparing Two Population Distributions

Suppose that we wish to compare consumer ratings (on a scale of 1 to 10) of two window cleaners. Six housewives are randomly selected from the consumer group and each rates one window treated with Cleaner A and

Table 18.1 *Consumer ratings for window cleaners*

	CLEANER	
HOUSEWIFE	A	B
1	10	7
2	7	5
3	8	7
4	5	2
5	7	6
6	9	6

one with Cleaner B. A 10 rating is "best." As you can see, we have used a paired-difference experiment (a randomized block design) to make the comparison. The cleaners A and B are treatments. We have asked each housewife to examine both cleaners in order to block out housewife-to-housewife variation.

The data for the paired-difference experiment are shown in table 18.1. Do the data present sufficient evidence to indicate a difference in consumer preference for the two cleaners?

No complicated statistical test is needed to answer the question. Indeed, we can use a rough-and-ready nonparametric test procedure, known as the *sign test*, that can almost be performed "by eye." That is, we note that the rating for Cleaner A exceeds the rating for B for all six housewives (thus the signs of the six differences are all positive). Assuming no difference between the cleaners, this result is equivalent to flipping a balanced coin six times and observing six heads (or tails). The probability of observing such an event, $(\frac{1}{2})^6 + (\frac{1}{2})^6 = \frac{1}{32}$, is quite small. Hence we would likely reject an hypothesis that the distributions of consumer preferences for the two cleaners are identical.

You will recall that we employed the same nonparametric statistical test as an alternative procedure for determining whether evidence existed to indicate a difference in the mean wear for the two types A and B of tires in the paired-difference experiment of section 9.5. Each pair of responses was compared and y (the number of times A exceeded B) was used as the test statistic. **This nonparametric test is called a sign test because y is the number of positive (or negative) signs associated with the differences.** The implied null hypothesis is that the two population distributions are identical and the resulting technique is completely independent of the form of the distribution of differences. Thus regardless of the distribution of differences, the probability that A exceeds B for a given pair is $p = .5$ when the null hypothesis is true (that is, when the distributions for A and B are identical). Then y possesses a binomial probability distribution, and a rejection region for y can be obtained by using the binomial probability distribution of chapter 6. In fact, the sign test uses exactly the same test procedure that was discussed in detail in section 6.8. Before reading further you may wish to review chapter 6 and, particularly, section 6.8.

We illustrate the use of the sign test with an example.

Example 18.1 In a market research experiment, a food-processing company undertook a study to determine the acceptability of a sugar substitute in their canned orange juice. Eleven families were given a liberal supply of both the current product (A) and the new one with the sugar substitute (B) and asked to use them over a four-week period and to state their preference. The results are shown in table 18.2.

Table 18.2 *Data for example 18.1*

FAMILY	PRODUCT		PREFERRED PRODUCT
	A	B	
1	−	+	B
2	−	+	B
3	+	−	A
4	−	+	B
5	0	0	no preference
6	−	+	B
7	−	+	B
8	+	−	A
9	−	+	B
10	−	+	B
11	−	+	B

Assume that the 11 families are representative of potential users of the company's product. Let y denote the number of families preferring the current product (A). Do the data present sufficient evidence to indicate a preference for one of the two orange drinks, the orange juice with a sugar substitute (B) or the current orange juice (A)? State the null hypothesis to be tested and use y as a test statistic.

Solution Because each pair of preferences A and B corresponds to a particular family, we see that the data were collected by using a paired-difference experiment. Consequently, the observed responses are paired as they appear in the data tabulation. Let y be the number of times that a family indicates a preference for Product A. Under the hypothesis that the two products are equally preferred, the probability p that A is preferred over B for a given family is $p = .5$. Or, equivalently, we wish to test an hypothesis that the binomial parameter p equals .5.

What should be done in case of ties between the paired responses? This difficulty is avoided by omitting tied pairs and thereby reducing the number n of pairs. In this example the effective sample size is 10, not 11.

Very large or very small values of y are certainly contradictory to the null hypothesis. Therefore, the rejection region for the test will be located by including the most extreme values of y that also provide an α that is feasible for the test.

Suppose that we wish α to be about .05 or .10. We would begin the se-

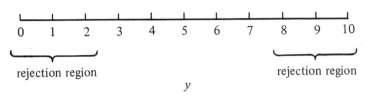

Figure 18.1 *Rejection region for example 18.1*

lection of the rejection region by including $y = 0$ and $y = 10$ and calculating the α associated with this region, using $p(y)$ [the probability distribution for the binomial random variable, chapter 6]. With $n = 10$ and $p = .5$, we refer to table 1 in the appendix and find that

$$\alpha = p(0) + p(10) = .002$$

Since the value of α is too small, the rejection region is expanded by including the next pair of y-values that are most contradictory to the null hypothesis, namely, $y = 1$ and $y = 9$. The value of α for this rejection region ($y = 0, 1, 9, 10$) is obtained from table 1 in the appendix.

$$\alpha = p(0) + p(1) + p(9) + p(10) = .022$$

This α is also too small, so we again expand the region. This time we include $y = 0, 1, 2, 8, 9, 10$. You may verify that the corresponding value of α is .11. Assuming that this value of α is acceptable to the experimenter (and to you), we will employ $y = 0, 1, 2, 8, 9, 10$ as the rejection region for the test. (See figure 18.1.)

From the data we observe that $y = 2$ and therefore we reject the null hypothesis. We conclude that sufficient evidence exists to indicate that for the population of buyers the two products are not equally preferred. In fact, it appears that a greater preference exists for Product B, the orange juice with the sugar substitute. The probability of rejecting the null hypothesis when it is true is only $\alpha = .11$, and therefore we are reasonably confident of our conclusion.

In example 18.1 we used the sign test as a rough and rapid tool for detecting a preference between two products. The rather large value of α does not disturb us. We can collect additional data (by allowing more families to conduct the experiment) if we are concerned about making a type I error in reaching our conclusion.

The values of α associated with the sign test can be obtained by using the normal approximation to the binomial probability distribution, as discussed in section 7.5. You can verify (by comparing exact probabilities with their approximations) that these approximations will be quite adequate for n as small as 10. This is due to the symmetry of the binomial probability distribution for $p = .5$.

For $n \geq 25$ you can conduct the sign test, saving time and effort, by using the z statistic (chapters 7 and 8) as a test statistic, where

$$z = \frac{y - np}{\sqrt{npq}} = \frac{y - .5n}{.5\sqrt{n}}$$

In using z you are testing the null hypothesis $p = .5$ against either the alternative $p \neq .5$, for a two-tailed test, or the alternatives $p > .5$ or $p < .5$, for a one-tailed test. The tests utilize the familiar rejection regions discussed in chapter 8.

Sign Test for Large Samples ($n \geq 25$)

Null hypothesis H_0: $p = .5$ (one treatment is not preferred to a second treatment, and vice versa).

Alternative hypothesis H_a: $p \neq .5$, for a two-tailed test. (Note: We use the two-tailed test as an example. Many analyses might require a one-tailed test.)

Test statistic: $z = \dfrac{y - .5n}{.5\sqrt{n}}$.

Rejection region: Reject H_0 if $z \geq z_{\alpha/2}$ or if $z \leq -z_{\alpha/2}$, where $z_{\alpha/2}$ is the z-value from table 3 of the appendix corresponding to an area of $\alpha/2$ in the upper tail of the normal distribution.

Example 18.2 A production superintendent claims that there is no difference in the employee accident rates for the day and evening shifts in a large manufacturing plant. The number of accidents per day is recorded for both the day and evening shifts for $n = 100$ days. It is found that the number of accidents per day for the evening shift exceeded the corresponding number of accidents on the day shift on 63 of the 100 days. Do these results provide sufficient evidence to indicate a difference in the distributions of accident rates for the two shifts?

Solution This is a paired-difference experiment, with $n = 100$ pairs of observations corresponding to the 100 days. To test the null hypothesis that the two distributions of accidents are identical, we use the test statistic

$$z = \frac{y - .5n}{.5\sqrt{n}}$$

where y represents the number of positive differences in daily accident rates. Then for $\alpha = .05$ we will reject the null hypothesis if $z \geq 1.96$ or $z \leq -1.96$. Substituting into the formula for z, we have

$$z = \frac{y - .5n}{.5\sqrt{n}} = \frac{63 - (.5)(100)}{.5\sqrt{100}} = \frac{13}{5} = 2.60$$

Since the calculated value of z exceeds $z_{\alpha/2} = 1.96$, we reject the null hypothesis. The data provide sufficient evidence to indicate a difference in the accident rate distributions for the day and evening shifts.

The sign test is most often employed for observations that have been randomly selected in pairs, using a paired-difference experiment. And the sign test is one of the few tests that can be employed when the only information available is that an A observation exceeds a B observation (or vice versa). However, the sign test can also be used to compare two population distributions where samples of equal size have been randomly, and independently, selected from the two populations. Then the pairs are formed by randomly matching each observation in sample A with an observation in sample B.

Some statisticians object to the use of the sign test for the independent random samples case because, in some situations, the rejection or nonrejection of the null hypothesis of "no difference between the two populations" would depend on the outcome of the random pairings. That is, one set of random pairings might lead to rejection of the null hypothesis; another might indicate nonrejection. This fact may disturb some users but the test is valid and the probability α of rejecting the null hypothesis when it is true will be exactly the value set by the experimenter.

The important point to note about the sign test is that sometimes it is one of the few tests you can use (see preceding comments) for comparing two populations. If all the observations can be ranked in order of their relative magnitudes, better tests (the rank sum tests of sections 18.4 and 18.5) are available for comparing two populations. But none is easier or faster than the sign test. Consequently, the sign test might be regarded as a "reader's test," the kind of test that you might use to evaluate (or eyeball) data in a technical report or a news article. It is not the best test for ranked observations but it is good, and it will enable you to evaluate data rapidly. If a difference between two population distributions is apparent when you use the sign test, it will be more apparent when using the rank sum tests.

Exercises

18.1. Using the tabulated binomial probabilities listed in table 1, find the significance levels between $\alpha = .01$ and $\alpha = .15$ that are available when using $n = 10$ paired observations. What are the corresponding rejection regions? Assume that you are using a one-tailed test.

18.2. Repeat the instructions in exercise 18.1 for sample sizes $n = 15, 20,$ and 25.

18.3. Much skepticism appears to exist about reported EPA mileage figures on new cars because the government tests only simulate actual driving conditions while ignoring such factors as wind, road conditions, traffic intensity, and driver differences. A private testing agency compared the EPA mileage rating of 11 new makes of automobiles with the mileage recorded on a test run under actual driving conditions. The results are shown in the table. (A indicates that the mileage obtained

Car Make	1	2	3	4	5	6	7	8	9	10	11
Test Results	E	O	E	A	E	E	E	E	E	A	E

under actual driving conditions was greater than the EPA figure; E indicates that the EPA mileage rating was greater; 0 indicates that the two mileage ratings were equal.)

Do the findings of the independent testing agency indicate that the EPA ratings are, on the average, higher than what you can expect to obtain in gasoline mileage while driving under normal driving conditions? Use a test with a level of significance as near as possible to 10%.

18.4.   Each of $n = 27$ prominent economists was asked whether he or she believes federal control of domestic oil companies would be in the best interests of the American economy. Fourteen of the economists indicated they favored controls for oil companies, 11 were opposed to controls, and 2 said they were not certain. Do these responses indicate that economists generally favor federal control of domestic oil companies as a policy in the best interests of the American economy? Test by using a level of significance as near as possible to 5%.

18.5.   The data in the table represent the amount spent on advertising (in thousands of dollars) on network television and on spot television by 10 of the top

ADVERTISER	NETWORK TELEVISION	SPOT TELEVISION
General Foods	$44,642	$49,259
Colgate-Palmolive	46,507	36,860
American Home Products	40,791	26,355
Lever Brothers	38,554	20,893
Philip Morris	36,685	11,491
Coca-Cola	15,527	16,944
Sears, Roebuck & Company	15,273	18,960
Gillette	27,479	16,320
General Mills	24,152	17,940
Kraftco	18,359	13,181

Source: Advertising Age, 21 June 1971.

national advertisers for 1970. Use the sign test, with a level of significance as near as possible to 5%, to determine whether these data suggest that the selected advertisers show a preference for one of the two television media sources over the other.

18.3 The Mann-Whitney U Test: Two Populations and Independent Random Samples

The sign test for comparing two population distributions ignores the actual magnitudes of the paired observations and thereby discards information that would be useful in detecting a departure from the null hypothesis. A statistical test that partially circumvents this loss by utilizing the relative magnitudes of the observations was proposed by Mann and Whitney and is equivalent to a test proposed independently by Wilcoxon.

Assume that you have independent random samples of sizes n_1 and n_2 from two populations, say A and B. The first step in finding the Mann-Whitney U statistic is to rank all $(n_1 + n_2)$ observations in order of magnitude, assigning a 1 to the smallest observation, a 2 to the second smallest, and so on. Ties in the observations can be handled by averaging the ranks that would have been assigned to the tied observations and assigning this average to each. Then calculate the sums of the ranks, T_A and T_B, for the two samples.

For example, suppose you have two random samples selected from populations A and B as shown in the table. The ranks corresponding to

A	B
28	33
31	29
27	35
25	30

these eight measurements are shown in the next table in parentheses. The rank sums $T_A = 12$ and $T_B = 24$ are also shown.

A		B	
28	(3)	33	(7)
31	(6)	29	(4)
27	(2)	35	(8)
25	(1)	30	(5)
$T_A = 12$		$T_B = 24$	

Since the ranks tend to be of the same relative magnitude as the original measurements, the rank sums T_A and T_B should not be too far apart for two identical population relative frequency distributions. On the other hand, the greater the difference in T_A and T_B, the greater the evidence to indicate a difference in the population relative frequency distributions for A and B. This information is used in the Mann-Whitney U statistic.

The formula for the Mann-Whitney U statistic can be given in terms of T_A or of T_B, one value of U being larger than the other, but the sum of the two U-values will always equal $n_1 n_2$. Since it is easier to construct a table of probabilities of U for only one tail of the U distribution, we will agree always to use the smaller value of U as a test statistic. The formulas for the two values of U, which we will denote as U_A and U_B, are as follows:

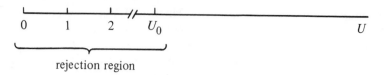

Figure 18.2 *Rejection region for a Mann-Whitney U test*

Formulas for the Mann-Whitney *U* Statistic

$$U_A = n_1 n_2 + \frac{n_1 (n_1 + 1)}{2} - T_A$$

$$U_B = n_1 n_2 + \frac{n_2 (n_2 + 1)}{2} - T_B$$

where $U_A + U_B = n_1 n_2$.

Table 8 of the appendix gives the probability that an observed value of *U* will be less than some specified value, say U_0. (For more comprehensive tables of the *U* statistic, see the source listed in the references.) This is the value of α for a one-tailed test. To find the value of α for a two-tailed test, double the tabulated probability. The smaller the value of *U*, the greater the evidence to indicate a difference between the population relative frequency distributions for A and B. The rejection region for the *U* statistic is shown in figure 18.2.

For example, suppose that $n_1 = 4$ and $n_2 = 5$. Then you would consult the third table of table 8, the table corresponding to $n_2 = 5$. The first few lines of table 8 for $n_2 = 5$ are shown in table 18.3.

Across the top of table 18.3 you see values of n_1. Values of U_0 are shown down the left side of the table. The entries give the probability that *U* will assume a small value—namely, the probability that $U \le U_0$. Since for this example $n_1 = 4$, you move across the top of the table to $n_1 = 4$.

Table 18.3 *An abbreviated version of table 8 of the appendix;*
$P(U \le U_0)$ *for* $n_2 = 5$

	n_1	1	2	$n_2 = 5$ 3	4	5
	0	.1667	.0476	.0179	.0079	.0040
	1	.3333	.0952	.0357	.0159	.0079
U_0	2	.5000	.1905	.0714	.0317	.0159
	3		.2857	.1250	.0556	.0278
	4		.4286	.1964	.0952	.0476
	5		.5714	.2857	.1429	.0754
	⋮		⋮	⋮	⋮	⋮

Move to the third row of the table corresponding to $U_0 = 2$ for $n_1 = 4$. Then you see the probability that U will be less than or equal to 2 is .0317. Similarly, moving across the row for $U_0 = 3$, you see that the probability that U is less than or equal to 3 is .0556. (This value is shaded in table 18.3.) So if you conduct a one-tailed Mann-Whitney U test with $n_1 = 4$ and $n_2 = 5$ and wish α to be near .05, you would reject the null hypothesis of equality of population relative frequency distributions when $U \leq 3$. The probability of a type I error for the test is $\alpha = .0556$.

Mann-Whitney *U* Test

Null hypothesis H_0: The relative frequency distributions for the two populations A and B are identical.

Alternative hypothesis H_a: The two population relative frequency distributions are not identical (a two-tailed test). Or the relative frequency distribution for population A is shifted to the right (or left) of the relative frequency distribution for population B (a one-tailed test).

Test statistic: U, the smaller of

$$U_A = n_1 n_2 + \frac{n_1 (n_1 + 1)}{2} - T_A$$

$$U_B = n_1 n_2 + \frac{n_2 (n_2 + 1)}{2} - T_B$$

Rejection region: For a given value of α,

1. for a two-tailed test reject H_0 if $U \leq U_0$, where $P(U \leq U_0) = \alpha/2$ [note that U_0 is the value such that $P(U \leq U_0)$ is equal to half of α];
2. for a one-tailed test reject H_0 if $U \leq U_0$, where $P(U \leq U_0) = \alpha$.

We illustrate the use of the Mann-Whitney U test with an example.

Example 18.3 A large accounting firm, with branch offices in a number of major cities, devised two different CPA study programs to assist junior accountants with their preparation for an upcoming CPA examination. To compare the effectiveness of the study programs, 8 large branch offices were selected by the firm and 50 junior accountants chosen from each office. The accountants from 4 offices were trained under Program A, the accountants from the other 4 under Program B. Following the completion of the CPA examination, the number of accountants passing the CPA exam was recorded. The results are shown in the table. Do these data

A	B
28	33
31	29
27	35
25	30

present sufficient evidence to indicate a difference in the population distributions for training programs A and B?

Solution The ranks are shown in the next table alongside the $(n_1 + n_2) = 8$ measurements. The rank sums for the two samples are also shown.

A		B	
28	(3)	33	(7)
31	(6)	29	(4)
27	(2)	35	(8)
25	(1)	30	(5)
$T_A = 12$		$T_B = 24$	

Then

$$U_A = n_1 n_2 + \frac{n_1(n_1 + 1)}{2} - T_A = (4)(4) + \frac{(4)(4 + 1)}{2} - 12 = 14$$

$$U_B = n_1 n_2 - U_A = 16 - 14 = 2$$

Since the question we wish to answer implies a two-tailed test and table 8 gives values of $P(U \le U_0)$ for specified sample sizes and values of U_0, we must double the tabulated value to find α. Suppose that we desire a value of α near .05. Checking table 8 for $n_1 = n_2 = 4$, we find $P(U \le 1) = .0286$. Using $U \le 1$ as the rejection region, α will equal $2(.0286) = .0572$, or, rounding to three decimal place, $\alpha = .057$. Because the smaller observed value of U is 2 (calculated above), U does not fall in the rejection region. Hence there is not sufficient evidence to show a difference in the population distributions for training programs A and B.

Example 18.4 Accounting errors are classified as those due to controllable effects (recording errors, computational errors, etc.) and those due to random effects (usually the result of the sampling). It is important that an accounting firm establish proper controls to eliminate all possible errors due to controllable effects while minimizing the impact of random factors. A governmental accounting office, responsible for auditing the books of governmental contractors, is evaluating a new method of auditing in an effort to reduce the number of errors found in its audits. To evaluate the quality of the new technique, 9 accounts were chosen for the experiment. Eighteen auditors were randomly assigned to the accounts, two to each, so that each account could be examined by using each of the

Table 18.4 *Data for example 18.4*

A		B	
125	(15)	89	(1)
116	(11)	101	(7)
133	(18)	97	(4)
115	(10)	95	(3)
123	(14)	94	(2)
120	(12)	102	(8)
132	(17)	98	(5.5)
128	(16)	106	(9)
121	(13)	98	(5.5)
$T_A = 126$		$T_B = 45$	

two (the new and the old) different auditing techniques, A and B. The data, the number of errors resulting from the use of the current auditing method A and the number of errors resulting from the new method B under study, are shown in table 18.4. Test the hypothesis of no difference in the distribution of the material (accounting) errors that result under the two different auditing techniques.

Solution   The data for this example are shown in table 18.4. The rank associated with each observation is given alongside in parentheses. Observe that although this is not a blocked (paired) experiment, each A observation exceeds its corresponding B measurement. And even though we do not have unique pairs, regardless of how the A and B measurements might be paired, the A measurement would exceed its B counterpart in nine of nine pairs. The sign test would then indicate rejection of the hypothesis of no difference in the population distributions of the number of material errors detected by the two auditing techniques. Certainly this procedure, which required less than 30 seconds to discuss, is more rapid than the calculations involved in the Student's *t* test of chapter 9. So if you can see something that is recognizable with the naked eye (the sign test), why use a microscope (a more powerful test)?

Although we have tested our hypothesis using the sign test, we will show that the same result can be quickly obtained using the more powerful Mann-Whitney U test. Thus for $n_1 = n_2 = 9$, we have

$$U_A = n_1 n_2 + \frac{n_1(n_1 + 1)}{2} - T_A = (9)(9) + \frac{(9)(10)}{2} - 126 = 0$$

From table 8 for $n_1 = n_2 = 9$, $P(U = 0) = .000$. The value of α for a two-tailed test and a rejection region of $U = 0$ and $U = 81$ is $\alpha = .000$. The calculated value of $U_A = 0$, falls in this rejection region. So we reject the null hypothesis, and with $\alpha = .000$ we feel very confident in our decision.

A simplified large-sample test can be obtained by using the familiar z statistic of chapter 8. When the population distributions are identical, it can be shown that the U statistic has expected value and variance of

$$E(U) = \frac{n_1 n_2}{2} \quad \text{and} \quad \sigma^2_U = \frac{n_1 n_2 (n_1 + n_2 + 1)}{12}$$

The distribution of

$$z = \frac{U - E(U)}{\sigma_U}$$

tends to normality, with mean 0 and variance equal to 1, as n_1 and n_2 become large. This approximation will be adequate when n_1 and n_2 are both larger than, say, 10. Thus for a two-tailed test with $\alpha = .05$, we would reject the null hypothesis if $|z| \geq 1.96$.

Mann-Whitney U Test for Large Samples

$(n_1 > 10$ and $n_2 > 10)$

Null hypothesis H_0: The relative frequency distributions for populations A and B are identical.

Alternative hypothesis H_a: The two population relative frequency distributions are not identical (a two-tailed test). Or the relative frequency distribution for population A is shifted to the right (or left) of the relative frequency distribution for population B (a one-tailed test).

Test statistic:
$$z = \frac{U - (n_1 n_2 / 2)}{\sqrt{n_1 n_2 (n_1 + n_2 + 1)/12}}$$

(Note: For the U statistic you can use either U_A or U_B.)

Rejection region: For a given value of α,

1. for a two-tailed test reject H_0 is $z \geq z_{\alpha/2}$ or if $z \leq -z_{\alpha/2}$;
2. for a one-tailed test place all of α in one tail of the z distribution and reject H_0 if $z \geq z_\alpha$ for an upper one-tailed test, or if $z \leq -z_\alpha$ for a lower one-tailed test.

Tabulated values of z are given in table 3 of the appendix.

What constitutes an "adequate" approximation is a matter of opinion. You will observe that by using the z statistic you will reach the same conclusion as when using the U test for example 18.4. Thus

$$z = \frac{0 - [(9)(9)/2]}{\sqrt{(9)(9)(9 + 9 + 1)/12}} = \frac{-40.5}{\sqrt{128.25}} = -3.58$$

This value of z falls in the rejection region ($|z| \geq 1.96$) and hence agrees with the sign test and the U test using the exact tabulated values for the rejection region.

Example 18.5 The U.S. Bureau of Census reports that all 50 states require taxing jurisdictions to evaluate the effectiveness of property valuations. This is done by an analysis of the ratios of assessed value (A) to sale value (S) for properties contained in the jurisdiction.* Ideally the A/S ratio is constant for all properties in a jurisdiction. When it is not, inequities may result in the property taxation structure.

A tax consultant wishes to compare the ratio of assessed value to sales value for properties in two sections of a city. An analysis of 24 property sales, $n_1 = 11$ from section 1 and $n_2 = 13$ from section 2, were randomly selected from the two sections of the city. The assumed value for each property was obtained from tax records and the ratio of assessed value to sale value was computed. The ratios for the two sections of the city are shown in table 18.5. Ranks of the ratios are shown in parentheses. Do

Table 18.5 *Data for example 18.5*

SECTION 1		SECTION 2	
.55	(9)	.49	(5)
.67	(17.5)	.68	(19)
.43	(1)	.59	(10.5)
.51	(6.5)	.72	(21)
.48	(3.5)	.67	(17.5)
.60	(12)	.75	(22.5)
.71	(20)	.65	(14.5)
.53	(8)	.77	(24)
.44	(2)	.62	(13)
.65	(14.5)	.48	(3.5)
.75	(22.5)	.59	(10.5)
		.51	(6.5)
		.66	(16)
$T_1 = 116.5$		$T_2 = 183.5$	

these data provide sufficient evidence to indicate a difference in the distributions of the ratios of assessed value to sale value for properties in the two sections of the city? Test a null hypothesis of no difference, using $\alpha = .05$.

Solution Since n_1 and n_2 are both larger than 10, we use the z test. Then we have that

$$E(U) = \frac{n_1 n_2}{2} = \frac{(11)(13)}{2} = 71.5$$

*U.S. Bureau of Census, *Trends in Assessed Valuations and Sales Ratios, 1956-1966*, State and Local Government and Special Studies, no. 54, March 1970.

$$\sigma^2_U = \frac{n_1 n_2 (n_1 + n_2 + 1)}{12} = \frac{(11)(13)(11 + 13 + 1)}{12} = 297.92$$

$$\sigma_U = \sqrt{297.92} = 17.3$$

The observed value of U is

$$U_2 = n_1 n_2 + \frac{n_2 (n_2 + 1)}{2} - T_2 = (11)(13) + \frac{(13)(14)}{2} - 183.5 = 50.5$$

We will test the null hypothesis that the distributions of ratios for the two sections of the city are identical against the alternative that they differ. Since this alternative implies a two-tailed test, we will reject the null hypothesis if $|z| \geq 1.96$ ($\alpha = .05$).

The observed value of the test statistic is

$$z = \frac{U - E(U)}{\sigma_U} = \frac{50.5 - 71.5}{17.3} = -1.21$$

Since z does not fall in the rejection region, there is not sufficient evidence to indicate a difference in the ratios of assessed value to sale value for properties located in the two sections of the city.

Why use the Mann-Whitney U test for example 18.5 rather than the parametric test employing the Student's t? Distributions of ratios often tend to be nonnormal and hence there is a good possibility that the requirements of the Student's t test are not met.

Parametric test procedures, using the z statistic (section 8.14) or the t statistic (section 9.4), are appropriate to test the hypothesis of no difference between the two population distributions if the assumptions underlying the tests are satisfied. If you have doubts on this point, the Mann-Whitney U test is a very good alternative.

Exercises

18.6.   The outputs from two filler machines in a food-processing plant were examined in order to compare their fill levels. The measurements given in the table represent the fluid ounces of content from five filled containers selected from the

Machine A	30.5	30.2	30.0	31.2	30.7
Machine B	30.9	31.0	31.5	31.4	31.3

output of each machine. Do these data present sufficient evidence to indicate a difference in the population of fill levels for the two machines? Use a level of significance of $\alpha = .10$.

a. Use the Mann-Whitney U test.
b. Use the Student's t test.

18.7.   The sales of new homes are intrinsically tied to interest rates. High interest rates mean that mortgage funds are in short supply and, when available, are very costly to the borrower. A real estate developer is interested in building a number of speculative homes in one of two communities. Knowing that interest rates vary from region to region, the contractor has recorded the interest rate for a conventional $35,000, thirty-year first mortgage from five financial institutions in Community A and from seven in Community B. These data are shown in the table.

Community A	8.6	8.9	8.6	8.7	8.5		
Community B	8.6	8.8	9.0	8.9	9.1	9.2	9.0

Do these data suggest that the average interest rate for a conventional $35,000, thirty-year first mortgage differs between the two communities? Test by using a 10% level of significance. What do these results suggest regarding which community the contractor should select for building the speculative houses?

18.8.   One researcher suggests that, among other reasons, consumer protection legislation has been ineffective because consumers do not possess sufficient understanding of the law to recognize illegal activities and do not have sufficient confidence that the legal system will punish violators.* Suppose a researcher wishes to examine consumer attitudes toward the cooling-off law, a provision that provides consumers time during which they may reconsider certain purchases made from direct-to-home salespeople. Since consumers in low-income areas are most likely to be in contact with direct-to-home salespeople, the researcher decided to compare the attitudes toward the cooling-off law by contacting n_1 = 10 residents from a low-income neighborhood and n_2 = 12 residents from a high-income neighborhood. Each respondent was asked to indicate, on a 10-point scale, the extent to which they agree or disagree with the following statement:

Consumer protection laws can help stop the unfair practices used in business.

The consumer responses are given in the table, with 1 indicating complete agreement with the statement and 10 indicating complete disagreement. Do these data suggest

Low-Income Neighborhood	8.5	9.5	9.0	7.5	8.0	8.5	10.0	5.0	6.5	8.5		
High-Income Neighborhood	4.0	3.5	5.5	6.5	7.0	6.0	2.0	5.0	4.5	5.5	3.0	6.5

that residents from low-income neighborhoods have less confidence in consumer protection laws than those from high-income neighborhoods? Use α = .05.

18.4 The Wilcoxon Rank Sum Test for a Paired Experiment

A rank sum test proposed by Wilcoxon can be used to analyze the paired-difference experiment of section 9.5 by considering the paired difference of the two treatments A and B. Under the null hypothesis of no difference

*D. H. Tootelian, "Attitudinal and Cognitive Readiness: Key Dimensions for Consumer Legislation," *Journal of Marketing*, July 1975.

in the distributions for A and B, you would expect (on the average) half of the differences in pairs to be negative and half to be positive. That is, the expected number of negative differences between pairs would be $n/2$ (where n is the number of pairs). Furthermore, it would follow that positive and negative differences of equal absolute magnitude should occur with equal probability. For example, the probability of a difference equal to $+2$ is the same as the probability of a difference equal to -2. If you were to order the differences according to their absolute values and rank them from smallest to largest, the expected rank sums for the negative and positive differences should be equal. Sizable differences in the rank sums of the positive and negative differences would provide evidence to indicate a difference between the distributions of responses for the two treatments A and B.

To carry out the Wilcoxon test, we calculate the n differences in pairs. Differences of 0 are eliminated and the number n of pairs is reduced accordingly. We rank the absolute values of the differences, assigning a 1 to the smallest, a 2 to the second smallest, and so on. Then we calculate the rank sum for the negative differences and the rank sum for the positive differences. The smaller of these two quantities, which we will designate as T, will be used as a test statistic to test the null hypothesis that the two population relative frequency histograms are identical. The smaller the value of T, the greater will be the weight of evidence favoring rejection of the null hypothesis.

To find the rejection region for T, we use table 9 of the appendix. An abbreviated version of table 9 is shown in table 18.6. Across the top of the table you see the number n of differences (the number of pairs). Values of α (denoted by the symbol P in Wilcoxon's table) for a one-tailed

Table 18.6 *An abbreviated version of table 9 of the appendix; critical values of T*

One-sided	Two-sided	$n = 5$	$n = 6$	$n = 7$	$n = 8$	$n = 9$	$n = 10$
$P = .05$	$P = .10$	1	2	4	6	8	11
$P = .025$	$P = .05$		1	2	4	6	8
$P = .01$	$P = .02$			0	2	3	5
$P = .005$	$P = .01$				0	2	3

One-sided	Two-sided	$n = 11$	$n = 12$	$n = 13$	$n = 14$	$n = 15$	$n = 16$
$P = .05$	$P = .10$	14	17	21	26	30	36
$P = .025$	$P = .05$	11	14	17	21	25	30
$P = .01$	$P = .02$	7	10	13	16	20	24
$P = .005$	$P = .01$	5	7	10	13	16	19

One-sided	Two-sided	$n = 17$
$P = .05$	$P = .10$	41
$P = .025$	$P = .05$	35
$P = .01$	$P = .02$	28
$P = .005$	$P = .01$	23

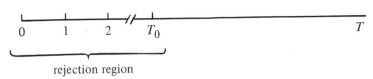

Figure 18.3 *Rejection region for the Wilcoxon rank sum test for a paired experiment*

test appear in the first column of the table. The second column gives values of α (P in Wilcoxon's notation) for a two-tailed test. Table entries are the critical values of T. You will recall that the critical value of a test statistic is the value that locates the upper (or lower) extreme of the rejection region.

For example, suppose you have $n = 7$ pairs and you are conducting a two-tailed test of the null hypothesis that the two population relative frequency distributions are identical. Checking the $n = 7$ column of table 18.6 and using the second row (corresponding to $P = \alpha = .05$ for a two-tailed test), you see the entry 2 (shaded in table 18.6). This is T_0, the critical value of T. As noted earlier, the smaller the value of T, the greater will be the evidence to reject the null hypothesis. Therefore, you will reject the null hypothesis for all values of T less than or equal to 2. The rejection region for the Wilcoxon rank sum test for a paired experiment is always of this form: reject H_0 if $T \le T_0$, where T_0 is the critical value of T. The rejection region is shown in figure 18.3.

Wilcoxon Rank Sum Test for a Paired Experiment

Null hypothesis H_0: The two population relative frequency distributions are identical.

Alternative hypothesis H_a: The two population relative frequency distributions are not identical (a two-tailed test). Or the relative frequency distribution for population A is shifted to the right (or left) of the relative frequency distributions for population B (a one-tailed test).

Test statistic: T, the smaller of the rank sum for positive differences and the rank sum for negative differences.

Rejection region: Reject H_0 if $T \le T_0$, where T_0 is the critical value given in table 9 of the appendix.

Example 18.6 Effective marketing requires knowledge of the seller's markets, because all marketing decisions are based to some degree on how wholesalers, distributors, and consumers will react to those decisions. To provide information useful in marketing decision making, a frozen foods manufacturer undertook a study to compare consumer opinion

Table 18.7 *Data for example 18.6*

RESPONDENT	LIKERT SCORE		DIFFERENCE	RANK
	A	B	(A − B)	
1	6.5	4.0	2.5	3
2	4.0	8.0	−4.0	4
3	5.5	6.5	−1.0	1.5
4	5.5	10.0	−4.5	5
5	7.0	8.0	−1.0	1.5
6	4.0	9.5	−5.5	6

of her line of frozen dinners (A) with those of her most active competitor (B). The study was conducted by having the purchasing manager of each of six nationwide supermarket chains respond on a 10-point Likert scale* according to their evaluation of the two products. In their analysis the respondents were asked to consider price, package design, variety, company promotion, and time before spoilage when scoring each product on a Likert scale. The results are shown in table 18.7. Do these data indicate that the population distributions of scaled responses differ significantly between Product A and Product B?

Solution The smaller rank sum is that for positive differences, and hence $T = 3$. From table 9 the critical value of T for a two-tailed test with $\alpha = .10$ is $T = 2$. Since the observed value of T exceeds the critical value of T, there is not sufficient evidence to indicate a significant difference in the population distributions of scaled responses associated with the two products. (Values of T less than or equal to the critical value imply rejection.)

When n is large (say 25 or more), T is approximately normally distributed, with a mean and a variance of

$$E(T) = \frac{n(n + 1)}{4} \quad \text{and} \quad \sigma_T^2 = \frac{n(n + 1)(2n + 1)}{24}$$

Then the z statistic is

$$z = \frac{T - E(T)}{\sigma_T}$$

*A Likert scale is an instrument that associates ordinal values with qualitative attributes. In this example the attribute is "measure of acceptability," for which the associated Likert scale would appear as follows:

Impression: very poor   acceptable   outstanding

| | | | | | | | | |

    1   2   3   4   5   6   7   8   9   10

The respondent is then asked to mark his or her response on the scale. The mark then defines an assumed measure of acceptability for the product in question.

and can be used as a test statistic. The test is conducted in exactly the same manner as for the large-sample Mann-Whitney U test. Thus for a two-tailed test and $\alpha = .05$, we reject the hypothesis of identical population distributions when $|z| \geq 1.96$.

Large-Sample Wilcoxon Rank Sum Test for a Paired Experiment ($n \geq 25$)

Null hypothesis H_0: The relative frequency distributions for populations A and B are identical.

Alternative hypothesis H_a: The two population relative frequency distributions are not identical (a two-tailed test). Or the relative frequency distribution for population A is shifted to the right (or left) of the relative frequency distribution for population B (a one-tailed test).

Test statistic: $z = \dfrac{T - [n(n + 1)/4]}{\sqrt{n(n + 1)(2n + 1)/24}}$.

Rejection region: For a given value of α,

1. for a two-tailed test reject H_0 if $z \geq z_{\alpha/2}$ or if $z \leq -z_{\alpha/2}$;
2. for a one-tailed test place all of α in one tail of the z distribution and reject H_0 if $z \geq z_\alpha$, for an upper one-tailed test, or if $z \leq -z_\alpha$, for a lower one-tailed test.

Tabulated values of z are given in table 3 of the appendix.

If any doubt exists about the applicability of the use of the t test for a paired experiment, the Wilcoxon test offers a very efficient alternative.

Exercises

18.9. Refer to exercise 18.5. Use the Wilcoxon rank sum test with $\alpha = .05$ to determine if the data suggest that the advertisers show a preference for one of the two television media sources over the other. Are your conclusions the same as those derived in exercise 18.5 when using the sign test? If they are different, explain the reason for the difference.

18.10. Two methods for controlling traffic were compared at each of $n = 12$ intersections with high accident rates. The traffic control methods, which we will call A and B, were each used for one week and the number of accidents occurring during this time was recorded. The order of use of each control method at each intersection was selected in a random manner. The recorded accidents are shown in the table. Do these data provide sufficient evidence to indicate a difference in the mean accident rates when using traffic control methods A and B?

INTERSECTION	TRAFFIC CONTROL METHOD		INTERSECTION	TRAFFIC CONTROL METHOD	
	A	B		A	B
1	5	4	7	2	3
2	6	4	8	4	1
3	8	9	9	7	9
4	3	2	10	5	2
5	6	3	11	6	5
6	1	0	12	1	1

a. Perform the analysis by using a sign test, with α as near as possible to .05.

b. Perform the analysis by using the Wilcoxon rank sum test for a paired experiment.

18.11.   Local public assessment offices are frequent users of statistical and computer-assisted procedures in order to maintain property valuations that are fair representations of market value. Such procedures allow the offices to rapidly update property valuations during times of inflation without incurring the expense of drastically increasing the size of the assessment force. A county assessment office has installed a computer-assisted property valuation program and would like to compare the efficiency of the new program with valuations generated by members of the current assessment staff. The assessment-sales ratio, the common measure of assessment accuracy used by all assessment offices, was computed for $n = 10$ recently sold properties from the county. Two assessment-sales ratios were computed for each property, one based on the computer assessment and the other based on the assessment provided by the assessor. These data are shown in the table.

ASSESSMENT METHOD	PROPERTY									
	1	2	3	4	5	6	7	8	9	10
computer model	.92	.98	.90	1.05	1.01	.95	1.10	1.05	.96	1.10
assessor	.88	.92	.95	.95	.93	.97	.97	.98	.92	.99

a. Do these data suggest that the mean assessment-sales ratio differs when using the computer model and when using the assessor to assess values of properties in the county? Use $\alpha = .10$.

b. Suppose an ideal assessment-sales ratio is 1.0, implying that the assessment is a perfectly accurate prediction of the sales price. Do the data provide sufficient evidence to indicate that the assessments generated by one of the two methods come closer to the ideal assessment-sales ratio than assessments generated by the other?

18.12.   As part of a collective-bargaining agreement, a corporation has agreed to offer a new health insurance plan to its employees. Two large insurance companies have each presented a health insurance plan to management; both are within the guidelines specified in the collective-bargaining agreement and would be equally costly to the corporation. To obtain workers' attitudes toward the two competing plans, the personnel officer randomly selected eight employees from the employee registry. Both insurance plans were explained in detail to each employee in the sample. The employees were then asked to rate each plan on a 10-point Likert scale (see example 18.6), with 1 indicating the plan is completely unacceptable and 10 indicating it is perfectly acceptable. The recorded responses are shown in the table. Do these data

INSURANCE PLAN	EMPLOYEE							
	1	2	3	4	5	6	7	8
A	8.0	7.5	7.0	8.5	9.5	9.0	8.5	8.0
B	5.0	4.0	9.0	7.5	8.0	7.0	6.0	8.0

suggest that the employees' attitudes differ significantly toward these two insurance plans? Use $\alpha = .10$.

18.5 The Runs Test: A Test for Randomness

Consider a production process in which manufactured items emerge in sequence and each is classified as either defective (D) or nondefective (N). We studied how we might compare the fraction defective over two equal time intervals by using the normal z test in chapter 8. We extended this to a test of an hypothesis about a constant fraction defective over two or more time intervals by using the chi-square test of chapter 17. The purpose of these tests was to detect a change or trend in the fraction defective p. Evidence to indicate an increasing fraction defective might show the need for a study to locate the source of difficulty. A decreasing value might suggest that a quality control program was having a beneficial effect in reducing the fraction defective.

Trends in a fraction defective (or other quality measures) are not the only indication of a lack of process control. A process may cause periodic runs of defectives, with the average fraction defective remaining constant, for all practical purposes, over long periods of time. For example, flashbulbs are manufactured on a rotating machine with a fixed number of positions for bulbs. A bulb is placed on the machine at a given position, the air is removed, oxygen is pumped into the bulb, and the glass base is flame-sealed. If a machine contains 20 positions, and several adjacent positions are faulty (too much heat in the sealing process), surges of defective bulbs will emerge from the process in a periodic manner. Tests to compare the fraction defective over equal intervals of time will not detect this periodic difficulty in the process. The periodicity, indicated by runs of defectives, is indicative of nonrandomness in the occurrence of defectives over time and can be detected by a **test for randomness**. The statistical test that we present, known as the **runs test**, is discussed in detail in the article by Wald and Wolfowitz (1940) listed in the references. Other practical applications of the runs test will follow.

As the name implies, the runs test studies a sequence of events where each element in the sequence may assume one of two outcomes, say success

S or failure F. The runs test is thus applied to a sequence of n_1 successes and n_2 failures. If we think of the sequence of items emerging from a manufacturing process as defective F or nondefective S, the observation of 20 items might yield the following:

SSSSSFFSSSFFFSSSSSSS

We notice the groupings of defectives and nondefectives and wonder whether this implies nonrandomness and, consequently, lack of process control.

Definition

A **run** is a maximal subsequence of like elements.

For example, the first 5 successes are a subsequence of 5 like elements and it is maximal in the sense that it includes the maximum number of like elements before encountering an F. (The first 4 elements form a subsequence of like elements, but it is not maximal because the fifth element could also be included.) Consequently, the 20 elements shown above are arranged in five runs, the first containing 5 S's, the second containing 2 F's, and so on.

A very small or very large number of runs in a sequence indicate nonrandomness. Therefore, we will use R (the number of runs in a sequence) as a test statistic to test the null hypothesis that the elements are arranged in random sequence. The rejection region when employing the alternative hypothesis that the sequence is nonrandom is $R \le k_1$ and $R \ge k_2$, as indicated in figure 18.4. A one-tailed test, using values of R in the upper tail, is employed for an alternative hypothesis that the expected value of R is larger than when randomness is present. This would imply an overmixing of the elements of the sequence. In contrast, a one-tailed test, using values of R in the lower tail, is used for an alternative hypothesis that the expected value of R is less than when randomness is present. This alternative would be used if you would not expect overmixing to occur and you could only expect a small number of runs (large groups of like elements). Or it would be used if overmixing was unimportant from a practical point of view.

To illustrate the selection of the alternative hypothesis, consider a test

Figure 18.4 *Rejection region for the runs test*

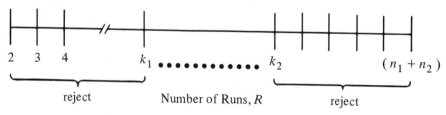

of the null hypothesis that the sequence of defective and nondefective flashbulbs emerging from a production line is random. If specific heat-sealing positions on the rotating sealing machine were operating improperly, defectives would be produced in a spaced and systematic manner, a condition that would result in a larger than expected number of runs (overmixing of defectives and nondefectives). To detect this situation you would use a one-tailed test, rejecting the null hypothesis of randomness for large values of R. Or a very small number of runs (undermixing) could occur if the heat supply (natural gas) to all sealing positions was inadvertently reduced. Then all sealing positions would produce defectives. The machine would produce, alternately, surges of defectives and nondefectives (depending on whether the gas supply was deficient or adequate), which would yield a smaller number of runs than expected. To detect this situation you would use a one-tailed test, rejecting the null hypothesis of randomness for small values of R. Finally, suppose you had no reason to expect either overmixing or undermixing but were simply looking for a lack of randomness (a troubleshooting check). Then you would place part of α in the upper tail of the R distribution and part in the lower tail and select a two-tailed rejection region, as shown in figure 18.4.

The rejection region for the runs test can be found by using table 10 of the appendix, which gives the probability that the number of runs R is less than or equal to some number a. An abbreviated version of table 10 is shown in table 18.8.

In the table the sample sizes (n_1, n_2) are shown at the left of the table and the number a is recorded across the top of the table. The table entries give the probability that $R \leq a$.

To illustrate the use of the table, suppose you have $n_1 = 4$ elements

Table 18.8 *An abbreviated version of table 10 of the appendix for the runs*

a

(n_1, n_2)	2	3	4	5	6	7	8	9
(2,3)	.200	.500	.900	1.000				
(2,4)	.133	.400	.800	1.000				
(2,5)	.095	.333	.714	1.000				
(2,6)	.071	.286	.643	1.000				
(2,7)	.056	.250	.583	1.000				
⋮	⋮	⋮	⋮	⋮				
(3,7)	.017	.083	.283	.583	.833	1.000		
(3,8)	.012	.067	.236	.533	.788	1.000		
(3,9)	.009	.055	.200	.491	.745	1.000		
(3,10)	.007	.045	.171	.455	.706	1.000		
(4,4)	.029	.114	.371	.629	.886	.971	1.000	
(4,5)	.016		.262	.500	.786	.929	.992	1.000
(4,6)	.010	.048	.190	.405	.690	.881	.976	1.000
(4,7)	.006	.033	.142	.333	.606	.833	.954	1.000
(4,8)	.004	.024	.109	.279	.533	.788	.929	1.000

of one type and $n_2 = 5$ of another. Furthermore, suppose you wish to conduct a one-tailed test of the hypothesis of randomness and you will reject H_0 if the number of runs is too small. Then the probability that you will observe $R \le 3$ runs is $\alpha = .071$, which is shaded in table 18.8.

On the other hand, suppose you wish to detect an overly large number of runs, rejecting the null hypothesis of randomness when R is large. Then for $n_1 = 4$ and $n_2 = 5$, the table gives $P(R \le 7) = .929$. Therefore, it follows that $P(R \ge 8) = 1 - P(R \le 7) = 1 - .929 = .071$. Then $\alpha = .071$ for this one-tailed test.

The two preceding examples explain how to find α if you conduct either a one-tailed test for undermixing or a one-tailed test for overmixing. A two-tailed test of nonrandomness to detect either undermixing or overmixing can be conducted by locating the rejection regions as shown in figure 18.4 and determining the critical values in the same manner as for the one-tailed tests.

Runs Test

Null hypothesis H_0: The sequence of elements, call them S's and F's, have been produced in a random manner.

Alternative hypothesis H_a: The elements have been produced in a nonrandom sequence (a two-tailed test). Or the process is nonrandom due solely to overmixing (an upper one-tailed test) or due solely to undermixing (a lower one-tailed test).

Test statistic: R, the number of runs.

Rejection region: For a given value of α,

1. for a two-tailed test reject H_0 if $R \le k_1$ or $R \ge k_2$ (see figure 18.4), where $P(R \le k_1) + P(R \ge k_2) = \alpha$ and k_1 and k_2 are obtained from table 10;

2. for a lower one-tailed test reject H_0 if $R \le k_1$, where $P(R \le k_1) = \alpha$ and k_1 is obtained from table 10; for an upper one-tailed test reject H_0 if $R \ge k_2$, where $P(R \ge k_2) = \alpha$ and k_2 is obtained from table 10.

Example 18.7   After a two-week study program to acquaint company personnel with U.S. tariff regulations, the manager of the international business division of a large manufacturing firm constructed a 20-question, true-false examination to test the effectiveness of the study program. The examination was constructed with correct answers running in the following sequence:

$$T F F T F T F T T T F T F F T F T F T T F$$

Does this sequence indicate a departure from randomness in the arrangement of T and F answers?

Solution The sequence contains $n_1 = 10$ true and $n_2 = 10$ false answers, with $R = 16$ runs. Nonrandomness can be indicated by either an unusually small or an unusually large number of runs, and, consequently, we will use a two-tailed test.

Suppose that we wish to use an α approximately equal to .05, with .025 or less in each tail of the rejection region. Then from table 10 with $n_1 = n_2 = 10$, we note $P(R \le 6) = .019$ and $P(R \le 15) = .981$. Then $P(R \ge 16) = .019$, and we will reject the hypothesis of randomness if $R \le 6$ or $R \ge 16$. Since $R = 16$ for the observed data, we conclude that evidence exists to indicate nonrandomness in the manager's arrangement of answers. His attempt to mix the answers was overdone.

A second application of the runs test is for detecting nonrandomness of a sequence of quantitative measurements over time. These sequences, known as **time series**, occur in many fields. For example, the measurement of a quality characteristic of an industrial product, the blood pressure of a human, and the price of a stock on the stock market all vary over time. Departures in randomness in a series, caused either by trends or periodicities, can be detected by examining the **deviations** of the time series measurements from their average. Negative and positive deviations could be denoted by S and F, respectively, and we could then test this time sequence of deviations for nonrandomness. We illustrate this application with an example.

Example 18.8 The management of a retailing firm feels that the firm's sales respond to a seasonal influence. If this were true, a better monthly sales forecast could be prepared by recognizing the effects of seasonality. The firm's sales over the last year are shown for each month in figure 18.5. The average sales $\bar{y}$ for the 12 months appears as shown. Note the deviations around $\bar{y}$. Do these sales data indicate a lack of randomness and thereby suggest a periodic relationship between sales and time of the year?

Figure 18.5 *Monthly sales for the firm*

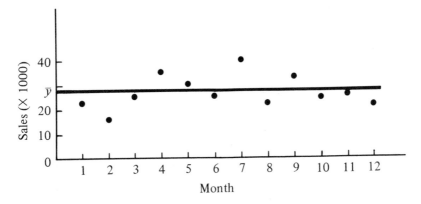

Solution If there is a seasonal trend in monthly sales, we would expect runs of high (or low) monthly sales corresponding to particular seasons. This would produce a smaller than expected number of runs and lead us to reject the hypothesis of randomness if $R \leq k_1$. That is, we will use a lower one-tailed test, placing α entirely in the lower tail of the R distribution. The sequence of negative (N) and positive (P) deviations, as indicated in figure 18.5, is

$$N \; N \; N \; P \; P \; N \; P \; N \; P \; N \; N \; N$$

Then $n_1 = 8$, $n_2 = 4$, and $R = 7$. Consulting table 10, $P(R \leq 7) = .788$. This value of R (i.e., $R = 7$) is not improbable, assuming the hypothesis of randomness to be true. Consequently, there is not sufficient evidence to indicate nonrandomness in this sequence of monthly sales. Therefore, there is insufficient evidence to support the notion of a seasonal (periodic) component in the firm's sales pattern.

The runs test can also be used to compare two population frequency distributions for a two-sample unpaired experiment. It provides an alternative to the Mann-Whitney U test of section 18.3. If the measurements for the two samples are arranged in order of magnitude, they will form a sequence. The measurements for samples 1 and 2 can be denoted as S and F, respectively, and we are once again concerned with a test for randomness. If all measurements for sample 1 are smaller than those for sample 2, the sequence will be S S S S ... S F F F ... F, giving $R = 2$ runs. A small value of R provides evidence of a difference in population frequency distributions, and the rejection region is $R \leq a$. This rejection region implies a one-tailed statistical test. An illustration of the application of the runs test to compare two population frequency distributions is left as an exercise.

As in the case of the other nonparametric test statistics studied in preceding sections of this chapter, the probability distribution for R tends to normality as n_1 and n_2 become large. The approximation is good when n_1 and n_2 are both greater than 10. Consequently, we may use the z statistic as a large-sample test statistic, where

$$z = \frac{R - E(R)}{\sigma_R}$$

$$E(R) = \frac{2n_1 n_2}{n_1 + n_2} + 1$$

$$\sigma_R^2 = \frac{2n_1 n_2 (2n_1 n_2 - n_1 - n_2)}{(n_1 + n_2)^2 (n_1 + n_2 - 1)}$$

$E(R)$ is the expected value of R and σ_R^2 is the variance of R. The rejection region for a two-tailed test with $\alpha = .05$ is $|z| \geq 1.96$.

Large-Sample Runs Test ($n_1 > 10$ and $n_2 > 10$)

Null hypothesis H_0: The sequence of elements, call them S's and F's, have been produced in a random manner.

Alternative hypothesis H_a: The elements have been produced in a nonrandom sequence (a two-tailed test). Or the process is nonrandom due solely to overmixing (an upper one-tailed test) or due solely to undermixing (a lower one-tailed test).

Test statistic: $z = \dfrac{R - \dfrac{2n_1 n_2}{n_1 + n_2} - 1}{\sqrt{\dfrac{2n_1 n_2 (2n_1 n_2 - n_1 - n_2)}{(n_1 + n_2)^2 (n_1 + n_2 - 1)}}}$.

Rejection region: For a given value of α,

1. for a two-tailed test reject H_0 if $z \geq z_{\alpha/2}$ or if $z \leq -z_{\alpha/2}$;
2. for a one-tailed test place all of α in one tail of the z distribution and reject H_0 if $z \geq z_a$, for an upper one-tailed test, or if $z \leq -z_\alpha$, for a lower one-tailed test.

Tabulated values of z are given in table 3 of the appendix.

Exercises

18.13. A filler machine in a food-processing plant is set so that the mean fill is 16 ounces. When the process is in control, the actual load per can should vary about this mean in a random manner, with a variance that remains stable over time. Suppose fill readings for the process, collected at one-hour intervals, were

Time Period	1	2	3	4	5	6	7	8
Fill Reading	15.87	16.14	15.98	16.03	16.05	16.01	16.08	15.94

as shown in the table. Do these data suggest that the process is behaving in a nonrandom manner? (Hint: Use the runs test to answer this question.)

18.14. Paper is produced in a continuous process. Suppose that a brightness measurement y is made on the paper once every hour and that the results are as shown in the table. Compute the average brightness measure $\bar{y}$. Then note the deviations about $\bar{y}$. Do these results indicate a lack of randomness and thereby suggest periodicity in the process and lack of control?

Time (hours)	1	2	3	4	5	6	7	8	9	10	11	12	13	14	15
Brightness, y	3.1	2.3	2.6	3.2	4.2	3.9	3.2	5.0	4.8	2.8	4.2	3.2	3.4	3.1	2.6

18.15. The so-called random walk hypothesis suggests that the movement of certain security prices is completely random and follows no discernible pattern over time. Listed in the table are the daily closing prices (rounded to the nearest dollar) over 20 consecutive market days for a certain security listed on the New York stock exchange. Do these data tend to support the notion that the daily closing prices

TIME (DAY)	PRICE	TIME (DAY)	PRICE
1	$21	11	$22
2	22	12	21
3	24	13	20
4	24	14	18
5	23	15	18
6	24	16	19
7	25	17	18
8	23	18	17
9	23	19	16
10	22	20	18

for this security follow a random walk? (Hint: Compute the mean security price for these data and note the runs of negative and positive deviations of daily prices from the 20-day average.)

18.16. Fifteen experimental batteries were selected at random from a pilot lot at Plant A, and 15 standard batteries were selected at random from production at Plant B. All 30 batteries were simultaneously placed under an electrical load of the same magnitude. The first battery to fail was an A, the second a B, the third a B, and so forth. The following sequence shows the order of failure for the 30 batteries:

A B B B A B A A B B B B A B A B B B B A A B A A A B A A A A

a. Using the large-sample U test, determine if there is sufficient evidence to conclude that the mean life for the experimental batteries is greater than the mean life for the standard batteries. Use $\alpha = .05$.

b. If, indeed, the experimental batteries have the greater mean life, what would be the effect on the expected number of runs? Using the large-sample runs test, determine whether there is a difference in the distributions of battery life for the two populations. Use $\alpha = .05$.

18.6 Rank Correlation Coefficient

In preceding sections we have used ranks to indicate the relative magnitude of observations in nonparametric tests for comparison of treatments. We will now employ the same technique in testing for a monotonic relation (one that is ever-increasing or ever-decreasing) between two ranked variables. A rank correlation provides a measure of the degree of linearity between the

Table 18.9 *Preference rankings and prices of television sets*

MANUFACTURER	PREFERENCE	PRICE
1	7	$449.50
2	4	525.00
3	2	479.95
4	6	499.95
5	1	580.00
6	3	549.95
7	8	469.95
8	5	532.50

ranking variables or a measure of the degree of monotonicity between the variables being observed. Thus a rank correlation coefficient is frequently referred to as a coefficient of agreement for preference data. Two common rank correlation coefficients are the Spearman r_s and the Kendall τ. We will present the Spearman r_s because its computation is identical to that for the sample correlation coefficient r of chapter 11. Kendall's rank correlation coefficient is discussed in detail in the text by Kendall and Stuart (1961) listed in the references.

Suppose an individual who is in the market for a 24-inch color television set ranks the standard models for each of the eight major television manufacturers. His preferences are listed in table 18.9 next to the manufacturer's suggested retail price for the set under consideration. Do the data suggest an agreement between the individual's preference rankings and the manufacturers' suggested retail prices? That is, is there a correlation between preferences and prices?

The two variables of interest are the preference rankings and the prices. The former is already in ranked form, and the prices may be ranked as shown in table 18.10. The ranks for tied observations are obtained by averaging the ranks that the tied observations would have, as for the Mann-Whitney U statistic. The Spearman rank correlation coefficient r_s is calculated by using the ranks as the paired measurements on the two variables x and y in the formula for r given in chapter 11.

Table 18.10 *Preference and price rankings of television sets*

MANUFACTURER	PREFERENCE RANKS, x_i	PRICE RANKS, y_i
1	7	1
2	4	5
3	2	3
4	6	4
5	1	8
6	3	7
7	8	2
8	5	6

Spearman's Rank Correlation Coefficient

$$r_s = \frac{SS_{xy}}{\sqrt{SS_x SS_y}}$$

where x_i and y_i represent the ranks of the ith pair of observations and

$$SS_{xy} = \sum_{i=1}^{n} (x_i - \bar{x})(y_i - \bar{y}) = \sum_{i=1}^{n} x_i y_i - \frac{\left(\sum_{i=1}^{n} x_i\right)\left(\sum_{i=1}^{n} y_i\right)}{n}$$

$$SS_x = \sum_{i=1}^{n} (x_i - \bar{x})^2 = \sum_{i=1}^{n} x_i^2 - \frac{\left(\sum_{i=1}^{n} x_i\right)^2}{n}$$

$$SS_y = \sum_{i=1}^{n} (y_i - \bar{y})^2 = \sum_{i=1}^{n} y_i^2 - \frac{\left(\sum_{i=1}^{n} y_i\right)^2}{n}$$

When there are no ties in either the x observations or the y observations, the expression for r_s algebraically reduces to the simpler expression shown in the next box.

$$r_s = 1 - \frac{6\sum d_i^2}{n(n^2 - 1)} \qquad \text{where} \qquad d_i = x_i - y_i$$

If the number of ties is small in comparison with the number of data pairs, little error will result in using this shortcut formula for calculating r_s. We illustrate the use of the formula with an example.

Example 18.9 Since the association between brand preference and price is an important aspect of pricing, one of the TV manufacturers wishes to know the degree of association, measured by the correlation r_s, between TV preference and price. Calculate r_s for the preference and price data shown in table 18.11.

Solution The values of d_i and d_i^2, $i = 1, 2, ..., 8$, are shown in table 18.11.
Substituting into the formula for r_s, we have

$$r_s = 1 - \frac{6\sum_{i=1}^{n} d_i^2}{n(n^2 - 1)} = 1 - \frac{6(144)}{8(64 - 1)} = -.714$$

Table 18.11 *Data for example 18.9*

MANUFACTURER	x_i	RANK, y_i	RANK, d_i	d_i^2
1	7	1	6	36
2	4	5	−1	1
3	2	3	−1	1
4	6	4	2	4
5	1	8	−7	49
6	3	7	−4	16
7	8	2	6	36
8	5	6	−1	1
				144

The Spearman rank correlation coefficient may be employed as a test statistic to test an hypothesis of no association between two populations. We assume that the n pairs of observations (x_i, y_i) have been randomly selected and therefore that the hypothesis of no association between the populations implies a random assignment of the n ranks within each sample. Each random assignment (for the two samples) represents a sample point associated with the experiment and a value of r_s could be calculated for each. Thus it is possible to calculate the probability that r_s assumes a large positive or negative value due solely to chance and thereby suggests an association between populations when none exists. The rejection region for a two-tailed test is shown in figure 18.6.

If the alternative hypothesis is that the correlation between x and y is negative, we reject H_0 for large negative values of r_s (in the lower tail of figure 18.6). Similarly, if the alternative hypothesis is that the correlation between x and y is positive, we reject H_0 for large positive values of r_s (in the upper tail of figure 18.6).

The critical values of r_s are given in table 11 of the appendix. An abbreviated version of table 11 is shown in table 18.12.

Across the top of table 18.12 (and table 11 of the appendix) are recorded the values of α that you might wish to use for a one-tailed test of the null hypothesis of no association between x and y. The number of ranked pairs n appears at the left side of the table. The table entries give the critical

Figure 18.6 *Rejection region for a two-tailed test of the null hypothesis of no association, using Spearman's rank correlation test*

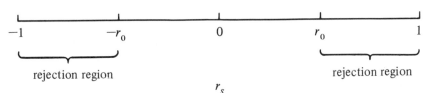

Table 18.12 *An abbreviated version of table 11 of the appendix for Spearman's rank correlation test*

n	$\alpha = 0.05$	$\alpha = 0.025$	$\alpha = 0.01$	$\alpha = 0.005$
5	0.900	—	—	—
6	0.829	0.886	0.943	—
7	0.714	0.786	0.893	—
8	0.643	0.738	0.833	0.881
9	0.600	0.683	0.783	0.833
10	0.564	0.648	0.745	0.794
11	0.523	0.623	0.736	0.818
12	0.497	0.591	0.703	0.780
13	0.475	0.566	0.673	0.745
14	0.457	0.545		
15	0.441	0.525		
16	0.425			
17	0.412			
18	0.399	$\vdots$	$\vdots$	$\vdots$
19	0.388			
20	0.377			

value r_0 for a one-tailed test. Thus $P(r_s \geq r_0) = \alpha$.

For example, suppose you have $n = 8$ ranked pairs and you know that if any correlation exists between the ranks, it must be positive. Then you would reject the null hypothesis of no association only for large positive values of r_s and would use a one-tailed test. Referring to table 18.12 and using the row corresponding to $n = 8$ and the column for $\alpha = .05$, you read $r_0 = .643$. Therefore, you would reject H_0 for all values of r_s greater than or equal to .643.

The test is conducted in exactly the same manner if you wish to test only for a large negative correlation. The only difference is that you would reject the null hypothesis if $r_s \leq -.643$. That is, you just use the negative of the tabulated value of r_0 as the lower-tail critical value.

To conduct a two-tailed test, you reject the null hypothesis if $r_s \geq r_0$ or if $r_s \leq -r_0$. The value of α for the test will be double the value shown at the top of table 11. For example, if $n = 8$ and you choose the .025 column, you will reject H_0 if $r_s \geq .738$ or if $r_s \leq -.738$. The α-value for the test is $2(.025) = .05$.

Spearman's Rank Correlation Test

Null hypothesis H_0: There is no association between the ranked pairs.

Alternative hypothesis H_a: There is an association between the ranked pairs (a two-tailed test). Or the correlation between the ranked pairs is positive (or negative) (a one-tailed test).

Test statistic: r_s.

Rejection region: For a given value of α,

         1. for a two-tailed test reject H_0 if $r_s \geq r_0$ or if $r_s \leq -r_0$, where r_0 is given in table 11 of the appendix; double the table value of α to obtain the value of α for the two-tailed test;
         2. for a one-tailed test reject H_0 if $r_s \geq r_0$, for an upper-tailed test, or if $r_s \leq -r_0$, for a lower-tailed test; the α-value for a one-tailed test is the value shown in table 11.

Example 18.10 Refer to example 18.9. Management realizes that the significance of a correlation depends on the true association and the sample sizes used to estimate the correlation coefficient. In example 18.9 the rank order correlation between TV preference and price was found to be $-.714$. Now management wishes to know the significance of this correlation tested against the null hypothesis of no association.

Solution The critical value of r_s for a one-tailed test with $\alpha = .05$ and $n = 8$ is .643. Let us assume that the correlation between the individual's preference rankings and the manufacturers' suggested retail prices could not possibly be positive. (That is, assume that a low preference rank means a television set is highly favorable and should be associated with a high price if the manufacturers' prices are indicators of quality, additional features, guarantee, and so forth.) The alternative hypothesis is that the population correlation coefficient is less than 0, and we are concerned with a one-tailed test. Thus α for the test is the table value of .05, and we will reject the null hypothesis if $r_s \leq -.643$.

The calculated value of the test statistic, $r_s = -.714$, is less than the critical value for $\alpha = .05$. Hence the null hypothesis of no association is rejected at the $\alpha = .05$ level of significance. It appears that some agreement does exist between preference and price.

Exercises

18.17. Eight applicants for some junior management positions were ranked in order of their suitability for employment (and prospects for success) by two experienced and highly trained personnel recruiters. The rankings are shown in the

	RANKINGS	
APPLICANT	Recruiter 1	Recruiter 2
1	7	8
2	4	6
3	1	3
4	3	1
5	6	5
6	8	7
7	5	4
8	2	2

table. Calculate r_s. Test an hypothesis of no association between the rankings of the two recruiting experts. What type of alternative hypothesis is appropriate for this test?

18.18. The profitability of a stock is typically measured by its return on equity, while its growth potential is measured by its earnings per share. But can a stock be profitable without showing growth potential, or can it show growth potential without also being profitable? To examine this issue an investor recorded the five-year average return on equity and the five-year average earnings per share for the 10 securities in his investment portfolio. The recorded results are shown in the table. Calculate

COMPANY	RETURN ON EQUITY (%)	EARNINGS PER SHARE (%)
Maytag	25.6	5.2
Hoover	15.3	10.9
Stanley Works	12.1	11.0
Magic Chef	11.4	7.4
Sunbeam	11.1	6.7
Simmons	9.6	5.4
Westinghouse Electric	9.4	2.0
Mohasco	8.6	6.3
Fedders	9.3	4.1
Tappan	7.1	2.8

Source: Reprinted by permission of *Forbes* magazine from the January 1975 issue, p. 168.

r_s. Do these data suggest that there is an association between profitability and growth potential for these securities? Test at the 5% level of significance.

18.19. Based on interviews with 2010 workers performing many different jobs, the Institute for Social Research constructed a "boredom chart" for a number of different occupations, with 100 being the average rating and with the higher the rating, the more boring the job.* The boredom measure for 12 different occupations is shown in the table. Also shown is a ranking of the 12 occupations in terms of the average salary earned by a member of each occupation, with 1 indicating the occupation

OCCUPATION	BOREDOM MEASURE	SALARY RANK
assembler	207	11
machine tender	169	9
monitor of continuous flow goods	122	10
accountant	107	4
engineer	100	6
computer programmer	96	7
delivery service courier	86	12
administrator	66	2
policeman	63	8
air traffic controller	59	3
professor	49	5
physician	48	1

*"Those Boring Jobs—Not All That Dull," *U.S. News and World Report*, 1 December 1975.

with the highest salary, 12 the occupation with the lowest salary. Calculate r_s. Do these data suggest that the occupations with the highest salaries are the least boring? Use $\alpha = .05$.

18.7 Summary

The nonparametric statistical tests presented in the preceding pages represent only a few of the many nonparametric statistical measures of inference available. A much larger collection of nonparametric test procedures, along with worked examples, is given in the tests by Siegel (1956) and Conover (1971), and an extensive bibliography of publications dealing with nonparametric statistical methods has been presented by Savage (1953). All are listed in the references at the end of the chapter.

We have indicated that nonparametric statistical procedures are particularly useful when the experimental observations are susceptible to ordering but cannot be measured on a quantitative scale. Since parametric statistical procedures usually cannot be applied to this type of data, you would use nonparametric methods to analyze the data.

A second application of nonparametric statistical methods is in testing hypotheses associated with populations of quantitative data when uncertainty exists about whether the assumptions for the population distributions are satisfied. Just how useful are nonparametric methods for this situation?

The Mann-Whitney U test can be used to compare two populations when the observations can be ranked according to their relative magnitudes and when the samples have been randomly and independently selected from the two populations.

The simplest nonparametric test, the sign test, is most often used for comparing two populations when the observations have been independently selected in pairs. If the differences between pairs can be ranked according to their relative magnitudes, you can use the Wilcoxon rank sum test (often called a signed ranks test) for comparing the two populations. This latter test utilizes more sample information than the sign test and consequently is more likely to detect a difference between the two populations if a difference exists.

Two additional nonparametric tests were presented in this chapter along with examples of their applications. The runs test, used to detect nonrandomness in a sequence of measurements, is a test that is most useful in detecting trends or periodicity in a sequence of observations collected over time (a time series). Spearman's rank correlation test is a nonparametric test for correlation between two variables when the observations associated with each variable can be ranked according to their relative magnitudes.

Supplementary Exercises

18.20. When would a statistician choose to employ a nonparametric statistical method instead of a parametric test like the t test or the z test?

18.21. The number of defective electrical fuses proceeding from each of two production lines A and B was recorded daily for a period of 10 days, with the results shown in the table. Assume that both production lines produced the same daily output.

| | LINE | |
DAY	A	B
1	172	201
2	165	179
3	206	159
4	184	192
5	174	177
6	142	170
7	190	182
8	169	179
9	161	169
10	200	210

Compare the number of defectives produced by A and B each day and let y denote the number of days when B exceeded A. Do these data present sufficient evidence to indicate that production line B produces more defectives, on the average, than A?

a. Use the sign test, with a level of significance as near as possible to .01.

b. Use the Wilcoxon test, with $\alpha = .01$.

18.22. A plywood manufacturer wished to compare the durability, under the presence of moisture, of two different brands of exterior veneer glue. The experiment consisted of applying both brands of glue A and B to each of ten 4 × 8, 3/4-inch plywood panels, with one brand being applied to a different 4 × 4 half of each panel. All ten panels were then immersed in a vat of water until one end peeled. The data shown in the table indicate which of the glued ends of each panel were noted to peel under the immersion experiment. In the case of panel 7, both glued

Panel	1	2	3	4	5	6	7	8	9	10
Glue	A	A	B	A	A	A	O	B	A	A

ends appeared to peel simultaneously. Do these results suggest that there is a difference between glues A and B in bonding strength under the presence of moisture? Use $\alpha = .05$.

18.23. An accounting firm is considering the use of a new format for its audit reports of subscribing firms. The new format consists of numerous figures, graphs, and charts instead of numerical and verbal descriptions of the firm's financial records. Twelve subscribing firms were issued audit reports using the new format. Of this group, 8 indicated they prefer the new format while 4 said they prefer the old format. Do these results offer sufficient evidence to indicate that the new report format is preferred over the old report format? Use a significance level as near .10 as possible.

18.24. In a study to determine the acceptability of a new package design, a firm decided to issue the product in its current package to only one store (Store B) and the product in its newly designed package to another store (Store A). In the past when both stores sold the product in its current package, weekly sales were almost identical. Over a period of 15 weeks, sales of the product showed the sales pattern given in the table. (0 implies that the sales in Store A and Store B were

Week	1 2 3 4 5 6 7 8 9 10 11 12 13 14 15
Store with Greatest Sales	B 0 A A B A B B A A A A 0 A A

equal for that week.) Do these results indicate a preference for the new package design? Test with a significance level as near as possible to .10.

18.25. An experiment was conducted by the advertising agent for a mail-order company to determine the effect of color versus black and white advertisements. To advertise a special sale, 1000 households were selected in each of eight cities across the nation. In each city the head of 500 households received a color advertising circular and another 500 received a black and white advertisement. Two months after all advertisements had been mailed, the number of mail orders placed by the recipients of each type of advertisement from each of the eight cities was found to be as shown in the table. Test the hypothesis that the two types of advertising are equally

| | TYPE OF ADVERTISEMENT | |
CITY	Color	Black and White
Atlanta	113	87
Boston	126	101
Chicago	89	90
Denver	105	82
Los Angeles	135	80
San Francisco	117	79
Seattle	175	113
St. Louis	71	93

effective, as measured by the number of orders placed by the recipients of each type of advertisement, against the alternative that the two types of advertisements are not equally effective. Use $\alpha = .05$.

18.26. Two makes of automobile tires were tested on the rear wheels of 12 different automobiles. The number of miles before tire failure was recorded for each tire. The results are shown in the table.

| | MAKE OF TIRE | |
AUTOMOBILE	A	B
1	17,500	22,000
2	26,450	23,000
3	25,000	24,000
4	28,000	32,000
5	16,500	19,650
6	24,000	33,000
7	27,500	25,000
8	26,000	37,000
9	22,400	31,500
10	28,300	25,900
11	18,900	31,000
12	24,500	30,000

a. Using the sign test, test the hypothesis that the two makes of tires have the same usable life, as measured by the miles before failure, against the alternative that the usable lives are not equally effective. Use $\alpha = .05$.

b. Suppose that make B is a fiberglass belted tire and make A is a conventional nylon cord tire, suggesting that make B has the capability of withstanding much more wear. Test the hypothesis of equal usable life against the alternative that make B is superior. Use $\alpha = .05$.

18.27.   Eight corporate advertising executives were asked to evaluate the comparative effectiveness to their firms of using television and radio advertisements versus advertisements in the printed media. Their responses, as listed on a 10-point Likert scale, are shown in the table. Use the Wilcoxon test to determine if the executives'

EXECUTIVE	RADIO AND TV	PRINTED MEDIA
A	5.0	4.0
B	6.2	3.0
C	7.5	5.0
D	8.0	4.0
E	4.5	6.0
F	6.6	5.0
G	6.0	6.0
H	5.0	5.5

opinions of advertising effectiveness differ significantly for the two advertising media. Use $\alpha = .05$.

18.28.   Use the sign test to test the hypothesis implied in exercise 18.27. Using a level of significance as near as possible to .05, is your conclusion here consistent with the conclusion reached by using the Wilcoxon test in exercise 18.27? Discuss any differences that may exist.

18.29.   The coded values for a measure of brightness in paper (light reflectivity) prepared by two different processes A and B are given for samples of size 9 drawn randomly from each of the two processes. Do the data present sufficient evidence

A	6.1	9.2	8.7	8.9	7.6	7.1	9.5	8.3	9.0
B	9.1	8.2	8.6	6.9	7.5	7.9	8.3	7.8	8.9

(with $\alpha = .10$) to indicate a difference in the populations of brightness measurements for the two processes?

a. Use the Mann-Whitney U test.

b. Use the Student's t test.

18.30.   The life in months of service before failure of the color television picture tube in 8 television sets manufactured by Firm A and 10 sets manufactured by Firm B are as given in the table. Using the U test to analyze the data, test

FIRM	LIFE OF PICTURE TUBE									
A	32	25	40	31	35	29	37	39		
B	41	39	36	47	45	34	48	44	43	33

to determine if the life in months of service before failure of the picture tube differs for the picture tubes manufactured by the two manufacturers. Use $\alpha = .10$.

18.31.   Refer to exercise 18.30. Test the hypothesis of equivalence of the population life distributions by using the runs test, with $\alpha = .10$. Compare your results here with the results of exercise 18.30. Explain any differences in your conclusions.

18.32.   In an effort to cut prices, neighborhood groups are forming buying clubs—small, informal cooperatives. The objective of these co-ops is to save money by eliminating part of the markup a retail store must include in its prices to cover overhead and allow for a profit. The disadvantage associated with buying clubs is their limited selection, which eventually causes many co-op members to lose interest in the co-op and to drop their membership. Interviews were conducted among 13 members of a local food co-op to examine this issue. Eight of the co-op members had been members for at least two months and 5 had been members for less than two months. Each was asked to respond on a 10-point scale to indicate the degree to which he or she was satisfied with the prices and selection offered by the co-op. These results are shown in the table. (A response of 1 indicates complete dissatisfaction

TIME OF MEMBERSHIP	MEASURE OF SATISFACTION							
at least 2 months	5.0	4.5	6.0	9.5	3.5	5.0	8.5	4.0
less than 2 months	8.0	9.0	7.5	9.5	10.0			

and 10 indicates complete satisfaction with the co-op.) Do these data suggest that those who have been members of the food co-op for at least two months are less satisfied with their membership than those who have been members for less than two months? Use $\alpha = .05$.

18.33.   The home office of a corporation selects its executives from the staff personnel of its two subsidiary companies, Company A and Company B. During the past three years, nine executives have been selected by the parent company from the subsidiaries, the first selected from B, the second from A, and so on. The following sequence shows the order in which the executives have been selected and the subsidiary firms from which they came:

$$B\ A\ A\ A\ B\ A\ A\ A\ B$$

Does this selection sequence provide sufficient evidence to imply nonrandomness in the selection of executives by the parent company from its subsidiaries?

18.34.   Items emerging from a continuous production process were classified as defective or nondefective. A sequence of items observed over time was as follows:

$$D\ N\ N\ N\ N\ N\ D\ D\ N\ N\ N\ N\ N\ D\ D$$

$$D\ N\ N\ N\ N\ D\ N\ N\ N\ D\ D\ N\ N\ N\ D\ D$$

a. Give the appropriate probability that $R \leq 11$, where $n_1 = 11$ and $n_2 = 23$.
b. Do these data suggest lack of randomness in the occurrence of defectives D and nondefectives N? Use the large-sample runs test.

18.35.   A quality control chart has been maintained for a certain measurable characteristic of items taken from a conveyor belt at a certain point in a production line. The measurements obtained today, in order of time, are

68.2	71.6	69.3	71.6	70.4	65.0	63.6	64.7
65.3	64.2	67.6	68.6	66.8	68.9	66.8	70.1

a. Classify the measurements in this time series as above or below the sample mean and determine (use the runs test) whether consecutive observations suggest lack of stability in the production process.

b. Divide the time period into two equal parts and compare the means, using the Student's t test. Do the data provide evidence of a shift in the mean level of the quality characteristics?

18.36.   If (as in the case of measurements produced by two well-calibrated measuring instruments) the means of two populations are equal, it is possible to use the Mann-Whitney U statistic for testing hypotheses about the population variances as follows:

i. Order the combined sample.
ii. Rank the ordered observations "from the outside in"; that is, rank the smallest observation as 1; the largest, 2; the next to smallest, 3; the next to largest, 4; and so forth. The final sequence of numbers induces an ordering on the population A items and the population B items. If $\sigma_A^2 > \sigma_B^2$, we would expect to find a preponderance of A's near the first of the sequence and thus a relatively small sum of ranks for the A observations.

a. Given the following measurements produced by well-calibrated precision instruments A and B, test to determine whether the more expensive instrument B is more precise than A. (Note that this would imply a one-tailed test.) Use the Mann-Whitney U test and α as near as possible to .05.

A	1060.21	1060.34	1060.27	1060.36	1060.40
B	1060.24	1060.28	1060.32	1060.30	

b. Test by using the F statistic of section 9.7.

18.37.   A large corporation selects college graduates for employment, using both interviews and a psychological achievement test. Interviews conducted at the home office of the company were far more expensive than the tests conducted on the campuses. Consequently, the personnel office was interested in determining whether the test scores were correlated with interview ratings and whether tests could be substituted for interviews. The idea was not to eliminate interviews but to reduce their number. To determine whether correlation was present, 10 prospects were ranked during interviews and then tested. The paired scores are shown in the table. Calculate

SUBJECT	INTERVIEW RANK	TEST SCORE
1	8	74
2	5	81
3	10	66
4	3	83
5	6	66
6	1	94
7	4	96
8	7	70
9	9	61
10	2	86

the Spearman rank correlation coefficient r_s. Rank 1 is assigned to the candidate judged to be the best.

18.38.   Refer to exercise 18.37. Do these data present sufficient evidence to indicate that the correlation between interview rankings and test scores is less than 0? If this evidence does exist, can we say that test scores could be used to reduce the number of interviews?

18.39.   A management scientist wished to examine the relationship between the image of buyers toward a particular manufacturer's product line and the distance between the buyer and the manufacturer's plant. Each of 12 buyers for regional variety stores rated the manufacturer's product line on a scale of 1 to 20. The data are given in the table. Calculate the Spearman rank correlation coefficient r_s.

BUYER	RATING	DISTANCE
1	12	75
2	7	165
3	5	2750
4	19	10
5	17	950
6	12	2230
7	9	680
8	18	60
9	3	1800
10	8	1150
11	15	700
12	4	510

18.40.   Refer to exercise 18.39. Do these data provide sufficient evidence to indicate a negative correlation between rating and distance?

18.41.   Business executives have often accused academicians of designing business curricula for middle and top management rather than for lower-level management, even though few students, if any, reach upper management within ten years of graduation. Banville and Domm examined this issue by noting management's assessment of the value of different courses of instruction offered in most typical business schools.* In their study a group of top managers and a group of lower-level

	RANKINGS	
COURSE	Top Management	Lower Management
general accounting	6	5
cost accounting	4	6.5
auditing	10	10
business economics	5	4
financial management	2	2
industrial marketing	8	9
managerial leadership	1	1
introduction to computers	7	3
statistics	9	8
business law	3	6.5

*G. R. Banville and D. Domm, "A Business Curriculum for Future Purchasing Officers," *Journal of Purchasing*, August 1973.

managers were separately asked to give consensus rankings of the relative importance of many different business courses. Listed in the table are the rankings given by each group for ten typical business courses included in the list studied by Banville and Domm. Do these data present sufficient evidence to indicate that the assessments of importance of business courses differ between top management and lower management?

18.42.   It is sometimes remarked that brand loyalists possess blind loyalty to a brand even though they may agree that a competing brand is more efficient in terms of the primary use of the product. To examine this theory nine housewives were contacted and asked to rank the seven leading brands of powdered laundry detergent in terms of the "cleaning power" they perceive the brands to possess. Five of the nine housewives were loyalists to Brand A while the other four possessed no brand loyalty among the five competing brands. The ranks given to Brand A by the nine housewives are listed in the table. (A rank of 1 implies greatest cleaning

Brand Loyalists	2	1	3	5	2
Brand Switchers	3	4	1	6	

power, 7 implies least cleaning power.) Do these data suggest that loyalists to Brand A rank it higher in "cleaning power" on the average, than those who possess no brand loyalty? Test at a 10% level of significance.

Experiences with Real Data

In a study noted earlier (see exercise 18.8), Tootelian criticized the viability of consumer protection laws in preventing abusive practices in the marketplace. Tootelian suggests that the major reason for the ineffectiveness of consumer protection laws is that they are often not oriented toward those who most need the protection. Most statutes require that consumers must take the initiative against those they believe have violated the law. The effectiveness of the statutes thus depends heavily on the assumption that consumers have sufficient understanding of consumer laws to recognize illegal activities (cognitive readiness) and have sufficient confidence that the legal system will punish violators (attitudinal readiness).

We have often heard those from other parts of the campus remark that business students are much more pragmatic and often more conservative in their interpretation of the problems of society than students in other fields. If this notion is true, we might expect that business students would possess greater attitudinal and cognitive readiness and thus have greater confidence in the efficacy of consumer protection laws than other students.

Construct an experiment to study this notion. Select random samples of n_1 = 10 students whose major is business and n_2 = 10 nonbusiness majors. Contact each chosen student personally and have each respond on a 10-point Likert scale to indicate the degree to which they agree with this statement:

Consumer protection laws can help stop the unfair practices in business.

Design the Likert scale as shown here, where a response of 1 indicates complete agreement with the statement and 10 indicates complete disagreement with it.

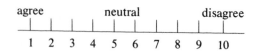

Have each respondent provide his or her response by placing a mark on the scale which you have provided for each in ballot form. Then carefully record the data obtained from each of the 20 respondents, making sure that you are obtaining an accurate representation of the numerical scalar response recorded by each respondent. That is, if an individual has placed a mark between 5 and 6, attempt to accurately define that mark in decimal form as 5.2, 5.3, or so on. Perform an analysis of your recorded data by using the Mann-Whitney U test. Do your data provide sufficient evidence to indicate that students of business possess greater attitudinal and cognitive readiness toward consumer legislation than nonbusiness students? Compare your findings with those obtained by other members of your class. Combine your data with those obtained by several other members of your class and use the large-sample Mann-Whitney U test to test the hypothesis.

References

BRADLEY, J. V. *Distribution-Free Statistical Tests.* Englewood Cliffs, N.J.: Prentice-Hall, 1968.

CONOVER, W. J. *Practical Nonparametric Statistics.* New York: Wiley, 1971.

KENDALL, M. G., and A. STUART. *The Advanced Theory of Statistics.* Vol. 2. New York: Hafner Publishing, 1961.

PIERCE, A. *Fundamentals of Nonparametric Statistics.* Belmont, Calif.: Dickenson Publishing, 1970.

SAVAGE, I. R. "Bibliography of Nonparametric Statistics and Related Topics." *Journal of the American Statistical Association* 48 (1953): 844–906.

Selected Tables in Mathematical Statistics. Chicago: Markham Publishing, 1970.

SIEGEL, S. *Nonparametric Statistics for the Behavioral Sciences.* New York: McGraw-Hill, 1956.

WALD, A., and J. WOLFOWITZ. "On a Test Whether Two Samples Are from the Same Population." *Annals of Mathematical Statistics* 2 (1940): 147–162.

appendix

Tables

Binomial

Table 1 *Binomial probability tables. Tabulated values are* $P(y \leq a) = \sum\limits_{y=1}^{a} p(y)$. *(Computations are rounded at the third decimal place.)*

(a) $n = 5$

p

a	0.01	0.05	0.10	0.20	0.30	0.40	0.50	0.60	0.70	0.80	0.90	0.95	0.99	a
0	.951	.774	.590	.328	.168	.078	.031	.010	.002	.000	.000	.000	.000	0
1	.999	.977	.919	.737	.528	.337	.188	.087	.031	.007	.000	.000	.000	1
2	1.000	.999	.991	.942	.837	.683	.500	.317	.163	.058	.009	.001	.000	2
3	1.000	1.000	1.000	.993	.969	.913	.812	.663	.472	.263	.081	.023	.001	3
4	1.000	1.000	1.000	1.000	.998	.990	.969	.922	.832	.672	.410	.226	.049	4

(b) $n = 10$

p

a	0.01	0.05	0.10	0.20	0.30	0.40	0.50	0.60	0.70	0.80	0.90	0.95	0.99	a
0	.904	.599	.349	.107	.028	.006	.001	.000	.000	.000	.000	.000	.000	0
1	.996	.914	.736	.376	.149	.046	.011	.002	.000	.000	.000	.000	.000	1
2	1.000	.988	.930	.678	.383	.167	.055	.012	.002	.000	.000	.000	.000	2
3	1.000	.999	.987	.879	.650	.382	.172	.055	.011	.001	.000	.000	.000	3
4	1.000	1.000	.998	.967	.850	.633	.377	.166	.047	.006	.000	.000	.000	4
5	1.000	1.000	1.000	.994	.953	.834	.623	.367	.150	.033	.002	.000	.000	5
6	1.000	1.000	1.000	.999	.989	.945	.828	.618	.350	.121	.013	.001	.000	6
7	1.000	1.000	1.000	1.000	.998	.988	.945	.833	.617	.322	.070	.012	.000	7
8	1.000	1.000	1.000	1.000	1.000	.998	.989	.954	.851	.624	.264	.086	.004	8
9	1.000	1.000	1.000	1.000	1.000	1.000	.999	.994	.972	.893	.651	.401	.096	9

Table 1 *Continued*

(c) $n = 15$

a	0.01	0.05	0.10	0.20	0.30	0.40	0.50	0.60	0.70	0.80	0.90	0.95	0.99	a
0	.860	.463	.206	.035	.005	.000	.000	.000	.000	.000	.000	.000	.000	0
1	.990	.829	.549	.167	.035	.005	.000	.000	.000	.000	.000	.000	.000	1
2	1.000	.964	.816	.398	.127	.027	.004	.000	.000	.000	.000	.000	.000	2
3	1.000	.995	.944	.648	.297	.091	.018	.002	.000	.000	.000	.000	.000	3
4	1.000	.999	.987	.836	.515	.217	.059	.009	.001	.000	.000	.000	.000	4
5	1.000	1.000	.998	.939	.722	.403	.151	.034	.004	.000	.000	.000	.000	5
6	1.000	1.000	1.000	.982	.869	.610	.304	.095	.015	.001	.000	.000	.000	6
7	1.000	1.000	1.000	.996	.950	.787	.500	.213	.050	.004	.000	.000	.000	7
8	1.000	1.000	1.000	.999	.985	.905	.696	.390	.131	.018	.000	.000	.000	8
9	1.000	1.000	1.000	1.000	.996	.966	.849	.597	.278	.061	.002	.000	.000	9
10	1.000	1.000	1.000	1.000	.999	.991	.941	.783	.485	.164	.013	.001	.000	10
11	1.000	1.000	1.000	1.000	1.000	.998	.982	.909	.703	.352	.056	.005	.000	11
12	1.000	1.000	1.000	1.000	1.000	1.000	.996	.973	.873	.602	.184	.036	.000	12
13	1.000	1.000	1.000	1.000	1.000	1.000	1.000	.995	.965	.833	.451	.171	.010	13
14	1.000	1.000	1.000	1.000	1.000	1.000	1.000	1.000	.995	.965	.794	.537	.140	14

Table 1 *Continued*

(d) $n = 20$

p π

r / a	0.01	0.05	0.10	0.20	0.30	0.40	0.50	0.60	0.70	0.80	0.90	0.95	0.99	a
0	.818	.358	.122	.002	.001	.000	.000	.000	.000	.000	.000	.000	.000	0
1	.983	.736	.392	.069	.008	.001	.000	.000	.000	.000	.000	.000	.000	1
2	.999	.925	.677	.206	.035	.004	.000	.000	.000	.000	.000	.000	.000	2
3	1.000	.984	.867	.411	.107	.016	.001	.000	.000	.000	.000	.000	.000	3
4	1.000	.997	.957	.630	.238	.051	.006	.000	.000	.000	.000	.000	.000	4
5	1.000	1.000	.989	.804	.416	.126	.021	.002	.000	.000	.000	.000	.000	5
6	1.000	1.000	.998	.913	.608	.250	.058	.006	.000	.000	.000	.000	.000	6
7	1.000	1.000	1.000	.968	.772	.416	.132	.021	.001	.000	.000	.000	.000	7
8	1.000	1.000	1.000	.990	.887	.596	.252	.057	.005	.000	.000	.000	.000	8
9	1.000	1.000	1.000	.997	.952	.755	.412	.128	.017	.001	.000	.000	.000	9
10	1.000	1.000	1.000	.999	.983	.872	.588	.245	.048	.003	.000	.000	.000	10
11	1.000	1.000	1.000	1.000	.995	.943	.748	.404	.113	.010	.000	.000	.000	11
12	1.000	1.000	1.000	1.000	.999	.979	.868	.584	.228	.032	.000	.000	.000	12
13	1.000	1.000	1.000	1.000	1.000	.994	.942	.750	.392	.087	.002	.000	.000	13
14	1.000	1.000	1.000	1.000	1.000	.998	.979	.874	.584	.196	.011	.000	.000	14
15	1.000	1.000	1.000	1.000	1.000	1.000	.994	.949	.762	.370	.043	.003	.000	15
16	1.000	1.000	1.000	1.000	1.000	1.000	.999	.984	.893	.589	.133	.016	.000	16
17	1.000	1.000	1.000	1.000	1.000	1.000	1.000	.996	.965	.794	.323	.075	.001	17
18	1.000	1.000	1.000	1.000	1.000	1.000	1.000	.999	.992	.931	.608	.264	.017	18
19	1.000	1.000	1.000	1.000	1.000	1.000	1.000	1.000	.999	.988	.878	.642	.182	19

Table 1 *Concluded*

(e) $n = 25$

a	0.01	0.05	0.10	0.20	0.30	0.40	0.50	0.60	0.70	0.80	0.90	0.95	0.99	a
0	.778	.277	.072	.004	.000	.000	.000	.000	.000	.000	.000	.000	.000	0
1	.974	.642	.271	.027	002	.000	.000	.000	.000	.000	.000	.000	.000	1
2	.998	.873	.537	.098	.009	.000	.000	.000	.000	.000	.000	.000	.000	2
3	1.000	.966	.764	.234	.033	.002	.000	.000	.000	.000	.000	.000	.000	3
4	1.000	.993	.902	.421	.090	.009	.000	.000	.000	.000	.000	.000	.000	4
5	1.000	.999	.967	.617	.193	.029	.002	.000	.000	.000	.000	.000	.000	5
6	1.000	1.000	.991	.780	.341	.074	.007	.000	.000	.000	.000	.000	.000	6
7	1.000	1.000	.998	.891	.512	.154	.022	.001	.000	.000	.000	.000	.000	7
8	1.000	1.000	1.000	.953	.677	.274	.054	.004	.000	.000	.000	.000	.000	8
9	1.000	1.000	1.000	.983	.811	.425	.115	.013	.000	.000	.000	.000	.000	9
10	1.000	1.000	1.000	.994	.902	.586	.212	.034	.002	.000	.000	.000	.000	10
11	1.000	1.000	1.000	.998	.956	.732	.345	.078	.006	.000	.000	.000	.000	11
12	1.000	1.000	1.000	1.000	.983	.846	.500	.154	.017	.000	.000	.000	.000	12
13	1.000	1.000	1.000	1.000	.994	.922	.655	.268	.044	.002	.000	.000	.000	13
14	1.000	1.000	1.000	1.000	.998	.966	.788	.414	.098	.006	.000	.000	.000	14
15	1.000	1.000	1.000	1.000	1.000	.987	.885	.575	.189	.017	.000	.000	.000	15
16	1.000	1.000	1.000	1.000	1.000	.996	.946	.726	.323	.047	.000	.000	.000	16
17	1.000	1.000	1.000	1.000	1.000	.999	.978	.846	.488	.109	.002	.000	.000	17
18	1.000	1.000	1.000	1.000	1.000	1.000	.993	.926	.659	.220	.009	.000	.000	18
19	1.000	1.000	1.000	1.000	1.000	1.000	.998	.971	.807	.383	.033	.001	.000	19
20	1.000	1.000	1.000	1.000	1.000	1.000	1.000	.991	.910	.579	.098	.007	.000	20
21	1.000	1.000	1.000	1.000	1.000	1.000	1.000	.998	.967	.766	.236	.034	.000	21
22	1.000	1.000	1.000	1.000	1.000	1.000	1.000	1.000	.991	.902	.463	.127	.002	22
23	1.000	1.000	1.000	1.000	1.000	1.000	1.000	1.000	.998	.973	.729	.358	.026	23
24	1.000	1.000	1.000	1.000	1.000	1.000	1.000	1.000	1.000	.996	.928	.723	.222	24

Poisson

Table 2 *Values of* $e^{-\mu}$

μ	$e^{-\mu}$	μ	$e^{-\mu}$	μ	$e^{-\mu}$	μ	$e^{-\mu}$
0.00	1.000000	2.60	.074274	5.10	.006097	7.60	.000501
0.10	.904837	2.70	.067206	5.20	.005517	7.70	.000453
0.20	.818731	2.80	.060810	5.30	.004992	7.80	.000410
0.30	.740818	2.90	.055023	5.40	.004517	7.90	.000371
0.40	.670320	3.00	.049787	5.50	.004087	8.00	.000336
0.50	.606531	3.10	.045049	5.60	.003698	8.10	.000304
0.60	.548812	3.20	.040762	5.70	.003346	8.20	.000275
0.70	.496585	3.30	.036883	5.80	.003028	8.30	.000249
0.80	.449329	3.40	.033373	5.90	.002739	8.40	.000225
0.90	.406570	3.50	.030197	6.00	.002479	8.50	.000204
1.00	.367879	3.60	.027324	6.10	.002243	8.60	.000184
1.10	.332871	3.70	.024724	6.20	.002029	8.70	.000167
1.20	.301194	3.80	.022371	6.30	.001836	8.80	.000151
1.30	.272532	3.90	.020242	6.40	.001661	8.90	.000136
1.40	.246597	4.00	.018316	6.50	.001503	9.00	.000123
1.50	.223130	4.10	.016573	6.60	.001360	9.10	.000112
1.60	.201897	4.20	.014996	6.70	.001231	9.20	.000101
1.70	.182684	4.30	.013569	6.80	.001114	9.30	.000091
1.80	.165299	4.40	.012277	6.90	.001008	9.40	.000083
1.90	.149569	4.50	.011109	7.00	.000912	9.50	.000075
2.00	.135335	4.60	.010052	7.10	.000825	9.60	.000068
2.10	.122456	4.70	.009095	7.20	.000747	9.70	.000061
2.20	.110803	4.80	.008230	7.30	.000676	9.80	.000056
2.30	.100259	4.90	.007447	7.40	.000611	9.90	.000050
2.40	.090718	5.00	.006738	7.50	.000553	10.00	.000045
2.50	.082085						

$$\text{est.} \mu = \bar{x} \pm 2(\sigma)$$
$$\sigma_{\bar{x}} = \sigma/\sqrt{n}$$
$$\text{est } \pi = \frac{r_1 + r_2}{n_1 + n_2} \quad \left\{ \begin{array}{l} \text{diff} \\ \text{of} \\ \text{two} \\ \text{props} \end{array} \right.$$

Table 3 *Normal curve areas*

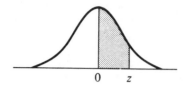

z	.00	.01	.02	.03	.04	.05	.06	.07	.08	.09
0.0	.0000	.0040	.0080	.0120	.0160	.0199	.0239	.0279	.0319	.0359
0.1	.0398	.0438	.0478	.0517	.0557	.0596	.0636	.0675	.0714	.0753
0.2	.0793	.0832	.0871	.0910	.0948	.0987	.1026	.1064	.1103	.1141
0.3	.1179	.1217	.1255	.1293	.1331	.1368	.1406	.1443	.1480	.1517
0.4	.1554	.1591	.1628	.1664	.1700	.1736	.1772	.1808	.1844	.1879
0.5	.1915	.1950	.1985	.2019	.2054	.2088	.2123	.2157	.2190	.2224
0.6	.2257	.2291	.2324	.2357	.2389	.2422	.2454	.2486	.2517	.2549
0.7	.2580	.2611	.2642	.2673	.2704	.2734	.2764	.2794	.2823	.2852
0.8	.2881	.2910	.2939	.2967	.2995	.3023	.3051	.3078	.3106	.3133
0.9	.3159	.3186	.3212	.3238	.3264	.3289	.3315	.3340	.3365	.3389
1.0	.3413	.3438	.3461	.3485	.3508	.3531	.3554	.3577	.3599	.3621
1.1	.3643	.3665	.3686	.3708	.3729	.3749	.3770	.3790	.3810	.3830
1.2	.3849	.3869	.3888	.3907	.3925	.3944	.3962	.3980	.3997	.4015
1.3	.4032	.4049	.4066	.4082	.4099	.4115	.4131	.4147	.4162	.4177
1.4	.4192	.4207	.4222	.4236	.4251	.4265	.4279	.4292	.4306	.4319
1.5	.4332	.4345	.4357	.4370	.4382	.4394	.4406	.4418	.4429	.4441
1.6	.4452	.4463	.4474	.4484	.4495	.4505	.4515	.4525	.4535	.4545
1.7	.4554	.4564	.4573	.4582	.4591	.4599	.4608	.4616	.4625	.4633
1.8	.4641	.4649	.4656	.4664	.4671	.4678	.4686	.4693	.4699	.4706
1.9	.4713	.4719	.4726	.4732	.4738	.4744	.4750	.4756	.4761	.4767
2.0	.4772	.4778	.4783	.4788	.4793	.4798	.4803	.4808	.4812	.4817
2.1	.4821	.4826	.4830	.4834	.4838	.4842	.4846	.4850	.4854	.4857
2.2	.4861	.4864	.4868	.4871	.4875	.4878	.4881	.4884	.4887	.4890
2.3	.4893	.4896	.4898	.4901	.4904	.4906	.4909	.4911	.4913	.4916
2.4	.4918	.4920	.4922	.4925	.4927	.4929	.4931	.4932	.4934	.4936
2.5	.4938	.4940	.4941	.4943	.4945	.4946	.4948	.4949	.4951	.4952
2.6	.4953	.4955	.4956	.4957	.4959	.4960	.4961	.4962	.4963	.4964
2.7	.4965	.4966	.4967	.4968	.4969	.4970	.4971	.4972	.4973	.4974
2.8	.4974	.4975	.4976	.4977	.4977	.4978	.4979	.4979	.4980	.4981
2.9	.4981	.4982	.4982	.4983	.4984	.4984	.4985	.4985	.4986	.4986
3.0	.4987	.4987	.4987	.4988	.4988	.4989	.4989	.4989	.4990	.4990

This table is abridged from Table 1 of *Statistical Tables and Formulas*, by A. Hald (New York: John Wiley & Sons, Inc., 1952). Reproduced by permission of A. Hald and the publishers, John Wiley & Sons, Inc.

Table 4 *Critical values of* t

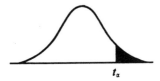

d.f.	$t_{.100}$	$t_{.050}$	$t_{.025}$	$t_{.010}$	$t_{.005}$
1	3.078	6.314	12.706	31.821	63.657
2	1.886	2.920	4.303	6.965	9.925
3	1.638	2.353	3.182	4.541	5.841
4	1.533	2.132	2.776	3.747	4.604
5	1.476	2.015	2.571	3.365	4.032
6	1.440	1.943	2.447	3.143	3.707
7	1.415	1.895	2.365	2.998	3.499
8	1.397	1.860	2.306	2.896	3.355
9	1.383	1.833	2.262	2.821	3.250
10	1.372	1.812	2.228	2.764	3.169
11	1.363	1.796	2.201	2.718	3.106
12	1.356	1.782	2.179	2.681	3.055
13	1.350	1.771	2.160	2.650	3.012
14	1.345	1.761	2.145	2.624	2.977
15	1.341	1.753	2.131	2.602	2.947
16	1.337	1.746	2.120	2.583	2.921
17	1.333	1.740	2.110	2.567	2.898
18	1.330	1.734	2.101	2.552	2.878
19	1.328	1.729	2.093	2.539	2.861
20	1.325	1.725	2.086	2.528	2.845
21	1.323	1.721	2.080	2.518	2.831
22	1.321	1.717	2.074	2.508	2.819
23	1.319	1.714	2.069	2.500	2.807
24	1.318	1.711	2.064	2.492	2.797
25	1.316	1.708	2.060	2.485	2.787
26	1.315	1.706	2.056	2.479	2.779
27	1.314	1.703	2.052	2.473	2.771
28	1.313	1.701	2.048	2.467	2.763
29	1.311	1.699	2.045	2.462	2.756
inf.	1.282	1.645	1.960	2.326	2.576

From "Table of Percentage Points of the *t*-Distribution." Computed by Maxine Merrington, *Biometrika*, Vol. 32 (1941), p. 300. Reproduced by permission of Professor D. V. Lindley.

Table 5 *Critical values of chi square*

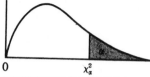

d.f.	$\chi^2_{0.995}$	$\chi^2_{0.990}$	$\chi^2_{0.975}$	$\chi^2_{0.950}$	$\chi^2_{0.900}$
1	0.0000393	0.0001571	0.0009821	0.0039321	0.0157908
2	0.0100251	0.0201007	0.0506356	0.102587	0.210720
3	0.0717212	0.114832	0.215795	0.351846	0.584375
4	0.206990	0.297110	0.484419	0.710721	1.063623
5	0.411740	0.554300	0.831211	1.145476	1.61031
6	0.675727	0.872085	1.237347	1.63539	2.20413
7	0.989265	1.239043	1.68987	2.16735	2.83311
8	1.344419	1.646482	2.17973	2.73264	3.48954
9	1.734926	2.087912	2.70039	3.32511	4.16816
10	2.15585	2.55821	3.24697	3.94030	4.86518
11	2.60321	3.05347	3.81575	4.57481	5.57779
12	3.07382	3.57056	4.40379	5.22603	6.30380
13	3.56503	4.10691	5.00874	5.89186	7.04150
14	4.07468	4.66043	5.62872	6.57063	7.78953
15	4.60094	5.22935	6.26214	7.26094	8.54675
16	5.14224	5.81221	6.90766	7.96164	9.31223
17	5.69724	6.40776	7.56418	8.67176	10.0852
18	6.26481	7.01491	8.23075	9.39046	10.8649
19	6.84398	7.63273	8.90655	10.1170	11.6509
20	7.43386	8.26040	9.59083	10.8508	12.4426
21	8.03366	8.89720	10.28293	11.5913	13.2396
22	8.64272	9.54249	10.9823	12.3380	14.0415
23	9.26042	10.19567	11.6885	13.0905	14.8479
24	9.88623	10.8564	12.4011	13.8484	15.6587
25	10.5197	11.5240	13.1197	14.6114	16.4734
26	11.1603	12.1981	13.8439	15.3791	17.2919
27	11.8076	12.8786	14.5733	16.1513	18.1138
28	12.4613	13.5648	15.3079	16.9279	18.9392
29	13.1211	14.2565	16.0471	17.7083	19.7677
30	13.7867	14.9535	16.7908	18.4926	20.5992
40	20.7065	22.1643	24.4331	26.5093	29.0505
50	27.9907	29.7067	32.3574	34.7642	37.6886
60	35.5346	37.4848	40.4817	43.1879	46.4589
70	43.2752	45.4418	48.7576	51.7393	55.3290
80	51.1720	53.5400	57.1532	60.3915	64.2778
90	59.1963	61.7541	65.6466	69.1260	73.2912
100	67.3276	70.0648	74.2219	77.9295	82.3581

Table 5 *Concluded*

$\chi^2_{0.100}$	$\chi^2_{0.050}$	$\chi^2_{0.025}$	$\chi^2_{0.010}$	$\chi^2_{0.005}$	d.f.
2.70554	3.84146	5.02389	6.63490	7.87944	1
4.60517	5.99147	7.37776	9.21034	10.5966	2
6.25139	7.81473	9.34840	11.3449	12.8381	3
7.77944	9.48773	11.1433	13.2767	14.8602	4
9.23635	11.0705	12.8325	15.0863	16.7496	5
10.6446	12.5916	14.4494	16.8119	18.5476	6
12.0170	14.0671	16.0128	18.4753	20.2777	7
13.3616	15.5073	17.5346	20.0902	21.9550	8
14.6837	16.9190	19.0228	21.6660	23.5893	9
15.9871	18.3070	20.4831	23.2093	25.1882	10
17.2750	19.6751	21.9200	24.7250	26.7569	11
18.5494	21.0261	23.3367	26.2170	28.2995	12
19.8119	22.3621	24.7356	27.6883	29.8194	13
21.0642	23.6848	26.1190	29.1413	31.3193	14
22.3072	24.9958	27.4884	30.5779	32.8013	15
23.5418	26.2962	28.8454	31.9999	34.2672	16
24.7690	27.5871	30.1910	33.4087	35.7185	17
25.9894	28.8693	31.5264	34.8053	37.1564	18
27.2036	30.1435	32.8523	36.1908	38.5822	19
28.4120	31.4104	34.1696	37.5662	39.9968	20
29.6151	32.6705	35.4789	38.9321	41.4010	21
30.8133	33.9244	36.7807	40.2894	42.7956	22
32.0069	35.1725	38.0757	41.6384	44.1813	23
33.1963	36.4151	39.3641	42.9798	45.5585	24
34.3816	37.6525	40.6465	44.3141	46.9278	25
35.5631	38.8852	41.9232	45.6417	48.2899	26
36.7412	40.1133	43.1944	46.9630	49.6449	27
37.9159	41.3372	44.4607	48.2782	50.9933	28
39.0875	42.5569	45.7222	49.5879	52.3356	29
40.2560	43.7729	46.9792	50.8922	53.6720	30
51.8050	55.7585	59.3417	63.6907	66.7659	40
63.1671	67.5048	71.4202	76.1539	79.4900	50
74.3970	79.0819	83.2976	88.3794	91.9517	60
85.5271	90.5312	95.0231	100.425	104.215	70
96.5782	101.879	106.629	112.329	116.321	80
107.565	113.145	118.136	124.116	128.299	90
118.498	124.342	129.561	135.807	140.169	100

From "Tables of the Percentage Points of the χ^2-Distribution." *Biometrika*, Vol. 32 (1941), pp. 188–189, by Catherine M. Thompson. Reproduced by permission of Professor D. V. Lindley.

Table 6 *Percentage points of the F distribution*

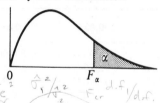

Degrees of Freedom ($\alpha = .05$)

ν_1	1	2	3	4	5	6	7	8	9
ν_2									
1	161.4	199.5	215.7	224.6	230.2	234.0	236.8	238.9	240.5
2	18.51	19.00	19.16	19.25	19.30	19.33	19.35	19.37	19.38
3	10.13	9.55	9.28	9.12	9.01	8.94	8.89	8.85	8.81
4	7.71	6.94	6.59	6.39	6.26	6.16	6.09	6.04	6.00
5	6.61	5.79	5.41	5.19	5.05	4.95	4.88	4.82	4.77
6	5.99	5.14	4.76	4.53	4.39	4.28	4.21	4.15	4.10
7	5.59	4.74	4.35	4.12	3.97	3.87	3.79	3.73	3.68
8	5.32	4.46	4.07	3.84	3.69	3.58	3.50	3.44	3.39
9	5.12	4.26	3.86	3.63	3.48	3.37	3.29	3.23	3.18
10	4.96	4.10	3.71	3.48	3.33	3.22	3.14	3.07	3.02
11	4.84	3.98	3.59	3.36	3.20	3.09	3.01	2.95	2.90
12	4.75	3.89	3.49	3.26	3.11	3.00	2.91	2.85	2.80
13	4.67	3.81	3.41	3.18	3.03	2.92	2.83	2.77	2.71
14	4.60	3.74	3.34	3.11	2.96	2.85	2.76	2.70	2.65
15	4.54	3.68	3.29	3.06	2.90	2.79	2.71	2.64	2.59
16	4.49	3.63	3.24	3.01	2.85	2.74	2.66	2.59	2.54
17	4.45	3.59	3.20	2.96	2.81	2.70	2.61	2.55	2.49
18	4.41	3.55	3.16	2.93	2.77	2.66	2.58	2.51	2.46
19	4.38	3.52	3.13	2.90	2.74	2.63	2.54	2.48	2.42
20	4.35	3.49	3.10	2.87	2.71	2.60	2.51	2.45	2.39
21	4.32	3.47	3.07	2.84	2.68	2.57	2.49	2.42	2.37
22	4.30	3.44	3.05	2.82	2.66	2.55	2.46	2.40	2.34
23	4.28	3.42	3.03	2.80	2.64	2.53	2.44	2.37	2.32
24	4.26	3.40	3.01	2.78	2.62	2.51	2.42	2.36	2.30
25	4.24	3.39	2.99	2.76	2.60	2.49	2.40	2.34	2.28
26	4.23	3.37	2.98	2.74	2.59	2.47	2.39	2.32	2.27
27	4.21	3.35	2.96	2.73	2.57	2.46	2.37	2.31	2.25
28	4.20	3.34	2.95	2.71	2.56	2.45	2.36	2.29	2.24
29	4.18	3.33	2.93	2.70	2.55	2.43	2.35	2.28	2.22
30	4.17	3.32	2.92	2.69	2.53	2.42	2.33	2.27	2.21
40	4.08	3.23	2.84	2.61	2.45	2.34	2.25	2.18	2.12
60	4.00	3.15	2.76	2.53	2.37	2.25	2.17	2.10	2.04
120	3.92	3.07	2.68	2.45	2.29	2.17	2.09	2.02	1.96
∞	3.84	3.00	2.60	2.37	2.21	2.10	2.01	1.94	1.88

Table 6 *Concluded*

10	12	15	20	24	30	40	60	120	∞	
241.9	243.9	245.9	248.0	249.1	250.1	251.1	252.2	253.3	254.3	1
19.40	19.41	19.43	19.45	19.45	19.46	19.47	19.48	19.49	19.50	2
8.79	8.74	8.70	8.66	8.64	8.62	8.59	8.57	8.55	8.53	3
5.96	5.91	5.86	5.80	5.77	5.75	5.72	5.69	5.66	5.63	4
4.74	4.68	4.62	4.56	4.53	4.50	4.46	4.43	4.40	4.36	5
4.06	4.00	3.94	3.87	3.84	3.81	3.77	3.74	3.70	3.67	6
3.64	3.57	3.51	3.44	3.41	3.38	3.34	3.30	3.27	3.23	7
3.35	3.28	3.22	3.15	3.12	3.08	3.04	3.01	2.97	2.93	8
3.14	3.07	3.01	2.94	2.90	2.86	2.83	2.79	2.75	2.71	9
2.98	2.91	2.85	2.77	2.74	2.70	2.66	2.62	2.58	2.54	10
2.85	2.79	2.72	2.65	2.61	2.57	2.53	2.49	2.45	2.40	11
2.75	2.69	2.62	2.54	2.51	2.47	2.43	2.38	2.34	2.30	12
2.67	2.60	2.53	2.46	2.42	2.38	2.34	2.30	2.25	2.21	13
2.60	2.53	2.46	2.39	2.35	2.31	2.27	2.22	2.18	2.13	14
2.54	2.48	2.40	2.33	2.29	2.25	2.20	2.16	2.11	2.07	15
2.49	2.42	2.35	2.28	2.24	2.19	2.15	2.11	2.06	2.01	16
2.45	2.38	2.31	2.23	2.19	2.15	2.10	2.06	2.01	1.96	17
2.41	2.34	2.27	2.19	2.15	2.11	2.06	2.02	1.97	1.92	18
2.38	2.31	2.23	2.16	2.11	2.07	2.03	1.98	1.93	1.88	19
2.35	2.28	2.20	2.12	2.08	2.04	1.99	1.95	1.90	1.84	20
2.32	2.25	2.18	2.10	2.05	2.01	1.96	1.92	1.87	1.81	21
2.30	2.23	2.15	2.07	2.03	1.98	1.94	1.89	1.84	1.78	22
2.27	2.20	2.13	2.05	2.01	1.96	1.91	1.86	1.81	1.76	23
2.25	2.18	2.11	2.03	1.98	1.94	1.89	1.84	1.79	1.73	24
2.24	2.16	2.09	2.01	1.96	1.92	1.87	1.82	1.77	1.71	25
2.22	2.15	2.07	1.99	1.95	1.90	1.85	1.80	1.75	1.69	26
2.20	2.13	2.06	1.97	1.93	1.88	1.84	1.79	1.73	1.67	27
2.19	2.12	2.04	1.96	1.91	1.87	1.82	1.77	1.71	1.65	28
2.18	2.10	2.03	1.94	1.90	1.85	1.81	1.75	1.70	1.64	29
2.16	2.09	2.01	1.93	1.89	1.84	1.79	1.74	1.68	1.62	30
2.08	2.00	1.92	1.84	1.79	1.74	1.69	1.64	1.58	1.51	40
1.99	1.92	1.84	1.75	1.70	1.65	1.59	1.53	1.47	1.39	60
1.91	1.83	1.75	1.66	1.61	1.55	1.50	1.43	1.35	1.25	120
1.83	1.75	1.67	1.57	1.52	1.46	1.39	1.32	1.22	1.00	∞

From "Tables of Percentage Points of the Inverted Beta (*F*)-Distribution," *Biometrika*, Vol. 33 (1943), pp. 73–88, by Maxine Merrington and Catherine M. Thompson. Reproduced by permission of Professor D. V. Lindley.

Table 7 *Percentage points of the F distribution*

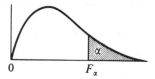

0 $\qquad F_\alpha$

Degrees of Freedom $(\alpha = .01)$

d.f. ν_1 / d.f. ν_2	1	2	3	4	5	6	7	8	9
1	4052	4999.5	5403	5625	5764	5859	5928	5982	6022
2	98.50	99.00	99.17	99.25	99.30	99.33	99.36	99.37	99.39
3	34.12	30.82	29.46	28.71	28.24	27.91	27.67	27.49	27.35
4	21.20	18.00	16.69	15.98	15.52	15.21	14.98	14.80	14.66
5	16.26	13.27	12.06	11.39	10.97	10.67	10.46	10.29	10.16
6	13.75	10.92	9.78	9.15	8.75	8.47	8.26	8.10	7.98
7	12.25	9.55	8.45	7.85	7.46	7.19	6.99	6.84	6.72
8	11.26	8.65	7.59	7.01	6.63	6.37	6.18	6.03	5.91
9	10.56	8.02	6.99	6.42	6.06	5.80	5.61	5.47	5.35
10	10.04	7.56	6.55	5.99	5.64	5.39	5.20	5.06	4.94
11	9.65	7.21	6.22	5.67	5.32	5.07	4.89	4.74	4.63
12	9.33	6.93	5.95	5.41	5.06	4.82	4.64	4.50	4.39
13	9.07	6.70	5.74	5.21	4.86	4.62	4.44	4.30	4.19
14	8.86	6.51	5.56	5.04	4.69	4.46	4.28	4.14	4.03
15	8.68	6.36	5.42	4.89	4.56	4.32	4.14	4.00	3.89
16	8.53	6.23	5.29	4.77	4.44	4.20	4.03	3.89	3.78
17	8.40	6.11	5.18	4.67	4.34	4.10	3.93	3.79	3.68
18	8.29	6.01	5.09	4.58	4.25	4.01	3.84	3.71	3.60
19	8.18	5.93	5.01	4.50	4.17	3.94	3.77	3.63	3.52
20	8.10	5.85	4.94	4.43	4.10	3.87	3.70	3.56	3.46
21	8.02	5.78	4.87	4.37	4.04	3.81	3.64	3.51	3.40
22	7.95	5.72	4.82	4.31	3.99	3.76	3.59	3.45	3.35
23	7.88	5.66	4.76	4.26	3.94	3.71	3.54	3.41	3.30
24	7.82	5.61	4.72	4.22	3.90	3.67	3.50	3.36	3.26
25	7.77	5.57	4.68	4.18	3.85	3.63	3.46	3.32	3.22
26	7.72	5.53	4.64	4.14	3.82	3.59	3.42	3.29	3.18
27	7.68	5.49	4.60	4.11	3.78	3.56	3.39	3.26	3.15
28	7.64	5.45	4.57	4.07	3.75	3.53	3.36	3.23	3.12
29	7.60	5.42	4.54	4.04	3.73	3.50	3.33	3.20	3.09
30	7.56	5.39	4.51	4.02	3.70	3.47	3.30	3.17	3.07
40	7.31	5.18	4.31	3.83	3.51	3.29	3.12	2.99	2.89
60	7.08	4.98	4.13	3.65	3.34	3.12	2.95	2.82	2.72
120	6.85	4.79	3.95	3.48	3.17	2.96	2.79	2.66	2.56
∞	6.63	4.61	3.78	3.32	3.02	2.80	2.64	2.51	2.41

Table 7 *Concluded*

10	12	15	20	24	30	40	60	120	∞	d.f./d.f.
6056	6106	6157	6209	6235	6261	6287	6313	6339	6366	1
99.40	99.42	99.43	99.45	99.46	99.47	99.47	99.48	99.49	99.50	2
27.23	27.05	26.87	26.69	26.60	26.50	26.41	26.32	26.22	26.13	3
14.55	14.37	14.20	14.02	13.93	13.84	13.75	13.65	13.56	13.46	4
10.05	9.89	9.72	9.55	9.47	9.38	9.29	9.20	9.11	9.02	5
7.87	7.72	7.56	7.40	7.31	7.23	7.14	7.06	6.97	6.88	6
6.62	6.47	6.31	6.16	6.07	5.99	5.91	5.82	5.74	5.65	7
5.81	5.67	5.52	5.36	5.28	5.20	5.12	5.03	4.95	4.86	8
5.26	5.11	4.96	4.81	4.73	4.65	4.57	4.48	4.40	4.31	9
4.85	4.71	4.56	4.41	4.33	4.25	4.17	4.08	4.00	3.91	10
4.54	4.40	4.25	4.10	4.02	3.94	3.86	3.78	3.69	3.60	11
4.30	4.16	4.01	3.86	3.78	3.70	3.62	3.54	3.45	3.36	12
4.10	3.96	3.82	3.66	3.59	3.51	3.43	3.34	3.25	3.17	13
3.94	3.80	3.66	3.51	3.43	3.35	3.27	3.18	3.09	3.00	14
3.80	3.67	3.52	3.37	3.29	3.21	3.13	3.05	2.96	2.87	15
3.69	3.55	3.41	3.26	3.18	3.10	3.02	2.93	2.84	2.75	16
3.59	3.46	3.31	3.16	3.08	3.00	2.92	2.83	2.75	2.65	17
3.51	3.37	3.23	3.08	3.00	2.92	2.84	2.75	2.66	2.57	18
3.43	3.30	3.15	3.00	2.92	2.84	2.76	2.67	2.58	2.49	19
3.37	3.23	3.09	2.94	2.86	2.78	2.69	2.61	2.52	2.42	20
3.31	3.17	3.03	2.88	2.80	2.72	2.64	2.55	2.46	2.36	21
3.26	3.12	2.98	2.83	2.75	2.67	2.58	2.50	2.40	2.31	22
3.21	3.07	2.93	2.78	2.70	2.62	2.54	2.45	2.35	2.26	23
3.17	3.03	2.89	2.74	2.66	2.58	2.49	2.40	2.31	2.21	24
3.13	2.99	2.85	2.70	2.62	2.54	2.45	2.36	2.27	2.17	25
3.09	2.96	2.81	2.66	2.58	2.50	2.42	2.33	2.23	2.13	26
3.06	2.93	2.78	2.63	2.55	2.47	2.38	2.29	2.20	2.10	27
3.03	2.90	2.75	2.60	2.52	2.44	2.35	2.26	2.17	2.06	28
3.00	2.87	2.73	2.57	2.49	2.41	2.33	2.23	2.14	2.03	29
2.98	2.84	2.70	2.55	2.47	2.39	2.30	2.21	2.11	2.01	30
2.80	2.66	2.52	2.37	2.29	2.20	2.11	2.02	1.92	1.80	40
2.63	2.50	2.35	2.20	2.12	2.03	1.94	1.84	1.73	1.60	60
2.47	2.34	2.19	2.03	1.95	1.86	1.76	1.66	1.53	1.38	120
2.32	2.18	2.04	1.88	1.79	1.70	1.59	1.47	1.32	1.00	∞

Table 8 *Distribution function of U. $P(U \leq U_0)$; U_0 is the argument*
$n_1 \leq n_2; 3 \leq n_2 \leq 10$.

n_1	$n_2 = 3$		
	1	2	3
0	.25	.10	.05
1	.50	.20	.10
U_0 2		.40	.20
3		.60	.35
4			.50

n_1	$n_2 = 4$			
	1	2	3	4
0	.2000	.0667	.0286	.0143
1	.4000	.1333	.0571	.0286
2	.6000	.2667	.1143	.0571
3		.4000	.2000	.1000
U_0 4		.6000	.3143	.1714
5			.4286	.2429
6			.5714	.3429
7				.4429
8				.5571

n_1	$n_2 = 5$				
	1	2	3	4	5
0	.1667	.0476	.0179	.0079	.0040
1	.3333	.0952	.0357	.0159	.0079
2	.5000	.1905	.0714	.0317	.0159
3		.2857	.1250	.0556	.0278
4		.4286	.1964	.0952	.0476
5		.5714	.2857	.1429	.0754
U_0 6			.3929	.2063	.1111
7			.5000	.2778	.1548
8				.3651	.2103
9				.4524	.2738
10				.5476	.3452
11					.4206
12					.5000

Table 8 *Continued*

$n_2 = 6$

n_1	1	2	3	4	5	6
0	.1429	.0357	.0119	.0048	.0022	.0011
1	.2857	.0714	.0238	.0095	.0043	.0022
2	.4286	.1429	.0476	.0190	.0087	.0043
3	.5714	.2143	.0833	.0333	.0152	.0076
4		.3214	.1310	.0571	.0260	.0130
5		.4286	.1905	.0857	.0411	.0206
6		.5714	.2738	.1286	.0628	.0325
7			.3571	.1762	.0887	.0465
8			.4524	.2381	.1234	.0660
U_0 9			.5476	.3048	.1645	.0898
10				.3810	.2143	.1201
11				.4571	.2684	.1548
12				.5429	.3312	.1970
13					.3961	.2424
14					.4654	.2944
15					.5346	.3496
16						.4091
17						.4686
18						.5314

$n_2 = 7$

n_1	1	2	3	4	5	6	7
0	.1250	.0278	.0083	.0030	.0013	.0006	.0003
1	.2500	.0556	.0167	.0061	.0025	.0012	.0006
2	.3750	.1111	.0333	.0121	.0051	.0023	.0012
3	.5000	.1667	.0583	.0212	.0088	.0041	.0020
4		.2500	.0917	.0364	.0152	.0070	.0035
5		.3333	.1333	.0545	.0240	.0111	.0055
6		.4444	.1917	.0818	.0366	.0175	.0087
7		.5556	.2583	.1152	.0530	.0256	.0131
8			.3333	.1576	.0745	.0367	.0189
9			.4167	.2061	.1010	.0507	.0265
10			.5000	.2636	.1338	.0688	.0364
11				.3242	.1717	.0903	.0487
U_0 12				.3939	.2159	.1171	.0641
13				.4636	.2652	.1474	.0825
14				.5364	.3194	.1830	.1043
15					.3775	.2226	.1297
16					.4381	.2669	.1588
17					.5000	.3141	.1914
18						.3654	.2279
19						.4178	.2675
20						.4726	.3100
21						.5274	.3552
22							.4024
23							.4508
24							.5000

Table 8 *Continued*

	$n_2 = 8$							
n_1	1	2	3	4	5	6	7	8
0	.1111	.0222	.0061	.0020	.0008	.0003	.0002	.0001
1	.2222	.0444	.0121	.0040	.0016	.0007	.0003	.0002
2	.3333	.0889	.0242	.0081	.0031	.0013	.0006	.0003
3	.4444	.1333	.0424	.0141	.0054	.0023	.0011	.0005
4	.5556	.2000	.0667	.0242	.0093	.0040	.0019	.0009
5		.2667	.0970	.0364	.0148	.0063	.0030	.0015
6		.3556	.1394	.0545	.0225	.0100	.0047	.0023
7		.4444	.1879	.0768	.0326	.0147	.0070	.0035
8		.5556	.2485	.1071	.0466	.0213	.0103	.0052
9			.3152	.1414	.0637	.0296	.0145	.0074
10			.3879	.1838	.0855	.0406	.0200	.0103
11			.4606	.2303	.1111	.0539	.0270	.0141
12			.5394	.2848	.1422	.0709	.0361	.0190
13				.3414	.1772	.0906	.0469	.0249
14				.4040	.2176	.1142	.0603	.0325
15				.4667	.2618	.1412	.0760	.0415
U_0 16				.5333	.3108	.1725	.0946	.0524
17					.3621	.2068	.1159	.0652
18					.4165	.2454	.1405	.0803
19					.4716	.2864	.1678	.0974
20					.5284	.3310	.1984	.1172
21						.3773	.2317	.1393
22						.4259	.2679	.1641
23						.4749	.3063	.1911
24						.5251	.3472	.2209
25							.3894	.2527
26							.4333	.2869
27							.4775	.3227
28							.5225	.3605
29								.3992
30								.4392
31								.4796
32								.5204

Table 8 *Continued*

$n_2 = 9$

n_1	1	2	3	4	5	6	7	8	9
0	.1000	.0182	.0045	.0014	.0005	.0002	.0001	.0000	.0000
1	.2000	.0364	.0091	.0028	.0010	.0004	.0002	.0001	.0000
2	.3000	.0727	.0182	.0056	.0020	.0008	.0003	.0002	.0001
3	.4000	.1091	.0318	.0098	.0035	.0014	.0006	.0003	.0001
4	.5000	.1636	.0500	.0168	.0060	.0024	.0010	.0005	.0002
5		.2182	.0727	.0252	.0095	.0038	.0017	.0008	.0004
6		.2909	.1045	.0378	.0145	.0060	.0026	.0012	.0006
7		.3636	.1409	.0531	.0210	.0088	.0039	.0019	.0009
8		.4545	.1864	.0741	.0300	.0128	.0058	.0028	.0014
9		.5455	.2409	.0993	.0415	.0180	.0082	.0039	.0020
10			.3000	.1301	.0559	.0248	.0115	.0056	.0028
11			.3636	.1650	.0734	.0332	.0156	.0076	.0039
12			.4318	.2070	.0949	.0440	.0209	.0103	.0053
13			.5000	.2517	.1199	.0567	.0274	.0137	.0071
14				.3021	.1489	.0723	.0356	.0180	.0094
15				.3552	.1818	.0905	.0454	.0232	.0122
16				.4126	.2188	.1119	.0571	.0296	.0157
17				.4699	.2592	.1361	.0708	.0372	.0200
18				.5301	.3032	.1638	.0869	.0464	.0252
19					.3497	.1942	.1052	.0570	.0313
U_0 20					.3986	.2280	.1261	.0694	.0385
21					.4491	.2643	.1496	.0836	.0470
22					.5000	.3035	.1755	.0998	.0567
23						.3445	.2039	.1179	.0680
24						.3878	.2349	.1383	.0807
25						.4320	.2680	.1606	.0951
26						.4773	.3032	.1852	.1112
27						.5227	.3403	.2117	.1290
28							.3788	.2404	.1487
29							.4185	.2707	.1701
30							.4591	.3029	.1933
31							.5000	.3365	.2181
32								.3715	.2447
33								.4074	.2729
34								.4442	.3024
35								.4813	.3332
36								.5187	.3652
37									.3981
38									.4317
39									.4657
40									.5000

Table 8 *Concluded*

$$n_2 = 10$$

n_1	1	2	3	4	5	6	7	8	9	10
0	.0909	.0152	.0035	.0010	.0003	.0001	.0001	.0000	.0000	.0000
1	.1818	.0303	.0070	.0020	.0007	.0002	.0001	.0000	.0000	.0000
2	.2727	.0606	.0140	.0040	.0013	.0005	.0002	.0001	.0000	.0000
3	.3636	.0909	.0245	.0070	.0023	.0009	.0004	.0002	.0001	.0000
4	.4545	.1364	.0385	.0120	.0040	.0015	.0006	.0003	.0001	.0001
5	.5455	.1818	.0559	.0180	.0063	.0024	.0010	.0004	.0002	.0001
6		.2424	.0804	.0270	.0097	.0037	.0015	.0007	.0003	.0002
7		.3030	.1084	.0380	.0140	.0055	.0023	.0010	.0005	.0002
8		.3788	.1434	.0529	.0200	.0080	.0034	.0015	.0007	.0004
9		.4545	.1853	.0709	.0276	.0112	.0048	.0022	.0011	.0005
10		.5455	.2343	.0939	.0376	.0156	.0068	.0031	.0015	.0008
11			.2867	.1199	.0496	.0210	.0093	.0043	.0021	.0010
12			.3462	.1518	.0646	.0280	.0125	.0058	.0028	.0014
13			.4056	.1868	.0823	.0363	.0165	.0078	.0038	.0019
14			.4685	.2268	.1032	.0467	.0215	.0103	.0051	.0026
15			.5315	.2697	.1272	.0589	.0277	.0133	.0066	.0034
16				.3177	.1548	.0736	.0351	.0171	.0086	.0045
17				.3666	.1855	.0903	.0439	.0217	.0110	.0057
18				.4196	.2198	.1099	.0544	.0273	.0140	.0073
19				.4725	.2567	.1317	.0665	.0338	.0175	.0093
20				.5275	.2970	.1566	.0806	.0416	.0217	.0116
21					.3393	.1838	.0966	.0506	.0267	.0144
22					.3839	.2139	.1148	.0610	.0326	.0177
23					.4296	.2461	.1349	.0729	.0394	.0216
24					.4765	.2811	.1574	.0864	.0474	.0262
U_0 25					.5235	.3177	.1819	.1015	.0564	.0315
26						.3564	.2087	.1185	.0667	.0376
27						.3962	.2374	.1371	.0782	.0446
28						.4374	.2681	.1577	.0912	.0526
29						.4789	.3004	.1800	.1055	.0615
30						.5211	.3345	.2041	.1214	.0716
31							.3698	.2299	.1388	.0827
32							.4063	.2574	.1577	.0952
33							.4434	.2863	.1781	.1088
34							.4811	.3167	.2001	.1237
35							.5189	.3482	.2235	.1399
36								.3809	.2483	.1575
37								.4143	.2745	.1763
38								.4484	.3019	.1965
39								.4827	.3304	.2179
40								.5173	.3598	.2406
41									.3901	.2644
42									.4211	.2894
43									.4524	.3153
44									.4841	.3421
45									.5159	.3697
46										.3980
47										.4267
48										.4559
49										.4853
50										.5147

Computed by M. Pagano, Department of Statistics, University of Florida.

Table 9 *Critical values of T in the Wilcoxon paired-difference test*

$$n = 5(1)50$$

One sided	Two-sided	$n = 5$	$n = 6$	$n = 7$	$n = 8$	$n = 9$	$n = 10$
$P = .05$	$P = .10$	1	2	4	6	8	11
$P = .025$	$P = .05$		1	2	4	6	8
$P = .01$	$P = .02$			0	2	3	5
$P = .005$	$P = .01$				0	2	3

One-sided	Two-sided	$n = 11$	$n = 12$	$n = 13$	$n = 14$	$n = 15$	$n = 16$
$P = .05$	$P = .10$	14	17	21	26	30	36
$P = .025$	$P = .05$	11	14	17	21	25	30
$P = .01$	$P = .02$	7	10	13	16	20	24
$P = .005$	$P = .01$	5	7	10	13	16	19

One-sided	Two-sided	$n = 17$	$n = 18$	$n = 19$	$n = 20$	$n = 21$	$n = 22$
$P = .05$	$P = .10$	41	47	54	60	68	75
$P = .025$	$P = .05$	35	40	46	52	59	66
$P = .01$	$P = .02$	28	33	38	43	49	56
$P = .005$	$P = .01$	23	28	32	37	43	49

One-sided	Two-sided	$n = 23$	$n = 24$	$n = 25$	$n = 26$	$n = 27$	$n = 28$
$P = .05$	$P = .10$	83	92	101	110	120	130
$P = .025$	$P = .05$	73	81	90	98	107	117
$P = .01$	$P = .02$	62	69	77	85	93	102
$P = .005$	$P = .01$	55	68	68	76	84	92

One-sided	Two-sided	$n = 29$	$n = 30$	$n = 31$	$n = 32$	$n = 33$	$n = 34$
$P = .05$	$P = .10$	141	152	163	175	188	201
$P = .025$	$P = .05$	127	137	148	159	171	183
$P = .01$	$P = .02$	111	120	130	141	151	162
$P = .005$	$P = .01$	100	109	118	128	138	149

One-sided	Two-sided	$n = 35$	$n = 36$	$n = 37$	$n = 38$	$n = 39$	
$P = .05$	$P = .10$	214	228	242	256	271	
$P = .025$	$P = .05$	195	208	222	235	250	
$P = .01$	$P = .02$	174	186	198	211	224	
$P = .005$	$P = .01$	160	171	183	195	208	

One-sided	Two-sided	$n = 40$	$n = 41$	$n = 42$	$n = 43$	$n = 44$	$n = 45$
$P = .05$	$P = .10$	287	303	319	336	353	371
$P = .025$	$P = .05$	264	279	295	311	327	344
$P = .01$	$P = .02$	238	252	267	281	297	313
$P = .005$	$P = .01$	221	234	248	262	277	292

One-sided	Two-sided	$n = 46$	$n = 47$	$n = 48$	$n = 49$	$n = 50$	
$P = .05$	$P = .10$	389	408	427	446	466	
$P = .025$	$P = .05$	361	379	397	415	434	
$P = .01$	$P = .02$	329	345	362	380	398	
$P = .005$	$P = .01$	307	323	339	356	373	

From "Some Rapid Approximate Statistical Procedures" (1964), 28, F. Wilcoxon and R. A. Wilcox. Reproduced with the kind permission of Lederle Laboratories, a division of American Cyanamid Company.

Table 10 *Distribution of the total number of runs R in samples of size*
   (n_1, n_2); $P(R \leq a)$

(n_1, n_2)	2	3	4	5	6	7	8	9	10
(2,3)	.200	.500	.900	1.000					
(2,4)	.133	.400	.800	1.000					
(2,5)	.095	.333	.714	1.000					
(2,6)	.071	.286	.643	1.000					
(2,7)	.056	.250	.583	1.000					
(2,8)	.044	.222	.533	1.000					
(2,9)	.036	.200	.491	1.000					
(2,10)	.030	.182	.455	1.000					
(3,3)	.100	.300	.700	.900	1.000				
(3,4)	.057	.200	.543	.800	.971	1.000			
(3,5)	.036	.143	.429	.714	.929	1.000			
(3,6)	.024	.107	.345	.643	.881	1.000			
(3,7)	.017	.083	.283	.583	.833	1.000			
(3,8)	.012	.067	.236	.533	.788	1.000			
(3,9)	.009	.055	.200	.491	.745	1.000			
(3,10)	.007	.045	.171	.455	.706	1.000			
(4,4)	.029	.114	.371	.629	.886	.971	1.000		
(4,5)	.016	.071	.262	.500	.786	.929	.992	1.000	
(4,6)	.010	.048	.190	.405	.690	.881	.976	1.000	
(4,7)	.006	.033	.142	.333	.606	.833	.954	1.000	
(4,8)	.004	.024	.109	.279	.533	.788	.929	1.000	
(4,9)	.003	.018	.085	.236	.471	.745	.902	1.000	
(4,10)	.002	.014	.068	.203	.419	.706	.874	1.000	
(5,5)	.008	.040	.167	.357	.643	.833	.960	.992	1.000
(5,6)	.004	.024	.110	.262	.522	.738	.911	.976	.998
(5,7)	.003	.015	.076	.197	.424	.652	.854	.955	.992
(5,8)	.002	.010	.054	.152	.347	.576	.793	.929	.984
(5,9)	.001	.007	.039	.119	.287	.510	.734	.902	.972
(5,10)	.001	.005	.029	.095	.239	.455	.678	.874	.958
(6,6)	.002	.013	.067	.175	.392	.608	.825	.933	.987
(6,7)	.001	.008	.043	.121	.296	.500	.733	.879	.966
(6,8)	.001	.005	.028	.086	.226	.413	.646	.821	.937
(6,9)	.000	.003	.019	.063	.175	.343	.566	.762	.902
(6,10)	.000	.002	.013	.047	.137	.288	.497	.706	.864
(7,7)	.001	.004	.025	.078	.209	.383	.617	.791	.922
(7,8)	.000	.002	.015	.051	.149	.296	.514	.704	.867
(7,9)	.000	.001	.010	.035	.108	.231	.427	.622	.806
(7,10)	.000	.001	.006	.024	.080	.182	.355	.549	.743
(8,8)	.000	.001	.009	.032	.100	.214	.405	.595	.786
(8,9)	.000	.001	.005	.020	.069	.157	.319	.500	.702
(8,10)	.000	.000	.003	.013	.048	.117	.251	.419	.621
(9,9)	.000	.000	.003	.012	.044	.109	.238	.399	.601
(9,10)	.000	.000	.002	.008	.029	.077	.179	.319	.510
(10,10)	.000	.000	.001	.004	.019	.051	.128	.242	.414

Table 10 *Concluded*

(n_1, n_2)	11	12	13	14	15	16	17	18	19	20
(2,3)										
(2,4)										
(2,5)										
(2,6)										
(2,7)										
(2,8)										
(2,9)										
(2,10)										
(3,3)										
(3,4)										
(3,5)										
(3,6)										
(3,7)										
(3,8)										
(3,9)										
(3,10)										
(4,4)										
(4,5)										
(4,6)										
(4,7)										
(4,8)										
(4,9)										
(4,10)										
(5,5)										
(5,6)	1.000									
(5,7)	1.000									
(5,8)	1.000									
(5,9)	1.000									
(5,10)	1.000									
(6,6)	.998	1.000								
(6,7)	.992	.999	1.000							
(6,8)	.984	.998	1.000							
(6,9)	.972	.994	1.000							
(6,10)	.958	.990	1.000							
(7,7)	.975	.996	.999	1.000						
(7,8)	.949	.988	.998	1.000	1.000					
(7,9)	.916	.975	.994	.999	1.000					
(7,10)	.879	.957	.990	.998	1.000					
(8,8)	.900	.968	.991	.999	1.000	1.000				
(8,9)	.843	.939	.980	.996	.999	1.000	1.000			
(8,10)	.782	.903	.964	.990	.998	1.000	1.000			
(9,9)	.762	.891	.956	.988	.997	1.000	1.000	1.000		
(9,10)	.681	.834	.923	.974	.992	.999	1.000	1.000	1.000	
(10,10)	.586	.758	.872	.949	.981	.996	.999	1.000	1.000	1.000

From "Tables for Testing Randomness of Grouping in a Sequence of Alternatives," F. Swed, and C. Eisenhart, *Annals of Mathematical Statistics*, Volume 14 (1943). Reproduced with the kind permission of the authors and of the Editor, *Annals of Mathematical Statistics*.

Table 11 *Critical values of Spearman's rank correlation coefficient*

n	α = 0.05	α = 0.025	α = 0.01	α = 0.005
5	0.900	—	—	—
6	0.829	0.886	0.943	—
7	0.714	0.786	0.893	—
8	0.643	0.738	0.833	0.881
9	0.600	0.683	0.783	0.833
10	0.564	0.648	0.745	0.794
11	0.523	0.623	0.736	0.818
12	0.497	0.591	0.703	0.780
13	0.475	0.566	0.673	0.745
14	0.457	0.545	0.646	0.716
15	0.441	0.525	0.623	0.689
16	0.425	0.507	0.601	0.666
17	0.412	0.490	0.582	0.645
18	0.399	0.476	0.564	0.625
19	0.388	0.462	0.549	0.608
20	0.377	0.450	0.534	0.591
21	0.368	0.438	0.521	0.576
22	0.359	0.428	0.508	0.562
23	0.351	0.418	0.496	0.549
24	0.343	0.409	0.485	0.537
25	0.336	0.400	0.475	0.526
26	0.329	0.392	0.465	0.515
27	0.323	0.385	0.456	0.505
28	0.317	0.377	0.448	0.496
29	0.311	0.370	0.440	0.487
30	0.305	0.364	0.432	0.478

From "Distribution of Sums of Squares of Rank Differences for Small Samples," E. G. Olds, *Annals of Mathematical Statistics.* Volume 9 (1938). Reproduced with the kind permission of the Editor, *Annals of Mathematical Statistics.*

Table 12 *Sine and cosine values for the functions* $x = \sin\left(\dfrac{2\pi t}{L}\right)$ *and*

$x = \cos\left(\dfrac{2\pi t}{L}\right)$, *for* $L = 4, 6,$ *and* 12

t	$\sin\left(\dfrac{2\pi t}{4}\right)$	$\cos\left(\dfrac{2\pi t}{4}\right)$	$\sin\left(\dfrac{2\pi t}{6}\right)$	$\cos\left(\dfrac{2\pi t}{6}\right)$	$\sin\left(\dfrac{2\pi t}{12}\right)$	$\cos\left(\dfrac{2\pi t}{12}\right)$
1	1	0	.866	.500	.500	.866
2	0	−1	.866	−.500	.866	.500
3	1	0	.000	−1.000	1.000	.000
4	0	1	−.866	−.500	.866	−.500
5	1	0	−.866	.500	.500	−.866
6	0	−1	.000	1.000	.000	−1.000
7	−1	0	.866	.500	−.500	−.866
8	0	1	.866	−.500	−.866	−.500
9	1	0	.000	−1.000	−1.000	.000
10	0	−1	−.866	−.500	−.866	.500
11	−1	0	−.866	.500	−.500	.866
12	0	1	.000	1.000	.000	1.000

Table 13 *Squares and roots*

Roots of numbers other than those given directly may be found by the following relations:

$$\sqrt{100n} = 10\sqrt{n}; \quad \sqrt{1000n} = 10\sqrt{10n}; \quad \sqrt{\tfrac{1}{10}n} = \tfrac{1}{10}\sqrt{10n};$$

$$\sqrt{\tfrac{1}{100}n} = \tfrac{1}{10}\sqrt{n}; \quad \sqrt{\tfrac{1}{1000}n} = \tfrac{1}{100}\sqrt{10n}; \quad \sqrt[3]{1000n} = 10\sqrt[3]{n};$$

$$\sqrt[3]{10,000n} = 10\sqrt[3]{10n}; \quad \sqrt[3]{100,000n} = 10\sqrt[3]{100n};$$

$$\sqrt[3]{\tfrac{1}{10}n} = \tfrac{1}{10}\sqrt[3]{100n}; \quad \sqrt[3]{\tfrac{1}{100}n} = \tfrac{1}{10}\sqrt[3]{10n}; \quad \sqrt[3]{\tfrac{1}{1000}n} = \tfrac{1}{10}\sqrt[3]{n}.$$

n	n^2	$\sqrt{n}$	$\sqrt{10n}$	n	n^2	$\sqrt{n}$	$\sqrt{10n}$
				30	900	5.477 226	17.32051
1	1	1.000 000	3.162 278	31	961	5.567 764	17.60682
2	4	1.414 214	4.472 136	32	1 024	5.656 854	17.88854
3	9	1.732 051	5.477 226	33	1 089	5.744 563	18.16590
4	16	2.000 000	6.324 555	34	1 156	5.830 952	18.43909
5	25	2.236 068	7.071 068	35	1 225	5.916 080	18.70829
6	36	2.449 490	7.745 967	36	1 296	6.000 000	18.97367
7	49	2.645 751	8.366 600	37	1 369	6.082 763	19.23538
8	64	2.828 427	8.944 272	38	1 444	6.164 414	19.49359
9	81	3.000 000	9.486 833	39	1 521	6.244 998	19.74842
10	100	3.162 278	10.00000	40	1 600	6.324 555	20.00000
11	121	3.316 625	10.48809	41	1 681	6.403 124	20.24846
12	144	3.464 102	10.95445	42	1 764	6.480 741	20.49390
13	169	3.605 551	11.40175	43	1 849	6.557 439	20.73644
14	196	3.741 657	11.83216	44	1 936	6.633 250	20.97618
15	225	3.872 983	12.24745	45	2 025	6.708 204	21.21320
16	256	4.000 000	12.64911	46	2 116	6.782 330	21.44761
17	289	4.123 106	13.03840	47	2 209	6.855 655	21.67948
18	324	4.242 641	13.41641	48	2 304	6.928 203	21.90890
19	361	4.358 899	13.78405	49	2 401	7.000 000	22.13594
20	400	4.472 136	14.14214	50	2 500	7.071 068	22.36068
21	441	4.582 576	14.49138	51	2 601	7.141 428	22.58318
22	484	4.690 416	14.83240	52	2 704	7.211 103	22.80351
23	529	4.795 832	15.16575	53	2 809	7.280 110	23.02173
24	576	4.898 979	15.49193	54	2 916	7.348 469	23.23790
25	625	5.000 000	15.81139	55	3 025	7.416 198	23.45208
26	676	5.099 020	16.12452	56	3 136	7.483 315	23.66432
27	729	5.196 152	16.43168	57	3 249	7.549 834	23.87467
28	784	5.291 503	16.73320	58	3 364	7.615 773	24.08319
29	841	5.385 165	17.02939	59	3 481	7.618 146	24.28992

Table 13 *Continued*

n	n²	√n	√10n	n	n²	√n	√10n
60	3 600	7.745 967	24.49490	100	10 000	10.00000	31.62278
61	3 721	7.810 250	24.69818	101	10 201	10.04998	31.78050
62	3 844	7.874 008	24.89980	102	10 404	10.09950	31.93744
63	3 969	7.937 254	25.09980	103	10 609	10.14889	32.09361
64	4 096	8.000 000	25.29822	104	10 816	10.19804	32.24903
65	4 225	8.062 258	25.49510	105	11 025	10.24695	32.40370
66	4 356	8.124 038	25.69047	106	11 236	10.29563	32.55764
67	4 489	8.185 353	25.88436	107	11 449	10.34408	32.71085
68	4 624	8.246 211	26.07681	108	11 664	10.39230	32.86335
69	4 761	8.306 624	26.26785	109	11 881	10.44031	33.01515
70	4 900	8.366 600	26.45751	110	12 100	10.48809	33.16625
71	5 041	8.426 150	26.64583	111	12 321	10.53565	33.31666
72	5 184	8.485 281	26.83282	112	12 544	10.58301	33.46640
73	5 329	8.544 004	27.01851	113	12 769	10.63015	33.61547
74	5 476	8.602 325	27.20294	114	12 996	10.67708	33.76389
75	5 625	8.660 254	27.38613	115	13 225	10.72381	33.91165
76	5 776	8.717 798	27.56810	116	13 456	10.77033	34.05877
77	5 929	8.774 964	27.74887	117	13 689	10.81665	34.20526
78	6 084	8.831 761	27.92848	118	13 924	10.86278	34.35113
79	6 241	8.888 194	28.10694	119	14 161	10.90871	34.49638
80	6 400	8.944 272	28.28427	120	14 400	10.95445	34.64102
81	6 561	9.000 000	28.46050	121	14 641	11.00000	34.78505
82	6 724	9.055 385	28.63564	122	14 884	11.04536	34.92850
83	6 889	9.110 434	28.80972	123	15 129	11.09054	35 07136
84	7 056	9.165 151	28.98275	124	15 376	11.13553	35 21363
85	7 225	9.219 544	29.15476	125	15 625	11.18034	35.35534
86	7 396	9.273 618	29.32576	126	15 876	11.22497	35.49648
87	7 569	9.327 379	29.49576	127	16 129	11.26943	35.63706
88	7 744	9.380 832	29.66479	128	16 384	11.31371	35.77709
89	7 921	9.433 981	29.83287	129	16 641	11.35782	35.91657
90	8 100	9.486 833	30.00000	130	16 900	11.40175	36.05551
91	8 281	9.539 392	30.16621	131	17 161	11.44552	36.19392
92	8 464	9.591 663	30.33150	132	17 424	11.48913	36.33180
93	8 649	9.643 651	30.49590	133	17 689	11.53256	36.46917
94	8 836	9.695 360	30.65942	134	17 956	11.57584	36.60601
95	9 025	9.746 794	30.82207	135	18 225	11.61895	36.74235
96	9 216	9.797 959	30.98387	136	18 496	11.66190	36.87818
97	9 409	9.848 858	31.14482	137	18 769	11.70470	37.01351
98	9 604	9.899 495	31.30495	138	19 044	11.74734	37.14835
99	9 801	9.949 874	31.46427	139	19 321	11.78983	37.28270

Table 13 *Continued*

n	n²	√n	√10n	n	n²	√n	√10n
140	19 600	11.83216	37.41657	**180**	32 400	13.41641	42.42641
141	19 881	11.87434	37.54997	181	32 761	13.45362	42.54409
142	20 164	11.91638	37.68289	182	33 124	13.49074	42.66146
143	20 449	11.95826	37.81534	183	33 489	13.52775	42.77850
144	20 736	12.00000	37.94733	184	33 856	13 56466	42.89522
145	21 025	12.04159	38.07887	185	34 225	13.60147	43.01163
146	21 316	12.08305	38.20995	186	34 596	13.63818	43.12772
147	21 609	12.12436	38.34058	187	34 969	13.67479	43.24350
148	21 904	12.16553	38.47077	188	35 344	13.71131	43.35897
149	22 201	12.20656	38.60052	189	35 721	13.74773	43.47413
150	22 500	12.24745	38.72983	**190**	36 100	13.78405	43.58899
151	22 801	12.28821	38.85872	191	36 481	13.82027	43.70355
152	23 104	12.32883	38.98718	192	36 864	13.85641	43.81780
153	23 409	12.36932	39.11521	193	37 249	13.89244	43.93177
154	23 716	12.40967	39.24283	194	37 636	13.92839	44.04543
155	24 025	12.44990	39.37004	195	38 025	13.96424	44.15880
156	24 336	12.49000	39.49684	196	38 416	14.00000	44.27189
157	24 649	12.52996	39.62323	197	38 809	14.03567	44.38468
158	24 964	12.56981	39.74921	198	39 204	14.07125	44.49719
159	25 281	12.60952	39.87480	199	39 601	14.10674	44.60942
160	25 600	12.64911	40.00000	**200**	40 000	14.14214	44.72136
161	25 921	12.68858	40.12481	201	40 401	14.17745	44.83302
162	26 244	12.72792	40.24922	202	40 804	14.21267	44.94441
163	26 569	12.76715	40.37326	203	41 209	14.24781	45.05552
164	26 806	12.80625	40.49691	204	41 616	14.28286	45.16636
165	27 225	12.84523	40.62019	205	42 025	14.31782	45.27693
166	27 556	12.88410	40.74310	206	42 436	14.35270	45.38722
167	27 889	12.92285	40.86563	207	42 849	14.38749	45.49725
168	28 224	12.96148	40.98780	208	43 264	14.42221	45.60702
169	28 561	13.00000	41.10961	209	43 681	14.45683	45.71652
170	28 900	13.03840	41.23106	**210**	44 100	14.49138	45.82576
171	29 241	13.07670	41.35215	211	44 521	14.52584	45.93474
172	29 584	13.11488	41.47288	212	44 944	14.56022	46.04346
173	29 929	13.15295	41.59327	213	45 369	14.59452	46.15192
174	30 276	13.19091	41.71331	214	45 796	14.62874	46.26013
175	30 625	13.22876	41.83300	215	46 225	14.66288	46.36809
176	30 976	13.26650	41.95235	216	46 656	14.69694	46.47580
177	31 329	13.30413	42.07137	217	47 089	14.73092	46.58326
178	31 684	13.34166	42.19005	218	47 524	14.76482	46.69047
179	32 041	13.37909	42.30829	219	47 961	14.79865	46.79744

Table 13 *Continued*

n	n^2	$\sqrt{n}$	$\sqrt{10n}$	n	n^2	$\sqrt{n}$	$\sqrt{10n}$
220	48 400	14.83240	46.90416	**260**	67 600	16.12452	50.99020
221	48 841	14.86607	47.01064	261	68 121	16.15549	51.08816
222	49 284	14.89966	47.11688	262	68 644	16.18641	51.18594
223	49 729	14.93318	47.22288	263	69 169	16.21727	51.28353
224	50 176	14.96663	47.32864	264	69 696	16.24808	51.38093
225	50 625	15.00000	47.43416	265	70 225	16.27882	51.47815
226	51 076	15.03330	47.53946	266	70 756	16.30951	51.57519
227	51 529	15.06652	47.64452	267	71 289	16.34013	51.67204
228	51 984	15.09967	47.74935	268	71 824	16.37071	51.76872
229	52 441	15.13275	47.85394	269	72 361	16.40122	51.86521
230	52 900	15.16575	47.95832	**270**	72 900	16.43168	51.96152
231	53 361	15.19868	48.06246	271	73 441	16.46208	52.05766
232	53 824	15.23155	48.16638	272	73 984	16.49242	52.15362
233	54 289	15.26434	48.27007	273	74 529	16.52271	52.24940
234	54 756	15.29706	48.37355	274	75 076	16.55295	52.34501
235	55 225	15.32971	48.47680	275	75 625	16.58312	52.44044
236	55 696	15.36229	48.57983	276	76 176	16.61235	52.53570
237	56 169	15.39480	48.68265	277	76 729	16.64332	52.63079
238	56 644	15.42725	48.78524	278	77 284	16.67333	52.72571
239	57 121	15.45962	48.88763	279	77 841	16.70329	52.82045
240	57 600	15.49193	48.98979	**280**	78 400	16.73320	52.91503
241	58 081	15.52417	49.09175	281	78 961	16.76305	53.00943
242	58 564	15.55635	49.19350	282	79 524	16.79286	53.10367
243	59 049	15.58846	49.29503	283	80 089	16.82260	53.19774
244	59 536	15.62050	49.39636	284	80 656	16.85230	53.29165
245	60 025	15.65248	49.49747	285	81 225	16.88194	53.38539
246	60 516	15.68439	49.59839	286	81 796	16.91153	53.47897
247	61 009	15.71623	49.69909	287	82 369	16.94107	53.57238
248	61 504	15.74902	49.79960	288	82 944	16.97056	53.66563
249	62 001	15.77973	49.89990	289	83 521	17.00000	53.75872
250	62 500	15.81139	50.00000	**290**	84 100	17.02939	53.85165
251	63 001	15.84298	50.09990	291	84 681	17.05872	53.94442
252	63 504	15.87451	50.19960	292	85 264	17.08801	54.03702
253	64 009	15.90597	50.29911	293	85 849	17.11724	54.12947
254	64 516	15.93738	50.39841	294	86 436	17.14643	54.22177
255	65 025	15.96872	50.49752	295	87 025	17.17556	54.31390
256	65 536	16.00000	50.59644	296	87 616	17.20465	54.40588
257	66 049	16.03122	50.69517	297	88 209	17.23369	54.49771
258	66 564	16.06238	50.79370	298	88 804	17.26268	54.58938
259	67 081	16.09348	50.89204	299	89 401	17.29162	54.68089

Table 13   *Continued*

n	n^2	$\sqrt{n}$	$\sqrt{10n}$	n	n^2	$\sqrt{n}$	$\sqrt{10n}$
300	90 000	17.32051	54.77226	**340**	115 600	18.43909	58.30952
301	90 601	17.34935	54.86347	341	116 281	18.46619	58.39521
302	91 204	17.37815	54.95453	342	116 964	18.49324	58.48077
303	91 809	17.40690	55.04544	343	117 649	18.52026	58.56620
304	92 416	17.43560	55.13620	344	118 336	18.54724	58.65151
305	93 025	17.46425	55.22681	345	119 025	18.57418	58.73670
306	93 636	17.49286	55.31727	346	119 716	18.60108	58.82176
307	94 249	17.52142	55.40758	347	120 409	18.62794	58.90671
308	94 864	17.54993	55.49775	348	121 104	18.65476	58.99152
309	95 481	17.57840	55.58777	349	121 801	18.68154	59.07622
310	96 100	17.60682	55.67764	**350**	122 500	18.70829	59.16080
311	96 721	17.63519	55.76737	351	123 201	18.73499	59.24525
312	97 344	17.66352	55.85696	352	123 904	18.76166	59.32959
313	97 969	17.69181	55.94640	353	124 609	18.78829	59.41380
314	98 596	17.72005	56.03570	354	125 316	18.81489	59.49790
315	99 225	17.74824	56.12486	355	126 025	18.84144	59.58188
316	99 856	17.77639	56.21388	356	126 736	18.86796	59.66574
317	100 489	17.80449	56.30275	357	127 449	18.89444	59.74948
318	101 124	17.83255	56.39149	358	128 164	18.92089	59.83310
319	101 761	17.86057	56.48008	359	128 881	18.94730	59.91661
320	102 400	17.88854	56.56854	**360**	129 600	18.97367	60.00000
321	103 041	17.91647	56.65686	361	130 321	19.00000	60.08328
322	103 684	17.94436	56.74504	362	131 044	19.02630	60.16644
323	104 329	17.97220	56.83309	363	131 769	19.05256	60.24948
324	104 976	18.00000	56.92100	364	132 496	19.07878	60.33241
325	105 625	18.02776	57.00877	365	133 225	19.10497	60.41523
326	106 276	18.05547	57.09641	366	133 956	19.13113	60.49793
327	106 929	18.08314	57.18391	367	134 689	19.15724	60.58052
328	107 584	18.11077	57.27128	368	135 424	19.18333	60.66300
329	108 241	18.13836	57.35852	369	136 161	19.20937	60.74537
330	108 900	18.16590	57.44563	**370**	136 900	19.23538	60.82763
331	109 561	18.19341	57.53260	371	137 641	19.26136	60.90977
332	110 224	18.22087	57.61944	372	138 384	19.28730	60.99180
333	110 889	18.24829	57.70615	373	139 129	19.31321	61.07373
334	111 556	18.27567	57.79273	374	139 876	19.33908	61.15554
335	112 225	18.30301	57.87918	375	140 625	19.36492	61.23724
336	112 896	18.33030	57.96551	376	141 376	19.39072	61.31884
337	113 569	18.35756	58.05170	377	142 129	19.41649	61.40033
338	114 244	18.38478	58.13777	378	142 884	19.44222	61.48170
339	114 921	18.41195	58.22371	379	143 641	19.46792	61.56298

Table 13 *Continued*

n	n²	√n	√10n	n	n²	√n	√10n
380	144 400	19.49359	61.64414	**420**	176 400	20.49390	64.80741
381	145 161	19.51922	61.72520	421	177 241	20.51828	64.88451
382	145 924	19.54482	61.80615	422	178 084	20.54264	64.96153
383	146 689	19.57039	61.88699	423	178 929	20.56696	65.03845
384	147 456	19.59592	61.96773	424	179 776	20.59126	65.11528
385	148 225	19.62142	62.04837	425	180 625	20.61553	65.19202
386	148 996	19.64688	62.12890	426	181 476	20.63977	65.26868
387	149 769	19.67232	62.20932	427	182 329	20.66398	65.34524
388	150 544	19.69772	62.28965	428	183 184	20.68816	65.42171
389	151 321	19.72308	62.36986	429	184 041	20.71232	65.49809
390	152 100	19.74842	62.44998	**430**	184 900	20.73644	65.57439
391	152 881	19.77372	62.52999	431	185 761	20.76054	65.65059
392	153 664	19.79899	62.60990	432	186 624	20.78461	65.72671
393	154 449	19.82423	62.68971	433	187 489	20.80865	65.80274
394	155 236	19.84943	62.76942	434	188 356	20.83267	65.87868
395	156 025	19.87461	62.84903	435	189 225	20.85665	65.95453
396	156 816	19.89975	62.92853	436	190 096	20.88061	66.03030
397	157 609	19.92486	63.00794	437	190 969	20.90454	66.10598
398	158 404	19.94994	63.08724	438	191 844	20.92845	66.18157
399	159 201	19.97498	63.16645	439	192 721	20.95233	66.25708
400	160 000	20.00000	63.24555	**440**	193 600	20.97618	66.33250
401	160 801	20.02498	63.32456	441	194 481	21.00000	66.40783
402	161 604	20.04994	63.40347	442	195 364	21.02380	66.48308
403	162 409	20.07486	63.48228	443	196 249	21.04757	66.55825
404	163 216	20.09975	63.56099	444	197 136	21.07131	66.63332
405	164 025	20.12461	63.63961	445	198 025	21.09502	66.70832
406	164 836	20.14944	63.71813	446	198 916	21.11871	66.78323
407	165 649	20.17424	63.79655	447	199 809	21.14237	66.85806
408	166 464	20.19901	63.87488	448	200 704	21.16601	66.93280
409	167 281	20.22375	63.95311	449	201 601	21.18962	67.00746
410	168 100	20.24864	64.03124	**450**	202 500	21.21320	67.08204
411	168 921	20.27313	64.10928	451	203 401	21.23676	67.15653
412	169 744	20.29778	64.18723	452	204 304	21.26029	67.23095
413	170 569	20.32240	54.26508	453	205 209	21.28380	67.30527
414	171 396	20.34699	64.34283	454	206 116	21.30728	67.37952
415	172 225	20.37155	64.42049	455	207 025	21.33073	67.45369
416	173 056	20.39608	64.49806	456	207 936	21.35416	67.52777
417	173 889	20.42058	64.57554	457	208 849	21.37756	67.60178
418	174 724	20.44505	64.65292	458	209 764	21.40093	67.67570
419	175 561	20.46949	64.73021	459	210 681	21.42429	67.74954

Table 13 *Continued*

n	n^2	$\sqrt{n}$	$\sqrt{10n}$	n	n^2	$\sqrt{n}$	$\sqrt{10n}$
460	211 600	21.44761	67.82330	**500**	250 000	22.36068	70.71068
461	212 521	21.47091	67.89698	501	251 001	22.38303	70.78135
462	213 444	21.49419	67.97058	502	252 004	22.40536	70.85196
463	214 369	21.51743	68.04410	503	253 009	22.42766	70.92249
464	215 296	21.54066	68.11755	504	254 016	22.44994	70.99296
465	216 225	21.56386	68.19091	505	255 025	22.47221	71.06335
466	217 156	21.58703	68.26419	506	256 036	22.49444	71.13368
467	218 089	21.61018	68.33740	507	257 049	22.51666	71.20393
468	219 024	21.63331	68.41053	508	258 064	22.53886	71.27412
469	219 961	21.65641	68.48957	509	259 081	22.56103	71.34424
470	220 900	21.67948	68.55655	**510**	260 100	22.58318	71.41428
471	221 841	21.70253	68.62944	511	261 121	22.60531	71.48426
472	222 784	21.72556	68.70226	512	262 144	22.62742	71.55418
473	223 729	21.74856	68.77500	513	263 169	22.64950	71.62402
474	224 676	21.77154	68.84766	514	264 196	22.67157	71.69379
475	225 625	21.79449	68.92024	515	265 225	22.69361	71.76350
476	226 576	21.81742	68.99275	516	266 256	22.71563	71.83314
477	227 529	21.84033	69.06519	517	267 289	22.73763	71.90271
478	228 484	21.86321	69.13754	518	268 324	22.75961	71.97222
479	229 441	21.88607	69.20983	519	269 361	22.78157	72.04165
480	230 400	21.90890	69.28203	**520**	270 400	22.80351	72.11103
481	231 361	21.93171	69.35416	521	271 441	22.82542	72.18033
482	232 324	21.95450	69.42622	522	272 484	22.84732	72.24957
483	233 289	21.97726	69.49820	523	273 529	22.86919	72.31874
484	234 256	22.00000	69.57011	524	274 576	22.89105	72.38784
485	235 225	22.02272	69.64194	525	275 625	22.91288	72.45688
486	236 196	22.04541	69.71370	526	276 676	22.93469	72.52586
487	237 169	22.06808	69.78539	527	277 729	22.95648	72.59477
488	238 144	22.09072	69.85700	528	278 784	22.97825	72.66361
489	239 121	22.11334	69.92853	529	279 841	23.00000	72.73239
490	240 100	22.13594	70.00000	**530**	280 900	23.02173	72.80110
491	241 081	22.15852	70.07139	531	281 961	23.04344	72.86975
492	242 064	22.18107	70.14271	532	283 024	23.06513	72.93833
493	243 049	22.20360	70.21396	533	284 089	23.08679	73.00685
494	244 036	22.22611	70.28513	534	285 156	23.10844	73.07530
495	245 025	22.24860	70.35624	535	286 225	23.13007	73.14369
496	246 016	22.27106	70.42727	536	287 296	23.15167	73.21202
497	247 009	22.29350	70.49823	537	288 369	23.17326	73.28028
498	248 004	22.31591	70.56912	538	289 444	23.19483	73.34848
499	249 001	22.33831	70.63993	539	290 521	23.21637	73.41662

Table 13 *Continued*

n	n²	√n	√10n	n	n²	√n	√10n
540	291 600	23.23790	73.48469	**580**	336 400	24.08319	76.15773
541	292 681	23.25941	73.55270	581	337 561	24.10394	76.22336
542	293 764	23.28089	73.62065	582	338 724	24.12468	76.28892
543	294 849	23.30236	73.68853	583	339 889	24.14539	76.35444
544	295 936	23.32381	73.75636	584	341 056	24.16609	76.41989
545	297 025	23.34524	73.82412	585	342 225	24.18677	76.48529
546	298 116	23.36664	73.89181	586	343 396	24.20744	76.55064
547	299 209	23.38803	73.95945	587	344 569	24.22808	76.61593
548	300 304	23.40940	74.02702	588	345 744	24.24871	76.68116
549	301 401	23.43075	74.09453	589	346 921	24.26932	76.74634
550	302 500	23.45208	74.16198	**590**	348 100	24.28992	76.81146
551	303 601	23.47339	74.22937	591	349 281	24.31049	76.87652
552	304 704	23.49468	74.29670	592	350 464	24.33105	76.94154
553	305 809	23.51595	74.36397	593	351 649	24.35159	77.00649
554	306 916	23.53720	74.43118	594	352 836	24.37212	77.07140
555	308 025	23.55844	74.49832	595	354 025	24.39262	77.13624
556	309 136	23.57965	74.56541	596	355 216	24.41311	77.20104
557	310 249	23.60085	74.63243	597	356 409	24.43358	77.26578
558	311 364	23.62202	74.69940	598	357 604	24.45404	77.33046
559	312 481	23.64318	74.76630	599	358 801	24.47448	77.39509
560	313 600	23.66432	74.83315	**600**	360 000	24.49490	77.45967
561	314 721	23.68544	74.89993	601	361 201	24.51530	77.52419
562	315 844	23.70654	74.96666	602	362 404	24.53569	77.58866
563	316 969	23.72762	75.03333	603	363 609	24.55606	77.65307
564	318 096	23.74868	75.09993	604	364 816	24.57641	77.71744
565	319 225	23.76973	75.16648	605	366 025	24.59675	77.78175
566	320 356	23.79075	75.23297	606	367 236	24.61707	77.84600
567	321 489	23.81176	75.29940	607	368 449	24.63737	77.91020
568	322 624	23.83275	75.36577	608	369 664	24.65766	77.97435
569	323 761	23.85372	75.43209	609	370 881	24.67793	78.03845
570	324 900	23.87467	75.49834	**610**	372 100	24.69818	78.10250
571	326 041	23.89561	75.56454	611	373 321	24.71841	78.16649
572	327 184	23.91652	75.63068	612	374 544	24.73863	78.23043
573	328 329	23.93742	75.69676	613	375 769	24.75884	78.29432
574	329 476	23.95830	75.76279	614	376 996	24.77902	78.35815
575	330 625	23.97916	75.82875	615	378 225	24.79919	78.42194
576	331 776	24.00000	75.89466	616	379 456	24.81935	78.48567
577	332 929	24.02082	75.96052	617	380 689	24.83948	78.54935
578	334 084	24.04163	76.02631	618	381 924	24.85961	78.61298
579	335 241	24.06242	76.09205	619	383 161	24.87971	78.67655

Table 13 *Continued*

n	n²	√n	√10n	n	n²	√n	√10n
620	384 400	24.89980	78.74008	660	435 600	25.69047	81.24038
621	385 641	24.91987	78.80355	661	436 921	25.70992	81.30191
622	386 884	24.93993	78.86698	662	438 244	25.72936	81.36338
623	388 129	24.95997	78.93035	663	439 569	25.74879	81.42481
624	389 376	24.97999	78.99367	664	440 896	25.76820	81.48620
625	390 625	25.00000	79.05694	665	442 225	25.78759	81.54753
626	391 876	25.01999	79.12016	666	443 556	25.80698	81.60882
627	393 129	25.03997	79.18333	667	444 889	25.82634	81.67007
628	394 384	25.05993	79.24645	668	446 224	25.84570	81.73127
629	395 641	25.07987	79.30952	669	447 561	25.86503	81.79242
630	396 900	25.09980	79.37254	670	448 900	25.88436	81.85353
631	398 161	25.11971	79.43551	671	450 241	25.90367	81.91459
632	399 424	25.13961	79.49843	672	451 584	25.92296	81.97561
633	400 689	25.15949	79.56130	673	452 929	25.94224	82.03658
634	401 956	25.17936	79.62412	674	454 276	25.96151	82.09750
635	403 225	25.19921	79.68689	675	455 625	25.98076	82.15838
636	404 496	25.21904	79.74961	676	456 976	26.00000	82.21922
637	405 769	25.23886	79.81228	677	458 329	26.01922	82.28001
638	407 044	25.25866	79.87490	678	459 684	26.03843	82.34076
639	408 321	25.27845	79.93748	679	461 041	26.05763	82.40146
640	409 600	25.29822	80.00000	680	462 400	26.07681	82.46211
641	410 881	25.31798	80.06248	681	463 761	26.09598	82.52272
642	412 164	25.33772	80.12490	682	465 124	26.11513	82.58329
643	413 449	25.35744	80.18728	683	466 489	26.13427	82.64381
644	414 736	25.37716	80.24961	684	467 856	26.15339	82.70429
645	416 025	25.39685	80.31189	685	469 225	26.17250	82.76473
646	417 316	25.41653	80.37413	686	470 596	26.19160	82.82512
647	418 609	25.43619	80.43631	687	471 969	26.21068	82.88546
648	419 904	25.45584	80.49845	688	473 344	26.22975	82.94577
649	421 201	25.47548	80.56054	689	474 721	26.24881	83.00602
650	422 500	25.49510	80.62258	690	476 100	26.26785	83.06624
651	423 801	25.51470	80.68457	691	477 481	26.28688	83.12641
652	425 104	25.53429	80.74652	692	478 864	26.30589	83.18654
653	426 409	25.55386	80.80842	693	480 249	26.32489	83.24662
654	427 716	25.57342	80.87027	694	481 636	26.34388	83.30666
655	429 025	25.59297	80.93207	695	483 025	26.36285	83.36666
656	430 336	25.61250	80.99383	696	484 416	26.38181	83.42661
657	431 649	25.63201	81.05554	697	485 809	26.40076	83.48653
658	432 964	25.65151	81.11720	698	487 204	26.41969	83.54639
659	434 281	25.67100	81.17881	699	488 601	26.43861	83.60622

Table 13 *Continued*

n	n²	√n	√10n	n	n²	√n	√10n
700	490 000	26.45751	83.66600	**740**	547 600	27.20294	86.02325
701	491 401	26.47640	83.72574	741	549 081	27.22132	86.08136
702	492 804	26.49528	83.78544	742	550 564	27.23968	86.13942
703	494 209	26.51415	83.84510	743	552 049	27.25803	86.19745
704	495 616	26.53300	83.90471	744	553 536	27.27636	86.25543
705	497 025	26.55184	83.96428	745	555 025	27.29469	86.31338
706	498 436	26.57066	84.02381	746	556 516	27.31300	86.37129
707	499 849	26.58947	84.08329	747	558 009	27.33130	86.42916
708	501 264	26.60827	84.14274	748	559 504	27.34959	86.48699
709	502 681	26.62705	84.20214	749	561 001	27.36786	86.54479
710	504 100	26.64583	84.26150	**750**	562 500	27.38613	86.60254
711	505 521	26.66458	84.32082	751	564 001	27.40438	86.66026
712	506 944	26.68333	84.38009	752	565 504	27.42262	86.71793
713	508 369	26.70206	84.43933	753	567 009	27.44085	86.77557
714	509 796	26.72078	84.49852	754	568 516	27.45906	86.83317
715	511 225	26.73948	84.55767	755	570 025	27.47726	86.89074
716	512 656	26.75818	84.61678	756	571 536	27.49545	86.94826
717	514 089	26.77686	84.67585	757	573 049	27.51363	87.00575
718	515 524	26.79552	84.73488	758	574 564	27.53180	87.06320
719	516 961	26.81418	84.79387	759	576 081	27.54995	87.12061
720	518 400	26.83282	84.85281	**760**	577 600	27.56810	87.17798
721	519 841	26.85144	84.91172	761	579 121	27.58623	87.23531
722	521 284	26.87006	84.97058	762	580 644	27.60435	87.29261
723	522 729	26.88866	85.02941	763	582 169	27.62245	87.34987
724	524 176	26.90725	85.08819	764	583 696	27.64055	87.40709
725	525 625	26.92582	85.14693	765	585 225	27.65863	87.46428
726	527 076	26.94439	85.20563	766	586 756	27.67671	87.52143
727	528 529	26.96294	85.26429	767	588 289	27.69476	87.57854
728	529 984	26.98148	85.32292	768	589 824	27.71281	87.63561
729	531 441	27.00000	85.38150	769	591 361	27.73085	87.69265
730	532 900	27.01851	85.44004	**770**	592 900	27.74887	87.74964
731	534 361	27.03701	85.49854	771	594 441	27.76689	87.80661
732	535 824	27.05550	85.55700	772	595 984	27.78489	87.86353
733	537 289	27.07397	85.61542	773	597 529	27.80288	87.92042
734	538 756	27.09243	85.67380	774	599 076	27.82086	87.97727
735	540 225	27.11088	85.73214	775	600 625	27.83882	88.03408
736	541 696	27.12932	85.79044	776	602 176	27.85678	88.09086
737	543 169	27.14774	85.84870	777	603 729	27.87472	88.14760
738	544 644	27.16616	85.90693	778	605 284	27.89265	88.20431
739	546 121	27.18455	85.96511	779	606 841	27.91057	88.26098

Table 13   *Continued*

n	n²	√n	√10n	n	n²	√n	√10n
780	608 400	27.92848	88.31761	**820**	672 400	28.63564	90.55385
781	609 961	27.94638	88.37420	821	674 041	28.65310	90.60905
782	611 524	27.96426	88.43076	822	675 684	28.67054	90.66422
783	613 089	27.98214	88.48729	823	677 329	28.68798	90.71935
784	614 656	28.00000	88.54377	824	678 976	28.70540	90.77445
785	616 225	28.01785	88.60023	825	680 625	28.72281	90.82951
786	617 796	28.03569	88.65664	826	682 726	28.74022	90.88454
787	619 369	28.05352	88.71302	827	683 929	28.75761	90.93954
788	620 944	28.07134	88.76936	828	685 584	28.77499	90.99451
789	622 521	28.08914	88.82567	829	687 241	28.79236	91.04944
790	624 100	28.10694	88.88194	**830**	688 900	28.80972	91.10434
791	625 681	28.12472	88.93818	831	690 561	28.82707	91.15920
792	627 264	28.14249	88.99428	832	692 224	28.84441	91.21403
793	628 849	28.16026	89.05055	833	693 889	28.86174	91.26883
794	630 436	28.17801	89.10668	834	695 556	28.87906	91.32360
795	632 025	28.19574	89.16277	835	697 225	28.89637	91.37833
796	633 616	28.21347	89.21883	836	698 896	28.91366	91.43304
797	635 209	28.23119	89.27486	837	700 569	28.93095	91.48770
798	636 804	28.24889	89.33085	838	702 244	28.94823	91.54234
799	638 401	28.26659	89.38680	839	703 921	28.96550	91.59694
800	640 000	28.28472	89.44272	**840**	705 600	28.98275	91.65151
801	641 601	28.30194	89.49860	841	707 281	29.00000	91.70605
802	643 204	28.31960	89.55445	842	708 964	29.01724	91.76056
803	644 809	28.33725	89.61027	843	710 649	29.03446	91.81503
804	646 416	28.35489	89.66605	844	712 336	29.05168	91.86947
805	648 025	28.37252	89.72179	845	714 025	29.06888	91.92388
806	649 636	28.39014	89.77750	846	715 716	29.08608	91.97826
807	651 249	28.40775	89.83318	847	717 409	29.10326	92.03260
808	652 864	28.42534	89.88882	848	719 104	29.12044	92.08692
809	654 481	28.44293	89.94443	849	720 801	29.13760	92.14120
810	656 100	28.46050	90.00000	**850**	722 500	29.15476	92.19544
811	657 721	28.47806	90.05554	851	724 201	29.17190	92.24966
812	659 344	28.49561	90.11104	852	725 904	29.18904	92.30385
813	660 969	28.51315	90.16651	853	727 609	29.20616	92.35800
814	662 596	28.53069	90.22195	854	729 316	29.22328	92.41212
815	664 225	28.54820	90.27735	855	731 025	29.24038	92.46621
816	665 856	28.56571	90.33272	856	732 736	29.25748	92.52027
817	667 489	28.58321	90.38805	857	734 449	29.27456	92.57429
818	669 124	28.60070	90.44335	858	736 164	29.29164	92.62829
819	670 761	28.61818	90.49862	859	737 881	29.30870	92.68225

Table 13 *Continued*

n	n^2	$\sqrt{n}$	$\sqrt{10n}$	n	n^2	$\sqrt{n}$	$\sqrt{10n}$
860	739 600	29.32576	92.73618	**900**	810 000	30.00000	94.86833
861	741 321	29.34280	92.79009	901	811 801	30.01666	94.92102
862	743 044	29.35984	92.84396	902	813 604	30.03331	94.97368
863	744 769	29.37686	92.89779	903	815 409	30.04996	95.02631
864	746 496	29.39388	92.95160	904	817 216	30.06659	95.07891
865	748 225	29.41088	93.00538	905	819 025	30.08322	95.13149
866	749 956	29.42788	93.05912	906	820 836	30.09983	95.18403
867	751 689	29.44486	93.11283	907	822 649	30.11644	95.23655
868	753 424	29.46184	93.16652	908	824 464	30.13304	95.28903
869	755 161	29.47881	93.22017	909	826 281	30.14963	95.34149
870	756 900	29.49576	93.27379	**910**	828 100	30.16621	95.39392
871	758 641	29.51271	93.32738	911	829 921	30.18278	95.44632
872	760 384	29.52965	93.38094	912	831 744	30.19934	95.49869
873	762 129	29.54657	93.43447	913	833 569	30.21589	95.55103
874	763 876	29.56349	93.48797	914	835 396	30.23243	95.60335
875	765 625	29.58040	93.54143	915	837 225	30.24897	95.65563
876	767 376	29.59730	93.59487	916	839 056	30.26549	95.70789
877	769 129	29.61419	93.64828	917	840 889	30.28201	95.76012
878	770 884	29.63106	93.70165	918	842 724	30.29851	95.81232
879	772 641	29.64793	93.75500	919	844 561	30.31501	95.86449
880	774 400	29.66479	93.80832	**920**	846 400	30.33150	95.91663
881	776 161	29.68164	93.86160	921	848 241	30.34798	95.96874
882	777 924	29.69848	93.91486	922	850 084	30.36445	96.02083
883	779 689	29.71532	93.96808	923	851 929	30.38092	96.07289
884	781 456	29.73214	94.02127	924	853 776	30.39737	96.12492
885	783 225	29.74895	94.07444	925	855 625	30.41381	96.17692
886	784 996	29.76575	94.12757	926	857 476	30.43025	96.22889
887	786 769	29.78255	94.18068	927	859 329	30.44667	96.28084
888	788 544	29.79933	94.23375	928	861 184	30.46309	96.33276
889	790 321	29.81610	94.28680	929	863 041	30.47950	96.38465
890	792 100	29.83287	94.33981	**930**	864 900	30.49590	96.43651
891	793 881	29.84962	94.39280	931	866 761	30.51229	96.48834
892	795 664	29.86637	94.44575	932	868 624	30.52868	96.54015
893	797 449	29.88311	94.49868	933	870 489	30.54505	96.59193
894	799 236	29.89983	94.55157	934	872 356	30.56141	96.64368
895	801 025	29.91655	94.60444	935	874 225	30.57777	96.69540
896	802 816	29.93326	94.65728	936	876 096	30.59412	96.74709
897	804 609	29.94996	94.71008	937	877 969	30.61046	96.79876
898	806 404	29.96665	94.76286	938	879 844	30.62679	96.85040
899	808 201	29.98333	94.81561	939	881 721	30.64311	96.90201

Table 13 *Concluded*

n	n²	√n	√10n	n	n²	√n	√10n
940	883 600	30.65942	96.95360	**970**	940 900	31.14482	98.48858
941	885 481	30.67572	97.00515	971	942 841	31.16087	98.53933
942	887 364	30.69202	97.05668	972	944 784	31.17691	98.59006
943	889 249	30.70831	97.10819	973	946 729	31.19295	98.64076
944	891 136	30.72458	97.15966	974	948 676	31.20897	98.69144
945	893 025	30.74085	97.21111	975	950 625	31.22499	98.74209
946	894 916	30.75711	97.26253	976	952 576	31.24100	98.79271
947	896 809	30.77337	97.31393	977	954 529	31.25700	98.84331
948	898 704	30.78961	97.36529	978	956 484	31.27299	98.89388
949	900 601	30.80584	97.41663	979	958 441	31.28898	98.94443
950	902 500	30.82207	97.46794	**980**	960 400	31.30495	98.99495
951	904 401	30.83829	97.51923	981	962 361	31.32092	99.04544
952	906 304	30.85450	97.57049	982	964 324	31.33688	99.09591
953	908 209	30.87070	97.62172	983	966 289	31.35283	99.14636
954	910 116	30.88689	97.67292	984	968 256	31.36877	99.19677
955	912 025	30.90307	97.72410	985	970 225	31.38471	99.24717
956	913 936	30.91925	97.77525	986	972 196	31.40064	99.29753
957	915 849	30.93542	97.82638	987	974 169	31.41656	99.34787
958	917 764	30.95158	97.87747	988	976 144	31.43247	99.39819
959	919 681	30.96773	97.92855	989	978 121	31.44837	99.44848
960	921 600	30.98387	97.97959	**990**	980 100	31.46427	99.49874
961	923 521	31.00000	98.03061	991	982 081	31.48015	99.54898
962	925 444	31.01612	98.08160	992	984 064	31.49603	99.59920
963	927 369	31.03224	98.13256	993	986 049	31.51190	99.64939
964	929 296	31.04835	98.18350	994	988 036	31.52777	99.69955
965	931 225	31.06445	98.23441	995	990 025	31.54362	99.74969
966	933 156	31.08054	98.28530	996	992 016	31.55947	99.79980
967	935 089	31.09662	98.33616	997	994 009	31.57531	99.84989
968	937 024	31.11270	98.38699	998	996 004	31.59114	99.89995
969	938 961	31.12876	98.43780	999	998 001	31.60696	99.94999
				1000	1000 000	31.62278	100.00000

From *Handbook of Tables for Probability and Statistics,* 2d ed. Edited by William H. Beyer (Cleveland: The Chemical Rubber Company, 1968). Reproduced by permission of CRC Press, Inc.

Table 14 Random numbers

Line/Col.	(1)	(2)	(3)	(4)	(5)	(6)	(7)	(8)	(9)	(10)	(11)	(12)	(13)	(14)
1	10480	15011	01536	02011	81647	91646	69179	14194	62590	36207	20969	99570	91291	90700
2	22368	46573	25595	85393	30995	89198	27982	53402	93965	34095	52666	19174	39615	99505
3	24130	48360	22527	97265	76393	64809	15179	24830	49340	32081	30680	19655	63348	58629
4	42167	93093	06243	61680	07856	16376	39440	53537	71341	57004	00849	74917	77758	16379
5	37570	39975	81837	16656	06121	91782	60468	81305	49684	60672	14110	06927	01263	54613
6	77921	06907	11008	42751	27756	53498	18602	70659	90655	15053	21916	81825	44394	42880
7	99562	72905	56420	69994	98872	31016	71194	18738	44013	48840	63213	21069	10634	12952
8	96301	91977	05463	07972	18876	20922	94595	56869	69014	60045	18425	84903	42508	32307
9	89579	14342	63661	10281	17453	18103	57740	84378	25331	12566	58678	44947	05585	56941
10	85475	36857	53342	53988	53060	59533	38867	62300	08158	17983	16439	11458	18593	64952
11	28918	69578	88231	33276	70997	79936	56865	05859	90106	31595	01547	85590	91610	78188
12	63553	40961	48235	03427	49626	69445	18663	72695	52180	20847	12234	90511	33703	90322
13	09429	93969	52636	92737	88974	33488	36320	17617	30015	08272	84115	27156	30613	74952
14	10365	61129	87529	85689	48237	52267	67689	93394	01511	26358	85104	20285	29975	89868
15	07119	97336	71048	08178	77233	13916	47564	81056	97735	85977	29372	74461	28551	90707
16	51085	12765	51821	51259	77452	16308	60756	92144	49442	53900	70960	63990	75601	40719
17	02368	21382	52404	60268	89368	19885	55322	44819	01188	65255	64835	44919	05944	55157
18	01011	54092	33362	94904	31273	04146	18594	29852	71585	85030	51132	01915	92747	64951
19	52162	53916	46369	58586	23216	14513	83149	98736	23495	64350	94738	17752	35156	35749
20	07056	97628	33787	09998	42698	06691	76988	13602	51851	46104	88916	19509	25625	58104
21	48663	91245	85828	14346	09172	30168	90229	04734	59193	22178	30421	61666	99904	32812
22	54164	58492	22421	74103	47070	25306	76468	26384	58151	06646	21524	15227	96909	44592
23	32639	32363	05597	24200	13363	38005	94342	28728	35806	06912	17012	64161	18296	22851
24	29334	27001	87637	87308	58731	00256	45834	15398	46557	41135	10367	07684	36188	18510
25	02488	33062	28834	07351	19731	92420	60952	61280	50001	67658	32586	86679	50720	94953

Abridged from *Handbook of Tables for Probability and Statistics*, 2d ed. Edited by William H. Beyer (Cleveland: The Chemical Rubber Company, 1968). Reproduced by permission of CRC Press, Inc.

Table 14 Continued

26	81525	72295	04839	96423	24878	66566	82651	14778	76797	14780	13300	87074	79666	95725
27	29676	20591	68086	26432	46901	89768	20849	81536	86645	12659	92259	57102	80428	25280
28	00742	57392	39064	66432	84673	32832	40027	61362	98947	96067	64760	64584	96096	98253
29	05366	04213	25669	26422	44407	37937	44048	63904	45766	66134	75470	66520	34693	90449
30	91921	26418	64117	94305	26766	39972	25940	22209	71500	64568	91402	42416	07844	09618
31	00582	04711	87917	77341	42206	35126	74087	99547	81817	42607	43808	76655	62028	76630
32	00725	69884	62797	56170	86324	88072	76222	36086	84637	93161	76038	65855	77919	88006
33	69011	65795	95876	55293	18988	27354	26575	08625	40801	59920	29841	80150	12777	48501
34	25976	57948	29888	88604	67917	48708	18912	82271	65424	69774	33611	54262	85963	03547
35	09763	83473	73577	12908	30883	18317	28290	35797	05998	41688	34952	37888	38917	88050
36	91567	42595	27958	30134	04024	86385	29880	99730	55536	84855	29080	09250	79656	73211
37	17955	56349	90999	49127	20044	59931	06115	20542	18059	02008	73708	83517	36103	42791
38	46503	18584	18845	49618	02304	51038	20655	58727	28168	15475	56942	53389	20562	87338
39	92157	89634	94824	78171	84610	82834	09922	25417	44137	48413	25555	21246	35509	20468
40	14577	62765	35605	81263	39667	47358	56873	56307	61607	49518	89656	20103	77490	18062
41	98427	07523	33362	64270	01638	92477	66969	98420	04880	45585	46565	04102	46880	45709
42	34914	63976	88720	82765	34476	17032	87589	40836	32427	70002	70663	88863	77775	69348
43	70060	28277	39475	46473	23219	53416	94970	25832	69975	94884	19661	72828	00102	66794
44	53976	54914	06990	67245	68350	82948	11398	42878	80287	88287	47363	46634	06541	97809
45	76072	29515	40980	07391	58745	25774	22987	80059	39911	96189	41151	14222	60697	59583
46	90725	52210	83974	29992	65831	38857	50490	83765	55657	14361	31720	57375	56228	41546
47	64364	67412	33339	31926	14883	24413	59744	92351	97473	89286	35931	04110	23726	51900
48	08962	00358	31662	25388	61642	34072	81249	35648	56891	69352	48373	45578	78547	81788
49	95012	68379	93526	70765	10592	04542	76463	54328	02349	17247	28865	14777	62730	92277
50	15664	10493	20492	38391	91132	21999	59516	81652	27195	48223	46751	22923	32261	85653
51	16408	81899	04153	53381	79401	21438	83035	92350	36693	31238	59649	91754	72772	02338
52	18629	81953	05520	91962	04739	13092	97662	24822	94730	06496	35090	04822	86774	88289
53	73115	35101	47498	87637	99016	71060	88824	71013	18735	20286	23153	72924	35165	43040
54	57491	16703	23167	49323	45021	33132	12544	41035	80780	45393	44812	12515	98931	91202
55	30405	83946	23792	14422	15059	45799	22716	19792	09983	74353	68668	30429	70735	25499
56	16631	35006	85900	98275	32388	52390	16815	69298	82739	38480	73817	32523	41961	44437
57	96773	20206	42559	78985	05300	22164	24369	54224	35083	19687	11052	91491	60383	19746
58	38935	64202	14349	82674	66523	44133	00697	35552	35970	19124	63318	29686	03387	59840
59	31624	76384	17403	53363	44167	64486	64758	75366	76554	31501	12614	33072	60332	92325
60	78919	19474	23632	27889	47914	02584	37680	20801	72152	39339	34806	08930	85001	87820
61	03931	33309	57047	74211	63445	17361	62825	39908	05607	91284	68833	25570	38818	46920
62	74426	33278	43972	10119	89917	15665	52872	73823	73144	88662	88970	74492	51805	99378
63	09066	00903	20795	95452	92648	45454	09552	88815	16553	51125	79375	97596	16296	66092
64	42238	12426	87025	14267	20979	04508	64535	31355	86064	29472	47689	05974	52468	16834
65	16153	08002	26504	41744	81959	65642	74240	56302	00033	67107	77510	70625	28725	34191

Table 14 *Concluded*

Line/Col.	(1)	(2)	(3)	(4)	(5)	(6)	(7)	(8)	(9)	(10)	(11)	(12)	(13)	(14)
66	21457	40742	29820	96783	29400	21840	15035	34537	33310	06116	95240	15957	16572	06004
67	21581	57802	02050	89728	17937	37621	47075	42080	97403	48626	68995	43805	33386	21597
68	55612	78095	83197	33732	05810	24813	86902	60397	16489	03264	88525	42786	05269	92532
69	44657	66999	99324	51281	84463	60563	79312	93454	68876	25471	93911	25650	12682	73572
70	91340	84979	46949	81973	37949	61023	43997	15263	80644	43942	89203	71795	99533	50501
71	91227	21199	31935	27022	84067	05462	35216	14486	29891	68607	41867	14951	91696	85065
72	50001	38140	66321	19924	72163	09538	12151	06878	91903	18749	34405	56087	82790	70925
73	65390	05224	72958	28609	81406	39147	25549	48542	42627	45233	57202	94617	23772	07896
74	27504	96131	83944	41575	10573	08619	64482	73923	36152	05184	94142	25299	84387	34925
75	37169	94851	39117	89632	00959	16487	65536	49071	39782	17095	02330	74301	00275	48280
76	11508	70225	51111	38351	19444	66499	71945	05422	13442	78675	84081	66938	93654	59894
77	37449	30362	06694	54690	04052	53115	62757	95348	78662	11163	81651	50245	34971	52924
78	46515	70331	85922	38329	57015	15765	97161	17869	45349	61796	66345	81073	49106	79860
79	30986	81223	42416	58353	21532	30502	32305	86482	05174	07901	54339	58861	74818	46942
80	63798	64995	46583	09785	44160	78128	83091	42865	92520	83531	80377	35909	81250	54238
81	82486	84846	99254	67632	43218	50076	21361	64816	51202	88124	41870	52689	51275	83556
82	21885	32906	92431	09060	64297	51674	64126	62570	26123	05155	59194	52799	28225	85762
83	60336	98782	07408	53458	13564	59089	26445	29789	85205	41001	12535	12133	14645	23541
84	43937	46891	24010	25560	86355	33941	25786	54990	71899	15475	95434	98227	21824	19585
85	97656	63175	89303	16275	07100	92063	21942	18611	47348	20203	18534	03862	78095	50136
86	03299	01221	05418	38982	55758	92237	26759	86367	21216	98442	08303	56313	91511	75928
87	79626	06486	03574	17668	07785	76020	79924	25651	83325	88428	85076	72811	22717	50585
88	85636	68335	47539	03129	65651	11977	02510	26113	99447	68645	34327	15152	55230	93448
89	18039	14367	61337	06177	12143	46609	32989	74014	64708	00533	35398	55408	13261	47908
90	08362	15656	60627	36478	65648	16764	53412	09013	07832	41574	17639	82163	60859	75567
91	79556	29068	04142	16268	15387	12856	66227	38358	22478	73373	88732	09443	82558	05250
92	92608	82674	27072	32534	17075	27698	98204	63863	11951	34648	88022	56148	34925	57031
93	23982	25835	40055	67006	12293	02753	14827	23235	35071	99704	37543	11601	35503	85171
94	09915	96306	05908	97901	28395	14186	00821	80703	70426	75647	76310	88717	37890	40129
95	59037	33300	26695	62247	69927	76123	50842	43834	86654	70959	79725	93872	28117	19233
96	42488	78077	69882	61657	34136	79180	97526	43092	04098	73571	80799	76536	71255	64239
97	46764	86273	63003	93017	31204	36692	40202	35275	57306	55543	53203	18098	47625	88684
98	03237	45430	55417	63282	90816	17349	88298	90183	36600	78406	06216	95787	42579	90730
99	86591	81482	52667	61582	14972	90053	89534	76036	49199	43716	97548	04379	46370	28672
100	38534	01715	94964	87288	65680	43772	39560	12918	86537	62738	19636	51132	25739	56947

Glossary

Acceptance region (chapters 6, 8, and 9)

In the theory of hypothesis testing, the set of values of the test statistic such that if the test statistic assumes one of these values, the null hypothesis is accepted.

Arithmetic mean (throughout the text)

The arithmetic mean of a set of n measurements y_1, y_2, y_3, ... , y_n is equal to the sum of the measurements divided by n.

Autocorrelation (chapters 12 and 15)

The internal correlation between members of a time series separated by a constant interval of time.

Autoregression (chapter 15)

The generation of a series of observations whereby the value of each observation is partly dependent on the values of those which have immediately preceded it. A regression structure where lagged response values assume the role of the independent variables.

Bayes's law (chapters 4 and 10)

If A is some event that occurs if and only if either B or $\bar{B}$ occurs, then

$$P(B|A) = \frac{P(A|B) P(B)}{P(A|B) P(B) + P(A|\bar{B}) P(\bar{B})} = \frac{P(AB)}{P(A)}$$

Binomial experiment (chapter 6)

An experiment consisting of n independent trials in which the outcome at each trial is a "success" with probability p or a "failure" with probability $1 - p$. We are interested in y, the number of successes observed during the n trials.

Binomial probability distribution (chapter 6)

A probability distribution giving the probability of y, the number of successes observed during the n trials of a binomial experiment:

$$p(y) = C_y^n p^y (1 - p)^{n-y} \qquad y = 0, 1, 2, ... , n$$

Box-Jenkins forecasting procedure (chapter 15)

A four-stage procedure for the development of a forecasting model which combines an autoregressive component involving past observed values with a moving-average computer involving current and part error terms.

Census (chapter 16)

A recording of every element contained in a population.

Central Limit Theorem (chapters 7, 8, and 9)

If random samples of n observations are drawn from a population with finite mean μ and standard deviation σ, then when n is large, the sample mean $\bar{y}$ will be approximately normally distributed with mean μ and standard deviation $\sigma/\sqrt{n}$. The approximation will become more and more accurate as n becomes large.

Cluster sample (chapter 16)

A cluster sample is obtained by first randomly selecting a set of m collections of sample elements, called clusters, from the population and then conducting a complete census within each cluster.

Coefficient of correlation (chapter 11)

A number between -1 and $+1$, which measures the linear dependence between two random variables. The limiting values -1 and $+1$ indicate perfect negative and perfect positive correlation, respectively, while a correlation of zero suggests a complete lack of association between the two variables.

Coefficient of determination (chapters 11 and 12)

A measure of the goodness of fit of a regression model that is equal to the square of the coefficient of correlation (simple or multiple, whichever is appropriate).

Combination (chapter 4)

The number of combinations of n objects taken r at a time is denoted by the symbol C_r^n, where

$$C_r^n = \frac{n!}{r!\,(n-r)!}$$

Complement of an event (chapter 4)

The complement of an event A is the collection of all sample points in the sample space that are not in A. The complement of A is denoted by the symbol $\bar{A}$, and $P(\bar{A}) = 1 - P(A)$.

Compound event (chapter 4)

An event composed of two or more simple events.

Conditional probability (chapter 4)

The probability of occurrence of an event A given that another event B has occurred is called the conditional probability of A given B and is denoted as $P(A|B)$. Computationally, $P(A|B) = P(AB)/P(B)$.

Confidence coefficient (chapters 8 and 9)

A probability associated with a confidence interval that expresses the probability that the interval will include the parameter value under study.

Confidence interval (chapters 8 and 9)

An interval computed from sample values. Intervals so constructed will straddle the estimated parameter $100(1 - \alpha)\%$ of the time in repeated sampling. The quantity $(1 - \alpha)$ is called the confidence coefficient.

Contingency table (chapter 17)

A two-way table for classifying the members of a group according to two or more identifying characteristics.

Continuous random variable (chapter 5)

A random variable defined over, and assuming the infinitely many values associated with, the points on a line interval.

Convenience sampling (chapter 16)

The selection of a sample that can be obtained simply and conveniently.

Correlogram (chapter 15)

A graph illustrating the autocorrelations between members of a time series (vertical axis) for different separations in time k (horizontal axis).

Cost of uncertainty (chapter 10)

The smallest expected opportunity loss in a decision-making problem involving uncertainty.

Critical value (chapters 6, 8, and 9)

In a statistical test of an hypothesis, the critical value is the value of the test statistic that separates the rejection and acceptance regions.

Cyclic effect (chapter 15)

A periodic movement in a time series that occurs as a result of stimuli from the economy and is generally not predictable.

Decision analysis (chapter 10)

The logical and quantitative analysis of all the factors that influence a decision.

Degrees of freedom (chapter 9 and throughout the text)

The number of linearly independent observations in a set of n observations. The degrees of freedom are equal to n minus the number of restrictions placed on the entire data set.

Delphi method (chapter 15)

A method for forecasting in the absence of a relevant data base in which a sequence of questionnaires are used to identify the factors that can best be of assistance in focusing attention on the resultant forecast.

Dependent variable (chapters 11, 12, and 15)

The predictand in a regression equation. The variable of interest in a regression equation which is said to be functionally related to one or more independent or predictor variables.

Design of an experiment (throughout the text)

The sampling procedure that enables the gathering of a maximum amount of information for a given expenditure.

Discrete random variable (chapter 5)

A random variable over a finite or a countably infinite number of points.

Econometric model (chapter 15)

An econometric model is a probabilistic model consisting of a system of one or more equations that describe the relationship among a number of economic and time series variables.

Empirical Rule (throughout the text)

Given a distribution of measurements that is approximately bell-shaped, the interval

$$(\mu \pm \sigma) \text{ contains approximately 68\% of the measurements.}$$

($\mu \pm 2\sigma$) contains approximately 95% of the measurements.

($\mu \pm 3\sigma$) contains approximately all the measurements.

Event (chapter 4)

A collection of sample points.

Exogenous variables (chapter 15)

Independent or predictor variables in an econometric model.

Expected value (chapters 5, 11, and 12)

Let y be a discrete random variable with probability distribution $p(y)$ and let $E(y)$ represent the expected value of y. Then

$$E(y) = \sum_{y} yp(y)$$

where the elements are summed over all values of the random variable y.

Exponential smoothing (chapters 14 and 15)

In time series analysis, a computational method that averages the first t time series values by increasingly weighting out the contribution of remote values. The exponentially smoothed value at time t is

$$S_t = \alpha \sum_{i=0}^{t-2} (1 - \alpha)^i y_{t-i} + (1 - \alpha)^{t-1} y_1$$

where $y_1, y_2, \ldots, y_t$ are the response values and α is the smoothing constant ($0 \le \alpha \le 1$).

Frame (chapter 16)

A list of sampling units.

Frequency distribution (chapter 3)

A specification of the way in which, probabilistically, the relative frequencies of members of a population are distributed according to the values of the variates they exhibit.

Frequency histogram (chapter 3)

A specification of the way in which the frequencies of members of a population are distributed according to the values of the variates they exhibit.

Historical analogy (chapter 15)

A method for forecasting the sales of a newly introduced product (or service) that uses the sales history of some previously introduced product as a guide.

Hypergeometric probability distribution (chapter 6)

Appropriate when sampling from a finite population of N elements of which k are identified as "successes" and $(N - k)$ are "failures." Then $p(y)$ gives the probability of observing y successes in a sample of n selected from the population.

Independent variable (chapters 11 and 12)

A nonrandom variable related to the response in a regression equation. One or more independent variables may be functionally related to the dependent variable. They are used in the regression equation to predict or estimate the value of the dependent variable.

Index number (chapter 14)

A quantity that shows the changes over time from some base period to a reference period of a variable process that is not directly observable in practice.

Intersection (chapter 4)

If A and B are two events in a sample space S, the intersection of A and B is the event composed of all sample points that are in both A and B and is denoted by AB.

Judgment sampling (chapter 16)

The selection of a sample which, according to the judgment and intuition of the sampler, accurately reflects the population.

Laspeyres index (chapter 14)

If the prices of a set of commodities in a base year are p_{01}, p_{02}, p_{03}, ... and q_{01}, q_{02}, q_{03}, ... are the quantities sold in the base period and p_{n1}, p_{n2}, p_{n3}, ... are the prices of the same commodities in a given year, the Laspeyres index is

$$L = \frac{\sum p_n q_0}{\sum p_0 q_0}$$

Least squares (chapters 11 and 12)

See Method of least squares.

Linear correlation (chapters 11 and 12)

A measure of the strength of the linear relationship between two variables y and x that is independent of their respective scales of measurement. Linear correlation is commonly measured by the coefficient of correlation.

Linear statistical model (chapter 12)

An equation of the form

$$y = \beta_0 + \beta_1 x_1 + \beta_2 x_2 + \cdots + \beta_k x_k + \epsilon$$

that relates a single dependent variable y to a set of independent variables $x_1, x_2, \ldots, x_k$, where $\beta_0, \beta_1, \ldots, \beta_k$ are unknown parameters and ϵ is a random error.

Median (chapter 3)

The median of a set of n measurements $y_1, y_2, y_3, \ldots, y_n$ is the value of y that falls in the middle when the measurements are arranged in order of magnitude.

Method of least squares (chapters 11, 12, and 15)

A technique for the estimation of the coefficients in a regression equation that chooses as the regression coefficients the values that minimize the sum of squares of the deviations of the observed values of y from those predicted.

Minimax decision (chapter 10)

In a decision analysis the minimax decision is the decision to select the action whose maximum opportunity loss is the smallest.

Mode (chapter 3)

The mode of a set of measurements $y_1, y_2, y_3, \ldots, y_n$ is the value of y that occurs with the greatest frequency. The mode is not necessarily unique.

Moving average (chapter 14)

An average computed from a time series by selecting process values from k

consecutive time periods, summing these values, and dividing by k. The moving average is then located at the middle of the span of the k values which contributed to it.

Multiple regression analysis (chapter 12)

A regression analysis in which the mean value of a dependent variable y is assumed to be related to a set of independent variables $x_1, x_2, \ldots, x_k$ by an expression of the form,

$$E(y|x_1, x_2, \ldots, x_k) = \beta_0 + \beta_1 x_1 + \beta_2 x_2 + \cdots + \beta_k x_k$$

Mutually exclusive events (chapter 4)

Two events A and B are said to be mutually exclusive if the event AB contains no sample points.

Nonparametric hypothesis (chapter 18)

A statistical hypothesis that does not involve population parameters but is concerned with the form of the population frequency distribution.

Nonresponse (chapter 16)

In a sample survey the failure to obtain information from a designated element for any reason.

Normal probability distribution (throughout the text)

A symmetric bell-shaped probability distribution of infinite range represented by the equation

$$f(y) = \frac{1}{\sigma \sqrt{2\pi}} \, e^{-(y-\mu)^2/2\sigma^2} \qquad (-\infty < y < \infty)$$

where μ is the mean and σ^2 is the variance of the distribution.

Null hypothesis (throughout the text)

In a statistical test of an hypothesis, the null hypothesis is a statement of the hypothesis to be tested.

One-tailed statistical test (chapters 8 and 9)

A statistical test of an hypothesis in which the rejection region is wholly located at one end of the distribution of the test statistic.

Operating characteristic curve (chapters 8 and 9)

A plot of the probability of accepting the null hypothesis when some alternative is true (the probability of a type II error) against various possible values of the alternative hypothesis.

Opportunity loss (chapter 10)

The opportunity loss L_{ij} for selecting action a_i given that the state of nature s_j is in effect is the difference between the maximum profit that could be realized if s_j occurs and the profit obtained by selecting action a_i.

Optimal decision (chapter 10)

In a decision analysis the optimal decision is a decision to select the action that maximizes the decision maker's objective.

Paired-difference test (chapter 9)

A test to compare two populations by using pairs of elements, one from each population, that are matched and hence nearly alike. Thus the test involves two samples of equal size, where the members of one sample can be paired against members of the other. Comparisons are made within the relatively homogeneous pairs (blocks).

Panel consensus method (chapter 15)

A method for forecasting in the absence of a relevant data base in which experts within the organization derive a forecast that represents a consensus of the opinions of the experts.

Parameter (throughout the text)

A numerical descriptive measure for the population.

Parametric hypothesis (throughout the text)

A statistical hypothesis about a population parameter.

Payoff table (chapter 10)

In a decision analysis the payoff table is a two-way table displaying the payoffs, either opportunity losses or profits, for selecting a particular action given that a specific state of nature is in effect.

Period of a cyclic or seasonal effect (chapters 14 and 15)

The number of time points between identifiable points of recurrence—between peaks and valleys—in a time series. The number of time points for one complete cycle or for a complete seasonal pattern to be exhibited.

Permutation (chapter 4)

An ordered arrangement of r distinct objects. The number of n objects selected in groups of size r is denoted by P_r^n, where

$$P_r^n = n(n - 1)(n - 2) \cdots (n - r + 1)$$

Point estimator (chapters 8 and 9)

A single number computed from a sample and used as an estimator of a population parameter.

Poisson probability distribution (chapter 6)

A model for finding the probability of count data resulting from any experiment, where the count y represents the number of rare events observed in a given unit of time or space.

Population (throughout the text)

A finite or infinite collection of measurements or individuals that comprises the totality of all possible measurements within the context of a particular statistical study.

Posterior probability (chapters 4 and 10)

The probability p_1 of an event at the outset of an experiment might be modified to p_2 in light of experimental evidence. The posterior probability p_2 is usually determined by employing Bayes's theorem.

Power of a statistical test (chapters 8 and 9)

The probability that the statistical test rejects the null hypothesis when some particular alternative is true. Power equals $1 - \beta$. The power is greatest when the probability of a type II error is least.

Prior probability (chapter 10)

The probability representing the likelihood of occurrence of an event before experimental evidence relevant to the event has been observed. The unconditional probabilities used in Bayes's theorem are prior probabilities.

Probability distribution (chapter 5)

A formula, table, or graph providing the probability associated with each value of the random variable if the random variable is discrete or providing the fraction of measurements in the population falling in specific intervals if it is continuous.

Probability of an event (throughout the text)

The probability of an event A is equal to the sum of the probabilities of the sample points in A. (*See also*, Subjective probability.)

Proportional allocation (chapter 16)

An allocation procedure that partitions the sample size among the strata proportional to the size of the strata when using stratified random sampling.

Qualitative variables (chapters 13 and 17)

Variables concerning qualitative or attribute data. The data are not necessarily representable in numerical form.

Quantitative variables (chapters 13 and 17)

Variables concerning quantitative or measurement data. The data can be represented on a numerical scale.

Quota sampling (chapter 16)

The selection of a predetermined number of elements from different sectors of the population.

Random sample (chapters 7, 16, and throughout the text)

Suppose that a sample of n measurements is drawn from a population consisting of N total measurements. If the sampling is conducted in such a way that each of the C_n^N samples has an equal probability of being selected, the sampling is said to be random and the result is said to be a random sample.

Random variable (chapters 4, 5, and throughout the text)

A numerical-valued function defined over a sample space.

Randomized block design (chapter 13)

An experimental design whereby treatments are randomly assigned within a set of blocks to eliminate bias.

Randomized response sampling (chapter 16)

A survey-sampling procedure for dealing with potentially sensitive or embarrassing material which requires that a question on the sensitive topic be paired with an innocuous question. The respondent then answers only one of the two questions which he or she has selected at random.

Range (chapters 3 and 16)

The range of a set of n measurements y_1, y_2, y_3, ... , y_n is the difference between the largest and smallest measurement.

Rank sum test (chapter 18)

A nonparametric statistical test proposed by Wilcoxon for comparing two population distributions by first ordering the combined observations from the samples selected from each population and then summing the ranks of the observations from one of the samples.

Ratio estimation (chapter 16)

An estimation procedure based on the relationship between two variables y and x which have been measured on the same set of sampled elements.

Regression analysis (chapters 11 and 12)

The process of fitting a regression equation to a set of data by using the method of least squares. Also includes the various statistical tests and estimates associated with the use of the fitted equation.

Regression equation (chapters 11, 12, and 15)

An equation expressing dependence of the mean of a dependent variable y on one or more independent or predictor variables x_1, x_2, x_3, ...

Rejection region (chapters 6, 8, and 9)

In the theory of hypothesis testing, the rejection region is the set of values such that if the test statistic assumes one of these values, the null hypothesis is rejected.

Run (chapter 18)

In a series of observations of attributes, the occurrence of an uninterrupted series of the same attribute is called a run. A run can be of length 1.

Sample (throughout the text)

Any subset of a population.

Sampling design (chapter 16)

A method for selecting a sample.

Sampling units (chapter 16)

Nonoverlapping collections of elements from the population.

Seasonal effect (chapters 14 and 15)

The rises and falls in a timeout series, which always occur at a particular time of year because of changes in the seasons.

Sign test (chapter 18)

A nonparametric statistical test of significance depending on the signs of differences between matched or unmatched pairs and not on the magnitudes of the differences.

Simple linear regression (chapter 11)

A regression analysis in which the mean value of a dependent variable y is assumed to be related to a single independent variable x by the expression $E(y|x) = \beta_0 + \beta_1 x$.

Simple random sample (chapter 8)

A simple random sample results when sampling is conducted in such a way that every possible sample of size n has an equal probability of being selected from a population.

Sinusoidal forecasting model (chapter 15)

A least squares forecasting model using sine and cosine functions of time as the independent or predictor variables in order to pick up cyclic effects over time which may exist within the time series.

Skewed distribution (chapter 3)

A frequency distribution that is not symmetric about its mean.

Smoothed statistic (chapters 14 and 15)

In time series analysis the smoothed statistic is an average of a set of process values, usually a moving average or an exponentially smoothed statistic, used to represent a time series after eliminating random fluctuations.

Smoothing constant (chapters 14 and 15)

The constant α $(0 \leq \alpha \leq 1)$ employed to weight out the contribution of remote response values in an exponential-smoothing scheme.

Standard deviation (chapter 3 and throughout the text)

The standard deviation of a set of n measurements $y_1, y_2, y_3, \ldots, y_n$ is equal to the positive square root of the variance of the measurements.

State of nature (chapter 10)

In a decision analysis the states of nature are the uncertain events over which the decision maker has no control.

Statistic (throughout the text)

A value computed from sample measurements, usually but not always as an estimator of some population parameter.

Stratified random sample (chapter 16)

A sample obtained by separating the population elements into nonoverlapping groups, called strata, and then selecting a simple random sample within each stratum.

Student's *t* distribution (chapters 9, 11, and 12)

The distribution of $t = (\bar{y} - \mu)/(s/\sqrt{n})$ for samples drawn from a normally distributed population and used for making inferences about population means when the population variance σ^2 is unknown and the sample size n is small.

Subjective probability (chapters 4 and 10)

A probability based partially or totally on the personal judgment and intuition of the decision maker, with little or no base in empirical evidence.

Symmetric distribution (chapter 3)

A frequency distribution for which the values of the distribution that are equidistant from the mean occur with equal frequency.

Systematic sampling (chapter 13)

A method of selecting a sample by a systematic method as opposed to random sampling, such as selecting each tenth name from a list or by sampling every third resident in every other block in an area sample.

Table of random numbers (chapter 16)

A table constructed so that each entry is equally likely to be any five-digit number from 00000 to 99999. The table of random numbers is used in selecting a random sample.

Tchebysheff's theorem (chapter 3)

Given a number k greater than or equal to 1 and a set of n measurements $y_1, y_2, \ldots, y_n$, at least $(1 - 1/k^2)$ of the measurements will lie within k standard deviations of their mean.

Test statistic (chapters 6, 9, and throughout the text)

A function of a sample of observations that provides a basis for testing a statistical hypothesis.

Time series (chapters 14 and 15)

Any sequence of measurements taken on a variable process over time. Usually illustrated as a graph whose vertical coordinate gives a value of the random response plotted against time on the horizontal axis.

Treatment (chapter 13)

A stimulus that is applied in order to observe its effect on the experimental situation. A treatment may refer to a physical substance, a procedure, or anything capable of controlled application according to the requirements of the experiment.

Two-stage cluster sample (chapter 16)

A two-stage cluster sample is obtained by choosing a simple random sample of clusters and then selecting a random sample of elements from each cluster.

Two-tailed statistical test (chapters 8 and 9)

A statistical test of an hypothesis in which the rejection region is separated by the acceptance region and is located in both ends of the distribution of the test statistic.

Type I error (chapters 6, 8, and throughout the text)

In the statistical test of an hypothesis, the error incurred by rejecting the null hypothesis when the null hypothesis is true.

Type II error (chapters 6, 8, and throughout the text)

In the statistical test of an hypothesis, the error incurred by accepting the null hypothesis when the null hypothesis is false and some alternative to the null hypothesis is true.

Unbiased estimator (chapter 9)

An unbiased estimator $\hat{\theta}$ of a parameter θ is an estimator for which the expected value of $\hat{\theta}$ is equal to θ. That is, $E(\hat{\theta}) = \theta$.

Union (chapter 4)

If A and B are two events in a sample space S, the union of A and B is the event containing all sample points in A or B or both.

Variance (chapter 3 and throughout the text)

The variance of a set of n measurements $y_1, y_2, \ldots, y_n$ is the average of the square of the deviations of the measurements about their mean.

Venn diagram (chapter 4)

A diagram portraying graphically each sample event as a sample point in a sample space S.

Reference

KENDALL, M. G., and W. R. BUCKLAND. *A Dictionary of Statistical Terms.* New York: Hafner, 1967.

Answers to Exercises

Chapter 2

2.1. a. -2 b. -1 c. -4 d. -1.5 e. $a + 1$ f. $-1 - a$

2.2. a. 1.5 b. $-8/3$ c. 0 d. $(x^2 - 1)/x$ e. $(a^2 - 2a)/(a - 1)$

2.3. $0, -1/2$

2.4. $-1, 1, 7$

2.5. y_1, y_2, y_3

2.6. a. 9 b. $1/2$ c. a^2 d. y^2

2.7. $2, 5, 8, 11$

2.8. $3, 6, 11, 18$

2.9. $y + 1$

2.10. a. -9 b. 65 c. $y_1 + y_2 + y_3 + y_4 - 8$ d. $3y + 12$

2.11. a. $0, 3, 8, 15$ b. $\displaystyle\sum_{i=1}^{4} (y_i^2 - 1)$ c. 26

2.12. a. $4, 1, 0, 1, 4$ b. $\displaystyle\sum_{i=1}^{5} (y - 3)^2$ c. 10

2.13. 30

2.14. $x_1 + x_2 + x_3 + x_4$

2.15. 38

2.16. 20

2.17. $2(y_1 + y_2 + y_3 + y_4) - 20$

2.18. $\displaystyle\sum_{i=1}^{n} y_i^2 - 3\sum_{i=1}^{n} y_i + 9n$

2.19. a. 3 b. 7 c. 11 d. -1 e. -5 f. $4a^2 + 3$ g. $-4a + 3$ h. $7 - 4y$

2.20. a. 0 b. 4 c. 49 d. $(x - 2)^2$ e. $(a - 3)^2$

2.21. a. 7 b. $a^2 + 2ab + b^2 - a - b + 1$

2.22. a. 0 b. undefined c. 3 d. $-(a^2 + 1)$

2.23. $1, (1 - a)$

2.24. a. $1/4$ b. -6 c. $(3/x^2) - (3/x) + 1$ d. $3x^2 - 21x + 37$

2.25. a. 2   b. 2   c. 2

2.26. 36

2.27. 18

2.28. 30

2.29. $5(x^2 + 6)$

2.30. $6x^2 + 55$

2.31. $3(3 + 2i)$

2.32. $10(y^2 + 1)$

2.33. $y_1 + y_2 - 3$

2.34. $\displaystyle\sum_{i=1}^{n} y_i - na$

2.35. $\displaystyle\sum_{i=1}^{n} y_i^2 - 2a \sum_{i=1}^{n} y_i + na^2$

2.36. 36

2.37. -29

2.38. 431

2.39. 610

2.40. 654

2.41. 649.077

2.46. a. 18   b. 76

2.47. 22

2.48. a. 27.7   b. 110.79

2.49. .1967

Chapter 3

3.1. $10.00 or $25.00 probably best

3.2. (I) overlapping class limits   (II) class intervals not of equal width   (III) does not include $15.00

3.3. b. $29/50$

3.4. b. $1/5$

3.8. mean = 2.8, median = 3, mode = 3

3.9. a. modal   b. median   c. mean

3.10. a. .68   b. .95   c. .025

3.11. a. about $3/4$ weigh 6.2 to 10.2 oz; about $8/9$ weigh 5.2 to 11.2 oz   b. about 68% weigh 7.2 to 9.2 oz; about 95% weigh 6.2 to 10.2 oz; almost all weigh 5.2 to 11.2 oz

3.12. The Empirical Rule probably does not apply.

3.13. $\bar{y} = 2$, $s^2 = 12/7$, $s = 1.309$

3.14. $\bar{y} = 3$, $s^2 = 24/5$, $s = 2.191$

3.15. $\bar{y} = -1$, $s^2 = 6.44$, $s = 2.54$

3.16. $s^2 = 2.62$, $s = 1.62$

3.17. b. mean $= 7.29$, median $= 7.5$, mode $= 7.5, 6.5$ c. $s^2 = 1.84$, $s = 1.35$

3.18. 7.5, 9.2; 6.5, 8.1

3.19. a. about $3/4$ from 4.59 to 9.99; about $8/9$ from 3.24 to 11.34   b. about 68% from 5.94 to 8.64; about 95% from 4.59 to 9.99; almost all from 3.24 to 11.34   c. .68, 1.00, 1.00

3.20. $\bar{y} = 2.01$, $s^2 = 1.36$, $s = 1.17$

3.21. $\bar{y} = 30.2$, $s = 7.51$

3.22. $\bar{y} = 7.46$, $s = 1.957$

3.23. a. $\bar{y} = 52.60$, $s = 22.68$ c. 16%   d. 2.5%

3.24. $\bar{y} = 28.001$, $s^2 = .0000185$

3.25. $\bar{y} = .03$, $s^2 = .0000025$

3.26. $\bar{y} = \$7.27$, $s = \$1.07$

3.27. a. $\bar{y} = .0224$, $s = .1414$ b. $\bar{y} = .0847$, $s = .5355$ c. $\bar{y} = 4.44$, $s^2 = 3.56$ d. $\bar{y} = 9.60$, $s = 5.84$

3.31. $\bar{y} = 6$, $s^2 = 6.8$, $s = 2.608$

3.32. $\bar{y} = 0$, $s^2 = 8.67$, $s = 2.944$

3.34. $\bar{y} = 9.28$, $s^2 = 31.38$, $s = 5.601$

3.37. 56%, 100%

3.41. range $= 19$; range $= 45$

3.42. a. 1.225, 1.35   b. 4.75, 5.60

3.43. a. 3.33   b. $\bar{y} = 22.8$, $s = 3.29$

3.44. range $= .5$; $s = .5/4 = .125$; actual $s = .1317$

3.45. .95

3.46. .84

3.47. .68

3.48. .16

3.49. a. about 68% between 7.85 and 9.25; about 95% between 7.15 and 9.95; almost all between 6.45 and 10.65   b. 2.5%

3.50. a. about 68% between 420 and 570; about 95% between 345 and 645; almost all between 270 and 720   b. .16

3.51. a. 251.36, 95.25   b. .68

3.52. $347

3.53. depends on classes used

3.54. $\bar{y} = -.72$, $s^2 = 31.38$

3.55. $\bar{y} = 2.69$, $s^2 = 184$

3.56. $10.16

3.57. 5.98 pounds sterling

3.58. 4752.24856

3.59. 10 degrees Celsius

3.60. 16%

Chapter 4

4.1. A: 4; B: 2, 4, 6; C: 1, 2; $P(A) = 1/6$; $P(B) = 1/2$; $P(C) = 1/3$

4.2. b. (bank 1 wins both), (bank 1 wins one and bank 2 wins one), (bank 1 wins one and bank 3 wins one) c. $P(A) = 1/2$

4.3. a. select 2 of 5 people c. $1/10$ d. $7/10, 3/10$

4.4. c. $1/4$ d. $3/4$

4.5. c. $P(A) = 1/36$; $P(B) = 1/6$; $P(C) = 15/36$; $P(D) = 1/36$; $P(E) = 0$; $P(F) = 15/36$

4.6. f. $P(A) = 1/7$; $P(B) = 3/7$; $P(C) = 5/7$; $P(D) = 5/7$; $P(AB) = 1/7$; $P(A \cup B) = 3/7$

4.8. $P(A) = 1/6$; $P(B) = 1/2$; $P(C) = 1/6$; $P(D) = 1/9$; $P(E) = 1/6$; $P(F) = 0$; $P(G) = 1/6$

4.9. $1/10$

4.10. $P(A_1) = .5$; $P(\bar{A}_1) = 1/2$; $P(A_1|B_1) = .73$; $P(A_2|B_1) = .27$; $P(A_1B_2) = .10$; $P(B_2|A_2) = .70$; $P(A_2|A_1) = 0$

4.11. $1/3$

4.12. .915

4.13. a. .06 b. .95 c. .931 d. .009 3. .18 f. .02

4.14. a. .45 b. .27 c. .78

4.15. a. .47 b. .2209 c. .2499 d. .2209

4.16. a. $1/2$ b. $1/3$

4.17. a. 1 b. .65 c. 1 d. 1

4.18. .0595; .077875

4.19. .0564

4.20. .972

4.21. .99

4.22. $1/4$

4.23. .8

4.24. .53846

4.25. 12

4.26. 6

4.27. 1000

4.28. 30

4.29. 5040

4.30. 151,200

4.31. 6

4.32. 15,504

4.33. 3,628,800

4.34. 5040

4.35. 5,720, 645, 482,000

4.36. 3; 9; 59,049

4.37. 380,204,032

4.38. 2,598,960

4.40. $P(A) = 1/3$; $P(B) = 2/3$; $P(C) = 1/3$; $P(D) = 1/6$; $P(E) = 2/3$

4.41. c. $P(A) = 3/4$

4.42. d. $P(A) = 1/8$

4.43. c. $P(A) = 3/10$

4.44. 2/3

4.45. a. 1/3; b. 4/9 c. 2/9

4.46. .6, .1

4.47. a. 5/6 b. 1/6

4.48. a. .144 b. .056 c. .944

4.49. $P(A) = .29$; $P(B) = .71$; $P(E) = .40$; $P(F) = .20$; $P(AF) = .048$; $P(BG) = .12$; $P(AB) = 0$; $P(A \cup F) = .44$; $P(B \cup F) = .76$; $P(A \cup B) = 1$; $P(A \cup C \cup D) = .44$

4.50. $P(A_1) = .475$; $P(B_3) = .25$; $P(A_1 B_4) = .125$; $P(B_1|A_3) = .5$; $P(A_2 \cup B_3) = .45$; $P(B_1 \cup B_4) = .625$; $P(B_2 B_4) = 0$

4.51. $P(A|B) = .50$; $P(A|C) = .50$; $P(B|C) = 1$; $P(AB) = .33$; $P(A|C) = .167$; $P(BC) = .33$; $P(A \cup B) = .67$; A and B are neither independent nor mutually exclusive; same for B and C.

4.52. $P(A) = .49$; $P(B) = .61$; $P(AB) = .38$; $P(A \cup B) = .72$; $P(A|B) = .623$

4.53. a. .36 b. .9172 c. .2432

4.54. a. .09 b. .10 c. .10 d. .20

4.55. a. 1/4 b. 1/12 c. 1/2

4.56. .189; .216

4.57. a. .73 b. .27

4.58. a. .81 b. .01 c. .90

4.59. .05

4.60. .58; yes

4.61. a. .10   b. .40

4.62. .68

4.63. .39

4.64. .49

4.65. .44

4.66. .41

4.67. .0001; .0198; .9801; .9801

4.68. .00039601; .03900798; .96059601; .99960399

4.69. .00314685; not independent

Chapter 5

5.1. a. discrete   b. discrete   c. continuous   d. discrete   e. continuous

5.2. a. continuous   b. discrete   c. discrete   d. continuous   e. continuous

5.3. $P(0) = .04; P(1) = .32; P(2) = .64$

5.5. $P(0) = 1/6; P(1) = 4/6; P(2) = 1/6$

5.7. a. .6; .24; .096   b. $p(y) = (.4)^{y-1} (.6)$

5.8. a. .25   b. .25

5.9. approximately .25

5.10. $E(y) = 1.0$

5.11. $E(y) = 2.0$

5.12. $E(\text{profit}) = \$100$

5.13. $4081.63

5.14. $E(y) = 1; V(y) = .8; 0; .05$

5.15. a. $E(y) = 5.68; \sigma^2(y) = 2.2176$ b. .04

5.16. 5

5.17. c. .064   d. .051   e. .064

5.18. $P(0) = .10; P(1) = .60; P(2) = .30$

5.19. $P(0) = .064; P(1) = .288; P(2) = .432; P(3) = .216$

5.21. a. discrete   b. continuous   c. continuous   d. discrete   e. discrete

5.22. $E(y) = 2.9167; V(y) = .9097$

5.23. $\sigma = .9538; 1/12$; yes; no

5.24. $E(y) = 1.2; \sigma^2(y) = .36$

5.25. $\sigma = .60$

5.26. $P(1) = .30; P(2) = .60; P(3) = .10; .7$

5.27. $1.75

5.29. 25

5.30. a. .50 b. .75 c. .40

5.31. $121.33

5.32. $P(0) = .064; P(1) = .288; P(2) = .432; P(3) = .216$

Chapter 6

6.1. a. binomial b. defective probabilities not equal c. continuous random variable d. binomial e. defective rate not constant

6.2. large n

6.3. a. .0016 b. .4096 c. .4096

6.5. a. .168 b. .528

6.6. a. .237 b. .896

6.7. a. .904 b. .004

6.9. $\mu = 1875; \sigma^2 = 468.75; \sigma = 21.65$; at least 75% between 1831.7 and 1918.3; at least 89% between 1810.05 and 1939.95; at least 94% between 1788.4 and 1961.6

6.10. a. 40 b. 4.90 c. no

6.11. a. 500 b. 15.81 c. no

6.12. $\mu = 250; \sigma^2 = 225$; at least 75% between 220 and 280; at least 89% between 205 and 295; at least 94% between 190 and 310

6.13. .175; .616

6.14. approximate values: $p(0) = .2865, p(1) = .3581$; actual values: $p(0) = .277, p(1) = .365$

6.15. a. .055 b. .832

6.16. .0067; .0337; .2650

6.17. .0166

6.18. a. $p(y) = C_y^2 C_{2-y}^3 / C_2^5$ b. .70

6.19. a. same as 6.18, part a b. .30 c. .10, .60, .30

6.20. a. $p(y) = C_y^2 C_{2-y}^2 / C_2^4$ b. 1/6 c. 1/6

6.21. b. .00002 c. .2355

6.22. a. .59 b. .168 c. .031 d. 1.000 e. -0

6.23. a. .917 b. .528 c. .188 d. 1.000 e. 0

6.24. a. .349 b. .028 c. .001 d. 1.000 e. 0

6.25. a. .736 b. .149 c. .011 d. 1.000 e. 0

6.27. b. $a = 1$ c. $a = 3$ d. $a = 2$

6.28. (25, 1)

6.29. $a = 4$

6.30. a. .062 b. .170

6.31. a. .376 b. .559

6.32. a. $H_0: p = .5$ b. .03125 c. .262208

6.33. yes, at $\alpha = .110$; H_0: $p = .5$; $y = 10$

6.34. a. .070 b. .383

6.35. a. binomial b. binomial c. p not constant d. binomial e. hypergeometric

6.36. a. .002 b. .630

6.37. a. $p(0) = .125$; $p(1) = .375$; $p(2) = .375$; $p(3) = .125$ c. $\mu = 1.5$; $\sigma = .866$ d. .75, 1.00

6.38. a. $p(0) = .729$; $p(1) = .243$; $p(2) = .027$; $p(3) = .001$ c. $\mu = .3$; $\sigma = .5196$

6.39. $\mu = 100$; $\sigma^2 = 90$; at least 75% between 81 and 119; at least 89% between 71.5 and 128.5; at least 94% between 62 and 138

6.40. a. .59049 b. .91854

6.41. a. .01 b. .9477

6.42. .122; .001; .000

6.43. .392; .008; .000

6.44. .677; .035; .000

6.45. .867; .107; .001

6.46. a. .575 b. 1.000 c. .380

6.48. $\mu = 2$; $\sigma = 1.34$; at least 75% from 0 to 4.68; at least 89% from 0 to 6.02; at least 94% from 0 to 7.36

6.49. a. .387 b. .651 c. .264

6.50. a. .873 b. .027 c. .127

6.51. a. .189 b. .193

6.52. .3

6.53. a. 1.25 b. .277 c. .127 d. no, lack of independence

6.54. .151

6.55. a. .000 b. .833

6.56. a. .008, larger b. .602, smaller

6.57. yes, if $\alpha \geq .09$

6.58. yes, if $\alpha \geq .055$; H_0: $p = .5$; $y = 8$

6.59. a. .60 b. .022

6.60. a. n large, p small b. counting occurrence of rare events

6.61. a. n too small b. Poisson c. Poisson d. p too large

6.62. .135

6.63. a. .0067 b. .1755 c. .265

6.64. a. .368 b. .2642 c. .0025

6.65. a. .2231 b. .5578

6.66. .19915

6.67. a. .0067 b. .4405 c. .9596 d. 5

Chapter 7

7.3. a. .4192 b. .2704 c. .1026 d. .6268 e. .1801 f. .3125 g. .4136 h. .9445

7.4. a. .3121 b. .1020 c. .7454 d. .8997 e. .7007 f. .0629

7.5. a. 1.645 b. −1.645 c. −.86 d. −1.12 e. 1.11 f. 1.645 g. 1.96

7.6. a. .2257 b. .6568 c. .0749 d. .7486 e. .3161 f. .0359

7.7. .0401

7.8. .0192; firm has questionable financial state

7.9. 11.918

7.10. a. .578 b. 10, 2.45, .5752

7.11. a. .630 b. .7995; they do not compare closely because of the skewness of b (20, .2)

7.12. a. .028 b. .3811

7.13. yes; $z = 2.4$

7.14. no; $z = -1.26$

7.15. a. .3849 b. .3159 c. .4452 d. .2734 e. .4265 f. .1628 g. .3227 h. .1586 i. .0730 j. .9500

7.16. a. 0 b. 1.1 c. .67 d. 2.575

7.17. a. .2266 b. .3830 c. .1056 d. .2172 e. .9332 f. .5596 g. .2578 h. .0668

7.18. a. .4332 b. .4772 c. .0401 d. .5 e. .8276 f. .1359

7.19. a. .1736 b. .2061 c. .5318 d. .1896 e. .1820 f. .0630

7.20. a. .0322 b. .5359

7.21. 85.36 minutes

7.22. a. .1056 b. .1056 c. .8944

7.23. 293.59

7.24. a. .618 b. .6156

7.25. a. .421 b. .3085

7.26. a. .444 b. .4599

7.27. a. .5339 b. .0668

7.28. a. .565 b. .957

7.29. almost 1

7.30. yes, $z = -3.265$

7.31. yes, $z = 1.89$

7.32. .9929

7.33. a. .5675

7.34. .8980

7.35. 206; 104

7.36. .9474

7.37. .1600

7.38. 383.5 hours

7.39. 1072 hours

7.40. a. 20   b. 4   c. .9512

7.41. .1314

7.42. a. .0228   b. .0022

7.42. yes; $z = -1.25$

7.44. .2865

7.45. serum effective; $z = 2.37$

7.46. $y \geq 26$

7.47. 183

7.48. yes; $z = -2.25$

Chapter 8

8.1. 117.5 ± 33.6

8.2. 2160 ± 162.6

8.3. $4.5 \pm .26$

8.4. .0918

8.5. (799.40, 872.60)

8.6. (8.22, 9.20)

8.7. (144.51, 150.39)

8.8. a. (16.18, 18.82)   b. reduce σ; reduce confidence

8.9. (1.115, 4.685)

8.10. $(-49.07, -28.93)$

8.11. $(-9.561, -6.239)$

8.12. $.3 \pm .065$

8.13. (.362, .483)

8.14. $.72 \pm .038$

8.15. (.798, .842)

8.16. (.183, .217)

8.17. $.04 \pm .106$

8.18. a. $.284 \pm .124$ b. (.182, .386)

8.19. $(-.142, .229)$

8.20. 400

8.21. 36

8.22. 34

8.23. 72

8.24. a. $H_0: p = .95$ b. $H_a: p < .95$ c. $z = -3.12$; reject H_0

8.25. no; $z = -1.6$

8.26. type I error: not building on site when speed ≥ 15; type II error: building on site when wind < 15; $\alpha = .10$ since type II error more serious

8.27. no, $z = -.615$

8.28. yes, $z = 2.86$

8.32. a. two-tailed b. one-tailed c. one-tailed d. two-tailed e. one-tailed

8.33. 1280 ± 28.4

8.34. .1587

8.35. (1256.641, 1303.359)

8.36. 11336 ± 174.5

8.37. (11110.89, 11561.11)

8.38. $.92 \pm .017$

8.39. (.898, .942)

8.40. (.055, .279)

8.41. (5.801, 7.199)

8.42. $.12 \pm .089$

8.43. (.608, .771)

8.44. $(-45.92, -38.08)$

8.45. -17 ± 5.28

8.46. $(-21.34, -12.66)$

8.47. $.06 \pm .072$

8.48. $(-.033, .153)$

8.49. $.112 \pm .157$

8.50. $(-.090, .314)$

8.51. 246, (4750, 4850)

8.52. 40,000; probability $\approx .95$

8.53. 6147

8.54. 384

8.55. 55

8.56. 35

8.57. (.1216, .2784); 385

8.58. a. $H_0: \mu = 1100$, $H_a: \mu < 1100$ b. $|z| > 1.96$ c. yes; $z = -2.85$

8.59. yes; $z = -2.53$

8.60. yes; $z = 6.45$

8.61. .13

8.62. no; $z = -1.00$

8.63. 400

8.64. yes; $z = -2.47$

8.65. no confidence in marketability

8.66. yes; $z = 4.00$

8.67. yes; $z = -7.577$

8.68. no; $z = 1.656$

8.69. $z = 3.14$; reject H_0

8.70. $z = 2.382$; yes

8.71. $z = 2.5$; reject H_0

8.72. yes; $z = 7.36$

8.73. no; $z = 1.428$; $\alpha = .01$

8.74. a. $H_0: p = .2$, $H_a: p > .2$ b. $z = 1.44$; $\alpha = .0749$

8.75. .1151

8.76. no; $z = 1.5$

8.77. no; $z = 1.835$

8.78. a. low α b. high α

8.79.

μ	13	13.5	14	14.5
β	.0033	.0427	.2358	.5103

8.80. a. 1.9576 b. no c. .1092 d. (1.817, 2.098) e. reduce confidence level
f. highly unlikely g. reduces width h. (45425, 52450) i. yes; $z = 2.886$

Chapter 9

9.2. no; $t = -2.50$

9.3. (22.853, 23.147)

9.4. (34.429, 39.851)

9.5. no; $t = 1.395$

9.6. no; $t = 1.956$

9.7. (2.116, 5.251)

9.8. no; $t = -1.436$

9.9. no; $t = .0225$

9.10. (13.701, 16.324)

9.11. yes; $t = -1.775$

9.12. yes; $t = 2.814$

9.13. yes; $t = -3.711$

9.14. yes, changes test from two-tailed to one-tailed

9.15. yes; $t = 3.354$

9.16. (23.537, 102.463)

9.17. a. approximately 29   b. not without reducing the confidence level

9.18. yes; $t = 3.643$

9.19. a. lower α b. higher α

9.20. no; $t = -.101$

9.21. $(-.672, .647)$

9.22. yes; $t = 5.275$

9.24. no; $X^2 = 32.294$

9.25. a. $(428.076, 447.924)$ b. $(554.276, 1457.497)$

9.26. yes; $X^2 = 14.8$

9.27. yes; $F = 3.268$

9.28. yes; $F = 2.486$

9.29. no; $F = 2.39$

9.30. $t = 1.042$; do not reject

9.32. no; $t = -1.341$

9.33. $(777.226, 812.774)$

9.34. 31, using $t_{.05,n-1} = z_{.05} = 1.645$, expecting n large

9.35. $(3.679, 4.221)$

9.36. yes; $t = 2.635$

9.37. $(7.030, 7.170)$

9.38. yes; $t = 2.108$

9.39. $(.00104, .01496)$

9.40. yes; $t = 2.8$

9.41. a. $(9.860, 12.740)$

9.42. n; $t = -1.95$

9.43. $(1.585, 3.282)$

9.44. 18

9.45. no; $t = -1.8$

9.46. no; $t = -1.567$

9.47. a. no; $t = -1.712$ b. $(-1.522, .0216)$

9.48. a. yes; $t = -2.497$ b. $(-1.704, -.092)$

9.49. $(-1.842, .092)$; yes

9.50. no; $t = .933$

9.51. yes; $t = 9.568$

9.52. a. $(21.174, 30.626)$ b. $(20.896, 30.882)$

9.53. $t = 2.425$

9.54. no; $t = -.434$

9.55. a. $(-4.722, 3.202)$ b. $(5.998, 13.122)$

9.56. no; $X^2 = 12.6$

9.57. (.00896, .05814), (.3786, .9645)

9.58. yes; $F = 2.407$

9.59. no; $F = 2.123$

9.60. no; $t = -1.125$

9.61. no; $X^2 = 3.504$

9.62. a. reduce variation   c. no; $t = 1.5267$ d. $(-288.95, 1488.95)$ e. $600
f. $(-288.95, 1488.95)$ g. 1.5%   h. $(-.7224\%, 3.722\%)$

Chapter 10

10.5.

	INVENTORY			
DEMAND	9	10	11	12
9	3.60	3.00	2.40	1.80
10	3.60	4.00	3.40	2.80
11	3.60	4.00	4.40	3.80
12	3.60	4.00	4.40	4.00

(table contains profits)

10.6. a. a_1: agree; a_2: not agree; s_1: downward float; s_2: no change; s_3: upward float

b.

	a_1	a_2
s_1	$21,000	0
s_2	6,000	0
s_3	$-1,500$	0

c.

	a_1	a_2
s_1	$ 0	$21,000
s_2	10	6,000
s_3	1,500	0

10.7. 11

10.8. yes; $4500

10.9. speculative restaurant; $E(\text{gain}|\text{restaurant}) = \$9375$; $E(\text{gain}|\text{savings account}) = \$1750$

10.10. yes; net profit $= \$8500$

10.11. b. no   c. at least $1/7$

10.12. at least $1/6$

10.13. b. variable-rate loan   c. up to $32.50

10.14. $.21

10.15. Yes, since the expected profit under certainty is $10,250.

10.16. $20,062.50

10.17. a. method B   b. up to $60

10.18. method B; posterior probabilities are .121, .611, and .268

10.19. a.

	GOOD	FAIR	POOR
New Design	$15,000	$5,000	−$2,000
Old Design	5,500	5,500	5,500

b. yes c. 2562.5

d. yes, E(profit) = $11,556; no; E(profit) = $3,209

e. EVSI = $1257.39; do not hire survey

10.20. a. inventory 9 b. accept offer c. accept offer d. buy competing firm e. either action

10.22. a. do not invest (.53 versus .55) b. do not invest (.59 versus .80)

10.27. a. profits b. time c. profits d. profits e. compatibility f. profits

10.28. a. utility to the state of the three situations b. assign them relative quantitative utilities

10.29. a. uncertainty b. demand/day

10.30. c. 60 d. 60 e. 90

10.31. a. A b. $37,000

10.32. $.59

10.33. a. .2, .5, .3 c. $65,500 d. $5,000

10.34. firm B

10.35. 5/7

10.36. the Japanese supplier; $1000 less expensive

10.37. a. tetra-essolan b. $128.00 c. $22.00

10.38. 3 homes

10.39. a.

	DEMAND			
	5	10	15	20
Old	4,000	8,500	22,000	44,500
New	0	7,000	28,000	63,000

b. new machine c. $17,500

10.40. buy new machine

10.41. buy Japanese motors; E(costs) are $15,000 versus $12,140.54

10.42. a. .013, .235, .573, .179 b. di-essolan c. $22.00−$6.39 = $15.61

10.43. a. .102, .392, .451, .055 b. di-essolan c. $22.00−$21.10 = $.90

10.44. use di-essolan if two or more defectives found

10.45. no; expected profit is −$250

10.46. a. do not publish; $E(P)$ = −$942,310 b. publish; $E(P)$ = $1,486,840

10.47. b. yes; list Angelacres c. yes; list Chancyhill

10.48. a. yes b. Blackacre; E(ut) = .53

Chapter 11

11.1. a. 5, −4

11.3. Probabilistic models have a random error component, deterministic models do not.

11.4. $\hat{y} = 3 - x$

11.5. $\hat{y} = 8.807 + .682x$

11.6. SSE = 2, $s^2 = .667$

11.7. SSE = 7.491, $s^2 = 1.498$

11.8. yes; $t = -4.90$

11.9. yes; $t = 2.95$

11.10. a. $\hat{y} = 3.5504 + .8836x$ c. reject H_0; $t = 3.33$

11.11. $.8836 \pm .5918$

11.12. $.732 \pm .466$

11.13. a. $\hat{y} = 4.3 + 1.5x$ c. 1.531   c. yes; $t = 3.83$

11.14. $2.000 \pm .984$

11.15. 12.560 ± 1.226

11.16. a. $\hat{y} = 24.7125 + 5.2702x$ c. 114.3 ± 4.9

11.17. 12.560 ± 3.377

11.18. 8.800 ± 2.263

11.19. a. $\hat{y} = 307.9 + 34.6x$ c. 2666.242   d. 515.5 ± 33.2 e. 515.5 ± 119.7

11.23. .762

11.24. a. .915   b. yes; $t = 7.17$

11.25. .764

11.26. 63.49%

11.27. a. $\hat{y} = 5344.220 - 430.358x$ b. $r = -.940$; yes, $t = -6.76$ c. 88.39%
        d. 1686.175 ± 671.425

11.31. a. 6; 3

11.32. 2/3; −5/3

11.33. a. $\hat{y} = 1.0 + .7x$ c. .367   d. yes; $t = 3.66$

11.34. a. $\hat{y} = 9.441 - 1.088x$ c. 2.747   d. yes; $t = -3.83$

11.35. $.700 + .451$

11.36. $-1.088 \pm .731$

11.37. 4.000 ± 1.611

11.38. 4.000 ± 4.555

11.39. $-.8635$

11.40. 74.56%

11.41. a. $\hat{y} = 83.074 - 1.185x$ c. 11.5137   d. yes; $t = -9.42$

11.42. 29.749 ± 2.898

11.43. SSR = 1021.053, SSE = 80.596, Total SS = 1101.649, R^2 = .9268

11.44. a. $\hat{y}$ = 4.445 + .827x   c. t = 11.80; reject H_0

11.45. .9725

11.46. a. $\hat{y}$ = .8 + 1.4x   c. .4; 1.333

11.47. 1.400 ± .272

11.48. 6.400 ± .636

11.49. t = 2.09; do not reject H_0

11.50. a. $\hat{y}$ = 1.731 + .445x   c. .0662

11.51. yes; t = 6.77

11.52. .9015

11.53. 4.846 ± .299

11.54. a. $\hat{y}$ = 46.000 − .316x   c. 19.033

11.55. −.3167 ± .1021

11.56. 30.167 ± 2.500

11.57. 30.167 ± 4.576
11.58. −.8716; .7597

11.59. a. $\hat{y}$ = 44.375 + 13.512x   c. 27.622   d. yes; t = 4.55

11.60. .8214; .6746

11.61. 79.506 ± 2.820

11.62. 79.506 ± 9.932

Chapter 12

12.1. $\hat{\beta}_0$ = 3.7097, $\hat{\beta}_1$ = 5.1613, $\hat{\beta}_2$ = 5.4839

12.2. F = 1.548; do not reject H_0

12.3. F = 1.548; t = 1.244

12.4. .0397 ± .0664

12.5. .1511 ± .1045

12.6. t = −3.007; reject H_0

12.7. a. E(y) = −2.491 + .099x_1 + .079x_2 + .086x_3 b. elevation and slope but not area

12.8. yes; t = 2.800

12.9. .099 ± .101268

12.10. a. 83.7058   b. .9654   c. E(y) = −51.034 + 57.600x_1 + 2.956x_2 − .934x_3 − 46.542x_4 + 46.355x_5 − 1.805x_6 d. advertising expense; utilities index   f. −1.805 ± 1.344

12.12. (1) $44,263   (2) $35,138   (3) $36,055   (4) $65,206   (5) $44,537

12.13. choose iii

12.14. a. yes; $F = 61.991$ and $R^2 = .776$ b. x_1, x_3, and x_5

12.21. b. $E(y) = 16.8166 + 1.0935x_1$, SSE $= 5.7848$ c. $F = 2.38$; do not reject H_0

12.25. a. yes; $t = -2.557$ b. yes; $F = 64.760$, $R^2 = .812$

12.27. $E(y) = 1.7125 - .1438x + .0135x^2$; reject H_0: $\beta_1 = 0$, $t = -3.02$; reject H_0: $\beta_2 = 0$, $t = 3.30$; $R^2 = .5732$

12.28. $E(y) = -1.7298 + 4.5055x - .5374x^2$; reject H_0: $\beta_1 = 0$, $t = 3.30$; reject H_0: $\beta_2 = 0$, $t = -2.69$; $R^2 = .7215$

12.29. yes; $F = 7.177$, $R^2 = .6146$

12.30. $E(y) = -12.4366 + 1.3251x_1 + 17.9296x_2 + .0075x_1^2 - 2.0789x_2^2$, with $F = 280.438$, $R^2 = .9825$

12.31. $E(y) = 12.89649 + 9.18561x_1 + 10.14408x_1^2 + .00055x_2^2$, with $F = 17.438$, $R^2 = .7658$

12.32. $E(y) = 134.0779 - 628.4045x_1 + 2.4161x_2 + 841.6853x_1^2 + .2286x_2^2 - 5.4993x_1 x_2$, with $F = 9.080$, $R^2 = .7049$

12.33. $F = .4114$; do not reject H_0: $\beta_3 = \beta_4 = \beta_5 = 0$

12.34. a. 30,102 units using the full model; 29,909 units using the reduced model

Chapter 13

13.2. $t = .858$, $F = .737$; do not reject H_0

13.3. a.

SOURCE	d.f.	SS	MS	F
treatment	2	163.33	81.67	.87
error	12	1130.00	94.16	
Totals	14	1293.33		

b. no; $F = .87$ c. no

13.4. a. 37.00 ± 9.46 b. 45.00 ± 18.92 c. -8.00 ± 13.37

13.5. a. yes; $F = 36.52$ b. -7.625 ± 2.395

13.6. a. yes; $F = 5.20$ b. $-.204 \pm .191$; $.022 \pm .191$; $-.348 \pm .191$

13.7. a. yes; $F = 5.70$ b. -9.535 ± 6.385

13.8. a. yes; $F = 5.29$ b. 1.578 ± 1.211

13.9. c. yes; $F = 23.57$

13.10. 2.875 ± 2.371

13.11. a. no; $F = 2.84$ b. yes, if $\alpha > .08$, since $F = 3.10$ c. 2.91 ± 2.29

13.14. $F = 2.00$; do not reject H_0

13.15. a. 225.00 ± 5.37 b. 8.33 ± 9.41

13.16. a.

SOURCE	d.f.	SS	MS	F
make	2	6.419	3.209	5.15
error	8	4.987	.623	
Totals	10	11.406		

b. yes; $F = 5.15$

13.17. a. $17.600 \pm .814$ b. 16.633 ± 1.051 c. $.967 \pm 1.330$

13.18. yes, if $\alpha > .06$, since $F = 3.02$

13.19. a. 48.600 ± 4.752 b. 56.200 ± 4.752

13.20. $F = 10.83$; reject H_0

13.22. no difference among sales areas (blocking ineffective); significant difference between promotional techniques

13.23. a. yes; $F = 6.47$ b. no; $F = 3.75$ c. $-1.20 \pm .65$

13.24. a. no; $F = 1.575$ b. If ineffective, it reduces the information available to perform the analysis.

13.25.

SOURCE	d.f.	SS	MS	F
dwellings	4	584.27	146.07	123.79
models	2	18.90	9.45	8.01
error	8	9.43	1.18	
Totals	14	612.60		

13.26. a. randomized block design   b. SSE $= 58.91$ c. $F = 23.57$; yes; reject H_0
d. No, the judge's response may have been influenced by site location considerations not included within the study.

13.27. a. yes; $F = 143.71$ b. no; $F = 2.81$ c. -2.55 ± 2.87

13.28. a. yes, difference exists in mean recognition time   b. $t = -2.73$; difference exists between layouts A and D

13.29.

SOURCE	d.f.	SS	MS	F
subject	3	.140	.047	6.62
layouts	4	.787	.197	27.70
error	12	.085	.0071	
Totals	19	1.012		

$F = 27.70$; reject the hypothesis of no difference in layout means.

Chapter 14

14.2. a. long-term, cyclic, random variation   b. long-term, random variation
c. seasonal, random variation   d. long-term, cyclic, random variation
e. long-term, cyclic, random variation

14.3. long-term, cyclic, random variation

14.4. b. seasonal, random variation

14.5. either determined by an examination of the time series over a long period of time or an analysis of the environmental conditions affecting its outcome; probably a cyclic effect

14.6. Seasonal effect and random variation both appear to be present.

14.7. The moving average is disadvantageous when the total number of response measurements is small since points of comparison are lost at each end of the time series.

14.9. All four components appear to be present.

14.10. Exponentially smoothed series with $\alpha = .1$ removed most of the seasonal component.

14.11. Long-term trend, seasonal effect, and random variation appear to be present.

14.13. a. housing 22.9%; food 36.1%; apparel 6.8%; transportation 16.3%; health 6.6%   b. housing $4056; food $3239; apparel $1004; transportation $1337; health $1576

14.14. a.

	1960	1965	1970	1975
Milk	112	116	151	177
Flour	105	115	132	190
Hamb.	132	139	200	226
Coffee	80	76	82	131
Potatoes	143	236	581	436

b. 100; 99; 103; 135; 171

14.15. 100; 111; 118; 165; 190

14.17. Long-term trend, seasonal effect, and random variation appear to be present. The time series is too short to determine if a cyclic effect is present.

14.20. Long-term trend, seasonal effect, and random variation appear to be present.

14.22. The 12-month moving-average model is the most effective model at removing the seasonal component.

14.23. a. The DJIA is not a true market average and may not be a reliable market indicator.   b. increase is 38.7% not 48%   c. The WPI measures wholesale price changes of producers; thus it may not be an appropriate measure in this case.   d. an effective use of index numbers

14.24. No, his income has risen only 42.8%. His "real" income in 1975 has actually dropped 20% since 1967.

14.25. b. 100, 110, 136, 139, 154

14.26. 100, 110, 137, 139, 151

14.27. 150

14.28. 150.5

14.29. a. 113, 122, 124   b. 120, 121, 130

14.30. 116, 121, 127

14.31.

	OCTOBER			
	10	11	12	13
Index	33.25	32.12	34.42	35.56

Chapter 15

15.2. b. $SSE_{lin} = 133,297.73$, $SSE_{quad} = 31,643.45$ c. linear model: $y_9 = 800.03$, $y_{10} = 866.70$; quadratic model: $y_9 = 1198.03$, $y_{10} = 1515.18$

15.3. Plots suggest a quadratic model should have been used instead of the linear model since the residuals from the linear model form a quadratic structure.

15.5. a. 25.485 b. 16.294; sinusoidal

15.6. They account for periodic effects in the times series.

15.7.

	Q1	Q2	Q3	Q4
Year 1	1.85	.55	2.05	3.75
Year 2	2.25	.55	2.45	4.55

15.9. $SSE_{lin} = 22.84$, $SSE_{sin} = 10.22$

15.10. $SSE = 4.55$

15.11. Use lagged terms, where the lag is equal to the length of the seasonal period.

15.12. a. $SSE = 88.7448$ b. $SSE = 17.3215$

15.13. Either temporarily shift α to a very large value or replace the smoothed statistics by the atypical occurrences.

15.14. a. double: $\hat{y}_{t+1} = 2.11s_t - 1.11\ s_t(2)$; triple: $\hat{y}_{t+1} = 3.3457s_t - 3.5802s_t(2) + 1.2346s_t(3)$ b. double: $\hat{y}_{t+2} = 2.22s_t - 1.22s_t(2)$; triple: $\hat{y}_{t+2} = 3.7037s_t - 4.1852s_t(2) + 1.4815s_t(3)$ c. double: $\hat{y}_{t+1} = 2.33s_t - 1.33s_t(2)$; triple: $\hat{y}_{t+1} = 4.11s_t - 4.88s_t(2) + 1.77s_t(3)$ d. double: $\hat{y}_{t+3} = 3s_t - 2s_t(2)$; triple: $\hat{y}_{t+3} = 6.67s_t - 9.33s_t(2) + 3.67s_t(3)$

15.15. Curtail production in the following months: Year 1—July, October; Year 2—May, September; Year 3—September. Increase production in the following months: Year 1—August; Year 2—March, July; Year 3—February.

15.17. $SSE = 31.28$

15.18.

	1974	1975	1976
Q1	.887	.971	1.011
Q2	.582	.581	.545
Q3	1.218	1.203	1.114
Q4	.778	.729	.734

$SSE = .1874$

15.19. Accuracy decreases as forecast lead time increases; $SSE = .2141$.

15.20. $\hat{y} = -728.021 + .37021x$, where $y = \ln(g)$ and $x = $ year; $\hat{g}_{1977} = \$3.64$

15.21. $\$3.64 \pm .14$

15.22. when the inherent pattern of the time series is not readily apparent

15.23. probably EWMA

15.26. Trend projections cannot detect changes in demand, attitudes toward stability of currencies, confidence in securities markets, and so forth.

15.27. a. sinusoidal model or EWMA b. multiple linear regression c. growth model or multiple exponential smoothing model d. EWMA

e. multiple linear regression or multiple exponential smoothing model

15.28. Econometric: uses predictive terms thought to be causally related to the response; time series: time series must have an inherent pattern over time; qualitative: very limited or no relevant data base available.

15.30. SSE = 37.60

15.31. SSE = 117.25

15.35. $\hat{y}_t = 2.464 + .888y_{t-1}$, with SSE = 82.51. Fit is better than that of the linear model, worse than that of the sinusoidal model.

15.37. Original series with Brown's multiple exponential smoothing model superimposed

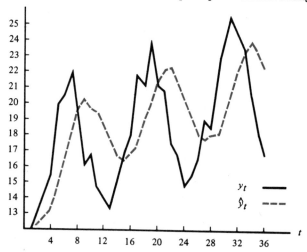

15.38. True values of the series with EWMA 1-month forecasts superimposed (Note that forecasts may vary with choice of initial values.)

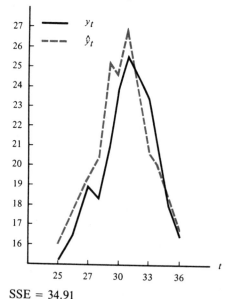

SSE = 34.91

15.39. True values of the series with EWMA 3-month forecasts superimposed

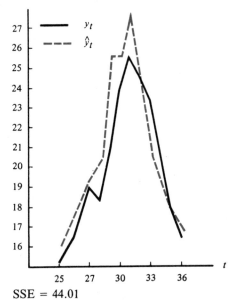

SSE = 44.01

15.40. True values of the series with forecasts superimposed

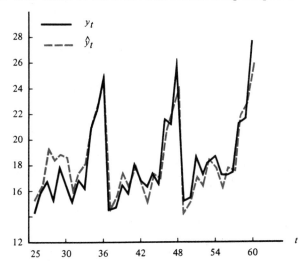

15.41. a. $\hat{\beta}_0 = .11526$, $\hat{\beta}_1 = .01051$, $\hat{\beta}_2 = -.08700$, $\hat{\beta}_3 = -.05866$ b. $R^2 = .9153$

15.42. a. $\hat{\beta}_0 = .11511$, $\hat{\beta}_1 = .01044$, $\hat{\beta}_2 = -.02953$, $\hat{\beta}_3 = -.04049$, $\hat{\beta}_4 = -.00288$, $\hat{\beta}_5 = -.00091$ b. about 2.73%

15.43.

	1966	1967	1968	1969
Q1	.281	.332	.380	.417
Q2	.403	.464	.519	.567
Q3	.506	.577	.644	.714
Q4	.278	.326	.370	.411

15.44. SSE (exercise 15.42) = .02396,; SSE (exercise 15.43) = .00460

15.45. True value of the series with forecasts superimposed

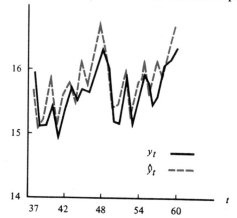

15.46. a. $\hat{y} = 2.5328 + .2231x$, where $y = \ln(g)$ and $x = $ year   b. $93,776

15.47. Combined model is $y_t = -8.1814 + .4803x_1 + .8977x_2$; SSE(exercise 15.37) = 442.2170; SSE(exercise 15.38) = 34.9053; SSE(exercise 15.47) = 13.9105.

Chapter 16

16.1. Random sampling; sample selected for every combination of n measurements from the population has an equal chance of being selected. We can calculate the probabilities of including certain observations in the sample.

16.2. Often too costly and time-consuming to sample entire population.

16.3. a. when the sample meets the conditions of the answer in exercise 16.1   b. when the interviewer selects people all from the same city block, or all from the same company, or all from the same law office, and soon; that is, when conditions of part a are not met.

16.4. a. unit-family get frame from phone book   b. unit-voter frame can be obtained from list of those who voted in last election   c. unit-employee frame from personnel list of all employees   d. unit-buyer frame from all invoices of last year

16.11. 57.5 ± 3.72

16.12. $6.8075 \pm .98997$

16.13. $.575 \pm .14905$

16.15. a. 21.9 ± 1.95187 b. $.7975.0 \pm 487.969$

16.16. a. 7.963 ± 3.1377 b. $.21500.1 \pm 8471.808$

16.17. a. $.36 \pm .118$ b. $.5 \pm .2887$

16.19. a. 2.45 ± 2.3856 b. 756.25 ± 1083.40

16.21. $.125 \pm .06729$

16.22. 18

16.23. 38, 26, 13

16.25. 72, with 21 from each of division 1 and 2, 18 from division 3, and 12 from division 4

16.26. 5 cars: 2 Fords, 2 Chevrolets, and 1 Plymouth

16.28. a. systematic   b. 2-stage cluster   c. ratio estimation   d. systematic   e. 2-stage cluster

16.29. use 2-stage cluster sampling

16.30. use ratio estimation

16.33. same answer as for exercise 16.1

16.34. a. nonrandom   b. random   c. nonrandom   d. random   e. random

16.35. a. saves time and dollars   b. saves time   c. saves time   d. saves time and dollars   e. saves time

16.37. $.26 \pm .121$

16.38. 27.5 ± 1.604

16.39. 13.05 ± 1.825

16.40. 1305 ± 182.5

16.41. $.493 \pm .1112$

16.42. 14.213 ± 1.503

16.43. 9686 ± 3080.625

16.44. 9991.739 ± 1056.609

16.45. Use randomized response sampling design; many acceptable designs within this type exist.

16.46. 165 (164.006 rounded up)

16.47. 8 males, 10 females (note: rounding upward)

16.50. .9882

16.51. tax revenue = $427,124.75

Chapter 17

17.2. yes; $x^2 = 95.06$ ($\chi^2_{.05} = 11.07$)

17.3. yes; $x^2 = 21.49$ ($\chi^2_{.05} = 9.49$)

17.4. yes; $x^2 = 19.297$ ($\chi^2_{.05} = 7.81$)

17.5. no; $x^2 = 1.342$ ($\chi^2_{.05} = 7.81$)

17.6. no; $x^2 = 1.886$ ($\chi^2_{.05} = 5.99$)

17.7. no; $x^2 = 1.651$ ($\chi^2_{.05} = 5.99$)

17.8. yes; $x^2 = 79.985$ ($\chi^2_{.10} = 14.68$)

17.9. yes; $x^2 = 11.414$ ($\chi^2_{.05} = 9.49$)

17.10. a. yes; $x^2 = 10.272$ ($\chi^2_{.05} = 3.84$)   b. yes; $z = 3.28$

17.11. no; $x^2 = 6.567$ ($\chi^2_{.10} = 7.78$)

17.12. no; $x^2 = 2.067$ ($\chi^2_{.10} = 4.61$)

17.13. yes; $x^2 = 24.480$ ($\chi^2_{.05} = 7.81$)

17.14. yes; $x^2 = 6.898$ ($\chi^2_{.05} = 5.99$)

17.15. no; $x^2 = 4.40$ ($\chi^2_{.05} = 9.49$)

17.16. yes; $x^2 = 9.654$ ($\chi^2_{.05} = 5.99$)

17.17. no; $x^2 = .938$ ($\chi^2_{.05} = 3.84$)

17.18. no; $x^2 = 4.703$ ($\chi^2_{.05} = 7.81$)

17.19. yes; $x^2 = 11.626$ ($\chi^2_{.05} = 9.49$)

17.20. no; $x^2 = 5.025$ ($\chi^2_{.05} = 5.99$)

17.21. no; $x^2 = 5.491$ ($\chi^2_{.05} = 5.99$)

17.22. yes; $x^2 = 6.489$ ($\chi^2_{.05} = 5.99$)

17.23. a. yes; $x^2 = 18.528$ ($\chi^2_{.05} = 3.84$)   b. yes; $z = 4.304$

17.24. yes; $x^2 = 43.560$ ($\chi^2_{.10} = 6.63$)

17.25. yes; $x^2 = 21.515$ ($\chi^2_{.05} = 7.81$)

17.26. no; $x^2 = 12.915$ ($\chi^2_{.05} = 15.51$)

17.27. yes; $x^2 = 54.112$ ($\chi^2_{.05} = 15.51$)

17.28. yes; $x^2 = 21.062$ ($\chi^2_{.10} = 7.78$)

17.29. yes; $x^2 = 38.862$ ($\chi^2_{.05} = 12.59$)

17.30. yes; $x^2 = 153.485$ ($\chi^2_{.05} = 15.51$)

Chapter 18

(Note: RR = rejection region.)

18.1. RR = 0, 1 and $\alpha = .011$; RR = 0, 1, 2 and $\alpha = .055$

18.2. $n = 15$, RR = 0, 1, 2, 3 for $\alpha = .018$ and RR = 0, ..., 4 for $\alpha = .059$; $n = 20$, RR = 0, ..., 5 for $\alpha = .021$; RR = 0, ..., 6 for $\alpha = .058$; RR = 0, ..., 7 for $\alpha = .132$; $n = 25$, RR = 0, ..., 7 for $\alpha = .022$; RR = 0, ..., 8 for $\alpha = .054$; RR = 0, ..., 9 for $\alpha = .115$

18.3. yes; $Y = 2$ with RR = 0, 1, 2 and $\alpha = .055$

18.4. no; $Y = 18$ with RR = 0, ..., 7 and 18, ..., 25 for $\alpha = .044$

18.5. do not reject the hypothesis of no preference; $Y = 7$ with RR = 0, 1, 9, 10 and $\alpha = .022$

18.6. a. yes; $U_0 = 2$ with $U_{.0754} = 5$ b. yes; $t = -2.937$ with $t_{.10} = 1.86$

18.7. yes; $U_0 = 4.5$ with $U_{.0745} = 8$; community A has lower borrowing rates

18.8. yes; $z = -3.396$

18.9. Reject hypothesis of no difference since $T = 6$ with $T_{.05} = 8$ for a two-tailed test. Conclusions are different because Wilcoxon test uses more information.

18.10. a. no; $Y = 8$ with RR = 0, 1, 2, 9, 10, 11 and $\alpha \approx .0656$ b. no; $T = 14.5$ with RR = 0, ..., 11 and $\alpha = .05$

18.11. a. yes; $T = 5$ with RR = 0, ..., 11 and $\alpha = .10$ b. no; RR = 0, 1, 2, 7, 8, 9 and $\alpha \approx .1798$

18.12. yes; $T = 3.5$ with RR = 0, ..., 4 and $\alpha = .10$

18.13. no; $R = 5$ with RR = 2, 7 and $\alpha = .086$

18.14. $\bar{y} = 34.4$; $R = 7$ with RR = 2, 3, 4, 11 and $\alpha = .071$; no evidence of lack of randomness

18.15. $\bar{y} = 20.9$; $R = 2$; $z = -4.287$; yes

18.16. a. $z = -1.805$; reject H_0 b. reduction of the expected number of runs; $z = -.37$; do not reject H_0

18.17. RR: $r_s \geq .643$; $\alpha = .05$; sample $r_s = .810$; reject H_0

18.18. RR: $r_s \geq .648$ (two-tailed); $\alpha = .05$; sample $r_s = .588$; reject H_0

18.19. RR: $r_s \geq .497$ (one-tailed); $\alpha = .05$; sample $r_s = .636$; reject H_0

18.21. a. $Y = 2$ with RR = 0, 1 and $\alpha = .011$; do not reject H_0 b. $T = 13$ with RR = 0, ..., 5 and $\alpha = .01$; do not reject H_0 (both tests are one-tailed)

18.22. sign test; $Y = 2$ with RR = 0, 1, 8, 9 and $\alpha = .0392$; no difference in bonding strengths

18.23. RR: $Y \geq 9$; $\alpha = .07$; do not reject H_0

18.24. RR: $Y \geq 9$; $\alpha = .133$; reject H_0

18.25. RR = 0, 1, 7, 8; $\alpha = .070$; do not reject H_0

18.26. a. RR = 0, 1, 11, 12; $\alpha = .0386$; do not reject H_0 b. RR = 0, 1, 2; $\alpha = .073$; do not reject H_0

18.27. RR: $T < 2$; sample $T = 4$; do not reject H_0

18.28. RR = 0, 1, 7, 8; $\alpha = .070$; sample $Y = 2$; do not reject H_0

18.29. a. $U_0 = 32$ with $\alpha = .094$; do not reject H_0 b. $t = .319$; do not reject H_0

18.30. $U_0 = 12.5$ with RR: $U < 21$ and $\alpha = .1012$; do not reject H_0

18.31. $R = 8$ with RR = 6, 7 and $\alpha = .10$; do not reject; difference in conclusions exist because the U test is more powerful.

18.32. $U_0 = 5.5$ with RR = 8, 9 and $\alpha = .05$; reject H_0

18.33. $R = 5$; $P(R < 2) = .024$; $P(R > 8) = 0$; do not reject H_0

18.34. a. $P(R \leq 11) = .0256$ b. yes; $z = -1.95$

18.35. a. $R = 7$; RR: $R < 6$ and $\alpha = .108$ (one-tailed); do not reject H_0 b. $t = .574$; do not reject H_0

18.36. a. $U = 3$; RR: $U \leq 3$ with $\alpha = .056$; reject H_0 b. $F = 4.91$; RR: $F > 9.12$ with $\alpha = .05$; do not reject H_0

18.37. $r_s = -.845$

18.38. RR: $r_s < -.564$ with $\alpha = .05$; reject H_0

18.39. $r_s = -.504$

18.40. RR: $r_s < -.497$ with $\alpha = .05$; reject H_0

18.41. $r_s = .766$; RR = $r_s > .564$ with $\alpha = .05$; reject H_0 and assume the two groups do provide similar rankings.

Index

geometric mean $r = \sqrt{bx \cdot y \cdot by x} = \sqrt{\dfrac{\Sigma xy}{\Sigma y^2} \cdot \dfrac{\Sigma xy}{\Sigma x^2}}$